A SHORT TABLE OF INTEGRALS

1. $\int u \, dv = uv - \int v \, du$

2. $\int \dfrac{dx}{x^2 + a^2} = \dfrac{1}{a} \tan^{-1} \dfrac{x}{a}$

3. $\int \sin \dfrac{x}{a} \, dx = -a \cos \dfrac{x}{a}$

4. $\int \cos \dfrac{x}{a} \, dx = a \sin \dfrac{x}{a}$

5. $\int \sin^2 x \, dx = \dfrac{x}{2} - \dfrac{1}{4} \sin 2x$

6. $\int \cos^2 x \, dx = \dfrac{x}{2} + \dfrac{1}{4} \sin 2x$

7. $\int \sin(ax + b) \, dx = -\dfrac{1}{a} \cos(ax + b)$

8. $\int \cos(ax + b) \, dx = \dfrac{1}{a} \sin(ax + b)$

9. $\int \sin px \sin qx \, dx = \dfrac{\sin(p - q)x}{2(p - q)} - \dfrac{\sin(p + q)x}{2(p + q)} \qquad p^2 \neq q^2$

10. $\int \cos px \cos qx \, dx = \dfrac{\sin(p - q)x}{2(p - q)} + \dfrac{\sin(p + q)x}{2(p + q)} \qquad p^2 \neq q^2$

11. $\int \sin px \cos qx \, dx = -\dfrac{\cos(p - q)x}{2(p - q)} - \dfrac{\cos(p + q)x}{2(p + q)} \qquad p^2 \neq q^2$

12. $\int x \sin ax \, dx = \dfrac{1}{a^2} \sin ax - \dfrac{x}{a} \cos ax$

13. $\int x \cos ax \, dx = \dfrac{1}{a^2} \cos ax + \dfrac{x}{a} \sin ax$

14. $\int e^{ax} \, dx = \dfrac{e^{ax}}{a}$

15. $\int x e^{ax} \, dx = \dfrac{e^{ax}}{a^2} (ax - 1)$

16. $\int e^{ax} \sin bx \, dx = \dfrac{e^{ax}}{a^2 + b^2} (a \sin bx - b \cos bx)$

17. $\int e^{ax} \cos bx \, dx = \dfrac{e^{ax}}{a^2 + b^2} (a \cos bx + b \sin bx)$

18. $\int_0^\infty e^{-ax} \, dx = \dfrac{1}{a} \qquad a > 0$

19. $\int_0^\infty x^{n-1} e^{-ax} \, dx = \dfrac{(n - 1)!}{a^n} \qquad n = 1, 2, 3, \ldots$

Other definite integrals between 0 and ∞ may be obtained from the Laplace transform with suitable numerical values for s.

Applied Circuit Analysis

Applied Circuit Analysis

Shlomo Karni
University of New Mexico

JOHN WILEY & SONS
New York • Chichester • Brisbane • Toronto • Singapore

Library of Congress Cataloging in Publication Data:

Karni, Shlomo, 1932–
    Applied circuit analysis.

    Includes index.
    1. Electric circuit analysis.   I. Title.
TK454.K36   1988   621.319′2   87–7514
ISBN 0-471-60498-4

Printed in the United States of America

10 9 8 7 6 5 4 3 2

Many characters in this book, such as t, for example, are real. Others, such as j, are purely imaginary. Still others are complex. All resemblance to similar characters which you encountered in the past and which you will undoubtedly encounter in the future is entirely intentional.

Contents

This section may be omitted at first without loss of continuity.

Preface

TO THE STUDENT

Applied Circuit Analysis is written to help you in your study of basic electrical circuits and systems. Although the expert guidance of an instructor is very beneficial, you, and only you, are the master of your own fate. On a smaller and more immediate scale, it means that you are responsible for successful progress in your studies.

Use this book daily. Start with Chapter 0 (no laughing matter) a day or two before classes begin; you'll be ready then. As you progress through the material, you'll find certain repetitions. There is no waste in them; they serve a useful purpose. They tie together several topics and improve your understanding from different but related points of view.

A few sections and problems are marked with an asterisk (*). These are slightly more advanced topics and they may be skipped at first. Be sure to get back to them at the end of the chapter—if not sooner.

Each topic is followed by solved examples. Do yourself a favor—solve each example on your own, then compare results. This is an invaluable method to check on your progress. If you don't quite get the example, go back and study again the previous topic.

The same goes for problems. They are listed by their numbers within the text, at the appropriate place for the particular topic (usually, but not always, after an example). Do them as assigned, or even beyond a class assignment. Some problems are solved in a later chapter. This is part of the learning process mentioned earlier, exposing several aspects of some topic. If you decide to read ahead to discover the solution, you *are* learning. Other problems are "leading." They take you one small step beyond the material that you've just learned and make you think about it.

TO THE INSTRUCTOR

Everyone agrees that *the* model undergraduate curriculum does not exist—and for very good reasons, indeed. On the other hand, several possible model curricula have been, and continue to be, developed. As a foundation in practically all such curricula, we find two semesters (or their equivalent) of "electrical circuits," "basic circuit analysis," "network analysis," or a similar synonymous topic. This foundation serves the more advanced topics, such as electronic circuits and devices, control systems, electric machinery, and nonlinear systems.

The scope of this book is aimed at the coverage of such a two-semester sequence. Let me make a few relevant remarks:

State Variables   The naturalness of capacitive voltages and inductive currents is easily accepted by students at an early stage. It is pedagogically sound and quite helpful to use, for instance, $v_C(t)$ as the unknown in the very first simple RC series circuit. An immediate beneficial "side effect" is the absence of ungainly integrals. These aspects become even more pronounced in the more general circuits, where integro-differential equations appear in both node analysis and loop analysis. Add to this the physical insights gained by state variables, and you have excellent reasons to present all three methods from an early start.

Computer Use   The use of computers by students depended in the past, as it does now and will depend in the future, on two main factors: (1) the availability of programs, from those in hand-held calculators to some inexpensive packages for personal computers, to the vast libraries at the computer center,† and (2) the *active* role played by the instructor in integrating such usage into the course. The instructor must be particularly careful here, or else the student is liable to lose sight of the forest for the trees. Perhaps a better metaphor is "to lose sight of who is wagging whom—the dog or the tail." The dog is always the analysis of circuits, and the tail is the computer that helps us in this analysis.

In addition to introducing certain standard programs (for example, the Gauss-Jordan elimination, a root finder, parts of CORNAP, etc.), the instructor should encourage, better yet, require, individual efforts from students to write short programs and to use PCs. Several problems throughout this text are formulated with this aspect in mind. Students, in general, are glad to be initiated into the mysteries of larger circuit analysis programs by this approach, and the learning benefits are rewarding.

Minimal Memorization   Memorization is antonymous to true learning. Students quickly develop fear and resentment of "cookbook" formulas. On the other hand, a simple logical explanation can make all the difference in the world. Even such a relatively difficult subject as convolution can be made palatable if we explain the idea

---

†*Two typical examples are the* Micro-Cap II *(Microcomputer Circuit Analysis Program) by Spectrum Software, Sunnyvale, CA, an interactive program for the HP 150 or the IBM PC; and the* I-G SPICE *by AB Associates, Tampa, FL, for larger mainframe computers (CRAY, CDC, IBM) and some minicomputers.*

of adding (superposition) of individual responses in order to get the total response. Similarly, the limits on the convolution integral are easily explained, not memorized. Or, consider the various parameters of two-port networks. Here, the cookbook formulas are disagreeable, at the very least. With a simple explanation (again, superposition) all the formulas and their derivations become natural and easy. This book is dedicated to the idea that a good explanation of the *why* makes the *how* much easier to understand.

Material, Majors, Nonmajors, etc.   As you examine this book, you'll undoubtedly recognize the best way to use the material for your particular class. Broadly speaking, selected material from the first twelve or thirteen chapters belongs in the first of the two semesters of circuit analysis; the rest is suitable for the second semester. Variations, of course, are possible: Chapter 3 on topology may be postponed until later, as well as the preliminary discussion of two-ports (Section 5-7) and of mutual inductance (Section 6-3).

With a few additional notes, this material can serve for the concurrent labs. Special effort has been made to phrase the suitable parts and problems in this way ("if we connect . . .", or "such an experiment . . . ," etc.)

For nonmajors, a one-semester (or its equivalent) selection can be made, deleting topics in Chapters 3, 5, 8, 13, 16, 18, 19, and 21. Here, again, your own dexterity and judgment will be guided by the course's requirements. Supplemental material may also be needed.

Finally, if you find an error, or a better way to solve a problem, please let me know. We are *all* students in the broad sense of the word.

ACKNOWLEDGMENTS

Valuable comments on the manuscript were received from several colleagues, among them Professors Carl Zimmer, Arizona State University; Gary Ford, University of California, Davis; Edward White, University of Virginia; D. F. Hunt, Polytechnic Institute of New York; Eugene Denman, University of Houston; Yogendra Kakad, University of North Carolina, Charlotte; Eddie Fowler, Kansas State University; James Holte, University of Minnesota; James Delansky, Pennsylvania State University; and Stewart Stanton, Montana State University. Special thanks go to my good friends, Professor Peter Dorato, University of New Mexico, and Professor Ben Leon, University of Kentucky.

SHLOMO KARNI

Applied Circuit Analysis

Chapter 0

Preliminaries

Don't let the number of this chapter fool you. The importance of the topics reviewed here is paramount. In addition to the usual prerequisites for this class (sophomore-junior standing with the associated courses in mathematics and physics), these topics are an absolute must. Your instructor will assume rightfully that you know them thoroughly, without the need of a review in class.

So, the burden of proof is on your shoulders. Go through each topic carefully. Solve all the listed problems to your complete satisfaction. Then check the given answers, and if you have any doubts or mistakes, review the topic in your favorite text. Be honest with yourself!

0-1 UNITS AND DIMENSIONS

The accepted International System of units (SI) adopted the following base and supplementary units:

Quantity	Unit name	Unit symbol
SI base units		
Length	meter	m
Mass	kilogram	kg
Time	second	s
Electric current	ampere	A
Thermodynamic temperature	kelvin	K
Amount of substance	mole	mol
Luminous intensity	candela	cd

SI supplementary units		
Plane angle	radian	rad
Solid angle	steradian	sr

Derived units with special names include, for instance,

Force	newton	N
Energy, work	joule	J
Power	watt	W

and others to be studied in this book.

The dimensions (units) of force are mass × acceleration, or

$$[\text{N}] = [\text{m} \cdot \text{kg/s}^2] \qquad (0\text{-}1)$$

Similarly, work is force × distance or

$$[\text{J}] = [\text{N} \cdot \text{m}] = [\text{m}^2 \cdot \text{kg/s}^2] \qquad (0\text{-}2)$$

A basic check on the correctness of an equation is the balance of dimensions (units) on both sides of the equality. For example, if someone "derives" the following equation for pressure:

$$p = \frac{f \cdot m \cdot A}{v^2}$$

where f is force, m mass, A area, and v velocity, the dimensions do not agree:

$$[\text{N/m}^2] \neq [\text{N} \cdot \text{kg} \cdot \text{m}^2/\text{m}^2 \cdot \text{s}^{-2}]$$

and therefore this equation is wrong.

In our studies throughout this book, many new equations will be derived. Check units often, even if they are given.

Probs. 0-1
through 0-7

0-2 FUNCTIONS AND PLOTTING

Forget for a moment that you may have access to a printer-plotter, capable of plotting functions. In all engineering disciplines, and certainly in electrical engineering, you must be able to visualize and sketch fairly accurately many common functions. The independent variable (abscissa) will be usually time, t, although at times a multiple of t, such as at, will be used. Another independent variable in our studies will be angular frequency.

Probs. 0-8
through 0-12

0-3 COMPLEX NUMBERS

This topic should be an old friend from your high school days. If you feel that you are a bit hazy, dust off your math book. A brief review is also provided in Appendix B at the end of this book. Whatever you do, don't neglect your fluency (nothing less!) in complex numbers.

Probs. 0-13
through 0-20

0-4 L'HÔPITAL'S RULE

This rule (often misspelled l'Hospital's) allows us to calculate limits of the form 0/0. It states that if

$$\lim_{t \to a} f(t) = 0, \qquad \lim_{t \to a} g(t) = 0 \tag{0-3}$$

then

$$\lim_{t \to a} \frac{f(t)}{g(t)} = \lim_{t \to a} \frac{f'(t)}{g'(t)} \tag{0-4}$$

provided the latter limit exists (finite or infinite). A similar rule holds if $t \to \infty$ in these limits. It may be necessary sometimes to repeat this process if f'/g' is, again, of the indeterminate form 0/0. For example,

$$\lim_{t \to 0} \frac{1 - \cos t}{t^2} = \lim_{t \to 0} \frac{\sin t}{2t} = \lim_{t \to 0} \frac{\cos t}{2} = \frac{1}{2}$$

It is important to remember that: (1) This rule *does not apply* if only one function goes to zero while the other does not. (2) The differentiation of $f(t)/g(t)$ is not done as a quotient—we simply differentiate numerator and denominator separately. (3) The same rule applies to the indeterminate form ∞/∞, that is, if

$$\lim_{t \to a} f(t) = \infty, \qquad \lim_{t \to a} g(t) = \infty \tag{0-5}$$

then

$$\lim_{t \to a} \frac{f(t)}{g(t)} = \lim_{t \to a} \frac{f'(t)}{g'(t)} \tag{0-6}$$

provided the latter limit exists.

Other indeterminate forms, such as $\infty - \infty$, or $0 \cdot \infty$, or 0^0, or ∞^0, or 0^∞, must be reduced to one of the forms 0/0 or ∞/∞ before l'Hôpital's rule is applied. For instance, to calculate

$$\lim_{t \to \infty} t^2 e^{-t} = ?$$

we recognize that this is of the form $\infty \cdot 0$, so we write it as

$$\lim_{t \to \infty} \frac{t^2}{e^t}$$

which is of the form ∞/∞. Now we can apply l'Hôpital's rule and obtain

$$\lim_{t \to \infty} \frac{t^2}{e^t} = \lim_{t \to \infty} \frac{2t}{e^t} = \lim_{t \to \infty} \frac{2}{e^t} = 0$$

and say, commonly, that the exponential e^{-t} decreases faster than the increase of the quadratic t^2.

Probs. 0-21 through 0-26

PROBLEMS

0-1 What are the dimensions (units) of $e = 2.71828\ldots$? Of $\pi = 3.14159\ldots$?

0-2 What are the units of k in e^{-kt}, where t is time?

0-3 The wavelength λ of a particle is given by

$$\lambda = \frac{h}{mv}$$

where m is the particle's mass and v its velocity. Show that h has the units $[\text{J}\cdot\text{s}]$.

0-4 The displacement $x(t)$ of a certain mass m is given as

$$x(t) = A \sin\left(\sqrt{\frac{k}{m}}\, t + B\right)$$

What are the units of A? Of B? Of $\sqrt{k/m}$? Of k?

0-5 The unit of angular velocity is rad/s. Relate it to rpm, revolutions per minute.

0-6 The density of water is 1 g/cm^3. What is it in kg/m^3?

0-7 The acceleration of a car is given as

$$a = 3 - 0.2t^2$$

What are the units of 3? Of 0.2?

0-8 Plot accurately on the same graph sheet, and for $t \geq 0$:
(a) $f_1(t) = 20e^{-0.1t}$ (b) $f_2(t) = 20e^{-t}$
(c) $f_3(t) = 20e^{-3t}$ (d) $f_4(t) = 20e^{-10t}$

You may use a calculator for repeated numerical evaluations, but no plotter-printer. Show points of intersection, asymptotes, etc.

0-9 Plot each function as indicated. Show important points.
(a) $g(t) = 10 \cos 5t$, $-\infty < t < \infty$
(b) $h(t) = 10 \sin 20t$ versus $(20t)$ as abscissa, $-\infty < t < \infty$
(c) $i(t) = 3 \cos[4t + (\pi/12)]$ versus $(4t)$
(d) $k(t) = 200e^{-t} \cos 5t$ versus t, $t \geq 0$

Hint: Use the two curves in Problems 0-8(b) and 0-9(a).

0-10 Calculate the first derivative, d/dt, of $f_1(t)$ in Problem 0-8(a), of $g(t)$ in Problem 0-9(a), and of $k(t)$ in Problem 0-9(d). Plot accurately each derivative versus t.

0-11 Calculate

$$\int_0^t f_2(x)\, dx$$

where $f_2(t)$ is in Problem 0-8(b). Plot the result for $t \geq 0$. Repeat for

$$\int_0^t i(x)\, dx$$

for $i(t)$ in Problem 0-9(c).

0-12 Plot $g(t)$ versus $h(t)$ of Problem 0-9. Your abscissa (horizontal axis) here is h, and your ordinate (vertical axis) is g. Such a plot is called *parametric*.
Hint: For a chosen value of $t = t_1$, calculate $h(t_1)$ and $g(t_1)$. Plot this point $(h(t_1), g(t_1))$ in the g-h plane. Repeat for $t = t_2, t = t_3, \ldots$.

Note: In Problems 0-13 through 0-20 you may *not* use the rectangular ↔ polar buttons on your calculator.

0-13 Given the four complex numbers

$$A = -3 + j2 \qquad B = 4e^{j\pi/10}$$
$$C = 1.6 - j2.4 \qquad D = 3e^{-j\pi/6}$$

with $j^2 = -1$. Plot each complex number on the complex plane.

0-14 Calculate

(a) $|A|$ (b) $|B|$ (c) $|A \cdot B|$ (d) $|A| \cdot |B|$

0-15 Calculate

$$E = A - B - D$$

and express it in its rectangular and its exponential forms. Plot E on the complex plane.

0-16 Calculate

$$G = B \cdot D$$

in rectangular and exponential forms. Plot G.

0-17 Calculate

$$H = \frac{A - D}{C}$$

in rectangular and exponential forms. Plot H.

0-18 Calculate C^2 and plot it.

0-19 Calculate $|C|^2$ and plot it.

0-20 Calculate $|C^2|$ and plot it.

0-21 Calculate

$$\lim_{t \to 0} t \ln t$$

0-22 Calculate

$$\lim_{t \to 0} \frac{\sin t}{t}$$

0-23 Calculate

$$\lim_{t \to \infty} t^n e^{-at}, \qquad a > 0, \qquad n = 1, 2, \ldots$$

0-24 What's wrong with the following result?

$$\lim_{s \to 1} \frac{4s^2 - 2s - 2}{s^2 - s} = \lim_{s \to 1} \frac{8s - 2}{2s - 1} = \lim \frac{8}{2} = 4$$

0-25 Calculate

$$\lim_{s \to \infty} K \frac{s^m + as^{m-1} + bs^{m-2} + \cdots}{s^n + ps^{n-1} + qs^{n-2} + \cdots}$$

where K, a, b, p, q are real coefficients and m and n are positive integers. Distinguish three cases: $m > n$, $m = n$, and $m < n$.

0-26 Calculate

$$\lim_{t \to \infty} e^{-at} \cos bt, \qquad a > 0, b > 0$$

and comment on your approach.

SOLUTIONS TO PROBLEMS

0-1 Both are pure, dimensionless numbers.†

0-2 Since e and its powers are dimensionless, k has the units of s^{-1}.

0-3 $[h] = [m \cdot kg \cdot m \cdot s^{-1}] = [J \cdot s]$

0-4 $[A] = [m], [B] = [rad], [\sqrt{k/m}] = [s^{-1}], [k] = [kg \cdot s^{-2}]$

0-5   1 rev/s $= 2\pi$ rad/s

$\therefore$ 60 rev/min $= 2\pi$ rad/s

$\therefore$ 1 rpm $= \dfrac{\pi}{30}$ rad/s

0-6 1 g/cm^3 $= 10^{-3}$ kg/(0.01)^3 m^3 $= 1000$ kg/m^3 $= 1$ ton/m^3

0-7 $[3] = [m \cdot s^{-2}], [0.2] = [m \cdot s^{-4}]$

0-8 See Prob. Fig. 0-8.

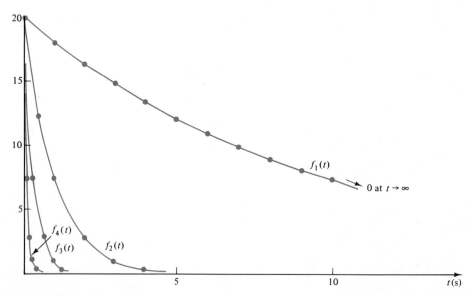

Problem 0-8

†One student dubbed them "non-dimensionalist!"

0-9 See Prob. Fig. 0-9. For (d), multiply the two curves point by point.

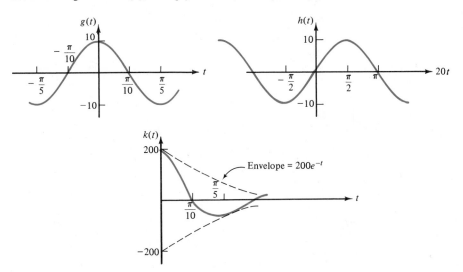

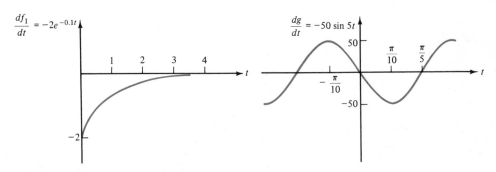

Problem 0-9

0-10 $f'_2(t) = -20e^{-t}$ $g'(t) = -50 \sin 5t$

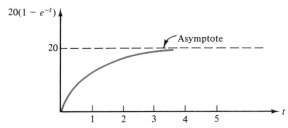

Problem 0-10

0-11 $\int_0^t 20e^{-x}\, dx = 20(1 - e^{-t})$.

The dummy variable x is needed because t is the upper limit on the integral.

Problem 0-11

0-13

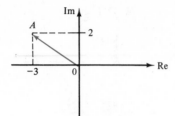

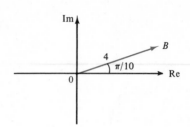

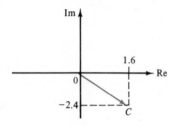

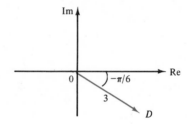

Problem 0-13

0-14 $|A| = \sqrt{(-3)^2 + (2)^2} = \sqrt{13} = 3.606$
$|B| = 4$
$|AB| = |A| \cdot |B| = 4\sqrt{13} = 14.424$

0-15 $E = -3 + j2 - 4\left(\cos\dfrac{\pi}{10} + j\sin\dfrac{\pi}{10}\right) - 3\left(\cos\dfrac{\pi}{6} - j\sin\dfrac{\pi}{6}\right)$

$= -3 + j2 - 4(0.951 + j0.309) - 3(0.866 - j0.5)$
$= -3 - 3.804 - 2.598 + j(2 - 1.236 + 1.5)$
$= -9.402 + j2.264$ in rectangular form.
$E = \sqrt{(-9.402)^2 + (2.264)^2}\, e^{j\phi}, \qquad \phi = \tan^{-1}(2.264/-9.402) = 2.904$ rad

0-16 $G = BD = (3.804 + j1.236)(2.598 - j1.5)$
$= (3.804)(2.598) + j(1.236)(2.598) - j(1.5)(3.804)$
$+ j(1.236)(-j1.5) = 9.883 + j3.211 - j5.706 + 1.854$
$= 11.737 - j2.50$

0-17 $H = \dfrac{-3 + j2 - (2.598 - j1.5)}{1.6 - j2.4} = \dfrac{-5.598 + j3.5}{1.6 - j2.4}$

$= \dfrac{-5.5981 + j3.5}{1.6 - j2.4}\dfrac{1.6 + j2.4}{1.6 + j2.4} = \dfrac{-17.357 - j7.835}{(1.6)^2 + (-2.4)^2}$

$= \dfrac{-17.357 - j7.835}{8.32} = -2.086 - j0.942$ in rectangular form

$H = \sqrt{(-2.086)^2 + (-0.942)^2}\, e^{j\phi}, \qquad \tan\phi = \dfrac{-0.942}{-2.086}$

0-18 $C^2 = C \cdot C = (1.6 - j2.4)^2 = 1.6^2 - 2(1.6)(j2.4) + (j2.4)^2$
$= 2.56 - j7.68 - 5.76 = -3.2 - j7.68$

0-19 $|C|^2 = (1.6)^2 + (-2.4)^2 = 8.32$, always a real, positive number

0-20 $|C^2| = |-3.2 - j7.68| = \sqrt{(-3.2)^2 + (-7.68)^2} = 8.32$ (ditto)

0-21 Since it is of the form $0 \cdot (-\infty)$, write it first as

$$\lim_{t \to 0} \frac{\ln t}{\dfrac{1}{t}} = \lim_{t \to 0} \frac{\dfrac{1}{t}}{\dfrac{-1}{t^2}} = \lim_{t \to 0} (-t) = 0$$

0-22 Apply l'Hôpital's rule immediately, since it is of the form $0/0$. Answer $= 1$.

0-23 Of the form $0 \cdot \infty$ Write as

$$\lim_{t \to \infty} \frac{t^n}{e^{at}} = \lim_{t \to \infty} \frac{nt^{n-1}}{ae^{at}} = \cdots = \lim_{t \to \infty} \frac{n!}{a^n e^{at}} = 0$$

0-24 The first step is O.K., since then we have $0/0$. The second step is wrong, since

$$\lim_{s \to 1} \frac{8s - 2}{2s - 1} = \frac{(8)(1) - 2}{(2)(1) - 1} = \frac{6}{1} = 6$$

is not an indeterminate form.

0-25 Use l'Hôpital's in all three cases, since all are indeterminate, ∞/∞. Answers:

$$m > n \qquad \lim = \infty$$
$$m = n \qquad \lim = K$$
$$m < n \qquad \lim = 0$$

0-26 Can't use l'Hôpital's rule, since $\cos bt$ never exceeds unity in magnitude. Therefore, with $|p| \leq 1$,

$$\lim_{t \to \infty} e^{-at} \cos bt = (0) \cdot (p) = 0$$

Chapter 1

Basic Units and Variables

1-1 INTRODUCTION

In the entire field of electrical engineering, circuit analysis is one of the most common and widely used disciplines. From microelectric devices to communication satellites to cross–country electric power distribution lines—circuit analysis is a major tool for electrical engineers. Many of the methods and models of circuit analysis are used in related fields, such as biomedical engineering. Finally, circuit analysis is an essential step in the *design* of circuits. We must master analysis before doing design.

    What is circuit analysis? First, let us define an electric *circuit* (or *network*) as a specific connection of components. Circuit analysis asks and answers the question, "What happens in this particular circuit?" To be a little more precise, let the box N in Figure 1-1 represent a certain circuit. To this circuit we apply an *input* (or *excitation*) and we are asked to calculate the *output* (*response*). Or, if we'd like to think of it in yet another way, this is a "cause-and-effect" problem: Given the circuit N and the cause (input), what is the effect (output)?

    In our first discussion, we are going to use some helpful analogies from other fields in order to make some points clearer. As a simple analogy here, consider a mechanical system (or network) consisting of a mass, m. The input to this system is a force, f, and the output is the acceleration a of the mass, found from the relationship $f = m \cdot a$.

Prob. 1-1

Figure 1-1   The problem of circuit analysis.

One comment is appropriate here. In mentioning the "cause-effect" explanation and in the example of the mass, we assume that the spatial dimensions of the mass are negligible: The mass is a *point* mass, a *lumped* element, and the same effect (acceleration) happens instantly throughout the entire mass. By contrast, consider a long chain, attached at one end to the wall. If we hold its free end and excite it with a jerk, the response will be a wavy motion which will not be instantaneous through the entire chain; it will travel down the length of the chain. The chain is not a lumped element, because of its *distributed* mass along its length.

In our studies of circuit analysis here we will deal only with lumped elements.   Prob. 1-2

1-2 SI SYSTEM OF UNITS

As mentioned in the previous chapter, it is an accepted practice, in scientific and engineering disciplines, to use the international system of units. This system, known by its French name, Le Système International d'Unités (SI), includes six base units: length (meter), mass (kilogram), time (second), electric current (ampere), temperature (kelvin), and luminous intensity (candela). The SI system also provides names and prefixes for decimal multiples and submultiples, as well as other derived and supplementary units.

For convenience, Table 1-1 lists the common SI decimal prefixes. This table appears also on the inside cover of this book, so that you don't have to tear out this page when you'll need to refer to it frequently. Boldface entries are the most common ones in electrical engineering.

Table 1-2 gives a listing of SI base units and derived units which are useful in our studies of circuit analysis. Do not worry if you don't recognize some of those units: you are here to learn them!

1-3 ELECTRIC CHARGE AND CURRENT

Electric charge is a quantity of electricity, and many phenomena in circuit analysis are related to electric charges, either in motion or stationary. It is an observable fact that electric charges are of two kinds, called positive and negative. The unit of measurement for electric charges is the *coulomb* (C). To get an idea of this unit,

TABLE 1-1 SI PREFIXES

Factor	Prefix	Symbol	Factor	Prefix	Symbol
10^{18}	exa	E	10^{-1}	deci	d
10^{15}	peta	P	10^{-2}	centi	c
10^{12}	tera	T	**10^{-3}**	**milli**	**m**
10^{9}	giga	G	**10^{-6}**	**micro**	**μ**
10^{6}	**mega**	**M**	**10^{-9}**	**nano**	**n**
10^{3}	**kilo**	**k**	**10^{-12}**	**pico**	**p**
10^{2}	hecto	h	10^{-15}	femto	f
10^{1}	deka	da	10^{-18}	atto	a

TABLE 1-2 SI BASE UNITS AND DERIVED UNITS

	Name	Symbol
Base unit:		
Length	meter	m
Mass	kilogram	kg
Time	second	s
Electric current	ampere	A
Derived unit:		
Frequency	hertz	Hz
Force	newton	N
Energy, work	joule	J
Power	watt	W
Electric charge	coulomb	C
Voltage	volt	V
Capacitance	farad	F
Resistance	ohm	Ω
Conductance	siemens, mho	S, ℧
Magentic flux	weber	Wb
Inductance	henry	H

consider the fact that one electron has a negative charge of 1.602×10^{-19} coulombs. In other words, 6.28×10^{18} electrons constitute one (negative) coulomb. By way of a simple analogy to coulombs, think of a quantity of water, say $1 \, cm^3$. Be careful, though, not to get carried away too far with such analogies: we can hold $1 \, cm^3$ of water in a cup (or in our hand), but there is no way of actually seeing one coulomb or one electron. We can only measure and describe *observable effects* of electric charges.

The symbol for electric charge is $q(t)$, where the lowercase letter indicates, in general, a function of time. The parenthetical (t) will emphasize this time dependence.

EXAMPLE 1-1 —————————————————————————————

A car battery, when freshly charged, has a charge of 108 kC. It begins to discharge according to

$$q(t) = 108 \times 10^3 \left(1 - \frac{t}{36{,}000} \right)$$

Plot $q(t)$ vs. t and find the time at which the charge is one third of its original value.

Solution. The plot of $q(t)$ is shown in Figure 1-2 from which we conclude that the battery will discharge to one-third its charge in $24 \times 10^3 \, s = 6\frac{2}{3}$ hours.

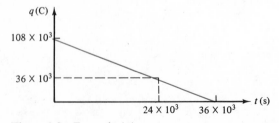

Figure 1-2 Example 1-1.

■ Prob. 1-3

We will be mostly interested in the flow of charges. For this purpose, we measure and define the *electric current* as

$$i(t) = \frac{dq(t)}{dt} \tag{1-1}$$

that is, the rate of change of charge with time. The current $i(t)$ is measured in *amperes* (A), often abbreviated as *amps*, and its units are coulombs/second (C/s).

When a suitable path is provided between the positive and the negative terminals of the battery in Example 1-1, an electric current will result. From physics, we can visualize this flow of charge as either negative charge flow towards the positive terminal, or positive charge flow towards the negative terminal. A positive charge flow in one direction is equivalent to a negative charge flow in the opposite direction. In electrical engineering, it is an accepted convention to measure current as the rate of flow of *positive* charges. (The convenient analogy here is the rate of flow of water, say in a pipe, in cm^3/s.)

EXAMPLE 1-2

In an electric conductor (analogous to a pipe of water), a hypothetical observer at point "X" counts a flow of 6 coulombs of positive charge per second from left to right and 4 coulombs of negative charge per second from right to left. See Figure 1-3. What is the current in the conductor?

Solution. The *net* flow of *positive* charges is 10 C/s from left to right, i.e., $i(t) = 10$ A, as shown. We can also say that the current is -10 A in the opposite direction.

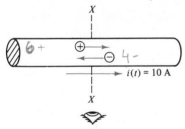

Eye of observer   **Figure 1-3**   Example 1-2.   ∎

This example brings to our attention two important aspects of the current $i(t)$. First, current is a "through" variable. We measure it at a point *in* the conductor, and since we have agreed to deal only with lumped elements ("point" elements without spatial variations), it is very convenient and brief to state that the current (in the particular element of the previous example) is 10 A.

The second important feature of $i(t)$ is the arrow showing its reference direction. Consider the following situation.

EXAMPLE 1-3

In a given element, as shown in Figure 1-4, the current is given by

$$i(t) = 10 \sin 100t$$

and is plotted versus time. It reaches its first maximum (10 A) at $t_1 = \pi/200 \approx 0.016$ s, and its first minimum (-10 A) at $t_2 = 3\pi/200 \approx 0.048$ s. Try now to answer this question: *Without* the reference arrow for $i(t)$, in which direction do 10 positive coulombs per second flow at $t = t_1$? And at $t = t_2$? We cannot answer without the reference arrow. We know that, at t_1, 10 A flow in the opposite direction to the 10 A at t_2—but without the arrow there is no way to tell. The arrow defines the positive direction.

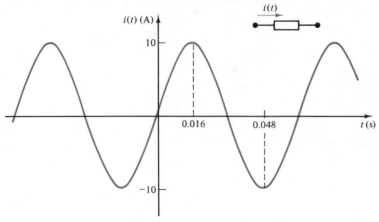

Figure 1-4   Example 1-3.   ■

The reference arrow for $i(t)$ is essential and *must be given*, or *is assumed*, with the following agreement: If the value of $i(t)$ at some time t_a is positive, $i(t_a) > 0$, then the flow of positive charges at that time is in the direction shown. If $i(t_a) < 0$, then, at $t = t_a$, positive charges flow in the opposite direction.

In summary, then, the current $i(t)$ is fully specified by its time function *and* by a reference direction.   Prob. 1-4

1-4 VOLTAGE OR POTENTIAL DIFFERENCES

Let us use a familiar analogy in order to introduce the idea of voltage (potential) difference. In our common gravitational field, let point B be at a height of 1000 meters above sea level, and point A at 1500 meters. See Figure 1-5. If we let a unit mass fall from A to B it will lose potential energy. On the other hand, if the unit mass is originally at B, then, in order to raise it to A, we must do work against the gravitational field. Such work is then stored as potential energy in m at point A.

In other words: in going from one point to another one in the gravitational field, energy is either *stored in m* or *recovered from m*. The two points are then at different

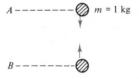

Figure 1-5   Analog to voltage difference.

drop = work = output
recovered

rise = work is required = input

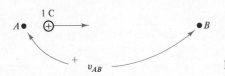

Figure 1-6   Voltage between A and B.

heights. In both cases, the amount of work depends on the *difference* of height between A and B (500 m here). The work *recovered from m* is equal in magnitude and opposite in sign to the work *stored in m*. If there is no height difference between points A and B, no work is done in going between these points (neglecting friction). Finally, keep in mind that the height difference between A and B exists regardless of whether the mass m is actually there, falling, or being raised.

We are ready now to translate these ideas into the electrical field. We replace the gravitational force field with an electric force field, and the mass m with an electric charge q. The *potential difference*, or *voltage*, between two points A and B in such a field is the amount of work (energy) per unit charge in moving it from A to B. More precisely,

$$v = \frac{dw}{dq}$$ (1-2)

where w is work (energy) in joules (J), q is charge in coulombs (C), and v, the voltage, is in volts (V). All quantities may be functions of time, as indicated by lowercase letters.

work

There is a voltage *drop* from point A to point B if work is recovered from the positive charge while moving from A to B. Conversely, there is a voltage *rise* from B to A if energy (work) is required to move the positive charge from B to A. The voltage rise is the negative of the voltage drop, and the energies are the negative of each other. Such a voltage can exist between the two points whether or not a charge is actually moved.

input ⟹ rise
output ⟹ drop

As in the case of current, the voltage is completely defined by its time function $v(t)$ and by a reference mark ($+$). This reference mark has the same purpose: if, at $t = t_1$, $v(t_1)$ is positive, then at $t = t_1$, there is a voltage drop from A to B, i.e., point A is at a higher potential than B. If, at $t = t_2$, $v(t_2) < 0$ then, at $t = t_2$, there is a voltage rise from A to B, i.e., point B is at a higher potential than A.

The voltage is *across* (or *between*) two points, whereas a current is *through* an element. It is sometimes convenient to use double subscripts as in Figure 1-6: by convention, v_{AB} means the voltage drop from A to B.

EXAMPLE 1-4 _____

An electron moves from point C to point D, and $v_{CD} = 6$ V. Determine the amount of energy, and whether spent or acquired by it.

Solution. From Equation 1-2 we have $(6)(-1.602 \times 10^{-19}) = -9.612 \times 10^{-19}$ J; in other words, the electron acquires 0.9612 attojoule (aJ). ∎

Probs. 1-5, 1-6, 1-7

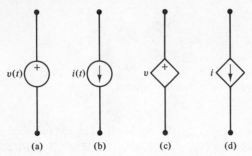

Figure 1-7   Voltage and current sources.

1-5 SOURCES

The input (excitation) applied to a circuit must come from a source of electric energy. We have two types of these sources:

A Voltage Source   A voltage source has a symbol shown in Figure 1-7(a). It is defined by the voltage $v(t)$ maintained always across the two terminals, together with its reference, no matter what is connected to those terminals.

EXAMPLE 1-5 _____

A car battery is a voltage source, where $v(t) = 12$ V. See Figure 1-8(a). This is a constant voltage, sometimes called a *dc* (direct current) *voltage*.   ∎

EXAMPLE 1-6 _____

The usual household voltage supply at the wall outlets in the U.S. is $v(t) = 170 \sin 377t$, as shown in Figure 1-8(b). We will discuss later the meanings of the numbers 170 and 377. At the moment, we note that this is a sinusoidal waveform, alternating between positive half cycles and negative half cycles. This voltage is called *ac* (alternating current). The terms "dc" and "ac" here have nothing to do with current; some terms simply become traditional, even if they are not quite accurate.

It is interesting to note the crossing points on the t axis of this sinusoidal voltage. Two such points are shown, at $\pi/377 \simeq \frac{1}{120}$ s and at $\frac{1}{60}$ s. Since it takes $\frac{1}{60}$ s for one complete cycle of the sinusoidal voltage, there are 60 cycles per second. Each second our household voltage source goes through 60 such cycles.

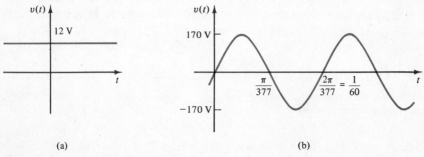

Figure 1-8   dc and ac voltages.   ∎   Prob. 1-8

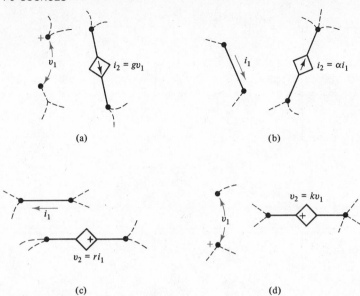

Figure 1-9 Dependent sources.

A Current Source. This source has a symbol shown in Figure 1-7(b), and is always defined by the specified waveform $i(t)$ and its reference direction, no matter what is connected to it. An electric welding machine and certain transistor circuits (some of which will be studied later) are examples of current sources. The terms dc and ac apply also to current sources.

Another distinction is important among sources. If a source (voltage or current) depends only on time, such a source is *independent*. For example, the two voltage sources in Figure 1-8 are independent. In some cases the waveform of a source depends on another current or voltage in the circuit. This is a *dependent* (*controlled*) source. A fairly common symbol for dependent sources is the diamond shape shown in Figure 1-7(c) and (d). Whether a circle or a diamond, there is no room for mistakes: The given expressions for v or i will tell us at once if the sources are independent or dependent.

In Figure 1-9 we see the four types of dependent sources: (a) a voltage-controlled current source, (b) a current-controlled current source, (c) a current-controlled voltage source and (d) a voltage-controlled voltage source. The constants g, α, r, and k are appropriate multipliers that will be discussed later.

Finally, let us make the observation that a voltage source (independent or dependent) has, by definition, a specified voltage across its two terminals regardless of the current through it. The current through it will be whatever is imposed by the circuit connected to that voltage source. In Figure 1-10(a), we show a given $v_s(t)$ unconnected to any circuit. It is obvious that $i = 0$ here. After we close the switch we will expect $i \neq 0$, and finding $i(t)$ will be part of our problem in circuit analysis.

By definition, a current source maintains a specified $i(t)$ through it regardless of the voltage across it. An unconnected source $i_s(t)$ is shown in Figure 1-10(c), and the

Prob. 1-9

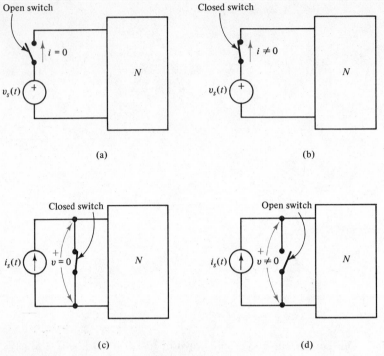

Figure 1-10   Unconnected and connected sources.

voltage across $i_s(t)$ is zero. When $i_s(t)$ excites a circuit, Figure 1-10(d), the voltage across $i_s(t)$ will be determined by the specific circuit N.

Two additional terms are introduced here and are illustrated in Figure 1-10:

1. An *open circuit* (*o.c.*) exists between two points if no conducting path exists between the points. In an open circuit the current is zero, $i = 0$. An open switch causes an open circuit in Figure 1-10.

2. A *short circuit* (*s.c.*) exists between two points if a perfect conducting path exists between the points. In a short circuit the voltage between them is zero, $v = 0$. These two points are electrically one point. A closed switch causes a short circuit in Figure 1-10.

1-6 POWER

If we take the product of voltage and current, Equations 1-1 and 1-2, we obtain the rate of change with time of work

$$v \cdot i = \frac{dw}{dq} \cdot \frac{dq}{dt} = \frac{dw}{dt} \qquad (1\text{-}3)$$

This rate of change is *power*, measured in *watts* (W), i.e., joules per second. We use the letter p for power which may vary with time

$$p(t) = v(t) \cdot i(t) \qquad (1\text{-}4)$$

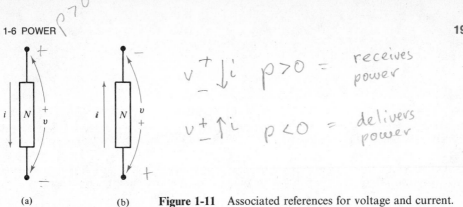

$v \overset{+}{\underset{-}{\Big|}} \downarrow i \quad p \gt 0 = $ receives power

$v \overset{+}{\underset{-}{\Big|}} \uparrow i \quad p \lt 0 = $ delivers power

(a)   (b)   **Figure 1-11** Associated references for voltage and current.

Let a two-terminal circuit N have a voltage $v(t)$ across it and a current $i(t)$ through it. The reference directions for v and i are arbitrary, and therefore we have four possible combinations. It is customary to adopt the following convention: if $p(t)$ is positive

$$p(t) = v(t)i(t) > 0 \qquad (1\text{-}5)$$

then the circuit N *receives* power as a consumer, or a *load*. If $p(t)$ is negative

$$p(t) = v(t)i(t) < 0 \qquad (1\text{-}6)$$

the circuit N *delivers* power as a source, or a supply. The references of $v(t)$ and $i(t)$ associated with this convention are shown in Figure 1-11: The ($+$) sign of $v(t)$ is at the *tail* of the arrow of $i(t)$. These will be called the *associated references* for v and i. These associated references define power received or power delivered according to Equations 1-5 and 1-6.

EXAMPLE 1-7

(a) A 12-V car battery (a dc voltage source) is connected to a network N_a, as shown in Figure 1-12(a). With the associated references given, the calculation shows that $i = 6$ A. What is the power in N_a, and is N_a delivering or receiving it?

Solution. With the reference as shown, we have

$$p = (12)(6) = 72 \text{ W}$$

and N_a is receiving 72 W (from the battery).

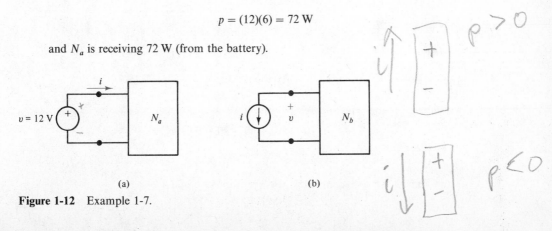

(a)   (b)

Figure 1-12 Example 1-7.

(b) In Figure 1-12(b), we are given, for $t > 0$,

$$i(t) = 10e^{-t}$$

and

$$v(t) = 2 \sin 4t$$

Calculate the power at $t_1 = \pi/8$ s and at $t_2 = 2\pi/5$ s, and determine if delivered or received by N_b.

Solution. Here we have the *opposite* of the associated references for N_b. Therefore

$$p(t) = -vi = -20e^{-t} \sin 4t, \qquad t > 0$$

and at $t = t_1$

$$p(t_1) = -20e^{-\pi/8} \sin \frac{\pi}{2} = -13.50 \text{ W}$$

so 13.50 W is *delivered* by N_b. At $t = t_2$

$$p(t_2) = -20e^{-2\pi/5} \sin \frac{8\pi}{5} = 5.41 \text{ W}$$

received by N_b.

Probs. 1-10,
1-11, 1-12,
1-13, 1-14,
1-15

The principle of conservation of energy implies that, in an entire network, the algebraic sum of all the powers in all the elements is zero,

$$\sum_k p_k = 0 \qquad\qquad (1\text{-}7)$$

as illustrated in the following example.

EXAMPLE 1-8 _____

In the network shown in Figure 1-13, the various elements (components) are symbolized by boxes. It is known that $p_1 = -100$ W, $p_2 = 30$ W, $p_3 = 40$ W, $p_4 = -20$ W, $p_5 = 60$ W. Calculate the power p_6 and determine if element 6 delivers or receives power.

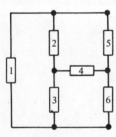

Figure 1-13   Example 1-8.

Solution. Using Equation 1-7, we write

$$-100 + 30 + 40 - 20 + 60 + p_6 = 0$$

and therefore

$$p_6 = -10 \text{ W}$$

and element 6 is delivering 10 W. (So are elements 1 and 4.)

Probs. 1-16,
1-17

1-7 IDEAL ELEMENTS AND MODELS

In studying elementary motion, we use Newton's law $f = ma$ (force = mass $\times$ acceleration). The mass is assumed to be lumped at one point; an analogous assumption is made for circuit elements. Such an assumption is valid under certain conditions, as we saw previously.

Another assumption made for the mass, in a particular problem, may be that it moves without friction. Again, such an assumption may be valid under certain conditions, and can be verified by actual observations (measurements or tests). If these tests confirm the predicted acceleration within an accepted accuracy, we say that our model of the mass is good. If not, we must modify the ideal model of the mass by introducing friction, for example.

The situation is similar with electrical elements. We assume ideal elements, calculate the response, and compare it with predicted or measured values. If the comparison is reasonable (within allowed engineering tolerances), the model of our circuit is good. Otherwise, we have to change the model and use nonideal elements.

For example, we use an ideal model for a dc source of 6 V; it maintains a constant voltage of 6 V across its terminals. In real life (e.g., a 6-V battery) such an ideal element does not exist—but it is useful for many calculations and applications. Another example of an ideal element is the switch, shown in Figure 1-10; ideally, it opens or closes *instantly* (in no time at all). Real life switches can be very fast, closing or opening in nanoseconds, but taking a *finite* amount of time. The use of ideal switches is justified by the simplified calculations which give correct results.

In our studies of circuit analysis, and, in fact, in the entire practice of electrical engineering, ideal elements are used, and modified as necessary.

PROBLEMS

1-1 Draw a block diagram, as in Figure 1-1 of the text, to describe as specifically as you can:
  (a) The headlights of a car
  (b) An electric toaster
  (c) A hi-fi record player
  In each case identify the input and the output.

1-2 The assumption of lumped elements is valid when the physical dimension of the element d is much smaller than the *wavelength* of the input signal λ. The wavelength is given by

$$\lambda = \frac{c}{f}$$

where c is the velocity of electrical signals, approximately equal to the velocity of light, $c = 3 \times 10^8$ meters per second. The *frequency* f is the rate of repetition of the signal, measured in *hertz* (Hz) or *cycles per second* (cps).
  (a) Check the validity of assuming lumped elements for audio signals, where the frequencies range from 20 Hz to 20 kHz.
  (b) Repeat for household electricity where $f = 60$ Hz and where some transmission lines are 5000 km long.
  (c) Repeat for computer circuits, where a typical signal has $f = 300$ MHz.

(d) Repeat for microwave signals, where, typically, $f = 10\,\text{GHz}$, and components may have a dimension of a few centimeters.

1-3 In the atmosphere above the ocean, there are ions (charge carriers) with an average density of $n_+ = 860/\text{cm}^3$ and $n_- = 700/\text{cm}^3$. Each ion has a charge ($+$ or $-$) of $1.602 \times 10^{-19}\,\text{C}$. Find the average charge density in C/m^3.

1-4 In a certain electric conductor, shown in the accompanying figure, the positive charge measured at $A-A'$ varies with time as

$$q(t) = 10e^{-2t}\,\text{C}, \qquad t > 0$$

Calculate and plot the current $i(t)$.

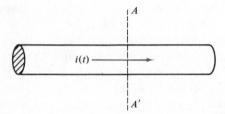

Problem 1-4

1-5 In an electric field, 3×10^6 electrons are moving from point X to point Y, and the work done on them is 1.8 nJ. Find the voltage between X and Y, with its reference polarity. (This situation is typical of electronic devices using beams of electrons, as, for example, in a TV picture tube.)

1-6 A common notation for voltage uses *double subscript*: v_{AB} is the voltage drop from A to B, with A having the ($+$) reference mark. Show that

$$v_{BA} = -v_{AB}$$

1-7 What is v_{YX} if 5 coulombs (positive charge) do work of 10 J in moving from Y to X?

1-8 Plot accurately vs. t the waveform of

$$v(t) = 170e^{-t}\sin 377t$$

and compare it with the household ac voltage shown in Figure 1-8 (maxima, points of intersection with the t axis).

1-9 Derive fully the units of the constant multipliers g, α, r, and k in the dependent sources (Figure 1-9). These units should be in terms of volts (V) and amperes (A).

1-10 In physics we often use the unit *electron-volt* (eV) for energy received or delivered by 1 electron as it moves through 1 volt of potential difference. How many eV are there in 1 J?

1-11 An electric heater of 2 kW is connected to a 110-V dc source. What current does it draw?

1-12 A dc motor delivers 5 hp at 90 percent efficiency and draws 18.84 A when connected to a dc voltage source. What is the voltage?

1-13 Household electric energy costs approximately 7 cents/kWh (kWh = kilowatthour). Calculate the cost of running a TV set, rated at 75 W, for 7 hours a day, for one week. Compare with a nightlight (5 W) burning continuously for 1 month.

1-14 A certain electrical element N is shown with its associated voltage and current references in Figure 1-11(a). It is known that

$$v(t) = 10 \sin 100t, \qquad t > 0$$
$$i(t) = 2 \cos 100t, \qquad t > 0$$

Calculate and plot vs. t the power $p(t)$. For the various ranges of time, state whether N receives or delivers power.

1-15 Repeat Problem 1-14 with $v(t) = 10 \sin 100t$ and $i(t) = 2 \sin (100t - \pi/3)$.

1-16 Since power is the rate of change of energy,

$$p(t) = \frac{dw}{dt}$$

we can calculate the total energy between two times t_1 and t_2 as

$$w \Big]_{t_1}^{t_2} = \int_{t_1}^{t_2} p(t)\, dt$$

Calculate the energy in the element in Problem 1-14 for:
(a) $t_1 = 0$ and $t_2 = \pi/400$ s
(b) $t_1 = 0$ and $t_2 = \pi/200$ s
(c) $t_1 = 0$ and $t_2 = \pi/100$ s

1-17 Repeat Problem 1-16 for the data in Problem 1-15.

Chapter 2

Resistance, Kirchhoff's Laws.
Node and Loop Analysis

2-1 THE RESISTOR. OHM'S LAW

The next element that we introduce is the resistor. The symbol for a resistor is shown
in Figure 2-1(a). Some common examples of resistors are the heating element in an
electric toaster and the filament in an incandescent light bulb, as well, of course, as the
numerous resistors in electronic devices.

As a result of laboratory experiments, Georg Simon Ohm formulated (1827) the
law relating voltage and current in a resistor. With the references for v and i as shown
in Figure 2-1(a), *Ohm's law* is

$$v(t) = Ri(t) \qquad (2\text{-}1)$$

where R is a positive constant, $R > 0$; it is called the *resistance* of the resistor and is
measured in ohms.† The capital Greek letter Ω is used to denote ohms. The units in

(a) (b)

Figure 2-1 (a) Resistor. (b) Its *v-i* characteristic.

† Certain devices have the characteristics of a negative resistance, $R < 0$. We'll have a chance to
study some of these later.

24

Equation 2-1 are $[V] = [\Omega][A]$. Ohm's law, then, states that voltage and current are proportional in a resistor, as shown by the v-i plot in Figure 2-1(b).

This v-i characteristic curve is a straight line passing through the origin. A resistor with such a characteristic is called *linear*.

EXAMPLE 2-1

A 6-V voltage source is connected at $t = 0$ to a resistor $R = 100\ \Omega$.
(a) Determine and plot versus time the voltage $v(t)$ across R.
(b) Determine and plot versus time the current $i(t)$ through R.

Solution. The voltage $v(t)$ is given by

$$v(t) = 0, \qquad t < 0$$
$$v(t) = 6, \qquad t > 0$$

and is plotted in Figure 2-2(b). The current is determined by Ohm's law, Equation 2-1, as

$$i(t) = 0, \qquad t < 0$$
$$i(t) = \tfrac{6}{100} = 0.06\ \text{A} = 60\ \text{mA}, \qquad t > 0$$

and is plotted in Figure 2-2(c).

[handwritten: $V = iR$ $i = \frac{V}{R}$]

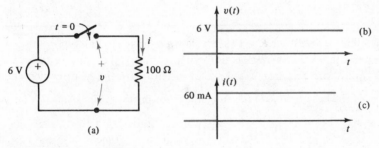

Figure 2-2 Example 2-1.

Probs. 2-1, 2-2, 2-3, 2-4

The model of a resistor (Figure 2-1) is ideal. It is also a lumped model (no spatial variations of v or i). The justification for using such a model is the fact that most "real life" resistors behave very closely to this model.

A good analogy to a resistor is the friction (a dashpot) in a mechanical system. There, force is proportional to velocity, in a relation similar to Equation 2-1.

We assumed that R is positive, $R > 0$. Again, this is a physical reality for many resistors. Another assumption is that R is constant with time. In some cases, there are *time-varying* resistors, denoted by $R(t)$, but in this book we will not discuss them.

The resistor is also *bilateral*, meaning that it obeys the same relation (Equation 2-1) if it is turned end for end. Certain other elements are not bilateral; their v-i relations are not the same if the element is turned. We will study some of them later.

In many cases, we will need to express $i(t)$ in terms of $v(t)$. For the resistor, we obtain from Equation 2-1

$$i(t) = \frac{1}{R} v(t) = Gv(t) \tag{2-2}$$

[handwritten bottom left:]
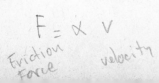
$F = \alpha\, V$
Friction force   velocity

where G is the _conductance_ of the resistor, measured in _siemens_ (S) or, more commonly, in _mhos_ (℧). The word "mho" is the reverse spelling of "ohm" and its symbol, ℧, is an inverted Ω. Who said that electrical engineers don't have a sense of humor?

Probs. 2-5, 2-6

2-2 POWER AND ENERGY. PASSIVITY

Let us calculate the power in a given resistor. With the associated reference directions as shown in Figure 2-1, we write

$$p(t) = v(t)i(t) = [Ri(t)]i(t) = R[i(t)]^2 \qquad (2\text{-}3)$$

or, using Equation 2-2,

$$p(t) = v(t)i(t) = v(t)[Gv(t)] = G[v(t)]^2 \qquad (2\text{-}4)$$

Since R (or G) is positive, we see that, at every instant, the power in the resistor is _positive_. Consequently, a resistor always _receives_ power, never delivers. For this reason, the resistor is called a _passive_ element (as contrasted with _active_). The power in the resistor is dissipated in the form of heat and light.

EXAMPLE 2-2 _____

The current $i(t)$ through a resistor $R = 10 \text{ k}\Omega$ is shown in Figure 2-3(a).
(a) Calculate and plot the instantaneous power $p(t)$.
(b) Calculate the total energy between $0 \le t \le 5$ s.

Solution. (a) The current $i(t)$ is given by the following expressions, for each interval of time:

$$i(t) = \begin{cases} 0 & \text{mA} & t < 0 \\ 10t & \text{mA} & 0 \le t \le 1 \\ 10 & \text{mA} & 1 \le t \le 3 \\ (40 - 10t) & \text{mA} & 3 \le t \le 5 \\ 0 & \text{mA} & t > 5 \end{cases}$$

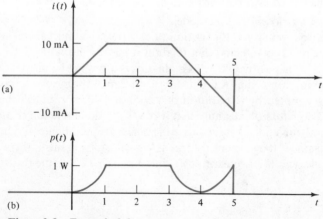

Figure 2-3 Example 2-2.

Be sure that you agree with these expressions! (Derive them if necessary—they are equations of straight lines.)

Consequently, $p(t) = Ri^2(t)$ yields

$$p(t) = \begin{cases} 0 & \text{W} & t < 0 \\ t^2 & \text{W} & 0 \leq t \leq 1 \\ 1 & \text{W} & 1 \leq t \leq 3 \\ (16 - 8t + t^2) & \text{W} & 3 \leq t \leq 5 \\ 0 & \text{W} & t > 5 \end{cases}$$

and its plot is shown in Figure 2-3(b).

(b) The total energy over the interval $0 \leq t \leq 5$ s is obtained as the integral of $p(t)$. Specifically, here we have three integrals, since $p(t)$ has three different expressions over this interval

$$w \Big]_{t=0}^{t=5} = \int_0^1 t^2\, dt + \int_1^3 dt + \int_3^5 (16 - 8t + t^2)\, dt = 3J \qquad \blacksquare$$

EXAMPLE 2-3

The voltage across a given resistor R is

$$v(t) = 170 \sin 377t \qquad \text{V}$$

(This is the household voltage in the U.S., mentioned in Chapter 1.)

(a) Calculate the total energy over one cycle of the voltage.

(b) Calculate the average power over one cycle.

Solution. (a) As shown in Chapter 1, a full cycle of $v(t)$ occurs between

$$377t_1 = 0 \qquad \therefore \quad t_1 = 0 \text{ s}$$

and

$$377t_2 = 2\pi \qquad \therefore \quad t_2 = \tfrac{1}{60}\, s$$

see Figure 2-4. Using Equation 2-4, we get

$$p(t) = G(170^2 \sin^2 377t) \qquad \text{W}$$

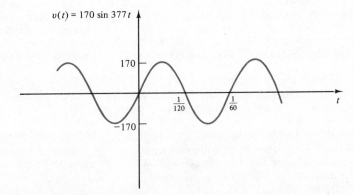

Figure 2-4 Example 2-3.

and

$$w\bigg]_{t_1}^{t_2} = \int_0^{1/60} 170^2 G \sin^2 377t \, dt = 170^2 G \int_0^{1/60} \sin^2 377t \, dt$$

$$= \frac{170^2}{(2)(60)} G \quad J$$

where we have used the integral

$$\int \sin^2 ax \, dx = \frac{x}{2} - \frac{1}{4a} \sin 2ax$$

(Go ahead, check it out in your favorite integral tables!)
(b) The average power is

$$P_{av} = \frac{1}{t_2 - t_1} w = 60 \frac{170^2 G}{(2)(60)} = \frac{170^2}{2} G \quad W$$

As a final question, consider here the following one: If we had a *constant* voltage, say $v(t) = V$, instead of the sinusoidal one, what value of V would produce the same average power? The answer is very simple: since the instantaneous power in this case is $p(t) = GV^2$, a constant, it is also the average power P_{av}. Therefore

$$GV^2 = \frac{170^2}{2} G$$

and

$$V = \frac{170}{\sqrt{2}} = 120 \quad V$$

This is called the *effective value* or the *RMS* (root-mean-square) value, and it is the one listed on the outlet ("120 volts").   ■   Prob. 2-7

We are ready now to consider our first problem in network analysis. Let us do it with an example.

EXAMPLE 2-4 _____

In Figure 2-5 we show the circuit model of a light bulb connected to a voltage outlet. In this circuit we recognize points where two (or more) elements are connected together. Such a point is called a *node*. The voltage source $v_s(t)$ is between nodes 1 and 5. To these nodes we connect the plug of the cord. The cord can be modeled as a resistor R_c. The on-off switch is between nodes 2 and 3. The bulb R_b is between nodes 3 and 4. There is no resistance ($R = 0$) between nodes 4 and 5; it is a short circuit.

The general problem of network analysis is: *Calculate the voltage and the current in every element of the given network.* Since this is our first "big" example, let us proceed slowly and methodically. As we learn more, we will be able to take some shortcuts.

Solution. We count the number of elements—the voltage source, R_c, the switch, R_b, and the short circuit 4—5. Altogether we have five elements. Our final solution will have to show, therefore, five voltages and five currents—altogether 10 unknowns. We need, therefore, 10 equations to solve for these 10 unknowns.

The next step is to show these 10 variables on the circuit diagram. Each element has a voltage across and a current through it. Each voltage and current must have a name

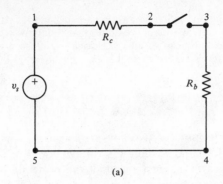

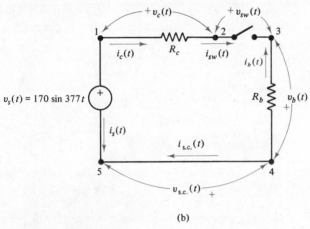

(b)

Figure 2-5 Example 2-4.

and a reference. The reference is either given, or we choose it arbitrarily. Let us discuss each element.

The voltage source: The voltage $v_s(t)$ is given by the function (170 sin 377t) and by the reference shown. It is, then, known. Its current, however, is unknown. Let us assign to it the variable $i_s(t)$ and the arbitrary reference shown.

The resistor R_c: The voltage and current are unknown. Let us assign the names $v_c(t)$ and $i_c(t)$, and the associated references as shown: The (+) of the voltage is at the tail of the arrow of the current.

The switch: The voltage v_{sw}, after closing, is zero (short circuit). The current, after closing, is shown with its arbitrarily chosen reference.

The resistor R_b: $v_b(t)$ and $i_b(t)$ are unknown. We choose (arbitrarily) the associated references as shown.

The short circuit 4—5: The voltage across any short circuit is zero by definition, $v_{s.c.}(t) = 0$. The current $i_{s.c.}(t)$ is unknown and is shown with a chosen reference.

Our problem is, then, reduced to seven unknowns: $i_s(t)$, $v_c(t)$, $i_c(t)$, $i_{sw}(t)$, $v_b(t)$, $i_b(t)$, and $i_{s.c.}(t)$. Let us see how many equations we can write. Ohm's law yields two:

$$v_c(t) = R_c i_c(t)$$

and

$$v_b(t) = R_b i_b(t)$$

Note that these two equations define the two resistors R_b and R_c, regardless of how they are connected in the circuit. We need five additional equations which will have to be related to the specific connections of all the elements in this particular circuit. We'll discuss these next, and return to complete this example later. ∎

2-3 KIRCHHOFF'S CURRENT LAW

In 1848, Gustav Kirchhoff formulated two laws related to circuits. These are called Kirchhoff's current law (KCL) and Kirchhoff's voltage law (KVL).

Kirchhoff's current law is a direct result of the principle of conservation of electric charge. Consider any node in a circuit, as shown in Figure 2-6. Reference directions for the currents are also shown. Since charge cannot accumulate or be destroyed at this node, the total rate of charge flow *into* the node must equal the total rate of charge flow *out* of the node. In other words,

$$i_1(t) + i_4(t) + i_5(t) = i_2(t) + i_3(t)$$

or

$$i_1(t) - i_2(t) - i_3(t) + i_4(t) + i_5(t) = 0$$

that is,

$$\sum i(t) = 0 \qquad \text{at a node} \tag{2-5}$$

This is Kirchhoff's current law (KCL): at every node, the instantaneous algebraic sum of all the currents is zero. By *algebraic* we mean that a ($+$) sign is assigned to currents entering the node and a ($-$) sign is assigned to currents leaving the node.

A convenient analogy here is a pipe system with water flowing in the pipes. The rate of flow of water, in cm^3/s, is analogous to $i(t)$.

Three important features are inherent to KCL, Equation 2-5:

1. It holds at every instant of time.
2. It holds regardless of the specific nature of the elements (resistors, sources, etc.) connected to the node.
3. Since multiplication by (-1) keeps Equation 2-5 valid, we don't have to memorize the convention of signs for it. We could have chosen ($+$) for currents leaving and ($-$) for currents entering. The important thing is to state the convention *clearly* and to stick with it.

EXAMPLE 2-5

In Figure 2-6, at $t = t_1$, we are given that:

$$i_1 = 2\,A, \qquad i_2 = -3\,A, \qquad i_3 = 1\,A, \qquad i_4 = -6\,A$$

(a) What is the *actual* direction of the currents (flow of positive charges) at $t = t_1$?
(b) What is $i_5(t_1)$?

Solution. (a) Since $i_1(t_1) > 0$ and $i_3(t_1) > 0$, these two currents are actually flowing at t_1 in the direction of their reference arrows. On the other hand, $i_2(t_1) < 0$, and its actual

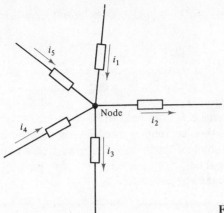

Figure 2-6   Kirchhoff's current law (KCL).

flow is opposite to its reference arrow. Similarly $i_4(t_1) < 0$ and flows opposite to its reference at t_1.

(b) Apply KCL at $t = t_1$, with the *given* references. We have

$$+2 - (-3) - 1 + (-6) + i_5(t_1) = 0$$

or

$$i_5(t_1) = 2 \text{ A}$$

It is important to recognize, and stick to, the signs ($+$ or $-$) when writing KCL. Currents entering a node, as shown by their given references, are assigned a ($+$) sign; currents referenced as leaving the node get a ($-$) sign. The value of each current, as in our example, may be positive or negative on its own; in the term $+(-6)$, the $+$ sign indicates that the current is entering the node, and the -6 indicates that the value of this current is -6 A; that is, the current is actually leaving the node and has a value of 6 A. ■

An extended concept of KCL at a node is KCL for a *cut set*. Consider a network N which can be separated (cut) into two parts N_1 and N_2, as shown in Figure 2-7. The two parts are connected by elements a, b, c, d whose currents are designated $i_a(t)$, $i_b(t)$, $i_c(t)$, $i_d(t)$. Imagine a plane P-P' perpendicular to the elements, "cutting" these elements. Since charge must be conserved on this imaginary plane, the total flow of

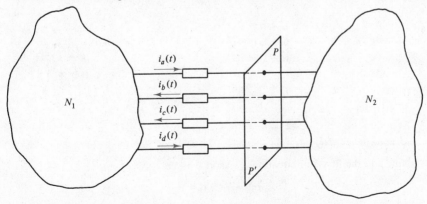

Figure 2-7   KCL for a cut set.

charge into the plane (from left to right) must equal that flowing out (from right to left). Here we have, then,†

$$i_a + i_d - i_b - i_c = 0 \qquad (2\text{-}6)$$

with the same convention of signs: $(+)$ for currents going into the plane one way, $(-)$ for those going the other way. Again, no memorization is needed: *We* choose a positive direction for a given problem.

This result is most amazing: Elements a, b, c, d need not be connected to a single common node—yet the algebraic sum of their currents is zero! The only restriction on these elements is: They must form a set which, if "cut" (by a closed surface, a plane, or with a pair of pliers), will cause the original network to separate into two parts N_1 and N_2. Such a set of elements is called (what else?) a *cut set*, and KCL reads

$$\sum i(t) = 0 \qquad \text{at a cut set} \qquad (2\text{-}7) \qquad \text{Prob. 2-8}$$

EXAMPLE 2-6 _____

In order to build up our confidence in cut sets, consider the network shown in Figure 2-8, where all the nodes are labeled by numbers $1, \dots, 8$. The cut set currents are, as before, $i_a, i_b, i_c,$ and i_d. The other currents are as shown $i_e, \dots, i_k$ with arbitrarily chosen references. Using $(+)$ for currents leaving and $(-)$ for currents entering, write KCL at every node.

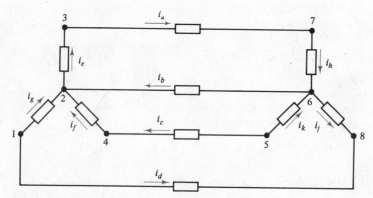

Figure 2-8   Example 2-6.

Solution.

At node 1: $\qquad\qquad i_g + i_d = 0$

At node 2: $\quad -i_g - i_f - i_b + i_e = 0$

At node 3: $\qquad\qquad -i_e + i_a = 0$

At node 4: $\qquad\qquad i_f - i_c = 0$

At node 5: $\qquad\qquad i_c + i_k = 0$

At node 6: $\quad i_b + i_j - i_h - i_k = 0$

At node 7: $\qquad\qquad -i_a + i_h = 0$

At node 8: $\qquad\qquad -i_j - i_d = 0$

Now add the first four equations (a valid mathematical operation!) to get

$$i_a - i_b - i_c + i_d = 0$$

† Don't forget: lowercase letters are functions of time, $i = i(t)$.

which is precisely KCL for the cut set *a-b-c-d*. This addition forms the part N_1 of the entire network. Why did i_e, i_f and i_g disappear in this addition? Because each one appears twice, entering a node and (obviously) leaving a node *inside* N_1.

If we add the last four equations, we'll get

$$-i_a + i_b + i_c - i_d = 0$$

the same equation (within a multiplier of -1) for the same cut set. This addition forms the part N_2 of the total network.

Finally, let us observe that a node is just a special case of a cut set. This cut set consists of all the elements connected to that node, and one part (N_1 or N_2) of the network is just that node. See Figure 2-9.

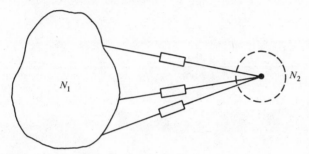

Figure 2-9   A special case of a cut set.

2-4 KIRCHHOFF'S VOLTAGE LAW

Kirchhoff's voltage law is a statement of the principle of conservation of energy. Let us remember that a voltage between two points A and B is the work (energy) per unit of positive charge, acquired by the charge or expended by the charge in moving from A to B. If work is stored by the charge (the charge acquires energy) then B is positive with respect to A; there is a voltage rise from A to B or, alternatively, a voltage drop from B to A. Let us remember and review here the useful analogy of a unit of (positive) mass moving between two points of different heights: If point B is higher than A (B is positive with respect to A), the mass will acquire energy as it moves from A to B, and we say that there is a rise from A to B, or a drop from B to A.

Consider now a unit of positive charge ($+1$ coulomb) moving from A to B, then from B to C, from C to D, and from D back to A. See Figure 2-10. This trip starts at A and ends at A. Let us also assume that there is a voltage *drop* from A to B, called $v_1(t)$ with its shown reference; a voltage *rise* from B to C, called $v_2(t)$; a voltage *rise* from C to D, called $v_3(t)$; and a voltage *drop* from D to A, called $v_4(t)$. The total energy acquired by the charge is therefore $v_1(t) + v_4(t)$. The total energy spent by the charge is $v_2(t) + v_3(t)$. Since the charge ended up where it started, we must have

$$v_1(t) + v_4(t) = v_2(t) + v_3(t) \tag{2-8}$$

or

$$v_1(t) - v_2(t) - v_3(t) + v_4(t) = 0 \tag{2-9}$$

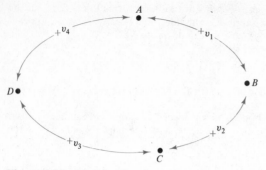

Figure 2-10 Kirchhoff's voltage law (KVL).

More compactly: Around a closed path, called a *loop*, the algebraic sum of all the voltages is zero,

$$\sum v(t) = 0 \qquad \text{around a loop} \tag{2-10}$$

where a voltage drop is given a $(+)$ sign and a voltage rise a $(-)$ sign. This is *Kirchhoff's voltage law* (KVL).

To continue the analogy used earlier: In going around a closed path, the algebraic sum of energy differences is zero; energy drops are counted as $(+)$, energy rises as $(-)$.

Three important features relate to KVL, Equation 2-10:

1. It holds at every instant of time.
2. It holds regardless of the specific nature of the elements around the loop. In fact, there may be no element between, say, C and D; it can be an open circuit.
3. Since multiplication by -1 is valid in Equation 2-10, we are free to choose $(+)$ for a voltage drop and $(-)$ for a voltage rise, or vice versa. The important thing is to state our choice in advance and stay with it.

Finally, let us turn our attention to the striking resemblance between KCL and KVL, repeated here

$$\sum i(t) = 0 \tag{2-5}$$
$$\sum v(t) = 0 \tag{2-10}$$

What is true (mathematically) in one equation for i holds true in the other for v. Such a resemblance runs throughout circuit analysis (and design) and is called *duality*. We say that KVL is dual to KCL, and a loop is dual to a cut set. We will have many opportunities in our future studies to explore duality and to benefit from it.

EXAMPLE 2-4 (concluded) —————————————————————————————————

Recall (turn back a couple of pages, please) that we had the following equations, defining the various elements:

Voltage source: $\qquad\qquad\qquad\qquad v_s(t) = 170 \sin 377t, \qquad t > 0$

Switch: $\qquad\qquad\qquad\qquad\qquad v_{sw}(t) = 0, \qquad\qquad\qquad t > 0,$

Wire 4–5: $\qquad\qquad\qquad\qquad\qquad v_{s.c.}(t) = 0, \qquad\qquad\qquad t > 0,$

Resistor R_c: $\qquad\qquad\qquad\qquad\quad v_c(t) = R_c i_c(t)$

Resistor R_b: $\qquad\qquad\qquad\qquad\quad v_b(t) = R_b i_b(t)$

We need now the equations that describe the connections of these elements, i.e., KCL and KVL. First, KCL with $+$ for currents leaving, $-$ for entering:

At node 1: $\qquad\qquad\qquad\qquad\qquad\qquad i_c + i_s = 0$

At node 2: $\qquad\qquad\qquad\qquad\qquad\qquad -i_c + i_{sw} = 0$

At node 3: $\qquad\qquad\qquad\qquad\qquad\qquad -i_{sw} - i_b = 0$

At node 4: $\qquad\qquad\qquad\qquad\qquad\qquad i_b + i_{s.c.} = 0$

At node 5: $\qquad\qquad\qquad\qquad\qquad\qquad -i_s - i_{s.c.} = 0$

From these equations we see that

$$i_c(t) = i_{sw}(t) = -i_b(t) = i_{s.c.}(t) = -i_s(t)$$

and a single unknown, $i_c(t)$, will do.

Next, apply KVL around the single loop defined by $1 \to 2 \to 3 \to 4 \to 5 \to 1$; it reads, with $(-)$ for a voltage rise, $(+)$ for a voltage drop:

$$v_c(t) + v_{sw}(t) - v_b(t) + v_{s.c.}(t) - v_s(t) = 0$$

In this KVL equation, substitute the proper expressions for each voltage

$$R_c i_c + 0 - R_b(-i_c) + 0 - 170 \sin 377t = 0$$

i.e.,

$$(R_c + R_b)i_c(t) = 170 \sin 377t$$

$$\therefore \quad i_c(t) = \frac{1}{R_c + R_b} \cdot 170 \sin 377t$$

With $i_c(t)$ known, every current in every element is known (the equations of KCL) and the two resistive voltages are known (Ohm's law).

In hindsight, we could have solved this circuit faster—but that's what hindsight is. We took carefully each step, namely:

1. Assign current and voltage variables, as needed.
2. Apply KCL and KVL.
3. Apply the equations which define each element.

As we progress, we will always take these steps, with some accelerated pace. ■ Prob. 2-9

2-5 NODE VOLTAGES AND NODE ANALYSIS

From the previous examples, it appears that we need a systematic method to solve networks, i.e., to find all the voltages, currents (and, therefore, power and energy) in all the elements. Such a method should be not only systematic but also efficient, and with a minimum number of unknowns and equations to solve.

Indeed, we will learn three such methods of solution, two of them in this chapter and the third one later on. Each method has its own advantages, and we will be able to make an intelligent choice among them for the particular problem at hand. Let us start with the method called *node analysis*, and consider a specific (though quite general) example.

EXAMPLE 2-7

Solve the network shown in Figure 2-11. For simplicity, we omitted the specific switches which accompany, as usual, the two sources, a dc (constant) current source, and a dc voltage source. Do not worry about the "unrealistic" values of the resistors: they are conveniently *scaled* values, pretty much like scaled distances on a map.†

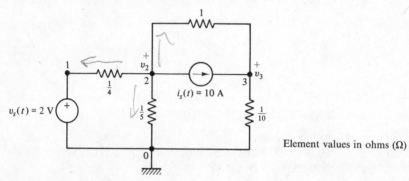

Element values in ohms (Ω)

Figure 2-11 Example 2-7.

Solution. Let us do an initial counting of variables: There are six elements; therefore, we are looking for 12 unknowns (one voltage and one current per element). The two sources, voltage $v_s(t)$ and current $i_s(t)$, are known—so we have 10 unknowns. This is still a huge number of unknowns (and equations) for such a simple network. We ought to be able to do better!

The first reduction will happen if we agree that, for a resistor, only the current *or* the voltage is needed, but not both. The other variable (voltage or current) is immediately available from Ohm's law, $v = Ri$ or $i = Gv$. Therefore, we have now six unknowns: four voltages across the resistors, the current in $v_s(t)$, and the voltage across $i_s(t)$.

Next, we introduce more convenient variables. Remember that a voltage between two points is a relative quantity, that is, point A is at a higher (or lower) voltage than point B. A good analogy here, we saw, is a geographical height, where we take the sea level as a convenient reference point (0 meter). Any other point is at a height above ($+$) or below ($-$) this reference. Also, the height difference between any two points can be calculated easily in terms of their heights with respect to sea level.

The same idea is used with *node voltages*. We choose (arbitrarily!) one node, and call it a *reference node* (or a *datum* node). In our example, let it be the node numbered "0." The reference node is usually marked by the symbol ⊥. Next, we seek out all the other nodes in the network, $n - 1$ in number if n is the total number of nodes. For each node, we ask the question: "Is the voltage of this node with respect to the reference node known or unknown?" If known, we do nothing (except note that it *is* known); if unknown, we assign to it a variable with a chosen reference.

† For full details on scaling, study Appendix C.

In our example, let us use the double subscript notation, v_{xy} = voltage *drop* from x to y, and seek out the $n - 1$ nodes:

Node 1: $\qquad\qquad\qquad\qquad v_{10} = v_s = 2 \text{ V} \qquad$ known

Node 2: $\qquad\qquad\qquad\qquad v_{20} \equiv v_2 \qquad\qquad$ unknown

Node 3: $\qquad\qquad\qquad\qquad v_{30} \equiv v_3 \qquad\qquad$ unknown

Thus, the only *two* unknowns in this network are the node voltages $v_2(t)$ and $v_3(t)$, as shown. Are they sufficient to solve the entire network? Yes, because if we calculate them, then the voltage drop across the $\frac{1}{4}$-Ω resistor is $v_{12} = v_s - v_2$; across the $\frac{1}{5}$-Ω resistor, the voltage drop is $v_{20} = v_2$; across the 1-Ω resistor the voltage drop is $v_{23} = v_2 - v_3$, which is also the voltage drop across the current source i_s; finally, the voltage drop across the 0.1-Ω resistor is $v_{30} = v_3$.

Now we have a reasonable number of unknowns, namely, two. We need two simultaneous equations for them. *For node voltages as unknowns, we write KCL at every unknown node*. Here, let us use the convention

$\qquad$ + for a current leaving a node

$\qquad$ − for a current entering a node

then we write

KCL at node 2: $\qquad\qquad 4(v_2 - 2) + 5v_2 + 10 + 1(v_2 - v_3) = 0$

KCL at node 3: $\qquad\qquad\qquad -10 + 10v_3 - 1(v_2 - v_3) = 0$

Rearrange these two equations as

$$10v_2 - v_3 = -2$$
$$-v_2 + 11v_3 = 10$$

In matrix notation†, these equations are

$$\begin{bmatrix} 10 & -1 \\ -1 & 11 \end{bmatrix} \begin{bmatrix} v_2 \\ v_3 \end{bmatrix} = \begin{bmatrix} -2 \\ 10 \end{bmatrix}$$

It is not a bad idea to review now an efficient method for solving these equations, called Gauss's elimination method (see Appendix A). Here it applies as follows:

1. Divide the first equation by 10 to get

$$v_2 - 0.1v_3 = -0.2$$

2. Add this equation to the second one, to get rid of v_2

$$0v_2 + 10.9v_3 = 9.8$$

This last equation yields immediately

$$\therefore \quad v_3 = \frac{9.8}{10.9} = 0.9 \text{ V}$$

Substitute back v_3 into the first equation to get v_2

$$v_2 - (0.1)(0.9) = -0.2$$
$$\therefore \quad v_2 = -0.11 \text{ V}$$

† Study Appendix A if necessary

After v_2 and v_3 are found, the complete solution of the network is (see Figure 2-12):

$$v_{12} = 2 - (-0.11) = 2.11 \text{ V} \qquad \therefore \quad i_{12} = 4(2.11) = 8.44 \text{ A}$$
$$v_{20} = -0.11 \text{ V or } v_{02} = 0.11 \text{ V} \qquad \therefore \quad i_{02} = 5(0.11) = 0.55 \text{ A}$$
$$v_{23} = -0.11 - 0.9 = -1.01 \text{ V or } v_{32} = 1.01 \text{ V} \qquad \therefore \quad i_{32} = 1.01 \text{ A}$$
$$v_{30} = v_3 = 0.9 \text{ V} \qquad \therefore \quad i_{30} = 10(0.9) = 9 \text{ A}$$

At node 1, KCL yields the current through the voltage source, 8.44 A, as shown.

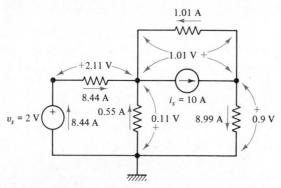

Figure 2-12   Example 2-7 (continued).

It would be interesting and instructive to check the power balance in this network. Since all voltages and currents are constant, the instantaneous powers will be constant. We remember that $p(t) = v(t)i(t)$, with the associated references of v and i dictating the sign and meaning of the power.

The voltage source:	$p = -(2)(9.44) = -16.88$ W	delivered to the network
The $\frac{1}{4}$-Ω resistor:	$p = +(2.11)(8.44) = +17.81$ W	received
The $\frac{1}{5}$-Ω resistor:	$p = +(0.11)(0.55) = +0.061$ W	received
The current source:	$p = -(1.01)(10) = -10.1$ W	delivered
The 1-Ω resistor:	$p = +(1.01)(1.01) = +1.02$ W	received
The 0.1-Ω resistor:	$p = +(0.9)(8.99) = +8.09$ W	received

As expected, $\Sigma p = 0$; that is, the total power delivered (by the sources) is equal to that received (dissipated) by the resistors.   ■

To summarize the general method of node analysis, let us repeat:

1. Choose one node as reference. The choice is arbitrary, although it is efficient to choose a node of one voltage source.
2. The remaining $(n - 1)$ node voltages are either known voltages or unknown voltages. Label the unknowns with variables $v_1, v_2, \ldots, v_p$ and with references. The number of these unknowns is $(n - 1)$ or less.

3. Write KCL at each unknown node, obtaining as many equations as unknowns. These equations, in matrix form, read

$$
\begin{bmatrix}
g_{11} & g_{12} & g_{13} & \cdots & g_{1p} \\
g_{21} & g_{22} & g_{23} & \cdots & g_{2p} \\
& & \vdots & & \\
g_{p1} & g_{p2} & g_{p3} & \cdots & g_{pp}
\end{bmatrix}
\begin{bmatrix}
v_1 \\ v_2 \\ \vdots \\ v_p
\end{bmatrix}
=
\begin{bmatrix}
j_1 \\ j_2 \\ \vdots \\ j_p
\end{bmatrix}
\tag{2-11a}
$$

that is,

$$
\mathbf{Gv} = \mathbf{j} \tag{2-11b}
$$

Here, $p \le n - 1$ is the number of unknowns and the number of equations; $\mathbf{v}$ is the matrix of the unknown node voltages; $\mathbf{G}$ is the matrix of the coefficients of $\mathbf{v}$, called the *node conductance matrix*; and $\mathbf{j}$ is the matrix of the known quantities on the right-hand side. Equation 2-11 resembles Ohm's law in matrix form for the entire network. Solve Equation 2-11 for the unknowns in $\mathbf{v}$.

4. Since each element is connected between two nodes and all node voltages are known now, express the voltage across each element as a difference of two node voltages.

5. For a resistor, use Ohm's law ($i = Gv$) to find the current. For a voltage source, use KCL to find its current. This completes the solution of the network.

Probs. 2-10, 2-11, 2-12, 2-13

2-6 LOOP CURRENTS AND LOOP ANALYSIS

The second systematic and efficient method for solving networks is by *loop analysis*. It is a dual method to node analysis. To outline its steps, let us go back to the network in Example 2-7 and solve it by loop analysis in the following example. See Figure 2-13.

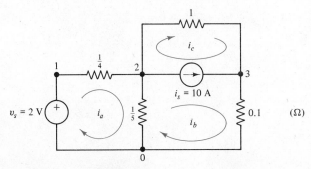

Figure 2-13 Example 2-8.

EXAMPLE 2-8 _____

Do quickly the initial count: six elements, therefore, six unknown element currents; less one known current (the current source $= 10$ A), for a total of five unknowns. We can do better! Let us define new unknown variables, called *loop currents*. A loop is a closed path, and we use here the three "obvious" loops, the window panes of the network.† The three loop currents $i_a(t)$, $i_b(t)$, and $i_c(t)$ are convenient new variables (just as node voltages were). Each is chosen with a name, i_a, i_b, and i_c, and with an *arbitrary* reference; in this case we chose them all clockwise (just as node voltages were chosen all $+$ with respect to reference).

As in the case of node voltages, where their sum or difference gave us the actual voltage across each element, so here the sum or difference of loop currents will yield the current in each element. Specifically,

In v_s: $\qquad\qquad\qquad\qquad\qquad i_s = i_{01} = i_a$

In $\frac{1}{4}$-Ω: $\qquad\qquad\qquad\qquad\quad i_{12} = i_a$

In $\frac{1}{5}$-Ω: $\qquad\qquad\qquad\qquad\quad i_{20} = i_a - i_b$

In $\frac{1}{10}$-Ω: $\qquad\qquad\qquad\qquad\quad i_{30} = i_b$

In 1-Ω: $\qquad\qquad\qquad\qquad\quad i_{32} = -i_c$

We have, at first count, three unknown loop currents, but the current source reduces this number further, because

$$i_{23} = 10 = i_b - i_c$$

or

$$i_c = i_b - 10$$

In a dual fashion, a voltage source between two nodes reduce the number of unknown node voltages.

We have, then, only two unknowns, i_a and i_b—not bad! What shall we use to get two equations in i_a and i_b? We already used KCL when we wrote the previous expressions for the current in each element. What remains? Of course, KVL (again, by duality to node analysis, where the final equations are KCL).

Let us write one KVL equation around the loop traced by i_a, and a second KVL equation around the loop formed by the $\frac{1}{5}$-, 1-, and $\frac{1}{10}$-Ω resistors. Why this loop and not the loop traced by i_b or i_c? For the dual reason that, in node analysis, we prefer not to write KCL at a node where a voltage source is connected: the current in the voltage source is unknown, on one hand, and, on the other, we don't need a KCL equation there. Here, in loop analysis, the voltage across the current source is unknown, but, at the same time, we don't need a third KVL equation.

The two KVL loop equations are, with $(+)$ for a voltage drop and $(-)$ for a voltage rise,

$$-2 + \tfrac{1}{4}i_a + \tfrac{1}{5}(i_a - i_b) = 0$$

for the first loop ($0 \rightarrow 1 \rightarrow 2 \rightarrow 0$), and

$$\tfrac{1}{5}(i_b - i_a) + 1(i_b - 10) + 0.1i_b = 0$$

for the second loop ($0 \rightarrow 2 \rightarrow 3 \rightarrow 0$).

Collect terms and arrange:

$$0.45i_a - 0.2i_b = 2$$
$$-0.2i_a + 1.3i_b = 10$$

In matrix form, these equations read

$$\begin{bmatrix} 0.45 & -0.2 \\ -0.2 & 1.3 \end{bmatrix} \begin{bmatrix} i_a \\ i_b \end{bmatrix} = \begin{bmatrix} 2 \\ 10 \end{bmatrix}$$

† Such a window pane is sometimes called a *mesh* and the current a *mesh current*.

Solve to get

$$i_a = 8.44 \text{ A} \qquad i_b = 8.99 \text{ A}$$

as before in Example 2-7.   ■

Probs. 2-16,
2-17, 2-18

The steps involved in loop analysis are:

1. Assign a loop current (arbitrary reference—clockwise or counterclockwise) in each window pane.
2. A current source can serve as a loop current, or can be the sum or difference of loop currents. Label the unknown loop currents by $i_a, i_b, \ldots$. There will be l (or less) such unknowns, if l is the number of window panes.
3. Write KVL around each loop. (Do not include a current source in KVL.) Obtain as many equations as unknowns. These equations, in matrix form, are

$$\begin{bmatrix} r_{11} & r_{12} & \cdots & r_{1q} \\ r_{21} & r_{22} & \cdots & r_{2q} \\ & & \vdots & \\ r_{q1} & r_{q2} & \cdots & r_{qq} \end{bmatrix} \begin{bmatrix} i_a \\ i_b \\ \vdots \\ i_q \end{bmatrix} = \begin{bmatrix} e_1 \\ e_2 \\ \vdots \\ e_q \end{bmatrix} \qquad (2\text{-}12a)$$

or, in matrix notation,

$$\mathbf{Ri} = \mathbf{e} \qquad (2\text{-}12b)$$

Here $q \le l$ is the number of unknowns; $\mathbf{i}$ is the matrix of unknown loop currents; $\mathbf{R}$ is the matrix of the coefficients, called the *loop resistance matrix*; and $\mathbf{e}$ is the matrix of the known quantities on the right-hand side. Equation 2-12 resembles Ohm's law in matrix form for the entire network.

These are, then, the two methods for network analysis. For a given network, we will choose the one with the fewer equations: it makes sense to opt for solving, say, four equations with four unknowns instead of seven equations with seven unknowns, right? Such an intelligent choice *must be done first* for every problem, unless, for some reason, the specific method of analysis is dictated to us (e.g., on a test, or by the availability of a particular computer program).

Probs. 2-14,
2-15

Another striking feature of these two methods is their duality. Review them carefully to recognize this duality: A node voltage is dual to a loop current; element voltages, as sums or differences of node voltages, are dual to element currents, as sums or differences of loop currents; all node voltages referenced $(+)$ with respect to ground are dual to all loop currents being clockwise; Equations 2-11 and 2-12 are dual of each other.

Keep in mind that the steps involved in node analysis and loop analysis, studied here, are quite general. They will apply later also, when we introduce other elements.

Probs. 2-19,
2-20, 2-21

Finally, if you have just a slight, lingering, uneasy feeling about one aspect of loop analysis, speak up! You are undoubtedly saying, "The identification and choice of node voltages is straightforward and ironclad. One node is datum, and the remaining nodes are clearly visible (with thick dots) in all cases. However, I am not so sure about loop currents. What if those window panes are not so obvious, as in the

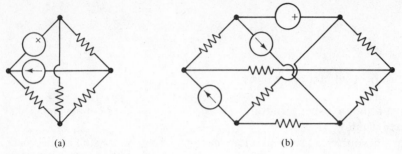

(a)   (b)

Figure 2-14   Where are the loops?

network in Figure 2-14(a)? Worse yet, what loops *do* I choose in Figure 2-14(b)?" (The small semicircular hooks in these circuits indicate wires that cross each other but without a node connection.) These are very valid questions, and, in fact, they will be studied in the next chapter.

*2-7 RECIPROCITY AND RECIPROCAL NETWORKS

Let us study an interesting property of certain linear circuits. Consider again loop analysis, applied to the network in Figure 2-15(a). Let $v_j(t)$ be the only voltage source applied, and let us write the loop equations (Equation 2-12) for this network

$$
\begin{bmatrix}
r_{11} & r_{12} & \cdots & r_{1q} \\
r_{21} & r_{22} & \cdots & r_{2q} \\
\multicolumn{4}{c}{\cdots\cdots\cdots\cdots\cdots} \\
r_{k1} & r_{k2} & \cdots & r_{kq} \\
\multicolumn{4}{c}{\cdots\cdots\cdots\cdots\cdots} \\
r_{q1} & r_{q2} & \cdots & r_{qq}
\end{bmatrix}
\begin{bmatrix}
i_1 \\ i_2 \\ \vdots \\ i_k \\ \vdots \\ i_q
\end{bmatrix}
=
\begin{bmatrix}
0 \\ 0 \\ v_j \\ \vdots \\ 0
\end{bmatrix}
\tag{2-13}
$$

Here the column matrix **e** contains a single nonzero entry in the jth loop, v_j. Don't let the picture fool you: loop k is *not* one big short circuit. It is partly hidden inside the box, and only a part of a wire is shown to identify i_k. Let us solve for the output, the kth loop current i_k, using Cramer's rule†

$$
i_k = \frac{1}{\det \mathbf{R}}
\begin{vmatrix}
r_{11} & \cdots & 0 & \cdots & r_{1q} \\
r_{21} & \cdots & 0 & \cdots & r_{2q} \\
 & & \vdots & & \\
\vdots & & v_j & & \vdots \\
 & & \vdots & & \\
r_{q1} & \cdots & 0 & \cdots & r_{qq}
\end{vmatrix}
\tag{2-14}
$$

†See Appendix A.

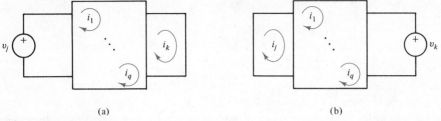

(a)   (b)

Figure 2-15   Reciprocity.

The denominator in Equation 2-14 is det $\mathbf{R}$, the determinant of the loop resistance matrix, that is, the determinant of the coefficients of the left-hand side in Equation 2-13. The numerator determinant is obtained by replacing the kth column of det $\mathbf{R}$ by the right-hand side of Equation 2-13.

If we expand Equation 2-14 by the kth column, we obtain a single term

$$i_k = \frac{\Delta_{jk}}{\det \mathbf{R}} v_j \tag{2-15}$$

where Δ_{jk} is the (j, k) cofactor; it is the smaller determinant obtained from det $\mathbf{R}$ by removing row j and column k, then multiplying the result by $(-1)^{j+k}$

$$\Delta_{jk} = (-1)^{j+k} \begin{vmatrix} r_{11} & r_{12} & \cdots & r_{1k} & \cdots & r_{1q} \\ r_{21} & r_{22} & \cdots & r_{2k} & \cdots & r_{2q} \\ \cdots\cdots & \cdots & \cdots & \cdots & \cdots & \cdots \\ r_{j1} & r_{j2} & & r_{jk} & \cdots & r_{jq} \\ \cdots\cdots & \cdots & \cdots & \cdots & \cdots & \cdots \\ r_{q1} & r_{q2} & \cdots & r_{qk} & \cdots & r_{qq} \end{vmatrix} \tag{2-16}$$

Now consider the same network in Figure 2-15(b), but with two changes: we remove v_j from loop j and insert v_k in loop k. When v_j is removed, it is replaced by a short circuit. When v_k is placed in loop k, we cut the wire in loop k and connect v_k to the two terminals of the cut. Calculate now the output i_j by Cramer's rule

$$i_j = \frac{\Delta_{kj}}{\det \mathbf{R}} v_k \tag{2-17}$$

where Δ_{kj} is the (k, j) cofactor.

If the network contains only linear resistors and no dependent sources, then $\mathbf{R}$, the loop resistance matrix, is *symmetric*, meaning that

$$r_{mn} = r_{nm} \tag{2-18}$$

This is easy to understand, because r_{mn} is the resistance in loop m common to loop n, and r_{nm} is the resistance in loop n common to loop m. Any dependent source may destroy this symmetry, $r_{mn} \neq r_{nm}$.

With $\mathbf{R}$ symmetric, we have

$$\Delta_{jk} = \Delta_{kj} \tag{2-19}$$

and, consequently, we can write from Equations 2-15 and 2-17

$$\frac{i_k}{v_j} = \frac{i_j}{v_k} \tag{2.20}$$

This equation expresses the *principle of reciprocity*, which may be stated as follows: If $v_j(t) = v_k(t)$, then $i_k(t)$ will be equal to $i_j(t)$ no matter what the structure and the element values are inside the box in Figure 2-15. A dual theorem holds for a single current input and measurement of a voltage output in a reciprocal network. Reciprocal networks, then, obey reciprocity. A network with dependent sources is *not* reciprocal, in general.

EXAMPLE 2-9 _____

For the network in Figure 2-16, the loop equations are (derive them in full!)

$$\begin{bmatrix} 3 + 2k & -2k - 2 \\ -2 & 9 \end{bmatrix} \begin{bmatrix} i_1 \\ i_2 \end{bmatrix} = \begin{bmatrix} v_1 \\ 0 \end{bmatrix}$$

and the network is not reciprocal since **R** is not symmetric. With $k = 0$ it becomes reciprocal.

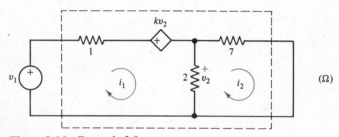

Figure 2-16 Example 2-9.

Probs. 2-22, 2-23

   It is important to recognize that the reciprocity principle applies only to a voltage input–current output or to a current input–voltage output. It does *not* apply to a voltage input–voltage output or to a current input–current output. Explain why not. *Hint*: Are the inner connections of the network—and hence det **R** and Δ_{jk}—unchanged with voltage input–voltage output?

PROBLEMS

2-1 The resistance, at a fixed temperature, is directly proportional to the length (l) of the resistor and inversely proportional to its cross section area (A)

$$R = \rho \frac{l}{A}$$

where ρ (rho) is the *resistivity*, in *ohm meters* ($\Omega \cdot m$), of the particular material of the resistor, and depends on the manufacturing process, the purity of the material, and temperature. Find the resistance of a copper-wound resistor ($\rho = 1.7 \times 10^{-8}$) of length 1 m and circular diameter of 1 mm. Copper, silver, and aluminum are examples of *conductors*, with ρ relatively small.

2-2 Materials with resistivity $10^{-4} < \rho < 10^{-7}$ $\Omega \cdot m$ are called *semiconductors*, and they provide the material for building transistors. Materials with very large resistivities ($\rho > 10^{10}$ $\Omega \cdot m$) are call *insulators*. Justify this name in terms of a current flowing in insulators. Mica, glass, and quartz are examples of insulators.

2-3 The voltage across a resistor $R = 50\,\Omega$ is given by

$$v(t) = 0, \qquad\qquad t < 0$$
$$v(t) = 100e^{-2t}, \qquad t > 0$$

Plot $v(t)$ and $i(t)$ for this resistor.

2-4 The current $i(t)$ in a resistor $R = 10\,k\Omega$ is shown. Sketch $v(t)$, the voltage across this resistor. Be sure to show all relevant values!

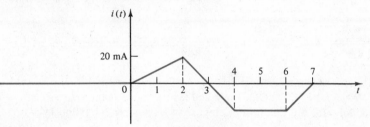

Problem 2-4

*2-5 (a) A constant voltage source is claimed to be a nonlinear resistor. Plot the v-i characteristic of such a voltage source ($v = V_0$, const.) and justify this claim.
(b) A constant current source ($i = I_0$, const.) is also a nonlinear resistor. Justify!
(c) An open circuit (o.c.) is considered a resistor. What is $R_{o.c.}$?
(d) A short circuit (s.c.) is considered a resistor. What is $R_{s.c.}$?

2-6 An electronic circuit element, called a tunnel diode, has a v-i characteristic as shown. Explain why, between the peak value of the current (I_p) and the valley current (I_v), the tunnel diode behaves like a *negative resistor*.

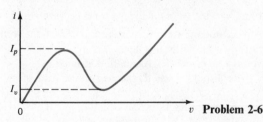

v **Problem 2-6**

2-7 Calculate the effective value of the current $i(t)$ as shown.

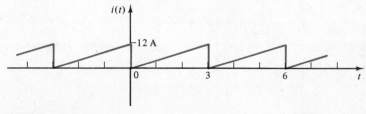

Problem 2-7

*2-8 In the figure we show a simple power transmission network, as found commonly. It is known that

$$i_a(t) = 140 \sin 377t$$

$$i_b(t) = 140 \sin (377t - 120°)$$

(a) Calculate $i_c(t)$ in the form $A \sin (377t + \phi)$.

(b) Plot all three currents vs. (377t) on the same graph.

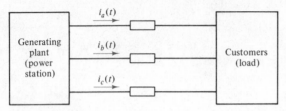

Problem 2-8

2-9 In the network shown, assign voltage and current variables (with references) to all the elements. Write KVL, KCL, Ohm's law, and reduce these equations to a single one in one unknown. Solve for this unknown. Compare with Example 2-4. (Hint: Duality!)

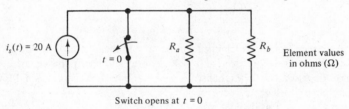

Switch opens at $t = 0$

Problem 2-9

2-10 (a) Use node analysis in Example 2-7 when $v_s(t) = 2e^{-t}$ V, and all other elements are the same.

(b) Calculate the total energy (from $t = 0$ to $t = \infty$) delivered (or received) by $v_s(t)$.

2-11 For the network shown, with a reference node chosen, there is one unknown node voltage, v_1. The other node voltage then is $(v_1 + 12)$. Verify this fact.

(a) What difficulty arises when you try to write a KCL equation at node v_1?

(b) To overcome this difficulty, write also KCL at the second node. Add these two equations.

(c) Show that step (b) is equivalent to writing KCL for the cut set consisting of the two current sources and the two resistors, shown in a dotted line.

(d) Solve completely and calculate the power in each element.

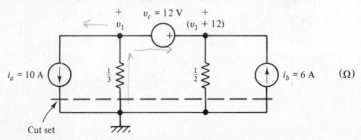

Problem 2-11

2-12 Solve by node analysis the network shown. Notice the dependent current source. Calculate the power in every element, and verify that $\sum p = 0$.

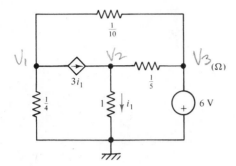

Problem 2-12

2-13 Solve by node analysis the network shown. Calculate and plot $p_a(t)$, the power of the source $v_a(t)$.

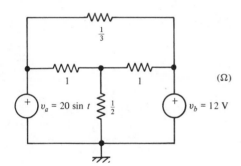

Problem 2-13

2-14 Let a network contain n nodes and n_v voltage sources. What is, at most, the number of equations needed for node analysis? Give a couple of examples, using both independent and dependent voltage sources.

2-15 Let a network contain l loops (window panes) and n_i current sources. What is, at most, the number of equations needed for node analysis? Give a couple of examples.

2-16 Solve Problem 2-9 by loop analysis and compare your results.

2-17 Solve Problem 2-11 by loop analysis and compare your results. *Hint*: Here assume $i_a = 10$ A as the loop current in the left loop, $i_b = 6$ A in the right loop, and one unknown loop current in the center loop.

2-18 Solve by loop analysis the network in Problem 2-13. This is a prime example of how an intelligent choice of the method (node analysis or loop analysis) can make all the difference in the world!

2-19 A transistor circuit is shown in part (a) of the figure, and its so-called "small-signal" equivalent model is shown in (b). From the equivalent model, calculate the output voltage v_{out} in terms of the input voltage v_{in}. *Note*: One resistor is given in kilohms.

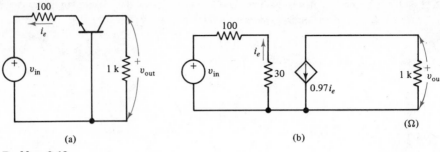

(a)   (b)

Problem 2-19

2-20 In the network shown, calculate v_{out} in terms of i_{in}.

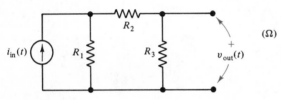

Problem 2-20

2-21 The network shown is known as a Wheatstone bridge. What must be the relation between R_1, R_2, R_3, and R_4 if $i_0 = 0$? The bridge is then said to be *balanced*. This circuit, with an ammeter in series with R_0, can be used to measure an unknown resistance, say R_1, if R_2, R_3, and R_4 are known.

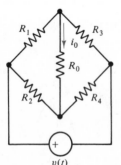

Problem 2-21

2-22 Carry out in full detail, on paper, the two experiments for testing reciprocity in Example 2-9 of Figure 2-16, showing that this network is not reciprocal.

2-23 Use the reciprocity theorem to calculate i_0 in the network shown. See Problem 2-21. Next, try to solve this problem directly, and see how much more difficult it is!

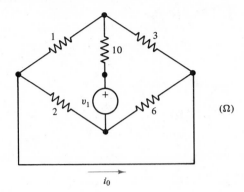

(Ω)

Problem 2-23

Chapter 3

Graphs and Networks

3-1 GRAPHS—DEFINITIONS AND PROPERTIES

Why are the two examples of Figure 2-14 in the previous chapter more complicated? Certainly not because of the nature of their elements. After all, a voltage source is always defined by its waveform $v(t)$, and its reference $(+)$, no matter how simple or complicated the network is. Similarly, a current source is always specified by $i(t)$ with its reference. A linear, constant resistor is always characterized by $v_R(t) = Ri_R(t)$, Ohm's law, no matter where the resistor is connected.

Yet these two networks, redrawn in Figure 3-1, are complicated because of their structure. Our previous "obvious" choice of loop currents may be adequate for simple networks, but it fails in more complicated networks. We want a *systematic* method for choosing and formulating loop equations and node equations. Furthermore, we want solid assurances that such equations can, indeed, be solved to yield unique answers.

To accomplish these goals, we study the *structure*, or *topology*, of the network in order to see clearly what approach to take for solving it.

Let us recognize the following important facts:

1. As mentioned, the v-i characteristic for an element (such as a resistor or a source) is inherent to the element *itself* and not to the way it is connected in a particular network. We can think of this v-i characteristic as the tag which is attached to the element by the manufacturer.
2. By contrast, Kirchhoff's laws do *not* depend on the types of elements; rather, they depend on the way that these elements are connected. For example, consider the networks shown in Figure 3-2(a) and (b). The "boxes" represent any type of element—a generalization of the two networks in Figure 3-1.

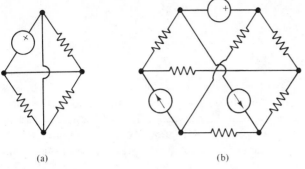

(a)   (b)

Figure 3-1   Two complicated networks.

With the current references shown in Figure 3-2(a), Kirchhoff's current law at node 2 reads: (+ leaving, − entering)

$$i_1(t) - i_2(t) + i_3(t) = 0 \qquad (3-1)$$

regardless of the nature of the elements connected to that node. Similarly, with the voltage references shown in Figure 3-2(b), Kirchhoff's voltage law

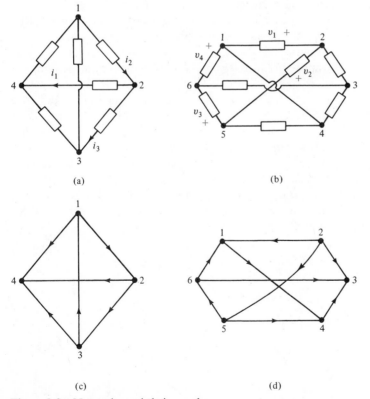

(a)   (b)

(c)   (d)

Figure 3-2   Networks and their graphs.

around the loop $1 \rightarrow 2 \rightarrow 5 \rightarrow 6 \rightarrow 1$ is ($+$ drop, $-$ rise)

$$-v_1(t) - v_2(t) + v_3(t) - v_4(t) = 0 \qquad (3\text{-}2)$$

no matter what those elements are.

To simplify our discussion further, we will draw the *graph* of a network as a connection of *branches*. A branch is simply a line segment, with its two end nodes, representing a circuit element. The graph of the network in Figure 3-2(a) is shown in Figure 3-2(c), and the graph of the network in Figure 3-2(b)—in Figure 3-2(d). Branches are interconnected only at the nodes; overlapping branches without a node do not form a connection there.

On each branch, let us show, for economy, only *one* reference arrow. It will be the reference for the current in the branch if it is a current source, or for the voltage drop if it is a voltage source, or (as in a resistor) for both current and voltage drop, in agreement with their associated references. The two graphs show these references. For all the resistors, arbitrary references were chosen, while the sources show their given references. A branch with a given reference is called *oriented*, and the graph with all references is also *oriented*.

As long as the interconnections of the branches are not changed, we can redraw a graph by stretching or shrinking branches, thinking of them as rubber bands. For example, the graph in Figure 3-3(a) can be redrawn as illustrated in Figure 3-3(b), with all nodes numbered $(1, \ldots, 4)$ and all branches labeled $(a, \ldots, f)$ for clarity. Quite obviously, the two graphs, Figure 3-3(a) and (b), are the same as far as KCL and KVL are concerned. Such graphs are called *isomorphic*—from the Greek, meaning "of the same structure."

Why do we bother with isomorphic graphs? Again, to simplify matters: The graph in Figure 3-3(b) has no branch intersections except at nodes, while the one in Figure 3-3(a) has such an "unpleasant" intersection in its center, where branch a lies on top of branch b. A graph that can be drawn (or redrawn isomorphically) without such overlapping intersections of branches is called *planar*. In other words, it can be drawn as a flat graph on a plane, so that no branches intersect at a point that is not a node. By contrast, a *nonplanar* graph cannot be drawn so: Some of its branches will intersect without a node, no matter how we redraw it. Such a graph is three-dimensional. The graph in Figure 3-2(d) is nonplanar (try it!).

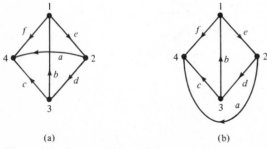

(a)   (b)

Figure 3-3 Isomorphic graphs.

It is a simple matter now to recognize the obvious loops in Figure 3-3(b). They are $1 \to f \to 4 \to a \to 2 \to e \to 1$, $1 \to f \to 4 \to c \to 3 \to b \to 1$, $3 \to c \to 4 \to a \to 2 \to d \to 3$, and $1 \to e \to 2 \to d \to 3 \to b \to 1$. We are still in the dark about loops in nonplanar graphs, as in Figure 3-2(d). Before tackling this problem, we need a few additional definitions and properties.

A branch and a node are *incident* to each other if the node is one of the two nodes of that branch. For example, in Figure 3-3, branches a, f, and c are incident to node 4, and branches b, c, and d are incident to node 3. The *degree* of a node is the number of branches incident to that node. So, in that figure, node 1 is of degree 3 (as it happens, every node in that graph has the same degree—but that's only a coincidence!). A node of degree zero is *isolated*.

A *subgraph* of a graph is a part (subset) of that graph. What's left after a subgraph has been picked is called its *complement*. Thus, a branch (with its two end nodes) is in the complement if, and only if, it is not in the subgraph. In Figure 3-4 we see a subgraph and its complement for the graph in Figure 3-3. In general, the choice of a subgraph (and, therefore, its complement) is not unique.

A *path* between two nodes of a graph is a sequence of branches, starting at the first node and ending at the second node, without tracing any branch twice and without going through any node more than once. To use our precise definitions, a path from node k to p is a subgraph in which node k is of degree one, node p is of degree one, and all other nodes are of degree two (in that subgraph). For example, a path from node 3 to node 4 is shown in solid lines in Figure 3-4. Verify to your complete satisfaction that it fits the precise definition of a path. (This is one of the several advantages of graph theory: preciseness and conciseness.) Also, satisfy yourself completely as to why $3 \to b \to 1 \to f \to 4 \to a \to 2 \to e \to 1 \to f \to 4$ is *not* a path from node 3 to node 4. A graph is said to be *connected* if there is at least one path between every two nodes. Such a graph is, in less precise terms, in one piece.

A *loop*, formerly used in an intuitive ("obvious") way, can be now defined precisely as a *closed path*; that is, it is a subgraph in which every node is of degree two. Some of the loops in Figure 3-4 are: $1 \to f \to 4 \to a \to 2 \to e \to 1$; $1 \to f \to 4 \to a \to 2 \to d \to 3 \to b \to 1$; $1 \to e \to 2 \to a \to 4 \to c \to 3 \to b \to 1$; $1 \to b \to 3 \to c \to 4 \to a \to 2 \to e \to 1$; and there are others.

This is where our original dilemma appears again: How do we know for sure which loops to use for KVL? How many of them are needed? How do we know that we have a solvable set of KVL equations?

Prob. 3-1

Prob. 3-2

Prob. 3-3

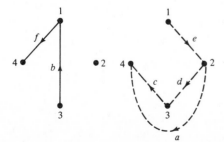

Figure 3-4 A subgraph (solid lines) and its complement (dotted lines).

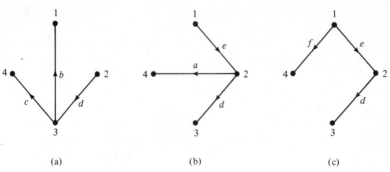

Figure 3-5   Trees for Figure 3-3.

In order to settle these and related questions, we need another important concept in graph theory. A *tree* of a connected graph is a connected subgraph containing all the nodes but no loops. It is important to satisfy *all* these requirements for a tree: (1) It must be a *connected* subgraph; (2) it must contain *all* the nodes; and (3) it must *not* have any loops. To build a tree (in a given graph), we follow these steps:

1. Draw all the original nodes (requirement 2).
2. Place enough branches between the nodes, from those given in the graph, until all the nodes are connected (requirement 1), without forming any loops (requirement 3).

Figure 3-5 shows three different trees for the graph in Figure 3-3 (there are more trees). The tree shown in Figure 3-5(a) "justifies" this name: With a bit of imagination, this subgraph looks like a tree (just add a trunk and some leaves!).

The branches of a tree are called (what else?) *tree branches*, and we designate their number by b_t. If n is the number of nodes in the graph or in the tree, we have

$$b_t = n - 1 \qquad (3\text{-}3)$$

Probs. 3-4,
3-5, 3-6

The complement of a tree is called, briefly, the *co-tree*. The three co-trees for Figure 3-5 are shown in Figure 3-6. The branches of the co-tree have also a special name: They are *links*, since they link the nodes of the tree. If the number of links is l and the number of the branches in the graph is b,

$$l + b_t = b \qquad (3\text{-}4)$$

or, using Equation 3-3, we get

$$l = b - n + 1 \qquad (3\text{-}5)$$

Equations 3-3 and 3-5 are simple and elegant. Moreover, they provide the basis for the systematic formulation of node equations and loop equations.

Prob. 3-7

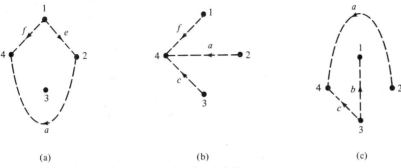

(a)   (b)   (c)

Figure 3-6   Co-trees for the trees in Figure 3-5.

Graph theory, with its tree analysis, counting links, etc., provides us with useful and essential information needed for circuit analysis. Specifically, it answers these questions:

1. Exactly how many loop equations are needed, and how to choose them so that they are guaranteed to have a solution.
2. Exactly how many node equations are needed, and how to choose them so that they are guaranteed to have a solution. Our intelligent choice will be, of course, to write the smaller of these two sets of equations when solving a network.

The emphasis here is on the words "exactly" and "guaranteed solution." Graph theory (topology) gives us absolute assurances for both. To illustrate these ideas, let us consider a simple example.

EXAMPLE 3-1 _____

In the circuit of Figure 3-7(a), we see the "obvious" three loop currents i_1, i_2, and i_3. However, there are three additional loops with their loop currents, labeled 4, 5, and 6. Why do we write three loop equations, and not four or five or six? Can we solve this network with *two* loop equations? We *must* have definitive answers to these questions before we embark on the actual writing of such equations.

Solution.   Suppose we write first the three "obvious" loop equations. They are

$$2i_1 - i_2 = 1$$

around loop 1,

$$-i_1 + 3i_2 - i_3 = 0$$

around loop 2, and

$$-i_2 + 3i_3 = 0$$

around loop 3. Their (correct) solution is $i_1 = 0.615$ A, $i_2 = 0.231$ A, $i_3 = 0.077$ A.

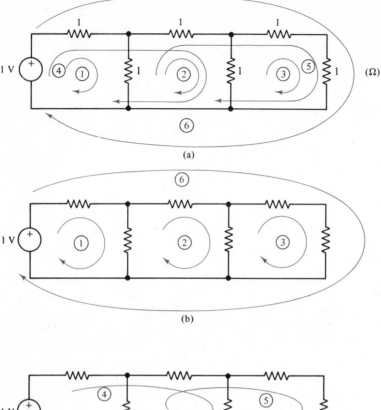

Figure 3-7 Example 3-1.

Let us now try four loops, as shown in Figure 3-7(b). We get the four loop equations (do it slowly!)

$$
\begin{aligned}
2i_1 - i_2 \qquad + i_6 &= 1 \qquad \text{(KVL loop 1)} \\
-i_1 + 3i_2 - i_3 + i_6 &= 0 \qquad \text{(KVL loop 2)} \\
-i_2 + 3i_3 + 2i_6 &= 0 \qquad \text{(KVL loop 3)} \\
i_1 + i_2 + 2i_3 + 4i_6 &= 1 \qquad \text{(KVL loop 6)}
\end{aligned}
$$

If we try to solve for the four unknowns, we find that all the answers are of the indeterminate form $\frac{0}{0}$. A second (or third) look at the four equations tells us that the fourth one is exactly the sum of the previous three—hence it is redundant, or, in more precise terms, it is linearly dependent on those three. Four (or more) loop equations for this network are, then, an overspecification.

How about less than three loop equations? The two loops in Figure 3-7(c) look promising: All the elements of the network are included in them, and some are common to both loops. Sounds good, right? Let us write these two loop equations as follows:

$$3i_4 + i_5 = 1 \quad \text{(KVL loop 4)}$$
$$i_4 + 3i_5 = 1 \quad \text{(KVL loop 5)}$$

These two equations *have* a unique solution

$$i_4 = i_5 = \tfrac{1}{4}\,\text{A}$$

Do we have the *correct* solution to our network? No! Even though KVL is satisfied with these answers around loop 4 and loop 5, it is *not* satisfied around loop 1, for instance:

$$(1)(\tfrac{1}{4}) - (1)(\tfrac{1}{4}) - 1 \neq 0$$

Nor is KVL satisfied around loops 2, 3, and 6. Here, two equations are an underspecification for this network.

How do we guarantee the correct number of equations? Their solution? These questions are answered by the topological equations developed in this chapter. ∎

3-2 NODE ANALYSIS REVISITED

At this stage, we can argue as follows: Even in the most complicated networks (or their graphs), the nodes are easily spotted. All that we need, then, is to do as in Chapter 2: Pick one node as a reference, and write KCL at the remaining $(n - 1)$ nodes. Isn't this simple and systematic enough?

The answer, if we are honest and critical, should be, "not quite." We don't know yet why the last nth KCL equation is not needed, or whether these $(n - 1)$ equations are always solvable.

To address the first question, consider again the graph in Figure 3-3, and write KCL for *every* node ($+$ leaving, $-$ entering). We get n equations,†

Node 1:	$-i_b + i_e + i_f = 0$	
Node 2:	$i_a + i_d - i_e = 0$	(3-6)
Node 3:	$i_b + i_c - i_d = 0$	
Node 4:	$-i_a - i_c - i_f = 0$	

We notice that every branch current appears exactly twice in these equations, once with a $(+)$ sign, once with a $(-)$ sign. The reason? Every branch is incident to precisely two nodes and is oriented *from* one and *into* the other, no more and no less. Therefore, any one KCL equation (say at node 3) is the negative of the sum of the remaining equations (here nodes 1, 2, and 4). To check, we add KCL at nodes 1, 2, and 4, with the result

$$-i_b - i_c + i_d = 0 \tag{3-7}$$

which is, indeed, the negative of KCL at node 3. It is, of course, the *same* as KCL at node 3, because a multiplication by -1 of Equation 3-7 yields KCL at node 3.

---

† Remember: $i_a = i_a(t)$, $i_b = i_b(t), \ldots$, etc.

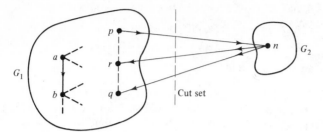

Figure 3-8   KCL at the nth node.

The same conclusion can be reached by physical reasoning. In Figure 3-8, we have enclosed in G_1 any $(n - 1)$ nodes, and in G_2 the nth node. All the branches connecting G_1 to G_2 form a cut set, which yields KCL at the nth node, looking at the cut set from the right of the cutting plane. See Figure 3-8. On the other hand, the sum of the currents in the cut set, on the left of the cutting plane, is equal to the sum of all KCL at the $(n - 1)$ nodes inside G_1: In all branches incident to nodes *inside* G_1, each current appears twice—once entering a node (for example, node b), and once leaving a node (node a). Such currents will add up to zero when we add the KCL equations for the $(n - 1)$ nodes in G_1. The only currents that will remain, then, are for branches which form the cut set—from node p to n, from node r to n, etc. The sum of these currents, therefore, is zero. However, this is not new—we reached this conclusion from the right side of the cutting plane. In summary, then, KCL at the nth node is embedded already in KCL at the previous $(n - 1)$ nodes; it is redundant.

Probs. 3-8, 3-9

We have, therefore, verified to our satisfaction that $(n - 1)$ KCL equations, with $(n - 1)$ node voltages as unknowns, are adequate for solving the network. If voltage sources are present, they will further reduce the number of unknowns and equations, as illustrated in the following example.

EXAMPLE 3-2 _____

In the graph of Figure 3-3, branch a is a current source $i_a(t) = 10$ A; branch c is a voltage source $v_c(t) = 6$ V; all other branches are 1-Ω resistors.

Solution.   In order to write the node equations, choose node 4 as reference. With respect to this node, v_1 and v_2 are the only unknowns, since $v_3 = 6$. The two node equations are (+ leaving, − entering)

At node 1:
$$\frac{v_1}{1} + \frac{v_1 - v_2}{1} - \frac{6 - v_1}{1} = 0$$

At node 2:
$$10 + \frac{v_2 - 6}{1} - \frac{v_1 - v_2}{1} = 0$$

that is,

$$3v_1 - v_2 = 6$$
$$-v_1 + 2v_2 = -4$$

yielding

$$v_1 = \tfrac{8}{5} \text{ V} \qquad v_2 = -\tfrac{6}{5} \text{ V} \qquad\qquad \blacksquare$$

The reason for choosing node 4 as reference in the previous example is obvious: Node 3 is at $+6$ V with respect to this reference. Node 3 is therefore known, and we don't need to write a KCL equation for it. Lucky us! If we had to write this equation, another unknown, the current in this voltage source, would be needed. We don't know this current, on one hand, and we don't need that equation, on the other. It is therefore *convenient*, although not mandatory, to make such a choice of reference. In the next example, we learn how to deal with a more complicated situation.

EXAMPLE 3-3 _____

In the same graph of Figure 3-3, let branch c be a voltage source, $v_c = 6$ V, and branch e also a voltage source, $v_e = 12$ V. All the other branches are resistors, as shown in Figure 3-9(a).

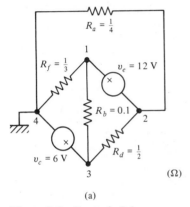

(a)   (b)

Figure 3-9 Example 3-3.

Solution. Here we can choose as reference *either* node 4 or node 2 (but obviously not both!). With node 4 as reference, node voltage v_1 is unknown; node voltage v_2 is then $(v_1 - 12)$ because there is a voltage drop of 12 V from node 1 to node 2, $v_e = v_1 - v_2 = 12$, and hence $v_2 = v_1 - 12$. Node voltage $v_3 = 6$.

It appears that there is one unknown, v_1. Let us write one KCL equation at that node, with $+$ for leaving, $-$ for entering:

$$3v_1 - 10(6 - v_1) + i_e = 0$$

But now we have i_e as another unknown! How to get rid of it? Since branch e is also incident to node 2, i_e will have, therefore, a minus sign there. Write KCL at node 2:

$$2(v_2 - 6) + 4v_2 - i_e = 0$$

Adding these equations, the current i_e *cancels* (that's the whole idea!)

$$13v_1 + 6v_2 - 72 = 0$$

and since $v_2 = v_1 - 12$, we get

$$19v_1 = 144$$
$$\therefore \quad v_1 = 7.58 \text{ V}$$ ■

Let us pause and review what we learned here:

1. The network has $n = 4$ nodes, so there are $n - 1 = 3$ node voltages. Two of these are known voltages (the sources), reducing the three node voltages to $3 - 2 = 1$, one unknown, v_1.
2. In writing KCL at node 1, we are forced to introduce another unknown, i_e, the current in a voltage source.
3. This voltage source is incident at node 2, where i_e will have a sign ($+$ or $-$) *opposite* to its sign at node 1.
4. Write, therefore, the auxiliary KCL equation at node 2, just to be able to add it to the previous KCL in order to cancel the terms $+i_e$ and $-i_e$.
5. From the graph's point of view, these two equations are:

$$i_f - i_b + i_e = 0$$

and

$$i_d + i_a - i_e = 0$$

Their sum is

$$i_a - i_b + i_d + i_f = 0$$

a KCL equation for the cut set $a - b - d - f$. This cut set leaves the branches e and c, the voltage sources, in the two separate parts of the graph G_1 and G_2, as shown in Figure 3-9(b); consequently, their currents, i_e and i_c, are of no concern to us.

In conclusion, then: The number of node equations (KCL) for node analysis is $(n - 1) - n_v$, where n_v is the number of the known voltage sources. Without voltage sources, KCL is written at every one of the $(n - 1)$ unknown nodes. With voltage sources, cut set equations are written.

Probs. 3-10, 3-11

EXAMPLE 3-4 _____

For the network shown in Figure 3-10(a), we draw its oriented graph in Fig. 3-10(b), labeling nodes and branches.

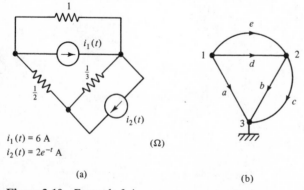

$i_1(t) = 6$ A

$i_2(t) = 2e^{-t}$ A

(Ω)

(a)   (b)

Figure 3-10   Example 3-4.

Solution. Since $n = 3$ and $n_v = 0$, we need $n - 1 = 2$ node equations. Choose node 3 as reference, and write KCL ($+$ leaving, $-$ entering):

At node 1: $\qquad\qquad +2v_1 + 6 + (v_1 - v_2) = 0$

At node 2: $\qquad\qquad -(v_1 - v_2) - 6 - 3(0 - v_2) + 2e^{-t} = 0$

Collect terms and rearrange:

$$3v_1 - v_2 = -6$$
$$-v_1 + 4v_2 = 6 - 2e^{-t}$$

Solve by determinants:

$$v_1 = \frac{\begin{vmatrix} -6 & -1 \\ 6 - 2e^{-t} & 4 \end{vmatrix}}{\begin{vmatrix} 3 & -1 \\ -1 & 4 \end{vmatrix}} = \frac{-18 - 2e^{-t}}{11} = -1.634 - 0.182e^{-t}$$

$$v_2 = \frac{\begin{vmatrix} 3 & -6 \\ -1 & 6 - 2e^{-t} \end{vmatrix}}{\begin{vmatrix} 3 & -1 \\ -1 & 4 \end{vmatrix}} = \frac{12 - 6e^{-t}}{11} = 1.1 - 0.55e^{-t}$$

With these node voltages known, the entire circuit is solved. For example, the voltage across the current source i_1 is:

$$v_d = v_1 - v_2 = -2.734 + 0.368e^{-t}$$

and so on. ∎

*3-3 FUNDAMENTAL CUT SETS

As we just realized, it is to our advantage to write cut set KCL equations when voltage sources are present. Again, node voltages are fairly easy to recognize by inspection, without too much fuss with the graph. How do we proceed in a systematic way to write cut set equations? Here the graph is absolutely a must—and a delight.

Recall how we build a tree: It has a maximum number of tree branches without forming any loops. The voltages of the tree branches are therefore *linearly independent*; review here Problem 3-8(e). In other words, since there are no loops in the tree, there is no relationship such as $v_a + v_b - v_c + v_d = 0$ among the voltages of the tree branches. Why? Because such a relationship is KVL for a loop—but there are no loops in the tree.

As a result, we can use tree branch voltages as unknowns for cut set analysis, and there are precisely $n - 1$ tree branches (unknowns) for this purpose. This is the same number of unknowns as in node analysis. Here, too, some of the tree branches may be n_v voltage sources, and they will reduce the number of unknowns to $n - 1 - n_v$, as in node analysis.

The systematic formulation of the cut set equations is, then, as follows:

1. Draw a *proper tree* of the graph. A proper tree, as distinguished from any other tree, must contain all the voltage sources.
2. Identify $n - 1$ *fundamental cut sets*, each such cut set consisting of exactly *one* tree branch and some links.
3. Write KCL for each fundamental cut set, using the tree branch voltages as unknowns.

The following example illustrates this method.

EXAMPLE 3-5

Let us rework Example 3-3 using fundamental cut sets.

Solution. The proper tree must contain the voltage sources (tree branches c and e). The last tree branch can be any one of those remaining, say d. This proper tree is shown in Figure 3-11(a) in solid lines, while the links are shown in dashed lines.

There are $n - 1 = 4 - 1 = 3$ tree branch voltages v_c, v_d, and v_e. However, $v_c = 6$ V, $v_e = 12$ V, and so we have only one unknown, v_d.

The fundamental cut set for tree branch d is d-b-f-a, as shown in Figure 3-11(b). Remember: A fundamental cut set (not just any old cut set) contains *one* tree branch and several links.

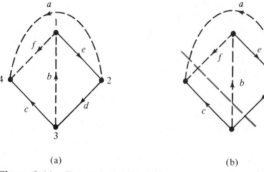

(a)   (b)

Figure 3-11   Example 3-5.

KCL for this fundamental cut set is $i_d - i_b + i_f + i_a = 0$ where we gave a $(+)$ sign to currents entering the cutting plane from one direction and a $(-)$ sign to those entering it from the opposite direction. In terms of branch voltages, this KCL equation reads

$$2v_d - 10v_b + 3v_f + 4v_a = 0$$

But KVL says that

$$v_b + v_e + v_d = 0 \qquad \therefore \quad v_b = -v_e - v_d = -12 - v_d$$

also,

$$v_f - v_c - v_d - v_e = 0 \qquad \therefore \quad v_f = 6 + v_d + 12 = 18 + v_d$$

and

$$v_a - v_c - v_d = 0 \qquad \therefore \quad v_a = 6 + v_d$$

As a result, the fundamental cut set equation becomes

$$2v_d - 10(-12 - v_d) + 3(18 + v_d) + 4(6 + v_d) = 0$$

or, finally,

$$v_d = -10.42 \text{ V}$$

With v_d known, the entire circuit is solved.

Probs. 3-12, 3-13

*3-4 FUNDAMENTAL LOOPS

Let us now consider the problem of formulating *systematically* loop (KVL) equations. Granted, with planar graphs, the loops are obvious—as are the node voltages in node analysis. In nonplanar graphs, the loops are not obvious.

We proceed in a dual manner to fundamental cut sets, starting with a proper tree. Here we will use the link currents as unknowns, and there will be $l = b - n + 1$ of them, as we proved in Equation 3-5. Links by themselves cannot form a cut set (one or more tree branches are needed). Therefore, there is no relationship of the form $i_a - i_b + i_c - i_d = 0$ among link currents, because such a relationship (KCL) requires that these links form a cut set. Hence, link currents are linearly independent.

As a result, link currents are valid variables to be used as unknowns in writing loop equations. Naturally, if there are n_i current sources, they will reduce the number of unknowns to $(l - n_i)$.

The systematic formulation of fundamental loop equations is then:

1. Draw a proper tree, saving all current sources to be links.
2. Identify l *fundamental loops*, each being formed by exactly *one* link and some tree branches.
3. Write KVL for each fundamental loop, using the link currents as unknowns.

EXAMPLE 3-6

Again, refer to the network in Example 3-3. Let us solve it now with fundamental loops.

Solution. A proper tree (solid lines) and the links (dotted lines) are shown in Figure 3-12(a). There are $l = b - n + 1 = 6 - 4 + 1 = 3$ unknown link currents: i_a, i_b, and i_f. Their corresponding fundamental loops are shown separately in Figure 3-12(b) for clarity. Remember: A fundamental loop (not just any old loop) consists of *one* link which creates it, and several tree branches.

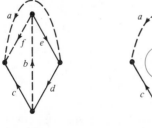

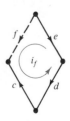

(a)   (b)

Figure 3-12   Example 3-6.

KVL for the first fundamental loop is

$$v_a - v_c - v_d = 0$$

for the second one,

$$v_b + v_e + v_d = 0$$

and, for the third one,

$$v_f - v_c - v_d - v_e = 0$$

where each loop was given, as loop current, the current of the link that creates it. In terms of the chosen unknowns, i_a, i_b, and i_f, we have

$$v_a = \tfrac{1}{4}i_a \qquad v_b = 0.1i_b \qquad v_d = \tfrac{1}{2}i_d = \tfrac{1}{2}(-i_a + i_b - i_f) \qquad v_f = \tfrac{1}{3}i_f$$

Also, $v_c = 6$ V, $v_e = 12$ V. The three fundamental loop equations become

$$\tfrac{1}{4}i_a - 6 - \tfrac{1}{2}(-i_a + i_b - i_f) = 0$$
$$0.1i_b + 12 + \tfrac{1}{2}(-i_a + i_b - i_f) = 0$$
$$\tfrac{1}{3}i_f - 6 - \tfrac{1}{2}(-i_a + i_b - i_f) - 12 = 0$$

or

$$\begin{bmatrix} 0.75 & -0.5 & 0.5 \\ -0.5 & 0.6 & -0.5 \\ 0.5 & -0.5 & 0.83 \end{bmatrix} \begin{bmatrix} i_a \\ i_b \\ i_f \end{bmatrix} = \begin{bmatrix} 6 \\ -12 \\ 18 \end{bmatrix}$$

The solution of which yields

$$i_a = -17.73 \text{ A}$$
$$i_b = -15.67 \text{ A}$$
$$i_f = 22.93 \text{ A}$$

As a quick check, we calculate here

$$v_d = \tfrac{1}{2}i_d = \tfrac{1}{2}(17.73 - 15.67 - 22.93) = -10.43 \text{ V}$$

in agreement (third decimal roundoff error) with the answer obtained in Example 3-5.

∎

EXAMPLE 3-7 ————————————————————————————

Let us turn our attention to one of the complicated networks in Figure 3-1(b), with its graph. They are repeated here for convenience. See Figure 3-13. Specific values are assigned here to the sources and to the resistors, so that we can solve this network.

Solution.   A quick count, first of all:

$$n = 6 \qquad n_v = 1$$

∴   Number of node (or cut set) equations needed is $n - 1 - n_v = 6 - 1 - 1 = 4$.

$$b = 9 \qquad n_i = 2$$

∴   Number of (fundamental or not) loop equations is

$$l - n_i = b - n + 1 - n_i = 9 - 6 + 1 - 2 = 2$$

It's *that* simple *and* definitive! This network can be solved with four KCL equations or with two KVL equations. Naturally, we opt for two equations.

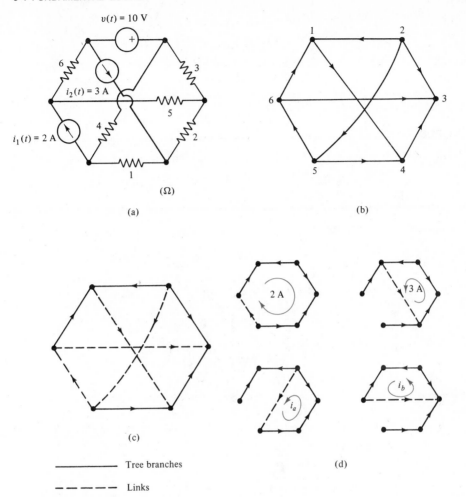

Figure 3-13 Example 3-7.

The proper tree is chosen as in Figure 3-13(c): voltage sources in it, current sources not in it. In Figure 3-13(d), we show separately the fundamental loops with their loop (link) currents. Each link current gives the fundamental loop its loop current and its reference.

KVL for the fundamental loop of i_a is ($+$ drop, $-$ rise):

$$4i_a + 1(i_a - 2) + 2(i_a - 2 + 3) + 3(i_a - 2 + 3 + i_b) = 0$$

and KVL for the fundamental loop of i_b is

$$5i_b + 3(i_b - 2 + 3 + i_a) + 10 + 6(i_b - 2) = 0$$

These two equations are

$$10i_a + 3i_b = -3$$
$$3i_a + 14i_b = -1$$

and their solution is

$$i_a = -0.298 \text{ A} \qquad i_b = -0.0076 \text{ A}$$

With i_a and i_b known, the entire network is solved.

Probs. 3-14, 3-15, 3-16

To conclude our discussion, let us consider another example, this time with dependent sources. We will see that the treatment is the same.

EXAMPLE 3-8 _____

In the network shown in Figure 3-14(a), we are given one independent source

$$i(t) = 3 \text{ A}$$

and one dependent source

$$v(t) = 2i_a(t)$$

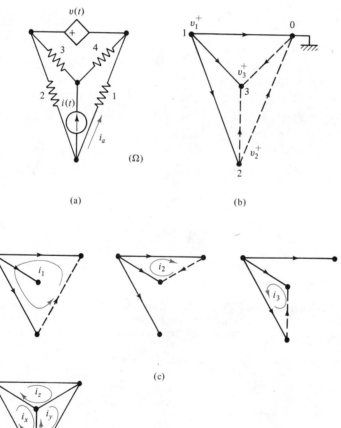

(a)

(b)

(c)

(d)

Figure 3-14   Example 3-8.

Solution. Let us solve the network twice—by node equations, and by loop equations. (Such duplication is not needed, of course, in real life. We are doing it here for practice.) A proper tree, including the voltage source but not the current source, is shown in Figure 3-14(b), together with the co-tree.

The number of node equations needed is

$$n - 1 - n_v = 4 - 1 - 1 = 2$$

Use node "0" as reference; then v_2 at node 2 and v_3 at node 3 are the unknowns. We know that

$$v_1(t) = v(t) = 2i_a(t) = 2v_2$$

KCL at node 2 reads ($+$ leaving, $-$ entering)

$$-\frac{v_1 - v_2}{2} + 3 + \frac{v_2}{1} = 0$$

and at node 3, we write

$$-\frac{v_1 - v_3}{3} - \frac{0 - v_3}{4} - 3 = 0$$

Rearranging these two equations and using $v_1 = 2v_2$, we get

$$\frac{v_2}{2} + 3 = 0$$

$$-\frac{2v_2}{3} + \frac{7}{12}v_3 = 3$$

$$\therefore \quad v_2 = -6 \text{ V}$$

$$\therefore \quad v_3 = -1.71 \text{ V}$$

The number of unknown loop currents needed is

$$b - n + 1 - n_i = 6 - 4 + 1 - 1 = 2$$

The three fundamental loops are shown in Figure 3-14(c), $i_3 = 3$, and the two unknowns are i_1 and i_2. Write KVL for each of these two loops ($+$ drop, $-$ rise):

$$1i_1 - 2i_a + 2(i_1 + 3) = 0$$
$$4i_2 + 3(i_2 + 3) + 2i_a = 0$$

But $i_a = i_1$, and therefore the two equations become

$$i_1 + 6 = 0$$
$$2i_1 + 7i_2 + 9 = 0$$
$$\therefore \quad i_1 = -6 \text{ A} \qquad i_2 = 0.428 \text{ A}$$

To verify our answers, we calculate

$$v_2 = 1i_1 = -6 \text{ V}$$
$$v_3 = -4i_2 = -1.71 \text{ V}$$

as before.

Do we have to be so "fancy," with proper trees, links, etc.? Not at all! The simple topological count can be done directly in the given circuit diagram, Figure 3-14(a), as follows:

$$b = \text{number of branches} = 6$$
$$n = \text{number of nodes} = 4$$
$$n_v = \text{number of voltage sources} = 1$$
$$n_i = \text{number of current sources} = 1$$

Therefore, the number of node equations needed is

$$n - 1 - n_v = 4 - 1 - 1 = 2$$

and the number of loop equations needed is

$$l - n_i = b - (n - 1) - n_i = 6 - 3 - 1 = 2$$

For node analysis, the unknown node voltages v_2 and v_3 are chosen as before, Figure 3-14(b), since $v_1 = 2v_2$. The two node equations are written as before. (Remember: We don't know the current through the voltage source—and we don't need KCL at node 1!)

For loop analysis, since the network is planar, the three mesh currents i_x, i_y, and i_z are shown in Figure 3-14(d). Again, the graph is not needed and we can draw them directly in Figure 3-14(a). Of these three, two are linearly dependent, since the current source is

$$3 = i_y - i_x$$

The two needed loop equations are written around any two loops which do not include the current source. (Remember: We don't know the voltage across the current source—and we don't need KVL around loop x or y!) Here, around loop z we write

$$2i_a + 4(i_z - i_y) + 3(i_z - i_x) = 0$$

Substitute for the controller i_a

$$i_a = -i_y = -3 - i_x$$

to get the first loop equation

$$-9i_x + 7i_z = 18$$

The second loop equation is written around the outside loop

$$-2i_y + 1i_y + 2i_x = 0$$

or

$$2i_x - 3 - i_x = 0$$

as the second loop equation, yielding immediately the answer

$$i_x = 3$$

Then

$$i_y = 6 \qquad i_z = 6.43$$

To check, we have here

$$v_2 = 1(-i_y) = -6$$
$$v_3 = 4(i_y - i_z) = -1.71$$

Probs. 3-17,
■ 3-18, 3-19, 3–

As a final remark, let us mention that for a *nonplanar* network, node analysis follows the same steps as always. Loop analysis, however, requires *fundamental* loops, not just any loops, in order to guarantee the proper equations and the correct solution.

In *all* cases, the first consideration must be economical: How many node equations $(n - 1 - n_v)$ versus how many loop equations $(b - n + 1 - n_i)$. *The lesser set is to be chosen for the given network*, formulated, and solved. A quick example will illustrate this important point again.

EXAMPLE 3-9 _____

In the network shown in Figure 3-15, we count

$$b = 6 \qquad n = 4 \qquad n_v = 3 \qquad n_i = 0$$

As a result, we need $6 - 4 + 1 = 3$ loop equations, with the three "obvious" meshes. How many node equations? The answer is $4 - 1 - 3 = 0$, no node equations at all! Surprised? Look again: The voltages of all three nodes with respect to a reference node are known; therefore, there is no need to solve any equations. The network *is* solved.

(Equally surprising is the number of people who will rush in to write the three loop equations.)

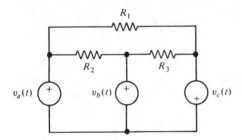

Figure 3-15   Example 3-9.   ■

PROBLEMS

3-1   Draw the oriented graph and one of its isomorphic graphs for each network shown in the figure. Is each graph planar? Nonplanar?

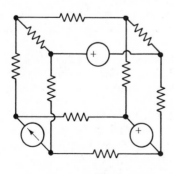

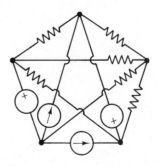

(a)   (b)

Problem 3-1

3-2 In some applications, we must count the total number of branches b in a graph. Devise a systematic, foolproof method for it. *Note*: Marking off each branch with a pencil while counting is *not* foolproof, especially in complicated graphs as shown in the figure. *Hint*: Every branch is incident to *exactly two* nodes. Illustrate your method for the graph shown.

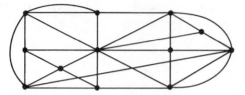

Problem 3-2

3-3 Enumerate *all* the loops in Figure 3-2(c) and (d).

3-4 Given that the graph in Figure 3-3 has a total of 16 trees, enumerate and draw them. Three are shown in Figure 3-5.

3-5 Prove that the number of branches in a path is one less than the number of nodes. *Hint*: Follow the definition of a path. Prove heuristically.

3-6 Prove Equation 3-3, relating the number of branches and nodes. *Hint*: Follow the construction of a tree. Prove heuristically or by mathematical induction.

3-7 Compare the properties of a tree with those of the co-tree. In particular: (1) Must a co-tree be a *connected* subgraph? (2) Must it contain all the original nodes? (3) Must it not contain closed loops? Give examples and counterexamples.

3-8 **(a)** Rewrite KCL for *all* the nodes, Equation 3-6, in matrix form as

$$\mathbf{A}_a \mathbf{i}_b = \mathbf{0}$$

where $\mathbf{A}_a$ is of order $(n \times b)$—a row for every node, a column for every branch—and is called the *incidence matrix* for the oriented graph. The matrix $\mathbf{i}_b$ is the column matrix, of order $(b \times 1)$, of the branch currents.
(b) What is typical of every column of $\mathbf{A}_a$? Explain why.
(c) As a consequence of (b), what happens if you add any $n - 1$ rows to the nth row?
(d) As a consequence of (b), can we delete any one row from $\mathbf{A}_a$ without loss of information? How can we restore that row, if needed later?
(e) In mathematical terms, we say that the KCL equation at the nth node is *linearly dependent* on the other $n - 1$ equations. Linear dependence means that such an equation is the sum, with some constant nonzero multipliers, of the other equations.

3-9 With an example of your choice, explain why we cannot delete *two* rows from $\mathbf{A}_a$. What is the sum of the $n - 2$ rows? Draw a graph for your example to illustrate your answers.

3-10 For the network shown:
(a) Draw its oriented graph. How many node equations are needed?
(b) Choose the unknown node voltages.
(c) Write and solve the node equations.
(d) From (c), calculate the current in the $\frac{1}{4}$-Ω resistor.

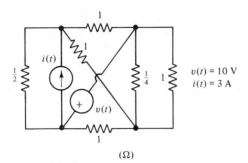

(Ω)   **Problem 3-10**

3-11 Repeat Problem 3-10 for the network shown.

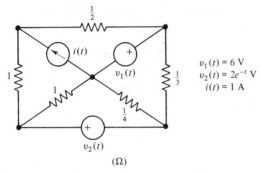

$v_1(t) = 6$ V
$v_2(t) = 2e^{-t}$ V
$i(t) = 1$ A

(Ω)

Problem 3-11

3-12 Solve Example 3-4 in the text by using fundamental cut sets. In drawing the proper tree, make sure that the current sources are links (and not tree branches). Why is this choice important?

3-13 Solve Problem 3-11 by using fundamental cut sets. See the note in Problem 3-12 about making a current source a link.

3-14 Consider the graph shown, with its tree (solid lines) and co-tree (dotted lines).
(a) Draw separately the l fundamental loops.
(b) For each fundamental loop, write KVL using branch voltages $(v_a, v_b, \ldots, v_f)$.
(c) Rewrite these l equations in matrix form as

$$\mathbf{B}_f \mathbf{v}_b = \mathbf{0}$$

——— Tree branch
– – – – Link

Problem 3-14

where $\mathbf{B}_f$ is of order $(l \times b)$—a row for every fundamental loop, a column for every branch—and is called the *fundamental loop matrix* for this graph. The matrix $\mathbf{v}_b$ is the column matrix of the branch voltages. *Note*: In parts (b) and (c), write the entries in alphabetical order; i.e., matrix $\mathbf{v}_b$ is

$$\mathbf{v}_b = \begin{bmatrix} v_a \\ v_b \\ v_c \\ v_d \\ v_e \\ v_f \end{bmatrix}$$

(d) What is typical of the left $(l \times l)$ submatrix in $\mathbf{B}_f$? Explain.

3-15 In the graph of Problem 3-14, let branch c be a current source, $i_c = 10$ A, and all the other branches resistors, $1\ \Omega$ each. Set up and solve the fundamental loop equations.

3-16 Repeat Problem 3-15, with one change: Branch d is a dependent current source, $i_d = 4i_a$. Remember: In the proper tree, current sources are links.

3-17 For the network shown:
  (a) How many node equations are needed?
  (b) How many loop equations are needed?
  (c) Choose between (a) and (b), and write these equations.
  (d) Solve (c), then calculate i_a.

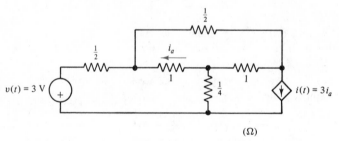

(Ω)

Problem 3-17

3-18 Repeat Problem 3-17 for the network shown.

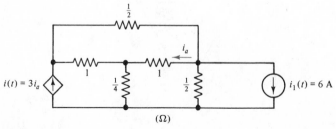

(Ω)

Problem 3-18

3-19 Repeat Problem 3-17 for the network shown.

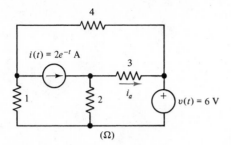

(Ω)   **Problem 3-19**

3-20 Use the fastest method of analysis (except copying!) to calculate the power of the current source i_2. Is this source delivering or receiving power?

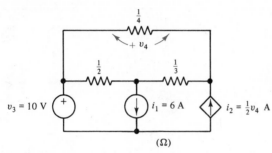

(Ω)

Problem 3-20

Chapter 4

Some Basic Resistive Circuits

4-1 SERIES CONNECTION

In many circuits, there are several ways to simplify our calculations and to take shortcuts (provided, of course, that we do so carefully). The first one is the *series connection*.

Elements (in particular, resistors) are connected *in series* if *the same current* flows through them. In Figure 4-1(a) we see a series connection of n resistors and a voltage source. Kirchhoff's current law (KCL) at each node verifies that

$$i(t) = i_1(t) = i_2(t) = \cdots = i_n(t) \tag{4-1}$$

and, therefore, all the elements are connected in series.

On the other hand, Kirchhoff's voltage law (KVL) around the loop reads

$$v(t) = v_1(t) + v_2(t) + \cdots + v_n(t) \tag{4-2}$$

Since we have Ohm's law for each resistor

$$v_k(t) = R_k i_k(t) \tag{4-3}$$

we substitute it into Equation 4-2 and use Equation 4-1:

$$
\begin{aligned}
v(t) &= R_1 i(t) + R_2 i(t) + \cdots + R_n i(t) \\
&= (R_1 + R_2 + \cdots + R_n) i(t)
\end{aligned}
\tag{4-4}
$$

This equation can be written as

$$v(t) = R_{\text{eq}} i(t) \tag{4-5}$$

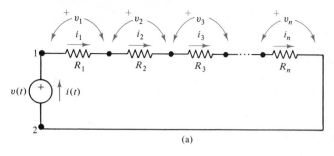

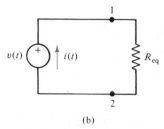

(a)

(b)

Figure 4-1   Series connection.

where the equivalent resistor

$$R_{eq} = R_1 + R_2 + \cdots + R_n = \sum_{p=1}^{n} R_p \qquad (4\text{-}6)$$

represents the equivalent resistance "seen" to the right of nodes 1 and 2. Therefore, the voltage source will have the *same* current $i(t)$ in the circuit of Figure 4-1(a) or (b).

In other words, we say: Resistors in series add up to give the equivalent total resistor.

4-2   PARALLEL CONNECTION

In a dual way to a series connection, we define a *parallel connection* of elements when they all have the *same voltage* across them. Figure 4-2(a) shows a parallel connection of *n* resistors, each one denoted by its conductance ($G = 1/R$), and a current source $i(t)$.

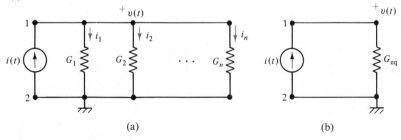

(a)   (b)

Figure 4-2   Parallel connection.

KCL at node 1 yields

$$i(t) = i_1(t) + i_2(t) + \cdots + i_n(t) \tag{4-7}$$

Since Ohm's law for each resistor is

$$i_k(t) = G_k v_k(t) \tag{4-8}$$

we have

$$\begin{aligned} i(t) &= G_1 v(t) + G_2 v(t) + \cdots + G_n v(t) \\ &= (G_1 + G_2 + \cdots + G_n) v(t) \end{aligned} \tag{4-9}$$

This equation can be written as

$$i(t) = G_{eq} v(t) \tag{4-10}$$

where the equivalent conductance

$$G_{eq} = G_1 + G_2 + \cdots + G_n = \sum_{p=1}^{n} G_p \tag{4-11}$$

represents the total conductance "seen" to the right of nodes 1 and 2 in either circuit of Figure 4-2(a) or (b). In other words, then, conductances in parallel add up to give the equivalent total conductance.

It is a common practice to use a shorthand notation for a parallel connection. For example, if R_1 is in parallel with R_2, we write $R_1 \| R_2$. The actual calculation is done, of course, with Equation 4-11.

Successive applications of series and parallel connections can simplify circuit analysis. This is illustrated in the following examples.

EXAMPLE 4-1 ⎯⎯⎯⎯⎯⎯⎯⎯⎯⎯⎯⎯⎯⎯⎯⎯⎯⎯⎯⎯⎯⎯⎯⎯⎯

Let us calculate the current in the voltage source, and all the currents and voltages in the circuit shown in Figure 4-3(a).

Solution. We do the calculations successively, with the appropriate nodes labeled for easy identification. The three resistors between nodes 4 and 2 are in parallel; so $10 \| 5 \| 4$ yield

$$G_{eq\,1} = \tfrac{1}{10} + \tfrac{1}{5} + \tfrac{1}{4} = 0.55 \, \mho$$

and

$$R_{eq\,1} = \frac{1}{G_{eq\,1}} = \frac{1}{0.55} = 1.82 \, \Omega$$

as shown in Figure 4-3(b). Next, the 2-Ω resistor is in series with 1.82 Ω, and so

$$R_{eq\,2} = 1.82 + 2 = 3.82 \, \Omega$$

shown in Figure 4-3(c). Here, there are now three resistors in parallel between nodes 3 and 2, and so $(3.82) \| (3) \| (6)$ yields

$$G_{eq\,3} = \frac{1}{3.82} + \frac{1}{3} + \frac{1}{6} = 0.762 \, \mho$$

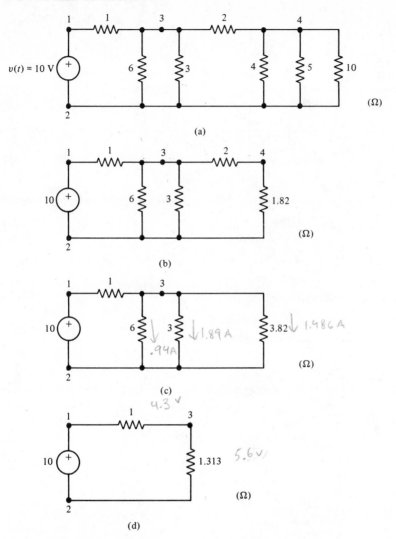

Figure 4-3 Example 4-1.

and

$$R_{eq\,3} = \frac{1}{0.762} = 1.313\ \Omega$$

Finally, in Figure 4-2(d), the two resistors are in series, so

$$R_{eq\,4} = 1 + 1.313 = 2.313\ \Omega$$

and consequently

$$i(t) = \frac{10}{2.313} = 4.324\ A$$

These calculations can be done quickly and effectively on your calculator. Symbolically, all these calculations may be summarized as

$$\{[(10\|5\|4) + 2]\|3\|6\} + 1 = 2.313 \ \Omega$$

To find the individual currents and voltages, we work "backwards," from Figure 4-3(d) to (c), to (b), and to (a):

In Figure 4-3(d) we have

$$v_{13} = (1)(4.324) = 4.324 \text{ V}$$
$$v_{32} = (1.313)(4.324) = 5.676 \text{ V}$$

with this v_{32}, we calculate in Figure 4-3(c) the currents in the 3.82-Ω, 3-Ω, and 6-Ω resistors, respectively,

$$i_{3.82 \ \Omega} = \frac{5.676}{3.82} = 1.486 \text{ A}$$

$$i_{3 \ \Omega} = \frac{5.676}{3} = 1.892 \text{ A}$$

$$i_{6 \ \Omega} = \frac{5.676}{6} = 0.946 \text{ A}$$

Moving, then, to Figure 4-3(b) we have

$$v_{42} = (1.486)(1.82) = 2.7 \text{ V}$$
$$v_{34} = (1.486)(2) = 2.972 \text{ V}$$

Finally, in Figure 4-3(a)

$$i_{10 \ \Omega} = \frac{2.7}{10} = 0.27 \text{ A}$$

$$i_{5 \ \Omega} = \frac{2.7}{5} = 0.54 \text{ A}$$

$$i_{4 \ \Omega} = \frac{2.7}{4} = 0.675 \text{ A}$$

The individual powers in the resistors are ($p = i^2 R$):

$p_1 = (4.324)^2 1 = 18.7 \text{ W}$	$p_4 = (0.675)^2 4 = 1.82 \text{ W}$
$p_6 = (0.946)^2 6 = 5.37 \text{ W}$	$p_5 = (0.54)^2 5 = 1.46 \text{ W}$
$p_3 = (1.892)^2 3 = 10.74 \text{ W}$	$p_{10} = (0.27)^2 10 = 0.73 \text{ W}$
$p_2 = (1.486)^2 2 = 4.42 \text{ W}$	

The total power in the resistors is 43.24 W. The power of voltage source is ($p = vi$):

$$p_v = -(10)(4.324) = -43.24 \text{ W}$$

delivered to all the resistors.   ■

EXAMPLE 4-2

The network shown in Figure 4-4 is a common one, known as a *ladder network* because of its shape. To find its equivalent resistance R_{eq} we start, again, at the right. The resistor R_7 is in series with G_8 (and everything in dotted lines), therefore

$$R_{aa'} = R_7 + \frac{1}{G_8 + \cdots}$$

The resistor G_6 is in parallel with $R_{aa'}$, therefore

$$G_{bb'} = G_6 + G_{aa'} = G_6 + \cfrac{1}{R_7 + \cfrac{1}{G_8 + \cdots}}$$

Again, R_5 is in series with $R_{bb'}$, so

$$R_{cc'} = R_5 + R_{bb'} = R_5 + \cfrac{1}{G_6 + \cfrac{1}{R_7 + \cfrac{1}{G_8 + \cdots}}}$$

Continuing in this fashion we obtain

$$R_{eq} = R_1 + \cfrac{1}{G_2 + \cfrac{1}{R_3 + \cfrac{1}{G_4 + \cfrac{1}{R_5 + \cdots}}}}$$

This expression is known as a *continued fraction*. Again, with numerical values for the R's and G's, it is easily calculated.

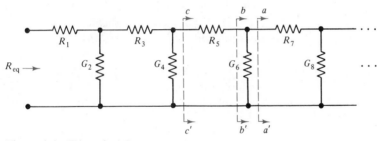

Figure 4-4 Example 4-2.

Probs. 4-1, 4-2, 4-3, 4-4

What if there are no series or parallel connections? To find the equivalent resistance (or conductance) then, we fall back on the basic methods, loop or node analysis. This is illustrated in the next two examples.

EXAMPLE 4-3 _____

Calculate R_{eq} at nodes 1–2 as shown in Figure 4-5. We recognize this network as a *bridged ladder* (the 5-Ω resistor bridges across the ladder), and no simple series or parallel combinations exist.

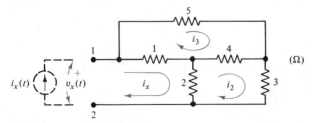

Figure 4-5 Example 4-3.

Solution.   To rephrase our problem, we are looking for a single resistor which, when connected to an arbitrary current source $i_x(t)$, shown in dotted lines, will have the same voltage $v_x(t)$ as the original network. Let us, therefore, calculate this $v_x(t)$.

Loop analysis is called for: two unknown loop currents, i_2 and i_3, as shown. (Node analysis requires three unknowns.) The two loop equations are

$$9i_2 - 4i_3 = 2i_x$$
$$-4i_2 + 10i_3 = i_x$$

Their solution is

$$i_2 = \tfrac{24}{74} i_x \qquad i_3 = \tfrac{17}{74} i_x$$

Therefore

$$v_x(t) = 1(i_x - i_3) + 2(i_x - i_2) = 2.122 i_x$$

and finally

$$R_{eq} = \frac{v_x}{i_x} = 2.122 \ \Omega \qquad\qquad \blacksquare$$

Obviously, we could have connected, instead of $i_x(t)$, a test voltage source $v_x(t)$, and solved for the resulting current $i_x(t)$. The main point is to realize the equivalence between the two circuits at nodes 1–2. This is emphasized in Figure 4-6. A second very important point to stress is also shown in the figure: The complicated network may contain resistors and *dependent* sources, but independent sources are not allowed. A little thought will make this restriction clear. Independent sources are (as their name implies) arbitrary. An independent voltage source, for example, may be 6 V, −10 sin 200*t* V, etc. It has no relationship to the resistors in that network. On the other hand, we are looking for an equivalent resistance which characterizes the

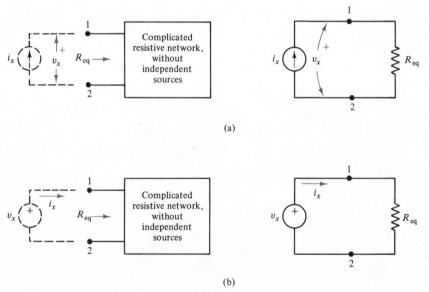

(a)

(b)

Figure 4-6   Equivalent resistance.

network and which depends very much on the values of those resistors. Consequently, we exclude independent sources.

Dependent sources, as expected, *are* related to the values of currents or voltages of resistors in that network, and, therefore, to the values of these resistors. We allow, then, dependent sources.

EXAMPLE 4-4 _____

Calculate R_{eq} at nodes 1–2 of the network shown in Figure 4-7, where a current-dependent current source is included with α, a constant, given.

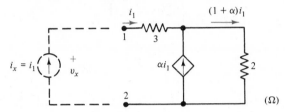

Figure 4-7   Example 4-4.

Solution.   Even though this dependent source is in parallel with the 2-Ω resistor, we don't know how to treat this combination. We must resort, therefore, to the general approach with a test current source $i_x (= i_1$ here). A simple calculation (KCL) shows that the current through the 2-Ω resistor is $(1 + \alpha)i_1$. Then KVL around the outer loop yields

$$-v_x + 3i_1 + 2(1 + \alpha)i_1 = 0$$

or

$$v_x = (5 + 2\alpha)i_1$$

and

$$R_{eq} = (5 + 2\alpha)\ \Omega$$

■ Probs. 4-5, 4-6, 4-7, 4-8

4-3 VOLTAGE DIVIDER AND CURRENT DIVIDER

Consider, again, the series connection of n resistors, as shown in Figure 4-8, and the voltage $v(t)$ across the entire series. We are interested in the voltage $v_k(t)$ across a single resistor R_k. Because of the series connection, we have

$$i(t) = \frac{v(t)}{R_1 + R_2 + \cdots + R_n} \tag{4-12}$$

Figure 4-8   Voltage divider.

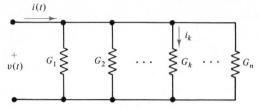

Figure 4-9 Current divider.

and across R_k

$$v_k(t) = R_k i_k(t) = R_k i(t) \tag{4-13}$$

Combining these two equations, we get

$$v_k(t) = \frac{R_k}{R_1 + R_2 + \cdots + R_n} v(t) \tag{4-14}$$

This relation is the voltage divider equation. In words, we say that the voltage across one of several resistors in series is a fraction of the total voltage; that fraction is the ratio of that resistor to the total resistance.

In a dual fashion, consider the parallel connection shown in Figure 4-9, with a total current $i(t)$. To find an individual current $i_k(t)$, we write

$$v(t) = \frac{i(t)}{G_1 + G_2 + \cdots + G_n} \tag{4-15}$$

and the current through G_k is

$$i_k(t) = G_k v_k(t) = G_k v(t) \tag{4-16}$$

that is,

$$i_k(t) = \frac{G_k}{G_1 + G_2 + \cdots + G_n} i(t) \tag{4-17}$$

This is the current divider equation, stating that the current through one of several conductances in parallel is a fraction of the total current, the fraction being the ratio of that conductance to the total conductance.

EXAMPLE 4-5 _____

Let us rework parts of Example 4-1 by using voltage and current dividers.

Solution. Specifically, in Figure 4-3(d), we have a voltage divider

$$v_{32} = \frac{1.313}{1.313 + 1} (10) = 5.676 \text{ V}$$

In Figure 4-3(c), we use current division

$$i_{3.82\ \Omega} = \frac{\dfrac{1}{3.82}}{\dfrac{1}{6} + \dfrac{1}{3} + \dfrac{1}{3.82}}(4.324) = 1.486\ \text{A}$$

$$i_{3\ \Omega} = \frac{\dfrac{1}{3}}{\dfrac{1}{6} + \dfrac{1}{3} + \dfrac{1}{3.82}}(4.324) = 1.892\ \text{A}$$

$$i_{6\ \Omega} = \frac{\dfrac{1}{6}}{\dfrac{1}{6} + \dfrac{1}{3} + \dfrac{1}{3.82}}(4.324) = 0.946\ \text{A}$$

and in Figure 4-3(b) a voltage divider

$$v_{34} = \frac{2}{2 + 1.82}(5.676) = 2.972\ \text{V}$$

■ Probs. 4-9, 4-10, 4-11

4-4 THE DELTA-WYE EQUIVALENCE

There are two connections of resistors which are fairly common, particularly in electric power systems. Shown in Figure 4-10, they are called "delta" (Δ) and "wye" (Y). Other names are "pi" (Π) for "delta" and "T" or "star" for "wye."

We define the equivalence between the Δ and the Y with respect to every two of the three outside nodes. Specifically, the resistance between nodes 1–2, 2–3, and 1–3 must be the same for both configurations. Therefore we have, between nodes 1–2,

$$R_b\|(R_a + R_c) = R_1 + R_3 \tag{4-18}$$

and between nodes 2–3,

$$R_c\|(R_a + R_b) = R_1 + R_2 \tag{4-19}$$

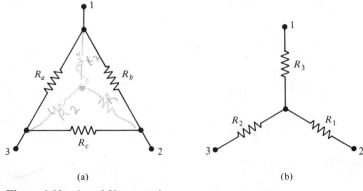

(a)   (b)

Figure 4-10 Δ and Y connections.

and between nodes 1–3,

$$R_a \| (R_b + R_c) = R_2 + R_3 \tag{4-20}$$

These three equations can be solved for R_a, R_b, and R_c when R_1, R_2, and R_3 are given or, vice versa, for R_1, R_2, and R_3 if R_a, R_b, and R_c are given. The results are

$$R_1 = \frac{R_b R_c}{R_a + R_b + R_c} \tag{4-21}$$

$$R_2 = \frac{R_a R_c}{R_a + R_b + R_c} \tag{4-22}$$

$$R_3 = \frac{R_a R_b}{R_a + R_b + R_c} \tag{4-23}$$

Alternately,

$$R_a = \frac{R_1 R_2 + R_1 R_3 + R_2 R_3}{R_1} \tag{4-24}$$

$$R_b = \frac{R_1 R_2 + R_1 R_3 + R_2 R_3}{R_2} \tag{4-25}$$

$$R_c = \frac{R_1 R_2 + R_1 R_3 + R_2 R_3}{R_3} \tag{4-26}$$

There are several mnemonic phrases that you can devise for these; for example, for Equations 4-21 to 4-23 use "the product of the two adjacent resistors divided by the sum," etc. In practice, it is better to use the actual given values, set up the equations of equivalence (Equations 4-18, 4-19, and 4-20), and solve them.

Probs. 4-11, 4-12

EXAMPLE 4-6 _____

Let us rework Example 4-3 using the Δ-Y equivalence.

    Solution.   Refer to Figure 4-11(a), where we show the equivalent Δ in dotted lines. Using Equations 4-24 to 4-26, we calculate

$$R_a = \frac{(1)(2) + (1)(4) + (2)(4)}{4} = 3.5 \, \Omega$$

$$R_b = \frac{(1)(2) + (1)(4) + (2)(4)}{2} = 7 \, \Omega$$

$$R_c = \frac{(1)(2) + (1)(4) + (2)(4)}{1} = 14 \, \Omega$$

The equivalent network is drawn in Figure 4-11(b), and we have

$$R_{eq} = 3.5 \| [(5 \| 7) + (14 \| 3)]$$

i.e.,

$$R_{eq} = \cfrac{1}{\cfrac{1}{3.5} + \cfrac{1}{\cfrac{1}{\frac{1}{5}+\frac{1}{7}} + \cfrac{1}{\frac{1}{14}+\frac{1}{3}}}} = 2.122\ \Omega$$

as before.

Alternately, we could convert the delta ($2\ \Omega$, $3\ \Omega$, $4\ \Omega$) into a Y, as shown in Figure 4-11(c). Using Equations 4-21 to 4-23 we get

$$R_1 = \frac{(2)(4)}{2 + 3 + 4} = 0.889\ \Omega$$

$$R_2 = \frac{(4)(3)}{2 + 3 + 4} = 1.333\ \Omega$$

$$R_3 = \frac{(2)(3)}{2 + 3 + 4} = 0.667\ \Omega$$

The resulting circuit is shown in Figure 4-11(d), where

$$R_{eq} = 0.667 + [(1 + 0.889)\|(5 + 1.333)]$$

i.e.,

$$R_{eq} = 0.667 + \cfrac{1}{\cfrac{1}{1.889} + \cfrac{1}{6.333}} = 2.122\ \Omega$$

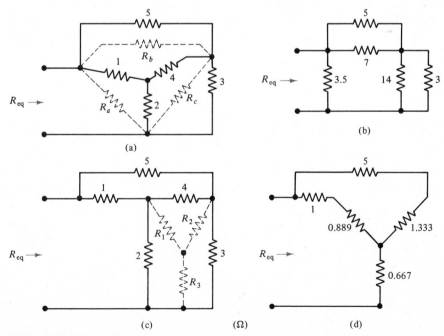

(a)

(b)

(c)

(Ω)

(d)

Figure 4-11 Example 4-6.

Probs. 4-13,
4-14, 4-15,
4-16, 4-17

4-5 THE OPERATIONAL AMPLIFIER

The operational amplifier (*op amp*, for short) is a very important and versatile circuit element. It is essentially a dependent source, and has earned itself a central place in many applications of electronic circuits. The adjective "operational" refers to the several mathematical operations (e.g., addition, multiplication, integration) that can be done with the help of op amp circuits. We will study them in this chapter and also later. Modern integrated circuit techniques provide us with cheap (less than 50 cents), reliable, and accurate op amps. In our study, we will concentrate on the fundamental operation, models, and circuit applications of op amps. The internal (actual) structure of op amps is not needed in order to understand and analyze their circuits. In this sense, then, we'll be doing the same thing as with resistors and sources: study, analyze, and apply the appropriate models.

The symbol of an op amp is shown in Figure 4-12(a). The op amp is a multiterminal device (unlike, say, a resistor, which has only two terminals). The two input terminals are labeled and called *inverting* and *noninverting terminals*, as shown. The reason for these names will become obvious soon. The output terminal is at the apex of the triangle. The two terminals marked with $+V$ and $-V$ are for the power supply to the op amp.

The circuit model of the op amp is shown in Figure 4-12(b). The relevant voltages are measured and referenced *with respect to a common reference* as v_1, v_2, and v_o, respectively. The two input currents and the output current are also shown, i_1, i_2, and i_o.

The op amp is a voltage-dependent voltage source and, as such, is shown in Figure 4-13(a). Here the input resistance R_i is very high (typically 10^5 Ω), the output resistance is low ($R_o \approx 10$ Ω), and A, the *open-loop gain*, very high ($A \approx 10^5$). The

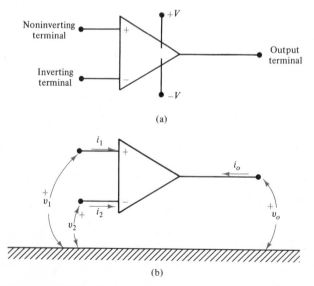

(a)

(b)

Figure 4-12 Op amp.

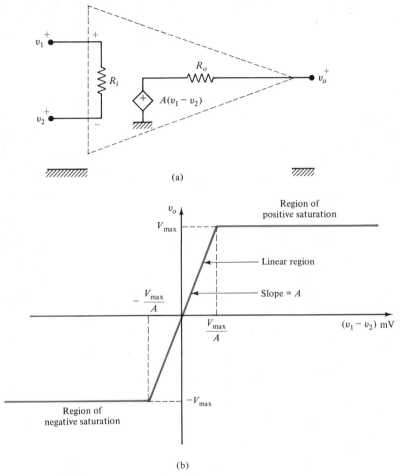

(a)

(b)

Figure 4-13   The op amp as a dependent source.

voltage source depends on the difference $(v_1 - v_2)$ of the input voltages; thus, of the op amp is a voltage-dependent voltage source.

In Figure 4-13(b) we see the output-input characteristic curve of the op amp. Over the linear region, v_o is linearly proportional to $(v_1 - v_2)$

$$v_o = A(v_1 - v_2) \qquad (4\text{-}27)$$

until saturation is reached. Typical values in this region are $V_{max} = 15$ V, and so $(v_1 - v_2)$ is typically a few millivolts $(1 \text{ mV} = 10^{-3} \text{ V})$. This explains the term "amplifier": The input voltage, a few millivolts, is amplified to several volts at the output. In the saturation regions, the output of the op amp remains $+V_{max}$ or $-V_{max}$. In almost all applications, and in our studies here, we will assume that the op amp remains in the linear (nonsaturated) region and therefore is in the same category—a linear element—as a resistor.

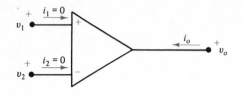

Figure 4-14   Ideal op amp.

An ideal model of an op amp, suitable for our calculations, is defined by the following relations:

$$R_i \to \infty \tag{4-28}$$

$$R_o \to 0 \tag{4-29}$$

$$A \to \infty \tag{4-30}$$

$$v_1 - v_2 = 0 \tag{4-31}$$

$$i_1 = i_2 = 0 \tag{4-32}$$

Note that Equation 4-28 implies Equation 4-32, while Equations 4-30 and 4-31 imply a finite, nonzero v_o. See Figure 4-14. We are ready now to analyze our first op amp circuit.

EXAMPLE 4-7 _____

(a) Calculate the output voltage v_σ in the circuit shown in Figure 4-15, with $v_a = 0$, $v_b = 0.5$ V. (b) Repeat, with $v_a = 1.5$ V, $v_b = 0.5$ V.

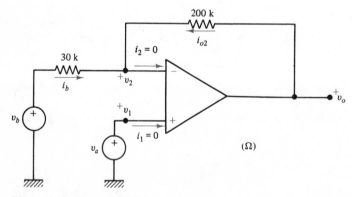

Figure 4-15   Example 4-7.

Solution.   (a) Since $v_a = 0$, we have $v_1 = 0$ (the noninverting terminal is grounded). Since $v_1 = v_2$ by Equation 4-31, we have here $v_2 = 0$. As a result

$$i_b = \frac{v_b - v_2}{3 \times 10^4} = \frac{0.5 - 0}{3 \times 10^4} = \frac{1}{60} \text{ mA}$$

The current in the 200-kΩ resistor, as shown, is

$$i_{02} = \frac{v_o - v_2}{2 \times 10^5} = \frac{v_o}{200} \text{ mA}$$

(as usual, the double subscript 02 indicates "from the output node 0 to node 2").

Apply KCL now at the inverting node 2, and remember that $i_2 = 0$, as in Equation 4-32. We get

$$i_b + i_{02} = 0$$

or

$$\frac{1}{60} \times 10^{-3} + \frac{v_o}{200} \times 10^{-3} = 0$$

i.e.,

$$v_o = -3.33 \text{ V}$$

(b) Using the same steps, we get here

$$v_2 = v_1 = v_a = 1.5 \text{ V} \qquad v_b = 0.5 \text{ V}$$

$$\therefore \quad i_b = \frac{0.5 - 1.5}{30 \times 10^3} = -\frac{1}{30} \text{ mA} \qquad \therefore \quad i_{02} = \frac{v_o - 1.5}{200} \text{ mA}$$

Since $i_b + i_{02} = 0$

$$\therefore \quad -\frac{1}{30} + \frac{v_o - 1.5}{200} = 0$$

$$\therefore \quad v_o = 8.17 \text{ V}$$

Note, for the moment in this example, that v_o is positive in one case and negative in the other, and also that amplification occurs in both cases. Also, the value of A ($\to \infty$) does not enter into the calculations. ■

Before we discuss several basic op amp circuits, let us make some general observations that will apply in all cases:

1. Even though topological considerations (Chapter 3) can be applied to the op amp, we won't do so, for the next reasons.
2. The equations of the op amp itself are very simple. They are essentially Equations 4-31 and 4-32, repeated here:

$$v_1 = v_2 \qquad\qquad (4\text{-}31)$$

and

$$i_1 = i_2 = 0 \qquad\qquad (4\text{-}32)$$

for the two input terminals.

3. The circuit that includes the op amp will dictate the other equations. These will invariably be some, or all, of:
   (a) KCL at the noninverting terminal
   (b) KCL at the inverting terminal
   (c) Ohm's law for one or more resistors between input and output
   (d) KVL in the input loop
   (e) KVL in the output loop

That's all, and it's that simple! Let us study now some standard op amp circuits.

EXAMPLE 4-8

The so-called *inverting amplifier* circuit is shown in Figure 4-16(a). The input voltage v_i is connected through a series resistor R_s, and there is a *feedback* resistor R_f from the output to the input.

Solution. As in the previous example, we have here

$$i_s = \frac{v_i - v_2}{R_s}$$

and

$$i_f = \frac{v_o - v_2}{R_f}$$

But

$$v_2 = v_1 = 0 \qquad \text{(datum)}$$

KCL at the inverting terminal yields

$$i_s + i_f = 0$$

that is,

$$\frac{v_i}{R_s} + \frac{v_o}{R_f} = 0$$

Therefore,

$$v_o = -\frac{R_f}{R_s} v_i \qquad\qquad (4\text{-}33)$$

The *closed-loop* voltage gain is defined as (v_o/v_i), and is here the finite ratio $(-R_f/R_s)$. It is not the same as the open-loop gain $(= A)$ in the absence of the feedback resistor.
   We can control the value of the closed-loop gain by properly choosing R_f and R_s. The output voltage v_o here has a negative polarity with respect to the input voltage v_i, hence the name *inverting* amplifier. The mathematical operation performed here is multiplication (sometimes called *scaling*) of v_i to get v_o.
   A shorthand symbol for this amplifier is shown in Figure 4-16(b), where $K = R_f/R_s$ and $v_o = -Kv_{\text{in}}$.

inverting

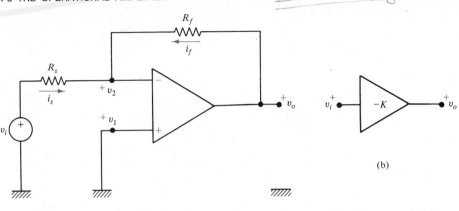

(a)

Figure 4-16 Inverting amplifier.

(b)

■ Prob. 4-18

EXAMPLE 4-9

Analyze the following op amp circuit, known as a *noninverting* amplifier. See Figure 4-17(a).

Solution. Here we have $i_s = i_1 = 0$. Consequently

$$v_i = v_1$$

In the output, we have a voltage divider (since $i_2 = 0$)

$$v_2 = \frac{R_2}{R_1 + R_2} v_o$$

But in the op amp

$$v_1 = v_2$$

non inverting

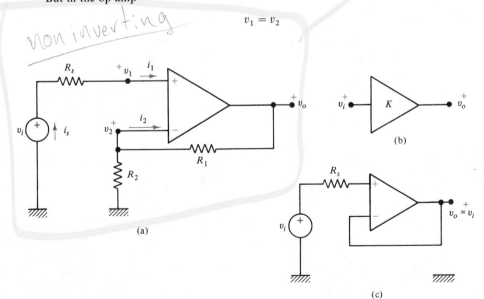

(a)

(b)

(c)

Figure 4-17 Noninverting amplifier.

Therefore

$$v_i = \frac{R_2}{R_1 + R_2} v_o$$

or

$$v_o = \frac{R_1 + R_2}{R_2} v_i = \left(1 + \frac{R_1}{R_2}\right) v_i \qquad (4\text{-}34)$$

with a closed-loop gain of $1 + (R_1/R_2)$. Again, we can choose values of R_1 and R_2 to vary this gain. Here the polarity of v_o is the same as v_i, and the amplifier is noninverting. The shorthand symbol for this amplifier is shown in Figure 4-17(b), with $K = (R_1 + R_2)/R_2$ and $v_o = Kv_i$.

If we choose $R_1 = 0$, we obtain

$$v_0 = v_i$$

and R_2 is of no consequence. Such a circuit is called a *voltage follower*, the output voltage being equal to the input, and is shown in Figure 4-17(c). Why do we need such an elaborate circuit if all we get is a voltage that is equal to itself? There is a very good reason: In many cases, we want to isolate electrically one part of a circuit without affecting voltages. An op amp has a very high input resistance ($R_i \to \infty$) and draws no input current ($i_1 = i_2 = 0$); this circuit, therefore, provides a good isolation, a *buffer*, between an input and an output. ■

Probs. 4-19, 4-20, 4-26

EXAMPLE 4-10 _____

Analyze the op amp circuit in Figure 4-18.

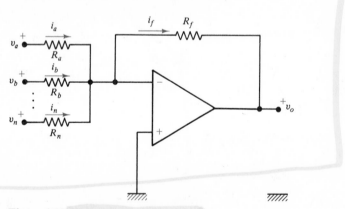

Figure 4-18 Adder (summer) circuit.

Solution. Applying KCL to the inverting terminal, we have

$$i_a + i_b + \cdots + i_n = i_f$$

i.e.,

$$\frac{v_a - v_2}{R_a} + \frac{v_b - v_2}{R_b} + \cdots + \frac{v_n - v_2}{R_n} = \frac{v_2 - v_o}{R_f}$$

But $v_2 = v_1 = 0$, and therefore

$$v_o = -\left(\frac{R_f}{R_a} v_a + \frac{R_f}{R_b} v_b + \cdots + \frac{R_f}{R_n} v_n\right) \tag{4-35}$$

This circuit, then, performs the operation of *addition* of several inputs; each input may be scaled by a factor R_f/R_k. In hi-fi circuits, such an adder may be used to mix (combine) several signals from several sources into one output.

Another interesting feature here is as follows: With each individual input alone, say, $v_a \neq 0$, $v_b = 0, \ldots, v_n = 0$, the circuit is an inverting amplifier obeying Equation 4-33. For several inputs, $v_a \neq 0$, $v_b \neq 0, \ldots, v_n \neq 0$, the total output is the *sum* of their individual outputs, Equation 4-35.

Probs. 4-21, 4-22, 4-23, 4-24, 4-25 ∎

PROBLEMS

4-1 Calculate the equivalent resistance of each network shown.

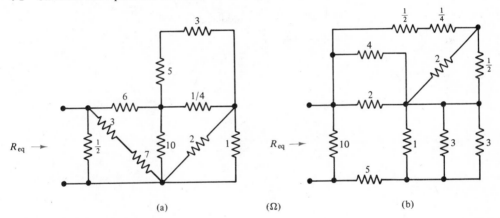

(a)

(Ω)

(b)

Problem 4-1

4-2 (a) Prove that the equivalent resistance of *two* resistors in parallel is

$$R_{eq} = \frac{R_1 R_2}{R_1 + R_2}$$

leading to the familiar rule, "product over sum." Prove quickly, without any algebra, that the same rule *is not true* for three (or more) resistors,

$$R_{eq} \neq \frac{R_1 R_2 R_3}{R_1 + R_2 + R_3}$$

(b) Prove the following geometrical construction to find R_{eq} for two resistors in parallel: Draw, to a proper scale, $AB = R_1$, as shown. Draw $CD = R_2$, parallel to AB. Connect B to C and D to A with straight lines. They intersect at P. Draw PQ parallel to AB. The length PQ is R_{eq} on the same scale.

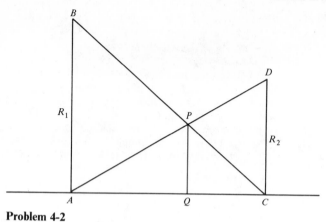

Problem 4-2

4-3 Calculate the equivalent resistance of the ladder network shown.

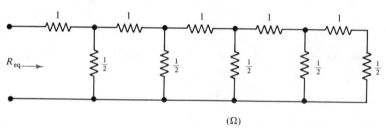

(Ω)

Problem 4-3

4-4 Calculate the equivalent resistance (R) of an *infinite* ladder network made up of 1-Ω resistors. *Hint*: Add two more resistors, as shown. What is then the equivalent resistance at $b - b'$?

$R = \dfrac{1 + \sqrt{5}}{2}$

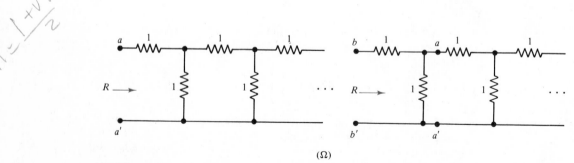

(Ω)

Problem 4-4

4-5 Calculate the equivalent conductance G_{eq} at nodes 1–2 of the network shown.

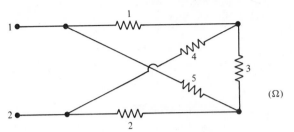

Problem 4-5

4-6 Repeat Problem 4-5 for the network shown; specifically here, your answer should be the same no matter what R_o is! See also Problem 2-21.

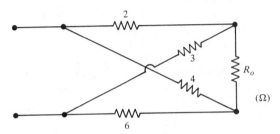

Problem 4-6

4-7 A certain circuit (using transistors) is modeled as shown. Here α and β are given constants. Find the equivalent resistance at nodes 1–2.

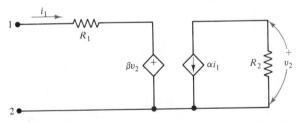

Problem 4-7

4-8 What is the equivalent resistance at nodes 1–2? What is it when $\alpha = -2$?

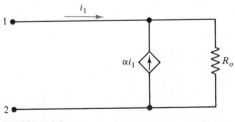

Problem 4-8

4-9 **(a)** In the network shown, what is the voltage v_{ab}?

   (b) If we connect a resistor $R_{ab} = 5\,\Omega$ across a–b, calculate the current in this resistor. Explain carefully why this current is not equal to v_{ab} of part (a) divided by $R = 5\,\Omega$.

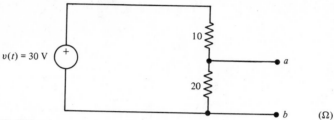

Problem 4-9

4-10 In the figure we show a proposed method for measuring unknown constant voltages. The voltage source V_o causes a current I_o (which is adjustable by means of the variable resistor R_o). Let us assume $I_o = 40\,\text{mA}$. The resistor AC is a uniform wire resistor R_w of total length l mm. The galvanometer G is a sensitive current-measuring instrument. The unknown dc voltage is connected to DE, and the sliding contact S is adjusted until G shows zero deflection (zero current). Calculate the unknown voltage v_{DE} in terms of the resistance of the portion SC. Discuss also the ranges of unknown voltages that can be measured by using different values of I_o (changing R_o).

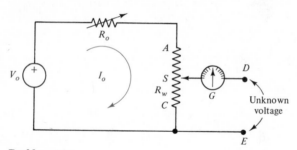

Problem 4-10

4-11 Verify, in full detail, the results given in Equations 4-21 through 4-26.

4-12 Rewrite Equations 4-24 to 4-26 using conductances. The results should resemble Equations 4-21 to 4-23.

4-13 Rework Problem 4-3 using the Δ-Y equivalence.

4-14 Rework Problem 4-5 using the Δ-Y equivalence.

4-15 Rework Problem 4-6 using the Δ-Y equivalence.

4-16 A dc ammeter is a measuring instrument for dc current. It consists of a moving coil to which a needle is attached. The deflection of the needle is directly proportional to the current flowing in the coil. This coil is rated by its current and voltage (for example, 10 mA and 100 mV). At this current rating, the needle is deflected to its full scale, and then the voltage drop across the coil is the rated one. In order to allow different ranges of currents to be measured in excess of the current rating of the coil, a resistor is connected in parallel ("shunt") with the coil by means of a selector switch, S. In the figure, R_3 is shunted across the coil. For our given example (10 mA, 100 mV), calculate the values of R_1, R_2, R_3, R_4 in order to be able to measure (full-scale reading) $I = 100$ mA, 1 A, 10 A, 20 A.

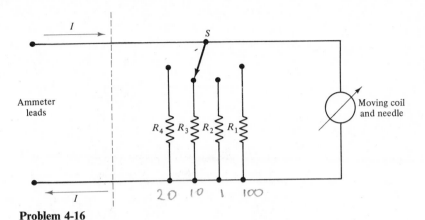

Problem 4-16

4-17 A dc voltmeter uses the same moving coil as the ammeter (Problem 4-16), but here a series resistor is connected in order to vary the range of voltages. Again, for the coil rated at 10 mA, 100 mV, calculate R_1, R_2, R_3, and R_4 in order to measure (full scale) 1 V, 5 V, 10 V, and 100 V.

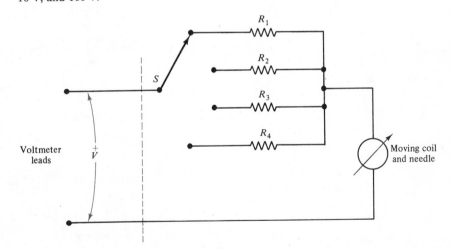

Problem 4-17

4-18 Derive the equation of an inverting amplifier (Equation 4-33) by using the op amp model of Figure 4-13(a) and appropriate limiting values ($R_o \to 0, A \to \infty$, etc.).

4-19 Derive the equation of a noninverting amplifier (Equation 4-34) by using the op amp model of Figure 4-13(a) with appropriate limiting values.

4-20 In the circuit shown, what is the output voltage v_o without the load resistor R_L? With the load resistor R_L? Use a buffer amplifier (voltage follower) to connect R_L without affecting v_o.

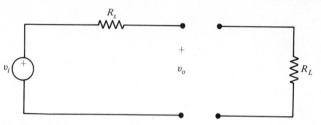

Problem 4-20

4-21 The circuit shown is proposed as an *ohmmeter* to measure unknown resistors R_x. The voltmeter's reading V is recorded. Find the value of R_x in terms of V. (*Important*: Notice the positive terminal of the voltmeter.)

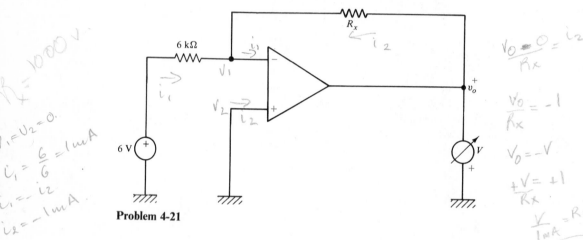

Problem 4-21

4-22 In the op amp circuit shown, calculate v_o in terms of the input voltages v_a and v_b.

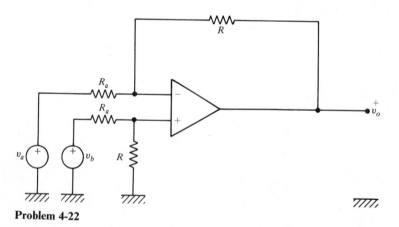

Problem 4-22

4-23 Analyze the op amp circuit shown to find the gain v_o/v_i. Is it an inverting or a noninverting amplifier?

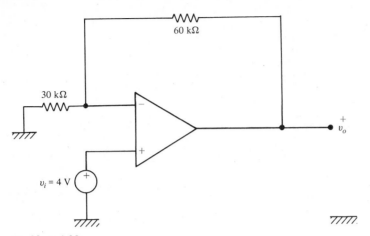

Problem 4-23

4-24 The circuit shown is proposed as a converter from a voltage source to a current source. Analyze it to verify this proposition. Here is, then, a "real-life" current source.

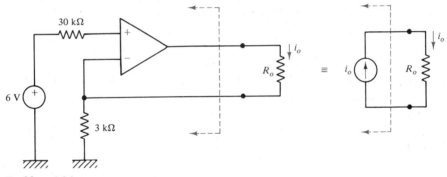

Problem 4-24

4-25 Analyze and calculate the gain in the following op amp circuit.

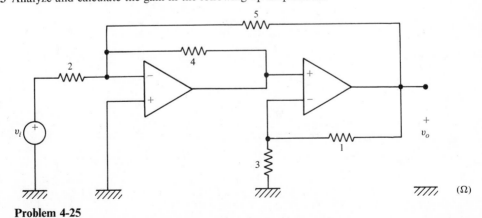

Problem 4-25

4-26 Find i_o in the op amp circuit shown.

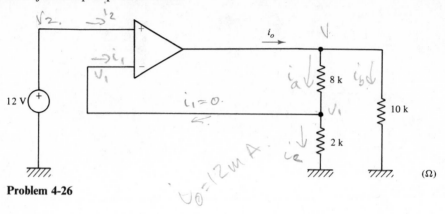

Problem 4-26

$$V_2 = V_1 = 12$$

$$i_o = i_a + i_b.$$

$$i_c = \frac{V_1}{2} = \frac{12}{2} = 6. \text{ mA}.$$

$$i_a = i_c = 6 \text{ m A}.$$

$$V = V_1 + 8 \, i_a$$

$$V = 12 + 48 = 60. \text{V}.$$

$$i_b = \frac{V}{10} = \frac{60}{10} = 6 \text{ m A}$$

$$i_o = 6 + 6 = 12 \text{ m A}.$$

Chapter 5

Network Theorems and Techniques

In this chapter we will study several theorems that are quite general. Their applications will include resistive networks first, and, in later chapters, networks with additional elements. Also, we will study more techniques of analysis.

5-1 LINEARITY AND SUPERPOSITION

In a previous chapter, we introduced a linear resistor by showing its i-v characteristic, Figure 2-1(b). Let us extend this concept and define a *linear element* as one whose output-input characteristic is a straight line passing through the origin. See Figure 5-1. The op amp (Figure 4-13) is another linear element, as long as it is not driven into saturation.

Let us call the input, a voltage or a current source, $e(t)$, where e stands for "excitation" (we can't use i for "input" because i is a standard notation for current). Likewise, let $r(t)$ be the response, a voltage or a current. Then the linear characteristic of Figure 5-1(b) has two implications:

1. If we multiply (scale) the input by a constant α, the output will be scaled by the same constant. Symbolically, we write it as follows:

$$\alpha e(t) \xrightarrow{N} \alpha r(t) \tag{5-1}$$

which is read as "the input $\alpha e(t)$ to N produces the output $\alpha r(t)$."

2. If we apply the sum of two inputs $e_1(t) + e_2(t)$, then the resulting output will be given by

$$[e_1(t) + e_2(t)] \xrightarrow{N} [r_1(t) + r_2(t)] \tag{5-2}$$

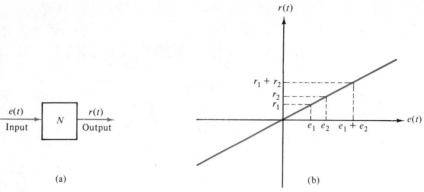

Figure 5-1 A linear element.

It makes sense, doesn't it? The total response is the sum of the individual responses.

Equations 5-1 and 5-2 may be combined into a single expression, as follows:

$$(\alpha e_1 + \beta e_2) \xrightarrow{N} (\alpha r_1 + \beta r_2) \tag{5-3}$$

where α and β are arbitrary constants. Equation 5-1 is obtained by letting $\beta = 0$, $\alpha \neq 0$. Equation 5-2 is derived with $\alpha = \beta = 1$.

The definition of *linearity* is given mathematically in Equation 5-3, and the property of adding individual responses in order to obtain the total response is called *superposition*. Briefly, we say that an element (or a network) is *linear* if it obeys the principle of superposition, and vice versa: It will obey superposition if it is linear.

EXAMPLE 5-1 _____

Consider a resistor, $R = 10\,\Omega$, as shown in Figure 5-2(a), excited by a dc voltage source $e_1(t) = 6$ V. The response is $r_1(t) = i_1(t) = 0.6$ A. In Figure 5-2(b), the input is $e_2(t) = 2$ V and the output $r_2(t) = i_2(t) = 0.2$ A.

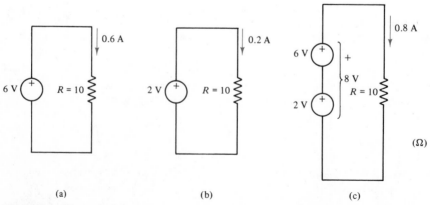

Figure 5-2 Example 5-1.

Solution. By superposition, with both sources as an input, and using Equation 5-2, we expect the output to be $r_1(t) + r_2(t) = 0.6 + 0.2 = 0.8$ A, as it is. This resistor obeys superposition and, therefore, is linear.

We note here that, in superposing the voltage sources, they are connected in series, since voltages add then. Current sources, when superposed, must be connected in parallel. ∎

EXAMPLE 5-2

Let the same resistor, $R = 10\,\Omega$, be connected across a voltage source given by $v(t) = 2e^{-t} + 4.3 \cos 20t$.

Solution. The current will be

$$i(t) = \frac{v(t)}{R} = \frac{2e^{-t} + 4.3 \cos 20t}{10} = 0.2e^{-t} + 0.43 \cos 20t = i_1(t) + i_2(t)$$

the superposition of $i_1(t)$, due to $v_1(t) = 2e^{-t}$, and of $i_2(t)$, due to $v_2(t) = 4.3 \cos 20t$. ∎

EXAMPLE 5-3

Consider now a resistor whose v-i characteristic is $v(t) = [i(t)]^3$, over the range, say, of $-4\,\mathrm{A} \le i \le 4\,\mathrm{A}$. This is shown in Figure 5-3(b), where the resistor is drawn in a box, to distinguish it.

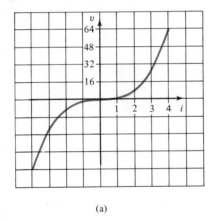

(a)

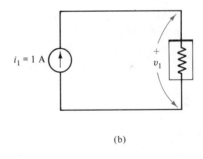

(b)

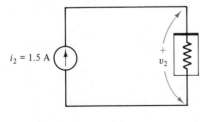

(c)

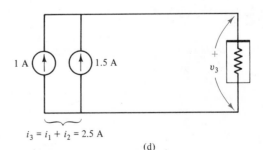

$i_3 = i_1 + i_2 = 2.5$ A

(d)

Figure 5-3 Example 5-3.

Solution. For an input $i_1 = 1$ A, the output is $v_1 = 1^3 = 1$ V. For a second input $i_2 = 1.5$ A, the output is $v_2 = (1.5)^3 = 3.375$ V. Now superpose both inputs, as in Figure 5-3(d); the total input is $i_3 = i_1 + i_2 = 2.5$ A, but the total output *is not* $1 + 3.375 = 4.375$ V. Rather, it is $i_3^3 = (2.5)^3 = 15.625$ V. Superposition is *not* obeyed here, and this resistor is *nonlinear*. ∎

In this book and in many electric circuits, both in theory and in practice, we deal only with linear elements. The reason is twofold: Many elements actually behave in a linear fashion, and the analysis of linear networks is considerably easier. There are, however, nonlinear elements whose characteristics cannot be linearized, or whose characteristics must be specifically nonlinear. Such elements and networks require nonlinear analysis, a much more difficult topic, studied in more advanced courses and texts.

EXAMPLE 5-4 ———————————————————————————————

The op amp adder (summer) circuit, in Example 4-10, is linear: Superposition is evident in the output. ∎

Probs. 5-1, 5-2, 5-3

The application of superposition to circuit analysis involves always these steps:

1. Consider one independent source (input) at a time.
2. Set all the other independent sources to zero. A current source is set to zero by replacing it with an open circuit, since then $i = 0$. A voltage source is zero when it is replaced by a short circuit, $v = 0$.
3. Dependent sources must *remain intact* because, by their very nature, they depend on a current or a voltage in the circuit.
4. Calculate the partial responses; then add them up to get the total response.

The following two examples illustrate this method.

EXAMPLE 5-5 ———————————————————————————————

Let us rework Example 2-7, repeated here, by using superposition. The entire network is shown in Figure 5-4(a).

Solution. In Figure 5-4(b) we retain only the independent voltage source, while the independent current source is made zero (o.c.). The partial responses v_2' and v_3' are calculated easily as follows: The total resistance between v_2' and reference is

$$\tfrac{1}{5} \| (1 + 0.1) = 0.169 \ \Omega$$

Then, by voltage division, we get

$$v_2' = \left(\frac{0.169}{\tfrac{1}{4} + 0.169} \right) 2 = \ = 0.807 \ \text{V}$$

as shown in Figure 5-4(c). Now v_3' can be calculated, again by voltage division in Figure 5-4(b)

$$v_3' = \left(\frac{0.1}{1 + 0.1} \right) v_2' = \left(\frac{0.1}{1.1} \right) 0.807 = 0.073 \ \text{V}$$

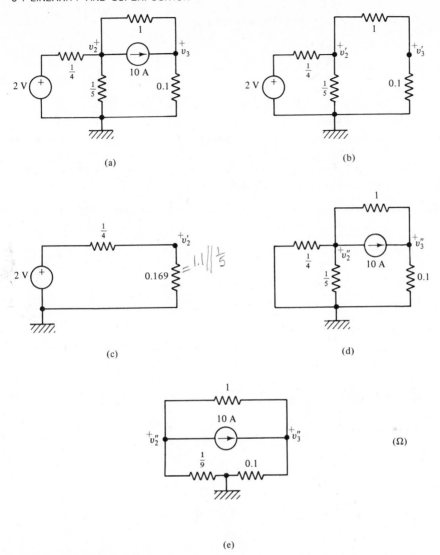

(a)

(b)

(c)

(d)

(Ω)

(e)

Figure 5-4 Example 5-5.

Next, we retain only the independent current source and make the independent voltage source zero, as shown in Figure 5-4(d). Here the two resistors $\frac{1}{4}\,\Omega$ and $\frac{1}{5}\,\Omega$ are in parallel, and the simplified circuit is redrawn in Figure 5-4(e). Using the current divider and Ohm's law, we obtain the partial responses

$$v_3'' = (0.1)\frac{4.737}{4.737 + 1}(10) = 0.826 \text{ V}$$

and

$$v_2'' = -\left(\frac{1}{9}\right)\frac{4.737}{4.737 + 1}(10) = -0.917 \text{ V}$$

Finally, by superposition, the total responses are

$$v_2 = v_2' + v_2'' = 0.807 - 0.917 = -0.11 \text{ V}$$
$$v_3 = v_3' + v_3'' = 0.073 + 0.826 = 0.9 \text{ V}$$

and the results agree with those obtained in Chapter 2. (Be sure that you agree totally with the minus sign in v_2''!)   ■

EXAMPLE 5-6 _____

Use superposition in the network shown in Figure 5-5(a) to calculate i_2. Notice the current-controlled (dependent) source.

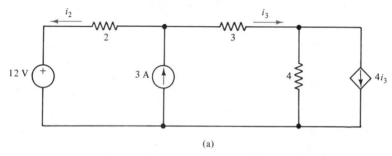

(a)

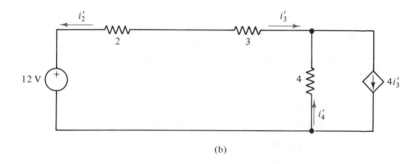

(b)

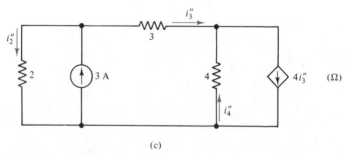

(c)

Figure 5-5 Example 5-6.

Solution. The first independent source, acting alone, but together with the dependent source, is shown in Figure 5-6(b). The reason for having to retain the dependent source, again, is: There *will* be a current i_3' in this circuit; therefore, the dependent source $4i_3'$ must

$$2\ddot{i}_2 + 3\ddot{i}_3$$
$$-2\ddot{i}_3 + 3\ddot{i}_3$$

be there too. In this circuit, we note that $i'_2 = -i'_3$. Also, at the top node of the 4-Ω resistor, we get by KCL that $i'_4 = 3i'_3$. Then we apply KVL to the single loop

$$-12 + 5i'_3 - 4(3i'_3) = 0$$

$$\therefore \quad i'_3 = -1.714 \text{ A} \qquad \therefore \quad i'_2 = -i'_3 = 1.714 \text{ A}$$

The independent current source is retained next, with the independent voltage source zero, but with the dependent source intact. See Figure 5-5(c). Here we write

$$i''_2 + i''_3 = 3 \qquad \text{(KCL)}$$
$$i''_4 = 3i''_3 \qquad \text{(KCL)}$$

and KVL around the resistive loop reads

$$2i''_2 + 4(3i''_3) - 3i''_3 = 0$$
$$\therefore \quad 2i''_2 + 9i''_3 = 0$$
$$\therefore \quad 2i''_2 + 9(3 - i''_2) = 0$$
$$\therefore \quad i''_2 = 3.86 \text{ A}$$

By superposition, then, the total current i_2 is $i_2 = i'_2 + i''_2 = 5.57$ A.

To check our result, let us solve this problem by conventional analysis. Here we don't need an elaborate, formal decision on loop analysis or node analysis. A quick, informal approach will do nicely. For the original network, Figure 5-5(a), we write

$$i_2 + i_3 = 3 \qquad \text{(KCL)}$$

and

$$-2i_2 + 3i_3 - 4(3i_3) = 12 \qquad \text{(KVL)}$$

The two equations are

$$i_2 + i_3 = 3$$
$$-2i_2 - 9i_3 = 12$$

and their solution yields

$$i_2 = 5.57 \text{ A}$$

Probs. 5-4,
5-5, 5-6, 5-7

Consider now the power in a linear resistor. With a current $i_1(t)$ through R, we have the instantaneous power

$$p_1(t) = i_1^2 R \qquad (5\text{-}4)$$

With a current $i_2(t)$, the instantaneous power is

$$p_2(t) = i_2^2 R \qquad (5\text{-}5)$$

By superposition, the total current is $i = i_1 + i_2$. Then the instantaneous power is

$$p(t) = i^2 R = (i_1 + i_2)^2 R = i_1^2 R + i_2^2 R + 2i_1 i_2 R \qquad (5\text{-}6)$$

but

$$p(t) \neq p_1(t) + p_2(t) \qquad (5\text{-}7)$$

because of the extra term $2i_2 i_2 R$. In other words, since the power-current relationship is *not* linear ($p = i^2 R$), superposition does not hold.

Prob. 5-8

5-2 LOOP AND NODE EQUATIONS REVISITED

We can make use of superposition when we want to write loop equations or node equations. The advantage in doing it so is that the equations may be written by inspection in their final form, without having to collect or rearrange terms. Sounds promising? Of course! It is best to learn this approach with specific examples, so let us consider them.

EXAMPLE 5-7 ——————————————————————————————————

Consider the network in Figure 5-6(a). Since the network is planar, we don't need graphs, trees, etc. A quick count tells us that we need three loop equations. The loop currents $i_1(t)$, $i_2(t)$, and $i_3(t)$ are shown, all chosen (conveniently, not necessarily) in one direction, clockwise.

Solution. As we studied earlier, a loop equation sums the voltages around the loop, and we use KVL, $\sum v = 0$, with a voltage drop being assigned a $(+)$ sign and a voltage rise a $(-)$ sign. Let us modify this convention now, and write, instead,

$$\sum v_R = \sum v_{\text{sources}}$$

that is, around a loop, the sum of all voltage drops in the resistors is equal to the sum of all voltage rises due to sources around that loop.

Now, apply the principle of superposition as follows: *The total voltage drop around a loop is the sum of the drops contributed by each loop current, acting alone, in that loop.* Specifically, around loop 1 we have:

1. With i_1 acting alone and $i_2 = 0$, $i_3 = 0$, the voltage drop is $(3 + 2 + 1)i_1 = 6i_1$. See Figure 5-6(b).
2. With i_2 acting alone, Figure 5-6(c), $i_1 = 0$, $i_3 = 0$, its voltage drop is across the 3-Ω resistor as shown. For loop 1, traced clockwise, it is a rise, a negative drop of $(-3i_2)$.
3. The loop current i_3, in this example, does not contribute any voltage to loop 1, that is, $(0 \cdot i_3)$. See Figure 5-6(d).
4. The total rise, due to the voltage sources around loop 1, is -6.

Consequently, KVL around loop 1 is

$$6i_1 - 3i_2 + 0 \cdot i_3 = -6$$

In a similar way we have for loop 2:

1. Loop current i_1 alone contributes $(-3i_1)$ around loop 2, in the direction of that loop. See Figure 5-6(e).
2. Loop current i_2 alone contributes $(4 + 3 + 1)i_2 = 8i_2$, as in Figure 5-6(f).
3. Loop current i_3 alone contributes $-4i_3$. See Figure 5-6(g).
4. The total rise due to sources is $-5e^{-t}$.

The second loop equation is then

$$-3i_1 + 8i_2 - 4i_3 = -5e^{-t}$$

Finally, around loop 3:

1. Loop current i_1 alone contributes $0 \cdot i_1$. See Figure 5-6(h).
2. Loop current i_2 alone contributes $(-4i_2)$ in the direction of that loop.
3. Loop current i_3 alone contributes $(10 + 4)i_3 = 14i_3$ in that loop.
4. The total rise of sources around the loop is $(v_1 - v_3)$.

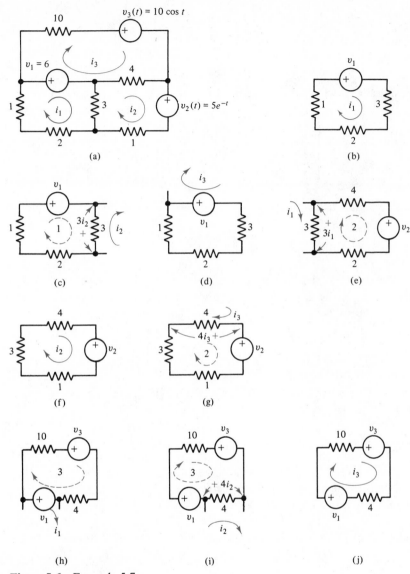

Figure 5-6 Example 5-7.

The third loop equation is, then:

$$0 \cdot i_1 - 4i_2 + 14i_3 = 6 - 10 \cos t$$

The three loop equations are repeated

$$6i_1 - 3i_2 + 0i_3 = -6$$
$$-3i_1 + 8i_2 - 4i_3 = -5e^{-t}$$
$$0i_1 - 4i_2 + 14i_3 = 6 - 10 \cos t$$

or, in matrix form

$$\begin{bmatrix} 6 & -3 & 0 \\ -3 & 8 & -4 \\ 0 & -4 & 14 \end{bmatrix} \begin{bmatrix} i_1 \\ i_2 \\ i_3 \end{bmatrix} = \begin{bmatrix} 6 \\ -5e^{-t} \\ 6 - 10\cos t \end{bmatrix}$$

ready for solution.   ■

This result agrees with the final form of Equation 2-12, rewritten here for our example as

$$\begin{bmatrix} r_{11} & r_{12} & r_{13} \\ r_{21} & r_{22} & r_{23} \\ r_{31} & r_{32} & r_{33} \end{bmatrix} \begin{bmatrix} i_1 \\ i_2 \\ i_3 \end{bmatrix} = \begin{bmatrix} e_1 \\ e_2 \\ e_3 \end{bmatrix}$$

or

$$\mathbf{Ri} = \mathbf{e} \qquad\qquad (5\text{-}8)$$

Several important observations can be made now:

1. The term $r_{11}i_1$ in the first equation, $r_{22}i_2$ in the second equation, and, in general, $r_{kk}i_k$ in the kth equation, represents the voltage contributed around the kth loop by its own loop current i_k acting alone. This term is always a voltage drop $(+r_{kk}i_k)$ in the kth loop because (obviously!) we choose the direction of the kth loop as that of i_k. We call r_{kk} the *self-resistance* of loop k.
2. The term $r_{12}i_2$ in the first equation is the voltage in loop 1 caused by i_2, flowing in a resistor common to loops 1 and 2. This term is a rise $(-r_{12}i_2)$ if i_2 flows in that resistor opposite to the direction of loop 1; it is a drop $(+r_{12}i_2)$ if i_2 flows in the resistor in the same direction as loop 1.
3. The general term $r_{pq}i_q$ is the voltage contributed to loop p by the loop current i_q flowing in a resistor common to loops p and q. This term is a rise $(-r_{pq}i_q)$ if i_q flows in that common resistor opposite to the direction of loop p; it is a drop $(+r_{pq}i_q)$ if i_q flows in the common resistor in the same direction as loop p. We call r_{pq} the *mutual resistance* between loops p and q.
4. The term e_p is the net voltage rise due to sources around loop p.

Let us practice this method with another example, this time using fundamental loops.

EXAMPLE 5-8

The network shown in Figure 5-7(a) has two independent sources, $i(t) = 10$ A, $v(t) = 2e^{-t}$ V.

Solution. The chosen tree is in Figure 5-7(b), and the four fundamental loops in Figure 5-7(c), (d), (e), (f). Loop current 1 is known, $i_1 = 10$ A—that's precisely the reason why the current source was chosen as a link and not a tree branch. Let us write, by inspection, the three loop equations around loops 2, 3, and 4. For loop 2, we have

$$-(1)10 + (3 + 5 + 1)i_2 + 5i_3 + 0 \cdot i_4 = 2e^{-t}$$

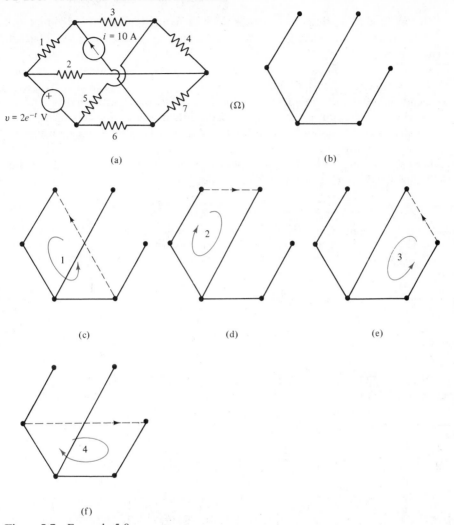

Figure 5-7 Example 5-8.

Around loop 3, we have

$$+(6)10 + 5i_2 + (4 + 5 + 6 + 7)i_3 - (6 + 7)i_4 = 0$$

And around loop 4,

$$-6(10) + 0 \cdot i_2 - (6 + 7)i_3 + (2 + 7 + 6)i_4 = 2e^{-t}$$

Finally,

$$\begin{bmatrix} 9 & 5 & 0 \\ 5 & 22 & -13 \\ 0 & -13 & 15 \end{bmatrix} \begin{bmatrix} i_2 \\ i_3 \\ i_4 \end{bmatrix} = \begin{bmatrix} 2e^{-t} + 10 \\ -60 \\ 2e^{-t} + 60 \end{bmatrix}$$

ready to be solved. ∎

This method of adding (by superposition) the contributions of individual loop currents is powerful and quick. With practice, the final equations, in matrix form, can be written almost by inspection!

A word of caution is in order here. Linear networks containing *dependent sources* obey superposition, of course. However, the method outlined above is not recommended (better yet: should be avoided) then. Why? Because it is hard to include directly the dependent sources with their coefficients in the coefficients of Equation 5-8. It is safer to go more slowly and to write the loop equations by the conventional methods we learned in previous chapters.

Probs. 5-9, 5-10, 5-11

The use of superposition in writing node equations goes, as expected, along dual lines. Let us illustrate with a couple of examples.

EXAMPLE 5-9

Consider the network in Figure 5-8(a), and ignore the fact that it can be solved with *one* loop equation. The two unknown node voltages v_1 and v_2 are shown, chosen conveniently (+) with respect to reference. This choice is dual to choosing all loop currents in one direction.

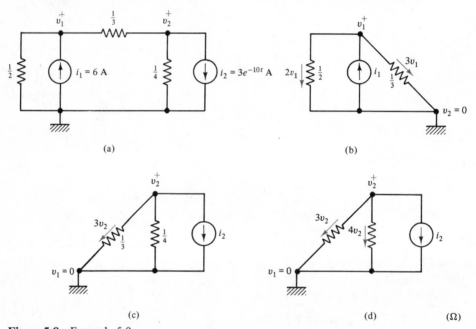

Figure 5-8 Example 5-9.

Solution. We write KCL at each node as $\sum i = 0$, with a (+) sign for a current leaving and a (−) sign for a current entering. Here we modify this convention and write, instead,

$$\sum i_R = \sum i_{sources}$$

that is, at a node, the sum of the currents leaving in the resistors is equal to the sum of the currents entering from the sources. Next, apply superposition: *The total current leaving a node is the sum of the currents contributed by each node voltage, acting alone, at that node.*

Specifically, we have at node 1:

1. Let v_1 act alone, $v_1 \neq 0$, $v_2 = 0$; see Figure 5-8(b). Since $v_2 = 0$, the $\frac{1}{4}$-Ω resistor and the current source i_2 do not affect the calculation. The total current leaving node 1 is $2v_1 + 3v_1 = 5v_1$.
2. With $v_2 \neq 0$, $v_1 = 0$, see Figure 5-8(c), the current source i_1 and the $\frac{1}{2}$-Ω resistor do not affect the calculation. The current leaving node 1 is $(-3v_2)$.
3. The total source entering node 1 is $i_1 = 6$ A. Consequently, KCL at node 1 reads

$$5v_1 - 3v_2 = 6$$

In a similar way, we have at node 2:

1. Node voltage $v_1 \neq 0$, acting alone, contributes a current $(-3v_1)$ leaving node 2. See Figure 5-8(b).
2. Node voltage $v_2 \neq 0$, acting alone, contributes two currents leaving, $+3v_2$ and $+4v_2$, that is, $+7v_2$. See Figure 5-8(d).
3. The total source entering node 2 is $-3e^{-10t}$.

Consequently, the KCL equation at node 2 is

$$-3v_1 + 7v_2 = -3e^{-10t}$$

The two equations are then

$$5v_1 - 3v_2 = 6$$
$$-3v_1 + 7v_2 = -3e^{-10t}$$

or, in matrix form

$$\begin{bmatrix} 5 & -3 \\ -3 & 7 \end{bmatrix} \begin{bmatrix} v_1 \\ v_2 \end{bmatrix} = \begin{bmatrix} 6 \\ -3e^{-10t} \end{bmatrix} \qquad \blacksquare$$

In a general notation, these equations are

$$\begin{bmatrix} g_{11} & g_{12} \\ g_{21} & g_{22} \end{bmatrix} \begin{bmatrix} v_1 \\ v_2 \end{bmatrix} = \begin{bmatrix} j_1 \\ j_2 \end{bmatrix}$$

or

$$\mathbf{Gv} = \mathbf{j} \qquad (5\text{-}9)$$

as in Equation 2-11.

The observations to be made here are:

1. The term $g_{kk}v_k$ in the kth equation is the current leaving node k with its own node voltage v_k acting alone. This term will be positive $(+g_{kk}v_k)$ if v_k is chosen $(+)$ with respect to reference. We call g_{kk} the *self-conductance* (in mho, $\mho$) of node k.
2. A term $g_{pq}v_q$ in the pth equation is the current contributed to node p by the node voltage v_q, through a conductance between nodes p and q. This term will be negative, $-g_{pq}v_q$, if node voltages v_p and v_q are chosen each with a $(+)$ with respect to reference. We call g_{pq} the *mutual conductance* between nodes p and q.
3. The term j_p is the net current, due to sources, entering node p.

Here, too, we remark that this is a very powerful and quick method. With practice, the individual contributions can be seen and written immediately. And here, too, we are cautioned to use conventional methods if *dependent sources* are present. No immediate by-inspection method is readily available in this case.

Probs. 5-12
5-13

5-3 THÉVENIN'S THEOREM

The ideas leading to Thévenin's theorem will be first introduced by a preliminary example.

EXAMPLE 5-10 _____

In the network shown in Figure 5-9(a), we want to find *only* i_R, the current through the resistor R. In order to be able to trace our calculations, we leave R as a lettered value, R ohms, with its two nodes labeled A and B.

Solution. Two loop equations will do, as shown. They are

$$4i_1 - 2i_R = 10$$
$$-2i_1 + (6 + R)i_R = -20$$

Their solution, by Cramer's rule for example, yields

$$i_R = \frac{\begin{vmatrix} 4 & 10 \\ -2 & -20 \end{vmatrix}}{\begin{vmatrix} 4 & -2 \\ -2 & 6+R \end{vmatrix}} = \frac{-60}{20 + 4R} = \frac{-15}{5 + R}$$

A simpler circuit for this expression of i_R is shown in Figure 5-9(b). As far as R is concerned, the complicated circuit in Figure 5-9(a) can be replaced by the simpler circuit in Figure 5-9(b), and R will not be able to tell the difference, that is, i_R (and v_R) will be the same.

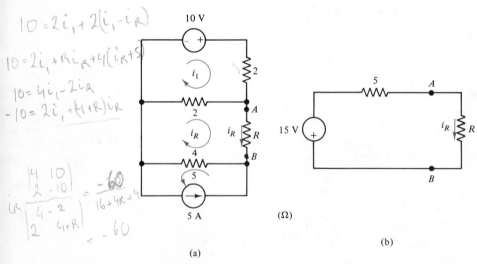

(a)

(b)

Figure 5-9 Example 5-10.

The theorem related to such a simplification of a circuit is attributed to L. C. Thévenin (a French engineer, 1857–1926) and, earlier, to H. von Helmholtz (a German professor of physics, 1821–1894). It will be called here *Thévenin's theorem* and is stated as follows: A linear network, consisting of resistors and sources (independent and dependent) and shown in Figure 5-10(a), can be replaced at its two terminals A–B, as far as an "observer network" is concerned, by a single voltage source v_{Th}, called the equivalent Thévenin voltage, in series with R_{Th}, the equivalent Thévenin resistance, as shown in Figure 5-10(b). There must be no electrical coupling (for example, with dependent sources) between the linear network and the observer. In other words, the only interaction between the linear network and the observer must be through the current i flowing in terminals A and B as shown.

The steps involved in calculating (or measuring in the laboratory) v_{Th} and R_{Th} are:

1. Disconnect the observer.
2. Calculate (or measure) the open-circuit (o.c.) voltage across the terminals A–B, as in Figure 5–10(c). This voltage is v_{Th}.
3. Set all the independent sources in the linear network to zero: Current sources are open-circuited ($i = 0$), and voltage sources are short-circuited ($v = 0$). *Dependent sources remain intact.* Calculate (or measure) the resistance at terminals A–B, as in Figure 5-10(d). This is R_{Th}.

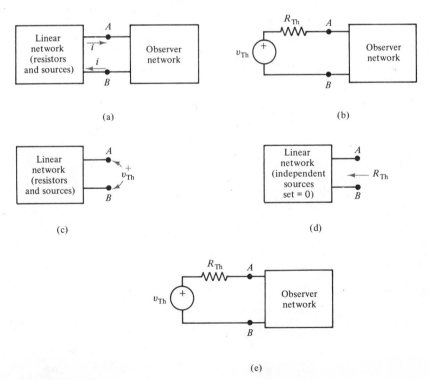

Figure 5-10   Thévenin's theorem.

4. Connect v_{Th} in series with R_{Th}, as shown in Figure 5-10(e). Reconnect the observer to A–B.

To illustrate these steps, let us rework Example 5-10.

EXAMPLE 5-10 (*continued*) ──────────────────────────────────

The entire network is shown in Figure 5-9(a). The observer is just R, and the linear network is everything else.

> *Solution.* In Figure 5-11(a), we show steps 1 and 2. The observer is disconnected, and we calculate v_{Th} as follows: make it part of a loop, then trace KVL for this loop. Here, the 10-V source causes a drop of 5 V across the 2-Ω resistor, while across the 4-Ω resistor there is a voltage of $(4)(5) = 20$ V as shown. KVL then reads
>
> $$-v_{Th} + 5 - 20 = 0$$
>
> or
>
> $$v_{Th} = -15 \text{ V}$$
>
> as shown earlier in Figure 5-9(b). To implement step 3, we set the two independent sources to zero, as shown in Figure 5-11(b). It is very simple to write by inspection
>
> $$R_{Th} = 4 + (2\|2) = 5 \,\Omega$$
>
> as obtained earlier. The final step 4 is shown in Figure 5-9(b), and from it we write
>
> $$i_R = \frac{v_{Th}}{R_{Th} + R} = \frac{-15}{5 + R}$$
>
> as before.

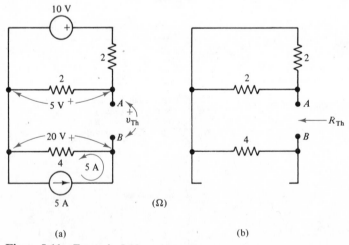

(a)   (b)

Figure 5-11 Example 5-10.

Probs. 5-14,
■ 5-15, 5-16

The next example illustrates the use of Thévenin's theorem when dependent sources are in the linear network.

EXAMPLE 5-11

Find the current i_R in the network shown, by Thévenin's theorem. Notice the current-dependent voltage source in the linear network.

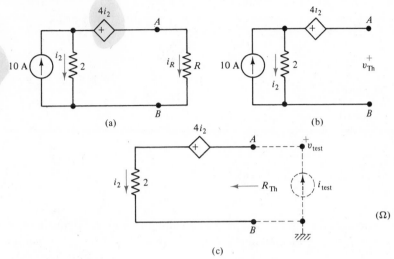

Figure 5-12 Example 5-11.

Solution. We open-circuit terminals A–B in Figure 5-12(b) and calculate v_{Th} as shown, using KVL

$$-v_{Th} - 4i_2 + 2i_2 = 0$$

$$\therefore \quad v_{Th} = -2i_2 = -20 \text{ V}$$

since $i_2 = 10$ A *in this circuit*.

To calculate R_{Th}, we set the independent source (10 A) to zero, but we leave the dependent source intact. See Figure 5-12(c). How do we find a resistance with a dependent source? Go back to basics: Connect to A–B an arbitrary current source i_{test}, and calculate the response voltage across it, v_{test}. Then

$$\boxed{R_{Th} = \frac{v_{test}}{i_{test}}} \qquad (5\text{-}10)$$

(We can, of course, connect v_{test} and calculate i_{test}—the result is the same.)

Here, we see that $i_2 = i_{test}$ (which is one good reason for choosing it as an input in this example!). Then, by KVL, we have

$$-v_{test} - 4i_2 + 2i_2 = 0$$

or, with $i_2 = i_{test}$,

$$v_{test} = -2i_{test}$$

that is,

$$R_{Th} = -2 \, \Omega \text{ (!)}$$

Finally, according to Thévenin, we get the answer

$$i_R = \frac{-20}{-2 + R}$$

Let us check this answer by a conventional method. One loop equation will solve the original network, with i_R being the loop current. We get

$$Ri_R - 2i_2 + 4i_2 = 0$$

But

$$i_2 = 10 - i_R \qquad \text{(KCL)}$$
$$\therefore \quad Ri_R - 2(10 - i_R) + 4(10 - i_R) = 0$$

or

$$i_R = \frac{-20}{-2 + R}$$

as before.   ∎

The following points are important to observe and repeat:

1. In finding v_{Th}, we work with a *simpler circuit*, after the observer has been open-circuited. The methods that we use for finding v_{Th} (KVL and KCL) apply to *that* circuit. In the previous example, Figure 5-12(b), the controller of the dependent source i_2 has a simple value, $i_2 = 10$ A, only because terminals A–B are open (and draw no current). In the original circuit, the controller is still i_2, but $i_2 \neq 10$ there (in fact, $i_2 = 10 - i_R$ there).

   In other words, the first step in Thévenin's theorem, open-circuiting terminals A–B, simplifies our calculations, and that's one good reason for using this theorem. A variable, say, i_2, may take on *different* values in two different circuits such as Figure 5-12(a) and Figure 5-12(b).

2. The Thévenin resistance cannot be calculated by series-parallel combinations when dependent sources are present. Instead, a test current (or voltage) must be connected to terminals A–B, and the resulting voltage (current) calculated. Then R_{Th} is given by Equation 5-10.

3. Typically (but not always), a *negative* Thévenin resistance may result with dependent sources.

4. Since the observer network has nothing to do with the calculation of the Thévenin circuit to its left, it can be arbitrary (linear, nonlinear, a single element, several elements, etc.).

Probs. 5-17, 5-18, 5-19

Finally, let us now show with an example the restriction on using Thévenin's theorem when there is an electrical coupling between the linear network and the observer.

EXAMPLE 5-12 ──

In the network shown in Figure 5-13, a current source is dependent on i_6. We *cannot* find the Thévenin equivalent of the circuit to the left of A–B, since there is coupling between it and the observer. The observer (to the right of A–B) contains the controlling current i_6 of the dependent source $5i_6$.

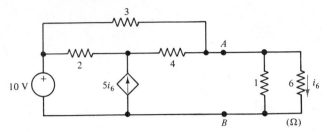

Figure 5-13 Example 5-12.

Probs. 5-20, 5-21

5-4 NORTON'S THEOREM

A dual circuit to the Thévenin equivalent was proposed by E. L. Norton (1898–   ). Instead of a voltage source v_{Th} in series with R_{Th}, the Norton equivalent circuit is a current source i_N in parallel with R_N, as in Figure 5-14(b). The two circuits must be equivalent at terminals A–B, meaning that an observer network connected to A–B of the Norton equivalent will draw the same current as if connected to the Thévenin equivalent.

In order to calculate i_N and R_N, then, let us find the Thévenin equivalent of the proposed Norton model. First, let us make the independent source i_N zero (open-circuit). The Thévenin resistance at terminals A–B of the Norton model is then

$$R_{Th} = R_N \qquad (5\text{-}11)$$

Next, the Thévenin voltage across terminal A–B of the Norton model is

$$v_{AB} = v_{Th} = i_N R_N = i_N R_{Th} \qquad (5\text{-}12)$$

or

$$i_N = \frac{v_{Th}}{R_{Th}} \qquad (5\text{-}13)$$

which is the current that will flow in the Thévenin circuit, Figure 5-14(a), if we short-circuit (s.c.) terminals A–B.

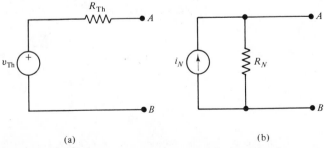

(a)   (b)

Figure 5-14 Thévenin's and Norton's equivalent circuits.

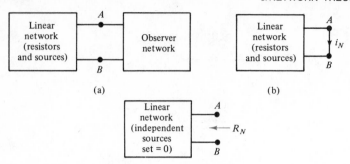

Figure 5-15 Steps in obtaining Norton's circuit.

In summary, to obtain the Norton equivalent circuit for a linear network, as far as an observer network is concerned, we do these steps:

1. Remove the observer.
2. Short-circuit terminals A–B, and calculate (or measure) the current there. This is i_N as in Figure 5-15(b).
3. Open terminals A–B, as in step 1. Set all independent sources to zero, but leave intact the dependent sources. Calculate R_N ($= R_{Th}$) at terminals A–B.
4. Connect the Norton source i_N, referenced as shown in Figure 5-14(b), in parallel with R_N. Across terminals A–B reconnect the observer network.

The equivalence of the two circuits, Thévenin's and Norton's, as shown in Figure 5-14, is sometimes called a *source transformation*: A voltage source in series with a resistor can be transformed (as far as an observer network is concerned) into a current source in parallel with a resistor, and vice versa.

EXAMPLE 5-13 _____

Find the current in R, as shown in Figure 5-16(a), using Norton's theorem.

Solution. In Figure 5-16(b) we remove the observer and short-circuit terminals A–B. The 4-Ω resistor carries no current then, and hence

$$i_{s.c.} = i_N = \tfrac{12}{2} = 6 \text{ A}$$

To calculate R_N ($= R_{Th}$), we look into the open-circuited terminals A–B while setting to zero the independent source, as in Figure 5-16(c). Then

$$R_N = 2\|4 = \tfrac{4}{3} \, \Omega$$

The final circuit is shown in Figure 5-16(d) and from it we write, using current division,

$$i_R = \frac{1/R}{1/R + \tfrac{3}{4}} (6) = \frac{8}{R + \tfrac{4}{3}}$$

This result can be quickly verified, either by conventional analysis or by Thévenin's method (do it!).

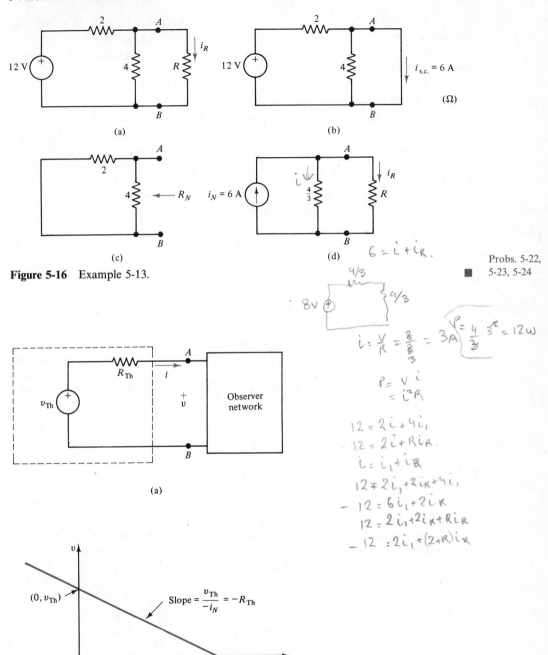

(a)

(b)

(c)

(d)

Figure 5-16　Example 5-13.

Probs. 5-22,
5-23, 5-24

$6 = i + i_R.$

$i = \dfrac{V}{R} = \dfrac{8}{\frac{8}{3}} = 3A$　$\dfrac{4}{3} 3^2 = 12w$

$P = V \cdot i$
$= i^2 R$

$12 = 2i + 4i_1$
$12 = 2i + R i_R.$
$i = i_1 + i_R$
$12 = 2 i_1 + 2 i_R + 4 i_1$
$12 = 6 i_1 + 2 i_R$
$12 = 2 i_1 + 2 i_R + R i_R$
$12 = 2 i_1 + (2 + R) i_R$

(a)

Observer
network

(b)

$\text{Slope} = \dfrac{v_{Th}}{-i_N} = -R_{Th}$

$(0, v_{Th})$

$(i_N, 0)$

Figure 5-17　*v-i* characteristics of the Thévenin circuit.

In conclusion, let us mention that, when we find the Thévenin (or Norton) equivalent circuit, all the individual sources and resistors inside the linear network lose their identity; we cannot calculate individual currents, voltages, power, etc., since everything is replaced by a single equivalent source and a single equivalent resistor. Such a loss is not great, because we are more interested in what is happening in the observer network, and *it* does not change.

It is also very instructive to derive the *v-i* characteristic curve of the Thévenin (or Norton) equivalent circuit, seen as a two-terminal network. We show it in Figure 5-17(a), where the dotted lines enclose this two-terminal network. KVL gives us the desired *v-i* relationship, $R_{Th}i + v = v_{Th}$. It resembles the equation $Ax + y = B$, with $A = R_{Th}$ and $B = v_{Th}$, both assumed constant.

Its plot in the *v-i* plane is a *straight line*, with the two intercepts and the slope as shown in Figure 5-17(b). When $i = 0$, the observer network is removed, and we have $v = v_{o.c.} = v_{Th}$ as expected. When $v = 0$, terminals A–B are short circuited and we have $i = i_{s.c.} = i_N$, as expected. For any other condition, with a finite, nonzero resistor in the observer network, the solution (i, v) will fall on that straight line.

5-5 MAXIMUM POWER TRANSFER

In many circuits, the observer network is a single resistor R. A question of practical interest arises in electronic circuits such as communication devices (transmitters and receivers): "What is the maximum power that can be transferred to a particular resistive load R?" This load, for example, is represented by the speakers of a hi-fi system. The situation is depicted in Figure 5-18 and is familiar to us from the previous discussion. In Figure 5-18(a), the original linear network is shown with the load (observer) resistor R. In Figure 5-18(b), we have replaced the linear network by its Thévenin's equivalent, and R still draws the same current i_R.

The power in R, which we wish to maximize, is given by

$$p = (i_R)^2 R = \left(\frac{v_{Th}}{R_{Th} + R}\right)^2 R \tag{5-14}$$

The Thévenin voltage and resistance, v_{Th} and R_{Th}, are fixed by the linear network and have nothing to do with the load R. To maximize p, then, we have to treat R as a variable. The maximum of p can be found by setting

$$\frac{dp}{dR} = 0 \tag{5-15}$$

which yields

$$v_{Th}^2 \frac{(R_{Th} + R)^2 - 2R(R_{Th} + R)}{(R_{Th} + R)^4} = 0 \tag{5-16}$$

$$\therefore \quad (R_{Th} + R)^2 = 2R(R_{Th} + R) \tag{5-17}$$

or

$$R = R_{Th} \tag{5-18}$$

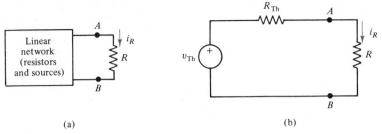

Figure 5-18 Maximum power transfer.

In other words, <u>maximum power transfer to the load R will occur when R is equal to R_{Th}</u>. This condition is also called *resistance match*; that is, the load resistance matches the source resistance. It is not very hard to show (by taking the second derivative) that this is, in fact, a maximum and not a minimum. A different argument, more physical, is suggested in Problem 5-25.

Prob. 5-25

EXAMPLE 5-13 (*continued*) ————————————————————

In the network of Figure 5-16, (a) What is the value of R for it to receive maximum power? (b) What is then this power? (c) What is then the power delivered by the 12-V source?

Solution.
(a) From Figure 5-16(d), we see that

$$R = R_{Th} = \tfrac{4}{3}\,\Omega$$

(b) From Figure 5-16(d), we have then

$$i_R = \frac{\tfrac{3}{4}}{\tfrac{3}{4} + \tfrac{3}{4}}\,6 = 3\text{ A} \qquad \therefore \quad p_R = (3)^2\tfrac{4}{3} = 12\text{ W}$$

(c) In Figure 5-16(a), we have then

$$v_{AB} = Ri_R = (\tfrac{4}{3})3 = 4\text{ V}$$

The current through the 4-Ω resistor is then

$$i_4 = \tfrac{4}{4} = 1\text{ A}$$

and the power dissipated in it is

$$p_4 = (1)^2 4 = 4\text{ W}$$

The current in the 2-Ω resistor is

$$i_2 = i_4 + i_R = 1 + 3 = 4\text{ A}$$

and the power dissipated in this resistor is

$$p_2 = (4)^2 2 = 32\text{ W}$$

The 12-V source must deliver

$$p_{12} = p_2 + p_4 + p_R = 32 + 4 + 12 = 48\text{ W}$$

of which only 25 percent ($= 12$ W) reaches the load.

Probs. 5-26,
5-27, 5-28 ∎

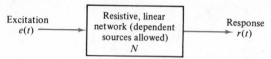

Figure 5-19   Input-output relations.

5-6 INPUT-OUTPUT RELATIONS. NETWORK FUNCTIONS

So far, we have answered, for resistive networks, the main problem posed in Chapter 1: Given an input (a source, an excitation) and a linear network, calculate the output (response). See Figure 5-19. In fact, in many circuits we had several inputs. We have also learned why the network N may not contain independent sources; besides, such a source can be considered a part of the input.

Let us confine our discussion now to a single input and a single output, as in Figure 5-19. Let us call the excitation $e(t)$, a voltage source or a current source. Also, let the response, a current or a voltage of some element, be $r(t)$. We have found, for our linear resistive networks, that the following relationship is true:

$$r(t) = He(t) \qquad (5\text{-}19)$$

that is, the response is proportional to the excitation through a multiplier H. This multiplier is called the *network function* for the particular input-output relation under consideration. Let us illustrate with several examples.

EXAMPLE 5-14 _____

Let the network be a resistor R, the excitation a current source $i(t)$, and the response the voltage $v(t)$.

Solution.   Then Equation 5-19 becomes

$$v(t) = Ri(t)$$

and the network function here is $H = R$, in ohms.   ∎

EXAMPLE 5-15 _____

If the resistor R is excited by a voltage source $v(t)$, the response is $i(t)$ and Equation 5-19 is here

$$i(t) = Gv(t)$$

where $H = G$ (mhos) in this case.   ∎

EXAMPLE 5-16 _____

In a voltage divider (Figure 4-8 and Equation 4-14) the input is $v(t)$ and the output is $v_k(t)$. Then Equation 5-19 reads

$$v_k(t) = \frac{R_k}{R_1 + R_2 + \cdots + R_n} v(t)$$

Here the appropriate network function is

$$H = \frac{R_k}{R_1 + R_2 + \cdots + R_n}$$

and is dimensionless (a pure number). ■

EXAMPLE 5-17

In an inverting op amp amplifier (Figure 4-16 and Equation 4-33) we write Equation 5-19 as

$$v_o(t) = -\frac{R_f}{R_s} v_i(t)$$

where $e(t) = v_i(t)$ and $r(t) = v_o(t)$. Here the network function is

$$H = -\frac{R_f}{R_s}$$

called earlier the closed-loop gain of the amplifier, and it is dimensionless. ■

EXAMPLE 5-18

The network function of a noninverting amplifier, as given in Figure 4-17 and Equation 4-34, is the dimensionless number

$$H = 1 + \frac{R_1}{R_2}$$ ■

From these examples, and in all our future ones, we recognize the following steps needed to calculate a network function H:

1. Identify the input $e(t)$.
2. Identify the output $r(t)$.
3. Analyze the given network with the given $e(t)$, and calculate $r(t)$.
4. Write $r(t)$ in terms of $e(t)$, as in Equation 5-19. Identify the appropriate H.

It is also important to recognize that the network function H depends exclusively on the parameters of the network (values of resistors, gains of amplifiers and dependent sources). In this sense, it is *characteristic of the network itself*. (In later studies, we will see that the network function has another parameter, or variable. Still, it will be characteristic of the network.)

EXAMPLE 5-19

For the network shown in Figure 5-20, calculate the network function if the output is (1) $i_a(t)$, (2) $v_c(t)$, (3) $v_a(t)$. Notice the (allowed) dependent source.

Solution. One loop equation will solve the entire network. Let i_x be this loop current, in a counterclockwise direction, KVL then reads, with $i_x = i_a - i$,

$$R_a i_a + R_b(K i_a + i_a - i) + R_c(i_a - i) = 0$$

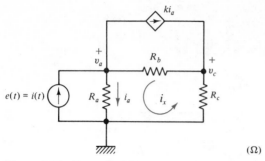

Figure 5-20   Example 5-19.

that is, for case (1)

$$r_1(t) = i_a(t) = \frac{R_b + R_c}{R_a + (k + 1)R_b + R_c} \, i(t) = H_1 e(t)$$

identifying the network function H_1. Similarly, for (2)

$$r_2(t) = v_c(t) = -i_x R_c = \frac{(KR_b + R_a) R_c}{R_a + (k + 1)R_b + R_c} \, i(t) = H_2 e(t)$$

and for (3) we have

$$r_3(t) = v_a(t) = R_a i_a(t) = \frac{R_a (R_b + R_c)}{R_a + (k + 1)R_b + R_c} \, i(t) = H_3 e(t)$$

For each input-output relation, we have calculated the appropriate network function. Each H shows its dependence only on the network parameters R_a, R_b, R_c, and k.

Dimensionally, H_1 is dimensionless (a pure number), since it relates a current output i_a to a current input i. Such a function is called a *current transfer function*. In the second case, H_2 has the units of ohms, relating an input current to an output voltage. It is a *transfer resistance function*; the word "transfer" reminds us that the output is measured at a different pair of nodes than the input. In the third case, H_3 is in ohms and is called the *driving-point resistance* or, simply, the *input resistance*, since input and output are at the *same* pair of terminals.

Finally, all three input-output relations and network functions may be put neatly together in matrix form, as follows

$$\begin{bmatrix} i_a \\ v_c \\ v_a \end{bmatrix} = \begin{bmatrix} H_1 \\ H_2 \\ H_3 \end{bmatrix} i$$

In general, this expression is

$$\mathbf{r}(t) = \mathbf{H} \, \mathbf{e}(t) \qquad\qquad (5\text{-}20)$$

which is the generalization of Equation 5-19. Here $\mathbf{r}(t)$ is the column matrix listing *several* outputs, $\mathbf{e}(t)$ is the column matrix of *several* possible inputs, and $\mathbf{H}$, the *matrix network function*, relates them according to Equation 5-20. This equation is a multi-input multi-output version of Equation 5-19. In the next section, and in later chapters, we'll have the opportunity to work with this relationship quite extensively. ■

Probs. 5-29, 5-30, 5-31

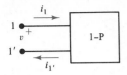

Figure 5-21 A one-port (1-P) network.

5-7 TWO-PORT (2-P) NETWORKS

What is an electrical port? It is a pair of terminals through which we can access a network, similar to the port of access on your personal computer, or to a port of entry through which we access a country. In Figure 5-21, we see a *one-port* (1-P) network. Terminals 1 and 1' are the port. Across 1-1' the voltage is v, and the currents *are equal*, as shown,

$$i_1 = i'_1 \tag{5-21}$$

This constraint on the currents *defines* a port: Two terminals constitute a port if, and only if, the currents are equal as shown.

From our previous studies and examples, we know that there is only one network function relating the two variables of 1-P, v and i. This network function is R_{dp}

$$v(t) = R_{dp}i(t) \tag{5-22}$$

that is, the driving-point (or input) resistance. Obviously, its inverse (reciprocal) is also a network function

$$i(t) = G_{dp}v(t) \tag{5-23a}$$

$$G_{dp} = R_{dp}^{-1} \tag{5-23b}$$

where G_{dp} is the driving-point (input) conductance.

A *two-port* (2-P) network, shown in Figure 5-22, has four port variables, a voltage and a current in each port. Note the standard references of v_1, i_1, v_2, and i_2. Again, the current restriction must be obeyed for a pair of terminals to qualify as a port. Two circuits are shown in Figure 5-23. In part (a), the current restriction is obeyed, as can be verified easily. The 2-P serves as a coupling network between a source v_s, with its internal resistance R_s, and a load R_L. In part (b), a slight

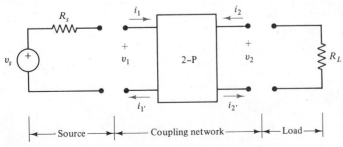

Figure 5-22 A two-port (2-P) network.

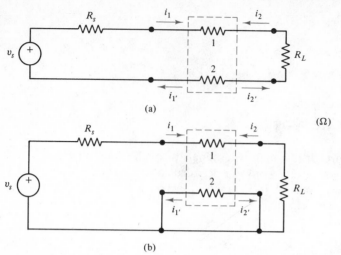

(a)

(b)

Figure 5-23   (a) A 2-P network. (b) A 4-T (4-terminal) network.

modification of the connections ruins the nature of the previous 2-P. Here we see that $i_1 = v_s/(R_s + 1 + R_L) = -i_{2'}$, but $i_{1'} = 0 = i_{2'}$. Therefore, $i_1 \neq i_{1'}$, $i_2 \neq i_{2'}$ and the network is not a 2-P. At best, we call it a *four-terminal* (4-T) network. In multiterminal networks, the port currents restriction does not hold.

Given, then, a 2-P network as shown in Figure 5-22. The internal structure of the 2-P can be quite complicated; as mentioned, the 2-P serves as a coupling network between two other networks (the load resistance is a 1-P!). We are interested in the *external* description of this 2-P, specifically, in the relationships among its port variables v_1, i_1, v_2, and i_2.

There are six different ways of relating two of these variables in terms of the other two. They are:

1. v_1 and v_2 in terms of i_1 and i_2 4. i_1 and v_2 in terms of v_1 and i_2
2. i_1 and i_2 in terms of v_1 and v_2 5. v_1 and i_1 in terms of v_2 and i_2
3. v_1 and i_2 in terms of i_1 and v_2 6. v_2 and i_2 in terms of v_1 and i_1

Let us explore these, one at a time, from the electrical input-output point of view, together with the associated mathematical relations.

1. The Open-Circuit Resistance Parameters   Just as in the case of a 1-P, the 2-P is assumed to be linear, containing only linear resistors and possibly dependent sources. Let the two inputs be $i_1(t)$ and $i_2(t)$, as shown in Figure 5-24(a). The outputs are, then, $v_1(t)$ and $v_2(t)$. Since the 2-P is linear, superposition must hold. The total response $v_1(t)$ is, therefore, the sum of the partial contributions of $i_1(t)$ alone and $i_2(t)$ alone. Mathematically, we write

$$v_1 = r_{11}i_1 + r_{12}i_2 \tag{5-24}$$

where r_{11} is (in ohms) a constant of proportionality, and the term $r_{11}i_1 = \hat{v}_1$ is the partial contribution due to i_1 alone. Similarly, r_{12} (ohms) is another constant and the

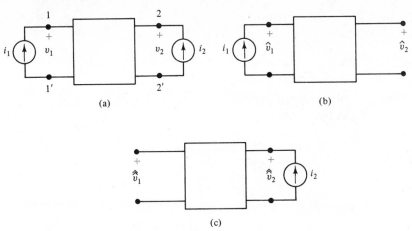

Figure 5-24 The o.c. parameters.

term $r_{12}i_2 = \hat{\hat{v}}_1$ is the partial response, due to i_2 alone, in the total response v_1. See Figure 5-24(b) and (c). At port 2, we have

$$v_2 = r_{21}i_1 + r_{22}i_2 \tag{5-25}$$

where $r_{21}i_1 = \hat{v}_2$ is the partial contribution of i_1 alone, and $r_{22}i_2 = \hat{\hat{v}}_2$ is the partial contribution of i_2 alone to the total response $v_2(t)$.

In matrix form, Equations 5-24 and 5-25 are

$$\begin{bmatrix} v_1 \\ v_2 \end{bmatrix} = \begin{bmatrix} r_{11} & r_{12} \\ r_{21} & r_{22} \end{bmatrix} \begin{bmatrix} i_1 \\ i_2 \end{bmatrix} \tag{5-26}$$

in the form of Equation 5-20

$$\mathbf{r}(t) = \mathbf{H}\,\mathbf{e}(t) \tag{5-20}$$

showing the response matrix $\mathbf{r}(t)$, the matrix network function $\mathbf{H}$, and the excitation matrix $\mathbf{e}(t)$. Be sure to recognize the difference between the four terms $r_{11}, r_{12}, r_{21}, r_{22}$ used here, and terms of the same form, r_{jk}, used in loop analysis in Chapter 2. In loop analysis, such terms represent self- and mutual resistances in a multiloop network. Here, the four terms serve only to relate v_1 and v_2 to i_1 and i_2 in the 2-P. The rest of the 2-P, inside the box, may contain many other loops that do not show explicitly in Equation 5-26.

What is the nature of the network functions r_{11}, r_{12}, r_{21}, and r_{22}? How can we calculate or measure them? The answer is, again, in the principle of superposition, shown in Figure 5-24(b) and (c). We can set up, in the laboratory on paper, these two experiments. First, with Figure 5-24(b), we have $i_1 \neq 0$ and $i_2 = 0$, port 2 being open-circuited. Then, again from Equation 5-24, 5-25, or 5-26, we have

$$\hat{v}_1 = r_{11}i_1 \qquad i_2 = 0 \tag{5-27a}$$
$$\hat{v}_2 = r_{21}i_1 \qquad i_2 = 0 \tag{5-27b}$$

In other words, r_{11} is the driving-point (input) resistance seen at port 1 with port 2 open-circuited; also r_{21} is the transfer resistance relating the open-circuit voltage output at port 2 to the input current i_1 at port 1. The second experiment, in Figure 5-24(c) and with Equation 5-24, 5-25, or 5-26, yields

$$\hat{v}_1 = r_{12}i_2 \qquad i_1 = 0 \tag{5-28a}$$

$$\hat{v}_2 = r_{22}i_2 \qquad i_1 = 0 \tag{5-28b}$$

that is, r_{12} is the transfer resistance relating the open-circuit voltage output at port 1 to the input current i_2 at port 2; and r_{22} is the driving-point resistance at port 2 with port 1 open-circuited.

Since these two experiments are done with port 1 or port 2 open-circuited, we call the parameters the *open-circuit resistance parameters*, and their matrix is designated as $\mathbf{R}_{o.c.}$.

$$\mathbf{R}_{o.c.} = \begin{bmatrix} r_{11} & r_{12} \\ r_{21} & r_{22} \end{bmatrix} \tag{5-29}$$

The letter $\mathbf{H}$ is a generic notation for network functions. Specific network functions are given specific letters, such as $\mathbf{R}_{o.c.}$ here.

EXAMPLE 5-20 _____

Find $\mathbf{R}_{o.c.}$ for the T two-port network shown in Figure 5-25.

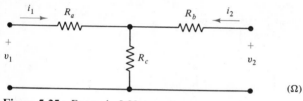

Figure 5-25   Example 5-20.

Solution.   With port 2 open-circuited and i_1 exciting port 1, we have, as in Equation 5-27,

$$\hat{v}_1 = (R_a + R_c)i_1 \qquad \therefore \quad r_{11} = R_a + R_c$$

and since there is no current in R_b,

$$\hat{v}_2 = R_c i_1 \qquad \therefore \quad r_{21} = R_c$$

With port 1 open-circuited and i_2 exciting port 2, we see that, with no current in R_a,

$$\hat{v}_1 = R_c i_2 \qquad \therefore \quad r_{12} = R_c$$

and

$$\hat{v}_2 = (R_b + R_c)i_2 \qquad \therefore \quad r_{22} = R_b + R_c$$

Consequently,

$$\mathbf{R}_{o.c.} = \begin{bmatrix} R_a + R_c & R_c \\ R_c & R_b + R_c \end{bmatrix}$$

Does this network look vaguely familiar? In Chapter 4, we saw it as a *3-terminal* (3-T) network (called a Y network). There, no restrictions were placed on the terminal currents. Here, the 3-T network was converted into a 2-P network.   ■

EXAMPLE 5-21 _____

Find $\mathbf{R}_{o.c.}$ for the 2-P network shown in Figure 5-26(a), representing a model of some electronic amplifier. There is a current-controlled current source in the 2-P network, with α a given constant.

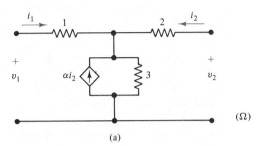

(a)

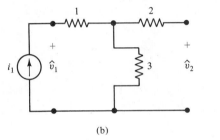

(b)

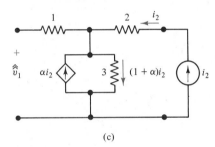

(c)

Figure 5-26   Example 5-21.

Solution.   We set up the two experiments carefully. With $i_2 = 0$, as in Figure 5-26(b), the source controlled by i_2 must also go to zero. Being a current source, it goes to zero by open-circuiting it. Then

$$\hat{v}_1 = (3 + 1)i_1 \qquad \therefore \quad r_{11} = 4$$
$$\hat{v}_2 = 3i_1 \qquad \therefore \quad r_{21} = 3$$

In the second experiment, $i_1 = 0$. Since $i_2 \neq 0$, the controlled source stays. The current in the 3-Ω resistor, by KCL, is $\alpha i_2 + i_2 = (1 + \alpha)i_2$. Therefore

$$\hat{\hat{v}}_1 = 3(1 + \alpha)i_2 \qquad \therefore \quad r_{12} = 3(1 + \alpha)$$

and

$$\hat{\hat{v}}_2 = 2i_2 + 3(1 + \alpha)i_2 \qquad \therefore \quad r_{22} = 5 + 3\alpha$$

Thus,

$$\mathbf{R}_{o.c.} = \begin{bmatrix} 4 & 3(1 + \alpha) \\ 3 & 5 + 3\alpha \end{bmatrix}$$

Probs. 5-32, 5-33, 5-34

The method outlined previously is very formal and somewhat tedious. We must set up two separate experiments, under some very specific conditions, and make sure that dependent sources are intact (if their controller is nonzero) or zero (if their controller is zero) in the particular experiment. All this may be quite lengthy and subject to errors. Let us outline another method for calculating $\mathbf{R}_{o.c.}$ which gives us all four parameters at once, without the need of two separate experiments and without ever worrying about the correct settings of dependent sources.

This second method, which we'll call the *informal method*, can be stated as follows: For a given 2-P, write *any* two valid KVL or KCL equations for that 2-P, then rearrange them as necessary in the form of Equation 5-26. That's all! No memorization is needed, no open circuit on one port or another, and the results usually fall out with little effort.

To illustrate this informal method, let us rework the two previous examples, then do another one.

EXAMPLE 5-22 ———————————————————————————————

Rework Example 5-20 by the informal method.

> *Solution.* In Figure 5-25, KCL at the center node tells us that the current through R_c is $(i_1 + i_2)$, referenced down. Then KVL in the left loop reads

$$-v_1 + R_a i_1 + R_c(i_1 + i_2) = 0$$

> and KVL in the right loop is

$$-v_2 + R_a i_1 + R_c(i_1 + i_2) = 0$$

Now rearrange these two equations into their desired form, with the v's on the left and the i's on the right (this is the required form—the only thing to memorize). We get

$$v_1 = R_a i_1 + R_c(i_1 + i_2) = (R_a + R_c)i_1 + R_c i_2$$
$$v_2 = R_c i_1 + (R_b + R_c)i_2$$

and we have the answer as before.   ∎

EXAMPLE 5-23 ———————————————————————————————

Rework Example 5-21 by the informal mnethod.

> *Solution.*   For the 2-P in Figure 5-26(a), *as given*, we do not set $i_1 = 0$ or $i_2 = 0$. Rather, by the informal method, we recognize that the current in the 3-Ω resistor is $(i_1 + i_2 + \alpha i_2)$, referenced down. Then KVL reads

$$v_1 = 1i_1 + 3(i_1 + i_2 + \alpha i_2)$$
$$v_2 = 2i_2 + 3(i_1 + i_2 + \alpha i_2)$$

A quick rearrangement yields the final answer

$$\begin{bmatrix} v_1 \\ v_2 \end{bmatrix} = \begin{bmatrix} 4 & 3 + 3\alpha \\ 3 & 5 + 3\alpha \end{bmatrix} \begin{bmatrix} i_1 \\ i_2 \end{bmatrix}$$ ∎

EXAMPLE 5-24 ———————————————————————————————

For the op-amp 2-P network shown in Figure 5-27, it may be risky to solve this problem by the formal method (a nagging question then: "What is i_0 when $i_2 = 0$?"). Using the informal method, we solve the entire network by inspection, as follows:

> *Solution*

> 1. Since the noninverting terminal draws no current, i_1 flows also through the 2-Ω resistor, as shown.

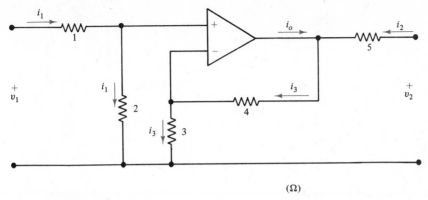

Figure 5-27 Example 5-24.

2. Since the inverting terminal draws no current, the 4-Ω resistor and the 3-Ω resistor are in series and carry the same current i_3 as shown.

3. The voltage drop across the 3-Ω resistor is $3i_3$, and, since there is zero voltage across the input terminals of the op amp, this is also the voltage across the 2-Ω resistor. Thus

$$3i_3 = 2i_1$$

or

$$i_3 = \tfrac{2}{3}i_1$$

4. KVL on the left reads

$$v_1 = 1i_1 + 2i_1 = 3i_1$$

5. KVL on the right reads

$$v_2 = 5i_2 + 4i_3 + 3i_3$$
$$= 5i_2 + 7(\tfrac{2}{3})i_1$$

The last two equations yield the final answer:

$$\begin{bmatrix} v_1 \\ v_2 \end{bmatrix} = \begin{bmatrix} 3 & 0 \\ \tfrac{14}{3} & 5 \end{bmatrix}$$

■

2. The Short-Circuit Conductance Parameters In a dual fashion to the previous derivation, we consider the 2-P shown in Figure 5-28(a). By superposition, the total response i_1 will be

$$i_1 = G_{11}v_1 + G_{12}v_2 \tag{5-30}$$

where $G_{11}v_1 = \hat{\imath}_1$ is the partial contribution due to v_1 alone, and G_{11}, in mhos, is a conductance. Also $G_{12}v_2 = \hat{\hat{\imath}}_1$ is the partial contribution due to v_2 alone. See Figure 5-28(b) and (c). At port 2, the total response is

$$i_2 = G_{21}v_1 + G_{22}v_2 \tag{5-31}$$

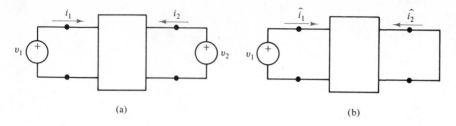

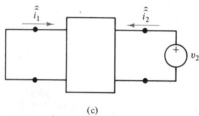

Figure 5-28   The s.c. parameters.

where $G_{21}v_1 = \hat{i}_2$ is the partial contribution of v_1 alone, and $G_{22}v_2 = \hat{i}_2$ is due to v_2 alone. In matrix form, then, we write

$$\begin{bmatrix} i_1 \\ i_2 \end{bmatrix} = \begin{bmatrix} G_{11} & G_{12} \\ G_{21} & G_{22} \end{bmatrix} \begin{bmatrix} v_1 \\ v_2 \end{bmatrix} \tag{5-32}$$

that is, in the form of Equation 5-20

$$\mathbf{r}(t) = \mathbf{H}\,\mathbf{e}(t) \tag{5-20}$$

The formal method of calculating (or measuring) G_{11}, G_{12}, G_{21}, and G_{22} proceeds with two experiments by superposition. First, with $v_1 \neq 0$, port 2 is short-circuited ($v_2 = 0$) as in Figure 5-28(b),

$$\hat{i}_1 = G_{11}v_1 \qquad v_2 = 0 \tag{5-33a}$$
$$\hat{i}_2 = G_{21}v_1 \qquad v_2 = 0 \tag{5-33b}$$

So, in other words, G_{11} is the driving-point conductance seen at port 1 with port 2 short-circuited. Then G_{21} is the transfer conductance relating the short-circuit current at port 2 to the input voltage v_1.

In the second experiment $v_2 \neq 0$ and port 1 is short-circuited ($v_1 = 0$). Then

$$\hat{i}_1 = G_{12}v_2 \qquad v_1 = 0 \tag{5-34a}$$
$$\hat{i}_2 = G_{22}v_2 \qquad v_1 = 0 \tag{5-34b}$$

where G_{12} is a transfer conductance and G_{22} a driving-point conductance. Since these calculations (or measurements) are done under short-circuit conditions, the parameters are called the *short-circuit conductance parameters*

$$\mathbf{G}_{s.c.} = \begin{bmatrix} G_{11} & G_{12} \\ G_{21} & G_{22} \end{bmatrix} \tag{5-35}$$

EXAMPLE 5-25

For the 2-P shown in Figure 5-29 (a π 2-P) calculate $\mathbf{G}_{s.c.}$. We have (verify carefully the two experiments!)

$$\mathbf{G}_{s.c.} = \begin{bmatrix} G_a + G_c & -G_c \\ -G_c & G_b + G_c \end{bmatrix}$$

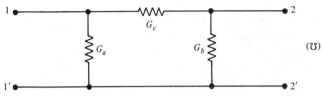

Figure 5-29 Example 5-25.

The informal method works equally well, and often better. Again, with dependent sources, we don't have to worry when such a source stays intact or becomes zero. The entire 2-P is considered, and the four parameters are obtained at once.

EXAMPLE 5-26

Find by the informal method the short-circuit parameters for the 2-P shown in Figure 5-30.

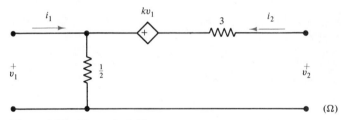

Figure 5-30 Example 5-26.

Solution. We write, by inspection, for the $\frac{1}{2}$-Ω resistor

$$i_1 + i_2 = 2v_1$$

and KVL on the right is

$$v_2 = 3i_2 - kv_1 + v_1$$

From the second equation we have

$$i_2 = \frac{k-1}{3}v_1 + \frac{1}{3}v_2 = G_{21}v_1 + G_{22}v_2$$

This equation is substituted into the first one to give us

$$i_1 = 2v_1 - i_2 = \frac{7-k}{3}v_1 - \frac{1}{3}v_2 = G_{11}v_1 + G_{12}v_2$$

Therefore

$$\mathbf{G}_{\text{s.c.}} = \begin{bmatrix} \dfrac{7-k}{3} & -\dfrac{1}{3} \\[2ex] \dfrac{k-1}{3} & \dfrac{1}{3} \end{bmatrix}$$

■

Another aspect of the informal method should become clear now. We can obtain $\mathbf{G}_{\text{s.c.}}$, or Equation 5-32, by an algebraic rearrangement of Equation 5-26. Let us repeat these equations here, for clarity

$$v_1 = r_{11}i_1 + r_{12}i_2$$
$$v_2 = r_{21}i_1 + r_{22}i_2 \tag{5-26}$$

and

$$i_1 = G_{11}v_1 + G_{12}v_2$$
$$i_2 = G_{21}v_1 + G_{22}v_2 \tag{5-32}$$

The informal method simply takes any two valid equations for the 2-P and rearranges them into a desired form. So, if we have Equation 5-26, the v's in term of the i's, we merely solve algebraically for the i's in term of the v's. Using Cramer's rule, we have then

$$i_1 = \frac{\begin{vmatrix} v_1 & r_{12} \\ v_2 & r_{22} \end{vmatrix}}{\begin{vmatrix} r_{11} & r_{12} \\ r_{21} & r_{22} \end{vmatrix}} = \frac{r_{22}}{\det \mathbf{R}_{\text{o.c.}}} v_1 + \frac{-r_{12}}{\det \mathbf{R}_{\text{o.c.}}} v_2 = G_{11}v_1 + G_{12}v_2 \tag{5-36}$$

$$i_2 = \frac{\begin{vmatrix} r_{11} & v_1 \\ r_{21} & v_2 \end{vmatrix}}{\det \mathbf{R}_{\text{o.c.}}} = \frac{-r_{21}}{\det \mathbf{R}_{\text{o.c.}}} v_1 + \frac{r_{11}}{\det \mathbf{R}_{\text{o.c.}}} v_2 = G_{21}v_1 + G_{22}v_2 \tag{5-37}$$

Here $\det \mathbf{R}_{\text{o.c.}} = r_{11}r_{22} - r_{12}r_{21}$ is the determinant of the o.c. parameters matrix. The results in Equations 5-36 and 5-37 give G_{11}, G_{12}, G_{21}, and G_{22}, provided, of course, that

$$\det \mathbf{R}_{\text{o.c.}} \neq 0 \tag{5-38}$$

These results can be summarized neatly in matrix notation. The system

$$\mathbf{v} = \mathbf{R}_{\text{o.c.}}\mathbf{i} \tag{5-26}$$

is the inverse of

$$\mathbf{i} = \mathbf{G}_{\text{s.c.}}\mathbf{v} \tag{5-32}$$

Specifically,

$$\mathbf{R}_{\text{o.c.}}^{-1} = \mathbf{G}_{\text{s.c.}} \tag{5-39}$$

where the inverse matrix $\mathbf{R}_{o.c.}^{-1}$ exists provided $\mathbf{R}_{o.c.}$ is nonsingular, satisfying Equation 5-38.

The same process applies, of course, if we have $\mathbf{G}_{s.c.}$ and want $\mathbf{R}_{o.c.}$. Then

$$\mathbf{R}_{o.c.} = \mathbf{G}_{s.c.}^{-1} \tag{5-40}$$

provided $\mathbf{G}_{s.c.}$ is nonsingular; that is, if

$$\det \mathbf{G}_{s.c.} \neq 0 \tag{5-41}$$

Two elementary 2-P's will illustrate these points.

EXAMPLE 5-27

The 2-P shown in Figure 5-31(a) has the open-circuit parameters

$$\mathbf{R}_{o.c.} = \begin{bmatrix} 1 & 1 \\ 1 & 1 \end{bmatrix}$$

Be sure to verify this, either formally or informally!

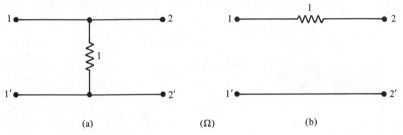

(a)   (Ω)   (b)

Figure 5-31 Example 5-27.

Solution. Since $\det \mathbf{R}_{o.c.} = 0$, the short-circuit matrix $\mathbf{G}_{s.c.}$ does not exist. What does it mean physically (electrically)? Try to set up (on paper!) the two experiments for calculating $\mathbf{G}_{s.c.}$ in a formal way, and see what happens.

In a dual way (yes, duality is prevalent in circuits), the 2-P in Figure 5-31(b) has

$$\mathbf{G}_{s.c.} = \begin{bmatrix} 1 & -1 \\ -1 & 1 \end{bmatrix}$$

(verify it!). Since $\det \mathbf{G}_{s.c.} = 0$, the open-circuit matrix $\mathbf{R}_{o.c.}$ does not exist. Verify here, too, what happens in the two formal experiments if you set them up to calculate $\mathbf{R}_{o.c.}$.

We see here another reason for having six different sets of parameters for 2-P's: A given 2-P may not have one (or more) set of parameters; it is comforting to have other parameters which describe this 2-P. ■ Probs. 5-35, 5-36, 5-37

3. The Hybrid Parameters These parameters are particularly useful in electronic transistor circuits, and are defined by

$$\begin{bmatrix} v_1 \\ i_2 \end{bmatrix} = \begin{bmatrix} h_{11} & h_{12} \\ h_{21} & h_{22} \end{bmatrix} \begin{bmatrix} i_1 \\ v_2 \end{bmatrix} \tag{5-42}$$

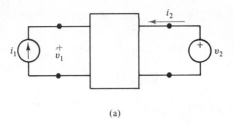

(a)

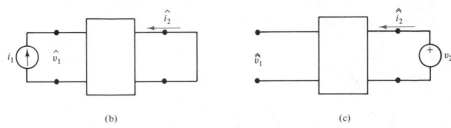

(b)   (c)

Figure 5-32 The hybrid parameters.

The proper identification of the inputs, $\mathbf{e}(t)$, and outputs, $\mathbf{r}(t)$, according to Equation 5-20, is

$$\mathbf{e}(t) = \begin{bmatrix} i_1 \\ v_2 \end{bmatrix} \qquad \mathbf{r}(t) = \begin{bmatrix} v_1 \\ i_2 \end{bmatrix} \tag{5-43}$$

This is shown in Figure 5-32(a). The use of superposition here gives, as in Figure 5-32(b),

$$\hat{v}_1 = h_{11} i_1 \qquad v_2 = 0 \tag{5-44}$$
$$\hat{i}_2 = h_{21} i_1 \qquad v_2 = 0 \tag{5-45}$$

with h_{11} being a driving-point resistance (with port 2 short-circuited) and h_{21} a dimensionless current transfer function. The second experiment, shown in Figure 5-32(c), yields

$$\hat{\hat{v}}_1 = h_{12} v_2 \qquad i_1 = 0 \tag{5-46}$$

and

$$\hat{\hat{i}}_2 = h_{22} v_2 \qquad i_1 = 0 \tag{5-47}$$

Here, h_{12} is a dimensionless voltage transfer function and h_{22} a driving-point conductance (with port 1 open-circuited).

Since the inputs are mixed in nature (one voltage and one current), as are the outputs, we call these parameters the *hybrid* (mixed) *parameters*

$$\mathbf{h} = \begin{bmatrix} h_{11} & h_{12} \\ h_{21} & h_{22} \end{bmatrix} \tag{5-48}$$

Here h stands for "hybrid." Again, the informal method is preferred when these are needed, particularly if another set of parameters is already known.

EXAMPLE 5-28 ───

Find the hybrid parameters for Example 5-26.

Solution. We have the short-circuit description

$$i_1 = \frac{7-k}{3} v_1 - \frac{1}{3} v_2$$

$$i_2 = \frac{k-1}{3} v_1 + \frac{1}{3} v_2$$

We must rearrange these to have v_1 and i_2 in term of i_1 and v_2.
The first equation yields immediately

$$v_1 = \frac{3}{7-k} i_1 + \frac{1}{7-k} v_2 = h_{11} i_1 + h_{12} v_2$$

When this expression is substituted into the second equation, the result is

$$i_2 = \frac{k-1}{7-k} i_1 + \left[\frac{k-1}{3(7-k)} + \frac{1}{3} \right] v_2 = h_{21} i_1 + h_{22} v_2$$

Please repeat this example by the formal method. It is instructive! ■

Probs. 5-38,
5-39, 5-40

4. The Inverse Hybrid Parameters As their name implies, these parameters are given
by

$$\begin{bmatrix} i_1 \\ v_2 \end{bmatrix} = \begin{bmatrix} g_{11} & g_{12} \\ g_{21} & g_{22} \end{bmatrix} \begin{bmatrix} v_1 \\ i_2 \end{bmatrix} \tag{5-49}$$

and the circuit is shown in Figure 5-33(a). The two experiments by superposition are
shown in Figure 5-33(b) and (c). The details of the derivation are fairly easy, and you
are encouraged to develop them step by step.

Probs. 5-41,
5-42,

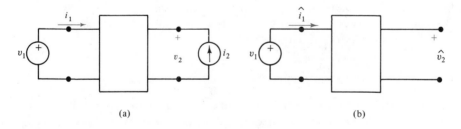

(a)   (b)

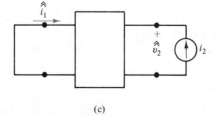

(c)

Figure 5-33 The inverse hybrid parameters.

From a mathematical point of view, we have

$$\mathbf{g} = \begin{bmatrix} g_{11} & g_{12} \\ g_{21} & g_{22} \end{bmatrix} = \mathbf{h}^{-1} \tag{5-50}$$

provided that $\mathbf{h}^{-1}$ exists, i.e.,

$$\det \mathbf{h} \neq 0 \tag{5-51}$$

5. The Chain Parameters The *chain parameters* (also called the *transmission parameters*) were of earlier use in 2-P networks. In those days, the concepts of network functions were not fully developed and, as a result, Equation 5-20 was not strictly followed. Nevertheless, relations among the 2-P voltages and currents were formulated (correctly!) as follows:

$$v_1 = Av_2 + B(-i_2) \tag{5-52}$$
$$i_1 = Cv_2 + D(-i_2) \tag{5-53}$$

or, in matrix form,

$$\begin{bmatrix} v_1 \\ i_1 \end{bmatrix} = \begin{bmatrix} A & B \\ C & D \end{bmatrix} \begin{bmatrix} v_2 \\ -i_2 \end{bmatrix} = \mathbf{T} \begin{bmatrix} v_2 \\ -i_2 \end{bmatrix} \tag{5-54}$$

where $\mathbf{T}$ is the chain (transmission) matrix. From the mathematical point of view (our informal method), this is certainly a valid formulation. It expresses the variables at port 1 in terms of the variables at port 2, as they are "transmitted" through the network—hence their name. The reason for the minus sign in i_2 is historical: In those days, the reference direction for i_2 was opposite to the one we use now.

It is easy to recognize that Equation 5-54 is not in the form of Equation 5-20: We can't have simultaneously v_2 and i_2 as inputs! As a result, the chain parameters are not necessarily network functions in the formal way. Nonetheless, we must emphasize that Equation 5-54 is a valid way of describing a 2-P; it has also certain advantages.

We can set up open-circuit or short-circuit experiments to help us calculate A, B, C, and D. For example, from Equation 5-52 we have

$$v_2 = \frac{1}{A} v_1 \qquad i_2 = 0 \tag{5-55}$$

yielding a valid network function, $1/A$, under open-circuit conditions in port 2. Once we have $1/A$, we can invert it, of course, to get A. (*Question*: Why can't we set up $v_1 = Av_2$, $i_2 = 0$ directly in Equation 5-52?)

Prob 5-43

EXAMPLE 5-29 _____

Calculate the chain parameters for each of the 2-P networks in Figure 5-31. These, as you may recall, did not have a $\mathbf{G}_{s.c.}$ and an $\mathbf{R}_{o.c.}$. To make it more general, let the 2-P in Figure 5-31(a) have a resistor R_a instead of 1 Ω. Also, in Figure 5-31(b), let it be R_b instead of 1 Ω.

Solution. For Figure 5-31(a), we write by inspection

$$v_1 = R_a(i_1 + i_2)$$
$$v_2 = R_a(i_1 + i_2)$$

Rearrange these in chain matrix form to read

$$\begin{bmatrix} v_1 \\ i_1 \end{bmatrix} = \begin{bmatrix} 1 & 0 \\ \dfrac{1}{R_a} & 1 \end{bmatrix} \begin{bmatrix} v_2 \\ -i_2 \end{bmatrix} \qquad \therefore \quad \mathbf{T}_a = \begin{bmatrix} 1 & 0 \\ \dfrac{1}{R_a} & 1 \end{bmatrix}$$

For Figure 5-31(b) we have

$$i_1 = -i_2$$
$$v_1 = R_b i_1 + v_2$$

or, in matrix form showing the chain parameters,

$$\begin{bmatrix} v_1 \\ i_1 \end{bmatrix} = \begin{bmatrix} 1 & R_b \\ 0 & 1 \end{bmatrix} \begin{bmatrix} v_2 \\ -i_2 \end{bmatrix} \qquad \therefore \quad \mathbf{T}_b = \begin{bmatrix} 1 & R_b \\ 0 & 1 \end{bmatrix} \qquad \blacksquare$$

EXAMPLE 5-30

Calculate the chain parameters of the 2-P shown in Figure 5-34. We recognize this 2-P as a chain connection of the 2-P in Figure 5-31(a) with Figure 5-31(b) to its right.

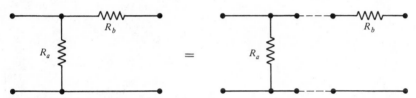

Figure 5-34   Example 5-30.

Solution.   The calculations are straightforward (do them!), with the final result

$$\mathbf{T} = \begin{bmatrix} A & B \\ C & D \end{bmatrix} = \begin{bmatrix} 1 & R_b \\ \dfrac{1}{R_a} & \dfrac{R_b}{R_a} + 1 \end{bmatrix}$$

A brief observation reveals that this result is the product of the two chain matrices

$$\mathbf{T}_a \cdot \mathbf{T}_b = \begin{bmatrix} 1 & 0 \\ \dfrac{1}{R_a} & 1 \end{bmatrix} \begin{bmatrix} 1 & R_b \\ 0 & 1 \end{bmatrix} = \begin{bmatrix} 1 & R_b \\ \dfrac{1}{R_a} & \dfrac{R_b}{R_a} + 1 \end{bmatrix}$$

In fact, it can be proved, in general, that the overall chain matrix of 2-P's connected in chain is the product of their individual chain matrices, done in the same order as the 2-P's. The order is important because we get a different overall 2-P if we interchange the order of the 2-P's; also matrix multiplication is not commutative, in general, and the order of multiplication is important.   ■

Probs. 5-44, 5-45

6. The Inverse Chain Parameters   Finally, the sixth possible form of relating the 2-P variables is with the *inverse chain* parameters. These are defined (as expected) by

$$\begin{bmatrix} v_2 \\ -i_2 \end{bmatrix} = \begin{bmatrix} A' & B' \\ C' & D' \end{bmatrix} \begin{bmatrix} v_1 \\ i_1 \end{bmatrix} \tag{5-56}$$

where the prime (') distinguishes these parameters from the previous ones.

The inverse chain parameters also do not show an input-output relation in the form of Equation 5-20. They are related to the chain parameters by

$$\mathbf{T'} = \begin{bmatrix} A' & B' \\ C' & D' \end{bmatrix} = \begin{bmatrix} A & B \\ C & D \end{bmatrix}^{-1} = \mathbf{T}^{-1} \tag{5-57}$$

when the inverse matrix exists.

Probs. 5-46, 5-47

To conclude this section, let us mention again the informal method of calculating any set of parameters from any other given set. All it takes is an algebraic rearrangement of terms. This is straightforward, but may be tedious. The good news is that it has been done, once and for all. Table 5-1 gives the complete interrelationships among the six sets of parameters.

TABLE 5-1   RELATIONSHIPS AMONG 2-P PARAMETERS

	$\mathbf{R}_{o.c.}$		$\mathbf{G}_{s.c.}$		$\mathbf{h}$		$\mathbf{g}$		$\mathbf{T}$		$\mathbf{T'}$	
$\mathbf{R}_{o.c.}$	r_{11}	r_{12}	$\dfrac{G_{22}}{\det \mathbf{G}}$	$\dfrac{-G_{12}}{\det \mathbf{G}}$	$\dfrac{\det \mathbf{h}}{h_{22}}$	$\dfrac{h_{12}}{h_{22}}$	$\dfrac{1}{g_{11}}$	$\dfrac{-g_{12}}{g_{11}}$	$\dfrac{A}{C}$	$\dfrac{\det \mathbf{T}}{C}$	$\dfrac{D'}{C'}$	
	r_{21}	r_{22}	$\dfrac{-G_{21}}{\det \mathbf{G}}$	$\dfrac{G_{11}}{\det \mathbf{G}}$	$\dfrac{-h_{21}}{h_{22}}$	$\dfrac{1}{h_{22}}$	$\dfrac{g_{21}}{g_{11}}$	$\dfrac{\det \mathbf{g}}{g_{11}}$	$\dfrac{1}{C}$	$\dfrac{D}{C}$	$\dfrac{\det \mathbf{T'}}{C'}$	
$\mathbf{G}_{s.c.}$	$\dfrac{r_{22}}{\det \mathbf{R}}$	$\dfrac{-r_{12}}{\det \mathbf{R}}$	G_{11}	G_{12}	$\dfrac{1}{h_{11}}$	$\dfrac{-h_{12}}{h_{11}}$	$\dfrac{\det \mathbf{g}}{g_{22}}$	$\dfrac{g_{12}}{g_{22}}$	$\dfrac{D}{B}$	$\dfrac{-\det \mathbf{T}}{B}$	$\dfrac{A'}{B'}$	
	$\dfrac{-r_{21}}{\det \mathbf{R}}$	$\dfrac{r_{11}}{\det \mathbf{R}}$	G_{21}	G_{22}	$\dfrac{h_{21}}{h_{11}}$	$\dfrac{\det \mathbf{h}}{h_{11}}$	$\dfrac{-g_{21}}{g_{22}}$	$\dfrac{1}{g_{22}}$	$\dfrac{-1}{B}$	$\dfrac{A}{B}$	$\dfrac{-\det \mathbf{T'}}{B'}$	
$\mathbf{h}$	$\dfrac{\det \mathbf{R}}{r_{22}}$	$\dfrac{r_{12}}{r_{22}}$	$\dfrac{1}{G_{11}}$	$\dfrac{-G_{12}}{G_{11}}$	h_{11}	h_{12}	$\dfrac{g_{22}}{\det \mathbf{g}}$	$\dfrac{-g_{12}}{\det \mathbf{g}}$	$\dfrac{B}{D}$	$\dfrac{\det \mathbf{T}}{D}$	$\dfrac{B'}{A'}$	
	$\dfrac{-r_{21}}{r_{22}}$	$\dfrac{1}{r_{22}}$	$\dfrac{G_{21}}{G_{11}}$	$\dfrac{\det \mathbf{G}}{G_{11}}$	h_{21}	h_{22}	$\dfrac{-g_{21}}{\det \mathbf{g}}$	$\dfrac{g_{11}}{\det \mathbf{g}}$	$\dfrac{-1}{D}$	$\dfrac{C}{D}$	$\dfrac{-\det \mathbf{T'}}{A'}$	
$\mathbf{g}$	$\dfrac{1}{r_{11}}$	$\dfrac{-r_{12}}{r_{11}}$	$\dfrac{\det \mathbf{G}}{G_{22}}$	$\dfrac{G_{12}}{G_{22}}$	$\dfrac{h_{22}}{\det \mathbf{h}}$	$\dfrac{-h_{12}}{\det \mathbf{h}}$	g_{11}	g_{12}	$\dfrac{C}{A}$	$\dfrac{-\det \mathbf{T}}{A}$	$\dfrac{C'}{D'}$	
	$\dfrac{r_{21}}{r_{11}}$	$\dfrac{\det \mathbf{R}}{r_{11}}$	$\dfrac{-G_{21}}{G_{22}}$	$\dfrac{1}{G_{22}}$	$\dfrac{-h_{21}}{\det \mathbf{h}}$	$\dfrac{h_{11}}{\det \mathbf{h}}$	g_{21}	g_{22}	$\dfrac{1}{A}$	$\dfrac{B}{A}$	$\dfrac{\det \mathbf{T'}}{D'}$	
$\mathbf{T}$	$\dfrac{r_{11}}{r_{21}}$	$\dfrac{\det \mathbf{R}}{r_{21}}$	$\dfrac{-G_{22}}{G_{21}}$	$\dfrac{-1}{G_{21}}$	$\dfrac{-\det \mathbf{h}}{h_{21}}$	$\dfrac{-h_{11}}{h_{21}}$	$\dfrac{1}{g_{21}}$	$\dfrac{g_{22}}{g_{21}}$	A	B	$\dfrac{D'}{\det \mathbf{T'}}$	
	$\dfrac{1}{r_{21}}$	$\dfrac{r_{22}}{r_{21}}$	$\dfrac{-\det \mathbf{G}}{G_{21}}$	$\dfrac{-G_{11}}{G_{21}}$	$\dfrac{-h_{22}}{h_{21}}$	$\dfrac{-1}{h_{21}}$	$\dfrac{g_{11}}{g_{21}}$	$\dfrac{\det \mathbf{g}}{g_{21}}$	C	D	$\dfrac{C'}{\det \mathbf{T'}}$	
$\mathbf{T'}$	$\dfrac{r_{22}}{r_{12}}$	$\dfrac{\det \mathbf{R}}{r_{12}}$	$\dfrac{-G_{11}}{G_{12}}$	$\dfrac{-1}{G_{12}}$	$\dfrac{1}{h_{12}}$	$\dfrac{h_{11}}{h_{12}}$	$\dfrac{-\det \mathbf{g}}{g_{12}}$	$\dfrac{-g_{22}}{g_{12}}$	$\dfrac{D}{\det \mathbf{T}}$	$\dfrac{B}{\det \mathbf{T}}$	A'	
	$\dfrac{1}{r_{12}}$	$\dfrac{r_{11}}{r_{12}}$	$\dfrac{-\det \mathbf{G}}{G_{12}}$	$\dfrac{-G_{22}}{G_{12}}$	$\dfrac{h_{22}}{h_{12}}$	$\dfrac{\det \mathbf{h}}{h_{12}}$	$\dfrac{-g_{11}}{g_{12}}$	$\dfrac{-1}{g_{12}}$	$\dfrac{C}{\det \mathbf{T}}$	$\dfrac{A}{\det \mathbf{T}}$	C'	

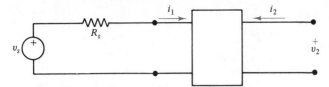

Figure 5-35 A singly terminated 2-P.

The following notation is used in this table, for the sake of brevity: det $\mathbf{R}$ is the determinant of $\mathbf{R}_{o.c.}$; det $\mathbf{G}$ is the determinant of $\mathbf{G}_{s.c.}$; det $\mathbf{h}$ is the determinant of $\mathbf{h}$, etc. Listings in the same row and in the same position are equivalent, so, for example,

$$\frac{r_{11}}{r_{21}} = \frac{-\det \mathbf{h}}{h_{21}} = \frac{1}{g_{21}} = A = \cdots$$

As mentioned (see Figure 5-22), a 2-P is used typically as a coupling network. Consider, first, a *singly terminated* 2-P, shown in Figure 5-35, where the 2-P, given by its $\mathbf{R}_{o.c.}$, has a termination only at port 1. This is a limiting case of Figure 5-22 with $R_L \to \infty$. We want to find the overall voltage transfer function H_v relating the output v_2 to the source v_s.

For the 2-P alone, we have

$$v_2 = r_{21}i_1 \qquad i_2 = 0 \tag{5-58}$$
$$v_1 = r_{11}i_1 \qquad i_2 = 0 \tag{5-59}$$

The termination at port 1 imposes the additional condition

$$v_s = R_s i_1 + v_1 \tag{5-60}$$

We eliminate i_1 and v_1 in these equations to get

$$v_2 = \frac{r_{21}}{r_{11} + R_s} v_s \qquad \therefore \quad H_v = \frac{r_{21}}{r_{11} + R_s} \tag{5-61}$$

A *doubly terminated* 2-P is the one shown in Figure 5-22; port 1 has the source and R_s, and port 2 is terminated in R_L. Here, too, a network function of interest is $H_v = v_2/v_s$. Again, the 2-P equations hold (but $i_1 \neq 0$, $i_2 \neq 0$), plus the two conditions at the ports

$$v_s = R_s i + v_1 \tag{5-62}$$

and

$$v_2 = -R_L i_2 \tag{5-63}$$

(Check that minus sign!) The algebraic details are left to you as a problem.

Probs. 5-48, 5-49, 5-50

PROBLEMS

5-1 Classify each of the following networks as linear or nonlinear.

(a) $e(t) \xrightarrow{\ N\ } \dfrac{d}{dt} e(t)$ (a differentiator)

(b) $e(t) \xrightarrow{\ N\ } \displaystyle\int_0^t e(x)\,dx$ (an integrator)

We will study models of such networks in a later chapter.

5-2 A network is described by the following output-input characteristic. (a) Is the network linear? (b) If the network is restricted to operate about a fixed point Q (called, appropriately enough, the *operating point*) with small signals Δe and Δr, is it linear then? The method of *small-signal analysis* about an operating point is very common in many electronic circuits.

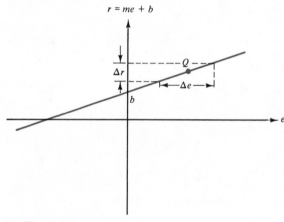

Problem 5-2

5-3 A spring, pulled by a force $f(t)$, is stretched a length $x(t)$. According to Hooke's law

$$f = Kx$$

where K is the spring's constant. Is the spring a linear element? If linear, are f and x limited to some range, and what happens beyond this range? Sketch the x vs. f characteristic and compare it with the characteristic of the op amp in Figure 4-13(b).

5-4 Use superposition to calculate v_o in the network shown. Verify your answer by another method. (Peeking in the solutions appendix is not acceptable!)

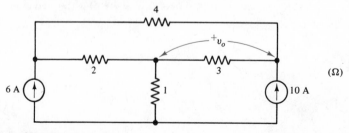

Problem 5-4

5-5 Solve for v_3 by superposition, then verify by another method.

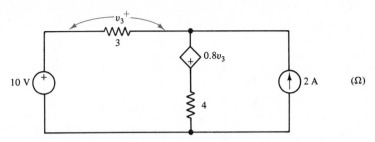

Problem 5-5

5-6 Use superposition to calculate the output voltage v_o in the op amp circuit shown. Name this circuit according to your answer.

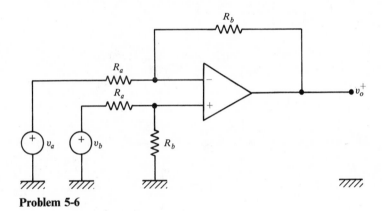

Problem 5-6

5-7 By superposition (and some informal calculations) find i_4 in the circuit shown.

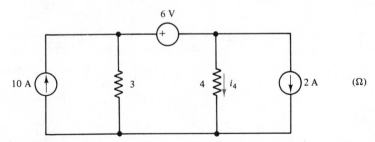

Problem 5-7

5-8 Find the current through the 6-V voltage source in Problem 5-7 using superposition. For each partial response (current), calculate the power in this source. Then calculate the total power in this source. Is superposition valid here? Why? Derive and explain in detail.

5-9 Using superposition and by inspection, write in their final form the loop equations for the network shown. Then solve these equations for the unknown loop currents (see Appendix A if necessary). Calculate the power of each current source, and state whether it receives or delivers this power.

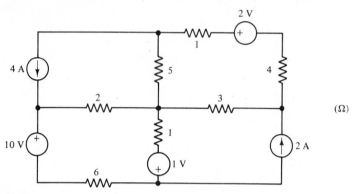

Problem 5-9

5-10 By superposition and inspection, formulate the loop equations of the network shown. Use the loop currents as given. Solve for these currents.

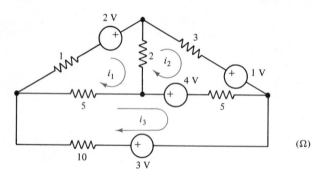

Problem 5-10

5-11 Carefully set up the loop equations for the network shown, using the same loops as in Problem 5-10. Inspect your results, the matrices **R** and **e**, compare with Problem 5-10,

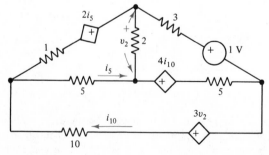

Problem 5-11

and see if you can relate them by inspection to the given network here. Solve for the loop currents and calculate the power in the source $4i_{10}$.

5-12 Use superposition to write by inspection the node equations for the network shown, in matrix form. Solve these equations (see Appendix A) and calculate the power of each source.

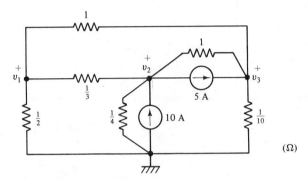

Problem 5-12

5-13 Set up carefully the node equations for the network shown. Check your result, $\mathbf{Gv} = \mathbf{j}$, and see if the matrices $\mathbf{G}$ and $\mathbf{j}$ could have been written by inspection.

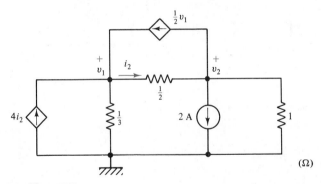

Problem 5-13

5-14 Use Thévenin's theorem to calculate i_4 in Problem 5-7.

5-15 Use Thévenin's theorem to calculate i_3 in Problem 5-10. Let the observer here be the 3-V source and the 10-Ω resistor.

5-16 (a) Calculate the current i_o in the *Wheatstone bridge* network shown, using Thévenin's theorem. Notice how much simpler it is by comparison with loop analysis!

(b) What is the relationship among R_1, R_2, R_3, and R_4 to make $i_o = 0$? Suggest a method to use this bridge in order to measure one unknown resistor (say, R_1) when R_2, R_3, and R_4 are known and adjustable.

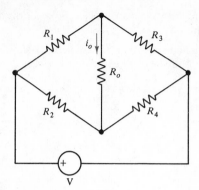

Problem 5-16

5-17 Calculate v_2 in Problem 5-13 by using Thévenin's theorem. Let the current source $i = 2$ A and the 1-Ω resistor be the observer network.

5-18 The circuit shown is a model of a common-base transistor. The terminals are e (emitter), b (base), and c (collector). The base is the common reference for voltages, hence the name. The parameters r_e, r_c, k_1, and k_2 are given. Note the two dependent sources. Calculate the current in the load resistor R_L using Thévenin's theorem, then verify by another (conventional) analysis.

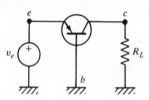
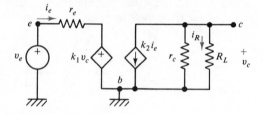

Problem 5-18

5-19 Calculate the input resistance at terminals A–B of the network shown. If $\alpha = -2$, what would you call this network?

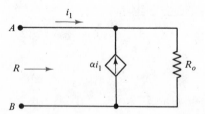

Problem 5-19

5-20 Solve Example 5-12 by a conventional method and show that your answer *is not* of the form

$$i_6 = \frac{v_{\mathrm{Th}}}{R_{\mathrm{Th}} + 6}$$

as it would be if Thévenin's theorem were applicable there.

*5-21 Prove Thévenin's theorem, using the following steps:
   1. In the original setup, shown in (a), let the voltage across the observer network be v_o and the current i_o.
   2. As far as the linear network is concerned, we can replace the observer with a voltage source v_o, with the current i_o still there, as shown in (b).
   3. Now apply superposition, since everything is linear in (b). The current i_o is contributed partly by v_o acting alone, and partly by the linear network acting alone.
   Develop these steps carefully to obtain the final Thévenin circuit of (c).

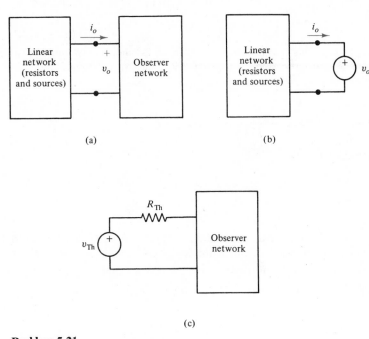

(a)   (b)

(c)

Problem 5-21

5-22 Use Norton's theorem to calculate i_4 in Problem 5-7.

5-23 Find the Norton equivalent circuit to the left of the 2-A source and the 1-Ω resistor in Problem 5-13.

5-24 Solve Problem 5-16 by using Norton's equivalent circuit.

5-25 Prove that the power delivered to a load resistance R is *maximum* (not minimum), as in Equations 5-15, 5-16, and 5-17. *Hint*: Consider two extreme cases, $R \to 0$ and $R \to \infty$, and sketch the plot of p as a function of R, remembering that R is passive.

5-26 Calculate the complete power balance (power in every element) for maximum power transfer in the circuit of Figure 5-9(a). Plot p_R vs. R fairly accurately, for $0 \le R \le \infty$.

5-27 Without using Equation 5-18, derive the expression for the power in R of Figure 5-12(a). Find the maximum of this power and the value of R then. Calculate then the complete power balance in the circuit. Can you relate your answers to Figure 5-12(b)? Explain.

5-28 What percentage of the available power from the Thévenin source, Figure 5-18(b), is dissipated in the load R when maximum power transfer happens?

5-29 In the network shown in Problem 5-13, let the input be the independent current source of 2 A. Use the results of Problem 5-13 to calculate: (1) The network function relating the output v_1 to the input. (2) The network function relating the output v_2 to the input. Write your results in matrix form, as in Equation 5-20.

5-30 Calculate the voltage transfer function v_c/v_e for the transistor circuit in Problem 5-18.

5-31 Rewrite your solution to Problem 5-12 in the form of Equation 5-20, that is,

$$\begin{bmatrix} v_1 \\ v_2 \\ v_3 \end{bmatrix} = \mathbf{H} \begin{bmatrix} 5 \\ 10 \end{bmatrix}$$

identifying fully the matrix network function **H**.

5-32 Calculate the o.c. resistance parameters for the π 2-P network shown by: (a) direct calculations, (b) converting the 2-P into a T 2-P, using the Δ-Y conversion (Chapter 4), shown in dotted lines.

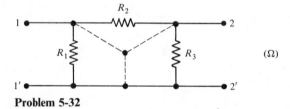

(Ω)

Problem 5-32

5-33 Calculate the o.c. resistance parameters of the 2-P in Problem 5-18 (without the source v_e and the load R_L). Port 1 is e–b; port 2 is c–b.

5-34 Calculate the o.c. resistance parameters of the 2-P shown. Comment on its suitability as a general model for $\mathbf{R}_{o.c.}$ of a 2-P.

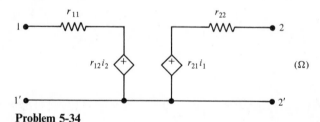

(Ω)

Problem 5-34

5-35 Calculate $\mathbf{G}_{s.c.}$ for the op amp 2-P shown. Does this 2-P have open-circuit parameters?

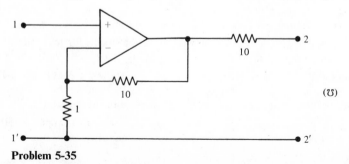

$(\mho)$

Problem 5-35

5-36 Calculate $\mathbf{G}_{s.c.}$ for the 2-P model shown.

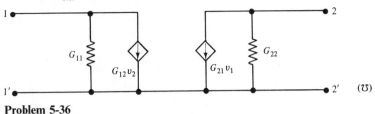

Problem 5-36

5-37 Calculate $\mathbf{R}_{o.c.}$ and $\mathbf{G}_{s.c.}$ for the 2-P shown.

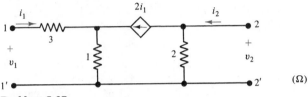

Problem 5-37

5-38 Find the hybrid parameters for the 2-P of Problem 5-18. (That is *the* reason for this model.) Excluded is the load resistance R_L.

5-39 Find the hybrid parameters for the 2-P in Problem 5-37: (a) by the formal method, (b) by the informal method.

5-40 Derive a general model for a 2-P using its hybrid parameters, similar to Problems 5-34 and 5-36.

5-41 Set up (on paper) the two experiments for the calculation of the **g** parameters. Draw a general model of a 2-P using these parameters (compare with Problems 5-34, 5-36, and 5-40).

5-42 Find the **g** parameters for the 2-P of Problem 5-35 by the informal method.

5-43 Set up three input-output experiments, under open-circuit or short-circuit conditions, to calculate the chain parameters B, C, D, similar to Equation 5-55 for A.

5-44 Calculate the chain parameters of the 2-P shown: (a) directly, (b) by using the chain parameters of the individual 2-P's.

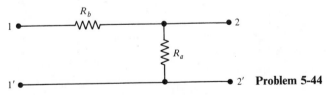

2' **Problem 5-44**

*5-45 Prove: The overall chain matrix of two 2-P's connected in chain is the product of their individual chain matrices, in the order of their connection. That is,

$$\begin{bmatrix} A & B \\ C & D \end{bmatrix}_{\text{total}} = \begin{bmatrix} A & B \\ C & D \end{bmatrix}_{\text{I}} \cdot \begin{bmatrix} A & B \\ C & D \end{bmatrix}_{\text{II}}$$

Hint: Write the 2-P chain equations for each 2-P; then relate v_2 and i_2 of 2-P I to v_1 and i_1 of 2-P II. Also relate v_1 and i_1 of the overall 2-P, and v_2 and i_2 of the overall 2-P, to 2-P I and 2-P II.

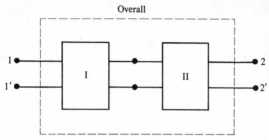

Overall

Problem 5-45

5-46 Find the inverse chain parameters of the 2-P in Problem 5-44: (a) directly, and (b) by inverting its chain matrix.

5-47 Find the inverse chain matrix of the 2-P in Problem 5-37.

5-48 Calculate the overall voltage transfer function of a doubly terminated 2-P, as given in Figure 5-22 and Equations 5-62 and 5-63 in two ways: (a) by eliminating the appropriate variables among the 2-P equations and Equations 5-62, 5-63; (b) by replacing the 2-P, v_s and R_s by its Thévenin equivalent, with R_L being the observer.

5-49 A singly terminated 2-P is shown, terminated with R_L. Use an appropriate set of parameters for the two-port (pay attention to the input!), together with the added condition in port 2, to obtain the current transfer function H_i

$$i_2 = H_i i_s$$

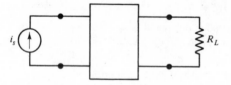

Problem 5-49

5-50 A common-emitter transistor circuit is shown in the figure, together with its equivalent 2-P model. Given typical values $r_e = 20\ \Omega$, $r_b = 100\ \Omega$, $r_c = 1\text{M}\ \Omega$, $\alpha = 0.98$, find: (a) the $\mathbf{R}_{\text{o.c.}}$ parameters of the transistor model, (b) the overall voltage transfer function of the circuit.

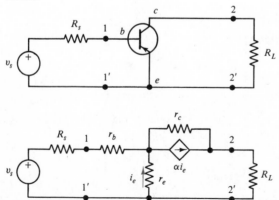

Problem 5-50

Chapter 6

Dynamic Elements

Our studies, so far, included resistive networks with independent and dependent sources. We saw in the previous chapters that, for all such networks, the output $r(t)$ is just a scaled version of the input $e(t)$ because the network function H is a constant number. This is shown in Figure 6-1 for an arbitrary form of $e(t)$ and $r(t)$. While this is an interesting and sometimes useful feature, it is also rather limited. If we want *waveshaping* circuits, that is, circuits that can change the shape of the waveform of an input, we need to have additional elements whose properties are different from those of a resistor. Why do we need waveshaping circuits? Because they are used almost everywhere—in your computer, in TV, in satellite communications, and in microwave ovens. A few of such waveforms are shown in Figure 6-2. We are ready to study two new elements which, together with resistors and dependent sources, allow us to do such exciting operations as waveshaping.

6-1 THE CAPACITOR

A capacitor (as you may remember from basic physics) is an element capable of storing electric energy. It is an observable physical experiment that when two conducting plates are separated by a nonconducting dielectric material, they form a capacitor, an element capable of storing electric charge and energy. The standard symbol for a capacitor is shown in Figure 6-3(a). A common capacitor is found in the electronic flash unit of a camera. A battery charges this capacitor; then, when you take a picture, the energy stored in the capacitor is used to fire the flashbulb.

A *linear, constant* capacitor is an ideal element (much like an ideal resistor), where the stored charge $q(t)$ is proportional to the voltage $v(t)$ across the capacitor

$$q(t) = Cv(t) \qquad (6\text{-}1)$$

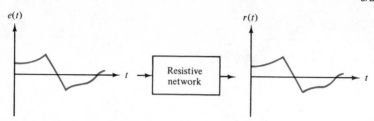

Figure 6-1   Input-output in a resistive network.

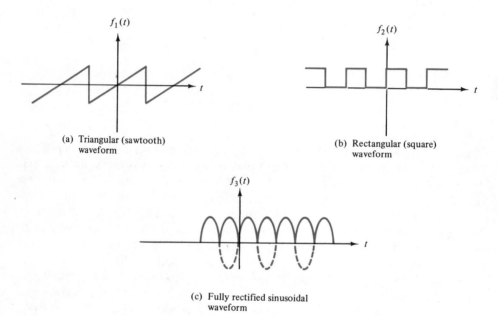

(a) Triangular (sawtooth) waveform

(b) Rectangular (square) waveform

(c) Fully rectified sinusoidal waveform

Figure 6-2   Some common waveforms.

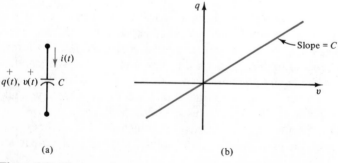

(a)   (b)

Figure 6-3   The capacitor.

See Figure 6-3(b) and compare it with the *v-i* characteristic of a linear resistor. The charge is measured in coulombs, and the voltage in volts. The positive constant C is the *capacitance* of the capacitor, and it is measured in *farads* (F). In many common circuits, small submultiples are used (1 μF = 10^{-6} F, 1 pF = 10^{-12} F, etc.). The unit farad is named after Michael Faraday (1791–1867), an English physicist.

In circuit analysis, we are interested mostly in voltages and currents. To obtain the *v-i* relation of a linear capacitor from its *q-v* relation, Equation 6-1, we remember the definition of current as the time rate of change of charge, in coulombs per second

$$i(t) = \frac{dq(t)}{dt} \qquad \text{C/s} \tag{6-2}$$

Using this relation in Equation 6-1, we obtain

$$i(t) = C \frac{dv(t)}{dt} \tag{6-3}$$

Therefore, the current in a capacitor is proportional to the derivative of the voltage across it, with the standard associated references as shown in Figure 6-3(a).   Prob. 6-1

EXAMPLE 6-1

Let the voltage across C be a constant dc value $v(t) = V_0$.

if $v(t) = dc$
capacitor = open circuit
$i(t) = 0$

  Solution. The current $i(t)$ is zero

$$i(t) = C \frac{dV_0}{dt} = 0$$

Here the capacitor acts as an open circuit (o.c.), $i = 0$. This makes sense: With a constant voltage V_0, the charge is also constant, $Q_0 = CV_0$, as in Equation 6-1. Since the charge is constant, there is no rate of change of charge—no current.

  Note the distinct difference by comparison with a resistor: There, zero current means zero voltage and vice versa. In a capacitor we can have zero current but a nonzero voltage!   ■

EXAMPLE 6-2

Let the voltage $v(t)$ across a capacitor ($C = 0.2$ F) be given as shown in Figure 6-4(a). It is required to plot the current $i(t)$.

  Solution. Since $v(t)$ has different expressions over the time axis, let us do it in the appropriate time segments:

  1. For $t < 0$:
$$v(t) = 0 \qquad \therefore \quad i(t) = 0.2 \frac{d}{dt}(0) = 0 \text{ A}$$

  2. For $0 < t < 2$:
$$v(t) = 50t \qquad \therefore \quad i(t) = 0.2 \frac{d}{dt}(50t) = 10 \text{ A}$$

  3. For $2 < t < 4$:
$$v(t) = 100 \qquad \therefore \quad i(t) = 0 \text{ A}$$

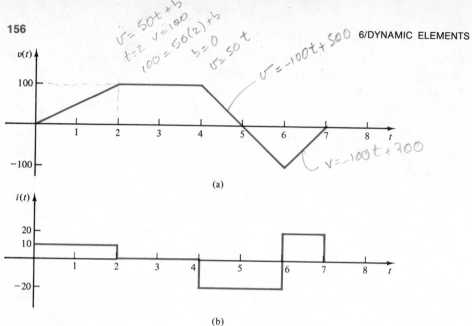

(a)

(b)

Figure 6-4 Example 6-2.

4. For $4 < t < 6$:

$$\frac{dv(t)}{dt} = -\frac{100}{1} = -100 \qquad \therefore \quad i(t) = (0.2)(-100) = -20 \text{ A}$$

5. For $6 < t < 7$:

$$\frac{dv}{dt} = \frac{100}{1} = 100 \qquad \therefore \quad i(t) = (0.2)(100) = 20 \text{ A}$$

6. For $t > 7$:

$$v(t) = 0 \qquad \therefore \quad i(t) = 0.2\frac{d}{dt}(0) = 0 \text{ A}$$

with the plot shown in Figure 6-4(b). *Question*: Why didn't we have to write the full expression for $v(t)$ in parts 4 and 5, as we did in the other parts?

Before continuing, let us pause and review the two waveforms in Figure 6-4. The waveform of $v(t)$ is continuous, while that of $i(t)$ is not; there are discontinuities—jumps—in $i(t)$ at $t = 0, 2, 4, 6,$ and 7 s. The waveform of $i(t)$ is essentially the first derivative of $v(t)$; so the capacitor acts as a waveshaper, a differentiator. ∎ Probs. 6-2, 6-3

The expression in Equation 6-3 gives the current in terms of voltage. It will be useful when we write later KCL equations. We want also the expression of $v(t)$ in terms of $i(t)$, that is, the inverse of Equation 6-3, in order to use it for KVL. To do so, we divide Equation 6-3 by C

$$\frac{dv}{dt} = \frac{1}{C}i(t) \tag{6-4}$$

then we integrate both sides between some arbitrary initial time t_0 and any later time $t > t_0$

$$\int_{t_0}^{t} \frac{dv}{dt} \, dt = \frac{1}{C} \int_{t_0}^{t} i(x) \, dx \tag{6-5}$$

Here we are using x as a dummy variable for the integrand on the right-hand side, since t is the upper limit and we don't want to confuse the variable of integration with the limit on the integral. Integrating Equation 6-5, we get

$$v(t) - v(t_0) = \frac{1}{C} \int_{t_0}^{t} i(x) \, dx \tag{6-6}$$

and finally,

$$v(t) = v(t_0) + \frac{1}{C} \int_{t_0}^{t} i(x) \, dx \tag{6-7}$$

This is the desired result. Intuitively, we could have predicted it immediately: Since the current is proportional to dv/dt, as in Equation 6-3, it is reasonable that the voltage will be proportional to the *integral* of the current. In integrating the current, we have to be careful with the limits of the integration, as well as with the appropriate constant of integration. These appear in Equations 6-5 and 6-7.

$\int_1^2 t^2 \, dt = \frac{t^3}{3} \big|_1^2 = \frac{8}{3} - \frac{1}{3} = \frac{7}{3}$

EXAMPLE 6-3 ───

If you are still a bit uneasy about the dummy variable of integration, try to work out the following simple problem in calculus: Integrate the function $f(t) = t^2$, (a) between $t = 1$ and $t = 2$ (answer: $\frac{7}{3}$), (b) between $t = 1$ and any $t > 1$.

Solution. In part (b) you'll have to use a dummy variable of integration, x, u, or any other letter, and write

$$\int_{1}^{t} x^2 \, dx = \frac{x^3}{3} \bigg]_{1}^{t} = \frac{1}{3} (t^3 - 1)$$

showing the need for the dummy variable. The final answer depends on t, the upper limit of the integration. ∎

Equation 6-7 makes sense also if we read it as follows: "The voltage at any time t is equal to the initial voltage $v(t_0)$ *plus* the voltage developed after the initial time till the time t." The integral in Equation 6-7 is the charge in the capacitor between t_0 and t, and charge divided by C is voltage, as in Equation 6-1.

In most applications, the initial time t_0 is conveniently chosen as $t_0 = 0$. That's when a switch is closed or is opened, and we start our time counting then. As a result, Equation 6-7 is rewritten as

$$v(t) = v(0) + \frac{1}{C} \int_{0}^{t} i(x) \, dx \tag{6-8}$$

We see, therefore, that the output (voltage is this case) of a capacitor at a given time $t = t_1$ depends on the input (current) at $t = t_1$ *and* on past values of the current

Prob. 6-4

$(t < t_1)$. This property defines the capacitor as a *dynamic* element. The capacitor has a *memory*, since past values of current affect present values of voltage. By contrast, a resistor is a *static* element: Its output $v(t)$ at $t = t_1$ depends *only* on the input $i(t)$ *at* $t = t_1$. It is called also an *instantaneous* or *memoryless* element.

EXAMPLE 6-4

A capacitor $(C = \frac{1}{2} \text{ F})$ has a current

$$i(t) = 2e^{-t} \qquad t \geq 0$$

$$-\int e^{-x} \, dx$$

Plot the voltage across it for $t > 0$.

$$-e^{-x}\big|_0^t = -e^{-t} + 1$$

Solution. From Equation 6-8 we write

$$v(t) = v(0) + 2 \int_0^t 2e^{-x} \, dx$$

$$= v(0) + 4(1 - e^{-t})$$

This is shown in Figure 6-5 for $v(0) = 0$ V and for $v(0) = -4$ V. Obviously, $v(0)$ *must be given* (or precalculated separately) for a unique answer to exist. The value of $v(0)$ is called the *initial value* of $v(t)$. Or else, an alternative initial condition may be given, $q(0)$, the initial charge on the capacitor. The two are related, of course, by Equation 6-1, $q(0) = Cv(0)$.

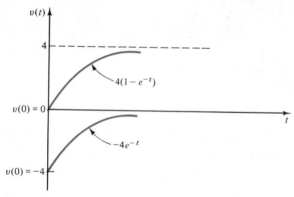

Figure 6-5 Example 6-4.

Probs. 6-5, 6-6

The principle of conservation of charge states that electric charge cannot change instantaneously. In other words, the plot of $q(t)$ must be continuous, without any discontinuities. See Figure 6-6(a). In Figure 6-6(b) we show a plot of $q(t)$ which changes instantaneously at $t = t_1$. Such a behavior is not allowed in a physical capacitor, since it violates conservation of charge at $t = t_1$. Also, if such a discontinuity in charge were allowed, it would imply an infinite current at $t = t_1$, since $i(t) = (d/dt)q(t)$. Obviously, infinite currents cannot exist in a physical system.†

As long as C does not change, the conservation of charge also means continuity of voltage, since $q(t) = Cv(t)$. Then we can say that the voltage across a constant

---

† In later chapters, we will use an idealized mathematical impulse function, which allows the description (on paper) of an infinitely large current during a very small time interval.

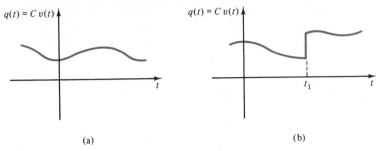

(a)   (b)

Figure 6-6 (a) Continuous $q(t)$. (b) Discontinuous $q(t)$ at $t = t_1$.

capacitor cannot change instantaneously. In particular, we write for the initial condition across a capacitor.

$$v(0^-) = v(0^+) \tag{6-9}$$

that is, the initial voltage "just before" $t = 0$ is also the initial voltage "just after" $t = 0$. These ideas are shown in Figure 6-7, where the intervals around $t = 0$ are enlarged in order to show clearly the points $t = 0^-$ and $t = 0^+$. Mathematically and physically, these intervals are infinitesimally small, of course. The need for such a distinctive "hair splitting" will become obvious in a later chapter. The important facts to remember here are the conservation of charge and of voltage across a capacitor, requiring continuous plots for $q(t)$ and $v(t)$. By contrast, we remember that the waveforms of $v(t)$ and $i(t)$ in a resistor *may* be discontinuous since $v(t) = Ri(t)$.

Prob. 6-7

Power and Energy   To calculate the instantaneous power $p(t)$ in a capacitor, we use the general expression

$$p(t) = v(t)i(t) \tag{6-10}$$

with the same convention as before: If the associated reference signs on v and i are maintained ($+$ of v at the tail of the arrow of i), then $p(t) > 0$ means that the element absorbs power, while $p(t) < 0$ means that the element delivers power.

We use Equation 6-3 in Equation 6-10 to get

$$p(t) = Cv(t)\frac{dv(t)}{dt} \quad \text{W} \tag{6-11}$$

as the instantaneous power for a capacitor.

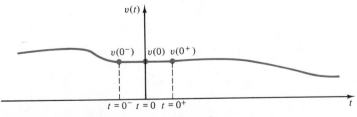

Figure 6-7   Continuity of a capacitive voltage.

EXAMPLE 6-5 ————————————————————————————————

For the waveforms of $v(t)$ and $i(t)$ in Example 6-2 (Figure 6-4), we can obtain the plot of $p(t)$ by simply multiplying the two curves, point by point.

Solution. For $t < 0$:

$$p(t) = v(t)i(t) = (0)(0) = 0$$

For $0 < t < 2$:

$$p(t) = (50t)(10) = 500t$$

For $2 < t < 4$:

$$p(t) = (100)(0) = 0$$

and so on. The plot of $p(t)$ is shown in Figure 6-8. We make the following observations:

1. Over $0 < t < 2$ and $5 < t < 6$, the capacitor receives power, since $p(t) > 0$ then.
2. Over $4 < t < 5$ and $6 < t < 7$, the capacitor delivers power, since $p(t) < 0$ then.
3. Over $2 < t < 4$, and for $t < 0$ and $t > 7$, the capacitor does not receive or deliver power, $p(t) = 0$ then.

Here, again, we contrast the capacitor with the resistor. The capacitor is capable of receiving, storing, and delivering power. The resistor only receives power (and dissipates it).

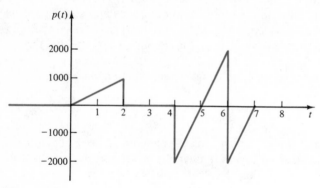

Figure 6-8   Example 6-5.   ■ Prob. 6-8

To calculate the energy absorbed in the capacitor between t_1 and t_2 we write, as always,

$$w = \int_{t_1}^{t_2} p(t)\, dt \tag{6-12}$$

and with Equation 6-11 this becomes

$$w = \int_{v_1}^{v_2} Cv\, dv \tag{6-13}$$

where $v_1 = v(t_1)$ and $v_2 = v(t_2)$. Integration of Equation 6-13 gives

$$w = \frac{C}{2}(v_2^2 - v_1^2) \quad \text{J} \tag{6-14}$$

EXAMPLE 6-6 _____

For the capacitor and $v(t)$ and $i(t)$ in Example 6-5, we calculate the energies for the various time intervals.

Solution. For $t < 0$:

$$w = 0 \qquad \text{(uncharged capacitor)}$$

For $0 < t < 2$:

$$w = \int_0^2 500t \ dt = \frac{0.2}{2}(100^2 - 0^2) = 1000 \text{ J}$$

and for $6 < t < 7$:

$$w = \frac{0.2}{2}[0^2 - (-100)^2] = -1000 \text{ J}$$

You should complete these calculations for the other intervals and show that energy is conserved (as it should be!) over all time. ■

Probs. 6-9, 6-10

6-2 THE INDUCTOR

An inductor is another dynamic element, capable of storing energy. In its classical form, it is a coil of wound wire, shown symbolically in Figure 6-9(a). Through many observable physical experiments (by A. M. Ampère, H. C. Oersted, and M. Faraday, among others), we conceive of a *magnetic flux* $\phi(t)$ associated with the current $i(t)$ in the inductor. In a simple experiment, demonstrating the magnetic effect of the flux, a compass needle shows a deflection when placed in the vicinity of a current-carrying wire. The flux created by the current causes the needle to deflect.

If you are worried about actually seeing the magnetic flux, then you should ask yourself also, "What about the electric charge in a capacitor?" In both cases, you don't have to worry. No one has ever seen (or, for that matter, will see) charge or flux. These are convenient physical abstractions; in terms of those abstractions, we *can* see and measure *their effects*, such as the deflection of the compass needle or the photographed trace of a charged particle.

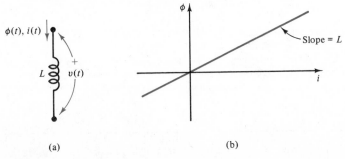

(a) (b)

Figure 6-9 The inductor.

A *linear, constant* inductor is an ideal element where the flux $\phi(t)$ is proportional to the current $i(t)$

$$\phi(t) = Li(t) \tag{6-15}$$

as shown in Figure 6-9(b). The unit of flux is the *weber* (Wb), and the current, as usual, is in amperes. The positive constant L is the *inductance* of the inductor, with the unit measured in *henrys* (H). The unit henry is named after the American scientist Joseph Henry (1797–1878). In common circuits, submultiples are often used (1 mH = 10^{-3} H, etc.). We say that the flux ϕ *links* ("embraces," "surrounds") the inductor L.

The voltage-current relation for the inductor is derived from an observable experiment: $v(t)$, as shown in Figure 6-9(a), is the time rate of change of the flux, that is,

$$v(t) = \frac{d\phi(t)}{dt} \quad \text{Wb/s} \tag{6-16}$$

Using this relation in Equation 6-15, we obtain

$$v(t) = L\frac{di(t)}{dt} \tag{6-17}$$

In a linear inductor, then, the voltage is proportional to the derivative of the current through it. The standard associated references for v and i are shown in Figure 6-9(a).

Do you begin to suspect duality between the capacitor and the inductor? You should! Their v–i relationships (Equations 6-3 and 6-17) are dual of each other: v in one relationship is replaced by i in the other. If we agree that charge and flux play dual roles, then duality exists in Equations 6-1 and 6-15, and in Equations 6-3 and 6-16.   Prob. 6-11

EXAMPLE 6-7 _____

Let the current through L be a constant value $i(t) = I_0$.

Solution.   Then the voltage $v(t)$ across the inductor is zero

$$v(t) = L\frac{dI_0}{dt} = 0$$

Here, the inductor acts as a short circuit (s.c.), $v = 0$. With a constant current I_0, the flux is also constant $\phi_0 = LI_0$ and its rate of change, the voltage, is zero.

Note here also the distinct difference by comparison with a resistor; there, I_0 would produce the voltage RI_0. In an inductor, we can have zero voltage but a nonzero current.

■

EXAMPLE 6-8 _____

Let the current $i(t)$ through an inductor ($L = 0.3$ H) be given as in Figure 6-10(a). We wish to plot the voltage $v(t)$ across L.

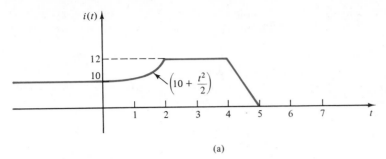

(a)

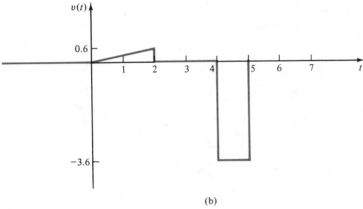

(b)

Figure 6-10   Example 6-8.

Solution.   For the different time segments we have

1. $t < 0$:

$$i(t) = I_0 = 10 \qquad \therefore \quad v(t) = 0.3 \frac{d}{dt} (10) = 0 \text{ V}$$

2. $0 < t < 2$:

$$i(t) = \frac{t^2}{2} + 10 \qquad \therefore \quad v(t) = 0.3 \frac{d}{dt} \left(\frac{t^2}{2} + 10 \right) = 0.3t \text{ V}$$

3. $2 < t < 4$:

$$i(t) = 12 \qquad \therefore \quad v(t) = 0 \text{ V}$$

4. $4 < t < 5$:

$$\frac{di}{dt} = -12 \qquad \therefore \quad v(t) = (0.3)(-12) = -3.6 \text{ V}$$

5. $t > 5$:

$$\frac{di}{dt} = 0 \qquad \therefore \quad v(t) = 0$$

and the waveform of $v(t)$ is plotted in Figure 6-10(b). As in Example 6-2, we did not need the analytical expression for $i(t)$ in the interval $4 < t < 5$, only its derivative.

While the waveform of $i(t)$ is continuous, that of $v(t)$ has "jump" discontinuities at $t = 2, 4,$ and 5. The inductor, like the capacitor, is a waveshaping element, differentiating the current.

Probs. 6-1
6-13

To obtain the i–v (current in terms of voltage) for an inductor, we start with Equation 6-17. It is the v–i relationship, useful for writing KVL. We need to invert it. Dividing by L, we get

$$\frac{di}{dt} = \frac{1}{L}\, v(t) \tag{6-18}$$

Next, we integrate between an initial time t_0 and any later time t

$$\int_{t_0}^{t} \frac{di}{dt}\, dt = \frac{1}{L} \int_{t_0}^{t} v(x)\, dx \tag{6-19}$$

with x a dummy variable of integration. Equation 6-19 then yields

$$i(t) - i(t_0) = \frac{1}{L} \int_{t_0}^{t} v(x)\, dx \tag{6-20}$$

or, finally,

$$i(t) = i(t_0) + \frac{1}{L} \int_{t_0}^{t} v(x)\, dx \tag{6-21}$$

If the initial time is $t_0 = 0$, we have

$$i(t) = i(0) + \frac{1}{L} \int_{0}^{t} v(x)\, dx \tag{6-22}$$

In words, the current at any time $t > 0$ is the initial current $i(0)$ plus the current that develops from $t = 0$ onward. The integral part in Equation 6-22 is the flux of the inductor for $t > 0$, and when divided by L it gives current, according to Equation 6-15.

The dynamic nature of the inductor is evident from Equation 6-21 or 6-22: The output (current) at time t_1 depends not only on the input (voltage) at that time but also on past values of the input. Like the capacitor, the inductor has a memory.

EXAMPLE 6-9 _____

An inductor $(L = \frac{1}{2}$ H) has a voltage across it

$$v(t) = 2e^{-t} \qquad t \geq 0$$

with the standard associated references. Also given is the initial current

$$i(0) = -4 \text{ A}$$

Calculate the current $i(t)$.

Solution. We write, according to Equation 6-22,

$$i(t) = -4 + 2 \int_{0}^{t} 2e^{-x}\, dx = -4e^{-t} \qquad t \geq 0$$

Its plot and the plot of $v(t)$ are shown in Figure 6-11.

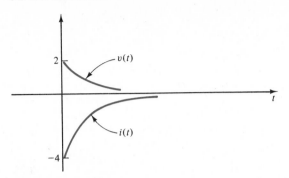

Figure 6-11   Example 6-9.   ■

As in the capacitor, the inductor has a natural initial condition associated with it, $i(0)$ or $\phi(0) = Li(0)$. This initial condition *must be given* (separately) if we want a unique answer for the current at $t > 0$, according to Equation 6-22. The inductor, just like the capacitor, is a dynamic element. Its i–v relationship involves an integration. The constant of integration is, in general, arbitrary. For a unique answer, this constant must be calculated or specified separately.

Just as the principle of conservation of charge applies to a capacitor, the *principle of conservation of flux* applies to the inductor. The flux $\phi(t)$ cannot have discontinuities. See Figure 6-6, where $q(t) = Cv(t)$ is now replaced by $\phi(t) = Li(t)$. A "jump" discontinuity in $\phi(t)$ would imply an infinitely large inductive voltage at this point, according to Equation 6-16. In physical circuits, only *finite* (although possibly very large) voltages can exist.†

As long as L does not change, the conservation of flux also means the continuity of current, since $\phi(t) = Li(t)$. The current through a constant inductor cannot change instantaneously. In particular

$$i(0^-) = i(0^+) \tag{6-23}$$

the initial current "just before" $t = 0$ is also the current "just after" $t = 0$, as shown in Figure 6-12.

Probs. 6-14, 6-15

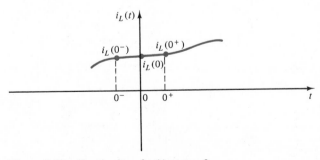

Figure 6-12   Continuity of $i_L(t)$ at $t = 0$.

† In later chapters, we will use an idealized mathematical impulse function, which allows the description (on paper) of an infinitely large voltage during a very small time interval.

6-3 MUTUAL INDUCTANCE

Consider an inductor L_1, with a current $i_1(t)$, as shown in Figure 6-13(a). The resulting flux is, according to Equation 6-15,

$$\phi(t) = L_1 i_1(t) \tag{6-24}$$

Now, let there be another inductor L_2 in the vicinity of L_1, as shown in Figure 6-13(b). This inductor carries no current; it is open-circuited. Being physically close to L_1, some of the flux $\phi(t)$ links L_2. We write $\phi(t)$ as the sum

$$\phi(t) = \phi_{11}(t) + \phi_{21}(t) \tag{6-25}$$

Here ϕ_{11} is the flux linking L_1 due to current i_1, and ϕ_{21} is the flux linking L_2 due to i_1. In the linear inductor L_2, this flux is proportional to i_1

$$\phi_{21} = M i_1 \tag{6-26}$$

where M is a constant, called the *mutual inductance* between L_1 and L_2. Its unit is also the henry (H).

We have now a similar observable experiment as before: L_2, having a varying flux $\phi_{21}(t)$ linking it, will develop a voltage v_2 across its terminals, given by Equation 6-16 and 6-26:

$$v_2(t) = \frac{d}{dt} \phi_{21}(t) = \frac{d}{dt} (M i_1) = M \frac{d i_1}{dt} \tag{6-27}$$

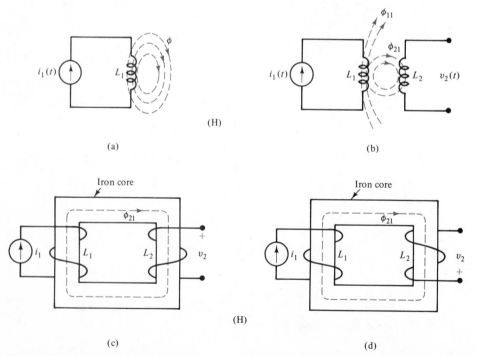

Figure 6-13   Mutual inductance.

At first, this is a startling observation: The inductor L_2, without any current, is just "sitting there," yet it develops a voltage v_2 across its terminals "out of nowhere!" However, the physical observation is firm. The voltage v_2 is *induced* in L_2 because of the mutual flux ϕ_{21} from i_1. At one time, such an observation was labeled "action at a distance": A current i_1 in L_1 causes a voltage v_2 at a distance from L_1. We prefer to use the more accurate terminology and speak of a voltage v_2 induced in L_2 because of a mutual inductance M (and a mutual flux ϕ_{21}) between L_1 and L_2. The phenomenon of mutual inductance and induced voltages is the basis for the operation of transformers, motors, and similar devices. Without mutual inductance, for example, our electric company would not be able to generate or distribute electricity!

We will return to mutual inductance in a later chapter. However, two important notes must be made at this point. First, from considerations of energy, it can be shown that, for two linear inductors L_1 and L_2, the mutual inductance M is given by

$$M = k\sqrt{L_1 L_2} \tag{6-28}$$

where $0 < k \leq 1$ is the *coefficient of coupling*, giving a measure of how closely (tightly) the two inductors are coupled. It depends not only on the physical closeness of L_1 and L_2 but also on the medium that provides for the mutual linkage. For example, if L_1 and L_2 are linked through air, the coefficient of coupling would be smaller (say, $k = 0.2$) than if L_1 and L_2 were wound on an iron core, as shown in Figure 6-13(c) and (d). An iron core provides a preferred path for the flux, and therefore a larger coefficient of coupling (say, $k = 0.98$).

Exactly the same observations and results are true when a current i_2 in L_2 induces an open-circuit voltage v_1 in L_1,

$$v_1 = M \frac{di_2}{dt} \tag{6-29}$$ Prob. 6-16

The second observation is related to the reference mark $(+)$ for the induced voltage v_2. If you remember the "right-hand rule" from your basic physics, you will be able to establish the direction of ϕ_{21} and, from it and the relative sense of the windings of L_1 and L_2, the reference of v_2. However, even if you remember it, this would be tedious or even impossible to apply every time. Just think of having to break the metal casing of a transformer (two coupled inductors) in order to check the sense of their windings!

Even without knowing the right-hand rule, we recognize the fact that when two inductors are mutually coupled as in Figure 6-13(c), the reference $(+)$ of v_2 will be reversed if the sense of windings of L_2 is reversed with respect to L_1, as in Figure 6-13(d).

To enable us to establish the $(+)$ mark on v_2, the following *dot convention* is used: A dot (or another mark) is shown at one terminal of L_1 and at one terminal of L_2. If a current i_1 enters the dot on L_1, then the induced voltage v_2 will be $(+)$ at the corresponding dot on L_2. This information is complete and totally equivalent to the right-hand rule and the relative sense of windings for L_1 and L_2. A pair of mutually coupled inductors, together with their dots and the references of i_1 and v_2, is shown in Figure 6-14. Let us practice with a couple of examples.

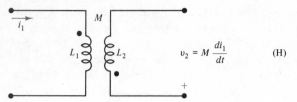

Figure 6-14   The dot convention.

EXAMPLE 6-10 _____

In the circuit of Figure 6-15(a), we have $L_1 = 2$ H, $L_2 = 3$ H, $k = 0.6$. If $i_1(t) = 10e^{-t}$ as shown, calculate the open-circuit induced voltage v_2 with its reference mark.

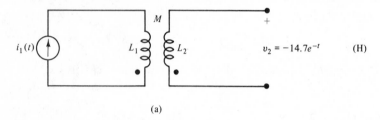

(a)

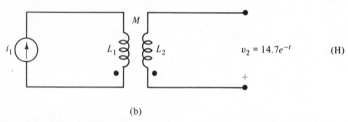

(b)

Figure 6-15   Example 6-10.

Solution.   First, we need M, the mutual inductance. As in Equation 6-28, we write

$$M = 0.6\sqrt{(2)(3)} = 1.47 \text{ H}$$

Next,

$$v_2 = M\frac{di_1}{dt} = 1.47\frac{d}{dt}(10e^{-t}) = -14.7e^{-t}$$

Finally, the dot convention: Since i_1 does *not* enter its dot, the induced voltage v_2 is *not* positive $(+)$ at its dot—its $(+)$ mark is at the *non*dotted terminal on L_2, as shown in Figure 6-15(a). If we wish now, we can show v_2 as in Figure 6-15(b).

   Let us be sure that we understand every step here. The expression (waveform) for v_2 is given by Equation 6-28. The second piece of information needed for v_2, its reference, is given by the dot convention. In using the dot convention here, we had to interpret it in its "negative" meaning, namely, a current *not* entering a dot induces a voltage which is *not* positive at the other dot. Or, if you will, a current entering a nondotted terminal induces a voltage which is positive at the other nondotted terminal. Finally, v_2 as shown in Figure 6-15(a) is identical to v_2 in Figure 6-15(b): The $(+)$ reference is reversed together with the minus sign in front of the expression.   ∎

EXAMPLE 6-11

For the two inductors shown in Figure 6-16, we are given $L_1 = 0.3$ H, $L_2 = 0.45$ H, $M = 0.22$ H. Also $i_2(t) = 10 \sin 377t$. Calculate the coefficient of coupling k, and the open-circuit induced voltage $v_1(t)$.

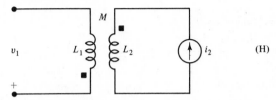

Figure 6-16 Example 6-11.

Solution. We have from Equation 6-28

$$k = \frac{M}{\sqrt{L_1 L_2}} = \frac{0.22}{\sqrt{(0.3)(0.45)}} = 0.6$$

Next, the induced voltage v_1 is

$$v_1 = 0.22 \frac{d}{dt}(10 \sin 377t) = 829 \cos 377t$$

According to the dot convention (here the dots are square) i_2, entering the square, induces v_1 which is referenced $(+)$ at the corresponding square on L_1, as shown in Figure 6-16. ∎

Probs. 6-17, 6-18, 6-19

Power and Energy Let us write the expressions for power and energy in an inductor by using duality on the corresponding capacitive expressions. The instantaneous power in an inductor is

$$p(t) = v(t)i(t) = Li(t)\frac{di(t)}{dt} \qquad \text{W} \tag{6-30}$$

and it can be positive, zero, or negative. The inductor may receive power ($p > 0$) or deliver power ($p < 0$).

EXAMPLE 6-12

For the inductor of Example 6-8, we have, for $t < 0$:

$$p(t) = (0)(10) = 0 \qquad \text{no power received or delivered}$$

For $0 < t < 2$ we have

$$p(t) = (0.3t)\left(\frac{t^2}{2} + 10\right) > 0 \qquad \text{power received by } L$$

For $2 < t < 4$:

$$p(t) = (0)(12) = 0 \qquad \text{no power}$$

For $4 < t < 5$:

$$p(t) = (-3.6)(-12t + 60) < 0 \qquad \text{power delivered by } L$$

For $t > 5$:

$$p(t) = 0 \qquad \text{no power}$$

Verify to your satisfaction the expression for $i(t)$ in the interval $4 < t < 5$, and the fact that $p(t) < 0$ there!   ■ Prob. 6-20

The energy absorbed in the inductor between t_1 and t_2 is

$$w\Big]_{t_1}^{t_2} = \int_{t_1}^{t_2} Li\frac{di}{dt}\,dt = \int_{i_1}^{i_2} Li\,di = \frac{L}{2}(i_2^2 - i_1^2) \qquad \text{J} \qquad (6\text{-}31)$$

Probs. 6-21
6-22, 6-23

To conclude and review our studies, we tabulate the equations for a resistor, a capacitor, and an inductor. Table 6-1 shows these relationships and stresses the duality between L and C. The subscripts R, C, and L are used for clarity.   Prob. 6-24

We have now all the basic "building blocks" of circuits: Sources (voltage, current) both independent and dependent, resistors, op amps, capacitors, and inductors (including mutual inductance). It is both amazing and comforting to know that practically all "real-life" circuits can be modeled, analyzed (and, later, designed) just with these components. That is what we'll be doing in the rest of the book.

TABLE 6-1 RELATIONSHIPS IN ELEMENTS

	R	C	L
Defining equation:	$v_R = Ri_R$	$q_C = Cv_C$	$\phi_L = Li_L$ (or Mi_L)
v-i:	$v_R = Ri_R$	$v_C = v_C(0) + \dfrac{1}{C}\displaystyle\int_0^t i_C(x)\,dx$	$v_L = L\dfrac{di_L}{dt}$
i-v:	$i_R = \dfrac{1}{R}v_R$	$i_C = C\dfrac{dv_C}{dt}$	$i_L = i_L(0) + \dfrac{1}{L}\displaystyle\int_0^t v_L(x)\,dx$
energy:	$w_R = R\displaystyle\int_{t_1}^{t_2} i_R^2(x)\,dx$	$w_C = \dfrac{C}{2}(v_2^2 - v_1^2)$	$w_L = \dfrac{L}{2}(i_2^2 - i_1^2)$

PROBLEMS

6-1 Check the units (dimensions) in Equation 6-3.

6-2 For the capacitor shown in Figure 6-3 we have $C = 0.3$ F and

$$v(t) = \begin{cases} 0 & t < 0 \\ te^{-4t} & t > 0 \end{cases}$$

Calculate and plot:
(a) The charge $q(t)$,
(b) The current $i(t)$. When does the charge reach its maximum?

6-3 The charge across a capacitor ($C = 0.1$ F) is given by

$$q(t) = \begin{cases} 10 & t < 0 \\ 10e^{-2t} & t > 0 \end{cases}$$

Calculate and plot $v(t)$ and $i(t)$.

6-4 Show that Equation 6-7 may be also written as

$$v(t) = \frac{1}{C} \int_{-\infty}^{t} i(x)\, dx$$

Hint:

$$\int_{-\infty}^{t} = \int_{-\infty}^{t_0} + \int_{t_0}^{t} \qquad \text{for any } t_0 < t$$

Why is Equation 6-7 more convenient than this equation?

6-5 The current through a capacitor ($C = 0.1$ F) is given by

$$i(t) = -20e^{-2t} \qquad t > 0$$

Also given is $v(0) = 100$ V.

(a) Show clearly the circuit diagram for this capacitor with the standard associated references for the voltage $v(t)$ and the current $i(t)$.

(b) Calculate $v(t)$ for $t > 0$.

(c) Plot $i(t)$ and $v(t)$ for $t > 0$ and compare with Problem 6-3.

6-6 In the circuit shown, the switch S closes at $t = 0$, the capacitor having no charge prior to $t = 0$.

(a) Plot $v_C(t)$, the voltage across the capacitor, for all t ($-\infty < t \le \infty$).

(b) Calculate and plot the current $i_C(t)$ vs. t.

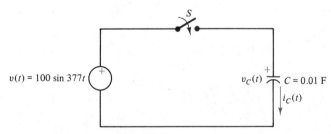

$v(t) = 100 \sin 377t$

$v_C(t)$ $C = 0.01$ F

$i_C(t)$

Problem 6-6

6-7 Starting with

$$i(t) = \frac{dq(t)}{dt}$$

(a) Integrate to obtain the expression of $q(t)$ in terms of $i(t)$. Don't forget initial conditions and the proper limits on the integral.

(b) Show that the integral in part (a) goes to zero as $t \to 0$ as long as i is not infinite. How is this result interpreted physically?

6-8 Calculate and plot the power $p(t)$ for Problem 6-6.

6-9 For Problems 6-6 and 6-8, calculate the energy:

(a) Between two successive times, t_1 and t_2, where $p(t_1) = p(t_2) = 0$ and $p > 0$ for $t_1 < t < t_2$.

(b) Between two successive times t_1 and t_2 where $p(t_1) = p(t_2) = 0$ and $p < 0$ for $t_1 < t < t_2$.

6-10 (a) Calculate the equivalent capacitance of two (uncharged) capacitors C_1 and C_2 connected in parallel. Extend to k capacitors $k = 3, 4, \ldots$.

(b) Repeat for a series connection. Contrast these results with their counterparts in resistors.

6-11 Check the dimensions (units) of Equation 6-17.

6-12 For the inductor shown in Figure 6-9, we have $L = 0.2$ H and

$$i(t) = \begin{cases} 0 & t < 0 \\ t^2 e^{-t} & t > 0 \end{cases}$$

Calculate and plot:

(a) The flux $\phi(t)$,

(b) The voltage $v(t)$. When do the current, the flux, and the voltage reach their maxima?

6-13 The flux in an inductor ($L = 0.4$ H) is given by

$$\phi(t) = \begin{cases} 0 & t < 0 \\ 20e^{-t} & t > 0 \end{cases}$$

Calculate and plot:

(a) The current

(b) The voltage

6-14 Consider a linear mechanical spring. Such a spring obeys Hooke's law

$$f(t) = Kx(t)$$

where $f(t)$ is the applied force and $x(t)$ is the elongation (extension) of the spring. Compare this expression with the q-v relation of a linear capacitor

$$v(t) = \frac{1}{C} q(t)$$

with $f \to v$, $x \to q$, $K \to 1/C$, where the arrow means "analogous (similar) to." *Note:* These are two different systems, one mechanical, the other electrical. Their equations exhibit analogies (similarities). Duality, on the other hand, exists among elements of a *single* system, say, electrical. Explore in full the analogy between: (1) the velocity across the spring (that is, the velocity of one terminal with respect to the fixed other terminal) and the current through the capacitor; (b) the energy stored in a spring and that stored in a capacitor.

6-15 Compare analogies between a linear spring and a linear inductor given by

$$i(t) = \frac{1}{L} \phi(t)$$

with $f \to i$, $K \to 1/L$, $x \to \phi$.

6-16 Calculate the mutual inductance between $L_1 = 2\,\text{H}$ and $L_2 = 3\,\text{H}$ with a coefficient of coupling
   (a) $k = 0.1$
   (b) $k = 0.9$
   (c) $k = 0$

6-17 Review, or study, the "right-hand rule" for establishing the reference of a mutually induced voltage.[1]

6-18 In Example 6-10, Figure 6-15(a), what is the voltage v_1 across L_1? Be sure to calculate its expression *and* show its reference.

6-19 Consider two linear inductors L_1 and L_2 coupled mutually, as shown. Let each inductor be excited by a current source. Calculate the total voltage across each inductor, v_1 and v_2. *Hint*: Use superposition.

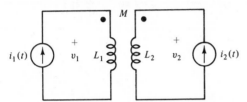

Problem 6-19

6-20 Calculate and plot the power $p(t)$ for the inductor in Example 6-9. Is the inductor receiving power or delivering it? From where?

6-21 Calculate and plot the energy absorbed in the inductor of Example 6-9. Verify that energy is conserved.

6-22 Calculate the equivalent inductance of two inductors (with zero initial conditions) connected:
   (a) In series
   (b) In parallel
   Extend your results to more than two inductors.

6-23 **(a)** The current through an inductor (L henrys), over the range $0 < t < 6$, is given by

$$i_L(t) = \tfrac{2}{3}t$$

   The inductor has no initial current (or energy). Plot $v_L(t)$. Calculate the energy stored in this inductor between $t_1 = 1\,\text{s}$ and $t_2 = 3\,\text{s}$.
   (b) Repeat, for the same inductor and zero initial current, if the current is given as

$$i_L(t) = \tfrac{1}{6}t^2 + \tfrac{1}{2}$$

   Draw conclusions!

*6-24 For the two inductors of Problem 6-19, calculate the instantaneous power $p = p_1 + p_2 = v_1 i_1 + v_2 i_2$ and the total energy between $t = 0$ and any $t > 0$. Assume no initial energy stored at $t = 0$.

REFERENCES

1. Sears, F. W., M. W. Zemansky, and H. D. Young, *University Physics* (6th ed.). Reading, Mass.: Addison-Wesley Publishing Co., 1983, chaps. 31–33.

Chapter 7

First-Order *RC* and *RL* Circuits

We are ready now to start our studies of dynamic circuits. As we remember, an all-resistive circuit is described by purely algebraic equations; therefore, its output waveform is identical to the input waveform, within a scaling multiplier.

Circuits with dynamic elements are described by more difficult equations, but they provide more exciting outputs.

7-1 THE OP AMP INTEGRATOR

One of the most useful circuits using a dynamic element is shown in Figure 7-1. An input voltage $v_i(t)$ is applied, and we are interested in the output voltage $v_o(t)$. Since the noninverting terminal is grounded, we have

$$v_1 = v_2 = 0 \tag{7-1}$$

Also, with their proper references, we write

$$i_R = \frac{v_i - v_2}{R} = \frac{v_i}{R} \tag{7-2}$$

and

$$i_C = C \frac{d}{dt}(v_o - v_2) = C \frac{dv_o}{dt} \tag{7-3}$$

KCL at the inverting terminal is

$$i_C + i_R = 0 \tag{7-4}$$

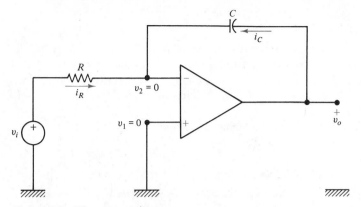

Figure 7-1 The op amp integrator.

that is,

$$C \frac{dv_o}{dt} + \frac{v_i}{R} = 0 \qquad (7\text{-}5)$$

In other words,

$$\frac{dv_o}{dt} = -\frac{1}{RC} v_i \qquad (7\text{-}6)$$

This is the desired input-output relation for this circuit. We see that the *first derivative* of the output appears in the equation. This is in contrast to the all-resistive networks studied in previous chapters. There, you will recall, the input-output relation was purely algebraic: The output (not its derivative) was equal to a constant times the input. See Equation 5-20. Here, not surprisingly, the capacitor's current introduces the first derivative of v_o.

Let us define the *order* $(=n)$ *of a network* as the highest derivative of the output which appears in the input-output equation for this network. In our op amp circuit, the order of the network is one $(n = 1)$.

To solve Equation 7-6 for $v_o(t)$, we integrate

$$v_o(t) = -\frac{1}{RC} \int_0^t v_i(x)\, dx \qquad (7\text{-}7)$$

provided the capacitor is initially uncharged. The output voltage $v_o(t)$ is, within the constant multiplier $-1/RC$, the integral of the input. If we choose, for example, $R = 1\ \text{M}\Omega$ and $C = 1\ \mu\text{F}$, this constant becomes -1. This circuit is called an *integrator*; it is extremely useful in waveshaping and in analog computers.

Probs. 7-1, 7-2

A similar result may be obtained by replacing the resistor R with an inductor L, and the capacitor C with a resistor R (Do it!). Such a circuit, however, is less practical; inductors, in real life, tend to be nonlinear, bulky, and more lossy (heat) than capacitors.

7-2 *RC* CIRCUIT—ZERO-INPUT RESPONSE

Let us consider a simple, but adequate, model of the flash unit in a camera. A battery (dc voltage source) charges a capacitor until the "ready" light glows, indicating that the capacitor is fully charged. Then, at $t = 0$, we press the shutter button: The switch S then connects C to the bulb R, and the current flowing through R causes the bulb to flash. See Figure 7-2(a).

We have for $t = 0^-$ (just prior to closing the switch on R)

$$v_C(0^-) = V_0 \tag{7-8}$$

as the initial condition of $v_C(t)$. We want to describe the behavior of the circuit for all $t > 0$, as shown in Figure 7-2(b); its tree and co-tree are shown in Figure 7-2(c). A quick preliminary decision tells us that we need only one node equation or one loop equation for this circuit. There are two excellent reasons for choosing node analysis here: (1) The node voltage $v_C(t)$ is a good variable since we know already its initial value $v_C(0^-)$; it would make a lot of sense to calculate $v_C(t)$ for all $t > 0$, as shown in Figure 7-2(d). A closely related reason is the choice of the capacitor as a tree branch. You recall (Chapter 3) that voltages of tree branches are independent variables. That is why we chose the capacitor as a tree branch. In this simple circuit, of course, we hardly need graph theory; however, these ideas will be used later in more complicated circuits. (2) If we decide to use loop analysis, the expression for the voltage across C will involve an integral of the loop current. We all agree that there is enough work dealing just with derivatives, so let us avoid integrals whenever possible.

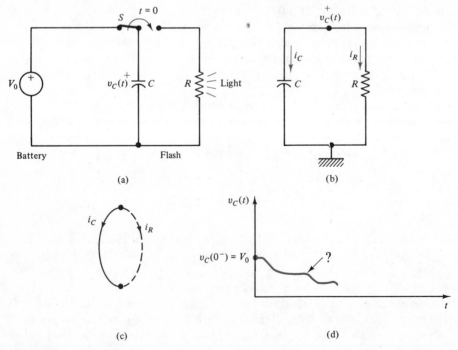

Figure 7-2 An *RC* circuit.

Node analysis, then, provides the following KCL equation in Figure 7-2(b), with $(+)$ for currents leaving the node:

$$i_C(t) + i_R(t) = 0 \qquad t > 0 \tag{7-9}$$

or

$$C\frac{dv_C(t)}{dt} + \frac{v_C(t)}{R} = 0 \qquad t > 0 \tag{7-10}$$

This is the equation for the unknown $v_C(t)$, including the unknown $v_C(t)$ and its first derivative. This is, therefore, a *first-order differential equation*, and as a result, the order of the circuit is $n = 1$. Also, the right-hand side of the equation is zero. We remember in previous chapters that the final form of a loop or a node equation contains, on the right-hand side, the sources (input) for that loop or that node. Here there is no source, no input for $t > 0$. We call Equation 7-10 a *zero-input* equation.†

A very important question comes up at this point (if you did not think of it yet, now is the time), "If the input is zero here, what excites the circuit? What will cause i_R to flow?" The answer is not difficult: The initial voltage $v_C(0^-)$ across the capacitor will discharge through R. This initial voltage makes things happen after $t = 0$. Why doesn't then $v_C(0^-)$ appear in Equation 7-10? Because the differential equation, describing KCL, holds for *all* $t > 0$; the value $v_C(0^-)$ is true *only* at one instant, $t = 0^-$. In fact, the number $v_C(0^-)$ is arbitrary. It can be specified without any relationship to Equation 7-10, because the capacitor C may be charged up to *any* V_0 for $t < 0$.

The complete statement of our problem, then, is: Solve the differential equation 7-10, with the given initial condition $v_C(0^-) = V_0$.

Before actually solving this problem, let us introduce another important concept for dynamic circuits. The capacitive voltage $v_C(t)$, if known for all $t > 0$, solves completely the circuit in Figure 7-2(b). That was the reason for choosing it as a node voltage. Let us reassure ourselves of this fact: If $v_C(t)$ is known, then $i_C(t) = C\, dv_C/dt$ is also known. Also, the voltage across the resistor is $v_C(t)$; it was zero for $t \le 0$. The current through the resistor is $i_R = v_C(t)/R = -i_C$. Thus, $v_C(t)$ solves the entire circuit.

The variable $v_C(t)$ is called a *state variable*. Its initial value, $v_C(0^-)$, is the *initial state* of the network. Here, the word "state" means "condition," as in "water in its liquid state," or "the state of the union." State variables are very useful in circuit analysis, in addition to loop currents and node voltages.

The solution of Equation 7-10 will be done by the most general method that is also applicable later in more complicated cases. Other methods, not quite as general, are mentioned in the problems for this section. Additional advantages of the general method are its minimal need of memorization and its close relationship to the so-called transform methods to be studied in later chapters of this book.

Probs. 7-3, 7-4

The general method for solving Equation 7-10 starts with the following question: "What *possible* function $v_C(t)$, when added to its first derivative, within

---

† Other names are *homogeneous* (in mathematics), or *force-free* (in physics).

multiplying constants (C and $1/R$), gives zero?" A brief survey of functions will tell us that the only possible candidate is the exponential function

$$v_C(t) = Ke^{st} \tag{7-11}$$

with suitable values of the constants K and s. Why? Because the derivative of e^{st} is still e^{st} (within a constant multiplier); therefore, it is the only hope for our problem. As an exercise, show that other functions, such as a polynomial in t, $v_C(t) = at^n + bt^{n-1} + \cdots$, are hopeless here.

Let us try, then, the (intelligently!) assumed solution, Equation 7-11, in Equation 7-10. We get

$$C(sKe^{st}) + \frac{1}{R} Ke^{st} = 0 \tag{7-12}$$

or

$$\left(Cs + \frac{1}{R} \right) Ke^{st} = 0 \tag{7-13}$$

In Equation 7-13, the product of two factors is zero. As a result, either factor (or both) must be zero. But we know that $v_C(t) = Ke^{st} \neq 0$ since a zero solution, $v_C(t) = 0$, while satisfying Equation 7-10, is not very interesting; it is a *trivial solution*. Therefore, we must have

$$Cs + \frac{1}{R} = 0 \tag{7-14}$$

which gives the value of the exponent s

$$s = -\frac{1}{RC} \tag{7-15}$$

Our solution, so far, is

$$v_C(t) = Ke^{-(1/RC)t} \tag{7-16}$$

and a quick substitution into the differential equation (Equation 7-10) proves that it is indeed a solution

$$C\left(-\frac{K}{RC} e^{-(1/RC)t} \right) + \frac{K}{R} e^{-(1/RC)t} = 0 \tag{7-17}$$

To determine K, we use the initial condition. At $t = 0^-$ we know that $v_C(0^-) = V_0$. Since $v_C(0^+) = v_C(0^-)$, as we learned in Chapter 6, we will simply designate it as $v_C(0)$.† At $t = 0$ we have then

$$V_0 = v_C(0) = Ke^{-(1/RC)0} = K \tag{7-18}$$

that is, $K = V_0$. The final answer is therefore

$$v_C(t) = V_0 e^{-(1/RC)t} \tag{7-19}$$

and its plot is shown in Figure 7-3, both for $V_0 > 0$ and for $V_0 < 0$.

---

†Later, we will make a distinction in the rare cases when $v_C(0^-) \neq v_C(0^+)$.

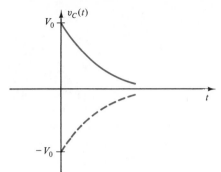

Figure 7-3 The zero-input response of an *RC* circuit.

Several important observations must be made about this solution:

1. It is the *zero-input* solution, meaning that no sources excite the circuit, just an initial condition. We see in Equation 7-19 that this solution, within the constant multiplier V_0, depends exclusively on R and C, the parameters of the network. Therefore, this solution is sometimes also called the *natural response*.

2. The order of the circuit is the order of its differential equation. Since a differential equation of order n requires n (arbitrary) initial conditions, the order of the circuit is also the number of those initial conditions.

3. The assumed solution Ke^{st} produced an algebraic equation for finding the exponent s. In our case, it was Equation 7-14. A little thought will convince us that if the differential equation is of order n, the algebraic equation for s will be of degree n. That is, if the differential equation is

$$a_n \frac{d^n v}{dt^n} + a_{n-1} \frac{d^{n-1} v}{dt^{n-1}} + \cdots + a_1 \frac{dv}{dt} + a_0 v = 0 \qquad (7\text{-}20)$$

then the algebraic equation for s will be

$$a_n s^n + a_{n-1} s^{n-1} + \cdots + a_1 s + a_0 = 0 \qquad (7\text{-}21)$$

This algebraic equation is called the *characteristic equation* for the particular circuit. The coefficients $a_n, a_{n-1}, \ldots, a_1, a_0$ in Equation 7-20 or 7-21 depend only on the element values of the circuit, and not, for example, on outside sources or initial conditions. In the characteristic equation 7-14, these coefficients are $a_1 = C$, $a_0 = 1/R$. These coefficients are characteristic to the network, hence the name for Equation 7-21. There will be n solutions (roots) of the characteristic equation $s_1, s_2, \ldots, s_n$. These are called the *characteristic values†* of the network. In our example, there is only one characteristic value, $s_1 = -1/RC$. We can appreciate now even more the reasons for assuming an exponential solution, Ke^{st}. Not only is it the most reasonable solution; by assuming it, we reduce the solution of a differential

† Or *eigenvalues* ("eigen" = "self" in German). Later, we'll call them *natural frequencies* for a good reason.

equation to solving an algebraic equation—a much easier, almost routine, task.

4. The arbitrary multiplying constant K (or, in the general case, $K_1, K_2, \ldots, K_n$) arises essentially because we integrate the differential equation. A first-order differential equation requires one integration to go from dv/dt to $v(t)$. The integration introduces an arbitrary constant of integration. To evaluate it, we need an additional piece of information, the initial condition. For a network of order n, we need n initial conditions to be able to evaluate the n constants $K_1, K_2, \ldots, K_n$.

5. The zero-input response in our *RC* circuit is a decaying exponential. A waveform that approaches zero ("dies away") as $t \to \infty$ is called *transient*.

Let us rewrite the transient voltage as

$$v_C(t) = V_0 e^{-t/RC} = V_0 e^{-t/\tau} \tag{7-22}$$

where $\tau = RC$ is necessarily in seconds (why?) and is called the *time constant* of this circuit. The time constant gives us a measure of how fast (or how slowly) the waveform $v_C(t)$ decays. After one time constant, $t = \tau = RC$, the value of $v_C(t)$ becomes

$$v_C(t)\Bigg]_{t=\tau} = V_0 e^{-1} = 0.3678 V_0 \tag{7-23}$$

that is, down to approximately 37 percent from its initial value. After five time constants, at $t = 5\tau$, the voltage will be

$$v_C(t)\Bigg]_{t=5\tau} = V_0 e^{-5} = 0.00674 V_0 \tag{7-24}$$

that is, less than 1 percent of its initial value; for all practical (engineering) purposes, v_C then is zero. See Figure 7-4(a). Also, for different values of R or C, we get different time constants for this circuit. Consider two time constants, τ_1 and τ_2. The circuit with the faster decay has the smaller time constant, $\tau_1 < \tau_2$, as shown in Figure 7-4(b); it takes τ_2 for $v_{C2}(t)$ to decay to 37 percent, while $v_{C_1}(t)$ takes only τ_1; clearly, $\tau_1 < \tau_2$.

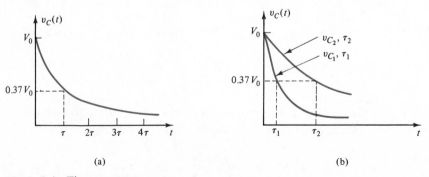

(a)   (b)

Figure 7-4 Time constants.

EXAMPLE 7-1 _____

For the circuit shown in Figure 7-2(b), we are given $V_0 = -2$ V, $R = 4$ MΩ, $C = 3$ μF. Calculate $v_C(t)$, the time constant τ, and the time it takes for V_0 to reach -1 V.

> *Solution.* In all our examples and problems, we will discourage memorization: It is unnecessary and risky (you may memorize the wrong equation). Rather, it is always better to rederive everything from basics. It is sincerely hoped that you, the student, will follow this advice. You will find it beneficial. After much practice, you will develop enough knowledge and confidence to take several shortcuts by remembering certain steps naturally and effortlessly.
>
> Here, we write KCL as

$$3 \times 10^{-6} \frac{dv_C}{dt} + \frac{1}{4 \times 10^6} v_C = 0$$

Multiply by 10^6 to clean it up

$$3 \frac{dv_C}{dt} + \frac{1}{4} v_C = 0$$

The characteristic equation is obtained by writing a multiplier s for every d/dt (see Equations 7-20 and 7-21):

$$\therefore \quad 3s + \tfrac{1}{4} = 0$$

The root of this equation is the characteristic value

$$s = -\tfrac{1}{12}$$

Therefore, the zero-input solution is

$$v_C(t) = Ke^{-(1/12)t}$$

At $t = 0^-$ we have

$$-2 = Ke^0 = K$$

Finally, then,

$$v_C(t) = -2e^{-t/12} = -2e^{-t/\tau}$$

The plot of this function is shown (in dashed lines) in Figure 7-3. The time constant of the circuit is $\tau = 12$ s. To reach -1 V, it will take t_1 and we must solve

$$v_C(t_1) = -1$$

or,

$$-1 = -2e^{-t_1/12}$$

$$\therefore \quad e^{-t_1/12} = 0.5$$

$$\therefore \quad \frac{t_1}{12} = 0.693$$

$$\therefore \quad t_1 = 8.32 \text{ s}$$

Probs. 7-5, 7-6, 7-7, 7-8

EXAMPLE 7-2 _____

Calculate the power and the energy balance in the *RC* circuit of Figure 7-2.

> *Solution.* At $t = 0^-$, the capacitor has the stored energy

$$w_C(0^-) = \tfrac{1}{2}CV_0^2 \qquad \text{J}$$

Thereafter, as $v_C(t)$ decreases, this energy is dissipated by the resistor. The voltage across the resistor is

$$v_R(t) = v_C(t) = V_0 e^{-t/RC}$$

$$\therefore \quad i_R(t) = \frac{V_0}{R} e^{-t/RC}$$

The total energy dissipated by R is

$$w_R \Big]_0^\infty = \int_0^\infty p_R(t) \, dt = \int_0^\infty v_R(t) i_R(t) \, dt$$

$$= \int_0^\infty \frac{V_0^2}{R} e^{-2t/RC} \, dt = \frac{CV_0^2}{2}$$

which establishes the energy balance between C and R.

Probs. 7-9,
■ 7-10

7-3 *RL* CIRCUIT—ZERO-INPUT RESPONSE

By duality, let us derive the zero-input response for the *RL* circuit shown in Figure 7-5(a). Here, the given initial current in L, $i_L(0^-) = I_0$, is the initial state, and has been established from a previous switching operation. The state variable $i_L(t)$ serves as a suitable loop current. We use it for similar reasons as $v_C(t)$ in the previous case: (1) $i_L(0^-)$ is given, and $i_L(t)$ for $t > 0$ will solve the entire network. (2) Link currents are independent variables (see Chapter 3); therefore, the inductor is chosen as a link in the graph, Figure 7-5(b). (3) If we chose to write one node equation here, the inductive current would involve the integral of v_L, and we want to avoid integrals when possible.

KVL around the loop provides the following equation

$$L \frac{di_L}{dt} + R i_L = 0 \qquad t > 0 \tag{7-25}$$

with $i_L(0^+) = i_L(0^-) = i_L(0) = I_0$, given. This is a first-order differential equation for the first-order *RL* circuit. It is a zero-input (homogeneous) equation.

Without unnecessary repetition, we assume the solution

$$i_L(t) = K e^{st} \tag{7-26}$$

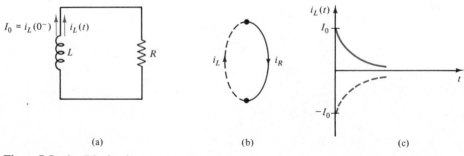

(a)   (b)   (c)

Figure 7-5   An *RL* circuit.

and obtain the characteristic equation

$$Ls + R = 0 \qquad (7\text{-}27)$$

whose solution is the characteristic value

$$s = -\frac{R}{L} \qquad (7\text{-}28)$$

Therefore,

$$i_L(t) = Ke^{-(R/L)t} = Ke^{-t/\tau} \qquad (7\text{-}29)$$

where the time constant is (in seconds)

$$\tau = \frac{L}{R} \qquad (7\text{-}30)$$

Using now the given initial condition in Equation 7-29, we get

$$I_0 = Ke^0 = K \qquad (7\text{-}31)$$

and so, finally,

$$i_L(t) = I_0 e^{-(R/L)t} \qquad (7\text{-}32)$$

is the zero-input response of the circuit shown, satisfying the initial condition. Its plot is shown in Figure 7-5(c) for $I_0 > 0$ and for $I_0 < 0$, and it is a transient waveform.

All our previous observations hold also here about the time constant and the rate of decay of $i_L(t)$. The entire derivation is, in fact, totally dual to the *RC* case. We did it in some detail in order to provide some solid practice.

EXAMPLE 7-3 ───────────────────────────────────────

The circuit in Figure 7-6 represents a certain magnetic relay. The switch is in position a for a long time; at $t = 0$, it moves instantly to position b. Such a switch is called a "make-before-break" switch, and, in moving from a to b, it makes contact with b before breaking the contact with a, without interrupting i. It will stay in position b as long as the current $i_L(t)$ does not go below 6 A, at which time the relay "trips." When will the relay trip?

Solution. For $t < 0$, the inductor acts as a short circuit (s.c.), from our studies in Chapter 6. Consequently,

$$i_L(0^-) = 10 \text{ A}$$

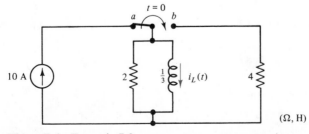

Figure 7-6 Example 7-3.

For $t > 0$, the two resistors combine in parallel across L; therefore

$$R_{eq} = 2\|4 = \tfrac{4}{3}\,\Omega$$

and, from Equation 7-32,

$$i_L(t) = 10^{-4t} \qquad t > 0$$

The tripping will happen at t_1, when

$$6 = 10e^{-4t_1}$$

$$\therefore \quad t_1 = -\tfrac{1}{4}\ln 0.6 = 128 \text{ ms}$$

Probs. 7-11
7-12

EXAMPLE 7-4 _____

In the network of Figure 7-7(a), the switch is opened, having remained closed for a very long time. Calculate $i_L(t)$ for $t > 0$ and the energy dissipated in the 5-Ω resistor during the first 10 ms after $t = 0$.

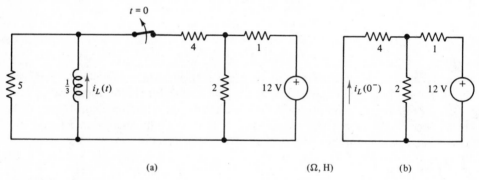

(a)   (Ω, H)   (b)

Figure 7-7 Example 7-4.

Solution. To calculate the initial current $i_L(0^-)$, we use Figure 7-7(b), with L acting as a short circuit. The calculation yields

$$i_L(0^-) = \frac{12}{(2\|4) + 1}\left(\frac{-\tfrac{1}{4}}{\tfrac{1}{4} + \tfrac{1}{2}}\right) = -1.71 \text{ A}$$

where the first fraction is the current in the 1-Ω resistor. The second fraction, with the minus sign, is the current division. The expression for $i_L(t)$, for $t > 0$, is then

$$i_L(t) = -1.71e^{-[5/(1/3)]t} = -1.71e^{-15t}$$

and this is also the expression for the current in the 5-Ω resistor after $t = 0$. Therefore, the energy dissipated is

$$w\Big]_0^{10^{-2}} = \int_0^{10^{-2}} 5(-1.71e^{-15t})^2\, dt = 0.126 \text{ J}$$

Compare this with the initial energy stored in L

$$w_L(0^-) = \tfrac{1}{2}Li_L^2(0^-) = \tfrac{1}{2}\cdot\tfrac{1}{3}(-1.71)^2 = 0.488 \text{ J}$$

that is, the resistor dissipates about 26 percent of the initial energy.

Probs. 7-13,
7-14

7-4 TOTAL RESPONSE OF A FIRST-ORDER CIRCUIT

Next, let us consider the circuit in Figure 7-8. The initial state is given, $v_C(0^-)$, and the switch opens at $t = 0$, exciting the circuit with the current source $i_s(t)$. As a first case, let this source be dc, $i_s(t) = I_s$, constant.

As before, with the capacitive voltage $v_C(t)$ as the state variable (C a tree branch), we write KCL

$$C \frac{dv_C}{dt} + \frac{1}{R} v_C = I_s \qquad t > 0 \tag{7-33}$$

This is a first-order differential equation, but its right-hand side is nonzero. As expected, it contains the source that excites the circuit. We call this equation *non-homogeneous* (it was homogeneous when the right-hand side was zero).

From the theory of differential equations, we know that the solution to Equation 7-33 consists of the sum

$$v_C(t) = v_{CH}(t) + v_{CP}(t) \tag{7-34}$$

where $v_{CH}(t)$, the *homogeneous solution,* is the solution to the homogeneous equation

$$C \frac{dv_{CH}}{dt} + \frac{1}{R} v_{CH} = 0 \tag{7-35}$$

and $v_{CP}(t)$ is a *particular solution,* depending on the right-hand side of Equation 7-33

$$C \frac{dv_{CP}}{dt} + \frac{1}{R} v_{CP} = I_s \tag{7-36}$$

In fact, if we add Equations 7-35 and 7-36 we'll get Equation 7-33, for which Equation 7-34 is the solution. You may ask, "Why isn't Equation 7-36 and *its* solution, v_{CP}, sufficient?" The answer is that we want *the most general* solution: Equation 7-36 and v_{CP} provide just a particular solution, and that's why we add to it the homogeneous solution.

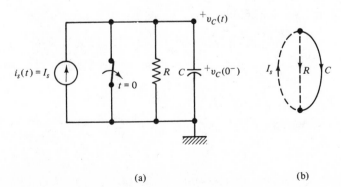

(a)   (b)

Figure 7-8   *RC* circuit with a source.

We know already the homogeneous solution from the previous sections. With the characteristic equation for Equation 7-35 being, again,

$$Cs + \frac{1}{R} = 0 \tag{7-37}$$

and its charcteristic value

$$s = -\frac{1}{RC} \tag{7-38}$$

we have

$$v_{CH}(t) = Ke^{-(1/RC)t} \tag{7-39}$$

exactly as before. Next, we must find a particular solution to Equation 7-36. Here, again, we use an intelligent argument: Since the right-hand side is a given constant, let us try for the particular solution a generalization of this number

$$v_{CP} = A \tag{7-40}$$

where A is an unknown constant. Substitute now this trial solution into its Equation 7-36, to get

$$C\frac{dA}{dt} + \frac{1}{R}A = I_s \tag{7-41}$$

or

$$0 + \frac{1}{R}A = I_s \tag{7-42}$$

From which we find

$$A = RI_s \tag{7-43}$$

The total solution to Equation 7-33 is therefore

$$v_C(t) = Ke^{-(1/RC)t} + RI_s \tag{7-44}$$

To determine K, we use now the given initial condition. At $t = 0^-$, $v_C(0^-)$ is given. Therefore,

$$v_C(0) = Ke^0 + RI_s = K + RI_s \tag{7-45}$$

from which

$$K = v_C(0) - RI_s \tag{7-46}$$

Substitute this value of K into Equation 7-44 to get the final answer

$$v_C(t) = [v_C(0) - RI_s]e^{-(1/RC)t} + RI_s \tag{7-47}$$

The plot of $v_C(t)$ is shown in Figure 7-9 for three cases: (a) $v_C(0^-) - RI_s > 0$; (b) $v_C(0^-) - RI_s < 0$; and (c) $v_C(0^-) = RI_s$.

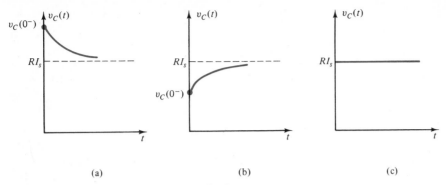

Figure 7-9   Total response of the *RC* circuit.

Several observations are in order here:

1. The total solution in Equation 7-47 consists of a *transient* part, the term that dies away with time

$$\lim_{t \to \infty} [v_C(0) - RI_s]e^{-(1/RC)t} = 0 \qquad (7\text{-}48)$$

There is another part in $v_C(t)$ that survives with time. That is the *steady-state*, nonzero part of $v_C(t)$

$$\lim_{t \to \infty} v_C(t) = RI_s \qquad (7\text{-}49)$$

and it is the asymptote for $v_C(t)$ in Figure 7-9(a) and (b). In case (c) there is no transient, and the entire solution is steady state.

2. Let us regroup the terms in Equation 7-47 as follows:

$$v_C(t) = v_C(0^-)e^{-(1/RC)t} + RI_s(1 - e^{-(1/RC)t}) \qquad (7\text{-}50)$$

The first term here, $v_C(0^-)e^{-(1/RC)t}$, is due only to $v_C(0^-)$, the initial state, without the source ($i_s = 0$); it is called the *zero-input* part, or the *natural response*, of the total solution. The second term, $RI_s(1 - e^{-t/RC})$, is due only to the input I_s without any initial state, $v_C(0^-) = 0$. It is called the *zero-state* part of the solution. We have here a general property of a linear system

$$\text{Total response} = \begin{pmatrix} \text{zero-input} \\ \text{response} \end{pmatrix} + \begin{pmatrix} \text{zero-state} \\ \text{response} \end{pmatrix} \qquad (7\text{-}51)$$

This is another manifestation of *superposition*, where the total response is the *sum* of the response due to initial conditions acting alone, *plus* the response due to sources acting alone.

3. Keep in mind the exact stage at which we used the initial condition to evaluate the arbitrary constant K. This was done only after we found the *total* response $v_C(t)$ in Equation 7-44. The reason is simple: $v_C(0^-)$ is the *total* initial voltage. We split our solution into two steps, the homogeneous and the particular, only for mathematical convenience. The circuit does not know of

this separation; it has physically one total $v_C(t)$. A common error, to be avoided, is to use the initial condition earlier in the homogeneous part (Equation 7-39). Don't rush!

4. The waveform of the total response may consist of a transient part and/or a steady-state part. To repeat, the transient part is the one that goes to zero as $t \to \infty$; the steady-state part is the one that remains nonzero as $t \to \infty$. Either part may be present or missing in the total response.

5. We must keep a clear distinction between the zero-input part and the zero-state part, on one hand, and the transient part and the steady-state part, on the other. These are two distinct ways to look at the total response. In the first one, we associate parts of the response with their respective causes; in the second one, we look at the behavior of the response for $t \to \infty$. These two views are, in general, unrelated. We *cannot* say, for example, that the zero-input part is always transient; nor can we say, in general, that the zero-state part is the steady state. Be sure to keep these concepts apart and to apply them properly in each case. Additional examples, in this chapter and later, will stress these ideas.

EXAMPLE 7-5 _____

At $t = 0$, a constant voltage source $v_s(t) = V_s$ is applied to the circuit shown in Figure 7-10. The initial voltage across C is given, $v_C(0^-)$.

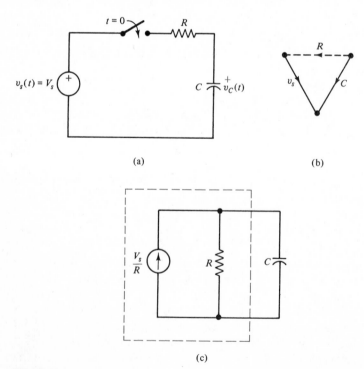

Figure 7-10 Example 7-5.

Solution. The graph for this circuit is shown in Figure 7-10(b), with the voltage source and the capacitor forming the tree, the resistor being a link. Again, one loop equation or one node equation will do. A loop current will require an integral for the expression of $v_C(t)$. Instead, we use the state variable $v_C(t)$ and write a node equation for it

$$C \frac{dv_C}{dt} + \frac{v_C - V_s}{R} = 0$$

or, after rearrangement,

$$C \frac{dv_C}{dt} + \frac{1}{R} v_C = \frac{V_s}{R}$$

This is precisely the previous equation for Figure 7-8, with $I_s = V_s/R$. We have rederived Norton's equivalent circuit for V_s in series with R, shown inside the dotted box in Figure 7-10(c)! Without further ado, then, we write from Equation 7-50

$$v_C(t) = v_C(0^-)e^{-t/RC} + V_s(1 - e^{-t/RC})$$

With convenient values $v_C(0^-) = -2$ V, $V_s = 12$ V, $R = 2 \,\Omega$, $C = 0.1$ F, we have

$$v_C(t) = -2e^{-5t} + 12(1 - e^{-5t}) = 12 - 14e^{-5t}$$

The transient part of $v_C(t)$ is $-14e^{-5t}$. The steady-state part of $v_C(t)$ is 12 V. This also makes sense from elementary considerations: After a very long time ($t \to \infty$) in Figure 7-10(a), the capacitor becomes fully charged and acts as an open circuit. Then there is no current through R, no voltage across R, and $v_C = V_s$. This is, in fact, the very same reasoning that we use in calculating initial conditions when they are not given. For all we know, our capacitor, after being fully charged, may be switched to another circuit. Its initial condition for that circuit will be $v_C(0^-) = 12$ V.

We note also that, just by looking at the final answer, $v_C(t) = 12 - 14e^{-5t}$, we cannot distinguish the zero-input and the zero-state parts. They were "blended" together from the previous step: There, the term $-2e^{-5t}$ is the zero-input part, and $12(1 - e^{-5t})$ is the zero-state part.   ■ Prob. 7-16

In this example, and in many others, a constant voltage (or current) is applied to the network at $t = 0$. This is shown in Figure 7-11(a) for a voltage source of 1 V and Figure 7-11(b) for a current source of 1 A. In the first case, the voltage across N, $v_N(t)$, is described by

$$v_N(t) = \begin{cases} 0 & t < 0 \\ 1 & t > 0 \end{cases} \tag{7-52}$$

Similarly, the current $i_N(t)$ in Figure 7-11(b) is given by

$$i_N(t) = \begin{cases} 0 & t < 0 \\ 1 & t > 0 \end{cases} \tag{7-53}$$

These two functions are called a *unit step function*, denoted by $u(t)$, defined as

$$u(t) = \begin{cases} 0 & t < 0 \\ 1 & t > 0 \end{cases} \tag{7-54}$$

and shown in Figure 7-11(c). The unit step function describes mathematically the operation of a switch. It will be used extensively in our studies.

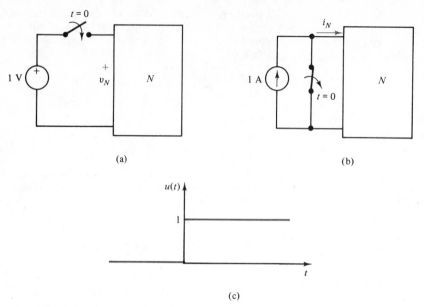

(a)   (b)

(c)

Figure 7-11   The unit step function.

EXAMPLE 7-6 ——

Calculate the zero-state response $v_C(t)$ in Figure 7-10 if $v_s(t) = u(t)$.

> *Solution.*   From Example 7-5 we have here $v_C(0^-) = 0$ (zero-state) and $V_s = 1$ for $t > 0$.
> Therefore,
>
> $$v_C(t) = 1 - e^{-t/RC}$$
>
> The zero-state response to a unit step input is called briefly the *step response* of the
> network. It is easily obtained in the laboratory.   ■   Prob. 7-15

To summarize, we list in an algorithmic way the step-by-step procedure for
solving a nonhomogeneous first-order differential equation of the general form

$$a_1 \frac{dr(t)}{dt} + a_0 r(t) = f(t) \tag{7-55}$$

where $r(t)$ is the unknown response, voltage or current, and $f(t)$ is a given source. The
coefficients a_1 and a_0 are constants of the network:

1. Find the homogeneous response $r_H(t)$ which satisfies

$$a_1 \frac{dr_H(t)}{dt} + a_0 r_H(t) = 0 \tag{7-56}$$

This leads to the characteristic equation

$$a_1 s + a_0 = 0 \tag{7-57}$$

TABLE 7-1 PARTICULAR SOLUTIONS

When $f(t)$ is	Use for $r_P(t)$
1. α_0, a constant	A, a constant
2. $\alpha_1 t$	$B_1 t + B_0$
3. $\alpha_n t^n$	$B_n t^n + B_{n-1} t^{n-1} + \cdots + B_1 t + B_0$
4. αe^{pt}	$B e^{pt}$
5. $\alpha \sin kt$	$B \sin (kt + D)$
6. $\alpha \cos kt$	$B \cos (kt + D)$

Notes:

1. When $f(t)$ contains a *sum* of terms, a corresponding *sum* must be assumed for $r_P(t)$; if a *product* of terms, use a corresponding product.
2. When the homogeneous solution already contains a term suggested for $r_P(t)$ in the second column, this term must be multiplied by t. If it appears q times in $r_H(t)$, multiply by t^q.

and the characteristic value

$$s_1 = -\frac{a_0}{a_1} \tag{7-58}$$

The homogeneous response is then

$$r_H(t) = K e^{s_1 t} \tag{7-59}$$

with K yet undetermined.

2. Find a particular solution r_P to

$$a_1 \frac{dr_P(t)}{dt} + a_0 r_P(t) = f(t) \tag{7-60}$$

by generalizing $f(t)$, according to Table 7-1. The particular solution does not have arbitrary constants.

3. The total response is

$$r(t) = r_H(t) + r_P(t) \tag{7-61}$$

4. Use the given initial condition for the total response to evaluate K.

EXAMPLE 7-7 ⎯⎯⎯⎯⎯⎯⎯⎯⎯⎯⎯⎯⎯⎯⎯⎯⎯⎯⎯⎯⎯⎯⎯⎯⎯⎯⎯⎯

A first-order RC circuit obeys the differential equation

$$2 \frac{dv}{dt} + 6v = 12e^{-t} \qquad t > 0$$

with $v(0) = -4$. Find the total response and classify it.

Solution.   We do it in the steps listed above:

1. The homogeneous response has the characteristic equation

$$2s + 6 = 0$$
$$\therefore \quad s_1 = -3$$

that is,

$$v_H(t) = Ke^{-3t}$$

2. For a particular solution, according to entry 4 in Table 7-1, we choose

$$v_P(t) = Be^{-t}$$

Substitute immediately this choice into the differential equation to get B:

$$2(-Be^{-t}) + 6(Be^{-t}) = 12e^{-t}$$
$$\therefore \quad 4Be^{-t} = 12e^{-t}$$
$$\therefore \quad B = 3$$
$$\therefore \quad v_P(t) = 3e^{-t}$$

3. The total response is

$$v(t) = Ke^{-3t} + 3e^{-t}$$

4. At $t = 0^-$

$$-4 = K + 3 \quad \therefore \quad K = -7$$
$$\therefore \quad v(t) = -7e^{-3t} + 3e^{-t} \qquad t > 0$$

This total response is entirely transient; it has no steady-state terms. Can you trace back and find the zero-input part and the zero-state part?   ■

EXAMPLE 7-8 _____

An *RC* first-order circuit obeys

$$\frac{dv}{dt} + v = 12e^{-t} \qquad t > 0$$

with $v(0) = -4$.

Solution.   Here the solution progresses as follows:

1. $$s + 1 = 0 \quad \therefore \quad s_1 = -1$$
$$\therefore \quad v_H(t) = Ke^{-t}$$

2. For a particular solution, we are tempted to use Be^{-t} as a generalization of $12e^{-t}$. However, as listed in note 2 in Table 7-1, $v_H(t) = Ke^{-t}$ is already of that form! We must therefore assume

$$v_P(t) = Bte^{-t}$$

Substituting this into the differential equation yields B

$$B(-te^{-t} + e^{-t}) + Bte^{-t} = 12e^{-t}$$
$$\therefore \quad B = 12$$

Therefore

$$v_P(t) = 12te^{-t}$$

3. The total solution is

$$v(t) = Ke^{-t} + 12e^{-t}$$

4. At $t = 0$:

$$-4 = K + 0 \qquad \therefore \quad K = -4$$
$$\therefore \quad v(t) = -4e^{-t} + 12te^{-t} \qquad t \geq 0$$

Just for fun, assume that you forgot to multiply the particular solution by t (we are all human beings, after all). Try $v_P = Be^{-t}$ in the differential equation and see how far you can go; not very far, because you'll get a nonsense like $0 = 12$. So, even if you forget to multiply by t, the system doesn't let you go too far!

The reason for the need of multiplying by t becomes clear now. In the particular solution we need something *new*. If the homogeneous solution is of the same form as $f(t)$, we get a new particular solution by multiplication with t.

When the input (right-hand side) is of the same form as the homogeneous solution, we say that a characteristic value (or a natural frequency) of the network is *excited*. In this example, the natural frequency $s_1 = -1$ is excited. We'll be considering this phenomenon later.   ∎

Probs. 7-22, 7-23

EXAMPLE 7-9

Solve the first-order differential equation

$$\frac{dv}{dt} + 3v = 2 \cos 10t \qquad t > 0$$

with $v(0^-) = 0$.

Solution.   The homogeneous part is

$$v_H(t) = Ke^{-3t}$$

(do it slowly, if you have to!). From Table 7-1, we must try

$$v_P(t) = B \cos (10t + D)$$

Substituting into the differential equation, we get

$$-10B \sin (10t + D) + 3B \cos (10t + D) = 2 \cos 10t$$

If we expand the sine and the cosine terms on the left, we get

$$(-10B \cos D - 3B \sin D) \sin 10t + (3B \cos D - 10B \sin D) \cos 10t = 2 \cos 10t$$

Comparing terms, we have two equations for B and D

$$-10B \cos D - 3B \sin D = 0$$
$$3B \cos D - 10B \sin D = 2$$

The first one yields

$$\tan D = -\tfrac{10}{3} \qquad \therefore \quad D = -73.3°$$

Then, from the second one we get

$$B = 0.192$$

The total response is therefore

$$v(t) = Ke^{-3t} + 0.192 \cos (10t - 73.3°)$$

At $t = 0^-$ we have

$$0 = K + 0.192 \cos (-73.3°) = K + 0.055$$
$$\therefore \quad K = -0.055$$

Finally,

$$v(t) = -0.055e^{-3t} + 0.192 \cos (10t - 73.3°)$$

The transient part is $-0.055e^{-3t}$. The steady-state part is $0.192 \cos (10t - 73.3°)$ and is oscillatory. This type of steady-state, resulting from a sinusoidal input, is extremely practical in many circuits. We will return to it in great detail in a later chapter. ■

Probs. 7-17,
7-18, 7-19,
7-20, 7-21

The same ideas apply to a first-order RL circuit with an input. You should study this case in detail by solving the problems listed here.

As a final note for these circuits, let us recognize that the steady-state part of the response in the RC and the RL case with a constant (dc) input may be obtained by inspection, or almost so. For dc, after a long time, an inductor acts as a short circuit, $v = L \, di/dt = L \cdot 0 = 0$, and a capacitor as an open circuit, $i = C \, dv/dt = C \cdot 0 = 0$. We made these observations in an earlier chapter. Therefore, in Figure 7-8, the steady-state part of $v_C(t)$, as $t \to \infty$, is RI_s; since the capacitor acts as an open circuit, the current I_s flows only through R, and the voltage across R is then RI_s, equal to the capacitor's voltage. This result is, of course, consistent with Equation 7-47.

A word of caution, though: *Don't* replace inductors with short circuits and capacitors with open circuits, except if all the sources are dc and you are only looking for the steady-state response. In all other cases, with time-varying sources, and when the transient, steady-state, zero-input, and zero-state parts of the solution are required, the methods of differential equations must be applied.

We conclude this chapter with an example that is both practical and uses several of the principles and methods studied so far.

EXAMPLE 7-10 _____

A series RC circuit, shown in Figure 7-12(a), is excited at $t = 0$ by a square pulse voltage

$$v_g(t) = \begin{cases} 20 & 0 < t < t_0 \\ 0 & \text{elsewhere} \end{cases}$$

as shown in Figure 7-12(b). Such a square pulse is very common in communications, radar, and computer circuits. It is also known as a *gate function* (because of its shape), hence the subscript g. The capacitor is initially uncharged, and we wish to calculate $v_R(t)$ in this circuit for $t > 0$.

Solution. The key to an easy solution is (yes!) superposition. The voltage $v_g(t)$ can be expressed as a difference of two step functions

$$v_g(t) = v_1(t) - v_2(t)$$

as shown in Figure 7-12(c). Here $v_1(t)$ is

$$v_1(t) = 20u(t)$$

where $u(t)$ is the unit step function (Equation 7-54). Similarly, $v_2(t)$ is the same as $v_1(t)$, but *delayed*: It starts t_0 s later.

By superposition, then, the voltage $v_R(t)$ will consist of the contribution of $v_1(t)$ *less* the contribution of $v_2(t)$. The whole beauty of the approach relies on the fact that the contribution of $v_2(t)$ is the same as the contribution of $v_1(t)$, *except* that it is delayed by t_0.

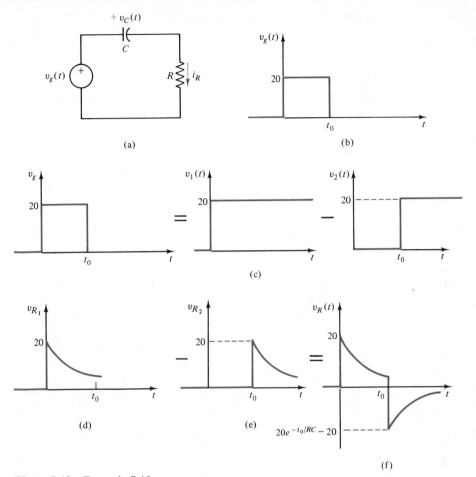

Figure 7-12   Example 7-10.

With $v_1(t)$ acting alone, we have

$$v_{R_1}(t) = Ri_{R_1}(t) = R\left(C\frac{dv_{C_1}}{dt}\right)$$

and KVL is

$$RC\frac{dv_{C_1}}{dt} + v_{C_1} = v_1 = 20 \qquad t > 0$$

The solution $v_{C_1}(t)$ of this equation is twenty times the step response in Example 7-6,

$$v_{C_1}(t) = 20 - 20e^{-t/RC}$$

Better yet, we ought to solve the differential equation directly and quickly, on its own, without the need to refer to, or memorize, previous results. Therefore

$$v_{R_1}(t) = v_1(t) - v_{C_1}(t) = 20e^{-t/RC}$$

Its plot is shown in Figure 7-12(d).

With $v_2(t)$ acting alone, the voltage across the resistor will be

$$v_{R_2}(t) = 20e^{-(t - t_0)/RC} \qquad t \geq t_0$$

that is, the same as $v_{R_1}(t)$ but delayed by t_0. So, in the expression of $v_{R_1}(t)$ we replace t by $(t - t_0)$. The plot of $v_{R_1}(t)$ is shown in Figure 7-12(e).

The final answer is therefore

$$v_R(t) = v_{R_1}(t) - v_{R_2}(t)$$

and is plotted in Figure 7-12(f).

Probs. 7-24, 7-25

*7-5 CALCULATING INITIAL CONDITIONS

So far, we have assumed that initial conditions are given to us for every circuit. This may be usually true, but in many cases we will have to calculate them separately. The situation is illustrated in Figure 7-13, where we show, in exaggeration, the three points on the time axis, $t = 0^-$, 0, and 0^+. A circuit has a past history that is completely summarized by the capacitive voltages and inductive currents at $t = 0^-$, $v_{C_1}(0^-)$, $v_{C_2}(0^-), \ldots, i_{L_1}(0^-), i_{L_2}(0^-), \ldots$.

At $t = 0$, the circuit is activated in a new configuration: A switch is closed or opened, and new elements are added or removed. We write differential equations that describe the behavior of the new circuit for all $t > 0$. To solve these equations completely, we need initial conditions such as $v_{C_1}(0^+)$, $i_{L_1}(0^+)$; that is, values of capacitive voltages and inductive currents just after $t = 0$. It is important to realize that these initial conditions do not depend on the differential equations; they must be obtained separately.

In Chapter 6, we discussed the conservation of charge which, in turn, implied the continuity of v_C; that is

$$v_C(0^-) = v_C(0^+) \tag{7-62}$$

and so v_C has no "jump," or discontinuity, at $t = 0$. Similarly, owing to conservation of flux, we found that i_L is continuous at $t = 0$,

$$i_L(0^-) = i_L(0^+) \tag{7-63}$$

To illustrate some exceptions to these rules, consider the following example.

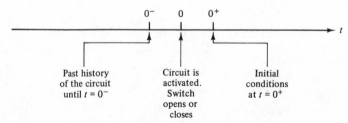

Figure 7-13 Initial conditions.

EXAMPLE 7-11

A dc voltage source, $v_s(t) = 1$ V, is applied at $t = 0$ to an uncharged capacitor, $v_C(0^-) = 0$. See Figure 7-14(a).

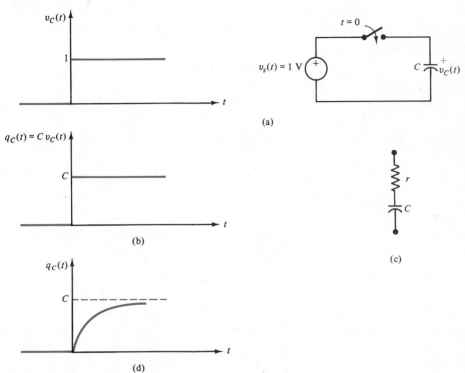

Figure 7-14   Example 7-11.

Solution. Clearly here, the capacitor's voltage is the unit step

$$v_C(t) = u(t) = \begin{cases} 0 & t < 0 \\ 1 & t > 0 \end{cases}$$

and we have $v_C(0^-) \neq v_C(0^+)$. What has happened? At $t = 0$, we changed instantly the charge across the capacitor from $q_C(0^-) = Cv_C(0^-) = 0$ to a finite value $q_C(0^+) = Cv_C(0^+) = C$, as shown in Figure 7-14(b). This step jump of the charge in zero time must be accompanied by an infinite current $i_C(t)$ at $t = 0$

$$i_C(t)\Big]_{t=0} = \frac{dq_C(t)}{dt}\Big]_{t=0} \to \infty$$

This current is called an *impulse*. At $t = 0$, the derivative (slope) of $q_C(t)$ is infinitely large.† In real life, of course, there are no infinite currents. What's the reason? A real-life capacitor, unlike our ideal model, is lossy, having a series resistance r, Figure 7-14(c). In such a case, we know from Example 7-5 that

$$q_C(t) = Cv_C(t) = C(1 - e^{-t/rC})$$

† We will study impulse functions in detail in Chapter 14.

and $q_C(t)$ is shown in Figure 7-14(d). The current $i_C(t)$ in this case is always finite. Also, we note that for $r \to 0$ the plot in Figure 7-14(d) approaches the one for the ideal capacitor, Figure 7-14(b).   ■ Prob. 7-26

EXAMPLE 7-12 _____

In a dual fashion, consider an inductor with $i_L(0^-) = 0$. It is excited at $t = 0$ with a current source $i_s(t) = 1$ A.

Solution.   The current $i_L(t)$ is therefore

$$i_L(t) = u(t)$$

and

$$i_L(0^-) \neq i_L(0^+)$$

The flux $\phi_L(t)$ is

$$Li_L(t) = \phi_L(t) = \begin{cases} L & t > 0 \\ 0 & t < 0 \end{cases}$$

and is accompanied by an impulse voltage across the inductor at $t = 0$

$$v_L(t) \bigg]_{t=0} = \frac{d\phi_L(t)}{dt} \bigg]_{t=0} \to \infty$$

Here, too, a real-life inductor will have some resistance in parallel with it, so that the flux will be continuous (without a step jump) at $t = 0$. Then the voltage will be finite there.

                                                                                                       ■

In conclusion, we say that if ideal capacitors form a loop, including possibly a voltage source, then we can expect discontinuities jumps in these capacitive voltages. An impulse current will accompany these voltage changes. Such a loop is called an *all-capacitive (all-C) loop*.

EXAMPLE 7-13 _____

Consider the circuit in Figure 7-15(a), where the all-*C* loop is marked with a dotted line. All capacitors are uncharged at $t = 0^-$.

Solution.   At $t = 0^+$ KVL must be satisfied; that is

$$-V_0 + v_{C_2}(0^+) + v_{C_1}(0^+) + v_{C_3}(0^+) = 0$$

or

$$-V_0 + \frac{q(0^+)}{C_2} + \frac{q(0^+)}{C_1} + \frac{q(0^+)}{C_3} = 0$$

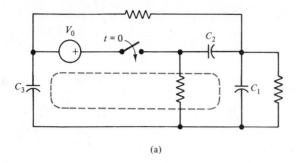

(a)

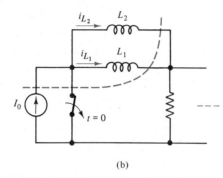

(b)

Figure 7-15 All-C loop and all-L node (cut set).

where $q(0^+)$ is the instantaneous charge, *just at $t = 0^+$*, across each capacitor. It is the same charge, because at $t = 0^+$ there is an impulse current flowing through the all-C loop. From this equation, we can solve for $q(0^+)$ and then

$$v_{C_1}(0^+) = \frac{q(0^+)}{C_1} \qquad v_{C_2}(0^+) = \frac{q(0^+)}{C_2} \qquad v_{C_3}(0^+) = \frac{q(0^+)}{C_3}$$

■ Prob. 7-29

Dually, if there is an *all-inductive node* (or a cut set) consisting only of ideal inductors and possibly a current source, then we can expect discontinuities in the inductive currents at $t = 0$.

EXAMPLE 7-14

In Figure 7-15(b), the switch opens at $t = 0$, and $i_{L_1}(0^-) = i_{L_2}(0^-) = 0$.

Solution. The all-L cut set is shown in a dotted line. For the instant $t = 0^+$, KCL reads

$$-I_0 + i_{L_1}(0^+) + i_{L_2}(0^+) = 0$$

or

$$-I_0 + \frac{\phi(0^+)}{L_1} + \frac{\phi(0^+)}{L_2} = 0$$

with an impulse voltage across L_1 and L_2.

■ Prob. 7-30

How do we calculate initial conditions in more complicated circuits, without all-C loops or all-L cut sets? It is easy, since then we know that $v_C(0^+) = v_C(0^-)$ and $i_L(0^+) = i_L(0^-)$. The steps to take are therefore:

1. Replace every voltage source $v_s(t)$ by a constant (dc) voltage source of value $v_s(0^+)$.
2. Replace every current source $i_s(t)$ by a dc current source of value $i_s(0^+)$.
3. Replace every capacitor by a dc voltage source, of value $v_C(0^+) = v_C(0^-)$. If the capacitor is initially uncharged, $v_C(0^-) = 0$; this amounts to a short circuit (s.c.).
4. Replace every inductor by a dc current source of value $i_L(0^+) = i_L(0^-)$. If the initial current in the inductor is zero, $i_L(0^-) = 0$; this amounts to an open circuit (o.c.).
5. In the resulting *resistive* network, calculate the necessary initial currents or voltages.

Prob. 7-36

EXAMPLE 7-15

In the network shown in Figure 7-16(a), we are given

$$i_L(0^-) = 0 \qquad v_C(0^-) = -1 \text{ V}$$

Let us calculate $i_R(0^+)$, $i_C(0^+)$, and $v_L(0^+)$.

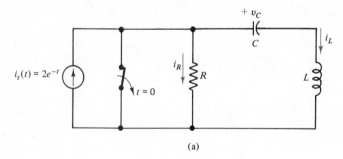

(a)

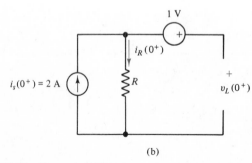

(b)

Figure 7-16 Example 7-15.

Solution. The entire network, valid *only* at $t = 0^+$, is shown in Figure 7-16(b). From it we write immediately

$$i_R(0^+) = 2 \text{ A} \qquad i_C(0^+) = 0$$

and from KVL

$$v_L(0^+) - 2R - 1 = 0$$

we obtain

$$v_L(0^+) = 2R + 1$$

Keep in mind that this method is good *only* for one "frozen" instant, $t = 0^+$. The simplified network in Figure 7-16(b) *is not* valid, of course, for all $t > 0$. ■

Probs. 7-31, 7-32

Finally, let us consider the evaluation of initial conditions for derivatives such as

$$\left.\frac{di_L(t)}{dt}\right]_{t=0^+} = i'_L(0^+) \tag{7-64}$$

or

$$\left.\frac{dv_C(t)}{dt}\right]_{t=0^+} = v'_C(0^+) \tag{7-65}$$

where the prime (') is the common notation for the first derivative. The general rule here is simple: Write KVL and/or KCL equations as needed, for all $t > 0$, then evaluate them for the one instant $t = 0^+$. To illustrate the method, let us use the previous example again.

EXAMPLE 7-16 _____

In the network of Figure 7-16(a), find $v'_C(0^+)$ and $i'_L(0^+)$.

Solution. For all $t > 0$, we have

$$i_C(t) = C\frac{dv_C(t)}{dt} = i_L(t)$$

Consequently

$$C\frac{dv_C(t)}{dt}\bigg]_{t=0^+} = i_L(0^+)$$

$$\therefore \quad v'_C(0^+) = \frac{1}{C}i_L(0^+) = 0$$

because $i_L(0^+) = 0$ in Example 7-15. Similarly, KVL for all $t > 0$ reads

$$v_C(t) + L\frac{di_L(t)}{dt} - Ri_R(t) = 0$$

For $t = 0^+$, we get

$$v_C(0^+) + Li'_L(0^+) - Ri_R(0^+) = 0$$

or

$$i'_L(0^+) = \frac{1}{L}[Ri_R(0^+) - v_C(0^+)] = \frac{1}{L}(2R + 1)$$

using the values of $i_R(0^+)$ and $v_C(0^+)$ obtained earlier. ■

Probs. 7-33, 7-34, 7-35

PROBLEMS

7-1 In the op amp integrator (Figure 7-1) interchange the position of R and C and prove that the resulting circuit is a *differentiator*.

7-2 Calculate the output $v_o(t)$ for a given input $v_i(t)$ in the circuit shown. The initial voltage across the capacitor is zero.

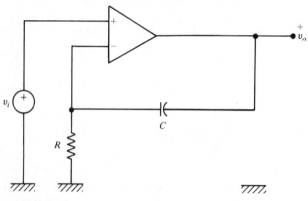

Problem 7-2

7-3 A method suitable for the solution of a first-order, zero-input (homogeneous) differential equation is by *separation of variables*. Specifically, in Equation 7-10,

$$\frac{dv_C}{dt} + \frac{1}{RC} v_C = 0$$

we separate the variables as

$$\frac{dv_C}{v_C} = -\frac{1}{RC} dt$$

then integrate both sides from $t = 0$ to any arbitrary t. Complete the steps and obtain the same answers as in the text.

*7-4 Yet another method for Problem 7-3 is by the *integrating factor*: Consider the general form of a first-order differential equation

$$\frac{dy(t)}{dt} + p(t)y(t) = f(t)$$

Prove that if we multiply it by the integrating factor

$$k(t) = e^{\int p(t)dt}$$

then the left-hand side becomes a total differential

$$\frac{d}{dt} [k(t)y(t)]$$

and hence both sides can be integrated between 0 and t. The final solution is

$$y(t) = e^{-\int p(t)dt} \int e^{\int p(t)dt} f(t)\, dt + K e^{-\int p(t)dt}$$

where K is the arbitrary constant of integration. Do all these steps in full detail, and compare the answer with the one in the text and in Problem 7-3.

7-5 In the circuit shown, switch S remains closed for a very long time ($-\infty < t \leq 0^-$). Then, at $t = 0$, it is opened.
(a) Calculate the initial voltage $v_C(0^-)$ across the capacitor.
(b) Calculate $v_C(t)$ for all $t \geq 0$.
Hint for (a): The capacitor acts as an open circuit for dc ($-\infty < t \leq 0^-$). In this problem you are given enough information to calculate the initial state rather than be given the initial state itself.

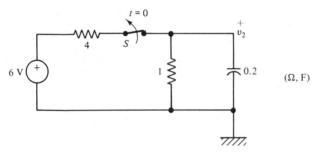

Problem 7-5

7-6 Calculate the state variable $v_C(t)$ in the network shown, in the following steps:
(a) Write KVL for the network. As a loop current, use the capacitor's current $C dv_C/dt$.
(b) Classify, with full justification, the response as being zero-input.
(c) Solve the differential equation.
(d) What is the time constant of this circuit?
Note the dependent source!

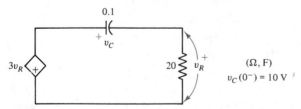

Problem 7-6

7-7 Prove the following property of the waveform of $v_C(t)$ in Figure 7-4(a): for any $t = t_0$, if we draw the tangent line at $v_C(t_0)$, this line will intersect the time axis at $t = t_0 + \tau$. This provides an easy graphical method for finding the time constant of such a circuit from a given plot of $v_C(t)$.

7-8 In the circuit shown, the switch has been in position a for a long time. At $t = 0$, it is moved to position b. Calculate and plot $i_1(t)$ for $t \geq 0$.

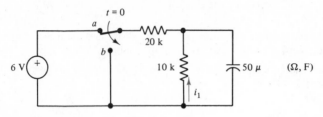

Problem 7-8

7-9 In the circuit shown, calculate the time constant τ for the zero-input response. *Hint*: Zero-input!

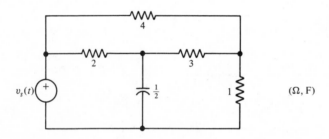

Problem 7-9

7-10 Repeat Problem 7-9, with a current source $i_s(t)$ instead of a voltage source.

7-11 Calculate the energy balance in the *RL* circuit of Figure 7-5.

7-12 Plot the zero-input current $i_L(t)$ of Figure 7-5 for a fixed value of R and for several values of L, showing how a decrease in L affects the plot. Next, keep L fixed and vary R, showing several plots.

7-13 In Example 7-4, find the time $t > 0$ when the 5-Ω resistor has dissipated:
(a) 50 percent of the initial energy in L
(b) 90 percent of the energy in L

7-14 In the figure we show a simple mechanical system, consisting of a mass M attached to a dashpot D (friction element). This is, for example, a model of a car (M) and a shock absorber (D). The force due to friction is proportional to the velocity, $f_D = D\dot{x}$. The mass is given an initial displacement, or an initial velocity, such as you do when you want to test the shock absorber.
(a) Show that the equation for the velocity $v(t)$ is a first-order differential equation, analogous to the zero-input RC; the analogous variables are voltage and velocity, C and M, $1/R$ and D, current and force.

(b) What is the solution for the velocity? The time constant of the system?

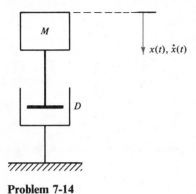

Problem 7-14

7-15 The switch is closed in position *a* for 0.6 s, then is moved to position *b*. Calculate and plot $v_C(t)$ for $t > 0$. The capacitor is uncharged before the switch closes on *a*.

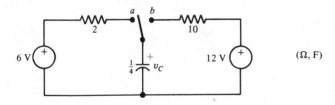

Problem 7-15

7-16 The experiment shown in Example 7-5 (Figure 7-10) can be easily set up in the laboratory, with several results measured or calculated. The constant voltage source, for example, can be a battery. Let $V_s = 12$ V, $R = 1$ kΩ, and C unknown (somebody rubbed off its value markings!). The initial voltage across C is zero. An oscilloscope is connected across C to record $v_C(t)$. For this experiment, plot a typical curve of $v_C(t)$, by comparison with Figure 7-9. Explain, with a full derivation, how to calculate the value of C from this plot. *Hint*: Draw a tangent to $v_C(t)$ at $t = 0$, calculate its intersection with the asymptote of $v_C(t)$, and, from it, the time constant τ.

7-17 Review Problem 7-4, and prove that the entries in Table 7-1 for the guessed particular solution are actually *derived* from the particular solution as obtained by the integrating factor,

$$y_P(t) = e^{-\int p(t)dt} \int e^{\int p(t)dt} f(t) \, dt$$

7-18 (a) Solve completely

1.
$$2\frac{dv}{dt} + 3v = 6e^{-4t} \qquad\qquad v(0^-) = -1$$

2.
$$\frac{dv}{dt} + 3v = 10te^{-4t} \qquad\qquad v(0^-) = 2$$

3.
$$\frac{dv}{dt} + 4v = 10te^{-4t} \qquad\qquad v(0^-) = 2$$

4.
$$3\frac{dv}{dt} + 10v = 4e^{-2t}\sin 50t \qquad v(0^-) = 0$$

5.
$$\frac{dv}{dt} - v = 1 \qquad\qquad v(0^-) = -3$$

(b) Plot the total response in each case. Comment about the physical (im)possibility of case 5.

7-19 (a) Calculate and plot the total response $i_L(t)$ for $t > 0$ in the network shown. Use the state variable $i_L(t)$ as a loop current. Given $i_L(0^-) = -4$ A.

(b) Use superposition to calculate just the zero-state response, then just the zero-input response, then add them. Each one can be written almost by inspection.

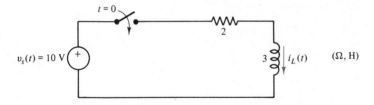

Problem 7-19

7-20 Calculate and plot the step response (see Example 7-6) of the network in Problem 7-19.

7-21 (a) In the network shown, the switch is closed to position *a* where it remains for 100 ms. Then it is moved to position *b*. Calculate and plot $i_L(t)$ for $t > 0$. Before position *a*, the inductor was unenergized.

(b) Calculate the energy balance in the *RL* circuit for position *b*.

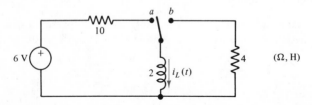

Problem 7-21

*7-22 A second-order network obeys the differential equation

$$\frac{d^2i}{dt^2} + 2\frac{di}{dt} + i = 10e^{-t}$$

(a) Write by inspection the characteristic equation.
(b) Calculate the characteristic values.
(c) Assume the proper form of the particular solution i_p and calculate it.

*7-23 Repeat Problem 7-22 if the right-hand source is $10te^{-t}$.

7-24 For the circuit shown in Problem 7-19, let

$$v_s(t) = \begin{cases} 10 \text{ V} & t > 2 \\ 0 & \text{elsewhere} \end{cases}$$

and zero initial condition. Calculate and plot $i_L(t)$.

7-25 Repeat Problem 7-24, with $i_L(0^-) = -4$ A.

*7-26 To get a better "feel" for the impulse current in Example 7-11, consider a limiting case first. Let $v_C(t)$ be given as shown in the figure. Obviously, as $\varepsilon \to 0$, $v_C(t) = u(t)$, the unit step.
(a) Draw carefully $q_C(t)$ for the given $v_C(t)$, with $\varepsilon \neq 0$.
(b) Draw carefully $i_C(t) = dq_C(t)/dt$, with $\varepsilon \neq 0$. Label the values of all the important points. In drawing it, remember that the derivative is the slope.
(c) Calculate the area under $i_C(t)$ from your plot in (b). Next, write this area as an integral. What is the electrical quantity represented by this integral?
(d) Let now $\varepsilon \to 0^+$ in (a), (b), and (c), and explain fully what happens to the current, to the charge, and to the voltage.

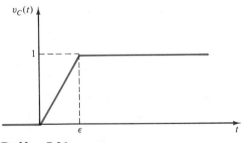

Problem 7-26

7-27 In each of the networks shown, predict by inspection whether or not the capacitive voltages will have a discontinuity at $t = 0$.

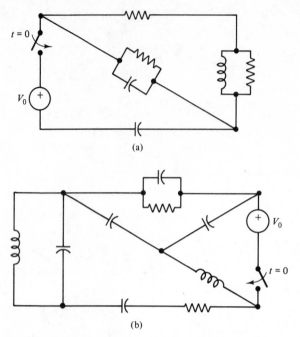

(a)

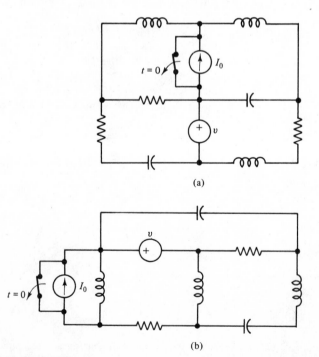

(b)

Problem 7-27

(a)

(b)

Problem 7-28

7-28 In each of the networks shown, predict by inspection whether or not the inductive currents will have a discontinuity at $t = 0$.

7-29 In Example 7-13, let $C_1 = 1$ F, $C_2 = 2$ F, $C_3 = 3$ F, $V_0 = 12$ V. Calculate all the initial values of the capacitive voltages and charges at $t = 0^+$.

7-30 In Example 7-14, let $L_1 = \frac{1}{2}$ H, $L_2 = 0.1$ H, $I_0 = 4$ A. Calculate all the inductive currents and fluxes at $t = 0^+$.

7-31 **(a)** In the network shown, the capacitive voltage and the inductive currents are zero for $t = 0^-$. Calculate the initial values at $t = 0^+$ of the capacitive current and the inductive voltages.

   (b) Repeat, if $v_C(0^-) = 3$ V ($+$ reference on the top plate) and the 1-H inductor carries $i_{L_1}(0^-) = 6$ A from right to left.

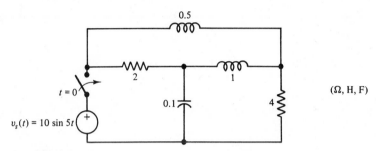

Problem 7-31

7-32 Rework Problem 7-31 if the source is $v_s(t) = 10 \cos 5t$.

7-33 In Example 7-16, calculate

$$v_C''(0^+) = \frac{d^2 v_C(t)}{dt^2}\bigg]_{t=0^+}$$

Hint: Use relationships that are valid for all $t > 0$, and obtain from them the required values.

7-34 For the RC network of Figure 7-8, calculate $v_C'(0^+)$, $v_C''(0^+)$. These values are useful also in the plotting of $v_C(t)$, giving the slope and the curvature of the plot at $t = 0^+$.

7-35 For the network shown, calculate the initial values at $t = 0^+$ of v_C, v_C' and v_C''.

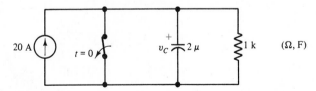

Problem 7-35

7-36 To help us in calculating initial values at $t = 0^-$, when a switch has been in the old position for a very long time, $-\infty < t \le 0^-$, let us establish rules similar to the rules in the text for $t = 0^+$:

   (a) How does a resistor behave if constant voltages or currents remain for a very long time?

(b) How does an inductor behave if a constant current flows through it for a very long time?

(c) How does a capacitor behave if a constant voltage is across it for a very long time? Based on your conclusions, draw the simplified resistive network and calculate $v_C(0^-)$ and $i_L(0^-)$ for the two networks shown.

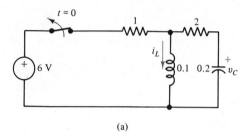

(a)

(Ω, H, F)

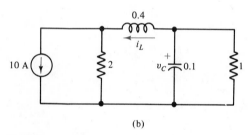

(b)

Problem 7-36

Chapter 8

Second-Order Circuits
Higher-Order Circuits

8-1 PRELIMINARY REMARKS AND EXAMPLE

In the previous chapter, we studied first-order circuits. Such a circuit, we saw, has one dynamic element (C or L), an input-output relation described by a first-order differential equation, and one initial condition. The initial energy stored in the dynamic element, $\frac{1}{2}Cv_C^2(0^-)$ or $\frac{1}{2}Li_L^2(0^-)$, is eventually dissipated in the resistors in the network, hence the decaying exponential forms of all voltages and currents in the zero-input solution.

A second-order circuit ($n = 2$), we fully expect, has two energy-storing elements with two arbitrary initial conditions, and a second-order differential equation describing its input-output relation. It will have a quadratic characteristic equation, with two characteristic values (natural frequencies). Also, the energy balance in such a circuit will be more complicated and more interesting.

EXAMPLE 8-1 ────────────────────────────────

Let us begin by considering the circuit shown in Figure 8-1(a), consisting of two capacitors and a resistor. Let the initial conditions be given as

$$v_1(0^-) = 2 \text{ V} \qquad v_2(0^-) = 0 \text{ V}$$

and let the resistor represent the (physical) resistance of the wires plus the lossy nature of both capacitors. The switch closes at $t = 0$. The tree and the link are shown in Figure 8-1(b). The two capacitors, having independent voltages, are chosen as tree branches.

Solution. Before writing any equations, we can decide qualitatively what will happen for $t > 0$. The charged capacitor will discharge some of its energy, partly in heat in the resistor and partly in charging the other capacitor until some balance is reached. Now let us write the necessary equations.

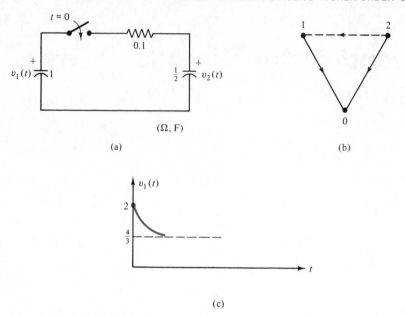

Figure 8-1   Example 8-1.

The two state variables (remember: This is a second-order circuit!) are $v_1(t)$ and $v_2(t)$, so let us use them. Another good reason against using an "obvious" loop current is an integral term, $v = (1/C) \int i \, dt$, in the loop analysis here. After $t = 0$, KCL at node 1 reads†

$$1 \frac{dv_1}{dt} + 10(v_1 - v_2) = 0$$

and at node 2

$$\frac{1}{2} \frac{dv_2}{dt} + 10(v_2 - v_1) = 0$$

with the convention sign ($+$) for currents leaving a node.

We have here two simultaneous first-order differential equations in the unknowns $v_1(t)$ and $v_2(t)$. Each equation is first-order because of the highest derivative in it; the equations are simultaneous because, just as in simultaneous algebraic equations, the unknowns appear simultaneously in both. Even before any further steps, we expect that these two first-order equations may be reduced to a single equivalent second-order differential equation, again because the order of the network is $n = 2$.

To obtain that single differential equation, we must eliminate v_1 or v_2 between the two. For example, we can solve the first equation for v_2

$$v_2 = 0.1 \frac{dv_1}{dt} + v_1$$

and take its first derivative

$$\frac{dv_2}{dt} = 0.1 \frac{d^2 v_1}{dt^2} + \frac{dv_1}{dt}$$

†Remember: Lowercase letters are functions of time, $v_1 = v_1(t)$, $v_2 = v_2(t)$.

Now substitute into the second equation

$$\frac{1}{2}\left(0.1\frac{d^2v_1}{dt^2} + \frac{dv_1}{dt}\right) + 10\left(0.1\frac{dv_1}{dt} + v_1\right) - 10v_1 = 0$$

The result is

$$\frac{d^2v_1}{dt^2} + 30\frac{dv_1}{dt} = 0$$

a second-order homogeneous differential equation for $v_1(t)$. (Try to get the single equation for v_2 by similar steps.)

The characteristic equation is

$$s^2 + 30s = 0$$

yielding two characteristic values,

$$s_1 = 0 \qquad s_2 = -30$$

Therefore, the total solution is (by superposition)

$$v_1(t) = K_1 e^{s_1 t} + K_2 e^{s_2 t} = K_1 + K_2 e^{-30t}$$

We need two initial conditions, $v_1(0^+)$ and $(dv_1/dt)_{0^+} = v_1'(0^+)$. The first one is given, $v_1(0^+) = v_1(0^-) = 2$ V, as studied in the previous chapter. In order to obtain the second one, we use the first original differential equation, evaluated at $t = 0^+$

$$1\left(\frac{dv_1}{dt}\right)_{0^+} + 10v_1(0^+) - 10v_2(0^+) = 0$$

or

$$v_1'(0^+) = \left(\frac{dv_1}{dt}\right)_{0^+} = 10v_2(0^+) - 10v_1(0^+) = -20$$

At $t = 0^+$ we have then from our solution

$$v_1(0^+) = 2 = K_1 + K_2 e^0 = K_1 + K_2$$
$$v_1'(0^+) = -20 = (-30K_2 e^{-30t})_{0^+} = -30K_2$$

where, in the last equation, we performed the differentiation

$$\frac{dv_1}{dt} = \frac{d}{dt}(K_1 + K_2 e^{-30t}) = -30K_2 e^{-30t}$$

The two equations for K_1 and K_2

$$K_1 + K_2 = 2$$
$$-30K_2 = -20$$

yield

$$K_1 = \tfrac{4}{3} \qquad K_2 = \tfrac{2}{3}$$

Finally,

$$v_1(t) = \tfrac{4}{3} + \tfrac{2}{3}e^{-30t} \qquad t \geq 0$$

Its plot is shown in Figure 8-1(c), showing a decay from the initial value, $v_1(0^+) = 2$, to the final steady-state value of $\tfrac{4}{3}$, with a time constant of $\tfrac{1}{30}$ second.

An alternate way of solving the second-order differential equation and extensions of this example are given in the problems listed here. ∎

Probs. 8-1, 8-2, 8-3

A much more elegant and easy way to solve Example 8-1 is outlined as follows:

EXAMPLE 8-1 (continued) ──────────────────────────────────

The two original simultaneous first-order differential equations in the two state variables $v_1(t)$ and $v_2(t)$ are repeated here

$$\frac{dv_1(t)}{dt} + 10(v_1 - v_2) = 0$$

$$\frac{1}{2}\frac{dv_2(t)}{dt} + 10(v_2 - v_1) = 0$$

with the given initial state $v_1(0^-) = 2$, $v_2(0^-) = 0$.

Without any substitutions or manipulations, we rewrite these two differential equations in matrix form

$$\begin{bmatrix} \dfrac{dv_1(t)}{dt} \\[2mm] \dfrac{dv_2(t)}{dt} \end{bmatrix} + \begin{bmatrix} 10 & -10 \\ -20 & 20 \end{bmatrix}\begin{bmatrix} v_1(t) \\ v_2(t) \end{bmatrix} = \begin{bmatrix} 0 \\ 0 \end{bmatrix}$$

or

$$\frac{d}{dt}\begin{bmatrix} v_1(t) \\ v_2(t) \end{bmatrix} = \begin{bmatrix} -10 & 10 \\ 20 & -20 \end{bmatrix}\begin{bmatrix} v_1(t) \\ v_2(t) \end{bmatrix} \qquad \begin{bmatrix} v_1(0_-) \\ v_2(0_-) \end{bmatrix} = \begin{bmatrix} 2 \\ 0 \end{bmatrix} \text{given}$$

That is,

$$\frac{d}{dt}\mathbf{x}(t) = \mathbf{A}\mathbf{x}(t) \qquad \mathbf{x}(0^-) \text{ given}$$

where

$$\mathbf{x}(t) = \begin{bmatrix} v_1(t) \\ v_2(t) \end{bmatrix}$$

is the column matrix of the state variables, called briefly the *state matrix* (or *state vector*).

We have, then, a *first-order matrix differential equation* for the unknown state matrix $\mathbf{x}(t)$, with the (expected!) initial condition matrix $\mathbf{x}(0^-)$ given.

In principle, this is as easy a problem as the *scalar* first-order differential equation

$$\frac{x(t)}{dt} = ax(t) \qquad x(0^-) \text{ given} \qquad\qquad ■$$

We will study in detail such first-order matrix differential equations in later chapters on state variable analysis. What makes them so attractive, among many reasons, are the facts that: (1) state variable equations are readily formulated by using KCL and KVL, as we saw in this preliminary example, and (2) state variables *always* yield only first-order differential equations and *never* an integral! (3) all the matrix calculations needed to solve the first-order matrix differential equation are standard and are readily available in technical computer libraries.

With such attractive promises of things to come, we must continue now with some additional important developments and concepts.

8-2 THE *LC* CIRCUIT AND ITS ZERO-INPUT RESPONSE

Let us now consider the zero-input response of the *LC* circuit shown in Figure 8-2(a). It is a second-order circuit, since the two dynamic elements can have two independent initial conditions, $v_C(0^-)$ and $i_L(0^-)$. Specifically here, we are given $v_C(0^-) \neq 0$ and $i_L(0^-) = 0$. At $t = 0$ the switch closes, and we must solve the network for all $t > 0$.

Again, before plunging into differential equations, let us try to understand what will happen in a qualitative way. The charged capacitor will discharge through the inductor, causing a current $i_L(t)$, and the inductor's energy will increase from its initial zero value. Since there is no resistor in the circuit, no energy will be dissipated in heat. The inductive energy will then flow back to the capacitor which, again, will discharge into the inductor, and so on. Some sort of oscillation will occur. The situation is analogous, as we saw in previous chapters, to the mechanical system shown in Figure 8-2(b): The mass m, analogous to the capacitor C, is given an initial displacement $x(0^-)$ and then released; the spring K, analogous to the inductor L, will extend and shrink in an oscillating fashion; here the lack of air friction and of internal spring friction is analogous to $R = 0$ in our circuit.

Let us derive and solve the equations for the *LC* circuit. The graph is shown in Figure 8-2(c), where, as always, the capacitor is chosen as a tree branch and the inductor as a link. The two state variables are $v_C(t)$ and $i_L(t)$. KCL at the top (or bottom) node is

$$C \frac{dv_C}{dt} + i_L = 0 \qquad (8\text{-}1)$$

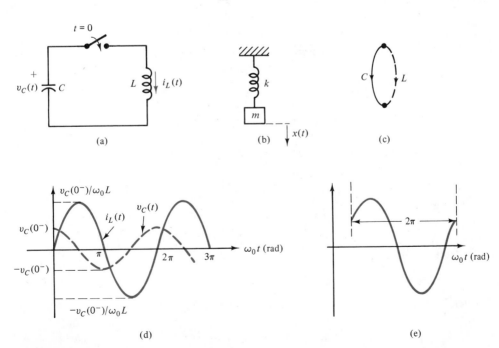

Figure 8-2 An *LC* circuit.

and KVL is

$$L \frac{di_L}{dt} - v_C = 0 \tag{8-2}$$

These are two simultaneous first-order differential equations. To obtain the single equivalent second-order differential equation, we differentiate Equation 8-2 once

$$L \frac{d^2 i_L}{dt^2} = \frac{dv_C}{dt} \tag{8-3}$$

and substitute into Equation 8-1, getting

$$LC \frac{d^2 i_L}{dt^2} + i_L = 0 \tag{8-4}$$

a second-order homogeneous (zero-input) equation.

The characteristic equation is

$$LCs^2 + 1 = 0 \tag{8-5}$$

or

$$-s^2 = \frac{1}{LC} = \omega_0^2 \tag{8-6}$$

where we introduce $\omega_0^2 = 1/LC$, a positive number, since $L > 0$, $C > 0$. The two characteristic values, or natural frequencies, are

$$s_1 = +j\omega_0 \qquad s_2 = -j\omega_0 \tag{8-7}$$

where $j = \sqrt{-1}$ is the unit imaginary number. Here s_1 and s_2 are pure imaginary and conjugate. The zero-input (here the total) solution for $i_L(t)$ is therefore

$$i_L(t) = K_1 e^{s_1 t} + K_2 e^{s_2 t} = K_1 e^{j\omega_0 t} + K_2 e^{-j\omega_0 t} \tag{8-8}$$

with K_1 and K_2 as the two arbitrary constants. While this expression is satisfactory in the mathematical sense, it conveys very little in the engineering, physical sense. Let us call on Euler's identity for an imaginary exponential function

$$e^{\pm j\theta} = \cos \theta \pm j \sin \theta \tag{8-9}$$

Then Equation 8-8 reads

$$i_L(t) = K_1(\cos \omega_0 t + j \sin \omega_0 t) + K_2(\cos \omega_0 t - j \sin \omega_0 t) \tag{8-10}$$

Collecting terms, we have

$$i_L(t) = K_3 \cos \omega_0 t + K_4 \sin \omega_0 t \tag{8-11}$$

where we use $K_3 = K_1 + K_2$ and $K_4 = j(K_1 - K_2)$ as the two new arbitrary constants. Don't be disturbed by j multiplying $(K_1 - K_2)$; just like K_1 and K_2, j is also a constant, and so is K_4. The *nature* of K_4, real or imaginary, has yet been determined. For now, it is just a constant.

Next, we need two initial conditions for $i_L(t)$. We have $i_L(0^-) = 0$. In order to calculate $i_L'(0^+) = (di_L/dt)_{0^+}$, use Equation 8-2 at $t = 0^+$

$$i_L'(0^+) = \left(\frac{di_L}{dt}\right)_{0^+} = \frac{1}{L} v_C(0^-) = \frac{1}{L} v_C(0^+) \tag{8-12}$$

We might remind ourselves here, from Chapter 7, that $v_C(0^+) = v_C(0^-)$ and $i_L(0^+) = i_L(0^-)$. Without these facts, we *cannot* relate $i_L(0^-)$ to $v_C(0^-)$ because, prior to $t = 0$, these are independent of each other.

With the two initial conditions established, we can calculate K_3 and K_4 in Equation 8-11. First, Equation 8-11 at $t = 0^+$ reads

$$i_L(0^+) = 0 = K_3 \cdot 1 + K_4 \cdot 0 \tag{8-13}$$

or

$$K_3 = 0 \tag{8-14}$$

Next, differentiate Equation 8-11 and use $t = 0^+$

$$i_L'(0^+) = \frac{v_C(0^+)}{L} = (\omega_0 K_4 \cos \omega_0 t)_{0^+} = \omega_0 K_4 \tag{8-15}$$

or

$$K_4 = \frac{v_C(0^+)}{\omega_0 L} \tag{8-16}$$

The final solution is therefore

$$i_L(t) = \frac{v_C(0^-)}{\omega_0 L} \sin \omega_0 t \tag{8-17}$$

From Equations 8-17 and 8-2 we get

$$v_C(t) = L \frac{di_L}{dt} = v_C(0^-) \cos \omega_0 t \tag{8-17a}$$

and its plot is shown in dotted lines in Figure 8-2(d). These plots of the behavior of $i_L(t)$ and $v_C(t)$ confirm our previous qualitative argument: At $t = 0$, the capacitor's energy is at its maximum, $\frac{1}{2} C v_C^2(0)$, and $i_L(0) = 0$. Then the capacitor discharges through L, v_C and its energy decrease, while i_L and its energy increase. At $\omega_0 t = \pi/2$, $v_C = v_L = 0$ and therefore $p_L = 0 = dw_L/dt$, because at this instant the inductor's energy is maximum. Then the process reverses itself: The inductor's current and energy decrease, while the capacitor's increase.

Several observations are in order here:

1. The current $i_L(t)$ is a pure sinusoid, oscillating between $v_C(0^-)/\omega_0 L$ and $-v_C(0^-)/\omega_0 L$. The positive peak of oscillation is the *amplitude* of the sinusoidal waveform, or its *maximum value* I_m

$$I_m = \frac{v_C(0^-)}{\omega_0 L} \tag{8-18}$$

2. The waveform of i_L is repetitive. The smallest "piece" of $i_L(t)$ which repeats itself is called a *cycle*. Such a cycle is shown in Figure 8-2(e). Any other cycle is also valid, provided its end is 2π radians from its beginning. The time for one cycle is called the *period* of the sinusoid, in seconds, and is denoted by T. We have therefore

$$\omega_0 T = 2\pi \tag{8-19}$$

or

$$T = \frac{2\pi}{\omega_0} \quad \text{s} \tag{8-20}$$

3. The number of cycles per second (cps) is called the *frequency* of the sinusoid, f. If one cycle takes T seconds, then, in 1 s there are

$$f = \frac{1}{T} \tag{8-21}$$

cycles. The unit for frequency is the *hertz* (Hz), named after the German physicist Heinrich Hertz (1857–1894). The quantity ω_0 has the units of radians per second, as seen in Equation 8-19, and is called the *radian frequency*. It is because of this term, when used in Equation 8-7, that we extend the name "natural frequencies" to all characteristic values, whether the response actually oscillates or not. Sinusoidal sources and responses are very common in all aspects of electrical engineering. We'll have much more to say and study about them in later chapters.

4. The current $i_L(t)$ is due only to initial conditions, without any source. It is the zero-input response, and it happens to be entirely steady-state. There is no transient part in $i_L(t)$. Here, again, we must make a clear distinction between the total response, consisting of the zero-input part plus the zero-state part, and the time behavior of the total response, being the transient part plus the steady-state part. Sometimes the zero-input part is the transient part (but not in the present case); in general, we cannot make such identifications.

5. Ideal lossless L and C are useful in models and in many approximations of real-life circuits. However, there may be some nonnegligible resistance in the circuit, as we saw in the previous chapters. We'll consider such cases later.

Probs. 8-4, 8-5

8-3 THE STEP RESPONSE OF THE *LC* CIRCUIT

As we recall, the step response is the zero-state (zero initial conditions) response to a unit step input. Specifically, let us consider the circuit of Figure 8-3(a), where the input is a unit step voltage $u(t)$ and L and C have zero initial conditions. The graph is shown in Figure 8-3(b). We can use here the state variables $v_C(t)$ and $i_L(t)$, as usual. In fact, it is always advisable to do so. However, if only for the sake of variety ("the spice of

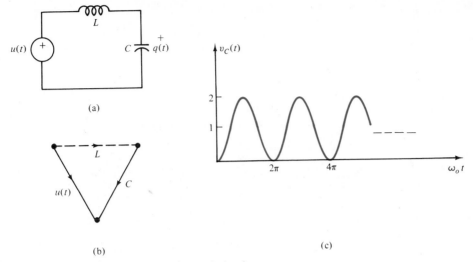

Figure 8-3 The step response of an *LC* circuit.

life"), let us use here $q(t)$, the charge across the capacitor, as the unknown. More seriously, there are learning advantages here, not just variety. We have

$$v_C(t) = \frac{1}{C} q(t) \tag{8-22}$$

and

$$i_L(t) = i_C(t) = \frac{dq(t)}{dt} \tag{8-23}$$

and so $q(t)$ is an acceptable variable. The appropriate initial conditions are

$$q(0^+) = Cv_C(0^+) = Cv_C(0^-) = 0 \tag{8-24}$$

and

$$i_L(0^+) = \frac{dq}{dt}\bigg]_{0^+} = q'(0^+) = 0 \tag{8-25}$$

To get the differential equation for the circuit, write KVL using Equations 8-22 and 8-23:

$$-u(t) + L \frac{d}{dt}\left(\frac{dq}{dt}\right) + \frac{1}{C} q(t) = 0 \tag{8-26}$$

or

$$L \frac{d^2q}{dt^2} + \frac{1}{C} q(t) = 1 \qquad t > 0 \tag{8-27}$$

One excellent reason for using $q(t)$, and not a loop current, emerges again: No integrals for the capacitive voltage!

This nonhomogeneous differential equation has the characteristic equation

$$Ls^2 + \frac{1}{C} = 0 \tag{8-28}$$

with the two natural frequencies

$$s_1 = j\omega_0 \qquad s_2 = -j\omega_0 \tag{8-29}$$

as expected, the same as before. The homogeneous solution is

$$q_H(t) = K_1 \cos \omega_0 t + K_2 \sin \omega_0 t \tag{8-30}$$

The particular solution, a generalization of the right-hand side of Equation 8-27, is a constant, $q_P = A$. Therefore

$$L \frac{d^2(A)}{dt^2} + \frac{1}{C} A = 1 \tag{8-31}$$

or

$$A = C \tag{8-32}$$

The total solution is then

$$q(t) = q_H + q_P = K_1 \cos \omega_0 t + K_2 \sin \omega_0 t + C \tag{8-33}$$

The first initial condition yields

$$q(0^+) = 0 = K_1 + C \tag{8-34}$$

The second initial condition yields

$$q'(0^+) = 0 = (-K_1 \omega_0 \sin \omega_0 t + K_2 \omega_0 \cos \omega_0 t)_{0^+} = K_2 \omega_0 \tag{8-35}$$

From Equations 8-34 and 8-35 we have

$$K_2 = 0 \qquad K_1 = -C \tag{8-36}$$

and so the total solution (Equation 8-33) becomes

$$q(t) = C(1 - \cos \omega_0 t) \tag{8-37}$$

and

$$v_C(t) = \frac{q(t)}{C} = 1 - \cos \omega_0 t \qquad t > 0 \tag{8-38}$$

is shown in Figure 8-3(c). It is the zero-state step response, and it is entirely steady-state.

Probs. 8-6, 8-7

8-4 THE *RLC* PARALLEL CIRCUIT AND ITS ZERO-INPUT RESPONSE

We consider the parallel *RLC* circuit in Figure 8-4. It is a second-order network ($n = 2$), with the two initial conditions $v_C(0^-)$ and $i_L(0^-)$ arbitrarily given. Again, we use the state variables $v_C(t)$ of the tree branch, and $i_L(t)$ of the link. Our preliminary

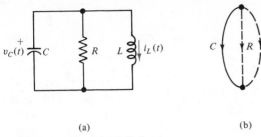

(a)   (b)

Figure 8-4   Parallel *RLC* circuit.

suspicion is that, due to the energy dissipated in R, there will be some *damping*. The details, of course, must be worked out fully.

KCL at the top node is, with $(+)$ for currents leaving,

$$C \frac{dv_C}{dt} + \frac{v_C}{R} + i_L = 0 \qquad (8\text{-}39)$$

where, again, we use the variable $i_L(t)$ to avoid the integral for the inductive term, $i_L = (1/L) \int_0^t v_C(x)\,dx + i_L(0^-)$. The second equation is obtained from KVL

$$v_C = L \frac{di_L}{dt} \qquad (8\text{-}40)$$

Differentiating Equation 8-40 and substituting into Equation 8-39 yields

$$LC \frac{d^2 i_L}{dt^2} + \frac{L}{R} \frac{di_L}{dt} + i_L = 0 \qquad (8\text{-}41)$$

the desired second-order differential equation. If we compare it with the LC (lossless) case in the previous section, we recognize the additional term $(L/R)\,di_L/dt$ here; in fact, with $R \rightarrow \infty$ (open circuited), we get Equation 8-4 as a special case of Equation 8-41. In the mechanical system of Figure 8-2(b), the added resistive term is analogous to the friction of a dashpot. The entire system would then model, for example, a shock absorber of a car: mass, spring and dashpot.

Prob. 8-8

Before we get into the solution of Equation 8-41, a word of caution and relief: Many steps will look quite complicated, because we will be working with letters (R, L, C) and not with numbers. There is no need to panic, though. These steps are quite logical and follow exactly the general pattern established so far. Also, with another sigh of relief, we don't have to memorize any of the final results. When we reach some conclusions, we'll draw them. Later, as each problem presents itself with numbers, not with letters, it will be easier to solve it *directly* on its own, rather than try to memorize previous formulas.

With these remarks in mind, let us proceed. The characteristic equation for the circuit is, after division by $LC = 1/\omega_0^2$,

$$s^2 + \frac{1}{RC} s + \omega_0^2 = 0 \qquad (8\text{-}42)$$

and this quadratic equation has two roots,

$$s_{1,2} = -\frac{1}{2RC} \pm \sqrt{\left(\frac{1}{2RC}\right)^2 - \omega_0^2} \tag{8-43}$$

the two natural frequencies (characteristic values) of the circuit. Their values will depend, in particular, on the square root in Equation 8-43. In passing, we notice that with $R \to \infty$, an open circuit, the characteristic equation and its two roots are reduced to the previous lossless LC case.

Let us define the *critical resistance*, R_{cr}, as the value of R which makes the square root in Equation 8-43 vanish; that is,

$$\left(\frac{1}{2R_{cr}C}\right)^2 - \frac{1}{LC} = 0 \tag{8-44}$$

or

$$R_{cr} = \frac{1}{2}\sqrt{\frac{L}{C}} \tag{8-45}$$

The *damping ratio* ζ is defined as the ratio of the critical resistance to the actual resistance

$$\zeta = \frac{R_{cr}}{R} \tag{8-46}$$

With these terms, the characteristic equation becomes

$$s^2 + 2\zeta\omega_0 s + \omega_0^2 = 0 \tag{8-47}$$

and the two characteristic values are

$$s_{1,2} = -\zeta\omega_0 \pm \omega_0\sqrt{\zeta^2 - 1} \tag{8-48}$$

We have three cases to consider:

1. $\zeta > 1$, or $R < R_{cr}$, the characteristic values are real and distinct.
2. $\zeta = 1$, or $R = R_{cr}$, the characteristic values are real and equal.
3. $\zeta < 1$, or $R > R_{cr}$, the characteristic values are complex and conjugate of each other, $s_2 = s_1^*$.

Let us stress that no memorization is required. The derivation using R_{cr}, ζ, and ω_0 was done primarily for two reasons: (1) these are commonly used engineering variables, and (2) the final results, Equations 8-47 and 8-48, are in their so-called standard forms. However, in each problem we should obtain our answers directly from scratch, as illustrated in the following examples.

Prob. 8-13

EXAMPLE 8-2 _____

In the parallel RLC circuit we are given†

$$R = 20\,\Omega \qquad L = 1\,\text{H} \qquad C = \tfrac{1}{25}\,\text{F}$$

†Remember that these are *scaled* values. See Appendix C.

Solution. The characteristic equation, quickly rederived, is Equation 8-42, which becomes here

$$s^2 + 1.25s + 25 = 0$$

Therefore, the two characteristic values are

$$s_{1,2} = -0.63 \pm j4.96$$

and we have case 3. Now verify formally, by using Equations 8-45 and 8-46,

$$R_{cr} = \frac{1}{2}\sqrt{\frac{L}{C}} = \frac{1}{2}\sqrt{25} = 2.5\,\Omega$$

$$\zeta = \frac{R_{cr}}{R} = \frac{2.5}{20} = 0.125 \qquad \zeta < 1 \qquad \blacksquare$$

EXAMPLE 8-3 ———————————————————————————————

Let now

$$R = 2.5\,\Omega \qquad L = 1\,\text{H} \qquad C = \tfrac{1}{25}\,\text{F}$$

and let us solve this problem on its own, without any memorization.

Solution. For KCL

$$i_R + i_L + i_C = 0$$

we have here

$$i_R = \frac{1}{2.5}\,v_C \qquad i_C = \frac{1}{25}\frac{dv_C}{dt}$$

and KVL is

$$v_C = v_L = 1\frac{di_L}{dt}$$

Therefore

$$\frac{1}{2.5}\left(1\frac{di_L}{dt}\right) + i_L + \frac{1}{25}\frac{d^2 i_L}{dt^2} = 0$$

For which, after multiplication by 25, we have the characteristic equation

$$s^2 + 10s + 25 = 0$$

or

$$(s + 5)^2 = 0$$

$$\therefore \quad s_1 = s_2 = -5$$

Evidently, this is case 2. Now confirm using Equations 8-45, 8-46, and 8-48

$$R_{cr} = \frac{1}{2}\sqrt{\frac{L}{C}} = \frac{1}{2}\sqrt{25} = 2.5\,\Omega = R$$

$$\zeta = \frac{R_{cr}}{R} = 1$$

$$\omega_0 = \frac{1}{\sqrt{LC}} = 5$$

$$s_{1,2} = -5, -5 \qquad\qquad\qquad \blacksquare$$

EXAMPLE 8-4 ────────────────────────────────

Let now

$$R = 2.44\,\Omega \qquad L = 1\,\text{H} \qquad C = \tfrac{1}{25}\,\text{F}$$

Solution. Here we get the characteristic equation

$$s^2 + 10.25s + 25 = 0$$

with the natural frequencies

$$s_1 = -4 \qquad s_2 = -6.25$$
$$\zeta = 1.025 \qquad \zeta > 1$$

clearly case 1. You should verify these results in full detail.

Probs. 8-9,
8-10, 8-11,
8-12

■

Now we are ready to write the solution to our problem, the zero-input response $i_L(t)$ for each case:

Case 1. When both characteristic values are real and distinct, we notice from Equation 8-48 or 8-43 that they are *real* and *negative*, because $\sqrt{\zeta^2 - 1} < \zeta$ when $\zeta > 1$. Then

$$s_1 = -\zeta\omega_0 + \omega_0\sqrt{\zeta^2 - 1} = -\alpha \qquad (8\text{-}49)$$

$$s_2 = -\zeta\omega_0 - \omega_0\sqrt{\zeta^2 - 1} = -\beta \qquad (8\text{-}50)$$

where α and β are positive numbers. Therefore, our solution is

$$i_L(t) = K_1 e^{-\alpha t} + K_2 e^{-\beta t} \qquad (8\text{-}51)$$

and is the sum of two decaying exponentials. This waveform is called *overdamped*.

EXAMPLE 8-5 ────────────────────────────────

In Example 8-4, let the two initial conditions be given as

$$i_L(0^-) = -1 \qquad v_C(0^-) = 0$$

Calculate and plot $i_L(t)$.

Solution. We have here

$$i_L(t) = K_1 e^{-6.25t} + K_2 e^{-4t}$$

The first initial condition yields

$$-1 = K_1 + K_2$$

For the second one, we write from Equation 8-40

$$1\left(\frac{di_L}{dt}\right)_{0^+} = v_C(0^+) = v_C(0^-) = 0 = (-6.25K_1 e^{-6.25t} - 4K_2 e^{-4t})_{0^+}$$

The two equations for K_1 and K_2 are therefore

$$K_1 + K_2 = -1$$
$$6.25K_1 + 4K_2 = 0$$

Their solution is

$$K_1 = 1.77 \qquad K_2 = -2.77$$

and

$$i_L(t) = 1.77e^{-6.25t} - 2.77e^{-4t} \qquad t > 0$$

In Figure 8-5 we show in dotted lines the two separate exponentials. The plot of $i_L(t)$ is their sum, shown in a solid line.

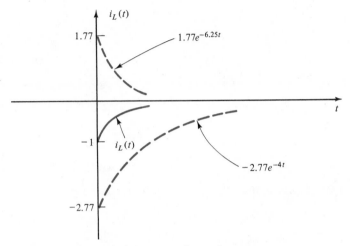

Figure 8-5   Example 8-5, an overdamped response.   ■

Case 2.   Here there are two equal, negative real characteristic values

$$s_1 = s_2 = -\zeta\omega_0 = -\omega_0 \tag{8-52}$$

where $\zeta = 1$. The solution is therefore

$$i_L(t) = K_3 e^{-\omega_0 t} + K_4 t e^{-\omega_0 t} \tag{8-53}$$

where the second term requires multiplication by t because of the repeated root, as we learned in Chapter 7. Just in case we forget to multiply by t, the result would be $i_L(t) = K_3 e^{-\omega_0 t} + K_4 e^{-\omega_0 t} = K_5 e^{-\omega_0 t}$, an obvious mistake since we need *two* arbitrary constants. So, we can't go very far with the mistake; it forces us to stop.

The two constants K_3 and K_4 are evaluated from the given initial conditions.

EXAMPLE 8-6 _____

In Example 8-3, let $i_L(0^+) = 0$ and $i'_L(0^+) = 6$ be given or precalculated.

Solution.   Then the first initial condition yields

$$0 = K_3 + K_4 \cdot 0 \qquad \therefore \quad K_3 = 0$$

and the second one

$$i'_L(0^+) = [K_4(-\omega_0 t e^{-\omega_0 t} + e^{-\omega_0 t})]_{0^+}$$

or

$$6 = K_4$$

and so

$$i_L(t) = 6te^{-5t} \qquad t > 0$$

The plot of $i_L(t)$ is shown in Figure 8-6. The waveform of $6te^{-5t}$ begins to rise, reaches a maximum (verify the values shown), and then decays. This behavior is typical to the term $K_4 te^{-\omega_0 t}$ in the critically damped case. With differential initial conditions, when $K_3 \neq 0$, the decaying exponential term $K_3 e^{-\omega_0 t}$ must be added to the plot.

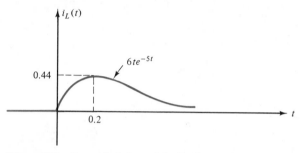

Figure 8-6   Example 8-6, a critically damped response.   ■ Prob. 8-14

Case 3.   Here, the two natural frequencies are complex conjugate

$$s_1 = -\zeta\omega_0 + j\omega_0\sqrt{1 - \zeta^2}$$
$$s_2 = s_1^* = -\zeta\omega_0 - j\omega_0\sqrt{1 - \zeta^2}$$

For convenience, let us introduce in this case the new variable

$$\omega_d = \omega_0\sqrt{1 - \zeta^2} \qquad\qquad (8\text{-}54)$$

Then the solution is

$$i_L(t) = K_7 e^{(-\zeta\omega_0 + j\omega_d)t} + K_8 e^{(-\zeta\omega_0 - j\omega_d)t}$$
$$= e^{-\zeta\omega_0 t}(K_7 e^{j\omega_d t} + K_8 e^{-j\omega_d t}) \qquad\qquad (8\text{-}55)$$

As in the *LC* case, we use Euler's identity for $e^{\pm j\theta}$ and obtain

$$i_L(t) = Ke^{-\zeta\omega_0 t} \sin(\omega_d t + \theta) \qquad\qquad (8\text{-}56)$$

where you should fill the missing steps. The two arbitrary constants are K and θ.

The waveform of $i_L(t)$ in Equation 8-56 is a sinusoid, oscillating at the *damped* radian frequency ω_d given by Equation 8-54. The subscript d reminds us of this fact. The amplitude of the oscillations is exponentially decaying, $Ke^{-\zeta\omega_0 t}$. This waveform is called *underdamped*.

EXAMPLE 8-7 _____

In Example 8-2, let $i_L(0^+) = 0$ and $i_L''(0^+) = -1$.

Solution.   Factor out $e^{-0.63t}$

$$i_L(t) = e^{-0.63t}(K_7 e^{j4.96t} + K_8 e^{-j4.96t})$$
$$= e^{-0.63t}(\hat{K}_7 \cos 4.96t + \hat{K}_8 \sin 4.96t)$$

If we let

$$\hat{K}_7 = K \sin \theta$$
$$\hat{K}_8 = K \cos \theta$$

then

$$\hat{K}_7 \cos 4.96t + \hat{K}_8 \sin 4.96t = K \sin (4.96t + \theta)$$

due to the trigonometric identity

$$\sin (x + y) = \sin x \cos y + \cos x \sin y$$

Consequently

$$i_L(t) = Ke^{-0.63t} \sin (4.96t + \theta)$$

To verify formally, use now Equation 8-54. In Example 8-2, we had

$$\zeta = 0.125$$

then

$$\omega_d = \omega_0 \sqrt{1 - \zeta^2} = (5)(0.992) = 4.96$$

and Equation 8-55 follows. Once more, we stress that memorization is not needed; the answer

$$i_L(t) = e^{-0.63t}(\hat{K}_7 \cos 4.96t + \hat{K}_8 \sin 4.96t)$$

is obtained directly from the problem, and is perfectly correct. The equivalent form, Equation 8-56, is somewhat easier to visualize and to plot.

To evaluate K and θ, use the initial conditions:

$$i_L(0^+) = 0 = K \sin \theta$$

and

$$i'_L(0^+) = K[-0.63e^{-0.63t} \sin (4.96t + \theta) + 4.96e^{-0.63t} \cos (4.96t + \theta)]_{0^+} = -1$$

To illustrate our approach without memorization, let us start with

$$i_L(t) = K_1 e^{s_1 t} + K_2 e^{s_2 t}$$

which is appropriate here. Differentiation yields

$$\frac{di_L}{dt} = K_1 s_1 e^{s_1 t} + K_2 s_2 e^{s_2 t}$$

For $t = 0^+$, these two equations became

$$K_1 + K_2 = 0$$
$$K_1 s_1 + K_2 s_2 = -1$$

We are leaving s_1 and s_2 in letter notation, rather than putting here $s_1 = -0.63 + j4.96$, $s_2 = -0.63 - j4.96$ and risking all kinds of numerical errors in the preliminary steps. The solution of the two constants is

$$K_1 = \frac{1}{s_2 - s_1} = \frac{1}{-j9.92} = j0.101$$

$$K_2 = \frac{1}{s_1 - s_2} = \frac{1}{j9.92} = -j0.101$$

where we've used $1/(-j) = j$ and $1/j = -j$. We also note that $K_2 = K_1^*$ in this example.

Now the solution is

$$i_L(t) = j0.101e^{(-0.63+j4.96)t} - j0.101e^{(-0.63-j4.96)t}$$

Factor $0.101je^{-0.63t}$ to get

$$i_L(t) = 0.101e^{-0.63t}j(e^{j4.96t} - e^{-j4.96t})$$

Use Euler's identity

$$\sin x = \frac{e^{jx} - e^{-jx}}{2j}$$

to get finally

$$i_L(t) = -0.202e^{-0.63t}\sin 4.96t \qquad t > 0$$

The plot of $i_L(t)$ is done best by first plotting separately $(-\sin 4.96t)$ and $(0.202e^{-0.63t})$, then multiplying the two curves point by point. See Figure 8-7. It is a damped sinusoid, oscillating at $\omega_d = 4.96$ rad/s within the envelope of $=0.202e^{-0.63t}$.

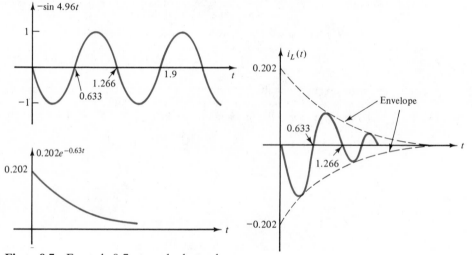

Figure 8-7   Example 8-7, an underdamped response.   ■   Prob. 8-15

These three cases are summarized as follows:

1. The overdamped response has two natural frequencies which are real, negative, distinct numbers. The waveform is the sum of two decaying exponentials.
2. The critically damped response has two real, negative, and equal natural frequencies. The waveform is the sum of a decaying exponential and of the same exponential waveform multiplied by t.
3. The underdamped response has two natural frequencies, complex and conjugate. The waveform is a sinusoid with a radian frequency equal to the imaginary part of the natural frequency; the amplitude decays exponentially with the real part of the natural frequency.

Prob. 8-16

To conclude this section, let us consider the following example.

EXAMPLE 8-8 _____

In the network of Figure 8-8(a), the switch, having been in position *a* for a very long time, is moved from *a* to *b* at $t = 0$. Calculate $v_C(t)$ for $t > 0$.

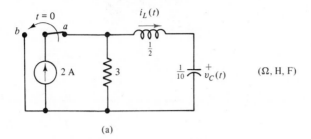

(a)

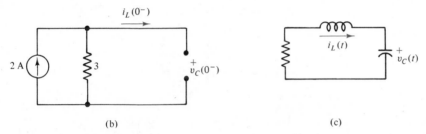

(b)   (c)

Figure 8-8 Example 8-8.

Solution. First, we must establish $v_C(0^-)$ and $i_L(0^-)$. Since the current source, for $-\infty < t < 0$, is constant, the inductor will behave, after that long a time, as a short circuit (s.c.) and the capacitor as an open circuit (o.c.). Refer to Chapter 7 and Problem 7-36 to review these ideas. Briefly, if a steady-state, constant current flows in *L*, then $L\, di/dt = 0$; with a steady-state, constant voltage across *C*, $C\, dv/dt = 0$. The equivalent resistive network, at $t = 0^-$, is therefore as shown in Figure 8-8(b); from it we see that

$$i_L(0^-) = 0 \qquad v_C(0^-) = 6$$

Since there are no all-capacitive loops or all-inductive nodes, these are also the initial conditions at $t = 0^+$

$$i_L(0^+) = 0 \qquad v_C(0^+) = 6$$

for the network after $t = 0$ in Figure 8-8(c). Using the state variables $v_C(t)$ and $i_L(t)$, we write for this network KCL

$$0.1 \frac{dv_C(t)}{dt} = i_L(t) \qquad t > 0$$

and KVL

$$\frac{1}{2} \frac{di_L(t)}{dt} + v_C(t) + 3i_L(t) = 0 \qquad t > 0$$

These equations reduce, as usual, to

$$\frac{1}{2}\left(0.1\,\frac{d^2 v_C}{dt^2}\right) + 3\left(0.1\,\frac{dv_C}{dt}\right) + v_C = 0$$

or

$$\frac{d^2 v_C(t)}{dt^2} + 6\,\frac{dv_C(t)}{dt} + 20 v_C(t) = 0$$

Do these steps in detail! The characteristic equation is

$$s^2 + 6s + 20 = 0$$

and the characteristic values are

$$s_{1,2} = -3 \pm j3.22$$

Therefore

$$v_C(t) = Ke^{-3t} \sin(3.32t + \theta)$$

To find $i_L(t)$, we write

$$i_L(t) = 0.1\,\frac{\ddot{v}_C}{dt} = 0.1\,Ke^{-3t}[3.32\cos(3.32t + \theta) - 3\sin(3.32t + \theta)]$$

Initial conditions at $t = 0^+$ yield

$$v_C(0^+) = K \sin\theta = 6$$
$$i_L(0^+) = 3.32\cos\theta - 3\sin\theta = 0$$

From the second equation, we get

$$\frac{\sin\theta}{\cos\theta} = \tan\theta = \frac{3.32}{3} \qquad \therefore \quad \theta = 47.9^0$$

The first equation then yields

$$K = \frac{6}{\sin\theta} = 8.1$$

Therefore

$$v_C(t) = 8.1e^{-3t}\sin(3.32t + 47.9^0) \qquad t > 0$$

Probs. 8-17
8-18, 8-19,
■ 8-20

8-5 THE STEP RESPONSE OF AN *RLC* SERIES CIRCUIT

We consider now the zero-state response of an *RLC* circuit to a unit step input. As before, it is an easy experiment to set up in the laboratory. Its results, experimentally and theoretically, are important.

A series *RLC* circuit, excited by a voltage source $v_s(t) = u(t)$, is shown in Figure 8-9(a). The two first-order differential equations in $i_L(t)$ and $v_C(t)$ for $t > 0$ are

$$C\,\frac{dv_C}{dt} = i_L(t) \qquad (\text{KCL}) \tag{8-57}$$

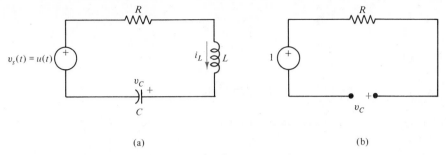

(a)   (b)

Figure 8-9 Step response of an *RLC* circuit.

and

$$L\frac{di}{dt} + Ri_L(t) + v_C(t) = 1 \qquad \text{(KVL)} \tag{8-58}$$

The difference between this case and its zero input is only the right-hand side of Equation 8-58, as expected. Differentiation of Equation 8-57 and substitution into Equation 8-58 give

$$LC\frac{d^2v_C}{dt^2} + RC\frac{dv_C}{dt} + v_C = 1 \tag{8-59}$$

the second-order, nonhomogeneous differential equation for $v_C(t)$.

The homogeneous (zero-input) solution is

$$v_H(t) = K_1 e^{s_1 t} + K_2 e^{s_2 t} \tag{8-60}$$

for the overdamped and the underdamped cases; it is

$$v_H(t) = K_3 e^{s_1 t} + K_4 t e^{s_1 t} \tag{8-61}$$

for the critically damped case. Here s_1 and s_2 are the natural frequencies, and the characteristic equation is, from Equation 8-59

$$s^2 + \frac{R}{L}s + \frac{1}{LC} = 0 \tag{8-62}$$

The particular solution, chosen as a generalization of the source on the right-hand side of Equation 8-59, is a constant

$$v_P(t) = A \tag{8-63}$$

Substitution in Equation 8-59 yields

$$LC\frac{d^2(A)}{dt^2} + RC\frac{d(A)}{dt} + A = 1 \tag{8-64}$$

that is,

$$A = 1 \tag{8-65}$$

The total solution is

$$v_C(t) = v_H(t) + v_P(t) = K_1 e^{s_1 t} + K_2 e^{s_2 t} + 1 \tag{8-66}$$

for the overdamped and underdamped cases. The evaluation of K_1 and K_2 is done now (not earlier). In this circuit, $v_C(0^+) = v_C(0^-) = 0$ and $i_L(0^+) = i_L(0^-) = 0$. Therefore

$$v_C(0^+) = K_1 + K_2 + 1 = 0 \tag{8-67}$$

and

$$i_L(0^+) = C v_C'(0^+) = C(K_1 s_1 + K_2 s_2) = 0 \tag{8-68}$$

These two equations provide the numerical values for K_1 and K_2. With these, Equation 8-66 is the final answer.

It is important to remember that no memorization is required (read over this last sentence: It is not self-contradictory!). The steps are standard, and they should be followed methodically for each problem with its specific numerical values.

In the zero-state response of Equation 8-66, we recognize the transient part

$$v_{tr}(t) = K_1 e^{s_1 t} + K_2 e^{s_2 t} \tag{8-69}$$

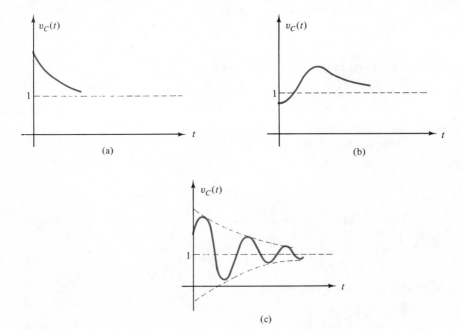

Figure 8-10   Typical plots of the zero-state step response.

because it approaches zero as $t \to \infty$ in all three cases (over-, under-, and critically damped). The steady-state part is nonzero as $t \to \infty$

$$v_{ss}(t) = 1 \qquad (8\text{-}70)$$

This part is easy to recognize by our previous method: For a dc (constant) input, an inductor becomes a short circuit as $t \to \infty$ and a capacitor an open circuit. See Figure 8-9(b). From this figure, we see immediately that $v_C(\infty) = 1$. Also, as a bonus, we see that $i_L(\infty) = 0$. These two values may be used as $v_C(0^-)$ and $i_L(0^-)$ for a new circuit which, at a new $t = 0$, includes the capacitor and the inductor in a new configuration.

Probs. 8-20 8-21, 8-22, 8-23, 8-24, 8-25

Typical plots of the zero-state step response are shown in Figure 8-10 for: (a) the overdamped, (b) critically damped, and (c) underdamped cases.

8-6 HIGHER-ORDER NETWORKS: LOOP, NODE, AND STATE EQUATIONS

Having solved in full detail first- and second-order networks, we face now the most general problem in network analysis: A given network consists of linear, constant resistors, capacitors, inductors (including mutuals), independent and dependent sources; also given, or precalculated, are initial conditions. At $t = 0$, it is excited by one or more sources of various waveforms. We must calculate the total response, that is, some or all of the currents and voltages of the elements.

From our studies of graphs and of resistive networks (Chapters 2 and 3), we know how to choose intelligently loop analysis or node analysis. In both cases, the resulting simultaneous equations for *RLC* circuits will contain derivatives, $i_C = C \, dv/dt$, $v_L = L \, di/dt$, as well as integrals, $v_C = (1/C) \int_0^t v(x) \, dx + v_C(0^-)$, $i_L = (1/L) \int_0^t i(x) \, dx + i_L(0^-)$. Therefore, loop analysis or node analysis will lead to *simultaneous integro-differential equations*. The direct solution of such equations is an extremely difficult task. It is somewhat comforting to be able to predict by inspection the order n of the network. This would be the order of the single differential equation which is equivalent to all the simultaneous equations, but it is much more difficult to actually *obtain* this equation! Just recall the relative difficulty of deriving it for second-order networks.

With dynamic elements (L, M, C), we can also formulate state equations using capacitive voltages and inductive currents as variables. These equations, as we saw, will contain only the unknowns and their first derivatives, but no integrals. This is already a significant relief; however, we still have n simultaneous first-order differential equations. Again, these are equivalent to a single nth-order differential equation. We know n in advance, but to obtain that equation is another matter.

Does all this sound hopeless? Definitely not. Rest assured that, before too long, we'll learn some powerful and relatively simple methods for solving such simultaneous equations. As an added promise, these methods will reduce the integro-differential equations to algebraic ones. Then we'll be able to handle them easily, just as in the all-resistive cases. For the time being, let us concentrate just on the correct *formulation* of these equations.

Loop Equations.   For these, we must use

$$v_R(t) = Ri_R(t)$$

$$v_L(t) = L\frac{di_L(t)}{dt} \tag{8-71}$$

$$v_C(t) = v_C(0^-) + \frac{1}{C}\int_{0^-}^{t} i_C(x)\, dx$$

where each branch current, i_R, i_L or i_C, is expressed in terms of the chosen loop currents. Before studying an example, it is well worth our while to review the derivation of the expressions in Equation 8-71.

There is no problem with $v_R(t) = Ri_R(t)$; this *is* the defining *v-i* relationship for the resistor. For an inductor, the defining relation is $\phi(t) = Li(t)$, flux being proportional to current. Then, since $v_L(t) = d\phi/dt$, we have $v_L(t) = L\, di_L(t)/dt$.

The *i-v* expression for the capacitor is obtained from its defining relation

$$q_C(x) = Cv_C(x) \tag{8-72}$$

Here we are using x as the variable for time. There should be no disagreement about that; *any* letter can be used, not just t. See Figure 8-11. Differentiating Equation 8-72 yields

$$i_C(x) = C\frac{dv_C(x)}{dx} \tag{8-73}$$

the *i-v* relationship. We need its inverse, the *v-i* relationship. From Equation 8-73 we write

$$\frac{dv_C(x)}{dx} = \frac{1}{C}i_C(x) \tag{8-74}$$

and then integrate both sides with respect to x ($=$ time), from $x = 0^-$ to *any* later time $x = t$

$$\int_{x=0^-}^{x=t} dv_C(x) = \frac{1}{C}\int_{x=0^-}^{x=t} i_C(x)\, dx \tag{8-75}$$

The left-hand side yields $v_C(t) - v_C(0^-)$, and therefore

$$v_C(t) = v_C(0^-) + \frac{1}{C}\int_{0^-}^{t} i_C(x)\, dx \tag{8-76}$$

which is the desired *v-i* relationship. It is clear now why we needed to use the dummy variable x (or y, or u, or ξ, or any other letter): It is the variable of integration, and t is the upper limit of the integral. The final answer, $v_C(t)$, is a function of t, as expected.

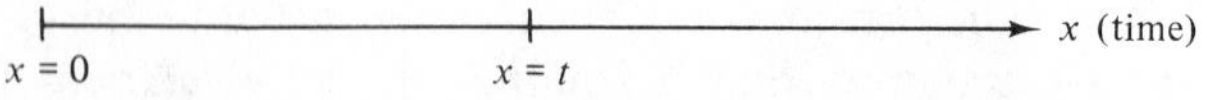

$x = 0$ $x = t$ x (time)

Figure 8-11   Dummy variable for time.

EXAMPLE 8-9 _____

Consider the network shown in Figure 8-12.

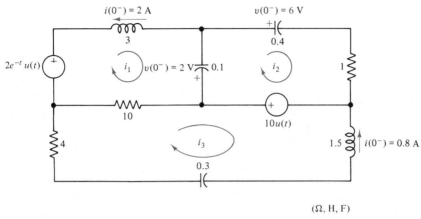

(Ω, H, F)

Figure 8-12 Example 8-9.

Solution. A quick count confirms the advantage of loop analysis (three unknowns) over node analysis (five unknowns). Initial conditions are indicated; where not indicated, assume they are zero. The sources are applied at $t = 0$, as indicated by the multiplying unit step function $u(t)$.

The three loop currents are conveniently chosen as shown $i_1(t)$, $i_2(t)$, and $i_3(t)$. KVL around the first loop reads, with the conventional $(+)$ for a drop and $(-)$ for a rise,

$$3\frac{di_1(t)}{dt} + \left\{\frac{1}{0.1}\int_0^t [i_1(x) - i_2(x)]\, dx - 2\right\} + 10[i_1(t) - i_3(t)] - 2e^{-t}u(t) = 0$$

Let us review each term carefully:

1. The voltage drop across the inductor, in the direction of the loop, is $+3di_1(t)/dt$. The initial current $i_1(0^-) = -2$ does not contribute to this drop. Should you forget what to do, try $L\, di_1(0^-)/dt$ and see what you get! The initial current will be used much later to evaluate arbitrary constants in the solution.
2. The voltage drop across the capacitor, in the direction of the loop, is written according to Equation 8-71. Here $i_C(x) = i_1(x) - i_2(x)$ and $v_C(0^-) = -2$, a rise in the direction of the loop.
3. The voltage drop across the resistor is $10(i_1 - i_3)$ in the first loop.
4. The voltage source is a rise in the first loop.

The second loop equation is

$$\left[6 + \frac{1}{0.4}\int_0^t i_2(x)\, dx\right] + 1i_2(t) - 10u(t) + \left\{2 + \frac{1}{0.1}\int_0^t [i_2(x) - i_1(x)]\, dx\right\} = 0$$

and you should confirm every term here. The third loop equation is

$$10[i_3(t) - i_1(t)] + 10u(t) + 1.5\frac{di_3(t)}{dt} + \frac{1}{0.3}\int_0^t i_3(x)\, dx + 4i_3(t) = 0$$

These are the three simultaneous integro-differential equations for this network

■ Prob. 8-26,

In a second example, let us practice with mutual inductance and with dependent sources.

EXAMPLE 8-10 _____

Solution. For the network shown in Figure 8-13 we write the first loop equation as

$$3i_1(t) + \left(4\frac{di_1}{dt} - 2\frac{di_2}{dt}\right) + \left\{4 + 100\int_0^t [i_1(x) + i_2(x)]\,dx\right\} - 6u(t) = 0$$

The mutually induced term needs a quick review: The inducing current i_2 does *not* enter its own dot; therefore, it induces a voltage $M\,di_2/dt$ which is *negative* at the other dot. In the first loop, therefore, this is a voltage rise.

Around the second loop we write

$$1i_2(t) + \left[3\frac{di_2(t)}{dt} - 2\frac{di_1(t)}{dt}\right] + \left\{4 + 100\int_0^t [i_2(x) + i_1(x)]\,dx\right\} + 2[-i_1(t)] = 0$$

Here the mutually induced term is $M\,di_1/dt$. The inducing current $i_1(t)$ *enters* its own dot; therefore, the induced voltage is *positive* at the other dot. In the second loop, this voltage is a rise.

Collecting terms and using the operators D and $1/D$ as in Problem 8-26, we have the final two integro-differential equations in matrix form

$$\begin{bmatrix} 3 + 4D + \dfrac{100}{D} & -2D + \dfrac{100}{D} \\[2ex] -2 - 2D + \dfrac{100}{D} & 1 + 3D + \dfrac{100}{D} \end{bmatrix} \begin{bmatrix} i_1(t) \\[1ex] i_2(t) \end{bmatrix} = \begin{bmatrix} -4 + 6u(t) \\[1ex] -4 \end{bmatrix}$$

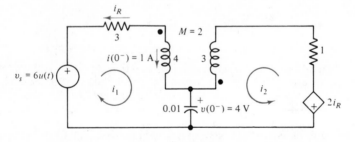

$(\Omega, \text{H}, \text{F})$

Figure 8-13   Example 8-10.

Probs. 8-2
8-28, 8-29

Node Equations.   Here we use the branch *i-v* relationships

$$i_R(t) = Gv_R(t)$$

$$i_C(t) = C\frac{dv_C(t)}{dt} \tag{8-77}$$

and

$$i_L(t) = i_L(0^-) + \frac{1}{L}\int_{0^-}^t v_L(x)\,dx$$

Be sure to review to your full satisfaction the last i-v relation for the inductor. It helps also to remember the duality properties between L and C.

EXAMPLE 8-11 ———————————————————————————————————

For the network shown in Figure 8-14, we need two node equations. By comparison, we need three loop equations; also its order is $n = 3$, so three state variable equations are required.

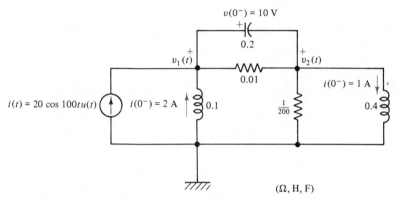

Figure 8-14   Example 8-11.

Solution.   The two node voltages are $v_1(t)$ and $v_2(t)$, as shown. With the convention $(+)$ for currents leaving and $(-)$ for entering, KCL at the first node is

$$-20 \cos 100t \, u(t) + \left[-2 + \frac{1}{0.1} \int_{0^-}^t v_1(x) \, dx\right] + 100[v_1(t) - v_2(t)]$$

$$+ 0.2 \frac{d}{dt}[v_1(t) - v_2(t)] = 0$$

where the inductive term consists of $i(0^-) = 2$ entering the node and $(1/0.1)\int_{0^-}^t v_1(x) \, dx$ *leaving* the node after $t = 0$. That last term is associated with the $(+)$ reference sign assigned to $v_1(t)$, the tail of the current arrow being at the $(+)$ of the voltage.

In a similar way, KCL at the second node reads

$$0.2 \frac{d}{dt}[v_2(t) - v_1(t)] + 100[v_2(t) - v_1(t)] + 200 \, v_2(t) + \left[1 + \frac{1}{0.4}\int_{0^-}^t v_2(x) \, dx\right] = 0$$

Here, the given initial current $i(0^-) = 1$ leaves the node, without any association to the chosen reference sign $(+)$ of v_2. The term $(1/0.4)\int_{0^-}^t v_2(x) \, dx$ is leaving the node because of the choice of reference $(+)$ of v_2. The final form of these two integro-differential equations is

$$\begin{bmatrix} 0.2D + 100 + \dfrac{10}{D} & -0.2D - 100 \\[2ex] -0.2D - 100 & 300 + 0.2D + \dfrac{2.5}{D} \end{bmatrix} \begin{bmatrix} v_1(t) \\ v_2(t) \end{bmatrix} = \begin{bmatrix} 20 \cos 100t \, u(t) + 2 \\ -1 \end{bmatrix}$$

Question: Where does the initial voltage across C enter into these equations? ∎

Such a general approach is also valid with dependent sources. With mutual inductances, node equations are very complicated; instead, we will use loop equations or state equations. Only in Chapter 14 will we learn how to handle mutual inductances in node analysis.

Probs. 8-30
8-31, 8-32

State Equations. We will later devote an entire chapter to the systematic formulation and solution of state equations. However, we have enough preliminary "feel" from our studies of *RL*, *RC*, and *RLC* circuits to study an example.

EXAMPLE 8-12 _____

Consider the network shown in Figure 8-15.

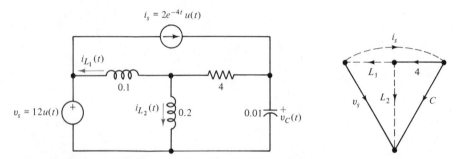

Figure 8-15 Example 8-12.

Solution. By inspection, its order is $n = 3$, and therefore the three unknowns will be $v_C(t)$, $i_{L_1}(t)$, and $i_{L_2}(t)$. Assume their initial values are $v_C(0^-) = -2$ V, $i_{L_1}(0^-) = 1$ A, $i_{L_2}(0^-) = 0$. As in our previous studies (and with further elaboration in the later chapter), we choose the tree and the co-tree as shown.

A fundamental cut set (KCL) equation is written with the capacitor as the only tree branch. It is

$$i_C(t) + i_{L_2}(t) + i_{L_1}(t) - i_s(t) = 0$$

or

$$i_C(t) = 0.01 \frac{dv_C}{dt} = -i_{L_1}(t) - i_{L_2}(t) + 2e^{-4t}u(t)$$

the first differential equation.

A fundamental loop (KVL) equation is written with L_1 as the only link. It is

$$0.1 \frac{di_{L_1}}{dt} + v_s(t) - v_C(t) + v_4(t) = 0$$

But

$$v_4(t) = 4i_4(t) = 4[i_{L_1}(t) + i_{L_2}(t)]$$

Consequently, the second state equation is

$$0.1 \frac{di_{L_1}}{dt} = -4i_{L_1} - 4i_{L_2} + v_C - 12u(t)$$

the second differential equation.

A fundamental loop equation with L_2 as the link yields

$$0.2 \frac{di_{L_2}(t)}{dt} - v_C(t) + 4[i_{L_1}(t) + i_{L_2}(t)] = 0$$

or

$$0.2 \frac{di_{L_2}(t)}{dt} = v_C(t) - 4i_{L_1}(t) - 4i_{L_2}(t)$$

the third differential equation.

After conveniently multiplying the first equation by 100, the second by 10 and the third by 5, we get the three simultaneous first-order differential equations in $v_C(t)$, $i_{L_1}(t)$, and $i_{L_2}(t)$ as follows:

$$\begin{bmatrix} \dfrac{dv_C(t)}{dt} \\[2mm] \dfrac{di_{L_1}(t)}{dt} \\[2mm] \dfrac{di_{L_2}(t)}{dt} \end{bmatrix} = \begin{bmatrix} 0 & -100 & -100 \\ 10 & -40 & -40 \\ 5 & -20 & -20 \end{bmatrix} \begin{bmatrix} v_C(t) \\ i_{L_1}(t) \\ i_{L_2}(t) \end{bmatrix} + \begin{bmatrix} 0 & 100 \\ -10 & 0 \\ 0 & 0 \end{bmatrix} \begin{bmatrix} 12u(t) \\ 2e^{-4t}u(t) \end{bmatrix}$$

This is the standard form of state equations, written as

$$\frac{d}{dt} \mathbf{x}(t) = \mathbf{A}\mathbf{x}(t) + \mathbf{B}\mathbf{e}(t)$$

Here $\mathbf{x}(t)$ is the state vector

$$\mathbf{x}(t) = \begin{bmatrix} v_C(t) \\ i_{L_1}(t) \\ i_{L_2}(t) \end{bmatrix}$$

containing the n unknown state variables. The vector $\mathbf{e}(t)$ is the matrix of the independent sources. Matrices $\mathbf{A}$ and $\mathbf{B}$ are constant matrices resulting from the formulation of these equations.

The solution of these equations will require the use of the given initial state vector, the three initial conditions

$$\mathbf{x}(0^-) = \begin{bmatrix} v_C(0^-) \\ i_{L_1}(0^-) \\ i_{L_2}(0^-) \end{bmatrix}$$

As mentioned, this is only a preliminary example. Since we'll devote an entire chapter to state equations, let us concentrate in the following chapters only on loop analysis and node analysis. ■

Probs. 8-33, 8-34

PROBLEMS

8-1 In Example 8-1, we solved the second-order differential equation

$$\frac{d^2 v_1}{dt^2} + 30 \frac{dv_1}{dt} = 0$$

Solve it by an alternate way, as follows: (a) Integrate it once, to obtain a non-homogeneous first-order differential equation. (b) Solve this equation, remembering to get both the homogeneous and the particular solutions. Verify that your answer agrees with the one in Example 8-1.

8-2 In Example 8-1:

(a) Obtain the differential equation for $v_2(t)$.

(b) Verify that the characteristic equation here is the same as for $v_1(t)$. It should be—after all, the circuit does not care how you obtain it!

(c) Solve completely for $v_2(t)$, having calculated the necessary initial condition $v_2'(0^+) = (dv_2/dt)_{0^+}$.

(d) Plot $v_2(t)$ vs. t and compare with Figure 8-1(c). Does the plot make sense? Compare the steady-state values of $v_1(t)$ and $v_2(t)$. Why are they what they are?

(e) Calculate the current through the resistor in terms of the two state variables $v_1(t)$ and $v_2(t)$.

(f) Calculate the energy balance in this circuit for $t \geq 0$. The initial energy stored in the capacitors must be equal to the energy dissipated in the resistor *plus* the final energy stored in the capacitors.

8-3 Repeat Example 8-1 if:

(a) $v_1(0^-) = v_2(0^-) = 2$ V

(b) $w_{C_1}(0^-) = \frac{1}{2}(1)(2)^2 = 2$ J $= w_{C_2}(0^-)$

8-4 For the LC circuit of Figure 8-2, let $L = 704$ mH, $C = 10$ μF, $v_C(0^-) = 10$ V, $i_L(0^-) = 0$.

(a) Plot accurately $i_L(t)$ vs. t. *Note*: In Figure 8-2(d) the plot is versus $\omega_0 t$. Identify clearly the zero points along the t axis.

(b) Calculate the (natural) frequency f of $i_L(t)$.

(c) Calculate the period T of $i_L(t)$.

(d) Obtain its state matrix differential equation.

8-5 Repeat the solution of the LC circuit, as in Section 8-2, this time with $v_C(0^-) \neq 0$ and $i_L(0^-) \neq 0$. *Hint*: The derivation is the same through, and including, Equation 8-11.

8-6 In the solution of the step response of an LC circuit (Section 8-3), check the dimensions (units) in Equation 8-33: On the left we have coulombs, and so K_1 and K_2 must be also in coulombs. What about C? Capacitance is in farads!

8-7 Calculate the current and the voltage for the inductor in Figure 8-3.

8-8 In the parallel RLC, zero-input circuit of Figure 8-4, use $\phi(t)$, the flux in the inductor, as a single unknown and obtain the second-order differential equation for it. What would be the two initial conditions for this case?

8-9 In the parallel RLC circuit, what corresponds to $\zeta = 0$? How does this justify the name damping factor?

8-10 For a *series RLC* circuit, derive:

(a) The differential equation for the zero-input response.

(b) The characteristic equation and the characteristic values in their standard forms (Equations 8-47 and 8-48).

(c) The expressions for the damping factor ζ and for the critical resistance, R_{cr}. Be careful: These are different from the parallel circuit.

(d) The classification of the characteristic values into three cases.

(e) Repeat (a), using $q_C(t)$ and $\phi_L(t)$ as state variables.

*8-11 Explore fully the values and the location of the two characteristic roots s_1 and $s_2 = s_1^*$ when $\zeta < 1$. Each root, being a complex number, has a real part and an imaginary part

$$s_1 = -\zeta\omega_0 + j\omega_0\sqrt{1 - \zeta^2}$$

$$s_2 = -\zeta\omega_0 - j\omega_0\sqrt{1 - \zeta^2}$$

Let a typical pair of roots be shown on the complex number plane as in the figure, where the real axis is designated as σ and the imaginary axis $j\omega$. Then $s = \sigma + j\omega$ in general. The two roots are shown with the symbol ×. (a) As ζ varies, $0 \le \zeta < 1$, show the location of s_1 and s_2. Prove that their location (locus) is on a semicircle centered at the origin with a radius $r = \omega_0$. (b) Extend your investigation of their locus to the remaining two cases, when $\zeta = 1$ and when $1 < \zeta < \infty$.

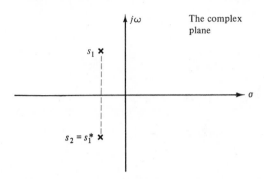

Problem 8-11

8-12 Another engineering term used with the RLC circuit, in addition to ω_0, ζ, and R_{cr}, is the *Q factor*, defined for the parallel RLC circuit as

$$Q = \frac{1}{\omega_0 RC}$$

Derive the relation between Q and ζ. Repeat for the series RLC circuit (Problem 8-10) where

$$Q = \frac{\omega_0 L}{R}$$

8-13 In the parallel RLC circuit, we have $R = 10\ \Omega$, $L = 0.2$ H. Find the range of values for C if the zero-input response is: (a) overdamped, (b) critically damped, (c) underdamped.

8-14 For the critically damped case (Equation 8-53 and Example 8-6) are there initial conditions $i_L(0^+)$ and $i_L'(0^+)$ which will make $K_4 = 0$? Prove your answer.

8-15 In a parallel RLC circuit, we have $R = 100\ \Omega$, $L = 1$ mH, $C = 0.1\ \mu$F. Given $i_L(0^+) = 1$ and $v_C(0^+) = -1$, calculate and plot accurately $v_C(t)$ for $t \ge 0$.

8-16 The *settling time* of a circuit is a convenient measure of how fast the transient voltage or current will decay to zero. For the zero-input RLC circuit, let the settling time be the time needed for the response to decay to 1 percent of its maximum value. Calculate the settling times for Examples 8-5, 8-6, and 8-7. Don't forget to calculate first the maximum value in each case.

8-17 The switch in the circuit shown has been in position a for a very long time. At $t = 0$, it is moved to position b. Calculate and plot $i_R(t)$ for $t \ge 0$.

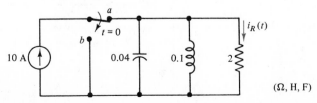

Problem 8-17

8-18 In the circuit shown, the switch has been closed for a long time. At $t = 0$, it is opened. Calculate and plot $i_L(t)$ for $t \geq 0$.

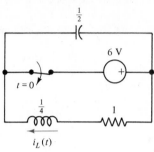

(Ω, H, F)   **Problem 8-18**

8-19 In the network shown, the switch is moved from a to b at $t = 0$, having been in a for a long time. Calculate $v_C(t)$ and $i_L(t)$ for $t \geq 0$.

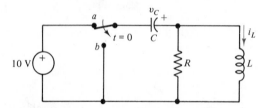

Problem 8-19

*8-20 From our studies of first- and second-order circuits, we recognize that their zero-input response waveforms depend on the natural frequencies. These natural frequencies are complex numbers of the general form

$$s = \sigma + j\omega$$

See Problem 8-11. In the RC and RL first-order circuits, we found that $s = -\alpha$, real and negative ($\omega \equiv 0$). The LC circuit has $s_{1,2} = \pm j\omega_0$, pure imaginary ($\sigma \equiv 0$), and the RLC circuit has either real negative or complex conjugate natural frequencies ($\sigma \neq 0$, $\omega \neq 0$).

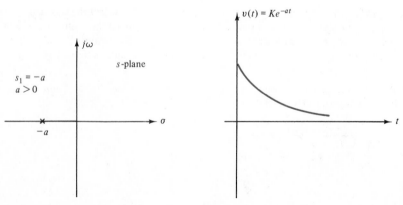

Problem 8-20

Prepare a table showing the waveform plotted vs. t, as related to the location of a natural frequency. The first entry is given as an illustration. Do it for the following cases:

(b)	$s_2 = -b$	where $b > a$	$a > 0, b > 0$
(c)	$s_3 = s_4 = -c$		$c > 0$
(d)	$s_5 = 0$		
(e)	$s_6, s_7 = \pm j\omega_e$		
(f)	$s_8, s_9 = \pm j\omega_f$		$\omega_f > \omega_e$
(g)	$s_{10}, s_{11} = -a \pm j\omega_e$		
(h)	$s_{12}, s_{13} = -b \pm j\omega_e$		
(i)	$s_{14}, s_{15} = -b \pm j\omega_f$		

What effect has the distance of s along the σ axis on the waveform? And the distance along the $j\omega$ axis?

8-21 In the network shown, the switch is opened at $t = 0$, after having been closed for a long time. Calculate and plot $i_L(t)$ for $t \geq 0$.

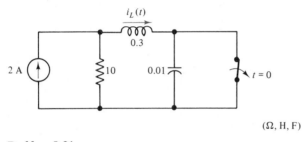

$(\Omega, \text{H}, \text{F})$

Problem 8-21

8-22 In the network shown, the switch is moved from position a to b at $t = 0$, after having been in position a for a long time. Calculate and plot $i_L(t)$ for $t \geq 0$.

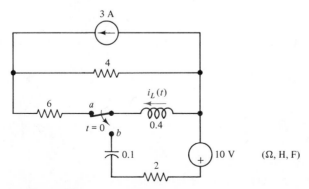

$(\Omega, \text{H}, \text{F})$

Problem 8-22

8-23 The two switches have been in positions *a* for a long time. At $t = 0$ they are moved to positions *b*. Calculate the energy balance in the new circuit for $t \geq 0$, as follows:
(a) The energy stored initially
(b) The total energy dissipated by the 2-Ω resistor

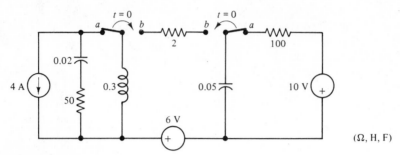

Problem 8-23

8-24 For the circuit shown, find:
(a) The characteristic equation
(b) The characteristic values
(c) The zero-input response $i(t)$ within arbitrary multiplying constants, and its classification as over-, under-, or critically damped

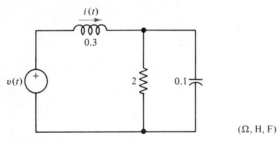

Problem 8-24

8-25 An *RLC* circuit is described by the equation

$$\frac{d^2 i}{dt^2} + 6 \frac{di}{dt} + 9i = 10e^{-t}$$

with $i(0^-) = -1$, $i'(0^-) = 4$.
(a) Calculate $i(t)$ completely. For the particular solution, refer to Table 7-1 for your choice.
(b) Identify the zero-state and the zero-input parts in your solution.
(c) Identify the transient part and the steady-state part in your solution.

8-26 In mathematical notation, there are two *operators* defined as

$$D \equiv \frac{d}{dt} \qquad \text{the differentiation operator}$$

and

$$\frac{1}{D} \equiv \int_{0^-}^{t} (\ ) \, dx \qquad \text{the integration operator}$$

So, for example, we write

$$L\frac{di(t)}{dt} = LDi(t)$$

and

$$\frac{1}{C}\int_{0^-}^{t} i(x)\, dx = \frac{1}{C}\frac{1}{D}i(t)$$

Use this notation to rewrite the final three loop equations in Example 8-9 in matrix form as

$$\mathbf{M}\begin{bmatrix} i_1(t) \\ i_2(t) \\ i_3(t) \end{bmatrix} = \mathbf{e}$$

where **M** contains the resistors, inductors, capacitors, D, and $1/D$. The column matrix **e** contains voltage sources and capacitive initial voltages.

8-27 For the network shown in the figure, write the loop equations in their final matrix form, as in Problem 8-26.

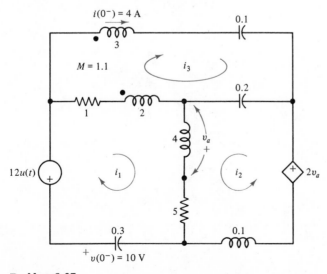

Problem 8-27

8-28 Write two fundamental loop equations for the network shown for the indicated graph.

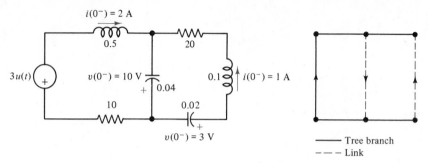

$(\Omega, \text{H}, \text{F})$

Problem 8-28

8-29 For the network shown, write fundamental loop equations. A chosen tree is shown, as well as the fundamental loops. All initial conditions are zero. Be very careful when you express each branch current in terms of the loop currents.

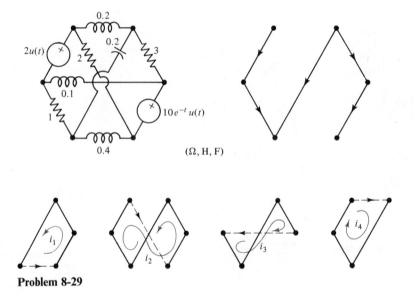

Problem 8-29

8-30 For the network shown, write two node equations in their final matrix form.

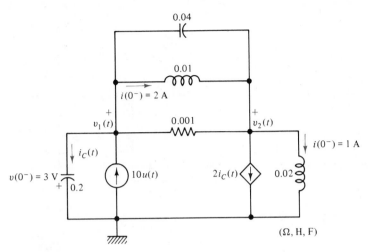

Problem 8-30

8-31 For the network shown, choose intelligently between loop and node equations. Write these equations.

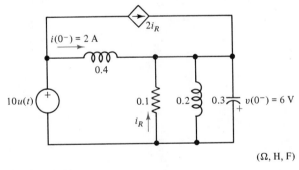

Problem 8-31

8-32 Write two node equations for the network shown. What is the form of the single differential equation for $v_1(t)$, if we could obtain it?

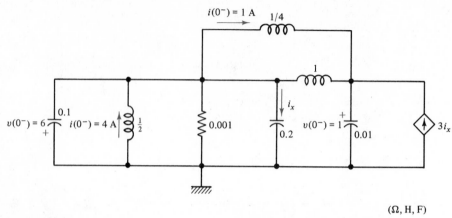

(Ω, H, F)

Problem 8-32

8-33 Write, as explicitly as you can by inspection, the single equivalent differential equation for $v_C(t)$ in Example 8-12.

8-34 Formulate the state equations for the network in Problem 8-31. Use the approach suggested in Example 8-12. For the tree, use the voltage source and the capacitor.

Chapter 9

Sinusoidal Steady-State Analysis

9-1 SINUSOIDAL SOURCES

In the previous two chapters, we learned about several responses of RLC circuits. In particular, we studied the case of a sinusoidal zero-input response in an LC circuit. Please review Section 8-2 to refresh your memory.

Sinusoidal sources and responses are very common in many circuits. For example, the household supply voltage in the United States is sinusoidal and is expressed as

$$v(t) = V_m \sin \omega t \qquad (9\text{-}1)$$

See Figure 9-1. Here, the amplitude of $v(t)$ is $V_m = 170$ V and the radian frequency is $\omega = 377$ rad/s, as studied in Chapter 8. We will review these concepts and expand our study in this chapter.

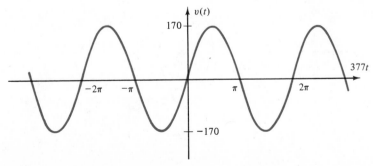

Figure 9-1 Plot of $v(t) = 170 \sin 377t$.

Before dealing with more general expressions and plots of sinusoidal waveforms, let us agree to deal only with the cosine function, rather than with both the cosine and the sine. We can do it because of the trigonometric identity

$$\sin x = \cos\left(x - \frac{\pi}{2}\right) \tag{9-2}$$

With this easy convention, we eliminate a lot of worry and extra work: Any sine function will be converted to a cosine, as in Equation 9-2. Later, we'll see other advantages of this convention. (Of course, by a similar convention, we could have decided to use only the sine function.)

Prob. 9-1

The plot of the function $f(t) = A_m \cos \omega t$ is shown in Figure 9-2(a), plotted versus ωt. In Figure 9-2(b) we see the same plot versus t. The difference between the two abscissas is merely a scaling factor. If we multiply t, in seconds, by ω, we get ωt in radians; points along ωt are angles, for example, $0, \pi/4, \pi, 2\pi, \ldots$. Conversely, if ωt is divided (scaled) by ω, the abscissa is in seconds, and the corresponding points are 0, $\pi/4\omega, \pi/\omega, 2\pi/\omega, \ldots$.

A *cycle* of a *periodic* (repetitive) function, you recall, is the smallest portion that repeats itself to produce the original function. The time duration of one cycle is called the *period* and is designated by T (in seconds). In Figure 9-2 we show one cycle and its period. We see that

$$\omega T = 2\pi \tag{9-3}$$

The *frequency* of the periodic waveform is the number of cycles per second (cps) and is

$$f = \frac{1}{T} \quad \text{Hz} \tag{9-4}$$

where the unit, hertz, is named after the German scientist Heinrich Hertz (1857–1894). For example, the U.S. residential voltage waveform is a sinusoid with $f = 60$ Hz, meaning that in 1 second the sinusoid goes through 60 cycles. From Equations 9-3 and 9-4 we obtain another useful relationship

$$\omega = \frac{2\pi}{T} = 2\pi f \tag{9-5}$$

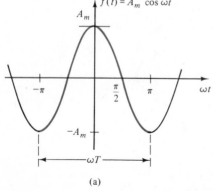
(a)

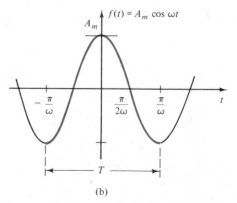
(b)

Figure 9-2   Plots of a cosine waveform.

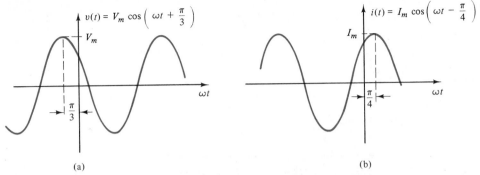

Figure 9-3 Sinusoidal waveforms with phase angles. (a) Positive θ. (b) Negative θ.

For example, the radian frequency of the household voltage is $\omega = (2\pi)(60) = 377$ rad/s as shown in Figure 9-1.

Other examples of sinusoidal frequencies are provided by commercial radio. Stations broadcasting on the AM band have a range of 540 to 1600 kHz (recall: 1 kHz $= 10^3$ Hz), while those broadcasting on the FM band have frequencies in the range of 88 to 108 MHz, where 1 MHz $= 10^6$ Hz. Probs. 9-2, 9-3, 9-4

The most general sinusoidal waveform is given by an extension of Figure 9-2. That is,

$$g(t) = A_m \cos(\omega t + \theta) \tag{9-6}$$

where $g(t)$ can be either a voltage or a current, and θ is the initial *phase angle* (in radians). In Figure 9-3 we show two plots: (a) when θ_1 is positive, $\theta_1 = \pi/3$, and (b) when θ_2 is negative, $\theta_2 = -\pi/4$. We see that, in the first case, the cosine waveform is shifted to the left by θ_1 radians along the ωt axis, and in the second case—to the right by θ_2 radians.

It is common in engineering practice to use degrees. Therefore, expressions like $v(t) = V_m \cos(\omega t + 60°)$, $i(t) = I_m \cos(\omega t - 45°)$ are acceptable and widely used, even though not strictly consistent in units.

If we compare $f(t) = A_m \cos \omega t$ in Figure 9-2(b) with $v(t) = V_m \cos(\omega t + 60°)$ in Figure 9-3(a), we see that a point on $v(t)$ is reached in time *ahead of* the corresponding point of $f(t)$. We say that $v(t)$ *leads* $f(t)$ by 60°. Similarly, $i(t)$ of Figure 9-3(b) *lags* (is behind) $f(t)$ by 45°, because a point on $i(t)$ arrives in time later than the corresponding point of $f(t)$. In Figure 9-3, $v(t)$ leads $i(t)$ by $60° - (-45°) = 105°$.

Let us pause for a summary:

1. A sinusoidal waveform, expressed as in Equation 9-6, is completely characterized by three parameters: Its amplitude, A_m, its radian frequency ω, and its phase angle θ.
2. Of two sinusoidal waveforms of the same frequency, one leads the other or lags behind it, depending on their phase angles. If both have the same phase angles, the waveforms are *in phase*. See Figure 9-4. Probs. 9-6, 9-7, 9-8

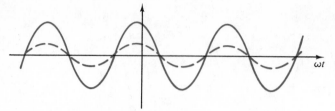

Figure 9-4   Two sinusoids in phase.

EXAMPLE 9-1 _____

In Figure 9-5(a) we have two sinusoidal waveforms $v_1(t)$ and $v_2(t)$. Write the cosine expressions for each one. Which one leads the other, and by what angle?

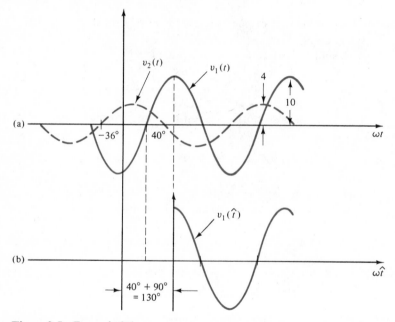

Figure 9-5   Example 9-1.

Solution.   Without any memorization or reference to Figure 9-3, let us redraw $v_1(t)$ as a cosine waveform without a phase angle. This is shown in Figure 9-5(b). Its expression is

$$v_1(\hat{t}) = 10 \cos \omega\hat{t}$$

where $\omega\hat{t}$ is the temporary abscissa. (Yes, we *do* have to remember the basic waveform of $\cos x$.) The relationship between the two abscissas is obviously

$$\omega\hat{t} = \omega t - 130°$$

Therefore

$$v_1(t) = 10 \cos (\omega t - 130°)$$

In a similar way we find

$$v_2(t) = 4 \cos (\omega t - 54°)$$

From the figure, or from these expressions, we see that $v_2(t)$ leads $v_1(t)$ by 76°; alternately, $v_1(t)$ lags behind $v_2(t)$ by 76°. ∎

9-2 FIRST-ORDER CIRCUITS: CLASSICAL SOLUTION

Let us consider the RL circuit shown in Figure 9-6. It is, for example, a simple model of an induction motor driven by a sinusoidal voltage source. We are interested in the zero-state response, so $i_L(0^-) = 0$. The differential equation for $i(t)$ is obtained by writing KVL:

$$L\frac{di(t)}{dt} + Ri(t) = V_m \cos \omega t \qquad t > 0 \tag{9-7}$$

From our previous studies, we know that the transient part of the current is

$$i_{tr}(t) = Ke^{-(R/L)t} \tag{9-8}$$

and it will vanish as t increases.

We wish to concentrate on the particular solution only, due to the sinusoidal right-hand side. According to our studies in Chapter 7 (Table 7-1), the particular solution will be of the form

$$i_P(t) = I_m \cos(\omega t + \theta) \tag{9-9}$$

Here, the amplitude I_m and the phase θ of the current are unknown and must be determined. However, ω is the same as in the voltage source, because differentiation, $L\,di/dt$, and multiplication by a constant, Ri, cannot change ω in Equation 9-9.

As usual, we must substitute the assumed solution, Equation 9-9, into Equation 9-7. To simplify the resulting calculations, we first rewrite Equation 9-9 as

$$i_P(t) = I_m \operatorname{Re}\{e^{j(\omega t + \theta)}\} \tag{9-10}$$

where we used Euler's identity

$$e^{jx} = \cos x + j\sin x = \operatorname{Re}\{e^{jx}\} + j\operatorname{Im}\{e^{jx}\} \tag{9-11}$$

with the notation "Re" meaning "the real part of ..." and "Im" meaning "the imaginary part of ... ."

With this notation, Equation 9-10 is substituted into Equation 9-7 to give

$$LI_m \operatorname{Re}\{j\omega e^{j(\omega t + \theta)}\} + RI_m \operatorname{Re}\{e^{j(\omega t + \theta)}\} = V_m \operatorname{Re}\{e^{j\omega t}\} \tag{9-12}$$

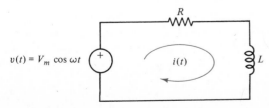

$v(t) = V_m \cos \omega t$

R

$i(t)$

L

Figure 9-6 An RL circuit.

As a second simplification, let us drop the "Re" altogether from Equation 9-12. Why can we do it? Because Equation 9-12 expresses an equality among the real parts of complex numbers; to satisfy this equality, it is certainly sufficient to equate the complex numbers themselves! Thus, Equation 9-12 reduces to

$$j\omega L I_m \, e^{j\omega t} e^{j\theta} + R I_m \, e^{j\omega t} \, e^{j\theta} = V_m \, e^{j\omega t} \tag{9-13}$$

The common mutiplying factor $e^{j\omega t}$ may be cancelled on both sides, and we have

$$I_m(j\omega L + R)\, e^{j\theta} = V_m \tag{9-14}$$

We recognize with delight that, by using Euler's identity, Equation 9-10, our original *differential* equation, Equation 9-7, has been reduced to an *algebraic* equation (no derivatives), Equation 9-13 or 9-14. This is a much simpler equation to solve!

Since Equation 9-14 has complex numbers, we equate magnitudes and then angles. Equating magnitudes on both sides, we remember that the amplitudes I_m and V_m are real, positive numbers. The result is

$$I_m \cdot |j\omega L + R| \cdot |e^{j\theta}| = V_m \tag{9-15}$$

and since $|e^{j\theta}| \equiv 1$, we get

$$I_m = \frac{V_m}{\sqrt{R^2 + (\omega L)^2}} \tag{9-16}$$

Next, equate angles on both sides of Equation 9-14,

$$0 + \tan^{-1} \frac{\omega L}{R} + \theta = 0 \tag{9-17}$$

to get

$$\theta = -\tan^{-1} \frac{\omega L}{R} \tag{9-18}$$

We have found the required I_m and θ. Therefore, Equation 9-10 reads

$$i_P(t) = I_m \cos(\omega t + \theta) = \frac{V_m}{\sqrt{R^2 + (\omega L)^2}} \cos\left(\omega t - \tan^{-1} \frac{\omega L}{R}\right) \tag{9-19}$$

This is the zero-state solution $i_P(t)$. As we see here, it is entirely steady-state, and it is drawn in Figure 9-7, together with the sinusoidal waveform of the input $v(t)$. The current $i(t)$ lags behind $v(t)$ by θ.

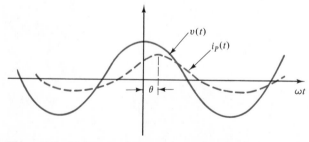

Figure 9-7   Voltage and current in an *RL* circuit.

Let us pause and review:

1. For a sinusoidal input, the zero-state particular solution is also a sinusoid of the *same* ω, but with a different amplitude and a different phase angle.
2. The amplitude of the current response is related to the amplitude of the voltage and to ω, R, and L.
3. The phase of the current response is related to ω, R, and L (but not to the amplitude of the input).
4. These relationships are fairly complicated, even for a simple first-order circuit. They involve a lot of manipulations.

What do we do in more complicated circuits? There ought to be an easier way to calculate the steady-state response for higher-order circuits with simultaneous integro-differential equations! Fortunately, there is such a way; it is called the phasor transform.

Probs. 9-10, 9-11, 9-12, 9-13

9-3 WHAT IS A TRANSFORM?

In general, we can say that a transform is *a simplified method* of doing something difficult. Many transform methods occur in everyday life, and we don't even think about them in this light. Here are a few:

1. Our languages (English, Japanese, etc.) are a transform. A language is a simplified method of communicating. Just think how difficult, or even totally impossible, it would be to communicate without a common language between you and your listener. A stranded traveler in a foreign country can attest to this difficulty.
2. The alphabet is a transform. The letter B represents in a simple way the sound associated with it. By itself, the symbol B means nothing. We have agreed, in advance, that it represents that particular sound.
3. Money is a transform. It is a *simplified* and *indirect* way to trade. The *difficult* and *direct* way of bartering may still be practiced, but just think how hard it is: You must have exactly what I may need (two sheep, for example), and, in return, I must have exactly what you need (a horse, for example), if we are to trade directly. With a monetary system, or a bead system, or a precious metal system, the values of the sheep and the horse are *represented* in dollars, or in ounces of gold, thereby simplifying trade. You don't have to give me sheep for my horse if it is not mutually agreeable. Instead, you give me the money transform of the value of the horse.
4. Of several mathematical transforms that are already familiar to us, one is the binary code. It is an indirect and easy way for computers (and even people) to handle mathematical operations. For example, the decimal multiplication $(12 \times 5)_{10}$ is transformed in binary code into $(1100 \times 101)_2$ by looking up in binary tables the binary transform for 12 and 5. The binary multiplication

yields 111100, and by looking it up in the binary tables, we see that it represents 60 decimal.

5. Another mathematical transform is the logarithmic one. Consider, for example, the calculation of $x = (9.5)^{1/3}$. The direct and difficult way to calculate x would be as follows:

$$x = (8 + 1.5)^{1/3} = \left[8\left(1 + \frac{1.5}{8} \right) \right]^{1/3}$$

Then we use the binomial expansion theorem on the inner parentheses:

$$x = 8^{1/3}\left(1 + \frac{1.5}{8} \right)^{1/3} = 2\left[1 + \frac{1}{3}\left(\frac{1.5}{8} \right) + \frac{1}{3}\left(\frac{-2}{3} \right)\frac{1}{2!}\left(\frac{1.5}{8} \right)^{2} + \cdots \right]$$

$$= 2.118$$

Quite difficult! We must first find, or guess, the nearest "nice" cube root (=8 here); then we must remember, or hunt down, the binomial expansion series, and worry about its convergence.

The logarithmic transform method is much simpler, though indirect. We use tables to represent the number 9.5 by its logarithm (base 10, say): log 9.5 = 0.978. Instead of the direct and difficult operation of raising to the power of $\frac{1}{3}$, the logarithmic transform requires a simpler step, multiplication by $\frac{1}{3}$. Therefore,

$$\log x = (\tfrac{1}{3})\,0.978 = 0.326$$

Finally, we look up in the logarithmic tables, and find the answer

$$x = \log^{-1}(0.326) = 2.118$$

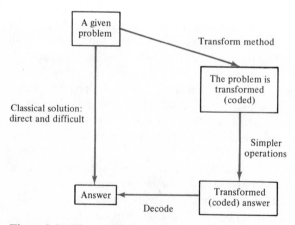

Figure 9-8   The transform method.

Let us summarize the features of a transform method:

1. It is a simpler, indirect way of performing difficult operations.
2. It is a code which must be understood by everyone who uses it. Otherwise, it is useless.
3. It has a set of rules, as well as "two-way" listings (tables) for encoding and decoding. These ideas are shown in Figure 9-8.

Probs. 9-14, 9-15, 9-16

A thorough review of complex numbers (Appendix B) is suggested here.

9-4 THE PHASOR TRANSFORM

The phasor transform method is particularly suitable for solving problems in the sinusoidal steady state, such as the one in the previous section. Encouraged by the simplified results there, let us develop the code and the operations of the phasor transform. Consider, again, the general sinusoidal waveform

$$f(t) = F_m \cos(\omega t + \theta) \tag{9-20}$$

where $f(t)$ is a generic notation for the time function (voltage, current, charge, flux, etc.). With Euler's identity, we can write

$$f(t) = \text{Re}\left\{F_m e^{j(\omega t + \theta)}\right\} \tag{9-21}$$

and let us agree to encode it as follows

$$f(t) \rightarrow F_m e^{j\theta} \tag{9-22}$$

This is our phasor transform (code): The sinusoidal time function $f(t)$ is transformed into (represented by) the complex number

$$\mathbf{F} = F_m e^{j\theta} \tag{9-23}$$

The arrow in Equation 9-22 reminds us that this is a code, a representation—*not* an equality.

In words, we drop from Equation 9-21 the "'Re" and $e^{j\omega t}$ (remember how it happened in Section 9-2?). The remaining complex number $\mathbf{F} = F_m e^{j\theta} = F_m \underline{/\theta}$ is the *phasor transform* of $f(t)$. Also, given the phasor $\mathbf{F}$, we decode it back into the time domain according to Equation 9-21; that is, we multiply $\mathbf{F}$ by $e^{j\omega t}$, then take the real part of the result, to obtain $f(t)$. Symbolically, we write

$$f(t) \leftrightarrow \mathbf{F} \tag{9-24}$$

to indicate this "two-way" transform, from the time domain into complex numbers, and back.

It is worth repeating that $f(t)$ and $\mathbf{F}$ are not equal to each other. They are transforms of each other. In a similar way, in the logarithmic transform, we represented 9.5 by 0.978; these two numbers are not equal to each other.

Let us try a few examples:

EXAMPLE 9-2

What is the phasor transform of $v(t) = 170 \sin 377t$, as in Equation 9-1?

> *Solution.* First, we convert it into its cosine equivalent, as agreed:
>
> $$v(t) = 170 \cos(377t - 90°) = \text{Re}\{170e^{j(377t - 90°)}\}$$
>
> and therefore its phasor transform is
>
> $$\mathbf{V} = 170\underline{/-90°} = -j170$$
>
> and we must remember (implicitly) $\omega = 377$ here, because $e^{j\omega t}$ had been cancelled. ∎

EXAMPLE 9-3

What is the current $i(t)$ if its phasor transform is given as

$$\mathbf{I} = 10\underline{/15°}$$

at $\omega = 4000$ rad/s?

> *Solution.* Using Equation 9-21, we write immediatly
>
> $$i(t) = 10 \cos(4000t + 15°)$$

The two phasors in Examples 9-2 and 9-3 are drawn in Figure 9-9, with convenient choices of scales for their lengths.

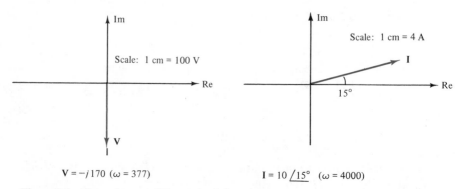

$$V = -j170 \ (\omega = 377) \qquad\qquad I = 10\underline{/15°} \ (\omega = 4000)$$

Figure 9-9   Two phasors of Examples 9-2 and 9-3.   ∎

There is a simple and elegant way to obtain the sinusoidal time function $f(t)$ from its phasor **F**. Let the given phasor be $\mathbf{F} = F_m\underline{/\theta}$, as in Equation 9-23. It is plotted as $\overrightarrow{OA}$ in Figure 9-10. Starting from this position, let the phasor rotate counterclockwise, in the positive sense, with a constant angular velocity ω rad/s. Then at $t = t_1$, say, it will be in the position $\overrightarrow{OB}$, at an angle $(\omega t_1 + \theta)$. According to Equation 9-21, $f(t_1)$ is $\overrightarrow{OB'}$, the projection of $\overrightarrow{OB}$ on the real axis! We plot $\overrightarrow{OB'}$ as the value of $f(t_1)$ versus ωt, and continue with other values of t, getting $\overrightarrow{OC'}, \overrightarrow{OD'}$, etc. The sinusoidal function $f(t) = F_m \cos(\omega t + \theta)$ is thus obtained point by point.

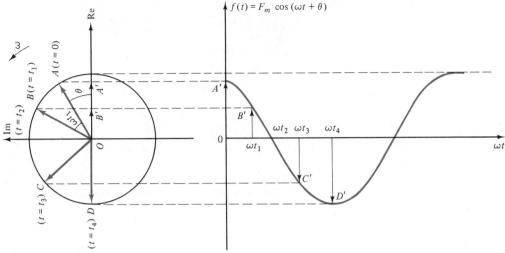

Figure 9-10   Graphical $f(t)$ from **F**.

9-5 DIFFERENTIATION, INTEGRATION, AND THEIR TRANSFORMS.

There are two relatively difficult operations in the classical time-domain loop or node analysis, differentiation and integration. Let us see how these are simplified in the phasor transform.

Let us write $f(t)$ as in Equation 9-21, repeated here

$$f(t) = \text{Re}\,\{F_m\,e^{j(\omega t + \theta)}\} \tag{9-21}$$

with its phasor transform, as before,

$$\mathbf{F} = F_m e^{j\theta} \tag{9-23}$$

Take now the first derivative of $f(t)$ in Equation 9-21

$$\frac{df(t)}{dt} = \text{Re}\,\{F_m \cdot j\omega \cdot e^{j(\omega t + \theta)}\} \tag{9-25}$$

where we used the fact that

$$\frac{d}{dt}\,\text{Re}\{\ \ \} = \text{Re}\,\frac{d}{dt}\{\ \ \}$$

(Prove this to yourself!) But the phasor transform of the function inside the braces of Equation 9-25 is, by our convention,

$$\frac{df(t)}{dt} \leftrightarrow j\omega F_m e^{j\theta} \tag{9-26}$$

With Equation 9-23, we can write Equation 9-26 as

$$\frac{df(t)}{dt} \leftrightarrow j\omega \mathbf{F} \tag{9-27}$$

In summary, we have shown that if

$$f(t) \leftrightarrow \mathbf{F} \qquad (9\text{-}24)$$

then

$$\frac{df(t)}{dt} \leftrightarrow j\omega\mathbf{F} \qquad (9\text{-}27)$$

Therefore, differentiation in the time domain of $f(t)$ is represented by multiplication by $j\omega$ of the corresponding phasor $\mathbf{F}$. There is no doubt that multiplication is a simpler operation than differentiation!

Probs. 9-1• 9-20

EXAMPLE 9-4 ——————————————————————————————

Let us calculate the steady-state solution of the RC circuit shown in Figure 9-11.

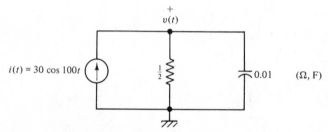

Figure 9-11   Example 9-4.

Solution.   The node equation for $v(t)$ is

$$0.01 \frac{dv(t)}{dt} + 2v(t) = 30 \cos 100t$$

Take the phasor transform of this equation, term by term, with $\omega = 100$ remembered

$$0.01(j100)\mathbf{V} + 2\mathbf{V} = 30\underline{/0^\circ}$$

As expected, the original differential equation has been transformed into an algebraic equation which is much easier to solve. We collect terms

$$(2 + j1)\mathbf{V} = 30\underline{/0^\circ}$$

and solve *algebraically* for $\mathbf{V}$:

$$\mathbf{V} = \frac{30\underline{/0^\circ}}{2 + j1} = \frac{30\underline{/0^\circ}}{2.24\underline{/26.6^\circ}} = 13.42\underline{/-26.6^\circ}$$

The time function $v(t)$ is therefore

$$\mathbf{V} \rightarrow v(t) = 13.42 \cos(100t - 26.6^\circ)$$

This is the end. Short and simple! If you are still in doubt, rework this example by the classical method of solution and compare with the RL circuit in Section 9-2.   ∎

Let us now consider integration. The time-domain expressions of interest to us are

$$v_C(t) = \frac{1}{C} \int_{0^-}^{t} i_C(x)\,dx \qquad (9\text{-}28\text{a})$$

and

$$i_L(t) = \frac{1}{L} \int_{0^-}^{t} v_L(x)\,dx \qquad (9\text{-}28\text{b})$$

the zero-state voltage across a capacitor, and the zero-state current through an inductor, respectively. Again, in the sinusoidal steady-state analysis, we are solving only for the zero-state response; therefore $v_C(0^-)$ and $i_L(0^-)$ are missing from these equations.

Use again the generic notation, and write either integral in Equation 9-28 as

$$\int_{0^-}^{t} f(x)\,dx = g(t) \qquad (9\text{-}29\text{a})$$

where the dummy variable x must be used. Also, we denote the entire integral by a single name, $g(t)$. Our problem is then to find $\mathbf{G}$, the phasor transfer of $g(t)$.

From Equation 9-29(a), we have

$$f(t) = \frac{dg(t)}{dt} \qquad (9\text{-}29\text{b})$$

and the phasor transform of Equation 9-29(b) is

$$\mathbf{F} = (j\omega)\mathbf{G} \qquad (9\text{-}30\text{a})$$

as we learned before. Consequently,

$$\mathbf{G} = \frac{1}{j\omega}\mathbf{F} \qquad (9\text{-}30\text{b})$$

that is,

$$\int_{0^-}^{t} f(x)\,dx \leftrightarrow \frac{1}{j\omega}\mathbf{F} \qquad (9\text{-}30\text{c})$$

In words, integration of $f(t)$ is represented by division by $j\omega$ of the corresponding phasor $\mathbf{F}$. Again, division is a simpler operation than integration. We also recognize that this result makes sense: Differentiation and integration are inverse operations of each other in the time domain, and so are multiplication and division by $j\omega$ in the phasor domain.

Probs. 9-21, 9-22, 9-23, 9-24, 9-25, 9-26

9-6 SIMULTANEOUS PHASOR NETWORK EQUATIONS

We are ready now to approach the bigger problem, the steady-state solution of simultaneous loop or node equations. First, let us illustrate it with one example for both cases.

EXAMPLE 9-5 _____

Consider the network shown in Figure 9-12(a). One node equation or two loop equations will solve it.

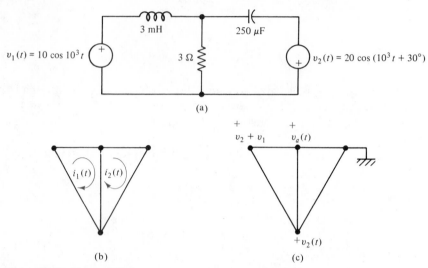

(a)

(b)   (c)

Figure 9-12 Example 9-5.

Solution. For now, let us do it both ways. First, by loop currents, as shown in Figure 9-12(b). The two loop equations are, with $(+)$ for a drop and $(-)$ for a rise,

$$3 \times 10^{-3} \frac{di_1}{dt} + 3[i_1(t) - i_2(t)] - 10 \cos 1000t = 0$$

$$\frac{1}{250 \times 10^{-6}} \int_0^t i_2(x)\,dx - 20 \cos(1000t + 30°) + 3[i_2(t) - i_1(t)] = 0$$

Notice that this is the zero-state formulation and solution; therefore, $v_C(0^-) = 0$ in the term for the capacitive voltage. The classical solution of these two simultaneous integro-differential equations is hopelessly difficult. Instead, let us use the phasor transform. It is worth repeating what we can and cannot do with phasors, and under what conditions: **1.** We can use phasors because only pure sinusoids at a single frequency ($\omega = 1000$) are involved. **2.** Phasors are complex numbers. They represent sinusoidal time-domain functions. **3.** We are getting only the steady-state part of the zero-state solution. Phasors cannot solve homogeneous integro-differential equations. **4.** All initial conditions are zero.

With this in mind, we take the phasor transform of the two loop equations. The results are

$$3 \times 10^{-3}(j1000)\mathbf{I}_1 + 3[\mathbf{I}_1 - \mathbf{I}_2] - 10\underline{/0°} = 0$$

$$\frac{1}{250 \times 10^{-6}} \frac{\mathbf{I}_2}{j1000} - 20\underline{/30°} + 3\mathbf{I}_2 - 3\mathbf{I}_1 = 0$$

As expected, these are *algebraic* equations. Collect terms and rearrange for solution

$$(3 + j3)\mathbf{I}_1 - 3\mathbf{I}_2 = 10$$

$$-3\mathbf{I}_1 + (3 - j4)\mathbf{I}_2 = 20\underline{/30°}$$

Solving by determinants, we get

$$\mathbf{I}_1 = \frac{\begin{vmatrix} 10 & -3 \\ 20\underline{/30^\circ} & 3-j4 \end{vmatrix}}{\begin{vmatrix} 3+j3 & -3 \\ -3 & 3-j4 \end{vmatrix}} = \frac{81.96-j10}{12-j3} = \frac{82.57\underline{/-6.96^\circ}}{12.37\underline{/-14.04^\circ}} = 6.68\underline{/7.08^\circ}$$

$$\mathbf{I}_2 = \frac{\begin{vmatrix} 3+j3 & 10 \\ -3 & 20\underline{/30^\circ} \end{vmatrix}}{\begin{vmatrix} 3+j3 & -3 \\ -3 & 3-j4 \end{vmatrix}} = \frac{51.96+j81.96}{12-j3} = \frac{97.04\underline{/57.63^\circ}}{12.37\underline{/-14.04^\circ}} = 7.84\underline{/71.67^\circ}$$

Therefore, the two loop currents are

$$\mathbf{I}_1 \rightarrow i_1(t) = 6.68\cos(1000t + 7.08^\circ)$$
$$\mathbf{I}_2 \rightarrow i_2(t) = 7.84\cos(1000t + 71.67^\circ)$$

For node analysis, see Figure 9-12(c) where a reference node was chosen and all other nodes have been labeled. There is only one unknown node voltage, $v_a(t)$. We write one KCL equation at this node ($+$ leaving, $-$ entering):

$$250 \times 10^{-6}\frac{dv_a(t)}{dt} + \frac{v_a(t) - v_2(t)}{3} + \frac{1}{3 \times 10^{-3}}\int_{0^-}^{t} [v_a(x) - v_2(x) - v_1(x)]\, dx = 0$$

Here, $i_L(0^-) = 0$ for the zero-state solution, and this term is missing from the expression for the inductive current. The phasor transform of this time-domain equation is an algebraic equation,

$$(250 \times 10^{-6})(j1000)\mathbf{V}_a + \frac{\mathbf{V}_a - 20\underline{/30^\circ}}{3} + \frac{1}{3 \times 10^{-3}}\frac{\mathbf{V}_a - 20\underline{/30^\circ} - 10\underline{/0^\circ}}{j1000} = 0$$

Collect terms in $\mathbf{V}_a$

$$(0.33 - j0.083)\mathbf{V}_a = 9.107 - j5.774$$

and solve algebraically

$$\mathbf{V}_a = \frac{9.107 - j5.774}{0.33 - j0.083} = 31.38\underline{/-18.34^\circ}$$

Therefore, the time-domain solution is

$$\mathbf{V}_a \rightarrow v_a(t) = 31.38\cos(1000t - 18.34^\circ)$$

It would be interesting to check the correctness of these two solutions by calculating, for example, the current in the capacitor using the node analysis. This current is found using the answer for $v_a(t)$ as

$$i_C(t) = C\frac{dv_C(t)}{dt} = 250 \times 10^{-6}\frac{dv_a(t)}{dt}$$
$$= -(250 \times 10^{-6})(31.38)1000\sin(1000t - 18.34^\circ)$$
$$= 7.845\cos(1000t + 71.66^\circ)$$

which agrees with the answer found by loop analysis.   ■   Probs. 9-27, 9-28

As can be seen, it is always a good habit to list a complex number in both forms, rectangular and polar. This is easily done on your hand-held calculator. Either form is then available when needed.

9-7 OHM'S LAW IN PHASOR NOTATION. IMPEDANCES AND ADMITTANCES

Before we extend these results to larger networks, let us make some general observations and introduce certain useful concepts. Our observations concern the v-i and i-v relationships of R, L, and C in the time domain and their respective phasor transforms.

Resistor R For any waveforms in general, and for sinusoids in particular, we have Ohm's law:

$$v_R(t) = Ri_R(t) \qquad (9\text{-}31)$$

Its phasor transform reads

$$\mathbf{V}_R = R\mathbf{I}_R \qquad (9\text{-}32)$$

Typical sinusoidal waveforms for $i_R(t)$ and $v_R(t)$ are shown in Figure 9-13(a), and we see that they are *in phase*. Their amplitudes (positive scalar numbers) are related by

$$V_{Rm} = RI_{Rm} \qquad (9\text{-}33)$$

The two phasors

$$\mathbf{I}_R = I_{Rm}\underline{/-\alpha} \qquad (9\text{-}34)$$

and

$$\mathbf{V}_R = V_{Rm}\underline{/-\alpha} \qquad (9\text{-}35)$$

are drawn as a *phasor diagram* in Figure 9-13(b), with convenient scales chosen for their lengths. The two phasors are collinear (parallel) because the two sinusoids are in phase.

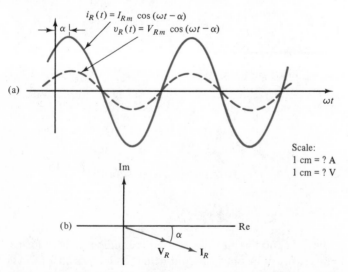

Figure 9-13   Resistor relations.

The i-v relationship for the resistor is

$$i_R(t) = Gv_R(t) \qquad G = \frac{1}{R} \tag{9-36}$$

and for pure sinusoids the phasor relationship is

$$\mathbf{I}_R = G\mathbf{V}_R \tag{9-37}$$

the algebraic inverse of Equation 9-32.

Nothing surprising about this, you say? Maybe so. However, we should recognize that Equations 9-32 and 9-37 are Ohm's law in the phasor transform. The fact that they resemble Ohm's law for *any* waveform in the time domain

$$v_R(t) = Ri_R(t) \qquad i_R(t) = Gv_R(t) \tag{9-38}$$

is only because of the very simple algebraic form of Equation 9-38.

Inductor L The time-domain v-i relation is

$$v_L(t) = L\frac{di_L}{dt}(t) \tag{9-39}$$

for all waveforms. For a sinusoidal case, the phasor transform of Equation 9-39 reads

$$\mathbf{V}_L = L \cdot j\omega \cdot \mathbf{I}_L = j\omega L\mathbf{I}_L \tag{9-40}$$

This phasor relation between complex numbers gives us, first, the relationship between the magnitude of the amplitudes V_{Lm} and I_{Lm} of the two sinusoids

$$V_{Lm} = \omega L I_{Lm} \tag{9-41}$$

and, next, between the phase angles

$$\not{\angle} \mathbf{V}_L = 90° + \not{\angle} \mathbf{I}_L \tag{9-42}$$

(The multiplier j in Equation 9-40 has an angle of 90°.)

The two sinusoidal waveforms are plotted in Figure 9-14(a), and we see that $v_L(t)$ leads $i_L(t)$ by 90°, or $i_L(t)$ lags $v_L(t)$ by 90°. The product ωL, relating their amplitudes in Equation 9-41, is in ohms, and more will be said about it soon. The two phasors

$$\mathbf{I}_L = I_{Lm}\underline{/\theta} \tag{9-43}$$

$$\mathbf{V}_L = V_{Lm}\underline{/\theta + 90°} \tag{9-44}$$

are drawn to scale in Figure 9-14(b) as the phasor diagram. They show clearly the phase difference between $v_L(t)$ and $i_L(t)$, as $\mathbf{V}_L$ is ahead of $\mathbf{I}_L$ by 90°.

In every phasor diagram, we are free to choose convenient scales for drawing lengths of phasors. Don't forget to mark your choice of scales, so that true values (V_m or I_m) can be read from the phasor diagram.

The i-v relation for the inductor, under zero-state conditions, is

$$i_L(t) = \frac{1}{L}\int_{0-}^{t} v_L(x)\,dx \tag{9-45}$$

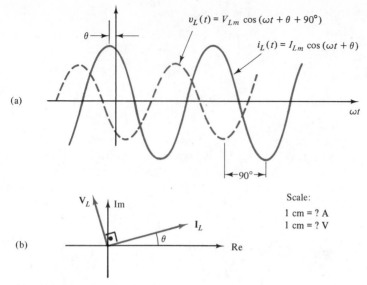

Figure 9-14 Inductor relations.

and the phasor transform of Equation 9-45 is

$$\mathbf{I}_L = \frac{1}{L}\frac{\mathbf{V}_L}{j\omega} = \frac{1}{j\omega L}\mathbf{V}_L \tag{9-46}$$

since integration is transformed into division by $j\omega$. Equations 9-40 and 9-46 are, of course, algebraic inverses of each other.

Capacitor C We should be able to write all the relationships by duality with L. However, let us rederive them for the added practice. The time domain, zero-state v-i relation is

$$v_C(t) = \frac{1}{C}\int_{0^-}^{t} i_C(x)\,dx \tag{9-47a}$$

for all waveforms. For sinusoids, the phasor transform of Equation 9-47a reads

$$\mathbf{V}_C = \frac{1}{C}\frac{\mathbf{I}_C}{j\omega} = \frac{1}{j\omega C}\mathbf{I}_C \tag{9-47b}$$

Again, equate magnitudes in Equation 9-47b

$$V_{Cm} = \left(\frac{1}{\omega C}\right)I_{Cm} \tag{9-47c}$$

to obtain the relation between the two amplitudes. Next, equate angles

$$\angle\,\mathbf{V}_C = -90° + \angle\,\mathbf{I}_C \tag{9-47d}$$

since $1/j$ contributes $-90°$.

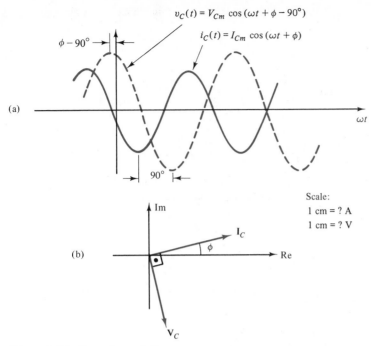

Figure 9-15   Capacitor relations.

The two sinusoidal waveforms are plotted in Figure 9-15(a), with $i_C(t)$ leading $v_C(t)$ by 90°. The same information is available in the phasor diagram, plotted to scale in Figure 9-15(b).

The *i-v* relationship for the capacitor is

$$i_C(t) = C\frac{dv_C(t)}{dt} \tag{9-48a}$$

and the immediate phasor transform of this equation yields

$$\mathbf{I}_C = j\omega C\mathbf{V}_C \tag{9-48b}$$

which is the algebraic inverse of Equation 9-47b.

To summarize, let us repeat the **V-I** phasor relations

$$\mathbf{V}_R = R\mathbf{I}_R \tag{9-32}$$

$$\mathbf{V}_L = j\omega L\mathbf{I}_L \tag{9-40}$$

$$\mathbf{V}_C = \frac{1}{j\omega C}\mathbf{I}_C \tag{9-47b}$$

Beautiful, aren't they? They all have the same general algebraic form, called *Ohm's law in phasor notation,*

$$\mathbf{V} = \mathbf{ZI} \tag{9-49}$$

where $\mathbf{Z}$ is a complex number called the *sinusoidal impedance*. Specifically, for a resistor,

$$\mathbf{Z}_R = R \tag{9-50}$$

for an inductor,

$$\mathbf{Z}_L = j\omega L \tag{9-51}$$

and for a capacitor,

$$\mathbf{Z}_C = \frac{1}{j\omega C} = -j\,\frac{1}{\omega C} \tag{9-52}$$

In each case, the impedance multiplies the phasor current $\mathbf{I}$ to give the phasor voltage $\mathbf{V}$. For the resistor, the impedance is a real number. For the inductor or the capacitor, it is a purely imaginary number.

In more general cricuits, the impedance—as a complex number—may be written in its rectangular form as

$$\mathbf{Z} = R + jX \tag{9-53}$$

where R is its *resistive part* and X is its *reactive part*. Thus, for a single resistor we can write

$$\mathbf{Z}_R = R + j\cdot 0 = R_R + jX_R \tag{9-54}$$

with the resistive part equal to the resistance R and with zero reactive part. For a single inductor, we have from Equation 9-51

$$\mathbf{Z}_L = 0 + j\omega L = R_L + jX_L \tag{9-55}$$

and so its resistive part is zero, while its reactive part is

$$X_L = \omega L \tag{9-56}$$

and is called the *reactance* of the inductor, in ohms. Similarly, for the capacitor, we write from Equation 9-52

$$\mathbf{Z}_C = 0 - j\,\frac{1}{\omega C} = R_C + jX_C \tag{9-57}$$

and therefore its resistive part is zero, while its *reactance* is

$$X_C = -\frac{1}{\omega C} \tag{9-58}$$

in ohms.

It is extremely important to repeat the difference between a phasor and an impedance. They are both complex numbers. A phasor is a convenient representation of a sinusoidal time function. An impedance does not represent any time function; it is just a complex number which multiplies a phasor current to give a phasor voltage.

As we saw, Equation 9-49 carries two pieces of information. First, the magnitude of $\mathbf{V}$ equals the magnitude of $\mathbf{ZI}$, that is

$$|\mathbf{V}| = |\mathbf{ZI}| = |\mathbf{Z}| \cdot |\mathbf{I}| \tag{9-59}$$

and this result expresses the relation between the amplitudes of voltages and currents in all three cases, as in Equations 9-33, 9-41, and 9-47c. Second, the angle of **V** equals the angle of **ZI**, which, in turn, equals the sum of the angles of **Z** and of **I**:

$$\measuredangle \mathbf{V} = \measuredangle \mathbf{Z} + \measuredangle \mathbf{I} \tag{9-60}$$

The three cases were obtained in Equations 9-32, 9-42, and 9-47d.

The inverse **I-V** phasor relations are

$$\mathbf{I}_R = G\mathbf{V}_R \tag{9-61}$$

$$\mathbf{I}_L = \frac{1}{j\omega L}\mathbf{V}_L \tag{9-62}$$

$$\mathbf{I}_C = j\omega C\mathbf{V}_C \tag{9-63}$$

and they are all of the general form

$$\mathbf{I} = \mathbf{YV} \tag{9-64}$$

The complex number **Y** is called the *sinusoidal admittance*. Quite obviously,

$$\mathbf{Y}_R = G = \frac{1}{R} = \frac{1}{\mathbf{Z}_R} \tag{9-65}$$

$$\mathbf{Y}_L = \frac{1}{j\omega L} = \frac{1}{\mathbf{Z}_L} \tag{9-66}$$

and

$$\mathbf{Y}_C = j\omega C = \frac{1}{\mathbf{Z}_C} \tag{9-67}$$

The admittance of each element is the reciprocal of its impedance; it is the complex number (not a phasor) which multiplies the phasor **V** to give the phasor **I**.

In general, the admittance, as a complex number, is written as

$$\mathbf{Y} = G + jB \tag{9-68}$$

where G is the *conductance*, and B the *susceptance*, both in mhos (℧). For an inductor, we have from Equation 9-66

$$\mathbf{Y}_L = 0 - j\frac{1}{\omega L} \tag{9-69}$$

that is,

$$G_L = 0 \qquad B_L = -\frac{1}{\omega L} \tag{9-70}$$

Similarly, for a capacitor,

$$\mathbf{Y}_C = 0 + j\omega C \tag{9-71}$$

with

$$G_C = 0 \qquad B_C = \omega C \tag{9-72}$$

Finally, the phasor Equation 9-64 means that

$$|\mathbf{I}| = |\mathbf{Y}| \cdot |\mathbf{V}| \qquad (9\text{-}73)$$

and

$$\measuredangle \mathbf{I} = \measuredangle \mathbf{Y} + \measuredangle \mathbf{V} \qquad (9\text{-}74)$$

Probs. 9-3
9-32, 9-33
9-34

Verify to your satisfaction these results for each element.

All these relations are summarized in Table 9-1. Then several examples follow.

TABLE 9-1 PHASOR RELATIONSHIPS FOR R, L, C

Element	Time relations	Phasor relations	Impedance	Admittance
R	$v_R(t) = Ri_R(t)$	$\mathbf{V}_R = R\mathbf{I}_R$	$\mathbf{Z}_R = R$	
	$i_R(t) = Gv_R(t)$	$\mathbf{I}_R = G\mathbf{V}_R$		$\mathbf{Y}_R = G$
L	$v_L(t) = L\dfrac{di_L(t)}{dt}$	$\mathbf{V}_L = j\omega L\mathbf{I}_L$	$\mathbf{Z}_L = j\omega L = jX_L$	
	$i_L(t) = \dfrac{1}{L}\displaystyle\int_{0^-}^{t} v_L(x)dx$	$\mathbf{I}_L = \dfrac{1}{j\omega L}\mathbf{V}_L$		$\mathbf{Y}_L = \dfrac{1}{j\omega L} = -jB_L$
C	$v_C(t) = \dfrac{1}{C}\displaystyle\int_{0^-}^{t} i_C(x)dx$	$\mathbf{V}_C = \dfrac{1}{j\omega C}\mathbf{I}_C$	$\mathbf{Z}_C = \dfrac{1}{j\omega C} = -jX_C$	
	$i_C(t) = C\dfrac{dv_C(t)}{dt}$	$\mathbf{I}_C = j\omega C\mathbf{V}_C$		$\mathbf{Y}_C = j\omega C = jB_C$

EXAMPLE 9-6

Following the previous approach and Example 9-5, let us formulate *directly*, and then solve the loop equation for the network shown in Figure 9-16(a), without the time-domain equation.

Solution. In Figure 9-16(b), we transformed the source and the current into their respective phasors and calculated the impedances of the elements. In particular, notice the minus sign in the impedance of the capacitor $\mathbf{Z}_C = 1/(j\omega C) = -j/(\omega C)$. Using Ohm's law (Equation 9-49) we write KVL in phasor form for this loop. The usual convention is ($+$) for a voltage drop and ($-$) for a rise:

$$-8\underline{/20^\circ} + 2\mathbf{I} + (j3)\mathbf{I} + (-j4)\mathbf{I} = 0$$

Collect terms

$$(2 + j3 - j4)\mathbf{I} = 8\underline{/20^\circ}$$

$$\therefore \quad (2 - j1)\mathbf{I} = 8\underline{/20^\circ}$$

$$\therefore \quad \mathbf{I} = \frac{8\underline{/20^\circ}}{2 - j1} = \frac{8\underline{/20^\circ}}{2.24\underline{/-26.56^\circ}} = 3.58\underline{/46.56^\circ}$$

Therefore, the steady-state current is

$$i(t) = 3.58 \cos(10t + 46.56^\circ)$$

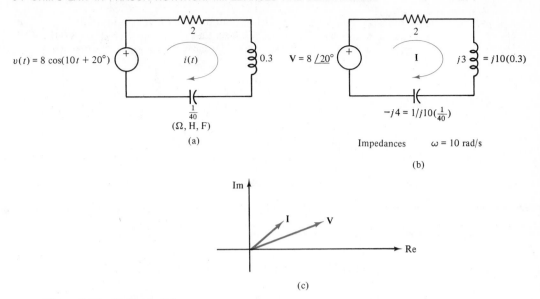

Figure 9-16   Example 9-6.

Compare, again, the amount of work here with that needed by the classical method in Chapter 7. The two phasors V and I are shown in Figure 9-16(c). ■

EXAMPLE 9-7

Write directly the phasor node equation for the network shown in Figure 9-17(a) and solve for $v(t)$.

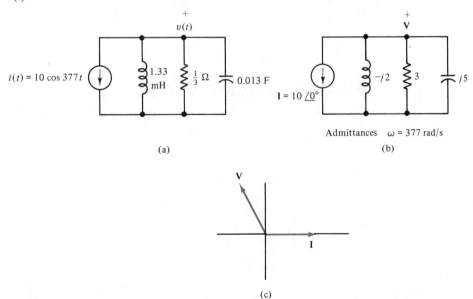

Figure 9-17   Example 9-7.

Solution. The circuit in Figure 9-17(b) shows the phasor transforms of $i(t)$ and $v(t)$, as well as the admittances of the elements. Check that minus sign in $\mathbf{Y}_L$! Now write KCL using Ohm's law of Equation 9-64. Use $(+)$ for currents leaving, $(-)$ for entering

$$10\underline{/0^\circ} + (-j2)\mathbf{V} + 3\mathbf{V} + (j5)\mathbf{V} = 0$$

$$\therefore \quad (3 + j3)\mathbf{V} = 10\underline{/180^\circ}$$

$$\therefore \quad \mathbf{V} = \frac{10\underline{/180^\circ}}{3 + j3} = \frac{10\underline{/180^\circ}}{4.24\underline{/45^\circ}} = 2.36\underline{/135^\circ}$$

Therefore, the zero-state, steady-state solution for $v(t)$ is

$$v(t) = 2.36 \cos(377t + 135^\circ)$$

The phasor diagram in Figure 9-17(c) shows $\mathbf{V}$ and $\mathbf{I}$.

Probs. 9-35
■ 9-36

9-8 GENERAL PHASOR NODE ANALYSIS

The steps to be taken are based on the previous examples:

1. Redraw the circuit with phasor notation for all voltages and currents. There will be $p \leq n - 1$ unknown phasor node voltages $\mathbf{V}_1, \mathbf{V}_2, \ldots, \mathbf{V}_p$, where n is the number of nodes. Label the elements (R, L, C) by their admittances.
2. Write KCL at each unknown node, using the I-V relation for each element connected to this node.
3. The resulting equations, in matrix form, read

$$\begin{bmatrix} y_{11} & y_{12} & y_{13} & \cdots & y_{1p} \\ y_{21} & y_{22} & y_{23} & \cdots & y_{2p} \\ \cdots\cdots\cdots\cdots\cdots\cdots\cdots\cdots \\ y_{p1} & y_{p2} & y_{p3} & \cdots & y_{pp} \end{bmatrix} \begin{bmatrix} \mathbf{V}_1 \\ \mathbf{V}_2 \\ \vdots \\ \mathbf{V}_p \end{bmatrix} = \begin{bmatrix} \mathbf{J}_1 \\ \mathbf{J}_2 \\ \vdots \\ \mathbf{J}_p \end{bmatrix} \qquad (9\text{-}75)$$

or

$$\mathbf{Y}_n \mathbf{V}_n = \mathbf{J}_n \qquad (9\text{-}76)$$

Here $\mathbf{Y}_n$ is the *node admittance matrix*, $\mathbf{V}_n$ is the column matrix of the p unknown phasor node voltages, and $\mathbf{J}_n$ is the column matrix of the known phasor sources. We did this development for purely resistive networks in Chapter 2, Equation 2-11. It is very instructive to compare it here.

EXAMPLE 9-8 _____

For the network in Figure 9-18(a), we have two unknowns, $v_a(t)$ and $v_b(t)$.

Solution. The circuit is redrawn in phasor notation and with admittances in Figure 9-18(b). Use $(+)$ for currents leaving and $(-)$ for currents entering to write KCL at node $\mathbf{V}_a$

$$3\mathbf{V}_a + (-j1)(\mathbf{V}_a - \mathbf{V}_b) = 0$$

and at node $\mathbf{V}_b$

$$(j2)\mathbf{V}_b + (-j1)(\mathbf{V}_b - \mathbf{V}_a) + (-j1.5)(\mathbf{V}_b - 1.2) = 0$$

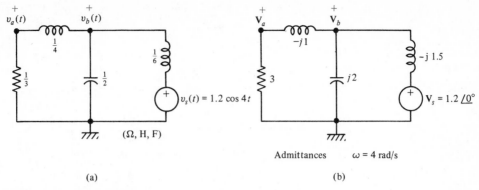

Figure 9-18   Example 9-8.

Collect and rearrange terms

$$(3 - j1)\mathbf{V}_a + j1\mathbf{V}_b = 0$$
$$j1\mathbf{V}_a - j0.5\mathbf{V}_b = -j1.8$$

In matrix form, these two equations are

$$\begin{bmatrix} 3 - j1 & j1 \\ j1 & -j0.5 \end{bmatrix} \begin{bmatrix} \mathbf{V}_a \\ \mathbf{V}_b \end{bmatrix} = \begin{bmatrix} 0 \\ -j1.8 \end{bmatrix}$$

Their solution is obtained by Cramer's rule (a good opportunity here to practice small determinants with complex numbers):

$$\mathbf{V}_a = \frac{\begin{vmatrix} 0 & j1 \\ -j1.8 & -j0.5 \end{vmatrix}}{\begin{vmatrix} 3 - j1 & j1 \\ j1 & -j0.5 \end{vmatrix}} = \frac{-1.8}{0.5 - j1.5} = \frac{1.8 \underline{/180°}}{1.58 \underline{/-71.6°}} = 1.14 \underline{/-108.4°}$$

$$\mathbf{V}_b = \frac{\begin{vmatrix} 3 - j1 & 0 \\ j1 & -j1.8 \end{vmatrix}}{0.5 - j1.5} = \frac{-1.8 - j5.4}{0.5 - j1.5} = \frac{5.7 \underline{/-108.4°}}{1.58 \underline{/-71.6}} = 3.6 \underline{/-36.8°}$$

Therefore

$$v_a(t) = 1.14 \cos (4t - 108.4°)$$
$$v_b(t) = 3.6 \cos (4t - 36.8°)$$

are the two steady-state node voltages. A good piece of advice is repeated here, if you haven't discovered it for yourself by now: When working with complex numbers, have them written in both rectangular and polar forms. It's fairly easy to do routinely; as soon as you get a number in one form, convert it into the other. You never know when you'll need it—and you'll have it ready then!

Prob. 9-37

EXAMPLE 9-9

Apply phasor node analysis to the network shown in Figure 9-19(a).

Solution. In Figure 9-19(b) we redraw the circuit with phasor notation and with admittances. Use (−) for currents entering and write KCL at the node $\mathbf{V}_a$ as

$$3\mathbf{V}_a + (-j2.5)(\mathbf{V}_a - \mathbf{V}_b) + j2(\mathbf{V}_a - \mathbf{V}_c) = 0$$

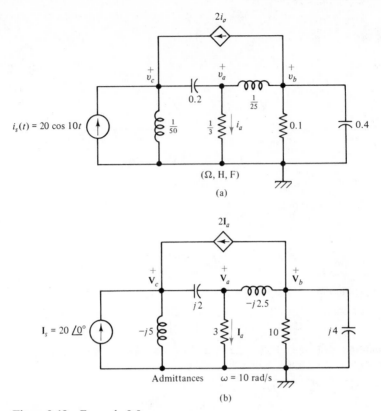

Figure 9-19 Example 9-9.

At the node $\mathbf{V}_b$, KCL is, with $\mathbf{I}_a = 3\mathbf{V}_a$,

$$10\mathbf{V}_b + j4\mathbf{V}_b + (-j2.5)(\mathbf{V}_b - \mathbf{V}_a) + 2(3\mathbf{V}_a) = 0$$

At node $\mathbf{V}_c$, KCL is

$$-20\underline{/0^\circ} + (-j5)\mathbf{V}_c + j2(\mathbf{V}_c - \mathbf{V}_a) - 2(3\mathbf{V}_a) = 0$$

Collect terms and arrange in the matrix form

$$\begin{bmatrix} 3 - j0.5 & j2.5 & -j2 \\ 6 + j2.5 & 10 + j1.5 & 0 \\ -6 - j2 & 0 & -j3 \end{bmatrix} \begin{bmatrix} \mathbf{V}_a \\ \mathbf{V}_b \\ \mathbf{V}_c \end{bmatrix} = \begin{bmatrix} 0 \\ 0 \\ 20 \end{bmatrix}$$

These are three simultaneous linear algebraic equations with complex coefficients. Cramer's rule for solving them is still valid, but it is long and very tedious for determinants larger than (2×2). Several hand-held calculators and most computer libraries have efficient routines to solve such equations. The solution here is

$$\mathbf{V}_a = -1.787 - j0.175 = 1.796\underline{/-174.4^\circ}$$
$$\mathbf{V}_b = 1.087 + j0.39 = 1.155\underline{/19.7^\circ}$$
$$\mathbf{V}_c = 1.542 + j3.21 = 3.56\underline{/64.3^\circ}$$

Therefore, the steady-state node voltages are

$$v_a(t) = 1.796 \cos(10t - 174.4°)$$
$$v_b(t) = 1.155 \cos(10t + 19.7°)$$
$$v_c(t) = 3.56 \cos(10t + 64.3°)$$

Probs. 9-38, 9-39, 9-40

A related comment to these examples involves the reciprocity theorem. Just as in Chapter 2 for purely resistive networks, we say here that a network is reciprocal if $\mathbf{Y}_n$ is symmetric, that is, if $y_{km} = y_{mk}$. The network in Example 9-8 is reciprocal, but the one in Example 9-9 is not. Typically, dependent sources may cause nonreciprocity.

9-9 GENERAL PHASOR LOOP ANALYSIS

The following steps are based on the previous examples:

1. Redraw the circuit with phasor notation for voltages and currents. There will be $q \leq l$ unknown phasor loop currents, $\mathbf{I}_1, \mathbf{I}_2, \ldots, \mathbf{I}_q$, where l is the number of links. Label the elements (R, L, C, M) by their impedances.
2. Write KVL around each loop, using the **V-I** relation for each element in this loop.
3. The resulting equations, in matrix form, read

$$\begin{bmatrix} z_{11} & z_{12} & \cdots & z_{1q} \\ z_{21} & z_{22} & \cdots & z_{2q} \\ \cdots\cdots\cdots\cdots\cdots\cdots \\ z_{q1} & z_{q2} & \cdots & z_{qq} \end{bmatrix} \begin{bmatrix} \mathbf{I}_1 \\ \mathbf{I}_2 \\ \vdots \\ \mathbf{I}_q \end{bmatrix} = \begin{bmatrix} \mathbf{E}_1 \\ \mathbf{E}_2 \\ \vdots \\ \mathbf{E}_q \end{bmatrix} \tag{9-77}$$

that is,

$$\mathbf{Z}_l \mathbf{I}_l = \mathbf{E}_l \tag{9-78}$$

The square matrix $\mathbf{Z}_l$ is the *loop impedance matrix*, and it is symmetric for a reciprocal network; $\mathbf{I}_l$ is the column matrix of the q unknown phasor loop currents, and $\mathbf{E}_l$ is the column matrix of known phasor sources. A special case of these equations was done for purely resistive networks in Chapter 2, Equation 2-12, and the two should be compared. *Question:* Why don't we need the phasor transform there?

EXAMPLE 9-10 —————————————————————————

A three-loop network is shown in Figure 9-20(a), including mutual inductance and a dependent source (but not the kitchen sink).

Solution. The phasor transformed circuit, with impedances, is shown in Figure 9-20(b). In particular, note the impedance due to M, the mutual inductance; you solved it, we hope, in Problem 9-34.

KVL around the first loop, with $(+)$ for a drop and $(-)$ for a rise, reads

$$(-j1)(\mathbf{I}_1 - \mathbf{I}_3) - \tfrac{1}{2}(\mathbf{I}_1 - \mathbf{I}_3) + 3\mathbf{I}_1 + j3\mathbf{I}_1 + j2.5\mathbf{I}_3 = 0$$

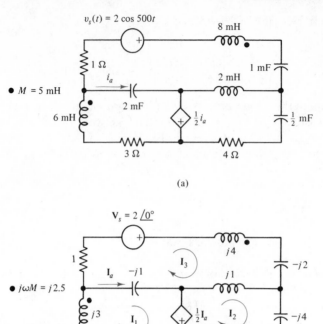

Figure 9-20   Example 9-10.

Here we use $\mathbf{I}_a = (\mathbf{I}_1 - \mathbf{I}_3)$. The mutually induced voltage is the last term in this equation. It is a drop across the $j3$ inductor, in the direction of the loop, because the inducing current $\mathbf{I}_3$ does not enter its own dot on the $j4$ inductor. Around the second loop we write

$$j1(\mathbf{I}_2 - \mathbf{I}_3) + (-j4)\mathbf{I}_2 + 4\mathbf{I}_2 + \tfrac{1}{2}(\mathbf{I}_1 - \mathbf{I}_3) = 0$$

Around the third loop we have

$$-2\underline{/0^\circ} + j4\mathbf{I}_3 + j2.5\mathbf{I}_1 + (-j2)\mathbf{I}_3 + j1(\mathbf{I}_3 - \mathbf{I}_2) + (-j1)(\mathbf{I}_3 - \mathbf{I}_1) + \mathbf{I}_3 = 0$$

Here, the term $j2.5\mathbf{I}_1$ is the mutually induced voltage in the $j4$ inductor due to $\mathbf{I}_1$.
Collect terms in matrix form

$$\begin{bmatrix} 2.5 + j2 & 0 & 0.5 + j3.5 \\ 0.5 & 4 - j3 & -0.5 - j1 \\ j3.5 & -j1 & 1 + j2 \end{bmatrix} \begin{bmatrix} \mathbf{I}_1 \\ \mathbf{I}_2 \\ \mathbf{I}_3 \end{bmatrix} = \begin{bmatrix} 0 \\ 0 \\ 2\underline{/0^\circ} \end{bmatrix}$$

The solution of these three linear equations yields

$$\mathbf{I}_1 = -0.316 - j0.444 = 0.545\underline{/-125.4^\circ}$$
$$\mathbf{I}_2 = -0.042 + j0.157 = 0.163\underline{/104.9^\circ}$$
$$\mathbf{I}_3 = 0.484 + j0.097 = 0.494\underline{/11.3^\circ}$$

Therefore

$$i_1(t) = 0.545 \cos (500t - 125.4°)$$
$$i_2(t) = 0.163 \cos (500t + 104.9°)$$
$$i_3(t) = 0.494 \cos (500t + 11.3°)$$

As a final note, this network is nonreciprocal, since $\mathbf{Z}_t$ is not symmetric. ■

Probs. 9-41, 9-42, 9-43

9-10 MORE ON PHASOR DIAGRAMS

The phasor transform method not only provides a powerful algebraic tool for solution; it can be used also in a graphical way, with amazing simplicity and results. After all, phasors are complex numbers. In a phasor diagram, they can be added and subtracted graphically. The results can be measured directly with a ruler and a protractor, and the answer is readily available from the scale used in that diagram.

Several such simple graphical solutions were given in some of the problems so far. Let us illustrate the method with a few additional examples.

EXAMPLE 9-11 _____

In the network shown in Figure 9-21 actual measurements with a voltmeter and an ammeter provided the following values:

$$|\mathbf{V}_R| = 60 \text{ V} \qquad |\mathbf{V}_L| = 50 \text{ V} \qquad |\mathbf{V}_C| = 100 \text{ V} \qquad |\mathbf{I}_R| = 10 \text{ A}$$

The absolute value signs remind us that voltmeters and ammeters give only scalar (pure number) readings.

It is required to draw a complete phasor diagram and, from it, to determine the source voltage $\mathbf{V}_s$.

Solution. In Figure 9-21(b) we draw to scale $\overrightarrow{OH} = \mathbf{I}_R = 10\underline{/0°}$. The choice of 0° for $\mathbf{I}_R$ is arbitrary, since nothing else was given. Since $\mathbf{V}_R$ is in phase with $\mathbf{I}_R$, we know that $\mathbf{V}_R = 60\underline{/0°}$, shown as $\overrightarrow{OA}$. The phasor $\overrightarrow{OH}$ is also $\mathbf{I}_L$ and $\mathbf{I}_C$, since

$$\mathbf{I}_R = \mathbf{I}_L = \mathbf{I}_C$$

in this circuit.

Next, we know that $\mathbf{V}_L$ leads $\mathbf{I}_L$ by 90°. Draw it to scale as $\overrightarrow{OB}$. Next, we know that $\mathbf{V}_C$ lags behind $\mathbf{I}_C$ by 90°; it is drawn to scale as $\overrightarrow{OD}$.

KVL around the loop tells us that

$$\mathbf{V}_s = \mathbf{V}_R + \mathbf{V}_L + \mathbf{V}_C$$

The phasor (complex number) addition is done by the familiar "head-to-tail" rule: Start with $\mathbf{V}_R = \overrightarrow{OA}$; at the head A of $\mathbf{V}_R$ draw $\mathbf{V}_L = \overrightarrow{AE} = \overrightarrow{OB}$, then at the head E of $\mathbf{V}_L$ draw $\mathbf{V}_C = \overrightarrow{OD} = \overrightarrow{EF}$. The sum $\mathbf{V}_R + \mathbf{V}_L + \mathbf{V}_C$ is therefore $\overrightarrow{OA} + \overrightarrow{AE} + \overrightarrow{EF} = \overrightarrow{OF} = \mathbf{V}_s$. Measuring the length of $\overrightarrow{OF}$ yields

$$|\mathbf{V}_s| = 78 \text{ V}$$

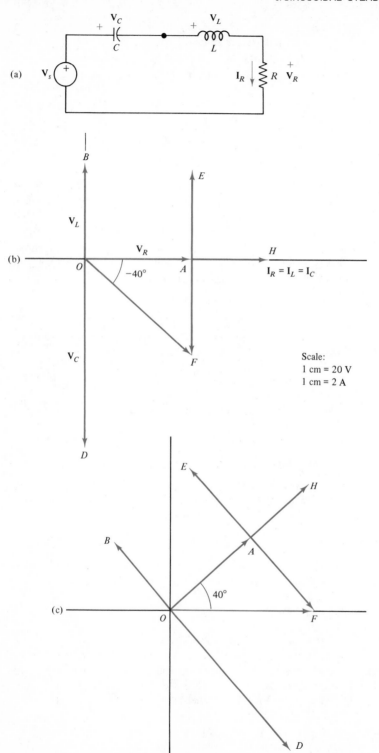

Figure 9-21 Example 9-11.

on the scale used here. The angle of V_s is measured as $-40°$. Thus

$$V_s = 78\underline{/-40°}$$

and

$$v_s(t) = 78 \cos (\omega t - 40°)$$

Take a minute or so to verify this answer analytically. You'll be pleasantly surprised.

What if we were told, in advance, that the phase angle of the source is $0°$? Our entire phasor diagram is still valid, but it must be rotated about O counterclockwise until $\overrightarrow{OF}$ is horizontal at $0°$. This is shown in Figure 9-21(c). Then we have $V_R = 60\underline{/40°}$, $V_L = 50\underline{/130°}$, and $V_C = 100\underline{/-50°}$, read directly from the phasor diagram. This illustrates the arbitrariness in choosing I_R at $0°$, unless given otherwise. In fact, the general rule is: *Unless otherwise given, choose the common phasor at $0°$ when possible.* Here, in the series circuit, the common phasor is $I_R = I_L = I_C$.

We recognize here also an important fact: The voltage across a single element in a circuit, V_C, can *exceed* in magnitude that of the total voltage applied, V_s. In other words, the length (magnitude) of $\overrightarrow{OD}$ can be greater than the length of $\overrightarrow{OF}$.

An unfortunately common mistake is to say that since the measured voltages across R, L, and C are 60, 50, and 100, then "obviously" the source voltage is $60 + 50 + 100 = 210$ V. The mistake here is, of course, in neglecting the phase relations among those voltages; while it *is* true that

$$V_s = V_R + V_L + V_C$$

in magnitude, however,

$$|V_s| = |V_R + V_L + V_C| \neq |V_R| + |V_L| + |V_C|$$

Probs. 9-44, 9-45

EXAMPLE 9-12

In Figure 9-22(a) we show an electric generator, a source of electric energy. The three generated voltages are sinusoidal. If

$$v_a(t) = V_m \cos \omega t$$
$$v_b(t) = V_m \cos (\omega t - 120°)$$

find $v_c(t)$ using a graphical phasor approach.

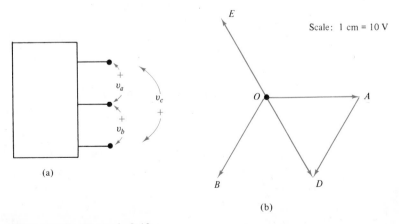

Scale: 1 cm = 10 V

(a)

(b)

Figure 9-22 Example 9-12.

Solution.   We have from KVL

$$v_a(t) + v_b(t) + v_c(t) = 0$$

or, in phasor notation,

$$\mathbf{V}_a + \mathbf{V}_b + \mathbf{V}_c = 0$$

That is,

$$\mathbf{V}_c = -(\mathbf{V}_a + \mathbf{V}_b)$$

The plot of $\mathbf{V}_a = V_m\underline{/0^\circ}$ is shown as $\overrightarrow{OA}$. The phasor $\mathbf{V}_b = V_m\underline{/-120^\circ}$ is drawn as $\overrightarrow{OB}$, of equal magnitude and lagging 120° behind $\mathbf{V}_a$. The sum $(\mathbf{V}_a + \mathbf{V}_b)$ is $\overrightarrow{OA} + \overrightarrow{AD} = \overrightarrow{OD}$, with $\overrightarrow{AD} = \overrightarrow{OB}$. Therefore $\mathbf{V}_c = -\overrightarrow{OD} = \overrightarrow{OE}$, and the geometry of the diagram tells us immediately that

$$\mathbf{V}_c = V_m\underline{/120^\circ}$$

In the time domain, then,

$$v_c(t) = V_m \cos{(\omega t + 120^\circ)} \qquad \blacksquare$$

We see that the phasor diagram provides a very clear picture of the relationships among voltages and currents, as well as a quick geometrical way of calculation. In our studies we will continue to use phasor diagrams whenever they are appropriate.

9-11 SERIES AND PARALLEL CONNECTIONS. VOLTAGE AND CURRENT DIVIDERS

Another useful extension from resistive networks (Chapter 2) is applicable to impedances. We say that several impedances are connected in series if the same current phasor flows through them. In Figure 9-23 we show n impedances, $\mathbf{Z}_1, \mathbf{Z}_2, \ldots, \mathbf{Z}_n$, connected in series. KVL, in phasor form, reads here

$$-\mathbf{V} + \mathbf{Z}_1\mathbf{I} + \mathbf{Z}_2\mathbf{I} + \cdots + \mathbf{Z}_n\mathbf{I} = 0 \qquad (9\text{-}79)$$

or

$$\mathbf{V} = (\mathbf{Z}_1 + \mathbf{Z}_2 + \cdots + \mathbf{Z}_n)\mathbf{I} = \mathbf{Z}_{eq}\mathbf{I} \qquad (9\text{-}80)$$

That is

$$\mathbf{Z}_{eq} = \sum_{k=1}^{n} \mathbf{Z}_k \qquad (9\text{-}81)$$

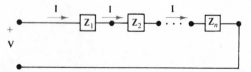

Figure 9-23   Series-connected impedances.

The rule is, then: Impedances in series add up to give the total equivalent impedance. In retrospect, the rule about series resistances adding up (Chapter 4) is a special case.

EXAMPLE 9-13 _____

Consider the series connection of R and L with a sinusoidal voltage source.

    Solution. The total impedance seen by the source is

$$\mathbf{Z} = R + jX_L$$

where X_L is the reactance of the inductor. Therefore the phasor current will be

$$\mathbf{I} = \frac{\mathbf{V}}{\mathbf{Z}} = \frac{\mathbf{V}}{R + j\omega L}$$

If $\mathbf{V} = V_m\underline{/0°}$, then

$$\mathbf{I} = \frac{V_m\underline{/0°}}{\sqrt{R^2 + \omega^2 L^2}\ \underline{/\tan^{-1}(\omega L/R)}} = I_m\ \left\lvert\ -\tan^{-1}\frac{\omega L}{R}\right.$$

as obtained by time-domain methods in Equations 9-16 and 9-18.
    The circuit and the phasor diagram are shown in Figure 9-24. From it we see that in such a circuit, $\mathbf{I}$ always lags $\mathbf{V}$. We refer to such circuits as *inductive.*

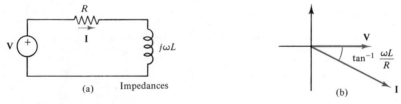

Figure 9-24 Example 9-13.   ■

EXAMPLE 9-14 _____

A series *RC* circuit with a sinusoidal source is shown in Figure 9-25.

    Solution. We write immediately

$$\mathbf{Z} = R + \frac{1}{j\omega C} = R - j\frac{1}{\omega C} = R + jX_C$$

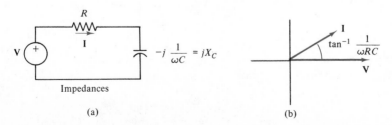

Figure 9-25 Example 9-14.

where X_C is the negative reactance of the capacitor. Therefore

$$I = \frac{V}{Z} = \frac{V_m\underline{/0°}}{[R^2 + (1/\omega^2 C^2)]^{1/2}\underline{/\tan^{-1}(-1/\omega CR)}} = I_m\underline{\bigg/\tan^{-1}\frac{1}{\omega RC}}$$

and I leads V. Such a circuit is called *capacitive*. ∎

In the general case, a total impedance can be written as

$$Z = R + jX_L \qquad X_L > 0 \tag{9-82}$$

for an inductive circuit, or as

$$Z = R + jX_C \qquad X_C < 0 \tag{9-83}$$

for a capacitive circuit. Since $V = ZI$, the angle of Z is added to the angle of I to give the angle of V. Therefore, in the inductive case, I lags behind V by

$$\theta_L = \tan^{-1}\frac{X_L}{R} \tag{9-84}$$

and in the capacitive case I leads V by

$$\theta_C = \tan^{-1}\frac{X_C}{R} \tag{9-85}$$

EXAMPLE 9-15 ──────────────────────────────────────

A series *RLC* circuit is shown in Figure 9-26(a). Calculate I and draw the phasors V and I.

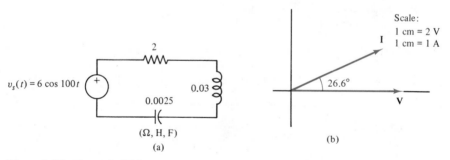

Figure 9-26 Example 9-15.

Solution. Here

$$Z = R + j\omega L + \frac{1}{j\omega C} = 2 + j3 + \frac{1}{j0.25} = 2 - j1$$

so the circuit is capacitive. We have

$$I = \frac{6\underline{/0°}}{2 - j1} = \frac{6\underline{/0°}}{2.24\underline{/-26.6°}} = 2.68\underline{/26.6°}$$

and the phasor diagram is shown in Figure 9-26(b).

Probs. 9-46,
9-47 ∎

EXAMPLE 9-16 _____

For the same circuit as in Figure 9-26, let the source be a unit step voltage

$$v_s(t) = u(t)$$

What is the impedance of the R, L, C series connection?

> *Solution.* Careful, now! If you are tempted to write $j\omega L$ or $1/(j\omega C)$, what *is* ω? There is no ω here. The source is not sinusoidal, and so the whole idea of sinusoidal impedance cannot be used. What can be done? The answer will have to wait until Chapters 13 and 14, where a more generalized impedance will be developed. In the meantime, the classical time-domain methods must be used to solve such problems. This was done in Chapter 8. ∎

Several admittances are said to be connected *in parallel* if they have the same voltage phasor **V** across them. In Figure 9-27 we show n such admittances. By a dual derivation of Equations 9-79 to 9-81, we get

$$\mathbf{Y}_{eq} = \sum_{k=1}^{n} \mathbf{Y}_k \tag{9-86}$$

the total equivalent admittance being the sum of these admittances. This result, when applied to pure resistors, becomes the special case of adding up conductances (Chapter 4).

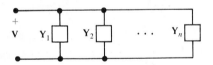

Figure 9-27 Parallel-connected admittances.

EXAMPLE 9-17 _____

A parallel connection of R ($=1/G$) and L is excited by a sinusoidal current source.

> *Solution.* The total admittance seen by the source is
>
> $$\mathbf{Y} = G + \frac{1}{j\omega L} = G + jB_L$$
>
> where B_L is the negative susceptance of the inductor. If $\mathbf{I} = I_m\underline{/0°}$, then the phasor voltage across the admittance will be according to Ohm's law
>
> $$\mathbf{V} = \frac{\mathbf{I}}{\mathbf{Y}} = \frac{I_m\underline{/0°}}{\sqrt{G^2 + B_L^2}\underline{/\tan^{-1}(-1/\omega LG)}} = V_m\underline{\left|\tan^{-1}\frac{1}{\omega LG}\right.}$$
>
> The circuit and the phasor diagram are shown in Figure 9-28, showing the **V** leading **I**, as it should in an inductive circuit.

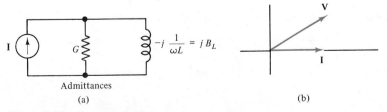

Admittances
(a)

(b)

Figure 9-28 Example 9-17.

∎ Prob. 9-48

Generally, a total admittance can be written as

$$Y = G + jB_L \qquad B_L < 0 \tag{9-87}$$

for an inductive circuit, and

$$Y = G + jB_C \qquad B_C > 0 \tag{9-88}$$

for a capacitive circuit. Since $I = YV$, the angle of Y is added to the angle of V to give the angle of I. In an inductive circuit, then, the voltage V leads the current by

$$\theta_L = \tan^{-1}\frac{B_L}{G} \tag{9-89}$$

while in a capacitive circuit V lags behind I by

$$\theta_C = \tan^{-1}\frac{B_C}{G} \tag{9-90}$$

There is no need to memorize equations like 9-84, 9-85, 9-89, or 9-90. Rather, solve each specific problem on its own, and the results will follow.

EXAMPLE 9-18 _____

For the circuit shown in Figure 9-29(a) we are given

$$R_1 = 1.1\ \Omega \qquad R_2 = 2\ \Omega \qquad L_3 = 100\ \mu\text{H}$$
$$L_1 = 150\ \mu\text{H} \qquad C_2 = 50\ \mu\text{F}$$

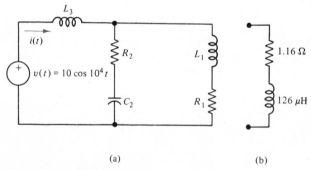

(a)   (b)

Figure 9-29   Example 9-18.

Solution.   The impedance of the $R_1 L_1$ branch is a series connection

$$Z_1 = 1.1 + j10^4(150)10^{-6} = 1.1 + j1.5 \quad \text{inductive}$$

The impedance of the $R_2 C_2$ branch is, again, a series connection

$$Z_2 = 2 + \frac{1}{j10^4(50)10^{-6}} = 2 - j2 \quad \text{capacitive}$$

The R_1L_1 and R_2C_2 branches are connected in parallel; therefore, their total admittance is

$$\mathbf{Y}_{12} = \mathbf{Y}_1 + \mathbf{Y}_2 = \frac{1}{1.1 + j1.5} + \frac{1}{2 - j2}$$
$$= (0.318 - j0.434) + (0.25 + j0.25)$$
$$= 0.818 - j0.184 = 0.838\underline{/-12.7^\circ} \quad \text{inductive}$$

How do we know that $\mathbf{Y}_{12}$ is inductive? Simple: The susceptance, the imaginary part of $\mathbf{Y}_{12}$, is negative. Alternatively, Ohm's law is $\mathbf{I} = \mathbf{YV}$; so if $\mathbf{V}$ is at an (arbitrary) angle α, the angle of $\mathbf{I}$ will be $(-12.7^\circ + \alpha)$, which is *lagging* behind $\mathbf{V}$. In an inductive circuit $\mathbf{I}$ lags behind $\mathbf{V}$.

Finally, the total impedance $\mathbf{Z}_s$ seen by the source is a series connection of L_3 and the rest

$$\mathbf{Z}_s = j10^4(100)10^{-6} + \frac{1}{\mathbf{Y}_{12}} = j1 + 1.19\underline{/12.7^\circ}$$
$$= j1 + 1.16 + j0.26 = 1.16 + j1.26 = 1.71\underline{/47.4^\circ}$$

and is inductive because, the reactive part of $\mathbf{Z}_s$ is positive. Alternatively, in $\mathbf{V} = \mathbf{ZI}$, the angle of $\mathbf{Z}$ is added to that of $\mathbf{I}$. Here, then, the phasor $\mathbf{V}$ leads $\mathbf{I}$ by 47.4°, clearly an inductive circuit.

As far as the source is concerned, the entire circuit in Figure 9-29(a) is equivalent to the simpler circuit in Figure 9-29(b). The current $i(t)$ in both cases is found from

$$\mathbf{I} = \mathbf{Y}_s\mathbf{V} = \frac{1}{1.71\underline{/47.4^\circ}} \; 10\underline{/0^\circ} = 5.8\underline{/-47.4^\circ}$$

therefore

$$i(t) = 5.8\cos(10^4t - 47.4^\circ)$$

Probs. 9-49, 9-50

If there are dependent sources, or if elements are not in simple series-parallel combinations, the more general approach is similar to the one learned in Chapter 4: Connect an arbitrary phasor source, and calculate the proper phasor response. We illustrate with an example similar to Example 4-4.

EXAMPLE 9-19

Find the equivalent impedance at terminals 1-2 of the network shown in Figure 9-30. Here $\mathbf{Z}_a$ and $\mathbf{Z}_b$ are known impedances, and α is a given constant for the dependent source.

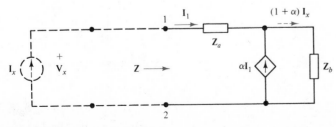

Figure 9-30 Example 9-19.

Solution. We connect an arbitrary current source $\mathbf{I}_x$ as shown in dotted lines, and calculate $\mathbf{V}_x$. Then, obviously, $\mathbf{V}_x = \mathbf{Z}\mathbf{I}_x$ and $\mathbf{Z}$ can be found.

We see that $\mathbf{I}_x = \mathbf{I}_1$, and also that the current through $\mathbf{Z}_b$ is $(1 + \alpha)\mathbf{I}_x$. KVL then yields

$$-\mathbf{V}_x + \mathbf{Z}_a\mathbf{I}_x + \mathbf{Z}_b(1 + \alpha)\mathbf{I}_x = 0$$

or

$$\mathbf{V}_x = [\mathbf{Z}_a + (1 + \alpha)\mathbf{Z}_b]\mathbf{I}_x$$

That is,

$$\mathbf{Z} = \mathbf{Z}_a + (1 + \alpha)\mathbf{Z}_b$$

■ Probs. 9-51
9-52, 9-53

Voltage and Current Dividers   As in the purely resistive case, we derive a useful relation for the phasor voltage divider. Specifically, let the network shown in Figure 9-31 consist of k impedances in series across a total voltage $\mathbf{V}$. Then

$$\mathbf{V}_k = \frac{\mathbf{Z}_k}{\Sigma\mathbf{Z}}\,\mathbf{V} \tag{9-91}$$

is the voltage across any impedance $\mathbf{Z}_k$.

In a dual fashion, the current divider equation for Figure 9-32 is

$$\mathbf{I}_k = \frac{\mathbf{Y}_k}{\Sigma\mathbf{Y}}\,\mathbf{I} \tag{9-92}$$

Derive both expressions in detail. It takes only a couple of simple steps.

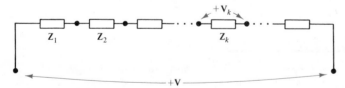

Figure 9-31   Voltage divider.

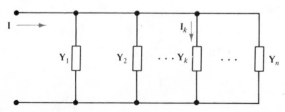

Figure 9-32   Current divider.

EXAMPLE 9-20 _____

In the network of Figure 9-26, Example 9-15, the voltage drop across the inductor is

$$\mathbf{V}_L = \frac{j3}{2-j}\, 6\underline{/0^\circ} = 8.04\underline{/116.6^\circ}$$

This checks with **I** as obtained there, since

$$\mathbf{V}_L = \mathbf{Z}_L \mathbf{I}_L = (j3)2.68\underline{/26.6^\circ} = 8.04\underline{/116.6^\circ}$$

Similarly,

$$\mathbf{V}_R = \frac{2}{2-j}\, 6\underline{/0^\circ} = 5.36\underline{/26.6^\circ}$$

and

$$\mathbf{V}_C = \frac{-j4}{2-j}\, 6\underline{/0^\circ} = 10.71\underline{/-63.4^\circ}$$ ∎

EXAMPLE 9-21 _____

In Example 9-7, Figure 9-17, we have

$$\mathbf{I}_R = -\frac{3}{-j2+3+j5}\, 10\underline{/0^\circ} = 7.07\underline{/135^\circ}$$

$$\mathbf{I}_L = -\frac{-j2}{3+j3}\, 10\underline{/0^\circ} = 4.71\underline{/45^\circ}$$

$$\mathbf{I}_C = -\frac{j5}{3+j3}\, 10\underline{/0^\circ} = 11.78\underline{/-135^\circ}$$

Here, the minus sign in each current divider equation accounts for the conventional reference of each current, the tail being at the $(+)$ of **V**, by comparison with the given reference for the total **I**.   ∎ Prob. 9-54

9-12 THÉVENIN'S AND NORTON'S THEOREMS

These two powerful theorems are also the extensions of their resistive cases studied in Chapter 5. Let us outline each one in turn here.

As far as an arbitrary observer network is concerned, Figure 9-33(a), the _linear_ network to the left at terminals _A-B_ can be replaced by its _Thévenin equivalent circuit_, as shown in Figure 9-33(b), consisting of a phasor Thévenin voltage source $\mathbf{V}_{Th}$ in series with a Thévenin impedance $\mathbf{Z}_{Th}$.

To obtain $\mathbf{V}_{Th}$, we disconnect the observer from terminals _A-B_, as in Figure 9-33(c). Leave the linear network untouched. The Thévenin voltage $\mathbf{V}_{Th}$ is the open-circuit voltage that appears across terminals _A-B_.

To calculate $\mathbf{Z}_{Th}$, take the linear network in Figure 9-33(c) and set all the _independent_ sources in it to zero: Independent current sources are open-circuited, and independent voltage sources are short-circuited, as in Figure 9-33(d). Dependent sources must remain untouched. Then find the impedance seen at terminals _A-B_, by series-parallel calculations (when possible), or by the general method studied earlier. This is $\mathbf{Z}_{Th}$. Connect $\mathbf{V}_{Th}$ and $\mathbf{Z}_{Th}$ to the observer, as in Figure 9-33(b).

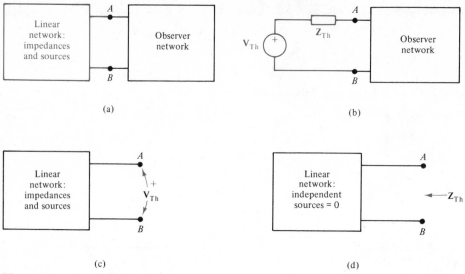

(a)   (b)

(c)   (d)

Figure 9-33 Thévenin's theorem.

The only restriction on the validity of Thévenin's theorem is that there must be *no coupling* between the linear network and the observer. Such a coupling can occur through mutual inductance (one inductor inside the observer, the other inside the linear network), or with a controlled source and its controller (one inside the linear network, the other inside the observer).

EXAMPLE 9-22

Let us find V_b across the capacitor in Example 9-8, Figure 9-18, by Thévenin's theorem.

Solution. Here, the capacitor is the observer, and everything else is in the linear network. As the first step, remove the observer. This is shown in Figure 9-34(b). The calculation of V_{Th} is then simply by a voltage divider

$$V_{Th} = \frac{\frac{1}{3} + j1}{\frac{1}{3} + j1 + j0.67} \, 1.2\underline{/0^\circ} = 0.744\underline{/-7.13^\circ}$$

Notice how easy this is: The original network is complicated, but Thévenin's theorem simplifies it.

In the next step, we set all independent sources to zero, resulting in Figure 9-34(c). The Thévenin impedance between A and B is then

$$Z_{Th} = (j0.67)\|(\tfrac{1}{3} + j1) = 0.414\underline{/82.9^\circ}$$

Be sure to check the numerical steps in detail. Finally, then, we draw Figure 9-34(d), replacing the observer and connecting it to the Thévenin equivalent network. Here V_b may be found again by a voltage divider equation

$$V_b = \frac{-j0.5}{-j0.5 + 0.414\underline{/82.9^\circ}} \, V_{Th} = 3.61\underline{/-36.67^\circ}$$

which agrees with Example 9-8.

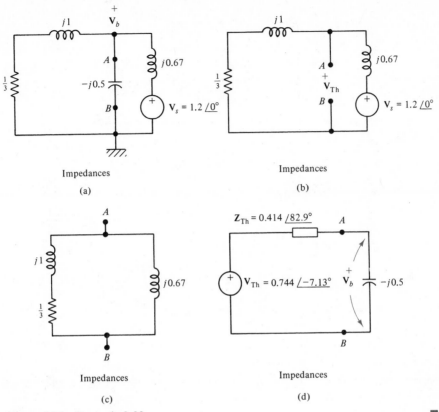

Figure 9-34   Example 9-22.

EXAMPLE 9-23

To illustrate Thévenin's theorem with dependent sources, consider the network in Figure 9-35(a). We wish to find the Thévenin equivalent circuit with respect to $\mathbf{Z}_o$ as the observer. Specifically, let $\mathbf{Z}_o = R_o$, a resistor.

Solution.   In Figure 9-35(b), the impedance $\mathbf{Z}_o$ is removed. The controller $\hat{\mathbf{I}}_1$ and the controlled source $\alpha\hat{\mathbf{I}}_1$ carry the "hat" (^) to remind us that these currents are not the same as in the original network in Figure 9-35(a), because of the removal of $\mathbf{Z}_o$. Still, the controller-controlled relation is maintained. Also, in Figure 9-35(b), $\hat{\mathbf{I}}_1$ flows as shown in $\mathbf{Z}_c$, since terminal A is open-circuited. We have therefore

$$\mathbf{V}_{Th} = -\mathbf{Z}_c\hat{\mathbf{I}}_1$$

To find $\hat{\mathbf{I}}_1$, write KVL around the outer loop

$$-\mathbf{V}_i - \mathbf{Z}_b(1 + \alpha)\hat{\mathbf{I}}_1 - \mathbf{Z}_a\hat{\mathbf{I}}_1 - \mathbf{Z}_c\hat{\mathbf{I}}_1 = 0$$

or

$$\hat{\mathbf{I}}_1 = \frac{\mathbf{V}_i}{\mathbf{Z}_a + \mathbf{Z}_b(1 + \alpha) + \mathbf{Z}_c}$$

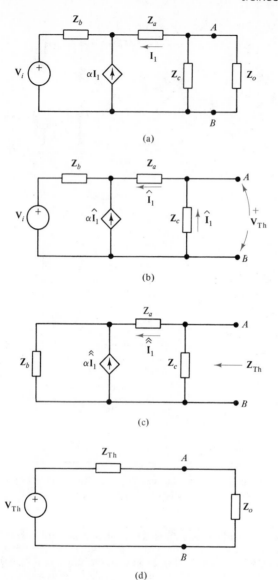

Figure 9-35 Example 9-23.

So

$$\mathbf{V}_{\text{Th}} = -\mathbf{Z}_c \hat{\mathbf{I}}_1 = \frac{-\mathbf{Z}_c \mathbf{V}_i}{\mathbf{Z}_a + \mathbf{Z}_b(1 + \alpha) + \mathbf{Z}_c}$$

To find $\mathbf{Z}_{\text{Th}}$, we set $\mathbf{V}_i$ to zero in Figure 9-35(b) and obtain the circuit of Figure 9-35(c). We recognize that $\mathbf{Z}_{\text{Th}}$ is the parallel combination of $\mathbf{Z}_c$ and the impedance found earlier in Example 9-19, Figure 9-30. Therefore

$$\mathbf{Z}_{\text{Th}} = \mathbf{Z}_c \| [\mathbf{Z}_a + (1 + \alpha)\mathbf{Z}_b]$$

With $\mathbf{V}_{\text{Th}}$ and $\mathbf{Z}_{\text{Th}}$ calculated, we connect the final circuit as in Figure 9-35(d). ■

EXAMPLE 9-24 _____

In Figure 9-35(a), find the Thévenin equivalent with respect to $\mathbf{Z}_o$ as the observer, when $\mathbf{Z}_o$ is an inductor ($\mathbf{Z}_o = j\omega L_o$) coupled mutually to another inductor, $\mathbf{Z}_b = j\omega L_b$. The mutual impedance is, as usual, $j\omega M$ with its given dot markings on L_o and L_b.

> *Solution.* Hold it! Thévenin's theorem cannot be applied here because of the coupling between the observer and the linear network. In fact, it is instructive to solve this problem by a conventional loop analysis and to recognize that $\mathbf{I}_o$, the current through $\mathbf{Z}_o$, *is not* of the form dictated by Figure 9-35(d)

Probs. 9-55, 9-56, 9-57, ■ 9-58

$$\mathbf{I}_o \neq \frac{\mathbf{V}_{Th}}{\mathbf{Z}_{Th} + \mathbf{Z}_o}$$

In summary, then: Thévenin's theorem can be applied provided the network seen by the observer consists of *linear* elements (R, L, C, M) plus independent and dependent sources. The observer is totally arbitrary, linear or nonlinear. There must not be any coupling between the linear network and the observer.

In a dual manner, we can replace the linear network by its *Norton's equivalent circuit*, as shown in Figure 9-36(b). The equivalent current source (phasor) $\mathbf{I}_N$ is the current in the short-circuit terminals *A-B* after removing the observer, as in Figure 9-36(c). The Norton impedance, $\mathbf{Z}_N$, *is the same* as the Thévenin impedance, $\mathbf{Z}_{Th}$, but here it is connected in parallel with $\mathbf{I}_N$, as shown in Figure 9-36(b).

To establish the equivalence between Norton and Thévenin, consider Figure 9-36(d) where the linear network is represented by its Thévenin equivalent circuit. Shorting terminals *A-B* here is the same as shorting them in Figure 9-36(c). Therefore

$$\mathbf{I}_N = \frac{\mathbf{V}_{Th}}{\mathbf{Z}_{Th}} \qquad\qquad (9\text{-}93)$$

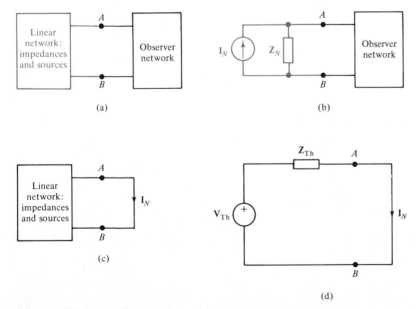

Figure 9-36 Norton's theorem.

On the other hand, let us find the Thévenin equivalent circuit seen by the observer in Figure 9-36(b): After the observer is removed, the open-circuit voltage across A-B is

$$\mathbf{V}_{Th} = \mathbf{Z}_N \mathbf{I}_N \qquad (9\text{-}94)$$

which is the same as in Equation 9-93 if

$$\mathbf{Z}_N = \mathbf{Z}_{Th} \qquad (9\text{-}95)$$

We have therefore established the equivalence between the Norton and Thévenin circuits.

The use of Norton's circuit may sometimes be easier, particularly in finding $\mathbf{Z}_N = \mathbf{Z}_{Th}$. Instead of having to connect a test source $\mathbf{I}_x$ and calculating $\mathbf{V}_x$, as in Example 9-19, Figure 9-30, we can short-circuit terminals A-B, as in Figure 9-36(c), calculate $\mathbf{I}_N$, and then find $\mathbf{Z}_N$ from Equation 9-93 or 9-94. The following example illustrates this method.

EXAMPLE 9-25 _____

In the network shown in Figure 9-37(a) the observer has already been removed.

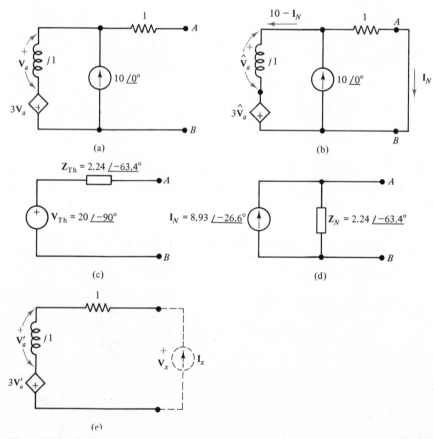

Figure 9-37 Example 9-25.

Solution. We find immediately $\mathbf{V}_{AB} = \mathbf{V}_{\text{Th}}$ as

$$\mathbf{V}_{\text{Th}} = \mathbf{V}_a - 3\mathbf{V}_a = -2\mathbf{V}_a = -2(j1)10\underline{/0°} = 20\underline{/-90°}$$

since there is no drop across the 1-Ω resistor.

Now short-circuit terminals *A-B*, as in Figure 9-37(b). Again, the "hat" ($\frown$) over the controller and the controlled source remind us that these are not the same as in Figure 9-37(a). In Figure 9-37(b) we write KVL around the outer loop as

$$1\mathbf{I}_N + 3\hat{\mathbf{V}}_a - \hat{\mathbf{V}}_a = 0 \qquad \therefore \quad \mathbf{I}_N = -2\hat{\mathbf{V}}_a$$

and

$$\hat{\mathbf{V}}_a = j1(10 - \mathbf{I}_N)$$

Consequently

$$-\frac{\mathbf{I}_N}{2} = j10 - j\mathbf{I}_N$$

or

$$\mathbf{I}_N = 8.93\underline{/-26.6°}$$

Finally

$$\mathbf{Z}_N = \mathbf{Z}_{\text{Th}} = \frac{\mathbf{V}_{\text{Th}}}{\mathbf{I}_N} = \frac{20\underline{/-90°}}{8.93\underline{/-26.6°}} = 2.24\underline{/-63.4°}$$

The Thévenin equivalent circuit is shown in Figure 9-37(c), while Norton's is shown in Figure 9-37(d).

To confirm our results, let us recalculate $\mathbf{Z}_{\text{Th}}$ in Figure 9-37(e) with a test current $\mathbf{I}_x$. As required, the independent current source is set to zero, but the dependent voltage source remains. We write KVL

$$-\mathbf{V}_x + 1\mathbf{I}_x + j1\mathbf{I}_x - 3(j1)\mathbf{I}_x = 0$$

or

$$\mathbf{V}_x = (1 - j2)\mathbf{I}_x$$

That is,

$$\mathbf{Z}_{\text{Th}} = 1 - j2 = 2.24\underline{/-63.4°}$$

There are no fixed rules about which method works better or faster. You will develop your own taste and preference as you solve more and more problems. The real beauty is in the variety of the possible approaches! ■

Probs. 9-59, 9-60

In concluding this discussion, let us recognize that the equivalence between Thévenin's and Norton's circuits is, in effect, a *source transformation*: To an observer, a (phasor) voltage source in series with an impedance is equivalent to a (phasor) current source in parallel with that impedance, and with Equation 9-93 relating them.

Probs. 9-61, 9-62

PROBLEMS

9-1 Convert the following waveforms into their cosine equivalent:

(a)
$$v_1(t) = 20 \sin \left(377t + \frac{\pi}{4} \right)$$

(b)
$$i_2(t) = 13.6 \sin \left(377t + \frac{\pi}{10} \right)$$

(c)
$$i_3(t) = 0.1 \sin \left(10^6 t + \frac{\pi}{2} \right)$$

(d)
$$v_4(t) = 4 \sin (t - \pi)$$

9-2 Assume that the speed of propagation of radio waves is the speed of light, i.e., $c = 3 \times 10^8$ m/s. The *wavelength*, in meters, of a sinusoidal signal is given by λ

$$\lambda = \frac{c}{f}$$

What is the range of wavelengths of the AM band? The FM band?

9-3 The International Telecommunication Union (ITU) has defined several frequency bands. Of these the very high frequency (VHF) range is 30 to 300 MHz, the ultra high frequency (UHF) is 300 to 3000 MHz, and the very low frequency (VLF) is 3 to 30 kHz. List the wavelength ranges for VHF, UHF, VLF.

9-4 A typical syllable of human speech has a duration of approximately $\frac{1}{8}$ s. What is the corresponding frequency? Radian frequency?

9-5 Two musical notes are said to be an *octave* apart if their frequencies are related by $f_2 = 2f_1$. If the note "middle A" has a frequency of 440 Hz, what is the frequency of a "high A" which is 2 octaves up?

9-6 Plot accurately on graph paper:
(a) $v(t) = 12 \cos (377t + 30°)$ vs. $377t$
(b) $i(t) = 10 \cos (377t - 15°)$ vs. $377t$

Which waveform lags and by what degree?

9-7 Plot accurately $i(t) = 10 \cos (50t - 15°)$ vs. $50t$ and compare with Problem 9-6(b). Can you say which of these two waveforms leads or lags? Why?

9-8 Plot accurately

(a)
$$v_1(t) = 100 \cos (1000t + 10°)$$

(b)
$$v_2(t) = 20 \cos (1000t - 170°)$$

and discuss the leading or lagging between v_1 and v_2.

9-9 The following identity is useful in our studies. Prove it:

$$A \sin x + B \cos x = \sqrt{A^2 + B^2} \cos \left(x - \tan^{-1} \frac{A}{B} \right)$$

Hint: Use the identity for the cosine

$$\cos (x \pm y) = \cos x \cos y \mp \sin x \sin y$$

and the right-angle triangle shown in the figure.

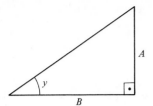

Problem 9-9

9-10 (a) In the parallel RC circuit shown, find the steady-state response $v(t)$ by the classical solution of the differential equation. Draw accurately $i(t)$ and $v(t)$.

(b) Could you have written the solution by inspection, considering the duality of this circuit and of the series RL circuit in the text? Explain fully.

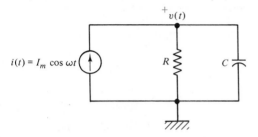

Problem 9-10

9-11 In the series RC circuit shown, use $v_C(t)$ as a state variable. Calculate the steady-state expression for $v_C(t)$ by solving classically the differential equation for it.

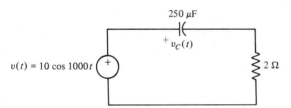

Problem 9-11

9-12 In the op amp circuit shown, find the steady-state output voltage $v_o(t)$ using the classical method of solution.

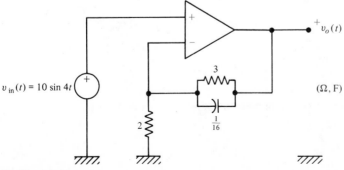

Problem 9-12

9-13 Calculate the steady-state inductive current $i_L(t)$. Note the dependent source!

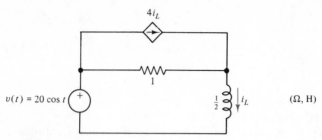

Problem 9-13

9-14 List and describe two additional transform methods, one nonmathematical.

9-15 List all the difficult, direct algebraic operations and their corresponding easy operations in the logarithmic transform.

9-16 Calculate longhand (directly)

$$x = \frac{(2.4)^2}{\sqrt{24}}$$

then use the logarithmic transform. *Note*: Calculators are *not* allowed here, except as a source of logarithm tables.

9-17 For each of the following waveforms: (a) plot it versus t; (b) write its phasor transform; (c) plot the phasor in the complex plane.

    (a) $v_1(t) = 25 \cos (100t - 80°)$

    (b) $i_2(t) = 40 \sin (377t + 25°)$

    (c) $v_3(t) = 170 \cos (t + 90°)$

    (d) $v_4(t) = 32 \cos (10t + 130°)$

    (e) $i_5(t) = -10.6 \cos (20t + 60°)$

Don't forget our convention to use only cosines.

9-18 Plot each phasor on the complex plane; then write the time function represented by it:

    (a) $\mathbf{I}_1 = 3.6\underline{/-14°}$ $(\omega = 10)$

    (b) $\mathbf{V}_2 = -3 + j4$ $(\omega = 377)$

    (c) $\mathbf{I}_3 = -1.2 - j3.0$ $(\omega = 10^4)$

    (d) $\mathbf{V}_4 = 43e^{j\pi/7}$ $(\omega = 200)$

    (e) $\mathbf{I}_5 = 0.058 - j0.067$ $(\omega = 1000)$

9-19 Prove the following properties of the phasor transform:

    (a) The transform of a sum of functions is the sum of the individual phasors

$$[f_1(t) + f_2(t)] \leftrightarrow \mathbf{F}_1 + \mathbf{F}_2$$

    (b) A constant multiplying a function also multiplies its phasor

$$\alpha f(t) \leftrightarrow \alpha \mathbf{F}$$

These properties qualify the phasor as a *linear* transform. Do these properties hold in the logarithmic transform?

9-20 Explain carefully why, despite the property listed in Problem 9-19, we *cannot* take the phasor transform of $f(t) = [10 \cos (377t + 40°) + 3 \cos (100t - 10°)]$.

9-21 In the example shown in Figure 9-11, replace the capacitor by an inductor $L = 0.01$ H. Write the time-domain node equation and solve it by phasors.

9-22 In the op amp circuit shown in Problem 9-12, find the steady-state voltage $v_o(t)$ using phasor methods.

9-23 In the network shown, calculate the steady-state voltage $v_L(t)$ using phasor methods. Reminisce of the classical solution to this problem.

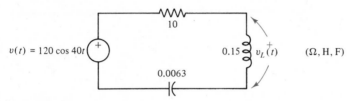

Problem 9-23

9-24 For each of the following equations, calculate the steady-state solution by the phasor method.

(a)
$$4 \frac{dv}{dt} + 3v = 8.5 \cos (2t + 110°)$$

(b)
$$\frac{1}{3} \frac{d^2i}{dt^2} + 6 \frac{di}{dt} + 500i = 0.2 \cos (30t - 14°)$$

(c)
$$\frac{d^2v}{dt^2} + 2 \frac{dv}{dt} + v = 16 \sin 3t$$

(d)
$$\frac{d^2i}{dt^2} + 7 \frac{di}{dt} + 12i = 20 \cos (t + 30°)$$

(e)
$$24 \frac{d^3y}{dt^3} - 15y = 31 \cos \left(\frac{t}{2} + 45° \right)$$

Hint:
$$\frac{d^2v}{dt^2} = \frac{d}{dt} \left(\frac{dv}{dt} \right), \text{ etc.}$$

9-25 Find the *complete* solutions of the differential equations in Problem 9-24 (a), (c), (d), given that

(a)
$$v(0^-) = 0$$

(c)
$$v(0^-) = 1 \qquad \left. \frac{dv}{dt} \right)_{0^-} = -3$$

(d)
$$i(0^-) = 0 \qquad \left. \frac{di}{dt} \right)_{0^-} = -2$$

9-26 Use phasor methods to write $f(t)$ as a *single* cosine waveform

$$f(t) = 10 \cos (377t - 30°) + 20 \cos (377t + 60°) - 40 \cos 377t$$

Next, try to do it directly in the time domain (using Problem 9-9), to really appreciate phasors.

9-27 Use the phasor method to calculate the steady-state $i_1(t)$ in the circuit shown. Do it by loop analysis, starting with the time-domain equations.

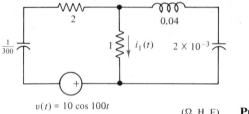

$v(t) = 10 \cos 100t$

$(\Omega, \text{H}, \text{F})$ **Problem 9-27**

9-28 Calculate the steady-state $v_C(t)$ in the circuit shown by phasor methods. Do it by node analysis and write first the time-domain equation(s).

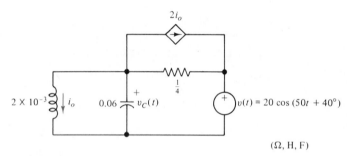

$(\Omega, \text{H}, \text{F})$

Problem 9-28

9-29 A single element is in the box, and the phasor diagram for it is shown to scale. Identify the element (R, L, or C) and its value (in Ω, H, F). Write the expressions for $v(t)$ and $i(t)$.

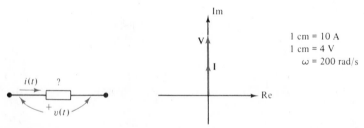

1 cm = 10 A
1 cm = 4 V
$\omega = 200$ rad/s

Problem 9-29

9-30 For the resistor shown, and the given references of $i_R(t)$ and $v_R(t)$, we are told that

$$i_R(t) = 4.3 \cos (100t + 45°)$$

Plot to scale $\mathbf{I}_R$ and $\mathbf{V}_R$. Be careful here!

Problem 9-30

9-31 The figure shows a simple model of a power transmission line, those overhead wires along highways and fields. The impedances are calculated for Europe, where the frequency is $f = 50$ Hz. (a) What is the value of the inductor in henrys? (b) Calculate the impedances of this model in the United States, where $f = 60$ Hz.

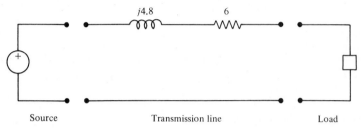

Source   Transmission line   Load

Problem 9-31

9-32 A certain filter network is shown in the figure. Calculate the impedance of each element when the filter is excited by a sinusoidal source at: (a) $f_1 = 100$ Hz, (b) $f_2 = 1$ kHz, (c) $f_3 = 1$ MHz.

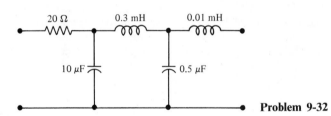

Problem 9-32

9-33 In each case, partial information is given for a single element. Calculate the missing details and draw to scale the phasor diagram:

(a) $\mathbf{V}_R = 10\underline{/115°}$ $v_R(t) = ?$ $R = 0.5\,\Omega$
     $\mathbf{I}_R = ?$ $i_R(t) = ?$ $f = 60$ Hz $\mathbf{Z}_R = ?$

(b) $\mathbf{V}_C = ?$ $v_C(t) = ?$ $C = 100\,\mu F$
     $\mathbf{I}_C = ?$ $i_C(t) = 0.6 \cos(314t - 10°)$ $f = ?$ $\mathbf{Z}_C = ?$
                                                    $X_C = ?$ $B_C = ?$

(c) $\mathbf{V}_L = ?$ $v_L(t) = ?$ $L = 0.3$ mH $f = 10$ kHz
     $\mathbf{I}_L = -4.2\underline{/60°}$ $i_L(t) = ?$ $\mathbf{Z}_L = ?$ $X_L = ?$ $B_L = ?$

(d) $\mathbf{V}_C = ?$ $v_C(t) = 1.1 \cos(2\pi 10^6 t - \theta)$ $C = ?$ $f = ?$
     $\mathbf{I}_C = 0.4\underline{/35°}$ $i_C(t) = ?$ $\mathbf{Z}_C = ?$ $B_C = ?$
                                                      $X_C = ?$ $\theta = ?$

9-34 Develop the phasor relations and the related impedances for two mutually coupled inductors as shown in the figure. In each case start with the time-domain equations, then apply to them the phasor transform.

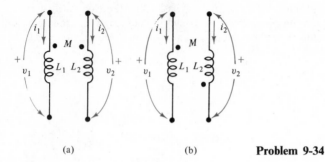

(a)   (b)   **Problem 9-34**

9-35 Calculate the individual branch currents I_R, I_L, and I_C in Example 9-7 and draw them to scale in the phasor diagram of Figure 9-17(c). On this phasor diagram, verify *graphically* that $I = I_R + I_L + I_C$.

9-36 Use phasor analysis to solve the circuit in Figure 9-6 in the text, confirming effortlessly the results in Equations 9-16, 9-18, and 9-19.

9-37 Complete Example 9-8, Figure 9-18, as follows: (a) Calculate the individual branch currents from the knowledge of V_a and V_b. (b) Plot the phasor diagram to scale, showing V_s, V_a, V_b and all the branch currents.

9-38 For Example 9-9, Figure 9-19, calculate all the branch currents (phasors) from the known V_a, V_b, and V_c. Draw these currents to scale on a phasor diagram and verify graphically that KCL is satisfied at each of the three nodes. This is an excellent check on the correctness of the entire solution.

9-39 Set up the phasor node equations, then solve for $v_1(t)$ in the network shown.

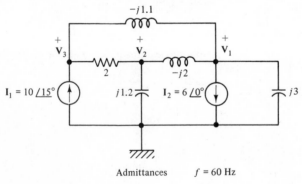

Admittances $f = 60$ Hz

Problem 9-39

9-40 For the op amp circuit shown, calculate the steady-state output voltage $v_o(t)$. Classify in words the operation of this circuit.

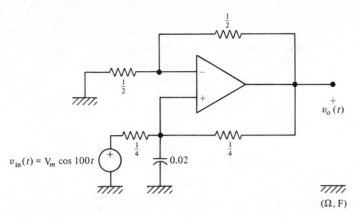

$v_{in}(t) = V_m \cos 100t$

(Ω, F)

Problem 9-40

9-41 Complete Example 9-10 as follows: Calculate the phasor voltage across the $j1$ inductor, the $-j4$ capacitor, and the 4-Ω resistor by using the solutions of $\mathbf{I}_1, \mathbf{I}_2,$ and $\mathbf{I}_3$. Plot these voltages to scale on a phasor diagram and verify graphically that, together with $\frac{1}{2}\mathbf{I}_a$, they satisfy KVL.

9-42 (a) Write the two loop equations for the network shown. (b) Solve for $\mathbf{I}_1$ and $\mathbf{I}_2$. (c) Calculate the time-domain expression for the voltage drop across the capacitor.

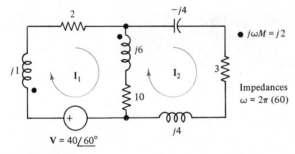

$V = 40\underline{/60°}$

Problem 9-42

9-43 Solve Example 9-8 by loop analysis. From your solution calculate $\mathbf{V}_a$ and $\mathbf{V}_b$ and compare with the answers in the text.

9-44 In Example 9-11, Figure 9-21, we are given the following measurements:

$$|\mathbf{V}_s| = |\mathbf{V}_R| = 60 \text{ V}$$

and there are no defects in the circuit such as open-circuited or short-circuited elements. Draw a voltage phasor diagram. What is $|\mathbf{V}_C|$? $|\mathbf{V}_L|$? Are those answers unique? Explain fully.

9-45 Consider an all-resistive network, similar to Example 9-11 and shown in the figure. It consists of a single input voltage source $\mathbf{V}_s$ and several resistors in an arbitrary connection. No dependent sources and no op amps are included. Can any one of the resistor's voltages exceed in magnitude $|\mathbf{V}_s|$, as happened in Example 9-11? Explain fully.

Problem 9-45

9-46 In Example 9-15, Figure 9-26, let the source be $v(t) = 6 \cos 150t$. Calculate $\mathbf{I}$, and draw the phasor diagram with $\mathbf{V}$ and $\mathbf{I}$. Is this circuit inductive or capacitive?

9-47 In a series RLC circuit excited by $v(t) = V_m \cos \omega t$, find the relationships among R, L, C, and ω for the circuit to be (a) inductive; (b) capacitive; (c) resistive, when $\mathbf{V}$ and $\mathbf{I}$ are in phase.

9-48 In a parallel RC circuit excited by a current source $\mathbf{I} = I_m \underline{/0°}$, calculate the total admittance and the voltage $\mathbf{V}$. Draw a phasor diagram.

9-49 Rework Example 9-18 step by step for

(a) $$v(t) = 10 \cos 100t$$

(b) $$v(t) = 10 \cos 10^6 t$$

9-50 In the network shown, individual elements are given by their impedances. Write the expression for the equivalent impedance of the ladder network as a continued fraction (see Example 4-2 in Chapter 4). Then simplify it to its polar form

$$\mathbf{Z}_{eq} = |\mathbf{Z}_{eq}| \underline{/\theta_z}$$

Is this an inductive or a capacitive circuit?

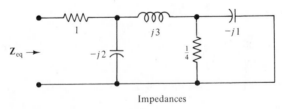

Impedances

Problem 9-50

9-51 Prove that the equivalent impedance of two impedances in parallel is

$$\mathbf{Z}_{eq} = \frac{\mathbf{Z}_1 \mathbf{Z}_2}{\mathbf{Z}_1 + \mathbf{Z}_2}$$

that is, the familiar rule, "product over sum." Prove also, without any fancy algebra, that this rule *cannot* be true for three (or more) impedances

$$Z_{eq} \neq \frac{Z_1 Z_2 Z_3}{Z_1 + Z_2 + Z_3}$$

Hint: Units (dimensions) of this equation. See also Problem 4-2.

9-52 Derive the delta-wye (Δ-Y) equivalence for impedances, by generalizing on Section 4-4 in Chapter 4. Specifically, prove that

$$Z_1 = \frac{Z_b Z_c}{Z_a + Z_b + Z_c} \qquad Z_a = \frac{P}{Z_1}$$

$$Z_2 = \frac{Z_a Z_c}{Z_a + Z_b + Z_c} \qquad Z_b = \frac{P}{Z_2}$$

$$Z_3 = \frac{Z_a Z_b}{Z_a + Z_b + Z_c} \qquad Z_c = \frac{P}{Z_3}$$

where $P = Z_1 Z_2 + Z_2 Z_3 + Z_1 Z_3$.

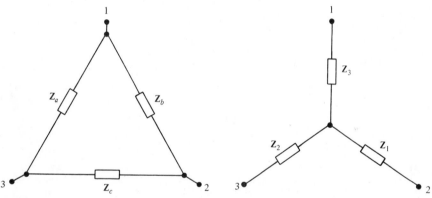

Problem 9-52

9-53 (a) Calculate **V** in the network shown by using node analysis.

(b) Repeat, by first converting the center Y into a Δ, then finding the total impedance seen by the source.

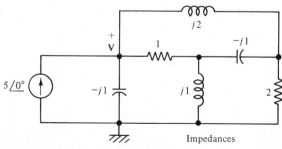

Impedances

Problem 9-53

9-54 Solve Example 9-8, Figure 9-18, by using combined impedances, admittances, and voltage and current dividers. Draw a phasor diagram showing V_s, V_a, and V_b.

9-55 Solve Problem 9-53 by Thévenin's theorem, the observer being the capacitor whose voltage is V.

9-56 Calculate I_2 in Problem 9-42 by making the 3-Ω resistor the observer.

9-57 In the figure shown, calculate I_2 by loop analysis. Show that the answer is not amenable to a Thévenin equivalent circuit with respect to $j\omega L_2$ as the observer, i.e., $I_2 \neq V_{Th}/(Z_{Th} + j\omega L_2)$.

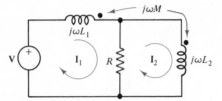

Problem 9-57

9-58 In the figure, we show the equivalent circuit of a transistor circuit (the so-called common-emitter circuit). Find its Thévenin equivalent circuit with respect to the load resistor R_L for a sinusoidal input $v_s(t)$.

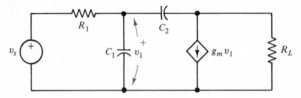

Problem 9-58

9-59 Solve Problem 9-53 by Norton's theorem, with the capacitor as the observer. Next, but not earlier, compare with your solution to Problem 9-55.

9-60 Solve Problem 9-58 by Norton's theorem.

9-61 **(a)** In the figure, write KVL for loop 2 and thereby prove that when you write a KVL equation in a loop adjacent to a current source, that equation performs automatically a source transformation.

　　　(b) Dually, prove that a KCL node equation adjacent to a voltage source performs automatically a source transformation. Do it by writing KCL at node 2.

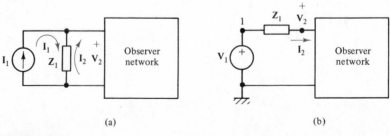

(a)　　　　　　　　　　　　　　　　　　　　　(b)

Problem 9-61

9-62 Find the steady-state current $i_R(t)$ by using phasors. *Scrambled hint*: O, no, it is supper!†

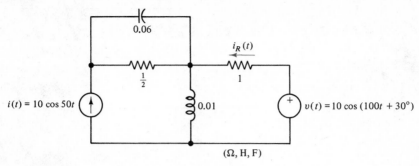

Problem 9-62

Chapter 10

Resonance and Frequency Response

10-1 RESONANCE AND *RLC* CIRCUITS

Resonance is a common phenomenon in many systems. Consider, for example, a company of soldiers marching to the periodic "left, right, left" of the drill sergeant. When they approach a foot bridge, the sergeant discontinues the rhythmic "left, right, left" and lets the soldiers march at their individual and unequal paces. Why? If they keep the uniform and periodic pace, the bridge may resonate and its vibrations may cause it to collapse.

Another common resonance occurs when you tune your radio receiver to a favorite station. Many simultaneous signals at different frequencies reach the antenna of the radio. A special filter circuit is tuned by you to resonate only to the particular frequency of your choice.

To develop these ideas, let us first do an example.

EXAMPLE 10-1 ────────────────────────────

Consider the *LC* circuit shown in Figure 10-1. Let us solve this problem in the classical way.

Solution. The second-order differential equation is, as in Chapter 8,

$$LC \frac{d^2 v_C(t)}{dt^2} + v_C(t) = V_m \cos \omega_1 t$$

The characteristic equation is

$$LCs^2 + 1 = 0$$

with the natural frequencies

$$s_1 = j\omega_0 \quad s_2 = -j\omega_0 \quad \omega_0^2 = \frac{1}{LC}$$

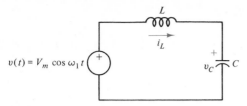

Figure 10-1   Example 10-1.

Therefore, the homogeneous solution is

$$v_{CH}(t) = A \cos{(\omega_0 t + \theta)}$$

The particular solution, as studied in Chapter 7 (Table 7-1) must be according to the source

$$v_{CP}(t) = B \cos{(\omega_1 t + \alpha)}$$

provided $\omega_1 \neq \omega_0$. Then the total solution is

$$v_C(t) = A \cos{(\omega_0 t + \theta)} + B \cos{(\omega_1 t + \alpha)}$$

with oscillations at the natural frequency ω_0 as well as at the input (source) frequency ω_1. The calculation of B and α can be done, of course, with phasors: We take the phasor transform, at ω_1, of the differential equation and obtain

$$LC(j\omega_1)^2 \mathbf{V}_C + \mathbf{V}_C = V_m \underline{/0°}$$

or

$$\mathbf{V}_C = \frac{V_m \underline{/0°}}{1 - \omega_1^2 LC} \qquad \omega_1^2 \neq \frac{1}{LC}$$

What happens if $\omega_1 = \omega_0 = 1/(LC)^{1/2}$? The homogeneous solution is still

$$v_C(t) = A \cos{(\omega_0 t + \theta)} = A \cos{(\omega_1 t + \theta)}$$

but the particular solution, according to Table 7-1, must be

$$v_{CP}(t) = Bt \cos{(\omega_1 t + \alpha)}$$

Notice the multiplication by t. This is necessary because the homogeneous solution is of the same form as the source. In more precise terms, the input $V_m \cos{\omega_1 t}$ excites a *natural frequency* of the circuit, $\omega_1 \Rightarrow \omega_0$. This frequency is the *resonant frequency* of the circuit. When this happens, the particular solution tends to grow very large

$$\lim_{t \to \infty} Bt \cos{(\omega_1 t + \alpha)} \to \infty$$

and that's the same reason for the march on the bridge. The smart drill sergeant breaks up the periodic input to the bridge, so as not to excite a natural frequency of the bridge, thus preventing resonance.

The same conclusion is reached by the phasor calculation. When $\omega_1 = 1/(LC)^{1/2} = \omega_0$, we see that

$$\mathbf{V}_C \to \infty$$

since the denominator is zero then.   ∎

The conclusion that we reached is important enough to be repeated. In fact, we will *define* resonance as the condition of a circuit having a natural frequency excited by an input. This frequency is the *resonant frequency*.

Prob. 10-1

10-2 SERIES *RLC* RESONANCE

To investigate further the phenomenon of resonance, let us consider a series *RLC* circuit excited by a sinusoidal voltage source

$$v(t) = V_m \cos \omega t \qquad (10\text{-}1)$$

as shown in Figure 10-2(a). This time, however, we will let ω, the sinusoidal frequency of the source, be a *variable* rather than a single fixed number. This way, we'll be able to study the behavior of the response as a function of ω. Such a dependence on ω is called a *frequency response*. All other parameters, V_m, R, L, and C, are kept constant.

First, the series impedance seen by the source is

$$\mathbf{Z} = R + j\omega L + \frac{1}{j\omega C} = R + j\left(\omega L - \frac{1}{\omega C}\right) \qquad (10\text{-}2)$$

The reactive term of $\mathbf{Z}$ is $X = \omega L - 1/(\omega C)$ and is dependent on ω. We plot it in Figure 10-2(b) as shown: The inductive reactance, $X_L = \omega L$ is a straight line through

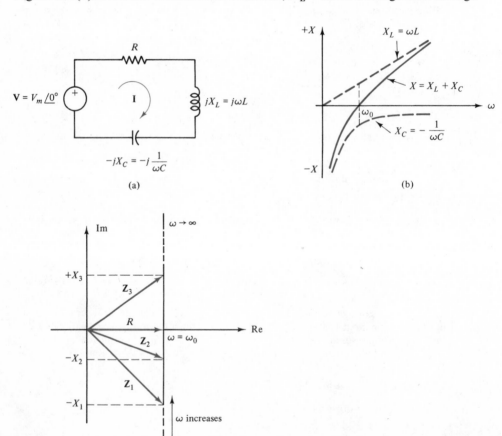

(a)

(b)

(c)

Figure 10-2 Series *RLC* circuit.

the origin with a slope $= L$. The capacitive reactance $X_C = -1/\omega C$ is a rectangular hyperbola. The total reactance X is shown in a solid line, and we see that it falls into three categories:

1. $X < 0$: Starting at $\omega = 0$ (dc) when the inductor is a short circuit ($\omega L = 0$) and the capacitor an open circuit, $1/\omega C \rightarrow \infty$, the total reactance is capacitive and the impedance is capacitive in nature. Typically, in this range, we have

$$\mathbf{Z}_1 = R - jX_1 \qquad \omega = \omega_1 \tag{10-3}$$

Such an impedance, a complex number (*not* a phasor), is drawn in Figure 10-2(c). As ω increases, X decreases in magnitude, and a typical impedance is

$$\mathbf{Z}_2 = R - jX_2 \qquad \omega_2 > \omega_1 \tag{10-4}$$

and is shown on the same plot. The locus of these capacitive impedances is the vertical line shown in Figure 10-2(c), at a fixed distance R from the origin.

2. $X = 0$: At a certain frequency, $\omega = \omega_0$, the total reactance is zero. Then

$$\omega_0 L - \frac{1}{\omega_0 C} = 0 \tag{10-5}$$

that is

$$\omega_0 = \frac{1}{\sqrt{LC}} \tag{10-6}$$

At this frequency, the impedance of the circuit is purely resistive

$$\mathbf{Z} = R \tag{10-7}$$

and is a real, positive number.

3. $X > 0$: As the frequency increases, $\omega > \omega_0$, the total reactance becomes inductive, $\omega L - 1/\omega C > 0$, and the impedance is inductive in nature. A typical impedance here is

$$\mathbf{Z}_3 = R + jX_3 \qquad \omega_3 > \omega_0 \tag{10-8}$$

and is shown in Figure 10-2(c). As ω becomes larger ($\omega \rightarrow \infty$), the inductor becomes an open circuit, $\omega L \rightarrow \infty$, while the capacitor is a short circuit, $1/\omega C \rightarrow 0$.

If we write $\mathbf{Z}$ in its polar form, we have

$$\mathbf{Z} = |\mathbf{Z}|\underline{/\theta_z} \tag{10-9}$$

where the magnitude of $\mathbf{Z}$ is

$$|\mathbf{Z}| = \sqrt{R^2 + X^2} \qquad 0 \leq \omega \leq \infty \tag{10-10}$$

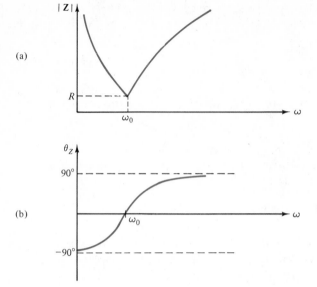

Figure 10-3   Magnitude and phase of **Z**.

and its angle is

$$
\theta_Z = \begin{cases} \tan^{-1}\left(\dfrac{-X}{R}\right) & 0 < \omega < \omega_0 \\[2ex] 0 & \omega = \omega_0 \\[2ex] \tan^{-1}\dfrac{X}{R} & \omega > \omega_0 \end{cases}
\tag{10-11}
$$

These two frequency response curves are plotted in Figure 10-3(a) and (b).   Prob. 10-2
    We are ready now to consider the frequency response of the current **I**. We write
for this circuit

$$
\mathbf{I} = \frac{\mathbf{V}}{\mathbf{Z}}
\tag{10-12}
$$

From this, the magnitude of **I** is

$$
|\mathbf{I}| = \frac{|\mathbf{V}|}{|\mathbf{Z}|} = \frac{V_m}{\sqrt{R^2 + X^2}}
\tag{10-13}
$$

and the phase angle of **I** is

$$
\theta_I = \theta_V - \theta_Z = \begin{cases} -\tan^{-1}\left(\dfrac{-X}{R}\right) & 0 < \omega < \omega_0 \\[2ex] 0 & \omega = \omega_0 \\[2ex] -\tan^{-1}\dfrac{X}{R} & \omega > \omega_0 \end{cases}
\tag{10-14}
$$

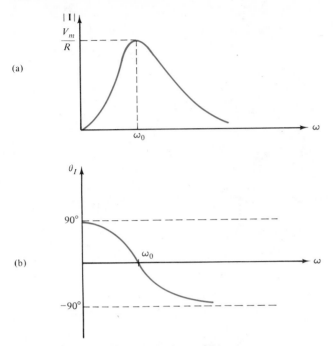

(a)

(b)

Figure 10-4   Magnitude and phase of **I**.

Here we used the given $\theta_V = 0°$ and Equation 10-11 for θ_Z. The frequency response plots for **I** are shown in Figure 10-4(a), (b). It is worthwhile to notice that $|\mathbf{I}|$ is proportional, by the constant multiplier V_m, to $1/|\mathbf{Z}|$. Thus, Figure 10-4(a) is essentially the inverse of Figure 10-3(a). Physically, Figure 10-4(a) also makes sense: At dc ($\omega = 0$), the capacitor is an open circuit and no current flows. As ω increases, the current increases in magnitude and reaches its maximum at the resonant frequency ω_0 because then $|\mathbf{Z}|$ is at its minimal value, $|\mathbf{Z}| = R$. As ω increases, $|\mathbf{Z}|$ increases and consequently, $|\mathbf{I}|$ decreases. As $\omega \to \infty$, the inductor acts as an open circuit and $|\mathbf{I}| = 0$ then.

It is very interesting to note the behavior of the phase of **I**. At $\omega \simeq 0$ it leads **V** by 90° (capacitive circuit); at resonance, it is *in phase* with **V**; as $\omega \to \infty$, it lags behind **V** by 90° (inductive circuit).

Probs. 10-3, 10-4

This discussion explains, in principle, the tuning of some radios. As you turn the knob, you vary the resonant frequency ω_0 by changing the value of a capacitor C. At the chosen ω_0, the current is maximum. This current is converted into sound energy in the speaker of the radio, and you hear your chosen station.

Before we summarize the important features of this discussion, a few additional aspects must be considered. First, the effect of R on the various responses. When $R = 0$, the circuit is lossless (LC), as analyzed in Example 10-1. The impedance is pure imaginary, $Z = \pm jX$. Its magnitude and phase plots are shown in Figure 10-5(a) and (b) for various values of R. The effects of R on the frequency response of **I** are shown in Figure 10-5(c) and (d). The fact that $|\mathbf{I}| \to \infty$ at ω_0 when $R = 0$ agrees with the same conclusion in Example 10-1.

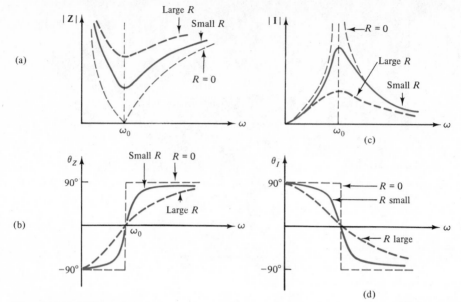

Figure 10-5   Effects of R.

Recall our discussion of the Q *factor*, or simply Q, of the circuit in Chapter 8. For the *RLC* series circuit, it is defined as

$$Q = \frac{\omega_0 L}{R} \tag{10-15}$$

and therefore the smaller R, the higher Q. This quality factor is the measure of the "slimness" of the plot of $|\mathbf{I}|$, Figure 10-5(c): The higher Q, the slimmer $|\mathbf{I}|$ around ω_0, hence a better selectivity, for example, of your radio receiver. With a low Q, the magnitude of $|\mathbf{I}|$ at ω_0 is not much different than at neighboring frequencies, and you'll receive a garble of two or more interfering stations.

Another convenient measure of the slimness of $|\mathbf{I}|$ is the bandwidth, also discussed in Chapter 8. Here, let us define the bandwidth B as

$$B = \omega_2 - \omega_1 \tag{10-16}$$

and it is shown in Figure 10-6. The two frequencies ω_2 and ω_1 are those at which $|\mathbf{I}|$ is down by a factor of $1/\sqrt{2}$ from its maximum. Look at your solution to Problem 10-3. These two frequencies are called the *half-power frequencies*, because then the power is half of that at resonance, power being proportional to the *square* of the current. The two half-power frequencies are

$$\omega_{1,2} = \pm \frac{R}{2L} + \sqrt{\left(\frac{R}{2L}\right)^2 + \omega_0^2} \tag{10-17}$$

Consequently, using the expression for Q in Equation 10-15, we obtain

$$B = \frac{\omega_0}{Q} \tag{10-18}$$

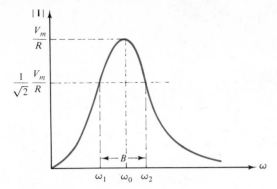

Figure 10-6 Bandwidth.

that is (as expected), the bandwidth is inversely proportional to Q: The higher the Q, the narrower the bandwidth. Also, from Equation 10-17 we get

$$\omega_1\omega_2 = \omega_0^2 \tag{10-19}$$

that is, the resonant frequency ω_0 is the geometrical average of ω_1 and ω_2.

Probs. 10-5, 10-6

10-3 ENERGY IN THE RESONANT CIRCUIT

The entire phenomenon of resonance is based physically on the interchange of energy between the inductor and the capacitor. It is important to understand this interchange.

In the series RLC circuit, let us take $\mathbf{I}$ as reference,

$$\mathbf{I} = I_m\underline{/0°} \tag{10-20}$$

Then the capacitor's voltage is

$$\mathbf{V}_C = \frac{1}{j\omega C}\mathbf{I} = \frac{I_m}{\omega C}\underline{/-90°} \tag{10-21}$$

The corresponding time-domain expressions are

$$i(t) = I_m\cos \omega t \tag{10-22}$$

and

$$v_C(t) = \frac{I_m}{\omega C}\cos(\omega t - 90°) = \frac{I_m}{\omega C}\sin \omega t \tag{10-23}$$

The electric energy in the capacitor is

$$w_C = \frac{1}{2}Cv_C^2 = \frac{I_m^2}{2\omega^2 C}\sin^2 \omega t \tag{10-24}$$

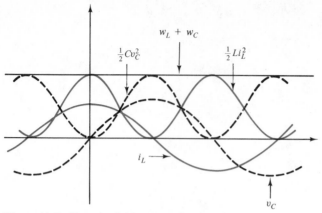

Figure 10-7 Energies in resonant circuit.

and the magnetic energy in the inductor is

$$w_L = \frac{1}{2} L i_L^2 = \frac{L I_m^2}{2} \cos^2 \omega t \tag{10-25}$$

These waveforms are plotted in Figure 10-7. We see that, at resonance, the capacitor delivers energy during one quarter cycle and the inductor receives it, then the roles are reversed during the next quarter cycle. The sum of these energies (Equations 10-24 and 10-25) is calculated with $L = 1/\omega_0^2 C$ and so

$$w_C + w_L = \frac{I_m^2}{2\omega_0^2 C} \sin^2 \omega_0 t + \frac{I_m^2}{2\omega_0^2 C} \cos^2 \omega_0 t = \frac{I_m^2}{2\omega_0^2 C} = \frac{L I_m^2}{2} \tag{10-26}$$

which is *constant*. This is amazing! At resonance, the outside world does not supply any energy to the inductor or to the capacitor; they simply exchange their energies internally. The only energy needed to be supplied from the outside is the loss (heat) in the resistor. A lossless *LC* circuit which starts with an initial stored energy will oscillate forever, and the energy distribution will follow the waveforms of Figure 10-7.

   The last two sections were rather lengthy with details and with formulas. There is no need to memorize any of these. In particular, do *not* memorize $\omega_0^2 = 1/LC$ as the panacea for all resonant circuits; it is not. A much better approach would be to understand the ideas well enough to rederive an equation when needed.   Prob. 10-7

10-4 INPUT-OUTPUT PHASOR RELATIONS

The input-output phasor relationship, used in Chapter 9 and here, can be written in general as

$$\mathbf{R} = \mathbf{H} \mathbf{E} \tag{10-27}$$

where **R** is the generic notation for the phasor response (output), and **E** for the phasor excitation (input). This relationship is the extension for phasors of Equation 5-19

$$r(t) = He(t) \tag{5-19}$$

which was obtained in Chapter 5 for purely resistive networks. The complex number **H** is the *network function* for the specific **R** and **E** under consideration. In the series *RLC* circuit we had

$$\mathbf{I} = \frac{1}{\mathbf{Z}} \mathbf{V} \qquad (10\text{-}28)$$

and therefore $\mathbf{R} = \mathbf{I}$, $\mathbf{E} = \mathbf{V}$ and $\mathbf{H} = 1/\mathbf{Z}$, the driving-point admittance of the circuit. In general, **H** is a complex number, written in its polar form as

$$\mathbf{H} = |\mathbf{H}|\underline{/\theta_H} \qquad (10\text{-}29)$$

From Equation 10-27, we recognize that **R** and **E** will be in phase when **H** is a real, positive number, that is, when

$$\theta_H = 0° \qquad (10\text{-}30)$$

As we saw in Section 10-2, that's when resonance occurs. We have, therefore, a *criterion for resonance*: Resonance is the condition of a circuit when the output phasor is in phase with the input phasor. Then Equation 10-30 holds.

Several examples will illustrate these points.

EXAMPLE 10-2 ———————————————————————————————

The circuit shown in Figure 10-8 is driven by a sinusoidal current source. Find the condition, or conditions, for resonance.

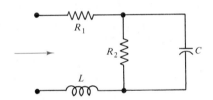

Figure 10-8 Example 10-2.

Solution. Here Equation 10-27 reads

$$\mathbf{V} = \mathbf{Z}\mathbf{I}$$

with $\mathbf{H} = \mathbf{Z}$, the driving-point impedance of the circuit. We have

$$\mathbf{Z} = R_1 + j\omega L + \frac{1}{G_2 + j\omega C}$$

where $G_2 = 1/R_2$. Clearing fractions in **Z**, we get

$$\mathbf{Z} = \frac{(R_1 G_2 + 1 - \omega^2 LC) + j(\omega L G_2 + \omega C R_1)}{G_2 + j\omega C}$$

At resonance $\theta_z = 0°$, which means that the angle of the numerator equals the angle of the denominator,

$$\tan^{-1} \frac{\omega L G_2 + \omega C R_1}{R_1 G_2 + 1 - \omega^2 LC} = \tan^{-1} \frac{\omega C}{G_2}$$

or

$$\frac{\omega(L G_2 + C R_1)}{R_1 G_2 + 1 - \omega^2 LC} = \frac{\omega C}{G_2}$$

One answer is $\omega = 0$ (dc). This is also obvious by inspecting the circuit: It acts then as a pure resistor $(R_1 + R_2)$. Canceling ω and solving yields

$$\omega^2 LC^2 = C - G_2^2 L$$

or

$$\omega = \sqrt{\frac{C - G_2^2 L}{LC^2}} \qquad C > G_2^2 L$$

as the second answer. For ω to be a (physical) real number, we must have $C > G_2^2 L$.

This example brings out the important, and often simple, check by inspection: If the network function $\mathbf{H}$ is purely resistive, $\theta_H = 0°$, at $\omega = 0$ or $\omega = \infty$, then there is resonance at those frequencies. Other nonzero resonant frequencies may exist. They require a detailed calculation. ∎

EXAMPLE 10-3

In the network shown in Figure 10-9, calculate the appropriate network function and value of C for resonance, assuming ω is fixed. Here the input is $\mathbf{E} = \mathbf{V}_s$ and the output is $\mathbf{R} = \mathbf{I}_2$.

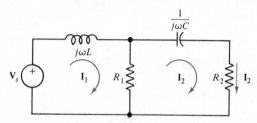

Figure 10-9 Example 10-3.

Solution. Having identified these, we proceed to calculate $\mathbf{I}_2$. The two loop equations are

$$\begin{bmatrix} j\omega L + R_1 & -R_1 \\ -R_1 & \dfrac{1}{j\omega C} + R_1 + R_2 \end{bmatrix} \begin{bmatrix} \mathbf{I}_1 \\ \mathbf{I}_2 \end{bmatrix} = \begin{bmatrix} \mathbf{V}_s \\ 0 \end{bmatrix}$$

and you should verify them to your satisfaction. Solving for $\mathbf{I}_2$, we have

$$\mathbf{I}_2 = \frac{\begin{vmatrix} j\omega L + R_1 & \mathbf{V}_s \\ -R_1 & 0 \end{vmatrix}}{\begin{vmatrix} j\omega L + R_1 & -R_1 \\ -R_1 & \dfrac{1}{j\omega C} + R_1 + R_2 \end{vmatrix}}$$

$$= \frac{R_1}{(L/C + R_1 R_2) + j[\omega L(R_1 + R_2) - R_1/\omega C]} \mathbf{V}_s$$

Therefore, we identify the appropriate network function here as a transfer admittance

$$\mathbf{H} = \frac{R_1}{(L/C + R_1 R_2) + j[\omega L(R_1 + R_2) - R_1/\omega C]}$$

It is a transfer function because it relates an output at one pair of terminals to an input at another pair of terminals; it is an admittance because, dimensionally, it multiplies a voltage to give a current.

Resonance will occur when $\theta_H = 0°$. Since the numerator's angle is zero, so must be the angle of the denominator. This will be true if the imaginary part in the denominator vanishes

$$\omega L(R_1 + R_2) - \frac{R_1}{\omega C} = 0$$

or

$$C = \frac{R_1}{\omega^2 L(R_1 + R_2)}$$

You should also verify by inspection that $\omega = 0$ or $\omega = \infty$ are not resonant frequencies.

■

EXAMPLE 10-4

In the circuit shown in Figure 10-10, calculate the voltage across the resistor R_2. Identify the network function and investigate resonance, and explain your conclusion.

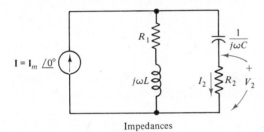

Impedances   **Figure 10-10**   Example 10-4.

Solution.   We have $\mathbf{V}_2 = R_2\mathbf{I}_2$ and $\mathbf{I}_2$ is obtained through the current divider equation. Therefore

$$\mathbf{V}_2 = R_2 \frac{R_1 + j\omega L}{R_1 + R_2 + j(\omega L - 1/\omega C)}\,\mathbf{I}$$

The network function here is a transfer impedance

$$\mathbf{H} = \frac{R_2(R_1 + j\omega L)}{R_1 + R_2 + j(\omega L - 1/\omega C)}$$

For resonance we must have, as the condition $\theta_H = 0°$,

$$\frac{\omega L}{R_1} = \frac{\omega L - 1/\omega C}{R_1 + R_2}$$

which has no physical (positive) solution for ω. (Verify this!) The conclusion is that $\mathbf{V}_2$ will never be in phase with $\mathbf{I}$. There is no resonance between these two.   ■

Probs. 10-8, 10-9, 10-10

We have studied two types of resonance: (1) the excitation of a natural frequency of a network by an input with the same natural frequency, and (2) the in-phase relationship between sinusoidal output and input. These resonances are different in most cases; they are almost the same if the Q of the circuit is high, that is, when the bandwidth is narrow. For a particular circuit, the statement of the problem should make clear which type of resonance is being considered.

10-5 FILTERS

A *filter* is a network that has certain desired frequency response characteristics. The general idea is shown in Figure 10-11. A specific filter is coupling a source to a load. Depending on its design, the filter will allow certain components of frequency to pass from the source to the load, and will stop other components of frequencies from passing to the load. In the series resonant circuit (Figure 10-4 or 10-6) we recognize a filter that *passes* the signal $|\mathbf{I}|$ at the frequency ω_0 and its neighborhood within that bandwidth. It *stops* (or *attenuates*) signals at all other frequencies, 0 to ω_1 and ω_2 to ∞, from reaching the load, the speakers of the radio in that example.

Since the input-output relations for a filter are general, as in Equation 10-27,

$$\mathbf{R} = \mathbf{HE} \tag{10-27}$$

we are interested in the frequency response characteristics of the filter, specifically its magnitude response $|\mathbf{H}|$ and its phase θ_H, as functions of ω. An *ideal* filter will pass without distortion signals of all the frequencies in a certain range, called the *passband*, and stop completely signals of all other frequencies, called the *stopband*. In addition, an ideal filter must be lossless so that the input energy of the signal is not dissipated in the filter and reaches the load.

In Figure 10-12 we show the magnitude characteristics of four major types of ideal filters. The ideal *lowpass* filter is shown in Figure 10-12(a): It passes, with a magnitude $|\mathbf{H}|$ scaled conveniently to 1, all signals from $\omega = 0$ (dc) to a *cutoff frequency* ω_c, and stops completely all signals of frequencies beyond ω_c. The ideal *highpass* filter characteristic is shown in Figure 10-2(b); no signals of frequencies less than the cutoff frequency are passed, and all signals beyond ω_c are passed. The ideal *bandpass* and *band elimination* filters are shown in Fig. 10-12(c) and (d).

These magnitude characteristics are ideal because we cannot achieve such perfectly square, sharp-cornered characteristics. Typically realistic characteristics are shown in dotted lines. The *RLC* series circuit (Figure 10-6) is therefore a bandpass filter. A filter whose magnitude characteristic is $|\mathbf{Z}|$, as shown in Figure 10-5(a), is a bandstop filter. Are you worried about the apparent "contradiction" here? It does not really exist. This *RLC* circuit is a bandpass filter, since its proper network function

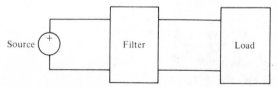

Figure 10-11   A filter.

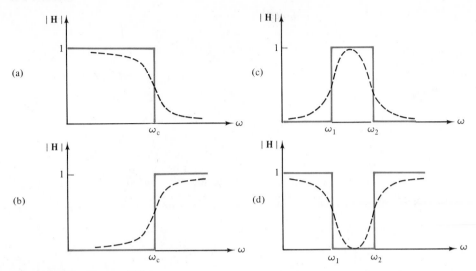

Figure 10-12 Types of filters.

is $1/\mathbf{Z}$, not $\mathbf{Z}$. This points out, again, the importance of identifying the correct network function for each case.

EXAMPLE 10-5

Plot the magnitude $|\mathbf{H}|$ of the filter shown in Figure 10-13(a) and classify it.

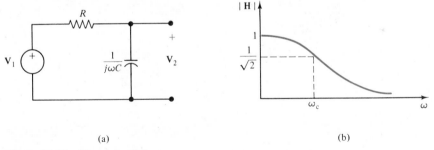

(a)   (b)

Figure 10-13 Example 10-5.

Solution. The input-output relation here is

$$\mathbf{V}_2 = \mathbf{H}\mathbf{V}_1$$

and therefore $\mathbf{H}$ is a dimensionless voltage transfer function. Before plunging into the mathematics of this problem, it is worthwhile to use some elementary physical arguments: At dc ($\omega = 0$) and low frequencies, the capacitor is an open circuit and therefore there is no current through C or R. At those frequencies, then, $|\mathbf{V}_2| \simeq |\mathbf{V}_1|$. As frequency increases, the capacitor's reactance decreases and so does $|\mathbf{V}_2|$ across it. Finally, at $\omega \to \infty$, the capacitor is a short circuit and $|\mathbf{V}_2| = 0$. It looks as if we have a (primitive) lowpass filter on our hands.

To confirm it, write

$$\mathbf{V}_2 = \frac{1/j\omega C}{1/j\omega C + R} \mathbf{V}_1$$

as the voltage divider equation here. Therefore

$$\mathbf{H} = \frac{1/j\omega C}{1/j\omega C + R} = \frac{1}{1 + j\omega RC}$$

and

$$|\mathbf{H}| = \frac{1}{\sqrt{1 + (\omega RC)^2}}$$

The plot of $|\mathbf{H}|$ is shown in Figure 10-13(b), and it is indeed a lowpass characteristic. A convenient way to measure ω_c, the cutoff frequency, is at the half-power point, as in the calculations of the bandwidth B. At ω_c, then, the magnitude is $1/\sqrt{2}$ of its maximum. Writing then

$$\frac{1}{\sqrt{2}} = \frac{1}{\sqrt{1 + (\omega_c RC)^2}}$$

we obtain here

$$\omega_c = \frac{1}{RC}$$
■

EXAMPLE 10-6

If we exchange the positions of R and C in the previous example, we obtain a highpass filter. See Figure 10-14.

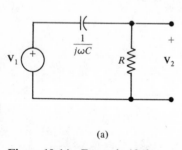

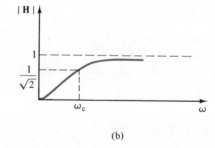

(a)   (b)

Figure 10-14   Example 10-6.

Solution. At dc and low frequencies, the capacitor prevents any current from flowing through R and therefore $|\mathbf{V}_2| \simeq 0$. At high frequencies, the capacitor is a short circuit and $|\mathbf{V}_2| \simeq |\mathbf{V}_1|$. Mathematically, we calculate

$$|\mathbf{H}| = \frac{\omega RC}{\sqrt{1 + (\omega RC)^2}}$$

and $\omega_c = 1/RC$ also. You should fill in the details of the calculations, as well as the plot of $|\mathbf{H}|$.
■

Probs. 10-11
10-12, 10-13

PROBLEMS

10-1 In the circuit shown, calculate the particular solution $i_P(t)$ by the classical way of differential equations. Plot it. Consider two cases:

(a) $\qquad\qquad\qquad\qquad\qquad\qquad\qquad \alpha \neq \dfrac{R}{L}$

(b) $\qquad\qquad\qquad\qquad\qquad\qquad\qquad \alpha = \dfrac{R}{L}$

When does resonance occur?

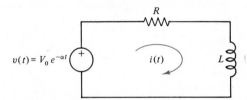

Problem 10-1

10-2 Prove that the polar plot of **Y** in the series RLC circuit (Figures 10-2, 10-3) is a circle, as shown. *Hint*: Write

$$\mathbf{Y} = \frac{1}{R \mp jX} = \frac{R \pm jX}{R^2 + X^2}$$

then set, as usual,

$$G = \frac{R}{R^2 + X^2} \qquad B = \frac{\pm X}{R^2 + X^2}$$

and relate G and B in the B-G plane.

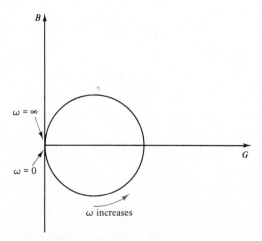

Problem 10-2

10-3 **(a)** In Figure 10-4(a), calculate the two frequencies at which the magnitude of **I** is $V_m/(R\sqrt{2})$, that is, $1/\sqrt{2}$ of its maximum value.
(b) In Figure 10-4(b), calculate the frequencies at which the phase of **I** is $\pm 45°$.

10-4 Based on Problem 10-2 and on Equations 10-13 and 10-14, draw the polar plot of **I** in the series *RLC* circuit.

10-5 **(a)** Prove that the two half-power frequencies (Equation 10-17) can be written as

$$\omega_{1,2} = \omega_0(\sqrt{\zeta^2 + 1} \pm \zeta)$$

where ζ is the damping factor studied in Chapter 8.
(b) Under what approximation can we write

$$\omega_1 = \omega_0 - \zeta\omega_0$$
$$\omega_2 = \omega_0 + \zeta\omega_0$$

(c) Prove that the bandwidth (Equation 10-18) can be written as

$$B = 2\zeta\omega_0$$

10-6 Prove that ω_0 is the arithmetic average of the two half-power frequencies if the abscissa is on a logarithmic scale, log ω, rather than a linear scale, ω. See Equation 10-19.

10-7 Provide a detailed derivation, with equations and figures, for the *parallel RLC* resonant circuit. Follow the steps in Section 10-2 for your derivation, noting as you go such features as duality.

10-8 In the circuit shown, the value of the inductor is

$$L = \frac{1}{\omega_1^2 C}$$

Calculate the output voltage across the resistor R. How does it vary with R?

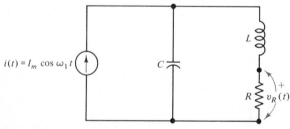

Problem 10-8

10-9 Calculate and name the network function **H** in the circuit shown. Investigate resonance. The two resistors are equal.

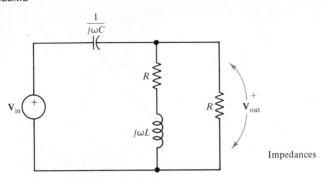

Problem 10-9

10-10 Design a series RLC circuit to the following specifications: resonance at $f_0 = 1.2$ MHz, bandwidth 4000 Hz, and an input impedance of magnitude 100 Ω at $f = 2f_0$.

10-11 For each of the filters shown, calculate **H**, plot $|\mathbf{H}|$, and classify the filter:
 (a) $\mathbf{Z}_1 = j\omega L$; $\mathbf{Z}_2 = R$.
 (b) $\mathbf{Z}_1$ is a parallel $R_1 C_1$ circuit; $\mathbf{Z}_2$ is a parallel $R_2 C_2$ circuit.
 (c) $\mathbf{Z}_1$ is a series RL circuit; $\mathbf{Z}_2$ is a parallel RC circuit. (*Note*: The same R in both circuits.)
 (d) $\mathbf{Z}_1$ is an inductor L; $\mathbf{Z}_2$ is a parallel RC circuit.

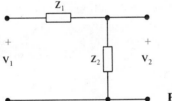

Problem 10-11

10-12 For the filter shown, we have $RC = 1$. **(a)** Calculate the voltage transfer function in $\mathbf{V}_2 = \mathbf{H}\mathbf{V}_1$ and prove that resonance occurs at $\omega_0 = 1/\sqrt{6}$. **(b)** Plot $|\mathbf{H}|$ and classify this filter.

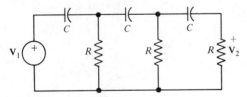

Problem 10-12

10-13 Calculate **H** for the filter shown, plot $|\mathbf{H}|$ and θ_H, and classify this filter. In order to understand the particular use of this filter, consider carefully its phase response, θ_H.

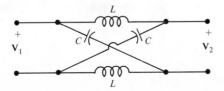

Problem 10-13

Chapter 11

Power in the Sinusoidal Steady State

The study of power accompanied our studies of the various elements from the earliest chapters. In this chapter, we discuss in detail power calculations for the sinusoidal steady state. This is an extremely important and practical topic, since the most common sources of electric energy operate in the sinusoidal steady state. Electric power companies provide the generation, transmission, and distribution of this energy.

11-1 INSTANTANEOUS AND AVERAGE POWER

Consider a one-port network whose sinusoidal impedance is $\mathbf{Z}$, with the associated $\mathbf{V}$ and $\mathbf{I}$ at its two terminals, as shown in Figure 11-1(a). Let us assume that

$$\mathbf{V} = V_m \underline{/\theta_v} \tag{11-1}$$

and

$$\mathbf{I} = I_m \underline{/\theta_i} \tag{11-2}$$

These two phasors are shown in Figure 11-1(c). The time-domain expressions for voltage and current are therefore

$$v(t) = V_m \cos(\omega t + \theta_v) \tag{11-3}$$

and

$$i(t) = I_m \cos(\omega t + \theta_i) \tag{11-4}$$

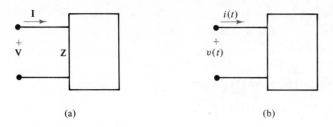

(a)   (b)

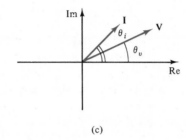

(c)

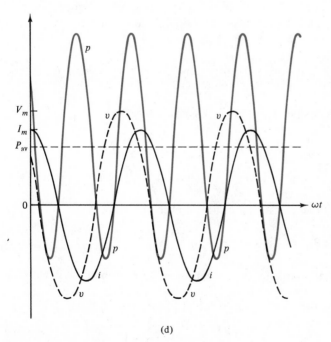

(d)

Figure 11-1   Instantaneous $v(t)$, $i(t)$, and $p(t)$.

referenced as in Figure 11-1(b). These are plotted in Figure 11-1(d). With these associated references, we remember that the instantaneous power in the circuit is

$$p(t) = v(t)i(t) \tag{11-5}$$

Here we obtain from Equations 11-3 and 11-4

$$p(t) = V_m I_m \cos(\omega t + \theta_v) \cos(\omega t + \theta_i) \qquad \text{W} \tag{11-6}$$

and the plot of $p(t)$ is also shown in Figure 11-1(d). In practice, this plot can be obtained point by point, by multiplying a value of $v(t)$, say, $v(t_1)$ by the corresponding value of $i(t)$, $i(t_1)$. Several observations can be made immediately:

1. The instantaneous power is zero whenever $v(t)$ *or* $i(t)$ is zero. At these instants, no power is delivered to or from the circuit.
2. The instantaneous power is positive whenever both v and i are of the same sign (both positive or both negative). Then, by our convention, power is delivered *to* the circuit.
3. The instantaneous power is negative whenever v and i are of opposite signs. Then power is delivered *by* the circuit.
4. The instantaneous power appears to oscillate at twice the frequency ω.
5. The *average* value of $p(t)$ appears to be positive; that is, the circuit *receives* a net (average) power.

   To support these observations more rigorously, let us use the trigonometric identity of the product of two cosine functions

$$\cos x \cos y = \tfrac{1}{2}[\cos(x - y) + \cos(x + y)] \tag{11-7}$$

in Equation 11-6. With $x = \omega t + \theta_v$ and $y = \omega t + \theta_i$ the result is

$$p(t) = \frac{V_m I_m}{2} \cos(\theta_v - \theta_i) + \frac{V_m I_m}{2} \cos(2\omega t + \theta_v + \theta_i) \tag{11-8}$$

We recognize immediately the average (constant) power

$$P_{av} = \frac{V_m I_m}{2} \cos(\theta_v - \theta_i) \qquad \text{W} \tag{11-9}$$

and the second term as a double-frequency term oscillating about P_{av} with the frequency 2ω.

   The angle $(\theta_v - \theta_i)$ is the angle between the phasors $\mathbf{V}$ and $\mathbf{I}$. Since

$$\mathbf{V} = \mathbf{ZI} \tag{11-10}$$

we have

$$\theta_v = \theta_z + \theta_i \tag{11-11}$$

or

$$\theta_v - \theta_i = \theta_z \tag{11-12}$$

When there will be no room for misunderstanding, we'll write simply

$$\theta \equiv \theta_z = \theta_v - \theta_i \tag{11-13}$$

and define the *power factor* (p.f.) of the circuit as

$$\text{p.f.} = \cos \theta \tag{11-14}$$

the cosine of the angle between $\mathbf{V}$ and $\mathbf{I}$. We know that $0 \le \cos \theta \le 1$. Then Equation 11-9 becomes

$$P_{av} = \frac{V_m I_m}{2} \cos \theta \tag{11-15}$$

The meaning of "power factor" can be interpreted as the percentage, from 0 to 1, of the maximum value of the average power $V_m I_m / 2$.

EXAMPLE 11-1 _____

In a single resistor, we have

$$v_R(t) = V_m \cos (\omega t + \theta_v)$$

and

$$i_R(t) = \frac{v_R(t)}{R} = I_m \cos (\omega t + \theta_i)$$

Solution. The phasor relationship is

$$\mathbf{V}_R = R\mathbf{I}_R \qquad \theta_i = \theta_v$$

that is, $\mathbf{V}$ and $\mathbf{I}$ are in phase, as in Figure 11-2(b). Therefore, the power factor is $\cos \theta = \cos 0° = 1$ and the average power is

$$P_{av} = \frac{V_m I_m}{2} = \frac{R I_m^2}{2} > 0$$

and the resistor receives power (and converts it into heat, light, etc.).

To review the time-domain relations, we have here

$$p_R(t) = v_R(t) i_R(t) = V_m I_m \cos^2 (\omega t + \theta_v)$$

Since

$$\cos^2 x = \tfrac{1}{2} + \tfrac{1}{2} \cos 2x$$

we have

$$p_R(t) = \frac{V_m I_m}{2} + \frac{V_m I_m}{2} \cos (2\omega t + \theta_v)$$

The average of $p_R(t)$ is indeed $V_m I_m / 2$. The plots of $v_R(t)$, $i_R(t)$, and $p_R(t)$ are shown in Figure 11-2(c). From it, we recognize that $v_R(t)$ and $i_R(t)$ are in phase, and that the resistor always receives power because, at every instant,

$$p_R(t) \ge 0$$

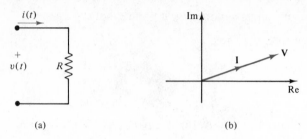

(a)   (b)

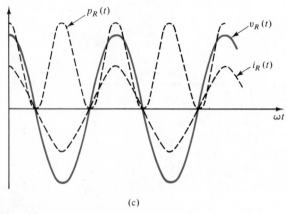

(c)

Figure 11-2   Example 11-1.

Probs. 11-1
■ 11-2

EXAMPLE 11-2

In a single capacitor C, we have

$$v_C(t) = V_m \cos (\omega t + \theta_v)$$

and

$$i_C(t) = C \frac{dv_C}{dt} = \omega C V_m \cos (\omega t + \theta_v + 90°)$$

$$= I_m \cos (\omega t + \theta_v + 90°)$$

Solution.   The phasor relationship is

$$\mathbf{V}_C = \frac{1}{j\omega C} \mathbf{I}_C \qquad \theta_i = \theta_v + 90°$$

as shown in Figure 11-3(b), with $\mathbf{I}_C$ leading $\mathbf{V}_C$ by 90°. The power factor here is

$$\cos \theta = \cos (\theta_v - \theta_i) = \cos (-90°) = 0$$

and

$$P_{av} = \frac{V_m I_m}{2} \cos \theta = 0$$

In the sinusoidal steady state, the capacitor is *lossless*, since the average power in it is zero.

The time-domain calculations yield

$$p_C(t) = v_C(t)i_C(t) = V_m I_m \cos(\omega t + \theta_v)\cos(\omega t + \theta_v + 90°) = \frac{V_m I_m}{2}\cos(2\omega t + 90°)$$

which is a sinusoid of double frequency with zero average, as in Figure 11-3(c). Over one-quarter cycle of $v(t)$ or $i(t)$, the capacitor receives power, $p_C(t) > 0$, and over the next quarter cycle it delivers back the same power, $p_C(t) < 0$.

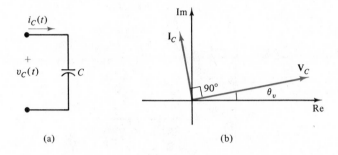

(a)   (b)

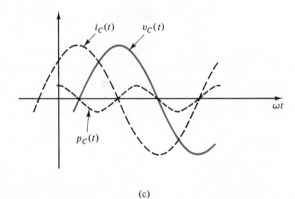

(c)

Figure 11-3   Example 11-2.   ■

EXAMPLE 11-3

In a single inductor L, we have

$$i_L(t) = I_m \cos(\omega t + \theta_i)$$

and

$$v_L(t) = L\frac{di_L(t)}{dt} = \omega L I_m \cos(\omega t + \theta_i + 90°)$$

Solution.   The phasor relationship is

$$\mathbf{V}_L = j\omega L \mathbf{I}_L \qquad \theta_v = \theta_i + 90°$$

as shown in Figure 11-4(b). The phasor $\mathbf{V}_L$ leads $\mathbf{I}_L$ by 90°. Here the power factor is

$$\cos \theta = \cos (\theta_v - \theta_i) = \cos 90° = 0$$

and

$$P_{\text{av}} = 0$$

Like the capacitor, an inductor is *lossless* in the sinusoidal steady state, receiving power, $p_L(t) > 0$, over a quarter cycle of $v(t)$ or $i(t)$ and returning it, $p_L(t) < 0$, over the next quarter cycle.

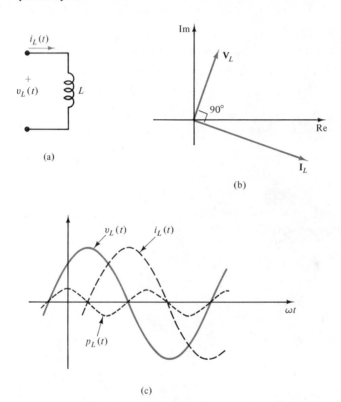

Figure 11-4   Example 11-3.   ■

In all these calculations, we must make sure not to forget the fundamental time-domain relationships. After all, only they have physical significance. Phasors are useful only to the extent that they allow us to interpret time-domain equations. By themselves, phasors are just a bunch of complex numbers without any physical, real-time meaning.

In specifying the power factor for a general circuit, we use the convention of capacitive and inductive circuits. Specifically, in a capacitive circuit, the phasor **I** *leads* the phasor **V**, and then we say that the power factor is *leading*. In an inductive circuit, **I** *lags* behind **V** and the power factor is *lagging*. See Figure 11-5.

Probs. 11-3
11-4

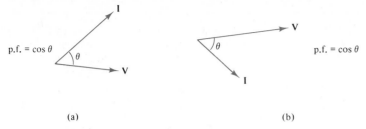

(a)   (b)

Figure 11-5   (a) Leading p.f. (b) Lagging p.f.

EXAMPLE 11-4 ——————————————————————————————

In a certain one-port network

$$\mathbf{V} = 10\underline{/-25°}$$

and

$$\mathbf{I} = 30\underline{/45°}$$

at $\omega = 1000$ rad/s. Calculate the power factor and the average power. Also, identify the circuit with its elements.

Solution.   We have here a capacitive circuit since $\mathbf{I}$ leads $\mathbf{V}$. Also

$$\theta = \theta_v - \theta_i = -25° - 45° = -70°$$

and therefore the power factor is

$$\cos{(-70°)} = \cos 70° = 0.342$$

leading. The average power is

$$P_{\text{av}} = \frac{(10)(30)}{2} \, 0.342 = 51.3 \text{ W}$$

The impedance of the circuit is

$$\mathbf{Z} = \frac{\mathbf{V}}{\mathbf{I}} = \frac{10\underline{/-25°}}{30\underline{/45°}} = 0.114 - j0.313 = R - jX_C$$

and consists of a resistor $R = 0.114 \, \Omega$ in series with a capacitor whose reactance is $-X_C = -0.313 \, \Omega = -1/\omega C$, and so $C = 1/(0.313)(1000) = 3.2$ mF. This answer, of course, is not unique. We could write also

$$\mathbf{Y} = \frac{\mathbf{I}}{\mathbf{V}} = \frac{30\underline{/45°}}{10\underline{/-25°}} = 1.03 + j2.82 = G + jB_C$$

and identify a parallel connection of a resistor of $1/1.03 = 0.98 \, \Omega$ and a capacitor whose susceptance is 2.82 ℧ $= \omega C$, or $C = 2.82$ mF.

We encounter here a basic property of circuit design, or circuit synthesis: In designing a circuit from given responses, there may be more than one answer. In circuit analysis, on the other hand, we have the circuit and calculate the responses; the answer here is unique.   ■

EXAMPLE 11-5 _____

In a certain one-port network, the following readings are made: V_m is measured with a voltmeter, $V_m = 65$ V; I_m is measured with an ammeter, $I_m = 2.4$ A; P_{av} is measured with a wattmeter, $P_{av} = 50$ W. Calculate the power factor and identify the circuit. Comment on the uniqueness of your answers.

Solution. We write from Equation 11-9

$$50 = \frac{(65)(2.4)}{2} \cos \theta$$

$$\therefore \quad \cos \theta = 0.64$$

Since

$$\cos \theta = \cos (-\theta)$$

we have either a capacitive circuit with a leading power factor *or* an inductive circuit with the same numerical lagging power factor.

In the capacitive case we write

$$|\mathbf{Z}| = \frac{|\mathbf{V}|}{|\mathbf{I}|} = \frac{65}{2.4} = 27.1 \, \Omega$$

and $\theta_z = \theta_v - \theta_i = -\cos^{-1} 0.64 = -50.21°$. Consequently

$$\mathbf{Z}_{RC} = 27.1\underline{/-50.21°} = 17.34 - j20.82 = R + j(-X_C)$$

In the inductive case

$$\theta_z = \theta_v - \theta_i = \cos^{-1} 0.64 = 50.21°$$
$$\therefore \quad \mathbf{Z}_{RL} = 27.1\underline{/50.21°} = 17.34 + j20.82 = R + jX_L$$

These possibilities are shown in Figure 11-6. Not only is the nature of the one-port network not unique (capacitive or inductive); within each category, the answers are not unique themselves, as we saw in Example 11-4.

As a final comment here, we are lucky that the power factor is $\cos \theta$ and not $\sin \theta$, for example. Had it been $\sin \theta$, then a capacitive circuit would have

$$\sin (-\theta) = -\sin \theta$$

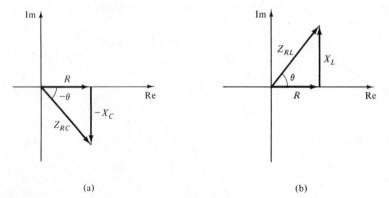

(a)   (b)

Figure 11-6 Example 11-5.

a negative power factor, meaning that an *RC* circuit could deliver power—an obvious absurdity. Actually, it is not simple luck. Our mathematical calculations must conform to the laws of nature, and therefore $\cos \theta$, and not $\sin \theta$, turns out to be the power factor.

Probs. 11-5,
11-6, 11-7

11-2 EFFECTIVE VALUES

To introduce the idea of effective values and their meanings, we turn back to Example 11-1. There we calculated the average power in a resistor with sinusoidal voltage and current. To repeat, we had

$$i_R(t) = I_m \cos \omega t \tag{11-4}$$

and

$$v_R(t) = Ri_R(t) = RI_m \cos \omega t = V_m \cos \omega t \tag{11-3}$$

Here we have assumed $\theta_i = 0°$ and, consequently $\theta_v = 0°$ also. The power factor is $\cos \theta = 1$. We found there

$$P_{av} = \frac{V_m I_m}{2} = \frac{RI_m^2}{2} \tag{11-16}$$

Now we pose the following question: What value of a *constant* (dc) current will produce the same average power in *R*? If we call this value I_{eff}, the effective value of the sinusoidal current, we have then

$$P_{av} = \frac{RI_m^2}{2} = RI_{eff}^2 \tag{11-17}$$

or

$$I_{eff} = \frac{I_m}{\sqrt{2}} \simeq 0.707 I_m \tag{11-18}$$

The situation is depicted in Figure 11-7(a) and (b). The sinusoidal current $I_m \cos \omega t$ produces the same average power in *R* as does the constant current I_{eff}.

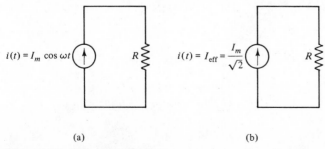

(a)

(b)

Figure 11-7 Effective value.

Since

$$I_m = \frac{V_m}{R} \tag{11-19}$$

and

$$P_{av} = \frac{V_m^2}{2R} = \frac{V_{eff}^2}{R} \tag{11-20}$$

we reach the same conclusion about the effective value of a sinusoidal voltage

$$V_{eff} = \frac{V_m}{\sqrt{2}} \simeq 0.707 V_m \tag{11-21}$$

Now we generalize this discussion to *any* periodic current or voltage. Such a waveform is shown in Figure 11-8(a). The instantaneous power in R is

$$p(t) = [i(t)]^2 R \tag{11-22}$$

Its average over one period (and repeating thereafter) is

$$P_{av} = \frac{1}{T} \int_0^T p(t)\, dt = \frac{1}{T} \int_0^T [i(t)]^2 R\, dt \tag{11-23}$$

Here the average of $p(t)$ is calculated as usual, by taking the sum (integral) of all the instantaneous values and averaging the sum by dividing by T. Another way to look at the average is geometrical: We find the area under $p(t)$ over one period (by

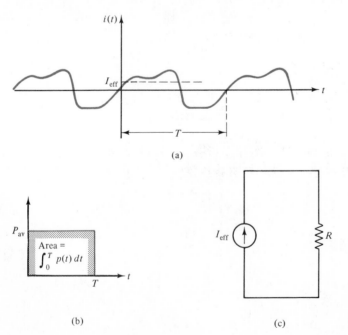

(a)

(b)   (c)

Figure 11-8 A periodic function.

integration); then we divide by the base T to find the height of a rectangle of the same area. See Figure 11-8(b).

We equate P_{av} in Equation 11-23 to the average power produced by the constant current I_{eff}

$$I_{eff}^2 R = \frac{1}{T} \int_0^T [i(t)]^2 R \, dt \tag{11-24}$$

From this we get

$$I_{eff} = \sqrt{\frac{1}{T} \int_0^T [i(t)]^2 \, dt} \tag{11-25}$$

Generically, the effective value of any periodic function $f(t)$ is defined as

$$F_{eff} = \sqrt{\frac{1}{T} \int_0^T [f(t)]^2 \, dt} \tag{11-26}$$

This general expression can be read as "the square root of the mean (average) of the squared function"; more briefly, it is called the root mean square (RMS) value of $f(t)$, or the effective value of $f(t)$. By extension, it is applicable to any periodic waveform, current or voltage.

We have found that the RMS value of a pure sinusoid is its amplitude divided by $\sqrt{2}$. This is the common value cited for sinusoids; for example, the outlet voltage in most U.S. households is 120 V effective; so the voltage waveform is $v(t) = 120\sqrt{2} \sin 2\pi(60)t = 170 \sin 377t$. If the rating of a lightbulb is, for example, "120 V, 100 W," it means that with an effective voltage of 120 V, the average power dissipated in the bulb is 100 W.

Using effective values, the expression for the average power (Equation 11-15) becomes

$$P_{av} = \frac{V_m}{\sqrt{2}} \frac{I_m}{\sqrt{2}} \cos \theta = V_{eff} I_{eff} \cos \theta \tag{11-27}$$

Several examples will illustrate these ideas.

EXAMPLE 11-6 _____

Calculate the average value and the effective value of the triangular voltage waveform shown in Figure 11-9. What is the average power dissipated in a 1-Ω resistor if this $v(t)$ is the voltage across it?

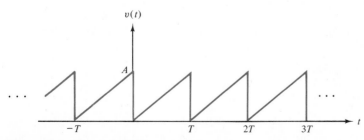

Figure 11-9   Example 11-6.

Solution.   The expression for $v(t)$ is

$$v(t) = \frac{A}{T} t \qquad 0 \le t < T$$

and it is periodic thereafter. Consequently, its average value is

$$V_{av} = \frac{1}{T} \int_0^T \frac{A}{T} t \, dt = \frac{A}{2} = 0.5\, A$$

which is also easy to recognize by inspection. The RMS (effective) value of $v(t)$ is

$$V_{eff} = \sqrt{\frac{1}{T} \int_0^T \left(\frac{A}{T} t\right)^2 dt} = \frac{A}{\sqrt{3}} \simeq 0.577\, A$$

The average power will be

$$P_{av} = \frac{V_{eff}^2}{R} = \frac{A^2}{3} \qquad W \qquad ■$$

EXAMPLE 11-7

A small electric motor is driven by a sinusoidal voltage source, the household outlet. The plate on the motor reads, "120 V, 600 W, 0.8 p.f. lagging." Calculate the impedance of this motor and draw a circuit model for it.

Solution.   From Equation 11-27 we have

$$600 = (120)(I_{eff})0.8$$

or

$$I_{eff} = 6.25 \text{ A}$$

If we take **V** as reference, we can write

$$\mathbf{V} = 120\sqrt{2}\underline{/0°}$$

and

$$\mathbf{I} = 6.25\sqrt{2}\underline{/-\cos^{-1} 0.8} = 6.25\sqrt{2}\underline{/-36.87°}$$

since this is a lagging power factor. From these two phasors, the impedance of the motor is

$$\mathbf{Z} = \frac{\mathbf{V}}{\mathbf{I}} = 19.2\underline{/36.87°} = 15.4 + j11.5 = R + jX_L$$

and we recognize the model as a series connection of $R = 15.4\ \Omega$ and $L = 11.5/(2\pi)60 = 31$ mH. ■

EXAMPLE 11-8

In the previous example, we can check our results by using Equation 11-17

$$P_{av} = I_{eff}^2 R = (6.25)^2 15.4 = 600 \text{ W}$$

since no average power is dissipated by the reactance.

The angle between $\mathbf{V}$ and $\mathbf{I}$ is

$$\theta = \theta_Z = \tan^{-1} \frac{11.5}{15.4}$$

and so the power factor is

$$\cos \theta = \cos\left(\tan^{-1} \frac{11.5}{15.4}\right) = 0.8$$

■

EXAMPLE 11-9

Find the average value and the effective value of $i(t)$ shown in Figure 11-10. Also calculate the heat energy dissipated by $R = 2\,\Omega$ carrying $i(t)$, between $t = 0$ and $t = 4$ s.

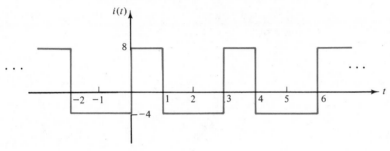

Figure 11-10 Example 11-9.

Solution. Here we have

$$i(t) = \begin{cases} 8 & 0 < t < 1 \\ -4 & 1 < t < 3 \end{cases}$$

for $i(t)$ over one period. Therefore

$$I_{av} = \frac{1}{T} \int_0^T i(t)\, dt = \frac{1}{3}\left[\int_0^1 8\, dt + \int_1^3 (-4)\, dt\right] = 0$$

This result can be recognized also by inspection. Whenever the total area under $f(t)$ is zero over a period, the average value F_{av} is zero. To calculate I_{eff} we write

$$I_{eff} = \sqrt{\frac{1}{3}\left[\int_0^1 (8)^2\, dt + \int_1^3 (-4)^2\, dt\right]} = 5.66\text{ A}$$

The energy dissipated is

$$w = \int_0^4 P_{av}\, dt = (5.66)^2(2)(4) = 256\text{ J}$$

■

Probs. 11-8, 11-9, 11-10, 11-11

11-3 REACTIVE POWER. COMPLEX POWER

In a previous section (Examples 11-2 and 11-3) we found that the instantaneous power in a pure capacitor is

$$p_C(t) = \frac{V_m I_m}{2} \sin 2\omega t \tag{11-28}$$

while for a pure inductor

$$p_L(t) = -\frac{V_m I_m}{2} \sin 2\omega t \qquad (11\text{-}29)$$

The average power in each case is zero. A capacitor and an inductor do not absorb and dissipate power like a resistor. The instantaneous power (Equation 11-28 or 11-29) is positive when energy is stored in L or in C, and is negative when that energy is removed from the element. The element is lossless since $P_{av} = 0$. Also, since $p_C(t) = -p_L(t)$ at every instant, we recognize the "give-and-take" interaction of energy between these elements. Their total instantaneous power is zero, as is their average power.

Let us derive the power relations in a general impedance $\mathbf{Z}$ using the geometry of complex numbers. An inductive circuit will have an impedance

$$\mathbf{Z} = R + jX_L \qquad (11\text{-}30)$$

with a lagging power factor, $\cos\theta$, as shown in Figure 11-11(a). A capacitive circuit has the impedance

$$\mathbf{Z} = R - jX_C \qquad (11\text{-}31)$$

and a leading power factor, $\cos\theta$, as shown in Figure 11-11(b).

In power calculations, it is convenient to draw phasor currents and voltages with their effective (rather than peak or amplitude) lengths. Let us adopt this convention and let the current through $\mathbf{Z}$ be the phasor $\mathbf{I}$

$$\mathbf{I} = I_{eff}\underline{/0^\circ} \qquad (11\text{-}32)$$

If we multiply each side of the *impedance triangle* in Figure 11-11(a) by I_{eff} (a positive number), we'll get the *voltage triangle* in (c): The horizontal side is $\mathbf{V}_R = V_R = I_{eff}R$, the voltage across R; the vertical side is (in magnitude) $V_L = I_{eff}X_L$, the voltage across X_L; and the hypotenuse is $V_Z = I_{eff}Z$, in magnitude the total voltage across $\mathbf{Z}$. For the capacitive case, the voltage triangle is shown in Figure 11-11(d), with $\mathbf{V}_R$ in phase with $\mathbf{I}$ and $\mathbf{V}_C$ lagging behind $\mathbf{I}$ by 90°. It is important to see that the voltage triangle is

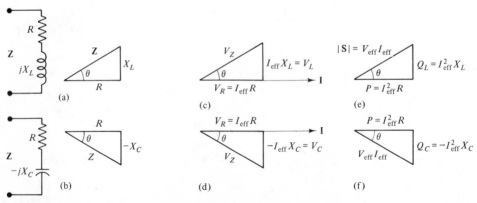

Figure 11-11   Impedance, voltage, and power triangles.

similar to the impedance triangle, since it is a scaled version of it: Every side is multiplied by I_{eff}. Thus, θ and $\cos\theta$ are preserved.

Now multiply the voltage triangle by I_{eff}. See Figure 11-11(e),(f). The horizontal side becomes

$$P = I_{eff}^2 R \qquad \text{W} \tag{11-33}$$

the average power dissipated in **Z**. For convenience, we dropped the subscript "average" from P, since there is no doubt about it. The vertical side becomes now

$$Q_L = I_{eff}^2 X \qquad \text{VAR} \tag{11-34}$$

the *reactive power* in the inductance. The units of Q_L are the same as P; however, to distinguish them, the units of Q_L are called volt-ampere reactive, or VAR. Reactive power is *not* dissipated or converted into another form like P. It is needed only to establish and maintain the electric and magnetic fields associated with capacitors and inductors. As an added way to distinguish between P and Q, the term *real power* is sometimes used for P.

The hypotenuse of the power triangle is the product $V_{eff}I_{eff}$, called the *volt-ampere* (VA) or *apparent power*. In complex number notation, we write

$$\mathbf{S} = P + jQ_L \tag{11-35}$$

directly from the power triangle. Also it is seen from the power triangle that

$$P = |\mathbf{S}|\cos\theta = V_{eff}I_{eff}\cos\theta \tag{11-36}$$

as expected, and

$$Q_L = |\mathbf{S}|\sin\theta = V_{eff}I_{eff}\sin\theta \tag{11-37}$$

The quantity **S** is called the *complex power*.

For the capacitive case, Figure 11-11(f), we have a *negative* reactive power

$$Q_C = -I_{eff}^2 X_C = -|\mathbf{S}|\sin\theta \tag{11-38}$$

in agreement with its opposite nature to Q_L. There is nothing mysterious about the negative reactive power of a capacitor. It does *not* mean that a capacitor provides power. All it means is that the reactive power (not the real power!) in a capacitor is of opposite sign to the reactive power in an inductor. The complex number equation for **S** in the capacitive case is

$$\mathbf{S} = P - jQ_C \tag{11-39}$$

For the general case, we can obtain the expression for **S** as follows. We write

$$\mathbf{V} = V_{eff}\underline{/\theta_v} = V_{eff}e^{j\theta_v} \tag{11-40}$$

and

$$\mathbf{I} = I_{eff}\underline{/\theta_i} = I_{eff}e^{j\theta_i} \tag{11-41}$$

in accordance with our new convention for phasor lengths. Then

$$\begin{aligned} P &= V_{eff}I_{eff}\cos\theta = V_{eff}I_{eff}\cos(\theta_v - \theta_i) \\ &= \text{Re}\left\{V_{eff}e^{j\theta_v}\cdot I_{eff}e^{-j\theta_i}\right\} \end{aligned} \tag{11-42}$$

where, as usual, Re stands for "the real part of ... ." Therefore, Equation 11-42 is rewritten together with Equations 11-40 and 11-41 as

$$P = \text{Re} (\mathbf{VI}^*) \tag{11-43}$$

where $\mathbf{I}^*$ is the complex conjugate of the phasor $\mathbf{I}$

$$\mathbf{I}^* = (I_{\text{eff}} e^{j\theta_i})^* = I_{\text{eff}} e^{-j\theta_i} \tag{11-44}$$

Therefore, the complex power $\mathbf{S}$ is given by

$$\mathbf{S} = \mathbf{VI}^* = P \pm jQ \tag{11-45}$$

where the $(+)$ sign is attributed to Q_L and the $(-)$ sign to Q_C.

EXAMPLE 11-10

In Example 11-7, we had

$$\mathbf{V} = 120\underline{/0^\circ}$$

and

$$\mathbf{I} = 6.25\underline{/-36.87^\circ}$$

Solution.

$$\mathbf{S} = \mathbf{VI}^* = (120\underline{/0^\circ})(6.25\underline{/36.87^\circ})$$
$$= 750\underline{/36.87^\circ} = 600 + j450$$

Consequently

$$P = \text{Re}\,\{\mathbf{S}\} = 600 \text{ W}$$

and, since the imaginary part of $\mathbf{S}$ is positive, we have an inductive circuit

$$Q_L = 450 \text{ VAR}$$

The apparent power to this circuit is

$$|\mathbf{S}| = (120)(6.25) = 750 \text{ VA}$$

as maintained by the source.   ■

EXAMPLE 11-11

Two loads are connected in parallel to a voltage source, as shown in Figure 11-12(a). The nameplate on load A reads, "3000 W, 220 V, 0.85 p.f. lagging." The nameplate on load B reads, "5000 VA, 220 V, 0.7 p.f. leading." Use the power triangles to calculate $\mathbf{V}$ and $\mathbf{I}$ and the power factor of the combined load.

Solution.   First, we use $\mathbf{V}$, the common phasor, as a reference at 0°

$$\mathbf{V} = 220\underline{/0^\circ}$$

The power triangle of load A is obtained as follows. We are given

$$P_A = 3000 \text{ W}$$

and

$$\cos \theta_A = 0.85 \text{ lagging (inductive)}$$
$$\therefore \quad \theta_A = 31.79^\circ$$

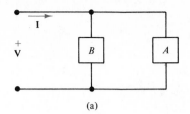

(a)

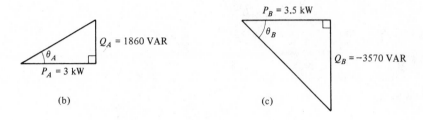

(b)   (c)

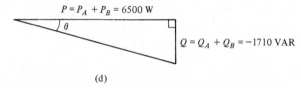

(d)

Figure 11-12   Example 11-11.

Then, since

$$\frac{Q_A}{P_A} = \tan \theta$$

we have

$$Q_A = P_A \tan \theta = (3000)(0.62) = 1860 \text{ VAR}$$

the inductive reactive power. The complete power triangle for A is drawn in Figure 11-12(b). For load B, we are given

$$|\mathbf{S}_B| = 5000 \text{ VA}$$

and

$$\cos \theta_B = 0.7 \text{ leading (capacitive)}$$
$$\therefore \quad \theta_B = 45.57°$$

Since

$$\frac{P_B}{|\mathbf{S}_B|} = \cos \theta_B$$
$$P_B = (5000)(0.7) = 3500 \text{ W}$$

Also, since

$$\frac{Q_B}{|S_B|} = \sin \theta_B$$

$$Q_B = -3570 \text{ VAR}$$

the capacitive reactive power. The power triangle for load B is shown in Figure 11-12(c).

The total power triangle at the source has therefore

$$P = 3000 + 3500 = 6500 \text{ W}$$

$$Q = 1860 - 3570 = -1710 \text{ VAR}$$

and is shown in Figure 11-12(d). From it, we obtain immediately

$$\theta = \tan^{-1} \frac{1710}{6500} = 14.74°$$

and the total power factor is

$$\cos \theta = \cos(-14.74°) = 0.967 \text{ leading}$$

As a final bonus, we can also calculate $\mathbf{I}$ as follows: The total apparent power at the source is

$$|S| = \sqrt{P^2 + Q^2} = 6721 \text{ VA} = V_{\text{eff}} I_{\text{eff}}$$

and therefore

$$I_{\text{eff}} = \frac{6721}{220} = 30.55 \text{ A}$$

Consequently, since we took $\mathbf{V}$ as reference

$$\mathbf{V} = 220\underline{/0°}$$

we have

$$\mathbf{I} = 30.55\underline{/14.74°}$$

because the power factor is leading.

This example illustrates how simple, yet powerful (pun intended) this method is using the ideas and the geometry of the power triangle. ■

Probs. 11-11
11-12, 11-13
11-14, 11-15

11-4 POWER FACTOR CORRECTION

The previous example will serve to illustrate the very common practice known as power factor correction. See Figure 11-13(a). An industrial plant is a typical load with an inherent lagging power factor, because electric motors are inductive circuits. If this power factor is low, say, 0.7, the load requires a large reactive power $+Q$ for a given P and θ, as shown in Figure 11-13(b). That's where the power company supplying this load becomes unhappy for two main reasons: (1) In order to maintain this power triangle, the power company must maintain a large apparent power $|S| = V_{\text{eff}} I_{\text{eff}}$. The customer (the load) pays for only a fraction of this apparent power, 70 percent in our example. The meter that runs up the electric bill measures only the energy

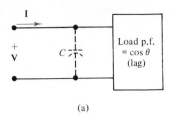

(a)

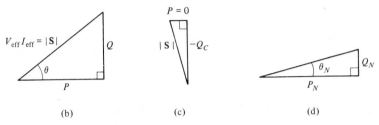

(b) (c) (d)

Figure 11-13 Power factor correction.

consumption in kilowatt hours (kWh), where $1 \text{ kWh} = (1000)(3600) \text{ watt} \cdot \text{second} = 3.6 \times 10^6$ joules, and the paid-up energy is

$$w \Big]_{t_1}^{t_2} = \int_{t_1}^{t_2} P \, dt = P(t_2 - t_1) \tag{11-46}$$

(2) With a large apparent power, the current in the transmission line is large. This causes heat losses in the line, $I_{\text{eff}}^2 R$. Perching birds enjoy it during cold weather, but they don't pay for it.

For these reasons, the power company can (and usually does) insist that the industrial plant *improve* its power factor to, say, 0.95 lagging. This is done by connecting an appropriate capacitor across the load. The power triangle for the capacitor is shown in Figure 11-13(c), assuming an ideal lossless capacitor. The drawing shows a horizontal P simply to clarify this point. In fact, $P_C = 0$ for the ideal capacitor, and the triangle becomes a single vertical line.

The new, overall power triangle for the load and the capacitor is shown in Figure 11-13(d). The new real power P_N is the same as before, $P_N = P$ since $P_C = 0$. The new reactive power is

$$Q_N = Q - Q_C \tag{11-47}$$

and its value is known from the specified new power factor, $\cos \theta_N$, in our example 0.95 lagging. We can therefore calculate the required reactive power of the capacitor, $-Q_C$, and the complete specifications for the capacitor.

EXAMPLE 11-12 _____

Assume an inductive load of the following specifications: 220 V, 3 kW, 0.7 p.f. The improved power factor is 0.95 lagging.

Solution. We have, from Figure 11-13(b),

$$Q = P \tan \theta = 3000 \tan (\cos^{-1} 0.7) = 3060 \text{ VAR}$$

In the new power triangle, we must have

$$Q_N = P_N \tan (\cos^{-1} 0.95) = (3000)(0.329) = 986 \text{ VAR}$$

The capacitor's reactive power is therefore

$$Q_C = 986 - 3060 = -2074 \text{ VAR}$$

Now, in Figure 11-13(c) we have

$$|Q_C| = |S_C| = V_{\text{eff}} I_{\text{eff}} = 220 I_{\text{eff}} \qquad \text{VA}$$

Therefore, the effective current in the capacitor is

$$I_{\text{eff}} = \frac{2074}{220} = 9.43 \text{ A}$$

The magnitude of the reactance of the capacitor is

$$\frac{|V_C|}{|I_C|} = |Z_C| = |X_C| = \frac{1}{\omega C} = \frac{V_{\text{eff}}}{I_{\text{eff}}} = \frac{220}{9.43} = 23.33 \ \Omega$$

and therefore, at $\omega = 2\pi(60) = 377$ rad/s, we get

$$C = \frac{1}{\omega |X_C|} = \frac{1}{(377)(23.33)} = 114 \ \mu\text{F}$$

The complete specifications for the capacitor are then: 114 μF, 220 V, 2074 VA. A practical capacitor will be 100 μF, 220 V, 2 kVA. ∎

Probs. 11-16 11-17, 11-18

11-5 MAXIMUM POWER TRANSFER

At times, we are interested in delivering to a load the maximum possible average power; this is possible if the impedance of the load can be adjusted or preselected. The situation is shown in Figure 11-14(a). Here we make use of Thévenin's theorem and replace the entire network seen by the load by its Thévenin's equivalent, as in Figure 11-14(b). We remember that as far as the load Z_o is concerned, the two configurations are the same: The voltage and current at terminals *A-B* remain unchanged.

We write

$$\mathbf{Z}_{\text{Th}} = R_{\text{Th}} + jX_{\text{Th}} \tag{11-48}$$

and

$$\mathbf{Z}_o = R_o + jX_o \tag{11-49}$$

allowing each reactive term to carry its sign, positive if inductive and negative if capacitive. In Figure 11-14(b) we have

$$\mathbf{I} = \frac{\mathbf{V}_{\text{Th}}}{\mathbf{Z}_{\text{Th}} + \mathbf{Z}_o} = \frac{\mathbf{V}_{\text{Th}}}{(R_{\text{Th}} + R_o) + j(X_{\text{Th}} + X_o)} \tag{11-50}$$

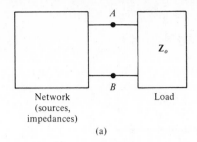

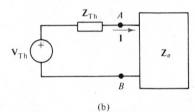

Figure 11-14   Maximum power transfer.

The average power P delivered to the load is then†

$$P = |\mathbf{I}|^2 R_o = \frac{|\mathbf{V}_{Th}|^2}{(R_{Th} + R_o)^2 + (X_{Th} + X_o)^2} R_o \tag{11-51}$$

and we wish to maximize this quantity. Let us remember that only the load is adjustable; in other words, $\mathbf{V}_{Th}$, R_{Th}, and X_{Th} are fixed by the network to the left of terminals A-B. Only R_o and X_o are adjustable, and they are independent variables.

Rather than rush into heavy calculus and set

$$\frac{\partial P}{\partial R_o} = 0 \tag{11-52}$$

and

$$\frac{\partial P}{\partial X_o} = 0 \tag{11-53}$$

for maximizing P with respect to these two variables, we should look again at Equation 11-51. The quantity $(X_{Th} + X_o)^2$ in the denominator is never negative. Therefore, it contributes to a larger denominator and, consequently, to a smaller P unless

$$X_o = -X_{Th} \tag{11-54}$$

With this condition satisfied, Equation 11-51 becomes

$$P = |\mathbf{V}_{Th}|^2 \frac{R_o}{(R_{Th} + R_o)^2} \tag{11-55}$$

---

† We are using effective (RMS) values for the phasors of all the sinusoidal voltages and currents.

and now we invoke Equation 11-52—a much easier task. The result is

$$\frac{\partial P}{\partial R_o} = |\mathbf{V}_{\text{Th}}|^2 \frac{(R_{\text{Th}} + R_o)^2 - 2R_o(R_{\text{Th}} + R_o)}{(R_{\text{Th}} + R_o)^4} = 0 \qquad (11\text{-}56)$$

From which we get

$$R_o = R_{\text{Th}} \qquad (11\text{-}57)$$

Combining Equations 11-54 and 11-57, we get the desired answer

$$\mathbf{Z}_o = R_o + jX_o = R_{\text{Th}} - jX_{\text{Th}} = \mathbf{Z}_{\text{Th}}^* \qquad (11\text{-}58)$$

For maximum power transfer, the adjustable load impedance must be equal to the *conjugate* of the Thévenin impedance. This power is then, from Equation 11-55,

$$P = |\mathbf{I}|^2 R_o = \left|\frac{\mathbf{V}_{\text{Th}}}{2R_{\text{Th}}}\right|^2 R_o = \frac{1}{4}\frac{|\mathbf{V}_{\text{Th}}|^2}{R_o} \qquad (11\text{-}59)$$

EXAMPLE 11-13 _____

In the network shown in Figure 11-15, what must be $\mathbf{Z}_o$ so that maximum average power is transferred to it? What is then this power?

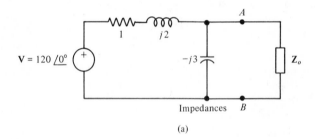

(a)

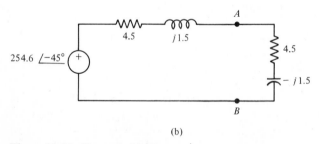

(b)

Figure 11-15   Example 11-13.

Solution.   We obtain $\mathbf{V}_{\text{Th}}$ by a voltage divider after removing $\mathbf{Z}_o$ as

$$\mathbf{V}_{\text{Th}} = 120\underline{/0°} \frac{-j3}{1 + j2 - j3} = 254.6\underline{/-45°}$$

and

$$\mathbf{Z}_{\text{Th}} = (-j3)\|(1 + j2) = 4.5 + j1.5$$

Therefore

$$\mathbf{Z}_o = \mathbf{Z}_{\mathrm{Th}}^* = 4.5 - j1.5$$

as shown in Figure 11-15(b). The average power to this load is then

Probs. 11-19,
11-20, 11-21,
11-22

$$P = \frac{1}{4} \frac{254.6^2}{4.5} = 3600 \text{ W}$$

PROBLEMS

11-1 Calculate the power factors for the circuits shown in (a) to (d).

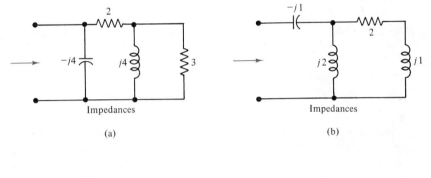

(a)   (b)

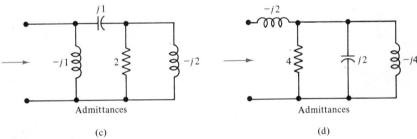

(c)   (d)

Problem 11-1

11-2 A one-port network has a current applied to it

$$i(t) = 3 \sin (10t + 45°)$$

The response steady-state voltage is

$$v(t) = 0.4 \cos (10t - 20°)$$

(a) Plot $i(t)$ and $v(t)$ accurately versus $(10t)$.
(b) Plot $p(t)$ by multiplying $i(t)$ by $v(t)$ point by point in (a).
(c) From the plot of $p(t)$ read a reasonably accurate value of the average power P_{av}.
(d) Calculate the power factor of this one-port network.
(e) Write the expressions for **V** and **I**.
(f) From (d) and (e) calculate P_{av} and compare with (c).

11-3 **(a)** In Problem 11-1, classify each power factor as leading or lagging.
     (b) For each circuit, draw an equivalent circuit having the same impedance but consisting of only two elements in series.

*11-4 An impedance $\mathbf{Z}_1$ has a lagging power factor of 0.8. A second impedance $\mathbf{Z}_2$ has a leading power factor of 0.8. What can you say about the power factor of

$$\mathbf{Z}_3 = \mathbf{Z}_1 + \mathbf{Z}_2$$

Justify fully your answer.

11-5 A load consists of the parallel connection of $R = 5\,\Omega$ and $L = 2$ mH. It is driven by a current source, $i_s(t) = 10 \cos(4000t + 15°)$.
(a) Calculate the average power received by the load.
(b) What is the power factor of the load?

11-6 A capacitor of susceptance $4\mho$ is connected in parallel with a conductance of $4\mho$ to a current source

$$i_s(t) = 3 \cos 10^6 t$$

(a) Calculate the average power received by the load.
(b) Calculate the power factor of the load.

11-7 For a one-port network we are given

$$v(t) = \sqrt{2}\,50 \cos(\omega t + 40°)$$
$$i(t) = \sqrt{2}\,10 \cos(\omega t - 65°)$$

with their standard associated references. Calculate P_{av} and state whether the one-port is delivering it or receiving it.

11-8 Calculate the effective value of the half-wave rectified sinusoidal voltage shown in the figure.

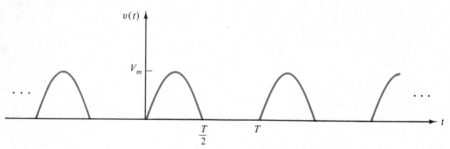

Problem 11-8

11-9 Calculate the effective value of the full-wave rectified sinusoidal voltage shown in the figure.

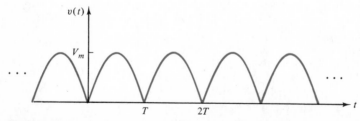

Problem 11-9

11-10 The current $i(t)$ through a resistor $R = 2\,\Omega$ is a periodic decaying exponential and is shown in the figure. Calculate the average power dissipated in R.

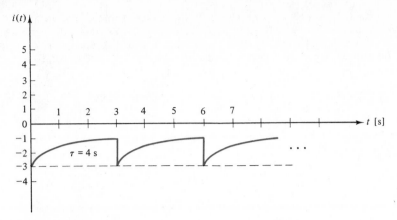

Problem 11-10

11-11 A load has the impedance $\mathbf{Z}_{load} = 10 + j2$. It is connected through a short transmission line whose impedance is $\mathbf{Z}_{line} = 0.8 + j0$ to a sinusoidal voltage source $V_{eff} = 120$ V. See the figure. Calculate: (a) the average power delivered to the load; (b) the average power in the line; (c) the power factor of the load; (d) the power factor seen by the source.

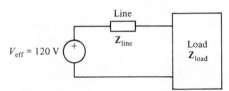

Problem 11-11

11-12 A resistor $R = 10 \, \Omega$ is connected in series with an inductor $L = 0.05$ H. In parallel with these is a capacitor $C = 2$ mF. This circuit is driven by a current source $i(t) = 10\sqrt{2} \cos 50t$.

(a) Calculate and draw the voltage triangles for the RL branch and for the C branch.

(b) Calculate and draw the power triangles.

(c) From (b) find the power factor of this circuit.

11-13 A series RL branch is given as $\mathbf{Z}_1 = 3 + j2$. A series RC branch is given as $\mathbf{Z}_2 = 2 - j3$. These are connected in parallel and a voltage source $\mathbf{V} = 50\underline{/0°}$ (effective value) excites them. Calculate and draw the power triangles of $\mathbf{Z}_1$, $\mathbf{Z}_2$, and of the total circuit. What is the total power factor?

11-14 For a one-port, we know that

$$\mathbf{V} = 110\underline{/0°} \qquad \omega = 500 \text{ rad/s}$$
$$\mathbf{I} = 30\underline{/-60°}$$

Calculate the complex power and the apparent power. From the reactive power and I_{eff}, identify the reactive element (L or C) and its value.

11-15 Prove the following useful expressions for P and Q:

(a) $P = \frac{1}{2}[\mathbf{VI^*} + \mathbf{V^*I}]$

(b) $Q = \dfrac{1}{2j}[\mathbf{VI^*} - \mathbf{V^*I}]$

11-16 Three loads are supplied by a generator (an ac voltage source), as shown: (1) Z_1 is a motor, rated at 50 kW, 220 V, 0.85 p.f. lagging. (2) Z_2 is an electric heater, purely resistive, rated at 8 kW. (3) Z_3 is an electric welding machine, modeled as $Z_3 = 3 + j3$. Calculate the capacitor needed in order to have the generator operate at 0.98 p.f. lagging. The entire system is in a European country where $f = 50$ Hz.

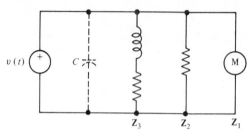

Problem 11-16

11-17 A generator (an ac voltage source) has a rating of 2 MVA ($= 2 \times 10^6$ VA). It supplies a load whose power factor is 0.85 lagging. In order to improve the power factor to 0.975 lagging, a capacitor is connected across the load. Calculate the rating of C in VAR if:
(a) The average power supplied by the generator is to remain constant.
(b) The apparent power supplied by the generator is to remain constant.

11-18 A load $Z = 10 + j8$ is supplied by a source of 120 V at 60 Hz. To improve the power factor to 0.95 lagging, a lossy capacitor is connected across the load. The lossy capacitor can be modeled as a series connection of a resistance r and a pure capacitance C. The power factor of this lossy capacitor is 0.087 leading. Calculate C.

11-19 Verify Equations 11-54 and 11-57 by applying Equations 11-52 and 11-53 to Equation 11-51.

11-20 Derive the condition for maximum power transfer for a purely resistive network and a purely resistive load R_o. Plot P versus R_o, where $0 \le R_o \le \infty$.

11-21 Assume that in Figure 11-14 the *magnitude* of the load impedance $|Z_o|$ can be adjusted but not its angle. Prove that maximum power transfer will occur when

$$|Z_o| = |Z_{Th}|$$

11-22 In the op amp circuit shown, calculate the output voltage v_o; then calculate the average power in an output resistor, $R_o = 10 \, \Omega$.

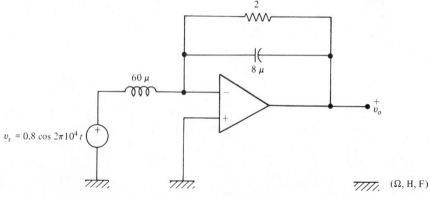

Problem 11-22

Chapter 12

Transformers

A transformer is an electrical device consisting of two (or more) mutually coupled inductors. In our study here, we'll review the phenomenon of mutual induction, and apply sinusoidal steady-state (phasor) methods to circuits with various transformers.

12-1 THE TRANSFORMER

A basic scheme of a transformer is shown in Figure 12-1(a). Two inductors (or "coils" in industrial lingo) are wound on a core of a ferromagnetic material such as iron, steel, and their alloys. A current $i_1(t)$ flows in one inductor. As we studied in Chapter 6, a magnetic flux ϕ_1 is established by this current, and in a linear inductor it is given by

$$\phi_1(t) = L_1 i_1(t) \qquad \text{webers} \tag{12-1}$$

This flux links, in part, its own inductor and, in part, the second inductor, as shown in Figure 12-1(a). We write

$$\phi_1(t) = \phi_{11}(t) + \phi_{21}(t) \tag{12-2}$$

where ϕ_{11} is the flux linking coil 1 and ϕ_{21} is the flux linking coil 2. The ferromagnetic properties of the core concentrate this flux and keep it within the path of the core, thus maximizing ϕ_{21}, the part of ϕ_1 which links the second inductor.

The direction of the flux lines is found by the familiar right-hand rule: Wrap your right hand's fingers in the direction of the flow of the current; your thumb then will point in the direction of the flux inside the inductor.

The time-varying flux $\phi_1(t)$ causes the voltage drop $v_1(t)$ across L_1

$$v_1 = \frac{d\phi_1}{dt} = L_1 \frac{di_1}{dt} \tag{12-3}$$

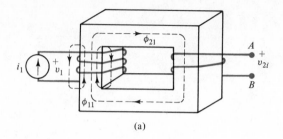

(a)

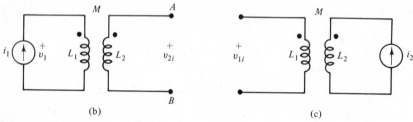

(b)   (c)

Figure 12-1   A transformer with one excitation.

with the standard associated references for voltage and current: The $(+)$ of the voltage is at the tail of the current's arrow. Also, the partial flux ϕ_{21}, created by i_1 and linking L_2, induces in L_2 a voltage v_{21}, as described in Chapter 6

$$v_{2i} = \frac{d\phi_{21}}{dt} = M\frac{di_1}{dt} \tag{12-4}$$

where M, in henrys, is the mutual inductance between L_1 and L_2. We are using the subscript i temporarily to remind us that this is an induced voltage in an open-circuited L_2.

The reference sign for v_2 is determined by the relative sense of the windings of L_1 and L_2, as follows: Let us mark with a dot the terminal of L_1 into which i_1 flows. Then we use the right-hand rule to determine the direction of the flux ϕ_1. Next, we repeat this experiment just for the second inductor L_2: We let i_2 flow into one arbitrarily marked terminal on L_2. By the right-hand rule we determine the direction of ϕ_2, the new flux created by i_2. If this direction is the same as ϕ_1 (caused by i_1), then the dot marked on L_2 is correct. If not, we mark the other terminal of L_2 with a dot.

The dot convention for *induced* voltages and their *inducing* currents, as studied in Chapter 6, is: The induced voltage v_{2i} will be $(+)$ at its dot on L_2 if the inducing current i_1 enters its own dot on L_1. This is shown in Figure 12-1(b). As a matter of practice, transformers carry the dot marks from the manufacturer, so we don't have to go through this experiment with every unit. The circuit diagram of Figure 12-1(a) is shown in Figure 12-1(b).

In Chapter 6 we gave also the relation between M, L_1, and L_2. It is

$$M = k\sqrt{L_1 L_2} \tag{12-5}$$

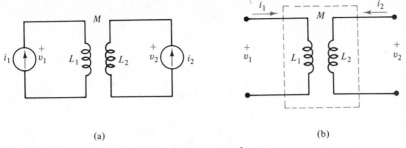

Figure 12-2 A transformer as a 2-port network.

where k is the *coefficient of coupling* between L_1 and L_2. It is a measure of how closely, or tightly, the two windings are linked. We stated there, without proof, the limits on k

$$0 < k \leq 1 \tag{12-6}$$

which sounds reasonable: $k = 0$ implies $M = 0$, and then there is no coupling at all; a weak coupling means a small k, and for a very tight coupling k approaches unity. It is very instructive to derive Equation 12-6 on the basis of energy considerations, and we will do it soon.

When both windings of this transformer carry currents, as shown in Figure 12-2(a), we use superposition, assuming (as we do throughout this book) that the transformer is linear. The total voltage across L_1 consists of the self-drop $L_1 \, di_1/dt$ due to i_1, plus the induced voltage $v_{1i} = \pm M \, di_2/dt$ due to i_2. We did not show the dots in this figure, and therefore we write ($\pm$) in front of this term. Then

$$v_1(t) = L_1 \frac{di_1}{dt} \pm M \frac{di_2}{dt} \tag{12-7}$$

Similarly, the total voltage v_2 consists of its own self-drop $L_2 \, di_2/dt$ plus the induced $v_{2i} = \pm M \, di_1/dt$

$$v_2(t) = \pm M \frac{di_1}{dt} + L_2 \frac{di_2}{dt} \tag{12-8}$$

Equations 12-7 and 12-8 are the basic equations of a two-winding transformer. If it is considered as a two-port network, we assume the variables v_1 and i_1 at one port and v_2 and i_2 at the other port, without actually worrying what type of sources (v or i) we have; these equations are always valid.

Prob. 12-1

EXAMPLE 12-1 _____

In the transformer circuit shown in Figure 12-3, write the two loop equations. The winding connected to a source is usually called the *primary winding* and the other one is the *secondary winding*. A third winding, if there is one, is called *tertiary*. Here the 1-Ω resistor represents the resistance of the primary winding and possibly the internal resistance of the voltage source. The output (load) resistance $R = 200 \, \Omega$ includes the resistance of the secondary winding.

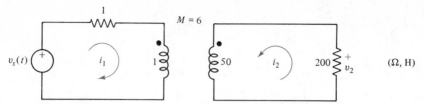

Figure 12-3 Example 12-1.

Solution. With the two loop currents as shown, we have the two loop equations

$$1i_1 + 1\frac{di_1}{dt} + 6\frac{di_2}{dt} = v_s(t)$$

$$6\frac{di_1}{dt} + 50\frac{di_2}{dt} + 200i_2 = 0$$

The output voltage is

$$v_2 = -200i_2 \qquad\qquad \blacksquare \quad \text{Prob. 12-2}$$

Now let us do some power and energy calculations for the transformer. (See also Problem 6-24.) For the transformer in Figure 12-2, we have the total instantaneous power

$$p(t) = p_1(t) + p_2(t) = v_1(t)i_1(t) + v_2(t)i_2(t) \qquad\qquad (12\text{-}9)$$

If we use Equations 12-7 and 12-8, this power is written as

$$p(t) = \left(L_1\frac{di_1}{dt} + M\frac{di_2}{dt}\right)i_1 + \left(L_2\frac{di_2}{dt} + M\frac{di_1}{dt}\right)i_2$$

$$= L_1 i_1\frac{di_1}{dt} + L_2 i_2\frac{di_2}{dt} + M\frac{d}{dt}(i_1 i_2) \qquad\qquad (12\text{-}10)$$

Here we use the ($+$) sign for the mutually induced voltages. A similar derivation holds for a ($-$) sign. Let us assume that the initial energy at $t = 0$ is zero, that is, $i_1(0^-) = i_2(0^-) = 0$; the transformer is taken fresh off the shelf. Then the energy delivered to it by the sources at any time $t > 0$ is the integral of $p(t)$,

$$w(t) = \int_0^t p(x)\,dx \qquad\qquad (12\text{-}11)$$

where, as usual, a dummy variable x must be used because t is the upper limit on the integral. Therefore

$$w(t) = \int_0^t\left[L_1 i_1\frac{di_1}{dx} + L_2 i_2\frac{di_2}{dx} + M\frac{d}{dx}(i_1 i_2)\right]dx \qquad\qquad (12\text{-}12)$$

The integration yields term, by term,

$$w(t) = \tfrac{1}{2}L_1 i_1^2(t) + \tfrac{1}{2}L_2 i_2^2(t) + M i_1(t)i_2(t) \qquad\qquad (12\text{-}13)$$

The first two terms are old friends: $\frac{1}{2}Li^2$ is the energy stored in an inductor, as we saw in Chapter 6. The third term accounts for the additional energy due to the mutual coupling.

Quite obviously, $w(t)$ cannot be negative. In other words, we must have

$$\tfrac{1}{2}L_1 i_1^2 + \tfrac{1}{2}L_2 i_2^2 + M i_1 i_2 \geq 0 \tag{12-14}$$

Let us sit back and think how to guarantee this inequality: The first two terms are never negative, each one being the product of a positive number (L_1 or L_2) and a *squared* real number. The third term, $M i_1 i_2$, will be positive if $i_1(t)$ and $i_2(t)$ are of the same sign, both positive or both negative. So the only way Equation 12-14 can go negative is if $i_1(t)$ and $i_2(t)$ are of opposite signs, that is, if

$$i_1(t) = -x i_2(t) \tag{12-15}$$

where x is any real, positive number. We divide Equation 12-14 by $i_2^2 > 0$ to get

$$\tfrac{1}{2}L_1 x^2 - Mx + \tfrac{1}{2}L_2 \geq 0 \tag{12-16}$$

Now divide by $\frac{1}{2}L_1$ and complete the square on the first two terms, to get

$$\left(x - \frac{M}{L_1}\right)^2 - \frac{M^2}{L_1^2} + \frac{L_2}{L_1} \geq 0 \tag{12-17}$$

Since the square in the parentheses is always nonnegative, Equation 12-17 will be satisfied if

$$-\frac{M^2}{L_1^2} + \frac{L_2}{L_1} \geq 0 \tag{12-18}$$

or finally,

$$M^2 \leq L_1 L_2 \tag{12-19a}$$

from which Equation 12-5 follows. Aha! The condition on the coefficient of coupling, $0 < k \leq 1$, is basically a result of the fact that the transformer, like the resistor, the inductor, or the capacitor, is a *passive* device, receiving (but not generating) energy.   Prob. 12-3

12-2 THE PERFECT TRANSFORMER. THE IDEAL TRANSFORMER

A *perfect* transformer is one whose coupling coefficient is maximum, that is, $k = 1$. This is called a perfect coupling and then we have

$$M = \sqrt{L_1 L_2} \tag{12-19b}$$

For such a transformer, let us calculate the voltage ratio v_1/v_2. We use Equations 12-7 and 12-8 together with Equation 12-19(b) to obtain

$$\frac{v_1}{v_2} = \frac{L_1 (di_1/dt) \pm \sqrt{L_1 L_2}\,(di_2/dt)}{\pm\sqrt{L_1 L_2}\,(di_1/dt) + L_2 (di_2/dt)} = \pm\sqrt{\frac{L_1}{L_2}} \tag{12-20}$$

Be sure to check the last equality.

In physical, linear inductors, the inductance is proportional to the square of the number of turns (windings) in it. That is,

$$L_1 \propto n_1^2 \qquad L_2 \propto n_2^2 \tag{12-21}$$

where n is the number of turns. Therefore, Equation 12-20 yields

$$\frac{v_1}{v_2} = \pm \sqrt{\frac{L_1}{L_2}} = \pm \sqrt{\frac{n_1^2}{n_2^2}} = \pm \frac{n_1}{n_2} = \pm a \tag{12-22}$$

where $a = n_1 : n_2$ is called *the turns ratio* of the transformer.

EXAMPLE 12-2

In a perfect transformer, we have

$$L_1 = 2\,\mathrm{H} \qquad M = 4\,\mathrm{H}$$

Calculate its turns ratio.

Solution. Since

$$M = \sqrt{L_1 L_2}$$

we have

$$4 = \sqrt{2L_2}$$

or

$$L_2 = 8\,\mathrm{H}$$

The turns ratio is

$$a = \frac{n_1}{n_2} = \sqrt{\frac{L_1}{L_2}} = \sqrt{\frac{2}{8}} = \frac{1}{2}$$

and so there are twice as many windings on the secondary as there are on the primary (500:1000, 2500:5000, etc.) ∎

In a similar way, let us calculate the current ratio i_1/i_2 in a perfect transformer. To do this, we must integrate the basic relation, Equation 12-7, between 0 and t; for simplicity, assume zero initial conditions. We get

$$i_1(t) = \frac{1}{L_1} \int_0^t v_1(x)\, dx \mp \frac{M}{L_1} i_2(t) \tag{12-23}$$

and since $M = \sqrt{L_1 L_2}$ and $a = \sqrt{L_1/L_2}$, the result is

$$i_1(t) = -\frac{1}{a} i_2(t) + \frac{1}{L_1} \int_0^t v_1(x)\, dx \tag{12-24}$$

An *ideal* transformer is a perfect transformer ($k = 1$) in which $L_1 \to \infty$, $L_2 \to \infty$, but their ratio is a finite constant. In an ideal transformer, then, Equation 12-24 yields

$$i_1(t) = -\frac{1}{a} i_2(t) \tag{12-25}$$

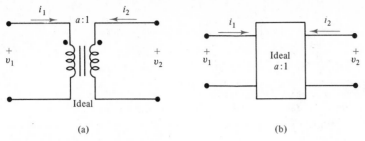

| (a) | (b) |

Figure 12-4 Ideal transformer.

Why do we bother with an ideal transformer? For several reasons. First, it is a good approximation (model) to a physical, iron-core transformer with a tight coupling between its primary and secondary windings. In such a transformer, the inductive reactances of L_1 and L_2 are usually very large by comparison with the load impedances in the sinusoidal steady state. Therefore, a simplified model like an ideal transformer can serve as a good approximation for a physical iron-core transformer. Another reason, closely related to the first one, is the simplicity and symmetry of the equations of the ideal transformer. From Equations 12-22 and 12-25, they are

$$\frac{v_1}{v_2} = a \tag{12-26}$$

$$\frac{i_1}{i_2} = -\frac{1}{a} \tag{12-27}$$

and so the ideal transformer, as shown in Figure 12-4(a), is completely identified by a single number, a, the turns ratio. The symbols of coils are still there, but all the inductances are infinitely large. Perhaps a better way would be to show it as a two-port network, Figure 12-4(b), *defined* by Equations 12-26 and 12-27.

EXAMPLE 12-3 _____

An ideal transformer $a:1$ is terminated on its secondary by a load resistance R_2, as shown in Figure 12-5. What is the resistance R_1 seen at the primary?

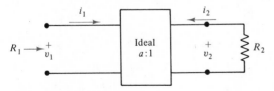

Figure 12-5 Example 12-3.

Solution. From Equation 12-26 we have

$$v_1 = av_2$$

But

$$v_2 = -R_2 i_2 \qquad \text{(notice the reference signs)}$$

Use now Equation 12-27 to get

$$v_1 = a(-R_2)(-ai_1) = a^2 R_2 i_1$$

Consequently,

$$R_1 = \frac{v_1}{i_1} = a^2 R_2$$

The ideal transformer can serve, therefore, as a device to scale the value of a resistance. A resistor of 10 Ω, connected to such a transformer with $a:1 = 5$, will look from the primary like a resistor of $(5)^2 10 = 250$ Ω.   ∎

Another common way to describe Equations 12-26 and 12-27 is to say that one quantity is *referred* from its side of the transformer to the other side. For example, Equation 12-26 yields

$$v_1 = av_2 = v_2' \tag{12-28}$$

and we say that v_2 (the secondary voltage) is referred to the primary side as $av_2 = v_2'$. A prime (') will be used to remind us that a quantity is referred. Similarly, from Equation 12-27, we can write

$$i_2 = -ai_1 = i_1' \tag{12-29}$$

where i_1' is the primary current referred to the secondary side. In Example 12-3, we referred the secondary resistance to the primary side as

$$R_2' = a^2 R_2 \tag{12-30}$$

A primary resistance R_1 will be referred to the secondary as

$$R_1' = \frac{1}{a^2} R_1 \tag{12-31}$$

EXAMPLE 12-4 _____

A hi-fi audio amplifier is shown in Figure 12-6. An 8-Ω speaker is to be connected to the amplifier; naturally, we wish to achieve maximum power transfer to the speaker. Calculate the turns ratio of the ideal transformer needed to perform the resistance match.

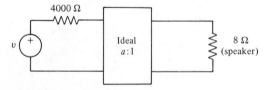

Figure 12-6 Example 12-4.

Solution. We know that, for maximum power transfer, the amplifier should see a resistance equal to its own resistance. The 8-Ω resistance must therefore be 4000 Ω when referred to the primary side. From Equation 12-30 then, we have

$$4000 = 8a^2$$

or

$$a = 22.4$$

An ideal transformer with $n_1 = 224$ and $n_2 = 10$, or any similar ratio, will do the job. In practice, a more common turns ratio of 20:1 will suffice.   ∎

Probs. 12-4, 12-5

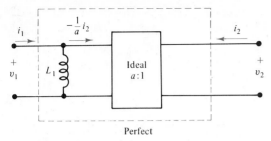

Figure 12-7   Perfect and ideal transformers.

   The relations between a perfect transformer and an ideal transformer can be explored further by comparing Equation 12-24 with Equation 12-25. With a finite L_1, the circuit in Figure 12-7 satisfies Equation 12-24 and therefore is a model for the perfect transformer.

Prob. 12-6

12-3 PHASOR ANALYSIS OF TRANSFORMERS

The sinusoidal steady state is a very common application of transformers: One or more sources at a single frequency are applied, and we are interested in the steady-state response. The two basic equations for mutual coupling (Equations 12-7 and 12-8) become in the phasor domain

$$\mathbf{V}_1 = j\omega L_1 \mathbf{I}_1 \pm j\omega M \mathbf{I}_2 \qquad (12\text{-}32)$$

and

$$\mathbf{V}_2 = \pm j\omega M \mathbf{I}_1 + j\omega L_2 \mathbf{I}_2 \qquad (12\text{-}33)$$

Here, again, the self-voltage drop in L_1 is $j\omega L_1 \mathbf{I}_1$, and the self-voltage drop in L_2 is $j\omega L_2 \mathbf{I}_2$. The induced voltage in L_1 due to $\mathbf{I}_2$ is $\pm j\omega M \mathbf{I}_2$, which is the phasor transform of $\pm M di_2(t)/dt$. The induced voltage in L_2 due to $\mathbf{I}_1$ is $\pm j\omega M \mathbf{I}_1$, the phasor transform of $\pm M di_1(t)/dt$.† The circuit diagram corresponding to Equations 12-32 and 12-33 is shown in Figure 12-8.

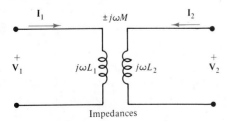

Figure 12-8   Phasor relations for Figure 12-2.

   † Let's not get tangled up in distinguishing between a *transform*, a mathematical operation, as in "phasor transform," and a *transformer*, an electrical device!

The phasor equations for a perfect transformer, obtained from Equations 12-22 and 12-24, are

$$\frac{\mathbf{V}_1}{\mathbf{V}_2} = a \tag{12-34}$$

and

$$\mathbf{I}_1 = -\frac{1}{a}\mathbf{I}_2 + \frac{1}{j\omega L}\mathbf{V}_1 \tag{12-35}$$

The ideal transformer is, again, defined by its turns ratio

$$\frac{\mathbf{V}_1}{\mathbf{V}_2} = -\frac{\mathbf{I}_2}{\mathbf{I}_1} = a \tag{12-36}$$

corresponding to Equations 12-26 and 12-27.

Prob. 12-7

Consider now the common use of a transformer as a coupling device between a source $\mathbf{V}_s$ and a general load $\mathbf{Z}_L$. The circuit is shown in Figure 12-9. Here, r_1 is the resistance of the primary winding, neglected in Figure 12-8 and Equations 12-32 and 12-33. Similarly, r_2 is the resistance of the secondary winding. The two currents $\mathbf{I}_1$ and $\mathbf{I}_2$ can serve as loop currents as we write the two loop equations

$$(\mathbf{Z}_s + r_1 + j\omega L_1)\mathbf{I}_1 \pm j\omega M\mathbf{I}_2 = \mathbf{V}_s \tag{12-37}$$

and

$$\pm j\omega M\mathbf{I}_1 + (r_2 + j\omega L_2 + \mathbf{Z}_L)\mathbf{I}_2 = 0 \tag{12-38}$$

The output voltage across the load is

$$\mathbf{V}_2 = -\mathbf{Z}_L\mathbf{I}_2 \tag{12-39}$$

This is all! It's *that* simple. From these three equations, we can calculate everything.

Let us simplify our notation as follows. Let

$$\mathbf{Z}_{11} = \mathbf{Z}_s + r_1 + j\omega L_1 \tag{12-40}$$

be the total self-impedance in loop 1, and let

$$\mathbf{Z}_{22} = r_2 + j\omega L_2 + \mathbf{Z}_L \tag{12-41}$$

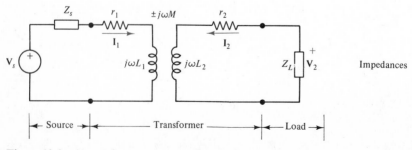

Figure 12-9 A transformer as a coupling device.

be the total self-impedance of loop 2. Also, let

$$\mathbf{Z}_{21} = \mathbf{Z}_{12} = \pm j\omega M \tag{12-42}$$

be the mutual impedance between loop 1 and loop 2. This notation is consistent with our general phasor loop analysis, as studied in Chapter 9.

The two loop equations then read

$$\mathbf{Z}_{11}\mathbf{I}_1 + \mathbf{Z}_{12}\mathbf{I}_2 = \mathbf{V}_s \tag{12-43}$$

and

$$\mathbf{Z}_{21}\mathbf{I}_1 + \mathbf{Z}_{22}\mathbf{I}_2 = 0 \tag{12-44}$$

The solution for the two loop currents is obtained easily as

$$\mathbf{I}_1 = \frac{\begin{vmatrix} \mathbf{V}_s & \mathbf{Z}_{12} \\ 0 & \mathbf{Z}_{22} \end{vmatrix}}{\begin{vmatrix} \mathbf{Z}_{11} & \mathbf{Z}_{12} \\ \mathbf{Z}_{21} & \mathbf{Z}_{22} \end{vmatrix}} = \frac{\mathbf{Z}_{22}}{\mathbf{Z}_{11}\mathbf{Z}_{22} - \mathbf{Z}_{12}^2}\,\mathbf{V}_s \tag{12-45}$$

and

$$\mathbf{I}_2 = \frac{\begin{vmatrix} \mathbf{Z}_{11} & \mathbf{V}_s \\ \mathbf{Z}_{21} & 0 \end{vmatrix}}{\begin{vmatrix} \mathbf{Z}_{11} & \mathbf{Z}_{12} \\ \mathbf{Z}_{21} & \mathbf{Z}_{22} \end{vmatrix}} = \frac{-\mathbf{Z}_{21}}{\mathbf{Z}_{11}\mathbf{Z}_{22} - \mathbf{Z}_{12}^2}\,\mathbf{V}_s = \frac{-\mathbf{Z}_{21}}{\mathbf{Z}_{22}}\,\mathbf{I}_1 \tag{12-46}$$

The total impedance seen by the source $\mathbf{V}_s$ is

$$\mathbf{Z}_{\text{total}} = \frac{\mathbf{V}_s}{\mathbf{I}_1} = \frac{\mathbf{Z}_{11}\mathbf{Z}_{22} - \mathbf{Z}_{12}^2}{\mathbf{Z}_{22}} = \mathbf{Z}_{11} - \frac{\mathbf{Z}_{12}^2}{\mathbf{Z}_{22}} = \mathbf{Z}_{11} + \frac{\omega^2 M^2}{\mathbf{Z}_{22}} \tag{12-47}$$

which is, reasonably enough, the series connection of $\mathbf{Z}_{11}$ and $\mathbf{Z}'_{22}$, the secondary total impedance referred by the transformer to the primary side

$$\mathbf{Z}'_{22} = \frac{\omega^2 M^2}{\mathbf{Z}_{22}} \tag{12-48}$$

We make several interesting observations about $\mathbf{Z}'_{22}$:

1. The actual impedance $\mathbf{Z}_{22}$ is referred to the primary as its *inverse*, and multiplied by $\omega^2 M^2$. If there were direct wire connections instead of the transformer, $\mathbf{Z}_{22}$ would *not* be inverted. The total impedance seen by the source would then simply be $\mathbf{Z}_L$.
2. It does not depend on the dot convention $(+M, \text{ or } -M)$, because of the *square* term $\omega^2 M^2$.

In order to explore a bit more the inversion of $\mathbf{Z}'_{22}$, let us write the actual impedance $\mathbf{Z}_{22}$ in its polar form

$$\mathbf{Z}_{22} = |\mathbf{Z}_{22}|\underline{/\alpha} \tag{12-49}$$

As usual, α is positive if $\mathbf{Z}_{22}$ is inductive, $\mathbf{Z}_{22} = R_{22} + jX_{22}$; if α is negative, $\mathbf{Z}_{22}$ is capacitive, $\mathbf{Z}_{22} = R_{22} - jX_{22}$. From Equations 12-48 and 12-49 we have

$$\mathbf{Z}'_{22} = \frac{\omega^2 M^2}{|\mathbf{Z}_{22}|/\alpha} = \frac{\omega^2 M^2}{|\mathbf{Z}_{22}|}\underline{/-\alpha} = \left(\frac{\omega M}{|\mathbf{Z}_{22}|}\right)^2 \mathbf{Z}^*_{22} \qquad (12\text{-}50)$$

where

$$\mathbf{Z}^*_{22} = |\mathbf{Z}_{22}|\underline{/-\alpha} \qquad (12\text{-}51)$$

is the *conjugate* of $\mathbf{Z}_{22}$. We see therefore that, apart from a positive multiplying scale factor $(\omega M/|\mathbf{Z}_{22}|)^2$, the secondary impedance $\mathbf{Z}_{22}$ is seen on the primary side as its conjugate: if $\mathbf{Z}_{22}$ is inductive, $\mathbf{Z}'_{22}$ is capacitive, and vice versa.

EXAMPLE 12-5 ⎯⎯⎯⎯⎯⎯⎯⎯⎯⎯⎯⎯⎯⎯⎯⎯⎯⎯⎯⎯⎯⎯⎯⎯⎯

In the circuit of Figure 12-9, we are given:

$$n_1 = 4000 \qquad\qquad n_2 = 1000$$
$$r_1 = 2 \qquad\qquad r_2 = 0.12$$
$$j\omega L_1 = j8 \qquad\qquad j\omega L_2 = j0.45$$
$$|\mathbf{V}_2| = 400 \text{ V(RMS)} \qquad k = 0.9, \text{ both dots on top}$$
$$\mathbf{Z}_s = 1 + j1 \qquad\qquad \mathbf{Z}_L = 2 - j1$$

Solution. We calculate

$$a = 4000:1000 = 4:1$$

$$\omega M = k\sqrt{(\omega L_1)(\omega L_2)} = 0.9\sqrt{(8)(0.45)} = 1.7$$

and according to Equations 12-40, 12-41, and 12-42,

$$\mathbf{Z}_{11} = 1 + j1 + 2 + j8 = 3 + j9 = 9.5\underline{/71.6°}$$
$$\mathbf{Z}_{22} = 0.12 + j0.45 + 2 - j1 = 2.12 - j0.55 = 2.19\underline{/-14.5°}$$
$$\mathbf{Z}_{12} = \mathbf{Z}_{21} = j1.7 = 1.7\underline{/90°}$$

The secondary impedance, referred to the primary side, is given by Equation 12-50 as

$$\mathbf{Z}'_{22} = \left(\frac{1.7}{2.19}\right)^2 2.19\underline{/14.5°} = 1.32\underline{/14.5°} = 1.28 + j0.33$$

and is inductive, whereas $\mathbf{Z}_{22}$ is capacitive. The total impedance seen by the source is then

$$\mathbf{Z}_{\text{total}} = \mathbf{Z}_{11} + \mathbf{Z}'_{22} = 3 + j9 + 1.28 + j0.33 = 10.26\underline{/65.36°}$$

The primary current is

$$\mathbf{I}_1 = \frac{\mathbf{V}_s}{\mathbf{Z}_{\text{total}}} = \frac{400\underline{/0°}}{10.26\underline{/65.36°}} = 39\underline{/-65.36°}$$

where we are using RMS values for all phasors. The secondary current is given by Equations 12-46 as

$$\mathbf{I}_2 = \frac{-j1.7}{2.19\underline{/-14.5°}} \, 39\underline{/-65.36°} = 30.3\underline{/-141°}$$

The voltage drop across the output load is given by Equation 12-39

$$\mathbf{V}_2 = -(2 - j1)30.3\underline{/-141^\circ} = 67.8\underline{/12.44^\circ}$$

The average power delivered to the load is

$$P_2 = (67.8)(30.3)\cos(12.44^\circ - 39^\circ) = 1840 \text{ W}$$

Here we had to reverse the sign of $\mathbf{I}_2$ (or $\mathbf{V}_2$) so that their associated references will be consistent with $P > 0$ being *delivered*; the addition of 180° to either angle of $\mathbf{V}_2$ or $\mathbf{I}_2$ reverses their sign. Here $180^\circ - 141^\circ = 39^\circ$ is the angle of $-\mathbf{I}_2$.

The total average power delivered by the source is

$$P_{\text{total}} = |\mathbf{I}_1|^2 R_{\text{total}} = (39)^2 4.28 = 6510 \text{ W}$$

The phasor diagram for this circuit is shown in Figure 12-10.

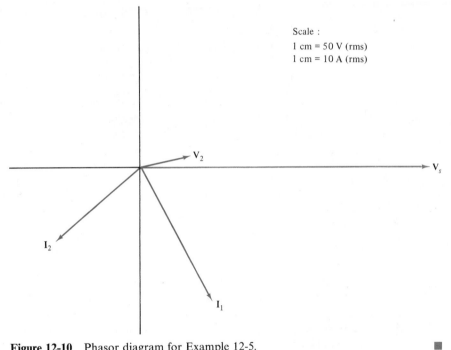

Figure 12-10   Phasor diagram for Example 12-5.   ■

At this point, if not earlier, we ought to ask the obvious question, "How do all these equations, such as Equations 12-45 and 12-46 for $\mathbf{I}_1$ and $\mathbf{I}_2$, and Equation 12-50 for the referred impedance, relate to an ideal transformer?" If our derivations are valid, they should also hold for the ideal transformer as a special case. Indeed, they do. To show this, we'll use the defining properties of an ideal transformer.

First, consider the input impedance of the loaded transformer in Figure 12-9. This input impedance is obtained from $\mathbf{Z}_{\text{total}}$ in Equation 12-47, with r_1 considered part of $\mathbf{Z}_s$ for convenience

$$\mathbf{Z}_{\text{in}} = \mathbf{Z}_{\text{total}} - \mathbf{Z}_s - r_1 = j\omega L_1 + \mathbf{Z}'_{22} = j\omega L_1 + \frac{\omega^2 M^2}{\mathbf{Z}_{22}} \qquad (12\text{-}52)$$

The ideal transformer has $k = 1$, $M^2 = L_1 L_2$. Therefore, we have

$$\mathbf{Z}_{\text{in}} = j\omega L_1 + \frac{\omega^2 L_1 L_2}{j\omega L_2 + r_2 + \mathbf{Z}_L} = \frac{j\omega L_1(r_2 + \mathbf{Z}_L)}{j\omega L_2 + r_2 + \mathbf{Z}_L} \qquad (12\text{-}53)$$

Next, in the ideal transformer $L_1 \rightarrow \infty$, $L_2 \rightarrow \infty$ but their ratio

$$\frac{L_1}{L_2} = \frac{n_1^2}{n_2^2} = a^2 \qquad (12\text{-}54)$$

is finite. With Equation 12-54, Equation 12-53 becomes

$$\mathbf{Z}_{\text{in}} = \frac{j\omega a^2 L_2(r_2 + \mathbf{Z}_L)}{j\omega L_2 + r_2 + \mathbf{Z}_L} = \frac{j\omega a^2(r_2 + \mathbf{Z}_L)}{j\omega + (r_2 + \mathbf{Z}_L)/L_2} \qquad (12\text{-}55)$$

Now let $L_2 \rightarrow \infty$, as required. Then

$$\mathbf{Z}_{\text{in}} = a^2(r_2 + \mathbf{Z}_L) = a^2 r_2 + a^2 \mathbf{Z}_L \qquad (12\text{-}56)$$

This result is in agreement with our previous results about referred resistances in Example 12-3. In a similar way, we can show that Equation 12-46, repeated here

$$\frac{\mathbf{I}_2}{\mathbf{I}_1} = \frac{-\mathbf{Z}_{21}}{\mathbf{Z}_{22}} \qquad (12\text{-}46)$$

becomes

$$\frac{\mathbf{I}_2}{\mathbf{I}_1} = -a \qquad (12\text{-}57)$$

for the ideal transformer ($k = 1$, $L_1 \rightarrow \infty$, $L_2 \rightarrow \infty$, $L_1 : L_2 = a^2$), as in Equation 12-36.   Prob. 12-8

In conclusion, then, we can develop several circuit models for a transformer. In Figure 12-11(a) we see the primary resistance r_1 and reactance $j\omega L_1$, an ideal transformer $a:1$, and the secondary resistance r_2 and reactance $j\omega L_2$. In Figure 12-11(b) all the secondary quantities are referred to the primary and combined when appropriate. For example, r_2 is referred as $r'_2 = a^2 r_2$ and combined in series with r_1; so is $(\omega L_2)' = a^2 \omega L_2$. The secondary voltage and current are referred as $\mathbf{V}'_2$ and $\mathbf{I}'_2$ where, as usual

$$\mathbf{V}'_2 = a\mathbf{V}_2 \qquad \mathbf{I}'_2 = -\frac{1}{a}\mathbf{I}_2 \qquad (12\text{-}58)$$

A more accurate circuit model is shown in Figure 12-11(c), where all the quantities are still referred to the primary. Ideally, if $\mathbf{I}_2 = 0$ (secondary is an open circuit) then $\mathbf{I}_1 = 0$ also. Practically, though, there flows a small exciting current then, $\mathbf{I}_1 = \mathbf{I}_\phi$. The additional parallel $r_c - L_\mu$ branch accounts for the fact that the iron core is not ideal and it requires a small magnetizing current $\mathbf{I}_\mu$ flowing through $j\omega L_\mu$. Also, the iron core develops heat losses in it, and these are accounted for by $\mathbf{I}_c$ flowing in the resistance r_c. The exciting current, $\mathbf{I}_\phi = \mathbf{I}_\mu + \mathbf{I}_c$, then, describes the actual, nonideal iron core of the transformer. In the following example we'll compare calculations using the equivalent circuits of Figure 12-11(b) and (c).

EXAMPLE 12-6

A transformer is rated at 50 kVA, $a = 400:2000$, with the following parameters:

$$r_1 = 0.015 \ \Omega \qquad\qquad r_2 = 0.45 \ \Omega$$

$$x_1 = \omega L_1 = 0.05 \ \Omega \qquad x_2 = \omega L_2 = 1.5 \ \Omega$$

$$r_c = 500 \ \Omega \qquad\qquad x_\mu = \omega L_\mu = 200 \ \Omega$$

The secondary load is rated at 45 kVA, 2000 V, 0.80 p.f. lagging. Calculate the primary voltage and current using Figure 12-11(b), then Figure 12-11(c).

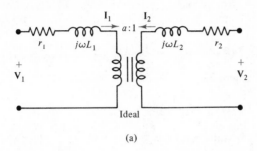

(a)

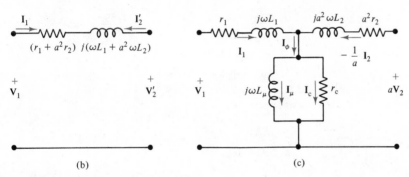

(b)   (c)

Figure 12-11   Models of a transformer.

Solution. Use Figure 12-11(b) and refer everything to the 400-V (primary) side, using magnitudes only:

$$r_1 = 0.015 \ \Omega \qquad\qquad r_2' = \left(\frac{400}{2000}\right)^2 0.45 = 0.018 \ \Omega$$

$$x_1 = \omega L_1 = 0.05 \ \Omega \qquad x_2' = \omega L_2' = \left(\frac{400}{2000}\right)^2 1.5 = 0.06 \ \Omega$$

$$V_2' = \left(\frac{400}{2000}\right)2000 = 400 \ \text{V}$$

$$I_2 = \frac{45 \times 10^3}{2000} = 22.5 \ \text{A} \qquad \therefore \quad I_2' = -\left(\frac{2000}{400}\right)22.5 = -112.5 \ \text{A}$$

Take now V_2' as reference

$$V_2' = 400\underline{/0°}$$

Then the phasor expression for $-I_2'$ is

$$-I_2' = 112.5\underline{/-36.8°} \qquad \cos^{-1} 0.8 = 36.8°$$

(Why did we write $-\mathbf{I}_2'$? *Hint*: References at the load!) From Figure 12-11(b) we write KVL as follows:

$$\mathbf{V}_1 = -\mathbf{I}_2'[(r_1 + r_2') + j(x_1 + x_2')] + \mathbf{V}_2'$$
$$= 112.5\underline{/-36.8°}(0.033 + j0.11) + 400\underline{/0°} = 410.4\underline{/1.07°}$$

and

$$\mathbf{I}_1 = -\mathbf{I}_2' = 112.5\underline{/-36.8°}$$

So the primary voltage is 410.4 V and the primary current is 112.5 A, both RMS.

With the model of Figure 12-11(c) we have, working back from load to source:

$$\mathbf{V}_2' = 400\underline{/0°} \qquad -\mathbf{I}_2' = 112.5\underline{/-36.8°}$$

as before. Now, the drop across $(r_2' + j\omega L_2')$ is

$$(0.018 + j0.06)112.5\underline{/-36.8°} = 7.05\underline{/36.5°} = 5.67 + j4.2$$

Therefore, the voltage across the parallel $L_\mu - r_c$ branch is

$$\mathbf{V}_\phi = \mathbf{V}_2' + 7.05\underline{/36.5°} = 405.67 + j4.2 = 405.7\underline{/0.59°}$$

The exciting current $\mathbf{I}_\phi$ is then

$$\mathbf{I}_\phi = \mathbf{I}_c + \mathbf{I}_\mu = \frac{405.7\underline{/0.59°}}{500} + \frac{405.7\underline{/0.59°}}{j200}$$
$$= 0.811\underline{/0.59°} + 2.029\underline{/-89.41°}$$
$$= 0.029 - j2.021 = 2.021\underline{/-89.18°}$$

The primary current is

$$\mathbf{I}_1 = -\mathbf{I}_2' + \mathbf{I}_\phi = 90.13 - j69.41 = 113.76\underline{/-37.6°}$$

The drop across the primary winding is

$$113.76\underline{/-37.6°}(0.015 + j0.05) = 5.94\underline{/35.7°} = 4.82 + j3.47$$

and therefore the primary voltage is

$$\mathbf{V}_1 = 405.7\underline{/0.59°} + 5.94\underline{/35.7°} = 410.5 + j7.67 = 410.6\underline{/1.07°}$$

The RMS primary voltage is 410.6 V, and the current is 113.76 A, by comparison with the values of 410.4 V and 112.5 A obtained with the simplified model. The agreement is quite good.

Probs. 12-ⁱ
■ 12-10

The phasor diagram for the transformer in Figure 12-11(c) is shown in Figure 12-12. Here, again, all the quantities are referred to the primary side. Conveniently, we take $\mathbf{V}_2'$, the load voltage, as reference (at 0°). Also, we reverse the reference on $-\mathbf{I}_2'$, so that $\mathbf{V}_2'$ and $\mathbf{I}_2'$ have the usual associated references for the load, that is, the (+) of $\mathbf{V}_2'$ is

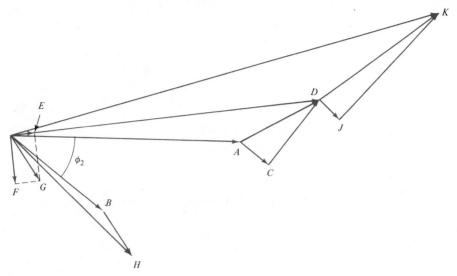

Figure 12-12   Phasor diagram of the transformer.

at the tail of the arrow of $\mathbf{I}'_2$. Then the power factor of the load is the usual cos ϕ_2 as shown. Phasors are not drawn to scale (lengths) in order to show more clearly their relationships. Here,

$$OA = \mathbf{V}'_2 = \text{load voltage}$$
$$OB = \mathbf{I}'_2 = \text{load current}$$
$$\cos \phi_2 = \text{p.f. of the load (shown lagging)}$$
$$AC = r'_2 \mathbf{I}'_2 = \text{voltage drop across } r'_2, \text{ in phase with } OB$$
$$CD = j\omega L'_2 \mathbf{I}'_2 = \text{voltage drop across } x'_2, \text{ leading } OB \text{ by } 90°$$
$$AD = \text{voltage drop across secondary winding impedance} = AC + CD$$
$$OD = \mathbf{V}_\phi = \text{voltage drop across the core branch} = \mathbf{V}'_2 + AD$$
$$OE = \frac{1}{r_c} \mathbf{V}_\phi = \mathbf{I}_c = \text{core heat loss current, in phase with } \mathbf{V}_\phi$$
$$OF = \frac{1}{jx_\mu} \mathbf{V}_\phi = \mathbf{I}_\mu = \text{core magnetizing current, lagging } \mathbf{V}_\phi \text{ by } 90°$$
$$OG = \mathbf{I}_\phi = \mathbf{I}_c + \mathbf{I}_\mu = \text{exciting current}$$
$$OH = \mathbf{I}_1 = \mathbf{I}'_2 + \mathbf{I}_\phi = OB + OG = OB + BH = \text{primary current}$$
$$DJ = r_1 \mathbf{I}_1 = \text{voltage drop across } r_1, \text{ in phase with } OH$$
$$JK = j\omega L_1 \mathbf{I}_1 = \text{voltage drop across } x_1, \text{ leading } OH \text{ by } 90°$$
$$DK = \text{voltage drop across primary winding impedance}$$
$$OK = \mathbf{V}_1 = \mathbf{V}_\phi + DK = \text{primary input voltage}$$

Look again: This complete phasor diagram is not complicated at all! Every phasor is accountable, and all that is needed is to use KCL, KVL, and voltage-current relations in resistors and inductors. The phasor diagram gives a grandstand view of all the quantities and their relationships. When drawn to scale (as it should be!) we can read off it directly the magnitudes and angles of interest.

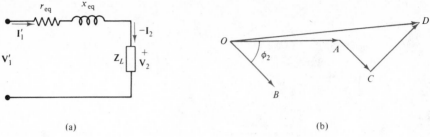

(a)   (b)

Figure 12-13   Regulation of a transformer.

12-4 VOLTAGE REGULATION AND EFFICIENCY

Of the several performance measures that are important in transformers, let us discuss two:

Voltage Regulation   This is a measure (usually in percent) of the change in the secondary output voltage with load. Specifically, for a given load at a constant power factor, the voltage regulation is

$$\text{Percent regulation} = \frac{|\mathbf{V}_{2\,o.c.}| - |\mathbf{V}_{2\,load}|}{|\mathbf{V}_{2\,load}|} \times 100 \qquad (12\text{-}59)$$

Let us use the simplified (and often quite accurate, as we saw) equivalent circuit of Figure 12-11(b), but this time referred to the secondary. It is shown in Figure 12-13(a). Here we use the notation

$$r_{eq} = r'_1 + r_2 \qquad (12\text{-}60)$$

and

$$x_{eq} = x'_1 + x_2 \qquad (12\text{-}61)$$

as the equivalent total resistance and reactance of the transformer windings. In Figure 12-13(b) we see the phasor diagram, with $OA = \mathbf{V}_2$, $OB = -\mathbf{I}_2$, the load voltage and current, respectively. Also, AC is the drop across r_{eq}, CD is the drop across x_{eq}, and $OD = \mathbf{V}'_1$. We have therefore $\mathbf{V}_{2\,o.c.} = \mathbf{V}'_1$ and

$$\text{Percent regulation} = \frac{|\mathbf{V}'_1| - |\mathbf{V}_2|}{|\mathbf{V}_2|} \times 100 = \frac{|OD| - |OA|}{|OA|} \times 100 \qquad (12\text{-}62)$$

at this particular power factor. A well-designed power transformer exhibits a voltage regulation of between 2 and 7 percent.

EXAMPLE 12-7 _____

For the transformer circuit of Figure 12-13, we are given $r_{eq} = 1.2\ \Omega$, $x_{eq} = 4.0\ \Omega$. Calculate the voltage regulation when the load is 15 kVA at the rated 1000 V, and 0.8 lagging power factor.

Solution. We have

$$|-I_2| = \frac{15 \times 10^3}{1000} = 15 \text{ A}$$

and

$$\phi_2 = \cos^{-1} 0.8 = 36.9°$$

also

$$|AC| = (15)(1.2) = 18 \text{ V}$$
$$|CD| = (15)(4) = 60 \text{ V}$$

See Figure 12-14. The voltage drop $|OD| - |OA|$ is approximately equal to $|AE|$, and we have

$$AE = AF + FE = AC \cos \phi_2 + CD \sin \phi_2$$
$$= (18)(0.8) + (60)(0.6) = 50.4 \text{ V}$$

Therefore

$$\text{Percent regulation} = \frac{50.4}{1000} \times 100 = 5.04 \text{ percent}$$

The geometrical approximation done here is very common in calculations of this type.

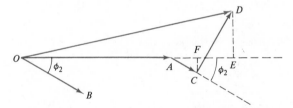

Figure 12-14   Example 12-7.   ■   Probs. 12-14, 12-15, 12-16

Efficiency   This is the ratio of output (useful) power to input power. That is

$$\eta = \frac{P_2}{P_1} = \frac{P_2}{P_2 + P_l} = \frac{|V_2| |I_2| \cos \phi_2}{|V_2| |I_2| \cos \phi_2 + P_l} \tag{12-63}$$

where P_l is the power loss in the transformer. As we have seen, it consists of the heat losses in the windings and in the core; that is,

$$P_l = |I_2|^2 r_{eq} + \frac{|V_1'|^2}{r_c'} \tag{12-64}$$

where the equivalent circuit of Problem 12-9 is used and all quantities are referred to the secondary. The core loss is constant for a given transformer, independent of the load.

EXAMPLE 12-8 _____

In the transformer of Example 12-7, it was determined that the core losses are 200 W. Calculate the efficiency.

    Solution.   We have the power output

$$P_2 = (15)(0.8) = 12 \text{ kW}$$

The heat loss in the windings is

$$|I_2|^2 \, r_{eq} = (15)^2 1.2 = 270 \text{ W}$$

Therefore

$$\eta = \frac{12{,}000}{12{,}000 + 270 + 200} \times 100 = 96.2 \text{ percent}$$
    ■   Prob. 12-17

    Finally, let us observe an interesting relationship between the core loss and the windings heat loss for maximum efficiency. Rewrite Equations 12-63 and 12-64 as

$$\eta = \frac{|V_2| \, |I_2| \cos \phi_2}{|V_2| \, |I_2| \cos \phi_2 + |I_2|^2 \, r_{eq} + W_c} \tag{12-65}$$

where W_c is the constant core loss $= |V_1'|^2/r_c$. If we divide throughout by $|I_2|$, the result is

$$\eta = \frac{|V_2| \cos \phi_2}{|V_2| \cos \phi_2 + |I_2| \, r_{eq} + W_c/|I_2|} \tag{12-66}$$

For a given load at a given $|V_2|$ and a given power factor $\cos \phi_2$, η will be maximum when the sum $|I_2| r_{eq} + W_c/|I_2|$ is minimum. Since the product of these two terms is constant, their sum is minimum when they are equal.† Therefore, maximum efficiency is obtained when

$$|I_2| r_{eq} = \frac{W_c}{|I_2|} \tag{12-67}$$

or

$$|I_2|^2 r_{eq} = W_c \tag{12-68}$$

Probs. 12-18
12-19, 12-20

that is, for a load that makes the windings losses equal to the (constant) core loss.

    In this chapter we have learned about various types of transformers and their circuit models. In addition, we have shown the advantages of phasor diagrams in connection with the analysis of transformers.

---

    † For a rectangle of a fixed area, the perimeter (and therefore half the perimeter) is minimum when the rectangle is a square. Or, if you want to do fancy work, set $d\eta/d|I_2| = 0$ and reach the same conclusion. Moral: If you can get that fly with a rolled-up newspaper, there is no need to bring out the big guns!

PROBLEMS

12-1 Consider the transformer in Figure 12-2(b) and its equations 12-7 and 12-8. Let $v_1(t)$ and $v_2(t)$ be the inputs, and let $i_1(t)$ and $i_2(t)$ be the unknown responses. These are, in fact, the two state variables for the circuit.

(a) Rearrange Equations 12-7 and 12-8 into the standard form of state equations in matrix form

$$\begin{bmatrix} \dfrac{di_1}{dt} \\[2mm] \dfrac{di_2}{dt} \end{bmatrix} = \mathbf{A} \begin{bmatrix} i_1(t) \\ i_2(t) \end{bmatrix} + \mathbf{B} \begin{bmatrix} v_1(t) \\ v_2(t) \end{bmatrix}$$

Note: You are *not solving* Equations 12-7 and 12-8 for $i_1(t)$ and $i_2(t)$, just rearranging them into the desired standard form. Identify fully the matrices **A** and **B**.

*(b) Consider now Figure 12-2(a). What are the unknown state variables? What is the standard form of the state equations?

12-2 **(a)** Write the state equation in matrix form for Example 12-1, i.e.,

$$\frac{d}{dt}\,\mathbf{x}(t) = \mathbf{A}\mathbf{x}(t) + \mathbf{B}\mathbf{e}(t)$$

where $\mathbf{x}(t)$ is the column matrix of the state variables and $\mathbf{e}(t)$ is the column matrix of the excitations (inputs). Identify fully the matrices **A** and **B**.

(b) Write the output matrix equation for Example 12-1 as

$$\mathbf{y}(t) = \mathbf{C}\mathbf{x}(t) + \mathbf{D}\mathbf{e}(t)$$

where **y** is the column matrix of the desired outputs. Identify fully the matrices **C** and **D**.

12-3 Derive Equation 12-19(a) from Equation 12-16 by using calculus: Find the minimum of Equation 12-16, prove that it is a minimum (and not a maximum), and from it obtain Equation 12-19(a).

12-4 Explain carefully, with references and a dot convention clearly shown, what is a negative turns ratio, for example, $-10{:}1$.

12-5 Two ideal transformers are connected in cascade as shown. Derive the *v-i* relations of the resulting 2-port network. Be careful with all your references.

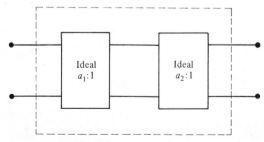

Problem 12-5

12-6 Obtain an alternate circuit for a perfect transformer and an ideal transformer, as in Figure 12-7, but with a finite L_2. *Hint*: Start with Equation 12-8 and integrate it.

*12-7 Equations 12-32 and 12-33 are the two loop equations for two coupled inductors. (a) By algebraic manipulations, invert them to yield the two phasor node equations (assuming V_1 and V_2 have one common reference, the bottom wire). (b) What condition must hold for L_1, L_2, and M for these node equations to exist? (c) From part (a), can you write the time-domain node equations? In this problem, we finally remove our long-standing "taboo" on node analysis for coupled inductors (see Chapter 6).

12-8 Derive in full detail Equation 12-35.

12-9 Another equivalent circuit for a transformer is shown below. It is obtained from Figure 12-11(c) by moving the core branch to the left, directly across V_1. Use this model to rework Example 12-6, and compare results.

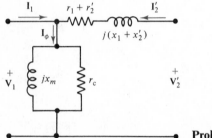

Problem 12-9

12-10 Yet another much simplified (but useful!) model for the transformer is obtained from Figure 12-11(b) by neglecting the total windings resistance, $r_1 \simeq r_2' \simeq 0$. Use such a model to rework Example 12-6, and compare results.

12-11 Draw to scale the complete phasor diagram for Example 12-6, using Figure 12-12 as a guide. From your completed phasor diagram read the power factor at the source.

12-12 Draw (not to scale but showing correct angles) the phasor diagram for Problem 12-9. Assume a lagging p.f. at the load.

12-13 Repeat Problem 12-12 for the transformer circuit of Figure 12-11(b).

*12-14 Use the geometry of the phasor diagram in Figure 12-13(b) to prove that the voltage regulation is (approximately) given by

$$\text{Percent regulation} = \left\{ \frac{|I_2|(r_{eq}\cos\phi_2 + x_{eq}\sin\phi_2)}{|V_2|} \right.$$
$$\left. + \frac{1}{2}\left[\frac{|I_2|(x_{eq}\cos\phi_2 - r_{eq}\sin\phi_2)}{|V_2|} \right]^2 \right\} \times 100$$

12-15 Calculate the voltage regulation of the transformer in Example 12-6.

12-16 Calculate the voltage regulation of Example 12-7, but this time the primary voltage is given by $|V_1'| = 1000$ V.

12-17 Use the circuit model and the results in Problem 12-9 to calculate the efficiency.

12-18 In order to determine the core loss W_c and the windings losses $|I_2|^2 r_{eq}$, two standard tests are run on the transformer:

    (a) *The open-circuit test.* The secondary (output) winding is kept open, the rated primary voltage is applied, and the input power is measured with a wattmeter.

    (b) *The short-circuit test.* The secondary winding is shorted, and a *reduced* primary voltage is applied such that the rated secondary current $|I_2|$ flows. Input power is measured.

    Using the equivalent circuit of Problem 12-9, prove that the power in test (a) is W_c, the core loss, and the power in test (b) is the windings loss. What additional assumptions, if any, must be made?

12-19 In the transformer of Problem 12-9, the open-circuit test was: 1500 V, 0.4 A, 0.7 p.f. The short-circuit test was: 50 V, 20 A, 0.4 p.f. Calculate the efficiency when the load is 10 kVA, 0.9 p.f. lagging.

12-20 Another type of efficiency in transformers is the so-called "all-day" efficiency, defined as

$$\eta_{ad} = \frac{\text{output in kWh}}{\text{input in kWh}}$$

over a 24-hour period. It measures the ratio of *energy* (not power) output to energy input. The reason for its usefulness is that most power distribution transformers are connected permanently with their primary to the main power line, while the secondary (output) is intermittently open, with no load connected, or closed, when the transformer is loaded. Thus, the core loss W_c is there continuously, but the windings losses are there only during a load period.

    An industrial transformer of 25 kW, 96 percent efficiency is loaded fully during 7 hours in a day. Calculate its all-day efficiency if the core loss is equal to the windings loss.

Chapter 13

Three-Phase Circuits

So far in our studies, we have considered typical electrical systems as shown in Figure 13-1. The source is a single-phase voltage source, connected by a single-phase transmission line to a single-phase load. We use the word "phase" to mean "circuit" here. In dc systems (for example, batteries connected to loads), single-phase operation is the only way to go. In ac sinusoidal systems, it is often more advantageous and more economical to use *polyphase* (many phases) circuits. Of these, the three-phase system is the most common, and we shall study it in some detail.

13-1 THREE-PHASE VOLTAGE SOURCES

A major factor in the use of three-phase systems is the economy of generating, operating, and transmitting ac power. These topics, by themselves, are the subject for several specialized courses and textbooks. For our purpose here, let us outline, just in principle, the method of generation of three-phase voltage sources.

In Figure 13-2(a) we see the basic principle of energy conversion: A prime mover, such as a gas or steam turbine, is driving an electric generator. The mechanical energy of the shaft is converted into electric energy, ready to be supplied by the generator. The elementary principle of the generated three-phase sources is shown in Figure 13-2(b). The fixed (nonrotating) frame of the generator, called the *stator*, has three inductive windings inside it. The shaft of the turbine turns the *rotor* of the generator, which is basically an electromagnet with a flux ϕ. As this flux sweeps past each winding on the stator, a voltage is induced in this winding due to $d\phi/dt$, the rate of change of the flux coupling this winding, just as we've studied in Chapters 6 and 12. Presto! We have three voltages available across the terminals of the three windings. These are shown in the conventional way in Figure 13-2(c), and are called phase voltage *a*, phase voltage *b*, and phase voltage *c*.

In a properly designed generator, operating in a normal steady state (no "blackouts" or "brownouts"), the three-phase voltages are sinusoids at the same radian frequency ω, with equal amplitudes, and 120° out of phase among themselves.

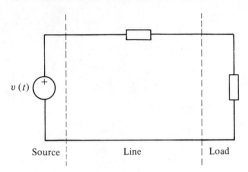

$v(t)$

| Source | Line | Load |

Figure 13-1 A single-phase circuit.

That is, if $v_a(t)$ is taken as reference, then

$$v_a(t) = \sqrt{2}\,V_{\text{eff}} \cos \omega t = V_m \cos \omega t$$

$$v_b(t) = \sqrt{2}\,V_{\text{eff}} \cos (\omega t - 120°) = V_m \cos (\omega t - 120°) \qquad (13\text{-}1)$$

$$v_c(t) = \sqrt{2}\; V_{\text{eff}} \cos (\omega t + 120°) = V_m \cos (\omega t + 120°)$$

These waveforms are shown in Figure 13-3(a), and you should complete this figure by filling in the missing points of zero crossings, etc. Their phasors are in Figure 13-3(b). As before, we'll use the effective (RMS) value for phasor lengths, rather than the amplitude V_m. Therefore,

$$\mathbf{V}_a = V_{\text{eff}}\underline{/0°}$$

$$\mathbf{V}_b = V_{\text{eff}}\underline{/-120°} \qquad (13\text{-}2)$$

$$\mathbf{V}_c = V_{\text{eff}}\underline{/-120°}$$

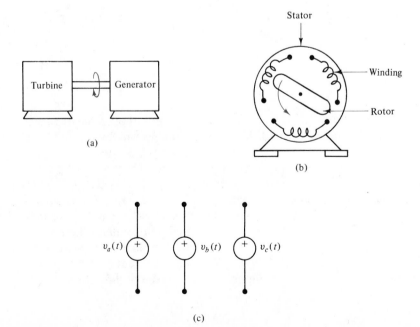

Stator

Winding

Rotor

Turbine   Generator

(a)

(b)

$v_a(t)$ $v_b(t)$ $v_c(t)$

(c)

Figure 13-2 A three-phase source.

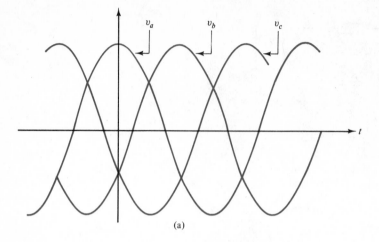

(a)

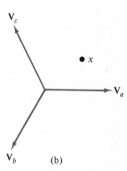

(b)

Figure 13-3 A balanced three-phase source (positive sequence).

A *balanced set of voltages* is one where they are sinusoids of the same frequency ω, of same amplitudes, and 120° apart. The set of voltages in Equation 13-1 is balanced.

In the three voltages that we have, phase *a* leads phase *b* by 120°, and phase *b* leads phase *c* by 120°. This is easy to verify in Figure 13-3(a): v_a starts at its maximum (V_m) at $t = 0$, v_b reaches its maximum 120° later, and v_c 120° after that. It is equally easy to verify from the phasor diagram: Place your finger anywhere on the diagram, say at point X as shown. Now let the entire phasor diagram rotate in the positive mathematical sense, that is, counterclockwise.† You see that phasor V_a passes your finger first, V_b follows 120° later, and V_c 120° after that. This sequence of voltages, *a-b-c*, as given in Equation 13-1 and in Figure 13-3, is called a *positive sequence*.

† Are you wearing a digital watch?

There is another possibility for the three balanced voltages, as follows:

$$v_a(t) = \sqrt{2}\,V_{\text{eff}} \cos \omega t$$
$$v_b(t) = \sqrt{2}\,V_{\text{eff}} \cos (\omega t + 120°) \qquad (13\text{-}3)$$
$$v_c(t) = \sqrt{2}\,V_{\text{eff}} \cos (\omega t - 120°)$$

Here v_b and v_c exchanged places with v_c and v_b in Equation 13-1. Thus, $v_a(t)$ leads $v_c(t)$ by 120°, and $v_c(t)$ leads $v_b(t)$ by 120°. This balanced sequence, a-c-b, is called a *negative sequence*.

EXAMPLE 13-1 _____

(a) Establish the sequence for

$$v_a(t) = 311 \cos (377t - 15°)$$
$$v_b(t) = 311 \cos (377t + 105°)$$
$$v_c(t) = 311 \cos (377t - 135°)$$

(b) Establish the sequence for

$$v_a(t) = 170 \cos (100t + 140°)$$
$$v_b(t) = 170 \cos (100t + 20°)$$
$$v_c(t) = 170 \cos (100t - 100°)$$

Solution. The phasor diagrams are shown in Figure 13-4. From these, we see that set (a) is a negative sequence and set (b) is a positive sequence.

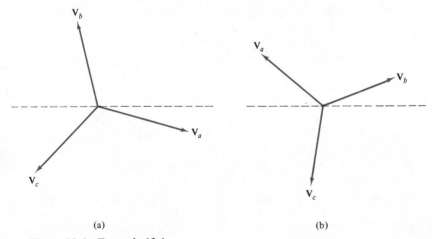

(a)   (b)

Figure 13-4 Example 13-1.   ■ Prob. 13-1

EXAMPLE 13-2 _____

In a negative sequence, we are given

$$v_b(t) = 50 \cos (10t + 30°)$$

Write the expressions for $v_a(t)$ and $v_c(t)$.

Solution. With the help of a quick phasor diagram (draw it!) we have

$$v_a(t) = 50 \cos(10t - 90°)$$
$$v_c(t) = 50 \cos(10t + 150°)$$

■ Prob. 13-2

Finally, we recognize in the balanced set of voltages the fact that their sum, at every instant, is zero

$$v_a(t) + v_b(t) + v_c(t) = 0 \tag{13-4}$$

which can be checked in Figure 13-3(a) or, equivalently, from the phasor diagram in Figure 13-3(b), where

$$\mathbf{V}_a + \mathbf{V}_b + \mathbf{V}_c = 0 \tag{13-5}$$

Verify Equation 13-5 graphically, and convince yourself again how much easier it is to deal with phasors than, say, actually adding the time functions in Equation 13-1.

In an *unbalanced* set of voltages, one or more of the features of a balanced set do not hold: The frequency may not be the same, the amplitudes may differ, or the relative phase angles may not be equal. As mentioned earlier, a sinusoidal steady-state normal operation of a three-phase generator involves always a balanced set, and most of our studies will have only balanced sets. However, during a transient operation (such as a shutdown due to a severe short circuit), unbalanced voltages appear. We will touch on this topic very briefly.

13-2 DELTA AND WYE CONNECTIONS

Now we have at our disposal the three balanced sources as in Figure 13-2(c). If we connect to each one of them separately a load through a transmission line, as in Figure 13-1, we'll have three single-phase circuits and a lot of wires (six lines between the sources and the loads). Quite obviously, that's not very economical. Of the many advantages of three-phase circuits, economy can be achieved by two possible connections of the voltage sources, that is, the windings in the stator of the generator.

The three sources may be connected in a *delta* (Δ) as shown in Figure 13-5(a). Three (instead of six) terminals, labeled *a*, *b*, and *c*, are then available for connecting to transmission lines and loads. The other possible connection is a *wye* (Y) connection, shown in Figure 13-5(b). Here, the three (−) terminals of the sources are connected together to form a *neutral point*, *n*, and the three other terminals *a*, *b*, and *c* are available for outside connections. The two distinct drawings of the Y-connected source are equivalent, of course; the one shaped like a Y gives this connection its name. An optional fourth terminal, the neutral *n*, is sometimes used in the Y source and is also shown. By its nature, a delta (Δ) connection has no neutral point.

Prob. 13-3

Three-phase loads can be connected in a similar way, either in a Δ or in a Y. These are shown in Figure 13-6(a) and (b), respectively. Here also, an optional neutral wire *N* is available in the Y.

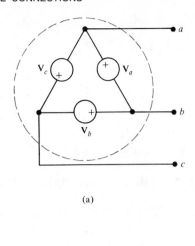

(a)

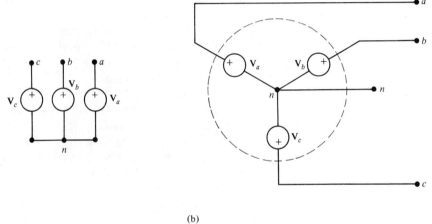

(b)

Figure 13-5 Δ-connected and Y-connected three-phase sources.

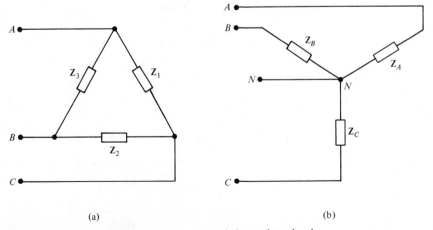

(a)   (b)

Figure 13-6 Δ-connected and Y-connected three-phase loads.

A load is *balanced* if all three impedances are equal. That is,

$$\mathbf{Z}_1 = \mathbf{Z}_2 = \mathbf{Z}_3 = \mathbf{Z}_\Delta \tag{13-6}$$

for a Δ-connected load, and

$$\mathbf{Z}_A = \mathbf{Z}_B = \mathbf{Z}_C = \mathbf{Z}_Y \tag{13-7}$$

for a Y-connected load. It is important to remember that when we equate complex numbers, as in Equation 13-6 or 13-7, magnitudes *and* angles must be equal; alternately, real parts must be equal *and* imaginary parts must be equal. Otherwise, the three-phase load is *unbalanced*.

Lastly, the three-phase transmission line connecting the source to the load is *balanced* if the impedances of the three lines are equal, each being $\mathbf{Z}_l$. If a neutral wire exists (between a Y-connected source and a Y-connected load) its impedance $\mathbf{Z}_0$ may be different from $\mathbf{Z}_l$. See Figure 13-7. When the three-phase ("3-ϕ" in common notation) source, the transmission line, and the load are all balanced, we say briefly that the *system* is balanced.

We are ready now to put together an entire system. Obviously, there are four possibilities: (1) a Y source and a Y load (with or without a neutral connection); (2) a Y source and a Δ load; (3) a Δ source and a Y load; (4) a Δ source and a Δ load. Before we embark on this mission, there are two important points to make.

The first one is a matter of reassurance. Faced with all these deltas, wyes, neutrals, etc., you may feel at bit discouraged at this point. The good news is that the analysis of a 3-ϕ is done precisely like any general analysis of networks: An intelligent choice of loop or node equations, followed by the formulation of these equations and their solution. We've done it in previous chapters. What will be new here? The general analysis, when applied to a balanced system, yields several simplified and interesting results. We will take the usual approach that we've adopted throughout our studies, namely, apply general principles of analysis without the need of memorization, and, in the process, we will learn the specialized results that hold for specific cases.

The second point is a matter of convenience and help. In order to keep track of the many voltages and currents in a 3-ϕ system, let us return to the use of *double subscripts* for voltages and currents, as introduced in an early chapter. We adopt the convention that v_{xy} (or $\mathbf{V}_{xy}$) is the voltage drop from x to y, as shown in Figure 13-8(a). Instead of the ($+$) sign as reference, the double subscript tells us that the ($+$) sign is at point x. The element connected between x and y can be a voltage source or an impedance. We have, of course,

$$v_{xy}(t) = -v_{yx}(t) \tag{13-8a}$$

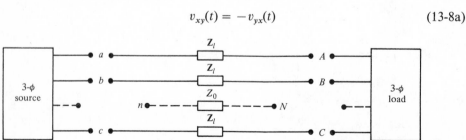

Figure 13-7   A three-phase transmission line.

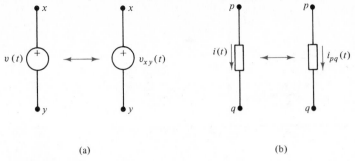

(a)   (b)

Figure 13-8   The double subscript notation.

and

$$\mathbf{V}_{xy} = -\mathbf{V}_{yx} \qquad\qquad (13\text{-}8b)$$

Also, from the basic definition of a voltage drop we can write

$$v_{xy}(t) = v_{xu}(t) + v_{uy}(t) \qquad\qquad (13\text{-}9a)$$

and

$$\mathbf{V}_{xy} = \mathbf{V}_{xu} + \mathbf{V}_{uy} \qquad\qquad (13\text{-}9b)$$

where u is any point between x and y. This is a very useful relationship, and it can be written "blindly" with full confidence. For example, $v_{AN} + v_{NB} + v_{BA} = 0$ is a correct KVL around the closed loop $A \to N \to B \to A$ regardless of where the intermediate points N and B are. As an application, this relationship holds in Figure 13-6(b).

The double subscript on a current, $i_{pq}(t)$, shows the positive reference of the current from p to q, as in Figure 13-8(b). It tells us that p is at the tail of the arrow. Here, also, a sign reversal occurs when the subscripts are switched

$$i_{pq}(t) = -i_{qp}(t) \qquad\qquad (13\text{-}10a)$$

and

$$\mathbf{I}_{pq} = -\mathbf{I}_{qp} \qquad\qquad (13\text{-}10b)$$

Here, too, we can write "blindly" a correct KCL as

$$\mathbf{I}_{pq} = \mathbf{I}_{qr} + \mathbf{I}_{qs} \qquad\qquad (13\text{-}11)$$

saying that the current from p to q is the sum of the current from q to r plus the current from q to s. For example, $\mathbf{I}_{CN} = \mathbf{I}_{NA} + \mathbf{I}_{NB}$ in Figure 13-6(b), assuming no neutral wire there.

13-3 THE Y-Y SYSTEM

We are ready now to apply our general methods of analysis. In order to stress their generality, we start with the *unbalanced* Y-source, Y-load (briefly: Y-Y) system in Figure 13-9. The source is three-phase, sinusoidal at a single frequency ω. Therefore,

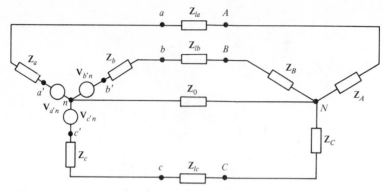

Figure 13-9 A Y-Y system.

we can use phasors. However, the amplitudes of the three phases are not equal, nor are the phase angles 120° apart. The internal impedances of the sources are unequal Z_a, Z_b, and Z_c. The transmission lines have impedances Z_{la}, Z_{lb}, and Z_{lc}. The unbalanced load is Z_A, Z_B and Z_C. The neutral line impedance is Z_0.

At a glance we recognize that it will take three loop equations to analyze this circuit. An additional moment of observation tells us that *one* node equation will do! If we let n, the neutral point of the source, be reference, then V_N, the voltage of the neutral of the load, is the only unknown. Not bad, eh? Write KCL at node N, with $(+)$ for currents leaving:

$$I_{NA} + I_{NB} + I_{NC} + I_{Nn} = 0 \qquad (13\text{-}12)$$

that is,

$$\frac{V_N - V_{a'n}}{Z_A + Z_{la} + Z_a} + \frac{V_N - V_{b'n}}{Z_B + Z_{lb} + Z_b} + \frac{V_N - V_{c'n}}{Z_C + Z_{lc} + Z_c} + \frac{V_N - 0}{Z_0} = 0 \qquad (13\text{-}13)$$

Solve this equation for the unknown V_N, and the entire problem is solved: All the currents and voltages can then be calculated.

Let us turn to the *balanced* case. Specifically, $V_{a'n}$, $V_{b'n}$, and $V_{c'n}$ are a balanced set (assume it to be a positive sequence). The three internal impedances of the sources are equal, as is the case of a usual 3-ϕ generator, $Z_a = Z_b = Z_c = Z_g$. All line impedances are equal $Z_{la} = Z_{lb} = Z_{lc} = Z_l$. The load is balanced, $Z_A = Z_B = Z_C = Z_Y$. With these conditions, Equation 13-13 reads

$$V_N\left(\frac{3}{Z_{\phi t}} + \frac{1}{Z_0}\right) = \frac{V_{a'n} + V_{b'n} + V_{c'n}}{Z_{\phi t}} \qquad (13\text{-}14)$$

where the total phase impedance $Z_{\phi t}$ is given by

$$Z_{\phi t} = Z_g + Z_l + Z_Y \qquad (13\text{-}15)$$

but the right-hand side of Equation 13-14 is zero, since the source is balanced. Consequently, the solution is

$$V_N = 0 \qquad (13\text{-}16)$$

Since $\mathbf{V}_n$ is also zero (reference), there is no current in the neutral line

$$\mathbf{I}_{Nn} = 0 \qquad (13\text{-}17)$$

and therefore, in a balanced Y-Y system, the neutral line nN is optional: It may be connected, it may be a short circuit ($\mathbf{Z}_0 = 0$), or we may ground n and N.

Continue now and calculate the *line currents* $\mathbf{I}_{aA}$, $\mathbf{I}_{bB}$, and $\mathbf{I}_{cC}$. As the name implies, a line current is a current in a transmission line connecting the source to the load. We have

$$\mathbf{I}_{aA} = \frac{\mathbf{V}_{a'n} - \mathbf{V}_N}{\mathbf{Z}_{\phi t}} = \frac{\mathbf{V}_{a'n}}{\mathbf{Z}_{\phi t}} \qquad (13\text{-}18)$$

because $\mathbf{V}_N = 0$. Similarly,

$$\mathbf{I}_{bB} = \frac{\mathbf{V}_{b'n}}{\mathbf{Z}_{\phi t}} \qquad (13\text{-}19)$$

and

$$\mathbf{I}_{cC} = \frac{\mathbf{V}_{c'n}}{\mathbf{Z}_{\phi t}} \qquad (13\text{-}20)$$

By assumption for the source, we have the phasors

$$\begin{aligned}
\mathbf{V}_{a'n} &= V_{\text{eff}}\underline{/0^\circ} \\
\mathbf{V}_{b'n} &= V_{\text{eff}}\underline{/-120^\circ} \\
\mathbf{V}_{c'n} &= V_{\text{eff}}\underline{/120^\circ}
\end{aligned} \qquad (13\text{-}21)$$

If we write

$$\mathbf{Z}_{\phi t} = Z_{\phi t}\underline{/\alpha} \qquad (13\text{-}22)$$

Then we have from Equations 13-18 to 13-20

$$\begin{aligned}
\mathbf{I}_{aA} &= I_{\text{eff}}\underline{/-\alpha} \\
\mathbf{I}_{bB} &= I_{\text{eff}}\underline{/-120^\circ - \alpha} \\
\mathbf{I}_{cC} &= I_{\text{eff}}\underline{/120^\circ - \alpha}
\end{aligned} \qquad (13\text{-}23)$$

where $I_{\text{eff}} = V_{\text{eff}}/Z_{\phi t}$. Therefore, the line currents form a *balanced* set of positive sequence, as shown in Figure 13-10. As a quick check here, we write KCL at N

$$\mathbf{I}_{Nn} = \mathbf{I}_{aA} + \mathbf{I}_{bB} + \mathbf{I}_{cC} = 0 \qquad (13\text{-}24)$$

because the three currents are balanced. This result agrees with Equation 13-17.

Take now the phase currents. A *phase current* is (what else?) the current in a phase of a load or of a source. In our load, the phase currents are $\mathbf{I}_{AN}$, $\mathbf{I}_{BN}$, and $\mathbf{I}_{CN}$. Whether the load is balanced or not, it is obvious that in the Y

$$\begin{aligned}
\mathbf{I}_{aA} &= \mathbf{I}_{AN} \\
\mathbf{I}_{bB} &= \mathbf{I}_{BN} \\
\mathbf{I}_{cC} &= \mathbf{I}_{CN}
\end{aligned} \qquad (13\text{-}25)$$

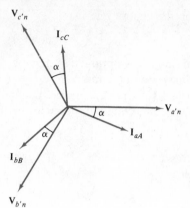

Figure 13-10   Phasor diagram for a balanced Y-Y system.

that is, in a Y connection, balanced or not, a phase current (phasor) is equal to a corresponding line current (phasor); this is simply due to the nature of the connections between the phases of the Y and the lines. Don't memorize this—just draw the picture when needed.

From Equation 13-23 and Figure 13-10, we can also conclude that the entire balanced 3-ϕ system can be solved by the three single-phase circuits, as shown in Figure 13-11. More efficiently, we need to solve only one such circuit, say for phase a, getting Equations 13-18 and 13-25. Since the system is balanced, the solutions for phases b and c form a balanced set with the solution for a. Knowing that the sequence is positive, we then have the entire solution for this 3-ϕ system.

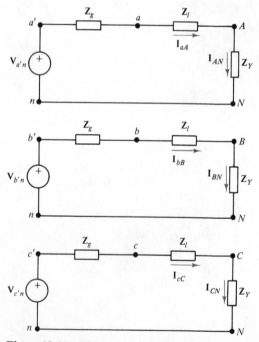

Figure 13-11   Three single-phase circuits.

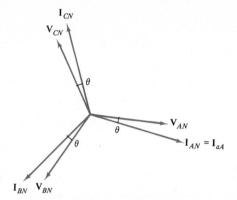

Figure 13-12 Phase currents and phase voltages, balanced Y load.

Let us consider next the *phase voltages* in the load. As the name indicates, they are the voltages across the phases of the load, namely, $\mathbf{V}_{AN}$, $\mathbf{V}_{BN}$, and $\mathbf{V}_{CN}$. We write, in our balanced Y load,

$$\mathbf{V}_{AN} = \mathbf{Z}_Y\mathbf{I}_{AN}$$
$$\mathbf{V}_{BN} = \mathbf{Z}_Y\mathbf{I}_{BN} \tag{13-26}$$
$$\mathbf{V}_{CN} = \mathbf{Z}_Y\mathbf{I}_{CN}$$

and recognize these phase voltages to be also balanced. If the angle of $\mathbf{Z}_Y$, a phase impedance, is θ (assume an inductive load, $\theta > 0$), then the phasor diagram of the phase voltages and phase currents is shown in Figure 13-12. (*Question*: How does θ differ from α in Figure 13-10?)

Finally, we consider the *line voltages*. A line voltage (or a *line-to-line voltage*) is the voltage between two lines. At our load, the line voltages are $\mathbf{V}_{AB}$, $\mathbf{V}_{BC}$, and $\mathbf{V}_{CA}$. At the source, they are $\mathbf{V}_{ab}$, $\mathbf{V}_{bc}$, and $\mathbf{V}_{ca}$. If there is a neutral line, we can speak of a *line-to-neutral* voltage. As mentioned, line quantities (line currents and line voltages) are important because they are external and easily accessible for measurements.

From Figure 13-9, or writing "blindly" using the double subscript notation, we have

$$\mathbf{V}_{AB} = \mathbf{V}_{AN} + \mathbf{V}_{NB}$$
$$\mathbf{V}_{BC} = \mathbf{V}_{BN} + \mathbf{V}_{NC} \tag{13-27}$$
$$\mathbf{V}_{CA} = \mathbf{V}_{CN} + \mathbf{V}_{NA}$$

as the line voltages for the Y load, balanced or not. When balanced, as in our present case, the phasor diagram is in Figure 13-13. We use a graphical construction to implement Equation 13-27. After all, that's one of the advantages of a phasor diagram! Note that $\mathbf{V}_{NB} = -\mathbf{V}_{BN}$ and it is drawn opposite to $\mathbf{V}_{BN}$ (at 180° to it). The phasor addition of $\mathbf{V}_{AN}$ and $\mathbf{V}_{NB}$ gives $\mathbf{V}_{AB}$. Because of the geometry of this construction, $\mathbf{V}_{AB}$ bisects the angle between $\mathbf{V}_{AN}$ and $\mathbf{V}_{NB}$, as shown. Also, if we drop a perpendicular from the tip of $\mathbf{V}_{NB}$ on $\mathbf{V}_{AB}$, it will bisect the length of $\mathbf{V}_{AB}$ (why?). Simple trigonometry therefore gives us

$$\frac{\frac{1}{2}V_{AB}}{V_{NB}} = \cos 30° = \frac{\sqrt{3}}{2} \tag{13-28}$$

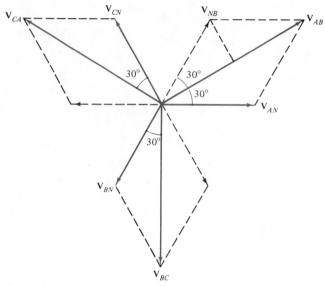

Figure 13-13   Line voltages and phase voltages, balanced Y.

or

$$V_{AB} = \sqrt{3}\, V_{NB} \tag{13-29}$$

that is,

$$V_l = \sqrt{3}\, V_{\text{ph}} \tag{13-30}$$

the magnitude of a line-to-line voltage in a balanced Y is $\sqrt{3}$ times the magnitude of a phase voltage. The other two line voltages are also shown in the phasor diagram.

In summary: In a Y, balanced or not, a line current *is* a phase current. When balanced, a line voltage is $\sqrt{3}$ times a phase voltage in magnitude, and there is a 30° angle between the phase voltage and the line voltage.

EXAMPLE 13-3 _____

A 3-ϕ balanced load is connected in Y, and $\mathbf{Z}_Y = 3 + j4$. The load is connected through a three-wire transmission line of negligible impedance to a 3-ϕ, Y-connected, positive-sequence generator, 60 Hz, of a line voltage 240 V. The internal impedance of the generator is also negligible. Calculate the phase currents, phase voltages, line currents, and line voltages at the load and at the generator, and draw a complete phasor diagram.

> *Solution.*   The system is shown in Figure 13-14(a). Notice the given external (easy to measure) line voltages of the generator. Let us go through the solution step by step, with minimal memorization or referral to "cookbook" formulas. It is safer and easier to re-derive previous results.
>
> We are given $V_{ab} = 240$ V. It is visually obvious from the figure that V_{ab} is larger than V_{an}—the distance ab is larger than an (don't take it too literally, though; this is only

a conveninent mnemonic help). By how much? by a multiplier of $\sqrt{3}$. Therefore, $V_{an} = 240/\sqrt{3} = 138$ V. Choose $\mathbf{V}_{an}$ as reference

$$\mathbf{V}_{an} = 138\underline{/0^\circ}$$

and, since the generator generates a positive sequence, we have

$$\mathbf{V}_{bn} = 138\underline{/-120^\circ}$$
$$\mathbf{V}_{cn} = 138\underline{/120^\circ}$$

Start the phasor diagram by drawing these to a convenient scale for voltages (1 cm = p volts). See Figure 13-14(b). Since there is no drop across the lines, these are also the phase voltages at the load $\mathbf{V}_{AN}$, $\mathbf{V}_{BN}$, and $\mathbf{V}_{CN}$, respectively. Add these labels to the phasor diagram.
  The phase current $\mathbf{I}_{AN}$ in the load is

$$\mathbf{I}_{AN} = \frac{\mathbf{V}_{AN}}{\mathbf{Z}_{AN}} = \frac{138\underline{/0^\circ}}{3 + j4} = 27.7\underline{/-53.1^\circ}$$

and is drawn to scale (1 cm = q amperes). Since the load is balanced, $\mathbf{I}_{BN}$ lags behind $\mathbf{V}_{BN}$ by 53.1°, as does $\mathbf{I}_{CN}$ behind $\mathbf{V}_{CN}$. Draw them, and we have the three phase currents in the load.
  From the figure of the circuit, the line currents $\mathbf{I}_{aA}$, $\mathbf{I}_{bB}$, and $\mathbf{I}_{cC}$ are equal to the respective phase currents in the load, $\mathbf{I}_{AN}$, $\mathbf{I}_{BN}$, and $\mathbf{I}_{CN}$. Label these on the phasor diagram.
  At the source, the phase currents are, again by inspection,

$$\mathbf{I}_{na} = \mathbf{I}_{aA}$$
$$\mathbf{I}_{nb} = \mathbf{I}_{bB}$$
$$\mathbf{I}_{nc} = \mathbf{I}_{cC}$$

Add these labels to the phasor diagram. The phase voltages at the source were already established. Finally, the line voltages at the source (equal to the line voltages at the load) are of magnitude 240 V each, as given. To establish their phasor representations, write (without memorization!)

$$\mathbf{V}_{ab} = \mathbf{V}_{an} + \mathbf{V}_{nb} = \mathbf{V}_{AN} + \mathbf{V}_{NB}$$

and do the geometrical addition. The length of $\mathbf{V}_{AB}$, on the scale of 1 cm = p volts, should be 240 V if you were reasonably accurate in your drawing. The remaining two line voltages $\mathbf{V}_{bc}$ and $\mathbf{V}_{ca}$ are

$$\mathbf{V}_{bc} = 240\underline{/-90^\circ}$$
$$\mathbf{V}_{ca} = 240\underline{/150^\circ}$$

They are not drawn in order not to clutter up our diagram.
  This is all. From here, we can get all the answers. For example: What is the time expression for $i_{an}(t)$? We read from the phasor diagram $\mathbf{I}_{na} = 27.7\underline{/-53.1^\circ}$ and therefore $\mathbf{I}_{an} = -\mathbf{I}_{na} = 27.7\underline{/-53.1^\circ} + 180^\circ = 27.7\underline{/126.9^\circ}$. Consequently, with $\omega = 2\pi(60) = 377$,

$$i_{an}(t) = 27.7\sqrt{2}\cos(377t + 126.9^\circ)$$

where $\sqrt{2}$ multiplies the RMS value of 27.7 to give the amplitude.
  As another question to answer, assume that $\mathbf{V}_{ab}$ was originally given at 0°, that is

$$\mathbf{V}_{ab} = 240\underline{/0^\circ}$$

Don't despair! The phasor diagram is still good. Just rotate it rigidly by 30° clockwise, to bring $\mathbf{V}_{ab}$ to the horizontal (0°) position. Now read $\mathbf{I}_{na} = 27.7\underline{/-83.1^\circ}$, $\mathbf{I}_{an} = 27.7\underline{/96.9^\circ}$, and therefore

$$i_{an}(t) = 27.7\sqrt{2}\cos(377t + 96.9^\circ)$$

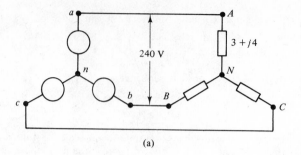

(a)

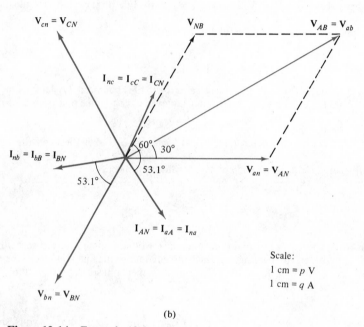

(b)

Figure 13-14   Example 13-3.

This example illustrates how relatively straightforward and simple a balanced 3-ϕ system can be. There is no need for "cookbook" formulas to be memorized. About the only caution to observe is to draw a neat, large, easy-to-read phasor diagram. ■

Probs.
13-4, 13-5,
13-6, 13-7

13-4 THE Y-Δ SYSTEM

Armed with the knowledge of the previous section, let us study the balanced Y-Δ system shown in Figure 13-15. We'll concentrate on the Δ-connected load. Our first observation is that phase voltages at the load are equal to line voltages, that is

$$\begin{aligned}
\mathbf{V}_{AB} &= \mathbf{V}_{ab} \\
\mathbf{V}_{BC} &= \mathbf{V}_{bc} \\
\mathbf{V}_{CA} &= \mathbf{V}_{ca}
\end{aligned} \qquad (13\text{-}31)$$

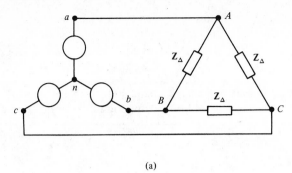

(a)

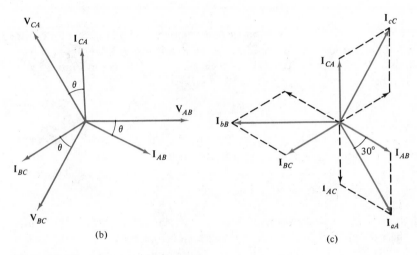

(b)

(c)

Figure 13-15 A balanced Y-Δ system.

by the very nature of the connection. Given, then, the line voltage, V_l, the phase voltages at the source are $V_{ph} = V_l/\sqrt{3}$, as in Equation 13-30. Any voltage, say, $\mathbf{V}_{AB}$, can be taken as reference

$$\mathbf{V}_{AB} = V_l\underline{/0°} \qquad (13\text{-}32)$$

and the other two follow (a positive sequence or, rarely, a given negative sequence); see Figure 13-15(b). Alternately, if the phase voltages of the source are

$$\begin{aligned}
\mathbf{V}_{an} &= V_{ph}\underline{/0°} \\
\mathbf{V}_{bn} &= V_{ph}\underline{/-120°} \\
\mathbf{V}_{cn} &= V_{ph}\underline{/120°}
\end{aligned} \qquad (13\text{-}33)$$

then we learned in the previous section that the line voltages (the phase voltages of the Δ load) are

$$\begin{aligned}
\mathbf{V}_{AB} &= \sqrt{3}\,V_{ph}\underline{/30°} \\
\mathbf{V}_{BC} &= \sqrt{3}\,V_{ph}\underline{/-90°} \\
\mathbf{V}_{CA} &= \sqrt{3}\,V_{ph}\underline{/150°}
\end{aligned} \qquad (13\text{-}34)$$

The phase currents in the load are $\mathbf{I}_{AB}$, $\mathbf{I}_{BC}$, and $\mathbf{I}_{CA}$, with

$$\mathbf{I}_{AB} = \frac{\mathbf{V}_{AB}}{\mathbf{Z}_{\Delta}} = \frac{V_l\underline{/0°}}{Z_{\Delta}\underline{/\theta}} \qquad (13\text{-}35)$$

and they form a balanced positive sequence, as in Figure 13-15(b).

The line currents are easy to obtain. From Figure 13-15(a), or "blindly," we write KCL

$$\mathbf{I}_{aA} = \mathbf{I}_{AB} + \mathbf{I}_{AC}$$

$$\mathbf{I}_{bB} = \mathbf{I}_{BA} + \mathbf{I}_{BC} \qquad (13\text{-}36)$$

$$\mathbf{I}_{cC} = \mathbf{I}_{CA} + \mathbf{I}_{CB}$$

as we learned in Equation 13-11. The phasor construction for $\mathbf{I}_{aA}$ is shown in Figure 13-15(c), where $\mathbf{I}_{AC}$ is opposite to $\mathbf{I}_{CA}$ and the geometry of adding $\mathbf{I}_{AB} + \mathbf{I}_{AC}$ is shown. Does it look familiar? Of course. We did the same thing for voltages in Figure 13-14. Therefore, for a balanced Δ

$$I_l = \sqrt{3}\, I_{\mathrm{ph}} \qquad (13\text{-}37)$$

in magnitude, the line current is $\sqrt{3}$ times the phase current. There is a 30° angle between a phase current phasor and a line current phasor.

In summary: In a Δ, balanced or not, a phase voltage *is* a line voltage, because of the connection itself. When balanced, a line current is $\sqrt{3}$ times a phase current in magnitude, and there is a 30° angle between them.

Again, there is no need to memorize. Just draw the circuit and from it recognize immediately these conclusions: A phase of the Δ is connected between two lines; a line current enters two phases of the Δ, hence it must be larger than either phase current by the ubiquitous factor of $\sqrt{3}$.

EXAMPLE 13-4 _____

In the circuit of Figure 13-15(a) let

$$\mathbf{Z}_{\Delta} = 9 + j12$$

and the source is a positive sequence generator with a line voltage of 220 V. Calculate the phase currents and voltages of the load, and the line currents.

Solution. Take $\mathbf{V}_{AB}$ as reference; the phase voltages are

$$\mathbf{V}_{AB} = 220\underline{/0°}$$

$$\mathbf{V}_{BC} = 220\underline{/-120°}$$

$$\mathbf{V}_{CA} = 220\underline{/120°}$$

as in Figure 13-15(b). The phase currents in the load are

$$\mathbf{I}_{AB} = \frac{\mathbf{V}_{AB}}{\mathbf{Z}_{AB}} = \frac{220\underline{/0°}}{9 + j12} = 14.7\underline{/-53.1°}$$

and therefore their positive sequence yields immediately

$$\mathbf{I}_{BC} = 14.7\underline{/-173.1°}$$

$$\mathbf{I}_{CA} = 14.7\underline{/66.9°}$$

The line current $\mathbf{I}_{aA}$ is (without memorization, just KCL at node A!)

$$\mathbf{I}_{aA} = \mathbf{I}_{AB} + \mathbf{I}_{AC} = 14.7\underline{/-53.1°} - 14.7\underline{/66.9°}$$

$$= 8.83 - j11.76 - (5.77 + j13.52)$$

$$= 3.06 - j25.28 = 25.47\underline{/-83.1°}$$

and the remaining two line currents follow a positive sequence, that is,

$$\mathbf{I}_{bB} = 25.47\underline{/-203.1°}$$

$$\mathbf{I}_{cC} = 25.47\underline{/36.9°}$$

Probs. 13-8,
13-9, 13-10,
■ 13-11

13-5 THE Δ-Y SYSTEM

The principles of analysis studied so far are applicable to this case also. Therefore, let us illustrate them with a specific example and draw the relevant conclusions.

EXAMPLE 13-5 _____

In the balanced circuit of Figure 13-16(a) let

$$\mathbf{Z}_Y = 3 + j4$$

and the source is a positive-sequence generator with a line voltage of 380 V. Calculate all the phase and line currents in the load and in the source.

Solution. We start with the phase voltages at the load. First

$$V_{\text{ph}} = \frac{380}{\sqrt{3}} = 220 \text{ V}$$

due to the Y-connected load. Then take $\mathbf{V}_{AN}$ as a convenient reference, getting

$$\mathbf{V}_{AN} = 220\underline{/0°}$$

$$\mathbf{V}_{BN} = 220\underline{/-120°}$$

$$\mathbf{V}_{CN} = 220\underline{/120°}$$

precisely as in Figure 13-13. (The same remark holds here, too: If $\mathbf{V}_{AB}$ is given at 0°, the entire phasor diagram must be rotated clockwise by 30°, and then $\mathbf{V}_{AN} = 220\underline{/-30°}$, etc.)
The line voltages at the load (and at the source) are then

$$\mathbf{V}_{AB} = \mathbf{V}_{ab} = 380\underline{/30°}$$

$$\mathbf{V}_{BC} = \mathbf{V}_{bc} = 380\underline{/-90°}$$

$$\mathbf{V}_{CA} = \mathbf{V}_{ca} = 380\underline{/150°}$$

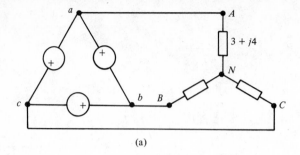

(a)

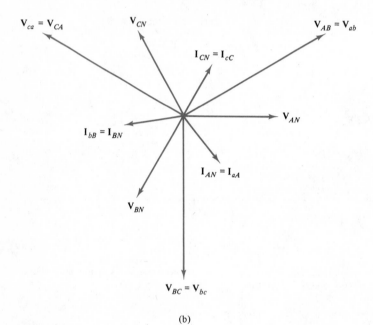

(b)

Figure 13-16   Example 13-5.

The load currents are

$$I_{AN} = \frac{220\underline{/0°}}{3 + j4} = 44\underline{/-53.1°}$$

$$I_{BN} = 44\underline{/-173.1°}$$

$$I_{CN} = 44\underline{/66.9°}$$

since they form a positive sequence. These are also the line currents, by inspection of the circuit

$$I_{aA} = I_{AN}$$

$$I_{bB} = I_{BN}$$

$$I_{cC} = I_{CN}$$

See Figure 13-16(b). The phase currents in the source are presented in Problem 13-12.   Probs. 13-1

13-13

13-6 THE Δ-Δ SYSTEM

This case, again, will be illustrated with an example, and we'll mention the important conclusions as we go along.

EXAMPLE 13-6 _____

In the balanced circuit of Figure 13-17(a) let

$$\mathbf{Z}_\Delta = 9 + j12$$

and the source is of positive sequence, 380 V line-to-line. Calculate the phase currents and phase voltages of the load and the line currents.

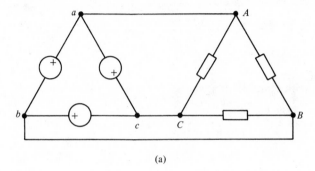

(a)

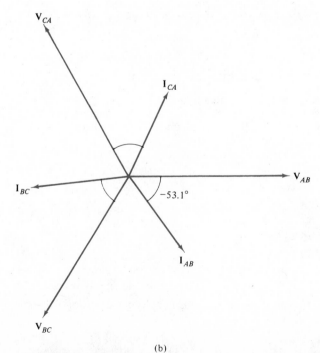

(b)

Figure 13-17 Example 13-6.

Solution.   Start conveniently with

$$\mathbf{V}_{ab} = \mathbf{V}_{AB} = 380\underline{/0°}$$

and therefore

$$\mathbf{V}_{bc} = \mathbf{V}_{BC} = 380\underline{/-120°}$$

$$\mathbf{V}_{ca} = \mathbf{V}_{CA} = 380\underline{/120°}$$

because of their positive sequence. In a Δ, line voltages *are* phase voltages in magnitude and angle. (Don't memorize it—just look at the circuit.)

The phase current $\mathbf{I}_{AB}$ in the load is

$$\mathbf{I}_{AB} = \frac{\mathbf{V}_{AB}}{\mathbf{Z}_{AB}} = \frac{380\underline{/0°}}{9 + j12} = 25.33\underline{/-53.1°}$$

and, without further ado, we write immediately

$$\mathbf{I}_{BC} = 25.33\underline{/-173.1°}$$

and

$$\mathbf{I}_{CA} = 25.33\underline{/66.9°}$$

being a positive sequence of currents.

The line currents are written with KCL at nodes *A*, *B*, and *C*

$$\mathbf{I}_{Aa} = \mathbf{I}_{CA} + \mathbf{I}_{BA}$$

$$\mathbf{I}_{Bb} = \mathbf{I}_{AB} + \mathbf{I}_{CB}$$

$$\mathbf{I}_{Cc} = \mathbf{I}_{AC} + \mathbf{I}_{BC}$$

We chose a different reference direction for our line currents just to practice the double subscript notation and to confirm how foolproof it is. The detailed calculations yield

$$\mathbf{I}_{Aa} = (10 + j23.3) + (-15.2 + j20.26) = -5.2 + j43.56 = 43.87\underline{/96.8°}$$

$$\mathbf{I}_{Bb} = (15.2 - j20.26) + (25.15 + j3.04) = 40.35 - j17.22 = 43.87\underline{/-23.2°}$$

$$\mathbf{I}_{Cc} = (-10 - j23.3) + (-25.15 - j3.04) = -35.15 - j26.3 = 43.87\underline{/-143.2°}$$

The last two, of course, could have been written by inspection after $\mathbf{I}_{Aa}$: They form a positive sequence. These line currents are not shown on the phasor diagram, to prevent clutter. It is worthwhile to check quickly the magnitude ratio of a line current to a phase current

$$\frac{43.87}{25.33} = 1.732 = \sqrt{3}$$

as expected.   ■ Prob. 13-14

13-7 AN UNBALANCED SYSTEM

To complete this discussion, let us go back to our introductory remarks for unbalanced systems. Here, no "nice" results can apply, no 120° in general, no factor of $\sqrt{3}$. The only way to approach an unbalanced system is by a general loop or node analysis.†

---

† In power systems analysis, there are several specialized methods for analyzing unbalanced systems. Here we are taking a broad, yet nonspecialized, approach.

What can get unbalanced? The load, for one. Although many loads are designed to be balanced, for example, a 3-ϕ ac motor, it can happen that one phase of such a motor becomes an open circuit. A transmission line is designed to be balanced; yet it, too, can become unbalanced. Finally, the 3-ϕ source may become unbalanced, for instance, during a severe short circuit.

EXAMPLE 13-7 ─────────────────────────────────

Let us take the Δ-Δ unbalanced system shown in Figure 13-18(a). The load and the line are unbalanced; the source is balanced, of positive sequence. We drew this circuit slightly differently, just to stress that there is nothing sacred about triangular-shaped drawings for 3-ϕ circuits.

Solution. As always, we must first make a choice of the method of analysis. For node analysis, if we let node c be reference (ground), nodes a and b have known voltages; nodes A, B, and C have unknown voltages; so we need three node equations. *Question*: What's wrong with using the formula that we learned in an earlier chapter, namely, the number of node equations = number of nodes $- 1 -$ number of voltage sources? Here, this would give $6 - 1 - 3 = 2$.

The number of unknown loop currents is three, as shown. The loop current $\mathbf{I}_\Delta$ inside the delta of the source is zero, as discussed earlier in this chapter. So, it's a toss-up between node analysis and loop analysis.

Let's do loop analysis. The three loop equations can be written by inspection (no exaggeration—you should be able to do it after Chapter 9). In matrix form they are

$$\begin{bmatrix} 6 - j1 & -2 - j1 & -3 + j4 \\ -2 - j1 & 6 + j6 & -3 - j4 \\ -3 + j4 & -3 - j4 & 10 + j3 \end{bmatrix} \begin{bmatrix} \mathbf{I}_1 \\ \mathbf{I}_2 \\ \mathbf{I}_3 \end{bmatrix} = \begin{bmatrix} 110\underline{/0°} \\ 110\underline{/-120°} \\ 0 \end{bmatrix}$$

If you don't quite see it, write carefully every loop equation, collect terms, and arrange in the final matrix form. The solution for $\mathbf{I}_1$, $\mathbf{I}_2$ and $\mathbf{I}_3$ is obtained by a Gauss-Jordan routine, such as the one given in Appendix A. The answers are

$$\mathbf{I}_1 = 6.6 - j10.4 = 12.3\underline{/-57.6°}$$

$$\mathbf{I}_2 = -13.1 - j15.2 = 20.1\underline{/-130.7°}$$

$$\mathbf{I}_3 = -4.3 - j14.2 = 14.8\underline{/-106.8°}$$

From these, all the required quantities are readily available. For example, the line currents are

$$\mathbf{I}_{aA} = \mathbf{I}_1 = 12.3\underline{/-57.6°}$$

$$\mathbf{I}_{bB} = \mathbf{I}_2 - \mathbf{I}_1 = -19.7 - j4.8 = 20.3\underline{/-166.3°}$$

$$\mathbf{I}_{cC} = -\mathbf{I}_2 = 13.1 + j15.2 = 20.1\underline{/49.3°}$$

The phase currents in the load are

$$\mathbf{I}_{AB} = \mathbf{I}_1 - \mathbf{I}_3 = 10.9 + j3.8 = 11.5\underline{/19.2°}$$

$$\mathbf{I}_{BC} = \mathbf{I}_2 - \mathbf{I}_3 = -8.8 - j1 = 8.9\underline{/-173.5°}$$

$$\mathbf{I}_{CA} = -\mathbf{I}_3 = 4.3 + j14.2 = 14.8\underline{/73.2°}$$

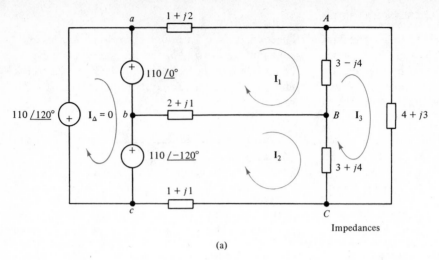

(a)

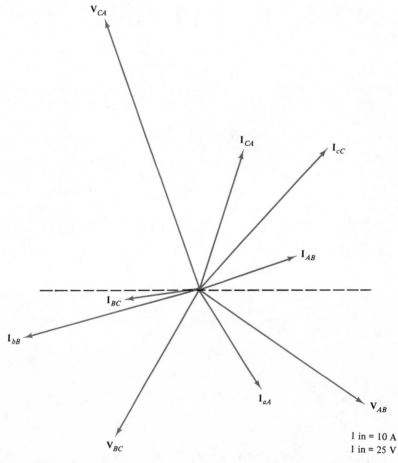

1 in = 10 A
1 in = 25 V

(b)

Figure 13-18 Example 13-7.

The phase voltages in the load are

$$\mathbf{V}_{AB} = (3 - j4)\mathbf{I}_{AB} = 57.5\underline{/-33.9°}$$

$$\mathbf{V}_{BC} = (3 + j4)\mathbf{I}_{BC} = 44.5\underline{/-120.3°}$$

$$\mathbf{V}_{CA} = (4 + j3)\mathbf{I}_{CA} = 74\underline{/110°}$$

The phasor diagram for this circuit is given in Figure 13-18(b), showing clearly the unbalanced nature of the system. Check on the correctness of this solution by creating a few cut sets and summing their currents.† ∎

Probs. 13-15, 13-16

13-8 POWER IN 3-φ SYSTEMS

In general, we can say that the total average power P_t in a 3-φ load is

$$P_t = P_1 + P_2 + P_3 \tag{13-38}$$

that is, the sum of the average powers in the three phases. Let us write

$$P_k = V_k I_k \cos \theta_k \tag{13-39}$$

as the average power in phase k, where, as usual, V_k is the effective phase voltage, I_k the effective phase current, and $\cos \theta_k$ the power factor of the phase. Then

$$P_t = \sum_{k=1}^{3} V_k I_k \cos \theta_k \quad \text{W} \tag{13-40}$$

Similarly, the total reactive power is

$$Q_t = \sum_{k=1}^{3} V_k I_k \sin \theta_k \quad \text{VAR} \tag{13-41}$$

while the volt-ampere for each phase is

$$S_k = V_k I_k \quad \text{VA}, \quad k = 1, 2, 3 \tag{13-42}$$

This is it, simple and general. It applies to an unbalanced load, where no additional simplifications can be made, as well as to a balanced load, where several simplifications can be made.

Prob. 13-17

Prior to that, however, let us backtrack a little and consider the *instantaneous power* in a 3-φ balanced system. If you recall from Chapter 11, that's how we started our discussion of power in a single-phase circuit.

In a *balanced* 3-φ load, connected in Y or in Δ, the three phase voltages are

$$v_1(t) = \sqrt{2}\, V_{\text{eff}} \cos \omega t$$

$$v_2(t) = \sqrt{2}\, V_{\text{eff}} \cos (\omega t - 120°) \tag{13-43}$$

$$v_3(t) = \sqrt{2}\, V_{\text{eff}} \cos (\omega t + 120°)$$

† Be involved. Do it!

assuming a positive sequence. For this balanced load, the corresponding phase currents are

$$i_1(t) = \sqrt{2}\,I_{\text{eff}}\cos{(\omega t - \theta)}$$

$$i_2(t) = \sqrt{2}\,I_{\text{eff}}\cos{(\omega t - 120° - \theta)} \qquad (13\text{-}44)$$

$$i_3(t) = \sqrt{2}\,I_{\text{eff}}\cos{(\omega t + 120° - \theta)}$$

where θ is the angle of the phase impedance. The total instantaneous power is

$$p_t(t) = \sum_{k=1}^{3} v_k(t)i_k(t) = 2V_{\text{eff}}I_{\text{eff}}[\cos{\omega t}\cos{(\omega t - \theta)}$$

$$+ \cos{(\omega t - 120°)}\cos{(\omega t - 120° - \theta)} + \cos{(\omega t + 120°)}\cos{(\omega t + 120° - \theta)}]$$

$$(13\text{-}45)$$

Now use the trigonometric identity

$$\cos{x}\cos{y} = \tfrac{1}{2}[\cos{(x - y)} + \cos{(x + y)}] \qquad (13\text{-}46)$$

exactly as we did in Chapter 9 for the single-phase power. With this identity, Equation 13-45 becomes

$$p_t(t) = V_{\text{eff}}I_{\text{eff}}[3\cos{\theta} + \cos{(2\omega t - \theta)} + \cos{(2\omega t - 120° - \theta)}$$

$$+ \cos{(2\omega t + 120° - \theta)}] \quad (13\text{-}47)$$

The second, third, and fourth terms in Equation 13-47 add up to zero

$$V_{\text{eff}}I_{\text{eff}}\cos{(2\omega t - \theta)} + V_{\text{eff}}I_{\text{eff}}\cos{(2\omega t - 120° - \theta)}$$

$$+ V_{\text{eff}}I_{\text{eff}}\cos{(2\omega t + 120° - \theta)} = 0 \quad (13\text{-}48)$$

because they are a *balanced set* (at a radian frequency of 2ω). Therefore, the total instantaneous power is

$$p_t(t) = 3V_{\text{eff}}I_{\text{eff}}\cos{\theta} \qquad (13\text{-}49)$$

which is constant with time, as shown in Figure 13-19.

This is an amazing result! In a single-phase circuit, we recall, the instantaneous power is sinusoidal (pulsating) at the double frequency 2ω, with an average value P. (See Figure 11-1.) In the 3-ϕ balanced case, these sinusoidal (2ω) terms cancel, and the

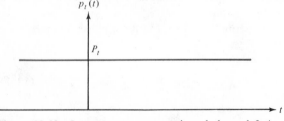

Figure 13-19 Instantaneous power in a balanced 3-ϕ load.

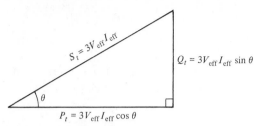

Figure 13-20 Power triangle for a balanced 3-ϕ load, with phase quantities.

instantaneous power is a smooth constant. That's another excellent reason for transmitting and distributing electric power by three-phase systems, as mentioned in the beginning of this chapter.

It does not take very long to recognize that, in Equation 13-49, the average power is also the instantaneous power†

$$p_t(t) = P_t = 3P_{ph} = 3V_{eff}I_{eff} \cos \theta \qquad (13\text{-}50)$$

and the total average power is three times the average power per phase, because the load is balanced. Equation 13-40 is reduced here to Equation 13-50. Similarly, the total reactive power in the balanced load is

$$Q_t = 3Q_{ph} = 3V_{eff}I_{eff} \sin \theta \qquad (13\text{-}51)$$

and the total volt-amperes in the balanced load are

$$S_t = 3S_{ph} = 3V_{eff}I_{eff} \qquad (13\text{-}52)$$

The power triangle for the balanced load is shown in Figure 13-20. Obviously, it is a triangle similar to the one for each phase by a scale factor of 3.

So far, we did not specify the connection (Y or Δ) of the load. Our calculations involved the *phases* of the load—which are valid for either a Y or a Δ connection. In an unbalanced load, we add the powers in the three different phases, and in the balanced load we multiply by 3 the power of one phase.

In either case, we must know (or measure) the phase voltage and the phase current, while the power factor of the phase is given. In most cases, such *phase* measurements are inconvenient, if not impossible. For example, a Δ-connected 3-ϕ motor is enclosed in its metal casing with only three terminals A, B, and C accessible for connecting to the line. Or, in a Y-connected load, the neutral N may not be accessible externally. Therefore, it would be nice to be able to express the average power in terms of *line* quantities (and the power factor, of course).

For an unbalanced load this is impossible, in general, because line voltages and currents bear no predicted relations to phase voltages and currents. We must simply use Equation 13-38. In a balanced load, the prospects are brighter. Consider each connection separately.

---

† The time average of a constant is that same constant.

Y-Connected Balanced Load   Here, as in Equation 13-50, we write

$$P_t = 3P_{\mathrm{ph}} = 3V_{\mathrm{ph}}I_{\mathrm{ph}} \cos \theta \qquad (13\text{-}53)$$

where we use, as required, effective values of voltage and current. The subscript "ph" is for phase quantities. In a balanced Y, we know the relations between phase and line quantities

$$V_{\mathrm{ph}} = \frac{V_l}{\sqrt{3}} \qquad (13\text{-}54)$$

and

$$I_{\mathrm{ph}} = I_l \qquad (13\text{-}55)$$

(Don't memorize, don't search frantically in previous pages; just sketch the circuit quickly or, better yet, close your eyes momentarily and visualize it.) Substitute these line quantities into Equation 13-53 to get

$$P_t = 3 \frac{V_l}{\sqrt{3}} I_l \cos \theta = \sqrt{3}\, V_l I_l \cos \theta \qquad (13\text{-}56)$$

Δ-Connected Balanced Load   Again,

$$P_t = 3V_{\mathrm{ph}}I_{\mathrm{ph}} \cos \theta \qquad (13\text{-}57)$$

but here

$$V_{\mathrm{ph}} = V_l \qquad (13\text{-}58)$$

and

$$I_{\mathrm{ph}} = \frac{I_l}{\sqrt{3}} \qquad (13\text{-}59)$$

So,

$$P_t = 3V_l \frac{I_l}{\sqrt{3}} \cos \theta = \sqrt{3}\, V_l I_l \cos \theta \qquad (13\text{-}60)$$

the same as Equation 13-56.

Another amazing and convenient result! The total average power of a balanced load does not depend on the connection of the load. Similarly, we have

$$Q_t = \sqrt{3}\, V_l I_l \sin \theta \qquad (13\text{-}61)$$

and

$$S_t = \sqrt{3}\, V_l I_l \qquad (13\text{-}62)$$

as shown in Figure 13-21.

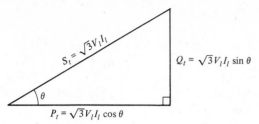

Figure 13-21 Power triangle for a balanced 3-ϕ load, with line quantities.

EXAMPLE 13-8

A 3-ϕ balanced load is rated at 1500 W, 0.8 p.f. lagging. It is fed from a line voltage of 220 V. Calculate the impedance per phase if: (a) it is Δ-connected, (b) it is Y-connected.

Solution. (a) We have

$$1500 = \sqrt{3}\,(220)I_l(0.8)$$

$$\therefore \quad I_l = 4.92 \text{ A}$$

$$\therefore \quad I_{ph} = \frac{1}{\sqrt{3}}(4.92) = 2.84 \text{ A}$$

$$\therefore \quad Z_{ph} = \frac{220}{2.84} = 77.44\ \Omega$$

$$\therefore \quad \mathbf{Z}_\Delta = 77.44\underline{/\cos^{-1}0.8} = 77.44\underline{/36.9^\circ} = 62 + j46.5$$

Check:

$$3I_{ph}^2 R = 3(2.84)^2 62 = 1500 \text{ W} = P_t$$

$$3I_{ph}^2 X = 3(2.84)^2 46.5 = 1125 \text{ VAR} = Q_t$$

$$3I_{pk}V_{ph} = 3(2.84)(220) = 1875 \text{ VA} = S_t$$

(b) Here also,

$$1500 = \sqrt{3}(220)I_l(0.8) \qquad \therefore \quad I_l = 4.92 \text{ A} = I_{ph}$$

$$\therefore \quad Z_{ph} = \frac{220/\sqrt{3}}{4.92} = 25.82\ \Omega$$

$$\therefore \quad \mathbf{Z}_Y = 25.82\underline{/36.9^\circ} = 20.6 + j15.5$$

which, by the way, confirms Problem 13-9.
Check:

$$3I_{ph}^2 R = 3(4.92)^2 20.6 = 1500 \text{ W} = P_t$$

$$3I_{ph}^2 X = 3(4.92)^2 15.5 = 1125 \text{ VAR} = Q_t$$

$$3I_{ph}V_{ph} = 3(4.92)\left(\frac{220}{\sqrt{3}}\right) = 1875 \text{ VA} = S_t$$

Probs. 13-18
■ 13-19, 13-20

13-9 POWER MEASUREMENTS

A *wattmeter* is an instrument that measures power. It consists of a current coil of a very low impedance and a voltage coil of a high impedance.† See Figure 13-22(a). When the wattmeter is connected, the current in each coil creates a flux. These fluxes create a mechanical torque between the coils. As a result, the voltage coil moves on a pivot, and a pointer attached to it indicates a reading on a scale.

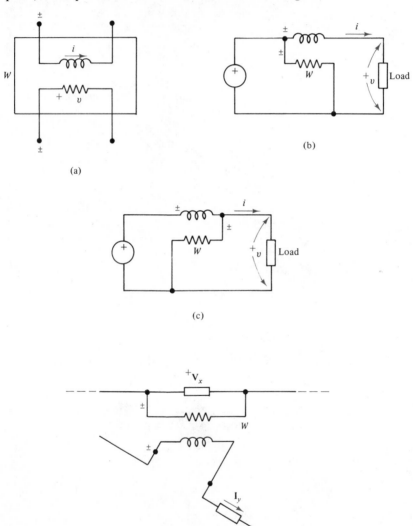

Figure 13-22   (a) Wattmeter. (b) Power measurement. (c) Power measurement. (d) Just a measurement.

† In technical jargon, the voltage coil is sometimes called the *pressure coil*.

In Figure 13-22(b), we show a typical connection of a wattmeter to measure the power delivered to the load in a single-phase circuit. The current coil is connected in series with the load, like an ammeter; its ($\pm$) marked terminal is at the tail of the load current. The flux in the current coil is therefore proportional to $i(t)$, the load current. The voltage coil is connected across the load in parallel, like a voltmeter; its ($\pm$) marked terminal is at the ($+$) reference of the load voltage. The flux in the voltage coil is therefore proportional to $v(t)$, the load voltage. The instantaneous torque is therefore proportional to the product $v(t)i(t)$, the instantaneous power. The reading of the pointer gives the average of this product if $v(t)$ and $i(t)$ are periodic

$$W = \frac{1}{T} \int_0^T v(t)i(t)\, dt \tag{13-63}$$

In a sinusoidal case, the reading is

$$W = P = V_{\text{eff}} I_{\text{eff}} \cos \theta \tag{13-64}$$

the average power delivered to the load.

A word or two about the two different connections in Figure 13-22(b) and (c). In the former, the current coil carries the true load current, but the voltage coil has the added voltage drop across the current coil. This added voltage is very small, since the impedance of the current coil is very small (that's the reason for designing it so!). In the latter, the voltage coil has the true load voltage, but the current coil has the added current in the voltage coil—again, a small value because of its high impedance. Thus, in either case, the wattmeter reading includes the small power loss in one or the other of its two coils. This loss is usually negligible. In precision meters, a special compensation is provided to give the correct reading. For our purposes, either connection will be valid.†

In Figure 13-22(d), the reading of the wattmeter will still be

$$W = \frac{1}{T} \int_0^T v_x(t)i_y(t)\, dt = V_x I_y \cos \beta \tag{13-65}$$

where V_x and I_y are the effective values of the sinusoidal $v_x(t)$ and $i_y(t)$, and β is the phase angle between them. As shown, v_x and i_y can be in different parts of the circuit; the reading of the wattmeter then is not power in a specific load, just the value given in Equation 13-65.

Probs. 13-21 13-22

For a 3-ϕ unbalanced load, it seems that we need three wattmeters, one per phase. The sum of their readings, $W_1 + W_2 + W_3$, will be the total power. However, such connections may not be possible because phase currents in a Δ and phase voltages in a Y may not be accessible. We must do with wattmeters connected in the lines leading to the load.

---

† The ac wattmeter, installed on your house by your friendly electric power company, operates essentially on the same principle. Here, the torque is produced on a revolving aluminum disk (which seems to run forever!) Through a gear train, the disk drives the register of the meter. Average power is integrated (added) by the continuous revolutions; this is the energy used, as indicated on the meter.

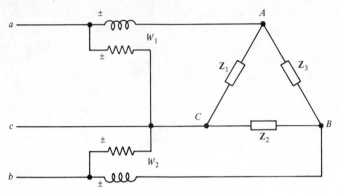

Figure 13-23 Two-wattmeter power measurement.

Consider, then, an unbalanced Δ load, as shown in Figure 13-23, with *two* wattmeters W_1 and W_2. The average readings of the two wattmeters are

$$W_1 = \frac{1}{T} \int_0^T v_{AC}(t) i_{aA}(t)\, dt \tag{13-66a}$$

and

$$W_2 = \frac{1}{T} \int_0^T v_{BC}(t) i_{bB}(t)\, dt \tag{13-66b}$$

because of the nature of their connections. Neither reading, on its own, has a meaning as the power is some part of the load, since the voltage coils have line-to-line voltages, v_{AC} and v_{BC}, while the current coils have two "unrelated" line currents i_{aA} and i_{bB}. The sum of readings, however, makes sense. It is

$$W_1 + W_2 = \frac{1}{T} \int_0^T (v_{AC} i_{aA} + v_{BC} i_{bB})\, dt$$

$$= \frac{1}{T} \int_0^T [v_{AC}(i_{AC} + i_{AB}) + v_{BC}(i_{BC} + i_{BA})]\, dt$$

$$= \frac{1}{T} \int_0^T v_{AC} i_{AC}\, dt + \frac{1}{T} \int_0^T v_{BC} i_{BC}\, dt + \frac{1}{T} \int_0^T (v_{AC} i_{AB} + v_{BC} i_{BA})\, dt \tag{13-67}$$

The integrand of the last integral is rewritten by using the convention of double subscript notation,

$$v_{AC} i_{AB} + v_{BC} i_{BA} = v_{AC} i_{AB} - v_{BC} i_{AB}$$

$$= (v_{AC} - v_{BC}) i_{AB} = (v_{AC} + v_{CB}) i_{AB} = v_{AB} i_{AB} \tag{13-68}$$

Equation 13-67 then reads

$$W_1 + W_2 = \frac{1}{T} \int_0^T v_{AC} i_{AC}\, dt + \frac{1}{T} \int_0^T v_{BC} i_{BC}\, dt + \frac{1}{T} \int_0^T v_{AB} i_{AB}\, dt$$

$$= P_{Z_1} + P_{Z_2} + P_{Z_3} \tag{13-69}$$

giving, indeed, the sum of the three average powers in the three branches of the load.   Prob. 13-23

EXAMPLE 13-9 _____

In Example 13-7, Figure 13-18, we found

$$\mathbf{V}_{AC} = 74\underline{/-70°}, \qquad \mathbf{I}_{AC} = 14.8\underline{/-106.8°}$$

(Watch those double subscripts!)

Solution. We calculate

$$P_{Z_1} = (74)(14.8) \cos(-70° + 106.8°) = 877 \text{ W}$$

Similarly

$$\mathbf{V}_{BC} = 44.5\underline{/-120.3°} \qquad \mathbf{I}_{BC} = 8.9\underline{/-173.5°}$$

$$\therefore \quad P_{Z_2} = (44.5)(8.9) \cos(-120.3° + 173.5°) = 237 \text{ W}$$

and

$$\mathbf{V}_{AB} = 57.5\underline{/-33.9°} \qquad \mathbf{I}_{AB} = 11.5\underline{/19.2°}$$

$$\therefore \quad P_{Z_3} = (57.5)(11.5) \cos(-33.9° - 19.2°) = 397 \text{ W}$$

The total power in the load is then

$$P_t = 877 + 237 + 397 = 1511 \text{ W}$$

With the connections of the two wattmeters as shown in Figure 13-23, we have

$$\mathbf{V}_{AC} = 74\underline{/-70°} \qquad \mathbf{I}_{aA} = 12.3\underline{/-57.6°}$$

and W_1 reads therefore

$$W_1 = (74)(12.3) \cos(-70° + 57.6°) = 888 \text{ W}$$

Furthermore

$$\mathbf{V}_{BC} = 44.5\underline{/-120.3°} \qquad \mathbf{I}_{bB} = 20.3\underline{/-166.3°}$$

and so

$$W_2 = (44.5)(20.3) \cos(-120.3° + 166.3°) = 625 \text{ W}$$

The sum of their readings is

$$W_1 + W_2 = 888 + 625 = 1513 \text{ W}$$

close enough!   ∎

Our derivation in Equations 13-65 through 13-69 was done for a delta load. However, the two wattmeters read *line* voltages and currents. In other words, the load can be either delta or wye. Also, since it was assumed unbalanced, the method and the results obviously hold for a balanced load. In summary: The two-wattmeter method of Figure 13-23 applies to *any* 3-ϕ, three-wire load. Prob. 13-24

As a special case, let us derive Equations 13-65 through 13-69 for a *balanced* load. Due to the symmetry of our previous results for voltages and currents in balanced loads, we should expect some nice symmetry in W_1 and W_2 here also. Specifically, let the load be Y-connected, with $\mathbf{Z}_Y = Z\underline{/\theta}$ per phase, as shown in Figure 13-24. The phasor diagrams of Figures 13-12 and 13-13 apply here. The reading of W_1 is

$$W_1 = \frac{1}{T} \int_0^T v_{AB} i_{aA} \, dt = V_{AB} I_{AN} \cos(30° + \theta) = V_l I_l \cos(30° + \theta) \quad \text{(13-70)}$$

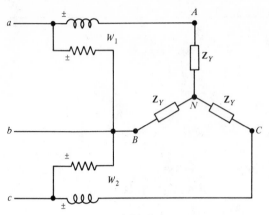

Figure 13-24 Balanced Y, 3-ϕ power measurement.

because there is a 30° angle between $\mathbf{V}_{AB}$ and $\mathbf{V}_{AN}$, and $\mathbf{I}_{AN}$ lags behind $\mathbf{V}_{AN}$ by θ°. Also, W_2 reads

$$W_2 = \frac{1}{T} \int_0^T v_{CB} i_{cC}\, dt = V_{CB} I_{CN} \cos(30° - \theta) = V_l I_l \cos(30° - \theta) \quad (13\text{-}71)$$

Be sure to satisfy yourself of these readings by carefully locating the appropriate phasors in Figures 13-12 and 13-13. In particular, don't forget that $\mathbf{V}_{CB} = -\mathbf{V}_{BC}$.

The sum $W_1 + W_2$, we know already, must be the total power. Use the trigonometric identity

$$\cos(x \pm y) = \cos x \cos y \mp \sin x \sin y \quad (13\text{-}72)$$

and add Equations 13-70 and 13-71. The result is

$$W_1 + W_2 = V_l I_l \left(\frac{\sqrt{3}}{2} \cos\theta - \frac{1}{2}\sin\theta + \frac{\sqrt{3}}{2}\cos\theta + \frac{1}{2}\sin\theta \right)$$

$$= \sqrt{3}\, V_l I_l \cos\theta = P_t \quad (13\text{-}73)$$

as in Equation 13-56 or 13-60. The result, then, is valid for a balanced Y *or* a balanced Δ load.

EXAMPLE 13-10 _____

In Example 13-5, Figure 13-16, we had

$$I_l = I_{\text{ph}} = 44 \text{ A}$$

Solution.   Therefore, quickly,

$$P_t = 3(44)^2 3 = 17{,}420 \text{ W}$$

$$\theta = \tan^{-1} \tfrac{4}{3} = 53.1° \qquad \cos\theta = 0.6$$

With the two wattmeters connected as in Figure 13-24, we have

$$W_1 = (380)(44) \cos (30° + 53.1°) = 2000 \text{ W}$$

and

$$W_2 = (380)(44) \cos (30° - 53.1°) = 15,380 \text{ W}$$

Their sum is

$$W_1 + W_2 = 17,380 \text{ W}$$

well within numerical errors of roundoff.

Probs. 13-25, 13-26, 13-27

Several additional conclusions can be reached from the readings of the two wattmeters (Equations 13-70 and 13-71):

1. If the load is purely resistive, $\theta = 0°$, both readings will be the same, $W_1 = W_2$.
2. If $\theta < 60°$, that is, the power factor is greater than 0.5, both readings are positive $W_1 > 0$, $W_2 > 0$.
3. If $\theta = 60°$ (p.f. = 0.5) the first reading is zero, $W_1 = 0$.†
4. If $\theta > 60°$, then $\cos (30° + \theta)$ is negative and W_1 will show a negative reading. To get an upscale reading, reverse the connection of the voltage coil or of the current coil. The power is still the sum of W_1 (negative) and W_2 (positive).

EXAMPLE 13-11

In Example 13-5, Figure 13-16, let the impedance per phase be

$$\mathbf{Z}_Y = 5\underline{/80°} = 0.868 + j4.924$$

Solution. Then

$$I_l = I_{ph} = 44 \text{ A}$$

and

$$W_1 = (380)(44) \cos (30° + 80°) = -5720 \text{ W}$$

$$W_2 = (380)(44) \cos (30° - 80°) = 10,750 \text{ W}$$

$$\therefore \quad P_t = -5720 + 10,750 = 5030 \text{ W}$$

Check:

$$P_t = 3(44)^2 0.868 = 5040 \text{ W}$$

with an acceptable error of 0.2 percent. Mind you, such errors as in this example and the previous one are due only to the numerical precision. The equations themselves are exact, of course.

Probs. 13-28, 13-29

---

† Assuming, of course, that this wattmeter is not broken.

13-10 DELTA-WYE TRANSFORMATION

We conclude this chapter with a discussion of an equivalence between a Δ load and a Y load. We must understand very clearly what is meant by such an equivalence. For example, when we say that three impedances in series are equivalent to a single impedance $\mathbf{Z}_t = \mathbf{Z}_1 + \mathbf{Z}_2 + \mathbf{Z}_3$, we mean that *at the accessible terminals, a* and *b*, voltage and current relations are the same. In converting these three impedances into a single equivalent one, we lose, for example, the ability to measure the voltage across $\mathbf{Z}_2$, because $\mathbf{Z}_2$ becomes inaccessible at the external terminals. See Figure 13-25(a).

A similar argument holds for the equivalence between a Δ load and a Y load, Figure 13-25(b). Given the Δ load, $\mathbf{Z}_1$, $\mathbf{Z}_2$, and $\mathbf{Z}_3$, we say that there is a Y load equivalent to it, $\mathbf{Z}_A$, $\mathbf{Z}_B$, and $\mathbf{Z}_C$ (shown in dotted lines), provided the voltages and currents at the accessible terminals *A*, *B*, and *C*—line voltages and line currents—remain the same. In converting the Δ to the Y (on paper more often than in real life), we lose $\mathbf{Z}_1$, $\mathbf{Z}_2$, and $\mathbf{Z}_3$, so we can't talk, for instance, of the phase current in $\mathbf{Z}_2$. Such a conversion, even if on paper only, is often very useful for two reasons: (1) usually we *are* interested only in line quantities, as illustrated amply in all our previous calculations and measurements; (2) subsequent calculations may be simplified by such a conversion.

For the Y to be equivalent to the Δ at the accessible terminals, the impedance between every two terminals must be the same for the Δ and for the Y. That is, between *A* and *B*, the Δ connection shows a total impedance of $\mathbf{Z}_1 + \mathbf{Z}_3$ in parallel with $\mathbf{Z}_2$, while the Y connection shows $\mathbf{Z}_A + \mathbf{Z}_B$. Therefore we write

$$\frac{(\mathbf{Z}_1 + \mathbf{Z}_3)\mathbf{Z}_2}{\mathbf{Z}_1 + \mathbf{Z}_2 + \mathbf{Z}_3} = \mathbf{Z}_A + \mathbf{Z}_B \tag{13-74}$$

Similarly, between *B* and *C*

$$\frac{(\mathbf{Z}_1 + \mathbf{Z}_2)\mathbf{Z}_3}{\mathbf{Z}_1 + \mathbf{Z}_2 + \mathbf{Z}_3} = \mathbf{Z}_B + \mathbf{Z}_C \tag{13-75}$$

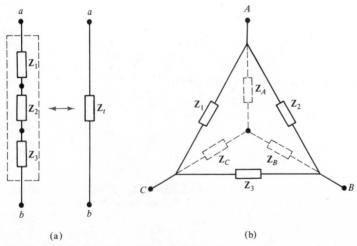

(a)   (b)

Figure 13-25 Equivalence of networks.

and between C and A

$$\frac{(\mathbf{Z}_2 + \mathbf{Z}_3)\mathbf{Z}_1}{\mathbf{Z}_1 + \mathbf{Z}_2 + \mathbf{Z}_3} = \mathbf{Z}_A + \mathbf{Z}_C \qquad (13\text{-}76)$$

These three equations can be solved for $\mathbf{Z}_A$, $\mathbf{Z}_B$, and $\mathbf{Z}_C$ by subtracting any two, then adding to the third one. We get

$$\mathbf{Z}_A = \frac{\mathbf{Z}_1\mathbf{Z}_2}{\mathbf{Z}_1 + \mathbf{Z}_2 + \mathbf{Z}_3}$$

$$\mathbf{Z}_B = \frac{\mathbf{Z}_2\mathbf{Z}_3}{\mathbf{Z}_1 + \mathbf{Z}_2 + \mathbf{Z}_3} \qquad (13\text{-}77)$$

$$\mathbf{Z}_C = \frac{\mathbf{Z}_1\mathbf{Z}_3}{\mathbf{Z}_1 + \mathbf{Z}_2 + \mathbf{Z}_3}$$

Conversely, if the Y is given (without a neutral wire!) and we want its Δ equivalent, we solve Equations 13-74, 13-75, and 13-76 for $\mathbf{Z}_1$, $\mathbf{Z}_2$, and $\mathbf{Z}_3$ to get

$$\mathbf{Z}_1 = \mathbf{Z}_A + \mathbf{Z}_C + \frac{\mathbf{Z}_A\mathbf{Z}_C}{\mathbf{Z}_B}$$

$$\mathbf{Z}_2 = \mathbf{Z}_A + \mathbf{Z}_B + \frac{\mathbf{Z}_A\mathbf{Z}_B}{\mathbf{Z}_C} \qquad (13\text{-}78)$$

$$\mathbf{Z}_3 = \mathbf{Z}_B + \mathbf{Z}_C + \frac{\mathbf{Z}_B\mathbf{Z}_C}{\mathbf{Z}_A}$$

Equations 13-77 and 13-78 constitute the Δ-Y, Y-Δ equivalence transformations. We met them earlier in Chapter 4, Section 4-4, for resistive networks only, and without the broader concepts of line voltages, currents, and neutrals. It is useful to be able to reestablish such a tie with the expanded implications.

EXAMPLE 13-12 _____

The Δ load in Example 13-7, Figure 13-18, is converted to a Y using our notation here.

Solution.

$$\mathbf{Z}_A = \frac{(3 - j4)(4 + j3)}{10 + j3} = 2.4\underline{/-32.9°} = 2 - j1.3$$

$$\mathbf{Z}_B = \frac{(3 + j4)(3 - j4)}{10 + j3} = 2.4\underline{/-16.7°} = 2.3 - j0.7$$

$$\mathbf{Z}_C = \frac{(3 + j4)(4 + j3)}{10 + j3} = 2.4\underline{/73.3°} = 0.7 + j2.3$$

The original Δ and its equivalent Y are shown in Figure 13-26. With the Y load, we need only *one* node equation to solve the system, by comparison with three loop equations there. This is the type of simplified calculations we mentioned! The unknown node

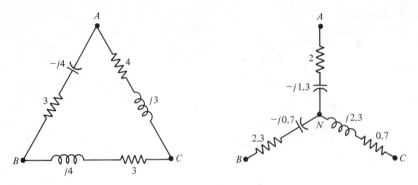

Impedances

Figure 13-26   Example 13-12.

voltage is $\mathbf{V}_N$, with node b of the source as reference. We have, with $(+)$ for currents leaving N,

$$\frac{\mathbf{V}_N - 110\underline{/0°}}{3 + j0.7} + \frac{\mathbf{V}_N}{4.3 + j0.3} + \frac{\mathbf{V}_N + 110\underline{/-120°}}{1.7 + j3.3} = 0$$

Solve for $\mathbf{V}_N$ and, from it, calculate the line currents and voltages. They are the same as in the original system.   Probs. 13-30, ■ 13-31

PROBLEMS

13-1   **(a)** Plot versus ωt and label carefully the waveforms of a negative sequence of three phase voltages.

  (b) Verify to your satisfaction that there are no additional possible three-phase sinusoidal sequences other than the positive one and the negative one.

13-2   In a positive sequence of voltages, we know that

$$\mathbf{V}_b = 120\underline{/-60°} \qquad \omega = 377$$

Draw the complete phasor diagram and from it write the expressions for $v_a(t)$, $v_b(t)$, and $v_c(t)$.

13-3   Explain in detail why, in a balanced Δ-connected source without any load, as in Figure 13-5(a), there is no circulating current in the loop formed by the delta. Such a current, if present, could damage the sources.

13-4   Repeat Example 13-3, but with a transmission line of impedance $\mathbf{Z}_l = 0.5 + j0.1$ per phase.

13-5   Repeat Example 13-3 with a negative sequence of voltages.

13-6   A positive-sequence Y-connected source, 380-V line voltage, is connected through a balanced transmission line, $\mathbf{Z}_l = 0.5 + j0.5$, to an unbalanced load connected in Y, with $\mathbf{Z}_{AN} = 10 + j8$, $\mathbf{Z}_{BN} = 10 - j8$, $\mathbf{Z}_{CN} = 10 + j0$. The neutral of the source (n) is connected to the neutral of the load (N) with a short circuit. Calculate the phase voltages and currents in the load and draw them on a phasor diagram. Use the phase currents as the unknowns.

13-7 In a Y-Y system, balanced or unbalanced, with or without a neutral connection, what is the sum of the line currents (and the neutral current, if a connection exists)? *Hint*: Think cut sets. Verify your answer for Problem 13-6. How is your answer applicable (or not) to a Δ-Δ, Δ-Y, Y-Δ system?

13-8 A balanced capacitive Δ load is given by

$$\mathbf{Z}_\Delta = 9 - j12$$

It is connected to a positive-sequence generator with a line voltage of 220 V. Calculate the phase currents and voltages of the load, and the line currents. Draw their phasor diagram to scale. Compare your results with Example 13-4. *Question*: Does the unmentioned connection of the generator (Y or Δ) affect your results? Explain why or why not.

13-9 A balanced Δ load draws the same line currents (in magnitude) as a balanced Y load. The 3-ϕ source is the same in both cases. Prove that

$$\mathbf{Z}_Y = \tfrac{1}{3}\mathbf{Z}_\Delta$$

13-10 Rework Example 13-4 with a negative sequence for the 3-ϕ source, that is

$$\mathbf{V}_{AB} = 220\underline{/0°}$$

$$\mathbf{V}_{BC} = 220\underline{/120°}$$

$$\mathbf{V}_{CA} = 220\underline{/-120°}$$

13-11 Draw a phasor diagram for a negative sequence of balanced phase voltages in a Y-connected load. From it, construct the phasors of the line voltages. Compare the relative position (30° leading or lagging) of a line voltage and a phase voltage in this case with the positive sequence (Figure 13-13).

13-12 Calculate the phase currents in the source of Example 13-5. *Suggestion*: Use the graphical approach on the phasor diagram. We know that

$$\mathbf{I}_{ba} + \mathbf{I}_{ca} = \mathbf{I}_{aA}$$

where $\mathbf{I}_{aA}$ is given. Also $I_{ba} = I_{ca} = I_{aA}/\sqrt{3}$ in magnitude, and there is a 30° angle there. Finally, don't forget that the phase currents in the source form a positive sequence.

13-13 In Example 13-5, add a balanced line impedance

$$\mathbf{Z}_l = 1 + j1$$

and solve completely.

13-14 In Example 13-6, add a balanced line impedance

$$\mathbf{Z}_l = 1 + j0.5$$

and solve completely. *Suggestion*: Use the result of Problem 13-9 and a single-phase equivalent circuit as in Figure 13-11.

13-15 Solve Example 13-7 by node analysis with $\mathbf{V}_A$, $\mathbf{V}_B$, and $\mathbf{V}_C$ as node voltages with respect to c. From your solution, calculate the load currents and compare with Example 13-7. (They'd better be the same.)

13-16 The Y-Y system in Figure 13-9 is given as follows: The source is balanced, positive sequence, 220 V per phase. The windings have equal impedances $\mathbf{Z}_a = \mathbf{Z}_b = \mathbf{Z}_c = j0.2$.

The line is balanced, $\mathbf{Z}_l = 0.6 + j0.4$. The load is $\mathbf{Z}_A = 2 + j0$; $\mathbf{Z}_B = 3 - j1$; $\mathbf{Z}_C = 1 + j2$. There is no neutral connection. Calculate the phase voltages and currents in the load.

13-17 Calculate the total average power, reactive power, and volt-amperes in the load of Example 13-7.

13-18 A small industrial park consists of two plants, each forming a balanced load. They are rated as follows:

> Load a: Δ, 300 kVA, 0.85 p.f. lagging
>
> Load b: Y, 400 kVA, 0.95 p.f. leading

The line voltage supplying the park is 13.8 kV. Calculate the line current and the p.f. of the industrial park as a single unit.

13-19 A 3-ϕ balanced load is rated at 300 kW, 0.8 p.f. lagging, 13.8-kV line voltage. It is proposed to improve its power factor to 0.95 lagging with a 3-ϕ balanced set of capacitors, connected in parallel with the load. Calculate the rating of each capacitor (in farads and volts) if the capacitors are connected: (a) in Δ, (b) in Y. A so-called *one-line circuit diagram* is shown in the figure, together with a full three-line diagram for the Δ-connected capacitors. The system operates at 60 Hz.

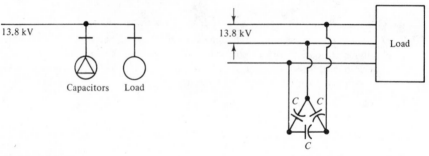

Problem 13-19.

13-20 A 3-ϕ balanced load is rated at 300 kW, 0.8 p.f. lagging, 2400-V line voltage. It is connected to the source by a balanced transmission line, $\mathbf{Z}_l = 0.2 + j1.1$. Calculate the line voltage at the source.

13-21 In certain cases, when measuring power, we must isolate the wattmeter, or scale the current and the voltage of the load to convenient values for the meter. Use a voltage coil transformer (turns ratio a_v:1) and a current coil transformer (turns ratio a_i:1) to draw a circuit diagram for such a connection. It will be a modification of Figure 13-22(b) or (c).

13-22 Based on Equation 13-65, describe in detail a laboratory experiment from which you can find the phase angle between two sinusoidal waveforms, using a voltmeter, an ammeter, and a wattmeter.

*13-23 Prove: In an n-wire system (n lines feeding a load), we need only $n - 1$ wattmeters to measure the total power to the load. Show the connections of these wattmeters in the lines.

13-24 Consider the measurement of power in a 3-ϕ unbalanced Y load without a neutral wire, as shown. The three wattmeters' current coils are in the lines, and their voltage coils form a Y of their own, with a neutral $\hat{N}$. Prove that $W_1 + W_2 + W_3$ is the total power P_t in the load. *Hint*: $v_{A\hat{N}} = v_{AN} + v_{N\hat{N}}$, etc.

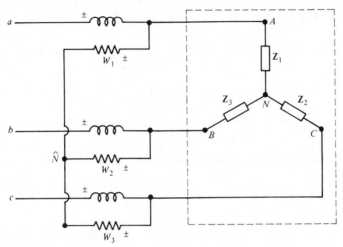

Problem 13-24.

13-25 Prove that in a balanced load with two wattmeters (Figure 13-24), the total reactive power Q_t is related to the difference $W_1 - W_2$. Derive this relationship.

13-26 Prove that in a balanced load with two wattmeters, the power factor of the load can be calculated from the two readings $(W_1 + W_2)$ and $(W_1 - W_2)$. Derive this relationship.

13-27 Prove that in a balanced load with two wattmeters, a reversal of the phase sequence will interchange the readings of W_1 and W_2.

13-28 In the derivations for the two-wattmeter method for a balanced load (Equations 13-70 through 13-73) the load was assumed inductive (lagging p.f.). Rework these derivations for a capacitive load (leading p.f.). Note carefully the individual readings of the two wattmeters, the condition for a zero reading in one, a negative reading, etc.

13-29 In a balanced 3-ϕ load, power is measured with two wattmeters. The ratio of the two readings is 2:1. What is the power factor of the load?

13-30 Use the Δ-Y transformation to verify your answer to Problem 13-9.

13-31 Complete the calculations in Example 13-12.

Chapter 14

The Laplace Transform

14-1 INTRODUCTION AND STATUS REPORT

It is time to sit back for a moment and take stock of our studies so far in the analysis of linear networks:

1. We know how to *formulate* and *solve* loop or node equations for a purely resistive network. These are algebraic equations with real coefficients.
2. We know how to *formulate* and *solve* the loop or node equations of a first-order circuit. Such an equation is a first-order differential equation, with one given initial condition. The classical solution is the sum of the homogeneous and particular solutions. The total response consists of the zero-input and the zero-state responses.
3. We know how to *formulate* and *solve* second-order circuits, either by two simultaneous first-order state differential equations or by loop or node analysis. The single equivalent differential equation must be obtained, then solved classically, as in first-order circuits.
4. We know how to formulate *but not solve* an nth-order circuit. The formulation involves simultaneous integro-differential loop or node equations, or n simultaneous first-order state differential equations, all with n given initial conditions. The handling of simultaneous differential (or integro-differential) equations by classical methods is hopelessly difficult, if not downright impossible.
5. If our inputs are pure sinusoidal of a single frequency, and if we are not interested in the zero-input solution (here, the transient solution), but want only the zero-state solution (here, the steady-state solution), we can use the phasor transform method to *solve* the network. This method transforms the

simultaneous integro-differential equations into algebraic equations with coefficients that are, in general, complex numbers. As such, the purely resistive network, case (1), may be considered a special case of this one.

Table 14-1 summarizes these achievements.

Obviously, we need to tackle the *solution* of case (4). Specifically, we need a powerful transform method that will: (1) make it easier to handle simultaneous integro-differential equations by transforming them into (what else?) simultaneous algebraic equations; (2) allow for various waveforms of inputs, such as e^{-at}, $e^{-bt} \cos \omega t$, etc., as well as dc and pure sinusoids; (3) handle easily initial conditions to yield the *total* response, zero-input plus zero-state.

Too good to be true? Not really. The Laplace transform, which we are about to learn, does all this—and more. Through its use, we can not only handle case (4), with cases (2), (3) and (5) as merely special cases; we can also learn a great deal about the general features of network analysis and, later, network design. Before going into this transform, we need to introduce two important functions.

14-2 THE UNIT STEP FUNCTION

In a typical circuit analysis problem, a source (voltage or current) is applied at $t = 0$. The initial conditions have been established just prior to that time, at $t = 0^-$. An example is shown in Figure 14-1(a), where a 6-V battery is applied to N. Such inputs were introduced in Chapters 7 and 8.

The operation of the switch can be described by the *unit step* function $u(t)$, defined as

$$u(t) = \begin{cases} 1 & t > 0 \\ 0 & t < 0 \end{cases} \tag{14-1}$$

TABLE 14-1 STATUS REPORT TO DATE

Problem	Formulation	Solution	Comments on solution
1. General resistive network	Yes	Yes	Complete
2. First-order network	Yes	Yes	Classical solution
3. Second-order network	Yes	Yes	Requires juggling of simultaneous differential equations
4. General nth-order network	Yes	No	Hopeless by classical methods
5. General nth order network	Yes	Partial	Phasors for sinusoidal steady state only

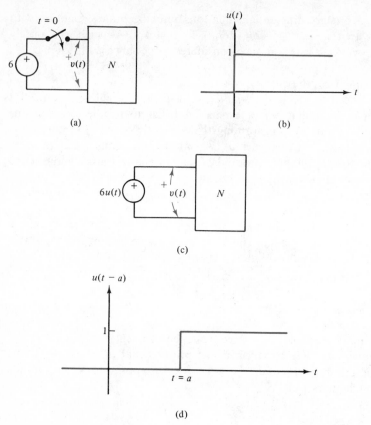

Figure 14-1   The unit step function.

and shown in Figure 14-1(b). This function has a discontinuity ("jump") *at* $t = 0$. In terms of $u(t)$, we can redraw the circuit of Figure 14-1(a) as in Figure 14-1(c): The voltage $v(t)$ across N is $6u(t)$, that is, zero for $t < 0$ and 6 for $t > 0$. The unit step function, then, serves as a mathematical-electrical switch.

A more general unit step function is defined as follows

$$u(t - a) = \begin{cases} 1 & t > a \\ 0 & t < a \end{cases} \tag{14-2}$$

that is, a switch activated at $t = a$, as shown in Figure 14-1(d). This is a *shifted* unit step. The previous case, $u(t)$, is a special case when $a = 0$.† In fact, we will define the most general unit step function, and the easiest to remember, as being equal to 1 when its argument is positive, and zero otherwise:

$$u(\cdot) = \begin{cases} 1 & \cdot > 0 \\ 0 & \cdot < 0 \end{cases} \tag{14-3}$$

The dot in the parentheses is the argument of the function.

† Don't be tempted to open the parentheses, $u(t - a) \neq u(t) - u(a)$. The argument of u must remain intact.

EXAMPLE 14-1

Plot the following step functions:
(a) $4u(t + 1)$ (c) $2u(-1 - t)$
(b) $u(-t + 2)$ (d) $-u(t - 3)$

Solution. We have from Equation 14-3:

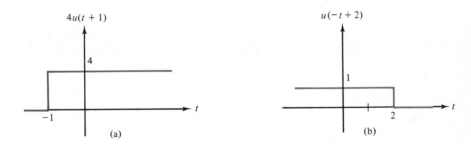

(a)
$$4u(t + 1) = \begin{cases} 4 & t + 1 > 0 \quad \therefore \quad t > -1 \\ 0 & \text{elsewhere} \end{cases}$$

(b)
$$u(-t + 2) = \begin{cases} 1 & -t + 2 > 0 \quad \therefore \quad t < 2 \\ 0 & \text{elsewhere} \end{cases}$$

(c)
$$2u(-1 - t) = \begin{cases} 2 & -1 - t > 0 \quad \therefore \quad t < -1 \\ 0 & \text{elsewhere} \end{cases}$$

(d)
$$-u(t - 3) = \begin{cases} -1 & t - 3 > 0 \quad \therefore \quad t > 3 \\ 0 & \text{elsewhere} \end{cases}$$

In each case, we applied the general definition of Equation 14-3. The plots are shown in Figure 14-2. The switching operation of case (b) might be described as "the source was connected to N until $t = 2$, then disconnected."

Question: What switching operation might be described by (d)?

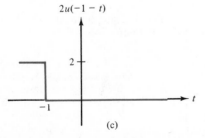

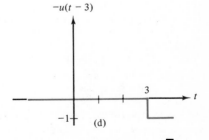

Figure 14-2 Example 14-1.   ■

The step function is also useful as a "building block" for many common waveforms in electrical engineering. Consider the *gate function* $p(t)$ shown in Figure 14-3(a); it is very common in communications and computer circuits. It serves to

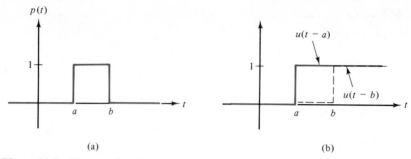

Figure 14-3   The gate function.

describe a source, or a signal, applied only from $t = a$ to $t = b$, with $b > a$. It is given by

$$p(t) = \begin{cases} 1 & a < t < b \\ 0 & \text{elsewhere} \end{cases} \tag{14-4}$$

In terms of step functions, its expression is very compact and elegant indeed

$$p(t) = u(t - a) - u(t - b) \tag{14-5}$$

where the two step functions are shown in Figure 14-3(b).

Probs. 14-1,
14-2, 14-3

EXAMPLE 14-2 ───────────────────────────────────────

Let us write the expression for $i(t)$ as shown in Figure 14-4(a). With "sectional" expressions, we can write, of course,

$$i(t) = \begin{cases} 0 & t < 0 \\ 10t & 0 < t < 1 \\ 10 & 1 < t < 3 \\ 25 - 5t & 3 < t < 5 \\ 0 & t > 5 \end{cases}$$

which is a bit clumsy and long. Instead, let us "build" this function, from left to right, using ramp functions (Problems 14-2 and 14-3—you solved them, didn't you?).

Solution.   Start with $10tu(t)$, as in Figure 14-4(b). This gives us the portion of $i(t)$ for $t < 0$ and for $0 < t < 1$; then it continues to climb past $t = 1$. At $t = 1$, we must flatten it to the constant value 10. To do this, add the shifted negative ramp function $-10(t - 1)u(t - 1)$ as shown. These two functions produce $i(t)$ correctly up to $t = 3$, and thereafter the constant 10 is maintained. We must "break" it at $t = 3$ to go down to $t = 5$. To do that, add another negative shifted ramp $-5(t - 3)u(t - 3)$; the slope of this ramp is easy to decide, because between $t = 3$ and $t = 5$, we have $\Delta i/\Delta t = 10/(-2) = -5$. The sum of the three functions will give $i(t)$ correctly up to $t = 5$, and thereafter it continues to fall. To flatten it to zero for $t > 5$, add a shifted ramp $5(t - 5)u(t - 5)$ as shown. Finally, then

$$i(t) = 10tu(t) - 10(t - 1)u(t - 1) - 5(t - 3)u(t - 3) + 5(t - 5)u(t - 5)$$

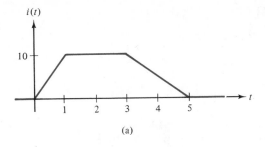

(a)

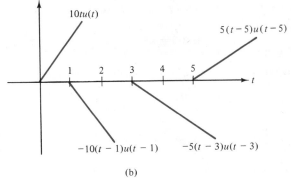

(b)

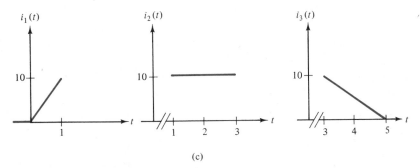

(c)

Figure 14-4   Example 14-2.

Another way to look at it is to say that $i(t)$ is the sum of the curve up to $t = 1$, *plus* the curve from $t = 1$ to $t = 3$, *plus* the curve from $t = 3$ on. These are, respectively,

$$i_1(t) = 10t[u(t) - u(t - 1)]$$
$$i_2(t) = 10[u(t - 1) - u(t - 3)]$$
$$i_3(t) = (25 - 5t)[u(t - 3) - u(t - 5)]$$

as shown in Figure 14-4(c). Here we have used gate functions as multipliers (in square brackets) to clip the appropriate functions.

You should verify that the sum $i_1 + i_2 + i_3$ is equal to $i(t)$ as obtained previously. The two methods are totally equivalent, and the choice is a matter of personal taste and visualization.   ■   Prob. 14-4

14-3 THE UNIT IMPULSE FUNCTION

If for no reason other than curiosity, we ought to ask, "if the unit step is the derivative of the unit ramp, as in Problem 14-2, what is the derivative of the unit step?" Actually, there are many good reasons to ask it. For instance, let the voltage across a capacitor be the unit step $u(t)$; the current will be $C \, du(t)/dt$, proportional to the derivative of $u(t)$.

Because of the discontinuity in the unit step function, let us obtain this derivative carefully, using a limiting process. In Figure 14-5(a), we show a new function, $g(t)$, which is continuous (without jumps). In Figure 14-5(b) we plot the derivative of $g(t)$, $dg(t)/dt$. This is easy. Up till $t = a$, $g(t) = 0$ and therefore its derivative is zero. Between $t = a$ and $t = a + \varepsilon$, $g(t)$ rises at a constant rate, and its derivative is $\Delta g/\Delta t = 1/\varepsilon$. For $t > a + \varepsilon$, $dg(t)/dt = 0$ again.

The derivative dg/dt is, then, a gate function, similar to Figure 14-3. In particular, note that the total area under the curve dg/dt is that of a rectangle (base × height)

$$\text{Area} = \varepsilon \cdot \frac{1}{\varepsilon} \equiv 1 \qquad (14\text{-}6)$$

and is always 1, *regardless* of ε. Since this is the case, let us now take the limit as $\varepsilon \to 0$. First of all, as $\varepsilon \to 0$, the function $g(t)$ becomes $u(t - a)$ as can be seen in Figure 14-5(c)

Prob. 14-5

$$\lim_{\varepsilon \to 0} g(t) = u(t - a) \qquad (14\text{-}7)$$

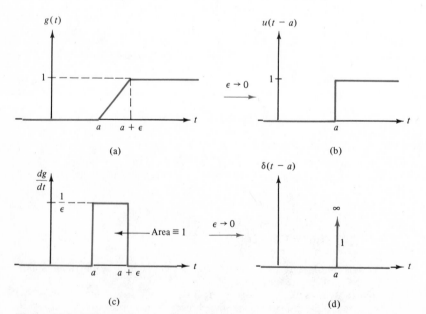

Figure 14-5 The unit impulse function.

As $\varepsilon \to 0$, the derivative $dg(t)/dt$ remains a gate function, a rectangle with a smaller and smaller base ε and a larger and larger height $1/\varepsilon$, but with a *constant* area as in Equation 14-6. Finally, this rectangle becomes

$$\lim_{\varepsilon \to 0} \frac{dg(t)}{dt} = \frac{d}{dt} u(t - a) = \delta(t - a) \tag{14-8}$$

the *unit impulse function*, shown in Figure 14-5(d): A very narrow-based rectangle ($\varepsilon = 0$), of infinite height, but with a finite area! To formalize it, we give the defining properties of the unit impulse function: It is zero, except at one point $t = a$

$$\delta(t - a) = 0 \qquad t \neq a \tag{14-9}$$

just as dg/dt is zero except between a and $a + \varepsilon$. Also

$$\delta(t - a) = \infty \qquad t = a \tag{14-10}$$

the impulse function is infinite ("blows up") at the one point; finally,

$$\int_{-\infty}^{\infty} \delta(t - a)\, dt = 1 \tag{14-11}$$

the total area, between $t = -\infty$ and $t = \infty$, under the impulse function is 1.

Phew! What a strain on our imagination—a function that is zero everywhere, except at one point where it is infinite, yet with a finite area. No wonder pure mathematicians refused (rightly) to accept this as a usual function. They had to develop a rigorous theory of so-called "generalized functions," or "distributions." Luckily for us, we don't need to worry about such pure mathematical rigor. The applications of the impulse function in electrical engineering are set on solid grounds.

Several examples ought to soften the initial shock, and, as we work more and more problems, we'll get used to these ideas.

EXAMPLE 14-3 ———————————————————————————————

In Figure 14-6, the network N is a single capacitor of C farads, with no initial voltage across it, $v_C(0^-) = 0$. We apply to it a voltage source $6u(t)$, a 6-V battery connected at $t = 0$.

Solution. The waveform of $v_C(t)$ is given by

$$v_C(t) = 6u(t) = \begin{cases} 0 & t < 0 \\ 6 & t > 0 \end{cases}$$

and is shown in Figure 14-6(b). The charge across C is therefore

$$q_C(t) = Cv_C(t) = 6Cu(t)$$

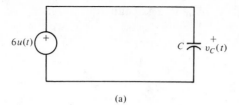

(a)

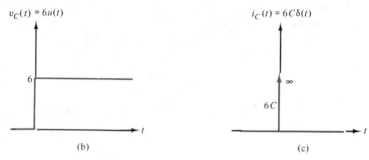

(b)   (c)

Figure 14-6   Example 14-3.

that is, zero for $t < 0$, then, at $t = 0$, jumping to the constant value of $6C$ coulombs. The current accompanying the charging of this capacitor is

$$i_C(t) = \frac{dq_C}{dt} = C\frac{dv_C}{dt} = C\frac{d}{dt}[6u(t)] = 6C\frac{du(t)}{dt} = 6C\,\delta(t)$$

where $du(t)/dt = \delta(t)$, the unit impulse at $t = 0$. The plot of $i_C(t)$ is shown in Figure 14-6(c). The area under the current is

$$\int_{-\infty}^{\infty} i_C(t)\,dt = \int_{-\infty}^{\infty} 6C\,\delta(t)\,dt = 6C\int_{-\infty}^{\infty} \delta(t)\,dt = 6C$$

the total charge. (The integral of the current is charge, as usual.) This area is usually marked next to the arrow of the impulse; it is sometimes called the *strength* of the impulse function. If not marked, it is assumed to be 1.

Do these answers make sense? Of course they do: The charge across the capacitor changes instantly, from zero at $t = 0^-$ to $6C$ at $t = 0^+$. Such an instantaneous change of charge Δq in zero time ($\Delta t = 0$) will require an *infinite* current $\Delta q/\Delta t$ at $t = 0$.

Now, in real life we know we don't have infinite currents, right? The circuit in Figure 14-6 is the usual ideal model. In a practical circuit, waveforms more similar to Figure 14-5 exist. The applied voltage is more like $g(t)$, since the switch takes ε seconds to close (ε may be very small, a few milliseconds perhaps). Consequently, the current i_C will be more like the gate function dg/dt, of a high (but finite) value, lasting for ε seconds, during which time the charge is transferred to C. Thereafter, with a constant charge, the current is zero again.

It is extremely important and useful to keep a balance between the ideal models with their mathematical descriptions and the practical circuits with their mathematical descriptions. These two approaches support and complement each other.   ■   Prob. 14-6

EXAMPLE 14-4

Plot the waveform of

$$v(t) = 10u(t) - 20u(t - 1) + 10u(t - 4)$$

Calculate and plot its derivative.

Solution. The derivative is, term by term,

$$\frac{dv}{dt} = 10\delta(t) - 20\delta(t - 1) + 10\delta(t - 4)$$

three impulse functions, at different times and of different strengths. The plots of $v(t)$ and dv/dt are in Figure 14-7.

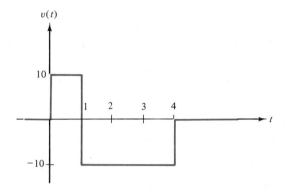

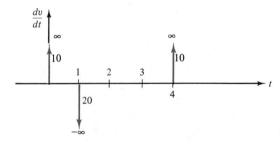

Figure 14-7   Example 14-4.   ■

Probs. 14-7, 14-8

We can continue the differentiation process and ask, "What is d^2g/dt^2 in Figure 14-5(a)?" To do it, we must differentiate the gate function dg/dt of Figure 14-5(b). We just did such a differentiation in Example 14-4, so we write, using Problem 14-5,

$$\frac{d^2g}{dt^2} = \frac{d}{dt}\frac{dg}{dt} = \frac{d}{dt}\left[\frac{1}{\varepsilon}u(t - a) - \frac{1}{\varepsilon}u(t - a - \varepsilon)\right]$$

$$= \frac{1}{\varepsilon}\delta(t - a) - \frac{1}{\varepsilon}\delta(t - a - \varepsilon) \qquad (14\text{-}12)$$

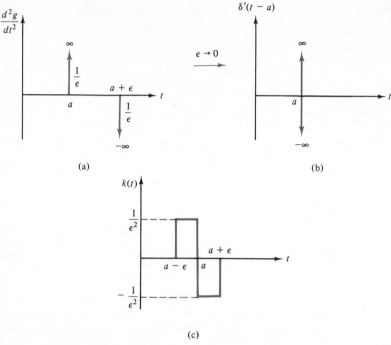

Figure 14-8   The unit doublet.

The plot of d^2g/dt^2 is shown in Figure 14-8(a). In the limit, as $\varepsilon \to 0$, we obtain the derivative of the unit impulse, called the *unit doublet*

$$\delta'(t - a) = \frac{d}{dt}\,\delta(t - a) \qquad (14\text{-}13)$$

Another formal approach to obtain the unit doublet is to let $\varepsilon \to 0$ for the function $k(t)$ shown in Figure 14-8(c).

   As a summary, we show in Figure 14-9 the ramp function, with its derivative the step function; the derivative of the step function is the impulse function, whose derivative is the doublet. In reverse, integration of the doublet yields the impulse, integration of the impulse yields the step function, and integration of the step function yields the ramp function.

   An important property of the impulse function is its *sampling property*, given by the definite integral I

Probs. 14-9
14-10

$$I = \int_{-\infty}^{\infty} f(t)\delta(t - a)\,dt = f(a) \qquad (14\text{-}14)$$

where $f(t)$ is an arbitrary function that is continuous at $t = a$. See Figure 14-10. Equation 14-14 says that when $f(t)$ is multiplied by $\delta(t - a)$, then integrated over the indicated range of time, the answer is $f(a)$, the value of $f(t)$ *at* the point $t = a$ where

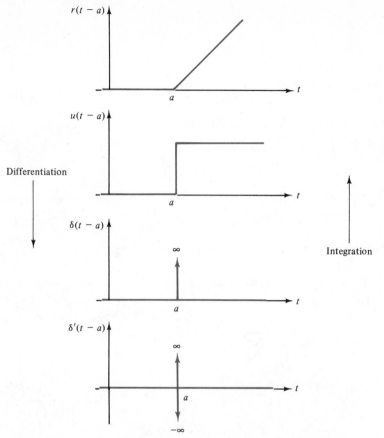

Figure 14-9 Relations among ramp, step, impulse, and doublet functions.

the impulse occurs. In this integral, then, $\delta(t - a)$ samples $f(t)$ and picks up only one value, $f(a)$. Another name for Equation 14-14 is the *sifting property* of $\delta(t - a)$.

To demonstrate the validity of Equation 14-14, we recognize that the integrand $f(t)\delta(t - a)$ is zero except between $t = a - \varepsilon$ and $t = a + \varepsilon$, because of $\delta(t - a)$. Therefore, Equation 14-14 can be written as

$$I = \int_{a-\varepsilon}^{a+\varepsilon} f(t)\delta(t - a)\, dt \tag{14-15}$$

By assumption, $f(t)$ is continuous (no jumps) at $t = a$. Therefore, over the small range of integration $a - \varepsilon \le t \le a + \varepsilon$, it takes on the value of $f(a)$. Therefore

$$I = \int_{a-\varepsilon}^{a+\varepsilon} f(a)\delta(t - a)\, dt = f(a) \int_{a-\varepsilon}^{a+\varepsilon} \delta(t - a)\, dt = f(a) \tag{14-16}$$

where $f(a)$, constant, comes out in front of the integral. The integral then expresses the area under the unit impulse, that is, 1.

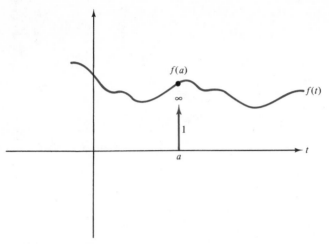

Figure 14-10   The sampling property of $\delta(t - a)$.

EXAMPLE 14-5 _____

We write *by inspection* the answers to the following fearsome-looking integrals:

Solution

(a)
$$\int_{-\infty}^{\infty} e^{-t}\delta(t - 2)\,dt = e^{-t}\bigg]_{t=2} = e^{-2} = 0.135$$

(b)
$$\int_{-\infty}^{\infty} \frac{t+1}{t^2+2}\,\delta(t-1)\,dt = \frac{t+1}{t^2+2}\bigg]_{t=1} = \frac{2}{3}$$

(c)
$$\int_{-3}^{10} t^2 e^{-2t}\delta(t+4)\,dt = 0,$$

because the integrand is zero over the range of integration; the impulse function occurs at $t = -4$, outside the range of integration.

(d)
$$\int_{0}^{2\pi} \sin \pi t \,\delta(t - \tfrac{1}{2})\,dt = \sin \pi t\bigg]_{t=1/2} = 1$$

(e)
$$\int_{-2\pi}^{0} \sin \pi t \,\delta(t - \tfrac{1}{2})\,dt = 0$$

for the same reason as (c).   ■   Prob. 14-1

EXAMPLE 14-6 _____

Let us review conservation of charge (and voltage) in a capacitor, as discussed in Chapter 6, Figure 6-6.

Solution.   There are two cases:
(a) The current in the capacitor has no impulse at $t = 0$. That is, $i(t)$ is smooth and well-behaved at $t = 0$, as in Figure 14-11(a). The interval 0^- to 0 to 0^+ is shown exaggerated, for clarity. Since

$$\frac{dq(t)}{dt} = i(t)$$

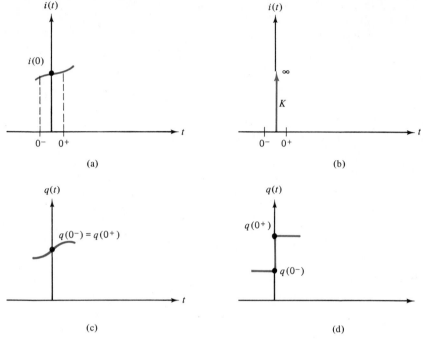

Figure 14-11 Continuous and discontinuous charge.

let us integrate between $t = 0^-$ and $t = 0^+$, to get

$$q(0^+) - q(0^-) = \int_{0^-}^{0^+} i(t)\, dt$$

The integral on the right is zero because it is the area of a rectangle of finite height, $i(0)$, and of zero base. Consequently,

$$q(0^+) = q(0^-)$$

and

$$v(0^+) = \frac{q(0^+)}{C} = \frac{q(0^-)}{C} = v(0^-)$$

charge and voltage do not change abruptly. See Figure 14-11(c).
(b) The current in the capacitor has an impulse of strength K at $t = 0$, as in Figure 14-11(b). Then

$$q(0^+) - q(0^-) = \int_{0^-}^{0^+} K\delta(t)\, dt = K \int_{0^-}^{0^+} \delta(t)\, dt = K$$

and therefore

$$q(0^+) = q(0^-) + K$$

showing a step discontinuity in charge (and the corresponding voltage), as in Figure 14-11(d).

Since true impulse functions do not happen in real-life circuits, we draw the familiar conclusion that charge and voltage in a capacitor cannot change instantly. However, we should keep in mind ideal models, where impulses *can* occur, and where charge *can* change abruptly.

The choice of $t = 0$ as the time of discontinuity is only convenient; it can be any other time, $t = a$, of course.

A dual derivation and conclusions apply to the flux and current in an inductor.

■   Prob. 14-1?

14-4 THE LAPLACE TRANSFORM

As mentioned in Section 14-1, the Laplace transform is used to handle the simultaneous integro-differential equations that arise in the analysis of linear, constant networks (loop, node, or state equations). This transform is named after P. S. Laplace (1749–1825), a French mathematician. To repeat the advantageous features of the Laplace transform, a flowchart is drawn in Figure 14-12, comparing the steps of the classical (direct) solution and of the Laplace transform (indirect) method of solution. As we saw, the direct solution is easy for first- and second-order circuits at the most. Beyond that, we find ourselves "stuck" in the very first step: How do we reduce, say, four simultaneous loop equations into a single differential equation, even if we know in advance (as we *do*) its order n? Subsequent steps, if we get there, are equally fraught with difficulties. With the Laplace transform, plain algebra takes over from the first step. With such pleasant prospects, let us proceed.

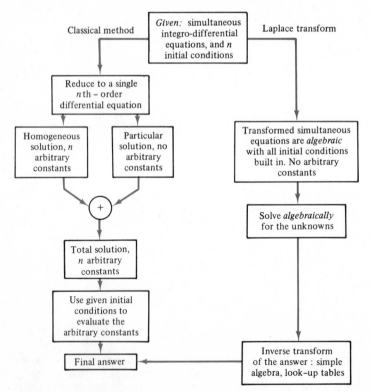

Figure 14-12   The flowchart of solutions.

As with any transform, we must start with the rule which defines that transform. The Laplace transform of a given time function $f(t)$ is defined as follows:

$$\mathcal{L}f(t) = \int_{0^-}^{\infty} f(t)e^{-st}\,dt = F(s) \tag{14-17}$$

We multiply $f(t)$ by e^{-st} and integrate with respect to t between $t = 0^-$ and $t = \infty$. The result of this definite integral will be only a function of s, not of t. Call it $F(s)$. The symbol $\mathcal{L}$ is read "the Laplace transform of ...." Conversely, the inverse Laplace transform of $F(s)$ is $f(t)$ and we write

$$\mathcal{L}^{-1}F(s) = f(t) \tag{14-18}$$

Compare with the phasor transform where we write symbolically

$$\mathcal{P}f(t) = \mathbf{F} \tag{14-19}$$

transforming a time function (sinusoid) to a complex number $\mathbf{F}$, and

$$\mathcal{P}^{-1}\mathbf{F} = f(t) \tag{14-20}$$

inverting $\mathbf{F}$ back into the time domain. (In Chapter 9, we used a double arrow to denote this transform.)

In the Laplace transform, the new variable s will be allowed in general to be a complex number. Furthermore, since e^{-st} is dimensionless, the unit of s is s^{-1}, that is, the same as frequency. We will write s as a complex number in its rectangular form

$$s = \sigma + j\omega \tag{14-21}$$

and call it the *complex frequency* variable. The Laplace transform, Equation 14-17, transforms $f(t)$, a function in the time domain, to $F(s)$, a function in the complex frequency domain.

What are the restrictions on $f(t)$ for it to have a Laplace transform? Unlike the phasor transform, which restricts $f(t)$ to be a pure sinusoid of a single frequency, the Laplace transform is much more general. First, let us consider the limits of the integral defining it in Equation 14-17.

The lower limit is $t = 0^-$, just before a switch closes or opens. That is nice for two reasons: (a) The entire past history of $f(t)$, $-\infty < t < 0$, is of no interest to us—and luckily so. Can you imagine the trouble if we had to know all this past history? In other words, the three *different* functions $f_1(t)$, $f_2(t)$, and $f_3(t)$ shown in Figure 14-13 have the *same* Laplace transform. (b) The natural initial conditions are given for $t = 0^-$; they completely summarize the past history of the network until then. They are the *initial state* of the network. At $t = 0$, we excite the network, write the necessary equations, and the Laplace transform takes over (with our firm guidance, of course). The initial state and the inputs for $t > 0$ will determine completely the response of the network for all $t > 0$.

As a result of remarks (a) and (b), let us agree on a unique $f(t)$ among the three possibilities shown in Figure 14-13. We will choose the second case and write $f(t)u(t)$ explicitly whenever necessary. Thus, there will be a unique (one-to-one) correspondence between a time function and its Laplace transform.

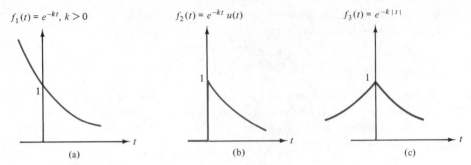

$f_1(t) = e^{-kt}, k > 0$ $f_2(t) = e^{-kt} u(t)$ $f_3(t) = e^{-k|t|}$

(a) (b) (c)

Figure 14-13 Three different functions, one Laplace transform.

The upper limit is $t = \infty$, a bit more troublesome. We must make sure that the integral converges (does not "blow up") at $t = \infty$. For this to happen, it is sufficient for $f(t)$ to be of *exponential order*. A function $f(t)$ is of exponential order if we can find a real constant c, such that

$$\lim_{t \to \infty} |f(t)e^{-ct}| = 0 \tag{14-22}$$

In other words, there should be an exponential e^{-ct} which, as $t \to \infty$, will cause $|f(t)e^{-ct}|$ to approach zero; that is, e^{-ct} must keep $f(t)$ and Equation 14-17 from diverging (blowing up) at the upper limit $t = \infty$. We are not interested if $f(t)e^{-ct}$ approaches zero from positive or from negative values—as long as it approaches zero. Hence the absolute value (magnitude) bars in Equation 14-22. Let us meet a few functions, of exponential order or not of exponential order.

EXAMPLE 14-7 _____

Consider $f(t) = e^{at}$, where a is a given real number (positive, negative, or zero).

Solution. We are looking for a constant c such that, as $t \to \infty$

$$|e^{at}e^{-ct}| = 0$$

That is,

$$|e^{(a-c)t}| = 0$$

Obviously, this will be true for any real c satisfying

$$c > a$$

Therefore e^{at} is of exponential order. We'll find its Laplace transform in short order. ■

EXAMPLE 14-8 _____

Consider $f(t) = \cos \beta t$. Is there a real number c such that, as $t \to \infty$

$$|\cos \beta t \, e^{-ct}| = 0?$$

Solution. The answer is "yes," because the magnitude of $\cos \beta t$ never exceeds 1, and therefore any

$$c > 0$$

will cause $|\cos \beta t \, e^{-ct}|$ to decay to zero. To satisfy yourself, plot $\cos \beta t \, e^{-0.001t}$ for large values of t. We conclude that $\cos \beta t$ (and similarly $\sin \beta t$) is of exponential order. ■

EXAMPLE 14-9

Consider the decaying sinusoidal function

$$f(t) = e^{-bt} \cos \beta t \qquad b > 0$$

and show that it is of exponential order by finding an appropriate c. Do it carefully! ■

EXAMPLE 14-10

The step function

$$f(t) = Ku(t - a)$$

is of exponential order, because

$$\lim_{t \to \infty} |Ku(t - a)e^{-ct}| = \lim_{t \to \infty} |Ke^{-ct}| = 0$$

for any $c > 0$. A suitable c has thus been found, and the function is of exponential order. ■

EXAMPLE 14-11

The impulse function

$$f(t) = K\delta(t - a)$$

is of exponential order. Why? Apply Equation 14-22, and exhibit a suitable c. ■

EXAMPLE 14-12

Consider now

$$f(t) = e^{t^2}$$

Solution. When we apply Equation 14-22, we have

$$\lim_{t \to \infty} |e^{t^2} e^{-ct}|$$

and we see that *there is no* constant c which will cause this limit to approach zero. The function e^{t^2} diverges (blows up) faster than any e^{-ct} which should keep it under control. The function e^{t^2} is *not* of exponential order and does *not* have a Laplace transform. We heave a small sigh of relief: After all, this function is not very common in electrical engineering. ■

Probs. 14-13, 14-14

We have seen now that all the waveforms that are common in engineering are of exponential order. Therefore they have Laplace transforms. Let us find several of these.

Start with the unit impulse at the origin, $f(t) = \delta(t)$. Use the definition of the Laplace transform, Equation 14-17:

$$\mathscr{L}\delta(t) = \int_{0-}^{\infty} \delta(t)e^{-st} \, dt = e^{-st} \bigg]_{t=0} = 1 \qquad (14\text{-}23)$$

where we evaluated the integral by the sampling property of $\delta(t)$; the impulse occurs at $t = 0$, within the range of integration.

This is a very nice result, indeed. The unruly, wild $\delta(t)$ has as its transform its strength, the constant 1. *Question*: What is the Laplace transform of $K\delta(t)$?

We enter our first entry in a table of Laplace transforms (Table 14-2). It looks like this:

$f(t)$	$F(s)$	Comments
$\delta(t)$	1	

In the left column, we have the time function $f(t)$. The center column has its Laplace transform, $F(s)$. Under "comments" we may list c, the exponent that guarantees the

TABLE 14-2 THE LAPLACE TRANSFORM

$$\text{Definition: } F(s) = \mathscr{L}\, f(t)u(t) = \int_{0-}^{\infty} f(t)e^{-st}\, dt$$

Functions		
$f(t)$	$F(s)$	Comments
$\delta(t)$	1	$c = 0$
$u(t)$	$\dfrac{1}{s}$	$c > 0$
$tu(t)$	$\dfrac{1}{s^2}$	$c > 0$
A, constant	$\dfrac{A}{s}$	$c > 0$
$t^n u(t) \quad n = 0, 1, 2, \ldots$	$\dfrac{n!}{s^{n+1}}$	$c > 0$
$e^{-at}u(t)$	$\dfrac{1}{s + a}$	$c > -a$
$t^n e^{-at}u(t)$	$\dfrac{n!}{(s + a)^{n+1}}$	$c > 0$
$\sin \beta t\, u(t)$	$\dfrac{\beta}{s^2 + \beta^2}$	$c > 0$
$\cos \beta t\, u(t)$	$\dfrac{s}{s^2 + \beta^2}$	$c > 0$
$e^{-at} \sin \beta t\, u(t)$	$\dfrac{\beta}{(s + a)^2 + \beta^2}$	$c > -a$
$e^{-at} \cos \beta t\, u(t)$	$\dfrac{s + a}{(s + a)^2 + \beta^2}$	$c > -a$
$\sinh \beta t\, u(t)$	$\dfrac{\beta}{s^2 - \beta^2}$	$c > |\beta|$
$\cosh \beta t\, u(t)$	$\dfrac{s}{s^2 - \beta^2}$	$c > |\beta|$

	Operations	
Time domain	s-domain	Comments
$af_1(t) + bf_2(t)$	$aF_1(s) + bF_2(s)$	Linearity
$\dfrac{df(t)}{dt}$	$sF(s) - f(0^-)$	Differentiation in the time domain
$\dfrac{d^2f(t)}{dt^2}$	$s^2F(s) - sf(0^-) - f'(0^-)$	
$\dfrac{d^nf(t)}{dt^n}$	$s^nF(s) - s^{n-1}f(0^-) - s^{n-2}f'(0^-)$ $- \cdots - f^{(n-1)}(0^-)$	
$\displaystyle\int_{0^-}^{t} f(x)\,dx$	$\dfrac{1}{s}F(s)$	Integration in the t-domain
$f(t-a)u(t-a)$	$e^{-as}F(s)$	Time shifting
$e^{-at}f(t)$	$F(s+a)$	Frequency shifting
$f(t) = f(t+T)$	$F(s) = \dfrac{F_1(s)}{1 - e^{-sT}}$ $F_1(s) = \displaystyle\int_{0}^{T} f(t)e^{-st}\,dt$	Periodic function
$f(at)$	$\dfrac{1}{a}F\!\left(\dfrac{s}{a}\right)$	Time scaling Frequency scaling
$t\,f(t)$	$-\dfrac{d}{ds}F(s)$	Multiplication by t
$f_1(t) * f_2(t)$ $= \displaystyle\int_{0^-}^{t} f_1(\lambda)f_2(t-\lambda)\,d\lambda$ $= \displaystyle\int_{0^-}^{t} f_1(t-\lambda)f_2(\lambda)\,d\lambda$	$F_1(s) \cdot F_2(s)$	Convolution in the time domain

existence of the Laplace transform, or any other relevant information. Table 14-2, found here and on the inside back cover, will be used like any transform table (logarithms, Morse code, etc.): For a given $f(t)$, we look up its $F(s)$ and use it, and for a given $F(s)$ we look up the corresponding $f(t)$.

Let us find the Laplace transform of the unit step function. Again, use Equation 14-17

$$\mathscr{L}u(t) = \int_{0^-}^{\infty} u(t)e^{-st}\,dt = \int_{0^+}^{\infty} 1e^{-st}\,dt = \left.\dfrac{e^{-st}}{-s}\right]_{0^+}^{\infty} = \dfrac{0-1}{-s} = \dfrac{1}{s} \qquad (14\text{-}24)$$

After setting up the defining integral, we always try to simplify the integrand. Over the range of our integral between 0^- and ∞, we have

$$\int_{0^-}^{\infty} u(t)e^{-st}\,dt = \int_{0^-}^{0^+} u(t)e^{-st}\,dt + \int_{0^+}^{\infty} 1e^{-st}\,dt$$

Since $u(t)$ is finite between 0^- and 0^+, the first integral on the right-hand side vanishes. Thus, the integrand in Equation 14-24 is $1e^{-st}$. This is a lot easier to integrate. The evaluation of the integral at $t = \infty$ yields $e^{-st}]_{t=\infty} = 0$ precisely because $u(t)$ is of exponential order. That's why we checked it first. Finally, at the lower limit we have $e^{-st}]_{t=0+} = 1$. Our second entry in Table 14-2 is then

$f(t)$	$F(s)$	Comments
$u(t)$	$\dfrac{1}{s}$	

Next, take the unit ramp function $tu(t)$. Notice that we multiply t by $u(t)$ to make a unique time function, as agreed. We write

$$\mathscr{L}tu(t) = \int_{0^-}^{\infty} tu(t)e^{-st}\,dt = \int_{0^-}^{\infty} te^{-st}\,dt \qquad (14\text{-}25)$$

where, again, we simplified the integrand. (When doing so, be sure that you do it legally. Some "simplifications" may be wrong!) To evaluate the last integral we can either look it up in a table of integrals as

$$\int xe^{ax}\,dx = \frac{xe^{ax}}{a} - \frac{e^{ax}}{a^2} \qquad (14\text{-}26)$$

or actually integrate it by parts. Just for the practice (and with the promise not to overdo it), let us do it. We have here $t = v$, $e^{-st} = u'$ and therefore

$$v' = 1 \qquad u = \frac{e^{-st}}{-s}$$

$$\therefore \int_{0^-}^{\infty} te^{-st}\,dt = \frac{te^{-st}}{-s}\bigg]_{0^-}^{\infty} - \int_{0^-}^{\infty} \frac{e^{-st}}{-s}\,dt$$

$$= \frac{te^{-st}}{-s}\bigg]_{0^-}^{\infty} + \frac{1}{s}\int_{0^-}^{\infty} e^{-st}\,dt = \frac{0-0}{-s} - \frac{1}{s^2}e^{-st}\bigg]_{0^-}^{\infty} \qquad (14\text{-}27)$$

$$= -\frac{1}{s^2}(0-1) = \frac{1}{s^2}$$

Here $te^{-st} = 0$ at the upper limit ($t = \infty$) because the function $tu(t)$ is of exponential order. Also $e^{-st} = 0$ at the upper limit from our previous derivation of $u(t)$.

Prob. 14-15

Another common waveform in electrical engineering is the decaying exponential $f(t) = e^{-at}u(t)$, $a > 0$. Its Laplace transform is

$$\mathscr{L}e^{-at}u(t) = \int_{0^-}^{\infty} e^{-at}u(t)e^{-st}\,dt = \int_{0^-}^{\infty} e^{-(a+s)t}\,dt$$

$$= \frac{e^{-(a+s)t}}{-(a+s)}\bigg]_{0^-}^{\infty} = \frac{0-1}{-(a+s)} = \frac{1}{s+a} \qquad (14\text{-}28)$$

Again, the integrand was simplified, $u(t) = 1$, and the exponents combined, prior to integrating. At the upper limit, $e^{-(s+a)t} = 0$ because we verified earlier in Example 14-7 that e^{-at} is a function of exponential order.

It is very instructive and often helpful to compare the Laplace transforms of $u(t)$ and $e^{-at}u(t)$,

$$\mathscr{L}u(t) = \frac{1}{s} \tag{14-24}$$

$$\mathscr{L}e^{-at}u(t) = \frac{1}{s+a} \tag{14-28}$$

If we let $a = 0$ in Equation 14-28, we obtain Equation 14-24. This is a powerful method of getting or verifying new transform pairs, $f(t)$ and $F(s)$, from existing pairs: Set an independent parameter (not s or t!) to a specific value.

EXAMPLE 14-13 _____

From the pair

$$\mathscr{L} \cos \beta t \, u(t) = \frac{s}{s^2 + \beta^2}$$

we set $\beta = 0$ to verify (again) that

$$\mathscr{L}u(t) = \frac{1}{s}$$

Probs. 14-16.
■   14-17

14-5 PROPERTIES OF THE LAPLACE TRANSFORM

Like any transform (phasor, logarithms, etc.), the Laplace transform has its unique properties. We must explore them, in order to be able to apply this transform to our integro-differential equations. In particular, we'll explore its linearity and its handling of derivatives and integrals.

Linearity   Let $f_1(t)$ and $f_2(t)$ have the respective transforms $F_1(s)$ and $F_2(s)$. Then, for any constants a and b, we have

$$\mathscr{L}[af_1(t) + bf_2(t)] = aF_1(s) + bF_2(s) \tag{14-29}$$

The proof is easy: Set up the defining integral for the left-hand side to get

$$\begin{aligned}
\mathscr{L}[af_1(t) + bf_2(t)] &= \int_{0^-}^{\infty} [af_1(t) + bf_2(t)]e^{-st} \, dt \\
&= \int_{0^-}^{\infty} af_1(t)e^{-st} \, dt + \int_{0^-}^{\infty} bf_2(t)e^{-st} \, dt \\
&= a\int_{0^-}^{\infty} f_1(t)e^{-st} \, dt + b\int_{0^-}^{\infty} f_2(t)e^{-st} \, dt \\
&= aF_1(s) + bF_2(s) \tag{14-30}
\end{aligned}$$

This result is based on two basic facts from calculus: (1) integration is linear (the integral of a sum is the sum of the integrals); and (2) a constant multiplier in the integrand can be brought out in front of the integral.

The Laplace transform is, therefore, linear (very much like superposition). By contrast, the logarithmic transform is *not* linear

$$\log (x + y) \neq \log x + \log y \tag{14-31}$$

but, then, to paraphrase a famous French proverb, "to each transform its own properties."† Where is the linearity of the Laplace transform useful? In taking the Laplace transform of an integro-differential equation in our circuit analysis, the left-hand side, and sometimes the right-hand side, will consist of a sum of several terms. We'll write the sum of their respective Laplace transforms.

EXAMPLE 14-14 _____

Calculate the Laplace transform of $f(t) = (2e^{-t} - 3e^{-4t})u(t)$.

    Solution.   We write using linearity

$$\mathscr{L}(2e^{-t} - 3e^{-4t})u(t) = 2\mathscr{L}e^{-t}u(t) - 3\mathscr{L}e^{-4t}u(t)$$

By looking up in Table 14-2, or Equation 14-28 we have

$$2\mathscr{L}e^{-t}(t) - 3\mathscr{L}e^{-4t}u(t) = \frac{2}{s + 1} - \frac{3}{s + 4} = \frac{-s + 5}{(s + 1)(s + 4)} = F(s)$$

The last term was obtained by combining the two fractions over the common denominator $(s + 1)(s + 4)$.   ■   Prob. 14-1

Transform of Time Derivatives In anticipation of the promised usefulness of the Laplace transform in differential equations, let us find the Laplace transform of $df(t)/dt$. Set up the defining integral from Equation 14-17

$$\mathscr{L}\frac{df(t)}{dt} = \int_{0-}^{\infty} \frac{df(t)}{dt} e^{-st} dt \tag{14-32}$$

and use integration by parts. Letting

$$\begin{aligned} u' &= f' &\quad \therefore \quad u &= f(t) \\ v &= e^{-st} &\quad \therefore \quad v' &= -se^{-st} \end{aligned} \tag{14-33}$$

we obtain

$$\mathscr{L}\frac{df(t)}{dt} = f(t)e^{-st}\Big]_{0-}^{\infty} - \int_{0-}^{\infty} -se^{-st}f(t)\, dt$$

$$= 0 - f(0^-) + s\int_{0-}^{\infty} f(t)e^{-st}\, dt = -f(0_-) + sF(s) \tag{14-34}$$

† In the logarithmic transform, $\log (a \cdot b) = \log a + \log b$. Don't even *think* of using this property in the Laplace transform! The Laplace transform of $[f_1(t) \cdot f_2(t)]$ *is not* $F_1(s) + F_2(s)$. For that matter, it is not $F_1(s) \cdot F_2(s)$, either.

where $f(t)e^{-st}$ at $t = \infty$ equals zero because $f(t)$ is assumed to be of exponential order. In the second integral with respect to t, the parameter s can be pulled in front as a multiplier. The final result is therefore

$$\mathcal{L}\frac{df(t)}{dt} = sF(s) - f(0^-) \tag{14-35}$$

The Laplace transform of the derivative of $f(t)$ is $sF(s)$, the transform of $f(t)$ multiplied by s, minus the initial value of $f(t)$. Two outstanding features of this transform are obvious: (1) it transforms a differentiation in the time domain into algebra (multiplication) in the frequency domain, and (2) the initial condition $f(0^-)$ is built into the transform from the start. See again Figure 14-12.

EXAMPLE 14-15 _____

As a quick verification for Equation 14-35, apply it to $f(t) = u(t)$, $F(s) = 1/s$.

Solution. We have then

$$\mathcal{L}\frac{df(t)}{dt} = \mathcal{L}\delta(t) = s \cdot \frac{1}{s} - u(0^-) = 1 - 0 = 1$$

as expected.   ■

We are ready for our first complete example in solving a network by the Laplace transform.

EXAMPLE 14-16 _____

For the *RL* network shown in Figure 14-14, KVL yields

$$2\frac{di(t)}{dt} + 4i(t) = 8u(t)$$

Here $i(t)$ is either the loop current or the state variable $i(t) = i_L(t)$. Also, $i(0^-) = 3$ is the given initial condition.

Solution. Take the Laplace transform of this equation, using the linearity property and Equation 14-35. The result is

$$2[sI(s) - 3] + 4I(s) = \frac{8}{s}$$

Be sure that every term is fully justified. Here $I(s)$ is the unknown transform of $i(t)$. Note how the differential equation in $i(t)$ is transformed into an algebraic equation in $I(s)$ with the initial condition incorporated right away.

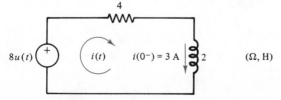

Figure 14-14   Example 14-16.

Solve algebraically for $I(s)$ by collecting terms and rearranging

$$(2s + 4)I(s) = \frac{8}{s} + 6$$

$$\therefore \quad I(s) = \frac{8}{s(2s + 4)} + \frac{6}{2s + 4}$$

This is the complete response, in the frequency domain. It consists of the zero-state response

$$\frac{8}{s(2s + 4)}$$

which is clearly due only to the input, $\mathscr{L}8u(t) = 8/s$, and not dependent on the initial state $i(0^-) = 3$; the second term

$$\frac{6}{2s + 4}$$

is the zero-input response, depending only on $i(0^-) = 3$. Trace back the origin of the number 6 in the numerator: It comes from $2i(0^-)$ in the original transformed equation.

If we had extensive Laplace transform tables[1,2,3] we would simply *look up* the time functions corresponding to these two terms, and add them up to give us the final answer. However, we can do equally well with a bit more of simple algebra. We simplify $I(s)$ further by dividing throughout by 2 (a legitimate algebraic operation) to get

$$I(s) = \frac{4}{s(s + 2)} + \frac{3}{s + 2}$$

and recognize immediately that the second term is listed in Table 14-2. For the first term, we'll have to break it up into partial fractions as follows:

$$\frac{4}{s(s + 2)} = \frac{2}{s} + \frac{-2}{s + 2}$$

You may remember a similar situation in calculus, when you had to evaluate an integral such as

$$\int \frac{4}{x(x + 2)} \, dx$$

Anyway, for now please verify that the sum of the partial fractions is correct, by recombining them

$$\frac{2}{s} + \frac{-2}{s + 2} = \frac{2s + 4 - 2s}{s(s + 2)} = \frac{4}{s(s + 2)}$$

We'll have a lot more to say about partial fractions later. The final result is then

$$I(s) = \frac{2}{s} + \frac{-2}{s + 2} + \frac{3}{s + 2}$$

Invert term by term (linearity!) using Table 14-2, to get

$$\mathscr{L}^{-1}I(s) = i(t) = \underbrace{2u(t) - 2e^{-2t}u(t)}_{\text{zero-state}} + \underbrace{3e^{-2t}u(t)}_{\text{zero-input}}$$

If we combine terms, we can write

$$i(t) = 2u(t) + e^{-2t}u(t)$$

showing the steady-state response, $2u(t)$, and the transient response, $e^{-2t}u(t)$.

This is it. Let us summarize these steps, keeping Figure 14-12 in front of us:

1. The original differential equation is transformed into an algebraic one. Initial conditions are incorporated immediately. There are no arbitrary constants to carry.
2. The unknown response in the s-domain is obtained easily by algebraic steps. It is the total response, consisting of the sum of the zero-state and zero-input parts. Each part is recognized easily, tracing it to its origin.
3. The inversion of the response to the time domain involves, again, some simple algebra and a look-up in the transform table.

Impressively simple and convincing, isn't it?

For the sake of comparison, as in Figure 14-12, let us solve this problem by the direct (classical) method, as we did in Chapter 7. From the given differential equation

$$2\frac{di}{dt} + 4i = 8 \qquad t \geq 0$$

we go first to the homogeneous solution

$$2\frac{di_H}{dt} + 4i_H = 0$$

The assumed solution $i_H(t) = Ke^{st}$ yields the characteristic equation of order $n = 1$ (equal to the order of the network)

$$2s + 4 = 0$$

Notice that the same characteristic polynomial $(2s + 4)$ appears as a "by-product" in the Laplace transform. It is the polynomial that multiplies $I(s)$ just before we solve algebraically for $I(s)$. This is a very important observation, and we will return to it again and again.

The characteristic value is the solution to the characteristic equation, that is

$$s = -2$$

and therefore

$$i_H(t) = Ke^{-2t}$$

with an arbitrary constant K.

Next we need the particular solution. Since the right-hand side is a constant $(= 8)$, let

$$i_P(t) = A$$

and substitute:

$$2\frac{d}{dt}(A) + 4A = 8$$

$$\therefore \quad A = 2$$

$$\therefore \quad i_P(t) = 2$$

Therefore, the total solution is

$$i(t) = i_H + i_P = Ke^{-2t} + 2$$

Now (and not earlier) use the initial condition

$$i(0^-) = 3 = K + 2 \qquad \therefore \quad K = 1$$

Finally, then,

$$i(t) = e^{-2t} + 2 \qquad t \geq 0$$

which can be written as

$$i(t) = e^{-2t}u(t) + 2u(t)$$

with $u(t)$ replacing the qualifier $t \geq 0$.

   The comparison speaks for itself. Even with a single (given) differential equation, the steps in the solution are long. Arbitrary constants must be carried to the end. There is no immediate way to separate the final answer into its zero-state and zero-input parts. Here, as in the Laplace transform, we must solve algebraically the characteristic equation. Here, this yields only the homogeneous solution (with arbitrary multiplying constants). In the Laplace transform, it yields a whole lot more, without any arbitrary constants.

   "So," you ask, "why do we even bother with the classical time-domain solution?" There are several aspects to the answer. Classical methods can be faster, particularly in low-order networks. In our example, we could have written the zero-input response by inspection, with some previous practice, as $i(0^-)e^{(-R/L)t}$. Similarly, we learned in Chapter 7 that the zero-state response is $V_0/R[1 - e^{-(R/L)t}] = 2(1 - e^{-2t})$. Some memorization is implied, but it comes naturally with practice. Finally, the Laplace transform, like any transform, is just a convenient tool. As engineers, we learn how to apply this tool most effectively.   ∎

EXAMPLE 14-17

Let us take a second-order ($n = 2$) *RLC* parallel circuit as shown in Figure 14-15. The initial conditions are given $v_C(0^-) = -1$ V, $i_L(0^-) = \frac{1}{2}$ A. Let us use state variable analysis, with $v_C(t)$ and $i_L(t)$ as the unknowns.

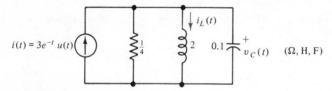

Figure 14-15   Example 14-17.

   Solution.   KCL at the top node reads

$$0.1\frac{dv_C(t)}{dt} + i_L(t) + 4v_C(t) = 3e^{-t}u(t)$$

and KVL for the *LC* loop is

$$v_C(t) - 2\frac{di_L(t)}{dt} = 0$$

   In the classical method of solution, we would have to obtain the single second-order differential equation by "juggling" these two equations. In higher-order circuits, and with *integro*-differential loop or node equations, this task is hopeless. With the Laplace transform, we transform each equation and continue algebraically.

   The first equation is transformed as follows:

$$0.1[sV_C(s) - (-1)] + I_L(s) + 4V_C(s) = \frac{3}{s+1}$$

Note in particular the first bracket: It is the transform of dv_C/dt, with $v_C(0^-) = -1$. The second equation is transformed as

$$V_C(s) - 2[sI_L(s) - \tfrac{1}{2}] = 0$$

the transform of zero being zero.

These two algebraic equations can be rearranged as

$$(0.1s + 4)V_C(s) + I_L(s) = \frac{3}{s+1} - 0.1$$

$$V_C(s) - 2sI_L(s) = -1$$

In matrix form, they are

$$\begin{bmatrix} 0.1s + 4 & 1 \\ 1 & -2s \end{bmatrix} \begin{bmatrix} V_C(s) \\ I_L(s) \end{bmatrix} = \begin{bmatrix} \dfrac{3}{s+1} \\ 0 \end{bmatrix} + \begin{bmatrix} -0.1 \\ -1 \end{bmatrix}$$

and are ready for solution. On the right-hand side, we show two separate matrices: The first one accounts for the zero-state response, since it contains only the sources; the second one causes the zero-input response since it contains only terms related to the initial conditions. ■   Prob. 14-19

The Laplace transform of the second derivative is handled in a similar way. We write

$$\mathscr{L} \frac{d^2 f(t)}{dt^2} = \mathscr{L} \frac{d}{dt}\left(\frac{df}{dt}\right) = \mathscr{L} \frac{d}{dt} g(t) \qquad (14\text{-}36)$$

where we let $g(t) = df/dt$ temporarily. Now, according to Equation 14-35,

$$\mathscr{L} \frac{dg(t)}{dt} = sG(s) - g(0^-) \qquad (14\text{-}37)$$

Also

$$G(s) = \mathscr{L}g(t) = \mathscr{L} \frac{df}{dt} = sF(s) - f(0^-) \qquad (14\text{-}38)$$

and

$$g(0^-) = \frac{df(t)}{dt}\bigg|_{t=0-} = f'(0^-) \qquad (14\text{-}39)$$

Therefore

$$\mathscr{L} \frac{d^2 f(t)}{dt^2} = s[sF(s) - f(0^-)] - f'(0^-) = s^2 F(s) - sf(0^-) - f'(0^-) \quad (14\text{-}40)$$

The second derivative of $f(t)$ is transformed into the multiplication of $F(s)$ by s^2, minus two terms involving the two initial conditions $f(0^-)$ and $f'(0^-)$. Equation 14-40 is entered in Table 14-2. Its extension to higher-order derivatives is also listed. The initial conditions are $f(0^-)$, the initial value of $f(t)$; $f'(0^-)$, the initial value of df/dt; down to $f^{(n-1)}(0^-)$, the initial value of $d^{n-1}f(t)/dt^{n-1}$.

EXAMPLE 14-8 _____

A certain third-order circuit obeys the differential equation

$$\frac{d^3 i(t)}{dt^3} + 2\frac{d^2 i(t)}{dt^2} + 6\frac{di(t)}{dt} + 4i(t) = 2u(t)$$

with $i(0^-) = -1$, $i'(0^-) = 3$, $i''(0^-) = -10$ given.

Solution. The Laplace transform is done term by term, using Table 14-2 as follows:

$$[s^3 I(s) - s^2(-1) - s(3) - (-10)] + 2[s^2 I(s) - s(-1) - 3]$$

$$+ 6[sI(s) - (-1)] + 4I(s) = \frac{2}{s}$$

Collecting terms and rearranging, we have

$$(s^3 + 2s^2 + 6s + 4)I(s) = \frac{2}{s} - s^2 + s - 10$$

ready for algebraic solution. The first term on the right, $2/s$, will account for the zero-state response; the others, for the zero-input response.   ∎

Transform of Integrals   We need the Laplace transform of the integral of $f(t)$, to handle the voltage across a capacitor in loop analysis or the current in an inductor in node analysis. Let us write

$$\mathcal{L}\left[\int_{0^-}^{t} f(x)\, dx\right] = \mathcal{L}y(t) = Y(s) \tag{14-41}$$

where we let temporarily

$$y(t) = \int_{0^-}^{t} f(x)\, dx \tag{14-42}$$

From Equation 14-42, we have immediately

$$\frac{dy(t)}{dt} = f(t) \tag{14-43}$$

Take the Laplace transform of Equation 14-43:

$$sY(s) - y(0^-) = F(s) \tag{14-44}$$

But, according to Equation 14-42,

$$y(0^-) = \int_{0^-}^{0^-} f(x)\, dx = 0 \tag{14-45}$$

Therefore, Equation 14-44 yields the final result

$$Y(s) = \mathcal{L}\left[\int_{0^-}^{t} f(x)\, dx\right] = \frac{1}{s} F(s) \tag{14-46}$$

The integral (from 0^- to t) of a time function is transformed to $F(s)$ divided by s. This also makes sense, because differentiation is transformed to multiplication by s. Integration is the inverse operation to differentiation, so it transforms to division by s.

EXAMPLE 14-19 ─────────────────────────

In Example 14-16, we found

$$\mathscr{L}^{-1}\frac{4}{s(s+2)}$$

by partial fractions. We can use Equation 14-46, instead.

Solution. Recognize that

$$\frac{4}{s(s+2)} = \frac{1}{s} \cdot \frac{4}{s+2}$$

and therefore its inverse Laplace is

$$\int_{0^-}^{t} f(x)\,dx$$

where

$$f(t) = \mathscr{L}^{-1}\frac{4}{s+2} = 4e^{-2t}u(t)$$

Therefore

$$\mathscr{L}^{-1}\frac{4}{s(s+2)} = \int_{0^-}^{t} 4e^{-2x}\,dx = -2e^{-2x}\Big]_{0^-}^{t} = (-2e^{-2t} + 2)u(t)$$

as obtained there. ■

Probs. 14-20, 14-21

You are justified if, by now, you are suspecting certain similarities between the Laplace transform and the phasor transform. In the phasor transform, differentiation in the time domain is transformed into multiplication by $j\omega$, and integration into division by $j\omega$. No initial conditions are involved with phasors, because the method is valid only for the zero-state response of pure sinusoids. Further implications of these similarities will be studied in the next chapter.

Another interesting comparison to be made is with the operators D and $1/D$ used in Chapter 8. The operator *notation*

$$Di(t) \equiv \frac{di(t)}{dt} \qquad D^2i(t) \equiv \frac{d^2i(t)}{dt^2} \tag{14-47}$$

and

$$\frac{1}{D}i(t) \equiv \int_{0^-}^{t} i(x)\,dx \tag{14-48}$$

is very similar to the algebraic multiplication or division by s.

14-6 ADDITIONAL PROPERTIES OF THE LAPLACE TRANSFORM

We conclude this chapter with several additional properties of the Laplace transform, as listed in Table 14-2. They are useful in various stages of the application of the Laplace transform, and we'll refer to them often.

Time Shifting   Consider a given original function $f(t)u(t)$, as shown in Figure 14-16(a), with its Laplace transform $F(s)$

$$\mathscr{L}f(t)u(t) = F(s) \tag{14-49}$$

If we shift this function by a units of time to the right, we obtain the *shifted* function $f(t - a)u(t - a)$. It is important to use the two step functions, $u(t)$ and $u(t - a)$, in these two plots. *Question*: What is the plot of just $f(t - a)$? of $f(t - a)u(t)$? Neither one is a truly shifted version of $f(t)u(t)$.

What is the Laplace transform of the shifted function? We set up the defining integral

$$\mathscr{L}f(t - a)u(t - a) = \int_{0^-}^{\infty} f(t - a)u(t - a)e^{-st}\, dt \tag{14-50}$$

and proceed, as usual, to simplify the integrand if possible. The step function $u(t - a)$ is zero until $t = a$, so the lower limit of the integral can be changed from 0 to a. Also, between $t = a$ and $t = \infty$, $u(t - a) = 1$. With these simplifications we have

$$\mathscr{L}f(t - a)u(t - a) = \int_{a}^{\infty} f(t - a)e^{-st}\, dt \tag{14-51}$$

In this integral we make the following substitution of variable, with the corresponding changes of limits

$$t - a = x \qquad \therefore\ dt = dx$$
$$\therefore\ \ t = a \rightarrow x = 0 \tag{14-52}$$
$$\therefore\ \ t = \infty \rightarrow x = \infty$$

and therefore

$$\mathscr{L}f(t - a)u(t - a) = \int_{0^-}^{\infty} f(x)e^{-(a+x)s}\, dx$$

$$= e^{-as}\int_{0^-}^{\infty} f(x)e^{-sx}\, dx$$

$$= e^{-as}F(s) \tag{14-53}$$

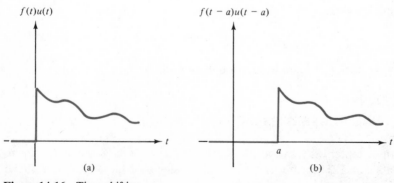

$f(t)u(t)$ $f(t - a)u(t - a)$

(a)   (b)

Figure 14-16   Time shifting.

The factor e^{-as} is a constant multiplier in the integrand, since integration is with respect to x. The last integral multiplying e^{-as} is, by definition, $F(s)$, the transform of the original nonshifted $f(t)u(t)$.

Time shifting transforms, then, into multiplication by e^{-as}. This is an easy visual sign to recognize.

EXAMPLE 14-20

Find the Laplace transform of the gate function

$$p(t) = u(t - a) - u(t - b)$$

as shown in Figure 14-3 (look it up!).

Solution. We write immediately the transform of the two shifted step functions

$$P(s) = \mathscr{L}p(t) = \frac{1}{s}e^{-as} - \frac{1}{s}e^{-bs} = \frac{e^{-as} - e^{-bs}}{s}$$ ∎

EXAMPLE 14-21

Find and plot the time function whose Laplace transform is

$$V(s) = \frac{10}{s}(1 - e^{-s})^2$$

Solution. Frightening? Not at all. Write it out fully as

$$V(s) = \frac{10}{s}(1 - 2e^{-s} + e^{-2s}) = \frac{10}{s} - \frac{20}{s}e^{-s} + \frac{10}{s}e^{-2s}$$

and invert term by term,

$$v(t) = \mathscr{L}^{-1}V(s) = 10u(t) - 20u(t - 1) + 10u(t - 2)$$

This waveform is shown in Figure 14-17

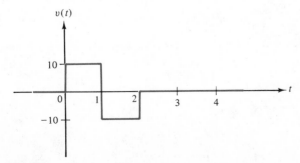

Figure 14-17 Example 14-21. ∎

Frequency Shifting   If the original function and its transform are given

$$\mathscr{L}f(t)u(t) = F(s) \tag{14-54}$$

let us calculate the transform of a new time function, $e^{-at}f(t)u(t)$. We write

$$\mathscr{L}e^{-at}f(t)u(t) = \int_{0^-}^{\infty} e^{-at}f(t)e^{-st}\,dt$$

$$= \int_{0^-}^{\infty} f(t)e^{-(s+a)t}\,dt = F(s+a) \tag{14-55}$$

where the last integral is recognized as the Laplace transform of $f(t)$, but with the argument $(s+a)$ instead of s. Thus, multiplying $f(t)$ by e^{-at} corresponds in the s-domain to replacing every s by $(s+a)$.

EXAMPLE 14-22 _____

Since

$$\mathscr{L}t^n u(t) = \frac{n!}{s^{n+1}}$$

we have

$$\mathscr{L}t^n e^{-at}u(t) = \frac{n!}{(s+a)^{n+1}} \qquad n = 0, 1, 2, \ldots$$

a most useful transform pair.   ■   Prob. 14-23

EXAMPLE 14-23 _____

Find

$$\mathscr{L}^{-1}\,\frac{s+2}{s^2 + 4s + 13}$$

Solution.   No such entry appears in Table 14-2, so let us do some algebraic enhancement. The numerator, $s + 2$, looks suspiciously like $(s + a)$ in Equation 14-55. Therefore, we complete the square in the denominator to get the same factor:

$$\frac{s+2}{s^2 + 4s + 13} = \frac{s+2}{(s+2)^2 + 9} = F(s+2)$$

where $s + a = s + 2$ is fully displayed. Hence, by Equation 14-55

$$\mathscr{L}^{-1}\,\frac{s+2}{(s+2)^2 + 9} = e^{-2t}\mathscr{L}^{-1}\,\frac{s}{s^2 + 9} = e^{-2t}\cos 3t\,u(t)$$

We'll use this property very often in our future calculations.   ■   Prob. 14-24

Periodic Functions   Let $f(t)u(t)$ be periodic for $t > 0$, with a period of T seconds. See Figure 14-18(a). Strictly speaking, a truly periodic function extends over the entire

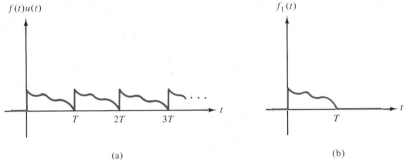

Figure 14-18 A periodic function.

time axis, $-\infty < t < \infty$, as we saw in previous chapters. However, with the Laplace transform, all our functions are zero for $t < 0$.

To find its Laplace transform, we consider its first cycle as a new function, $f_1(t)$, defined as

$$f_1(t) = \begin{cases} f(t) & 0 < t < T \\ 0 & \text{elsewhere} \end{cases} \tag{14-56}$$

and shown in Figure 14-18(b). Then we can write our periodic function as follows:

$$f(t)u(t) = f_1(t) + f_1(t - T)u(t - T) + f_1(t - 2T)u(t - 2T) + \cdots \tag{14-57}$$

which states that the periodic function is the sum of its first cycle *plus* the same cycle shifted by one period, *plus* the same cycle shifted by two periods, etc. Taking the Laplace transform of Equation 14-57, we get

$$F(s) = F_1(s) + e^{-Ts}F_1(s) + e^{-2Ts}F_1(s) + \cdots \tag{14-58}$$

in accordance with Equation 14-53. Rewrite this equation as

$$F(s) = F_1(s)(1 + e^{-Ts} + e^{-2Ts} + \cdots) = \frac{F_1(s)}{1 - e^{-Ts}} \tag{14-59}$$

where the geometric series in the parentheses is written in closed form using

$$1 + x + x^2 + \cdots = \frac{1}{1 - x} \tag{14-60}$$

We have therefore

$$\mathscr{L}f(t)u(t) = \frac{F_1(s)}{1 - e^{-Ts}} \tag{14-61}$$

The Laplace of a periodic function is given in terms of the Laplace transform of its first cycle

$$F_1(s) = \int_0^T f(t)e^{-st}\, dt \tag{14-62}$$

It is nice to note that, while $f(t)u(t)$ is given in an *open* form as an infinite series, Equation 14-57, its transform, is in *closed* form (no " $+\cdots$"), Equation 14-61.

EXAMPLE 14-24 _____

Let the periodic function be the square wave shown in Figure 14-19.

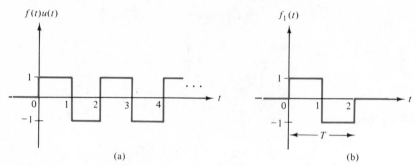

Figure 14-19 Example 14-24.

Solution. The Laplace transform of its first cycle is

$$F_1(s) = \int_0^2 f_1(t)e^{-st}\,dt = \int_0^1 1e^{-st}\,dt + \int_1^2 -1e^{-st}\,dt$$

where there are two separate expressions for $f_1(t)$ over $0 < t < T$. The direct evaluation of these integrals gives

$$F_1(s) = \frac{1 - 2e^{-s} + e^{-2s}}{s}$$

(check it!). As a quick and indirect review, we can write by inspection

$$\mathscr{L}^{-1}F_1(s) = \mathscr{L}^{-1}\frac{1}{s} - 2\mathscr{L}^{-1}\frac{1}{s}e^{-s} + \mathscr{L}^{-1}\frac{1}{s}e^{-2s}$$

$$= u(t) - 2u(t - 1) + u(t - 2)$$

using the time-shifting property. This is indeed the valid expression for $f_1(t)$—and we could have started here. Compare with Example 14-21.

Finally, using Equation 14-61

$$F(s) = \frac{1 - 2e^{-s} + e^{-2s}}{s(1 - e^{-2s})}$$

is the Laplace transform of the periodic square wave.

In calculating inverse transforms, a factor $(1 - e^{-Ts})$ in the denominator is a telltale for a periodic function. Simply ignore it, and the remaining is $F_1(s)$. Find its inverse, $f_1(t)$, and repeat $f_1(t)$ periodically with the *known* period T seen in $(1 - e^{-Ts})$.

■ Prob. 14-25

Time Scaling and Frequency Scaling It is often useful to scale (multiply) the independent variable, t, by a positive constant a. Then we find

$$\mathscr{L}f(at)u(t) = \frac{1}{a}F\left(\frac{s}{a}\right) \qquad a > 0 \qquad\qquad (14\text{-}63)$$

Thus, scaling t to at causes s to be scaled to s/a, and there is an extra division by a.

EXAMPLE 14-25

From

$$\mathscr{L} \sin t \, u(t) = \frac{1}{s^2 + 1}$$

we have

$$\mathscr{L} \sin \beta t \, u(t) = \frac{1}{\beta} \frac{1}{(s/\beta)^2 + 1} = \frac{\beta}{s^2 + \beta_2} \qquad \blacksquare$$

The result in Equation 14-63 may be also interpreted for scaling the frequency variable s. We have

$$\mathscr{L}^{-1}F\left(\frac{s}{a}\right) = af(at)u(t) \qquad a > 0 \qquad (14\text{-}64)$$

Multiplication by t From an original $f(t)u(t)$, we form a new function $tf(t)u(t)$. What is its Laplace transform?

Start with the original transform pair

$$F(s) = \int_{0^-}^{\infty} f(t)e^{-st} \, dt \qquad (14\text{-}65)$$

and differentiate both sides with respect to s. This is certainly a valid operation

$$\frac{d}{ds} F(s) = \frac{d}{ds} \int_{0^-}^{\infty} f(t)e^{-st} \, dt \qquad (14\text{-}66)$$

Furthermore, since the integral is with respect to t, and s is independent of t, we can move d/ds under the integral. In plain words, we are summing terms (integrating) over time, and we want then to differentiate this sum with respect to an independent variable s. We can certainly reverse the order: First differentiate each term with respect to s, then sum (integrate) over time. Thus

$$\frac{dF(s)}{ds} = \int_{0^-}^{\infty}\left[\frac{d}{ds} f(t)e^{-st}\right] dt$$

$$= \int_{0^-}^{\infty}\left[f(t)\frac{d}{ds} e^{-st}\right] dt$$

$$= \int_{0^-}^{\infty} f(t)(-te^{-st}) \, dt$$

$$= -\int_{0^-}^{\infty} [tf(t)]e^{-st} \, dt \qquad (14\text{-}67)$$

where, in the second step, $f(t)$ is a constant as far as d/ds is concerned. Our final result reads then

$$\mathscr{L} tf(t)u(t) = -\frac{dF(s)}{ds} \qquad (14\text{-}68)$$

Multiplication by t corresponds to differentiation with respect to s, with a minus sign added.

EXAMPLE 14-26 ——————————————————————————————

From

$$\mathscr{L}e^{-at}u(t) = \frac{1}{s+a}$$

we get

$$\mathscr{L}te^{-at}u(t) = -\frac{d}{ds}\frac{1}{s+a} = \frac{1}{(s+a)^2}$$

Continue in the same fashion on the last result

$$\mathscr{L}t^2e^{-at}u(t) = -\frac{d}{ds}\frac{1}{(s+a)^2} = \frac{2}{(s+a)^3}$$

and so on.

Probs. 14-27
■ 14-28, 14-29

Equipped with all these tools, we are ready now to tackle the analysis of a general network by the Laplace transform. We do it in the next chapters.

PROBLEMS

14-1 Write the expression for each function, as shown, using step functions and other functions as needed. *Hint*: In (c) and (d), multiply an appropriate gate function by an appropriate "regular" function.

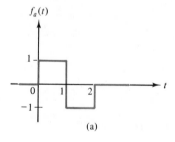

(a)

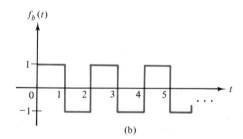

(b)

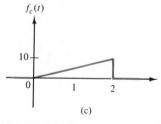

(c)

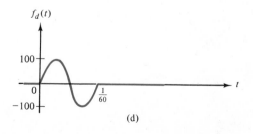

(d)

Problem 14-1

14-2 The *unit ramp* function, $r(t)$, is shown in the figure. Its expression is

$$r(t) = tu(t)$$

Show graphically that

$$\frac{dr(t)}{dt} = u(t)$$

or, alternately,

$$r(t) = \int_0^t u(x)\, dx$$

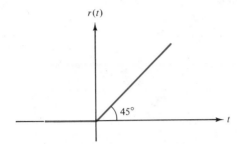

Problem 14-2

14-3 A shifted ramp function is shown. Its expression is

$$r(t-a) = (t-a)u(t-a)$$

(a) Verify carefully the correctness of this expression by plotting $(t-a)$ and $u(t-a)$ separately, then multiplying them.

(b) Plot the functions $(a > 0)$: (a) $tu(t-a)$, (b) $(t-a)u(t)$, and compare. Note in particular the *uniform* argument in the shifted ramp, from $tu(t)$ to $(t-a)u(t-a)$.

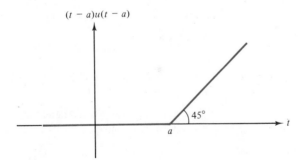

Problem 14-3

14-4 Write an expression for the "sawtooth" waveform $v(t)$ shown. (Such a waveform is used in your TV set.)

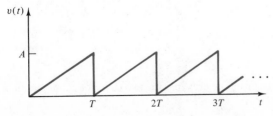

Problem 14-4

14-5 Write the expression for $dg(t)/dt$ in Figure 14-5(b) as a gate function using Equation 14-5. Don't forget its height, $1/\varepsilon$. Compare this expression with the fundamental definition of the derivative

$$\frac{dg(t)}{dt} = \lim_{h \to 0} \frac{g(t+h) - g(t)}{h}$$

14-6 Formulate *and* solve the dual problem to Example 14-3. Specifically, an inductor L is excited at $t = 0$ by a unit step current source. Draw carefully the circuit, showing the appropriate switch (closing or opening?). Write the expressions for $i_L(t)$, $\phi_L(t)$, and $v_L(t)$. Sketch their waveforms.

14-7 Plot the functions

$$f(t) = \frac{1}{\varepsilon} e^{-t/\varepsilon} u(t) \qquad \varepsilon > 0$$

and prove that

$$\lim_{\varepsilon \to 0} f(t) = \delta(t)$$

by showing how Equations 14-9, 14-10, and 14-11 are satisfied.

14-8 Repeat Problem 14-7 for the triangular-shaped function shown.

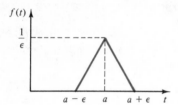

Problem 14-8

14-9 In Figure 14-9 of the text, what is the angle of the ramp function if the height of the step function is 2? If the strength of the impulse function is 0.88?

14-10 Extend Figure 14-9 one step upward; that is, what function, when differentiated, yields the ramp? Plot that function and give it its appropriate name.

14-11 Write the answer to each integral:

(a) $\displaystyle\int_{-\infty}^{\infty} \cos 4\pi t \,\delta(t + 1)\, dt$

(b) $\displaystyle\int_{0}^{\pi} \ln(t + 1)\delta(t - 1)\, dt$

(c) $\displaystyle\int_{0}^{\infty} e^t \delta(t + 1)\, dt$

(d) $\displaystyle\int_{-1}^{1} e^{-t}\delta(t)\, dt$

14-12 Derive in detail and with appropriate drawings the two cases for conservation of flux in a constant inductor. See Example 14-6.

*14-13 Another common definition for a function of exponential order is: There can be found two real constants, M and k, such that

$$|f(t)| < Me^{kt}$$

as $t \to \infty$. Prove that this is equivalent to our definition, Equation 14-22.

14-14 Check carefully each function to verify if it is or is not of exponential order, by exhibiting the appropriate c.

    (a) $f(t) = e^{j\beta t}$ Hint: $|f(t)| = ?$

    (b) $f(t) = \dfrac{d}{dt}\delta(t - a) = \delta'(t - a)$

    (c) $f(t) = t^n$ $n = 0, 1, 2, \ldots$ Hint: Use l'Hôpital's rule on t^n/e^{ct} when $t \to \infty$.

    (d) $f(t) = t^n e^{-at}$ $a > 0$

    (e) $f(t) = t \cos \beta t$

    (f) $f(t) = te^{-at} \sin \beta t$ $a > 0$

    (g) $f(t) = \cosh \beta t$ (review the definition of the hyperbolic cosine!)

    (h) $f(t) = t^t$

14-15 Calculate $\mathscr{L}t^n u(t)$, $n = 2, 3, \ldots$ and compare with the entry in Table 14-2. To evaluate the integral, look it up in a table of indefinite integrals.

14-16 Set $a = 0$ in Table 14-2 for $f(t) = e^{-at} \cos \beta t u(t)$ to verify a previous transform pair.

14-17 Complete Problem 14-15 by verifying in full detail the remaining entries in Table 14-2.

14-18 Use the linearity property of the Laplace transform on Euler's identity

$$e^{j\beta t} = \cos \beta t + j \sin \beta t$$

to obtain effortlessly $\mathscr{L} \cos \beta t$, $\mathscr{L} \sin \beta t$. Compare with the results in Table 14-2.

14-19 Complete Example 14-17 as far as calculating the zero-state response of $V_C(s)$.
    (a) Write your answer in the form

$$V_C(s) = \frac{\text{polynomial in } s}{\text{polynomial in } s}$$

    (b) What is the characteristic equation for this circuit? Review Example 14-16 to recall the method for obtaining the characteristic equation. **(c)** Find the characteristic values and, accordingly, classify the zero-input response as overdamped, underdamped, or critically damped.

14-20 Check the validity of Equation 14-46 on
    (a) $f_1(t) = u(t)$
    (b) $f_2(t) = e^{-at}$
    (c) $f_3(t) = \sin \beta t$
by doing two separate steps: (1) Integrate each $f(t)$ between 0^- and t. From Table 14-2, write the Laplace transform of the result. (2) Divide each $F(s)$ by s and compare with the result of step 1.

14-21 Find

$$\mathscr{L}^{-1}\frac{1}{s^2} = \mathscr{L}^{-1}\frac{1}{s} \cdot \frac{1}{s}$$

using Equation 14-46. Continue to

$$\mathscr{L}^{-1}\frac{2}{s^3} = \mathscr{L}^{-1}\frac{1}{s} \cdot \frac{2}{s^2}$$

and on to

$$\mathscr{L}^{-1}\frac{n!}{s^{n+1}} = \mathscr{L}^{-1}\frac{1}{s} \cdot \frac{n!}{s^n}$$

confirming, from a different approach, your result in Problem 14-15.

14-22 Derive the Laplace transform of $v(t)$ in Problem 14-4. Use shifted functions.

14-23 **(a)** Use the frequency-shifting property to obtain new transform pairs from the following ones:

(1) $\mathscr{L}u(t) = \dfrac{1}{s}$

(2) $\mathscr{L}\cos\beta tu(t) = \dfrac{s}{s^2 + \beta^2}$

(3) $\mathscr{L}e^{-bt}u(t) = \dfrac{1}{s + b}$

(b) Find $\mathscr{L}^{-1}$ for:

(1) $\dfrac{4}{(s + 1)^3}$

(2) $\dfrac{3}{(s - 1)^2}$

(3) $\dfrac{10}{(s + 2)^4}$

14-24 By completing the square, and with Equation 14-55, find the inverse Laplace transform of:

(a) $V(s) = \dfrac{s + 1}{s^2 + 2s + 15}$

(b) $I(s) = \dfrac{3}{s^2 + 4s + 20}$

(c) $V(s) = \dfrac{s - 1}{s^2 + 2s + 30}$

14-25 Plot the periodic function $f(t)u(t)$ given by its first cycle

$$f_1(t) = \begin{cases} t & 0 < t < 2 \\ 0 & \text{elsewhere} \end{cases}$$

and calculate its Laplace transform.

14-26 Prove the property of scaling (Equation 14-63); then demonstrate it on

$$\frac{2s}{s^2 + 36} = \frac{s/2}{(s/2)^2 + 3^2}$$

14-27 Use the differentiation property (Equation 14-68) to generate a new pair from

(a) $\mathcal{L} \cos \beta t u(t) = \dfrac{s}{s^2 + \beta^2}$

(b) $\mathcal{L} \sin \beta t u(t) = \dfrac{\beta}{s^2 + \beta^2}$

*14-28 Let a time function contain a parameter, say b, and let us stress it by writing $f(t, b)$. For example $f(t, b) = e^{-bt}u(t)$. Let its Laplace transform be written as $F(s, b)$. Prove that

$$\mathcal{L} \frac{\partial}{\partial b} f(t, b) = \frac{\partial}{\partial b} F(s, b)$$

thus generating new transform pairs. Illustrate on

(a) $f(t, b) = e^{-bt}u(t)$
(b) $f(t, b) = \sin bt\, u(t)$

*14-29 Among its many applications, a table of Laplace transform such as Table 14-2 can be used to evaluate definite integrals. Use Table 14-2 to evaluate *by inspection*

(a) $\displaystyle\int_0^\infty e^{-10t}\, dt$

(b) $\displaystyle\int_0^\infty e^{-4t} \sin 10t\, dt$

(c) $\displaystyle\int_0^\infty t^3 e^{-4t}\, dt$

by a proper choice of a value for s in each case. Don't you wish you had had Table 14-2 when you took calculus?

REFERENCES

1. F. E. Nixon, *Handbook of Laplace Transformation.* Englewood Cliffs, N.J.: Prentice Hall, Inc., 1960.
2. G. A. Korn, *Basic Tables in Electrical Engineering.* New York: McGraw Hill Book Co., 1965.
3. G. Doetsch, *Guide to the Applications of the Laplace and Z-Transforms.* New York: Van Nostrand Reinhold Co., 1971.

Chapter 15

Laplace Transform in Network Analysis

In this chapter, we study in detail the application of the Laplace transform to loop equations and node equations. A separate chapter is devoted to state variables.

15-1 LOOP ANALYSIS

In writing loop equations, we have three types of terms. For a resistor, the v–i relationship is

$$v_R(t) = Ri_R(t) \tag{15-1}$$

for an inductor, or a mutual coupling,

$$v_L(t) = L\frac{di_L(t)}{dt} \quad \text{or} \quad v_L(t) = M\frac{di_L(t)}{dt} \tag{15-2}$$

and for a capacitor

$$v_C(t) = \frac{1}{C}\int_{0-}^{t} i_C(x)\,dx + v_C(0^-) \tag{15-3}$$

Their respective Laplace transforms are, from the previous chapter,

$$V_R(s) = RI_R(s) \tag{15-4}$$

$$V_L(s) = L[sI_L(s) - i_L(0^-)] \quad \text{or} \quad V_L(s) = M[sI_L(s) - i_L(0^-)] \tag{15-5}$$

and

$$V_C(s) = \frac{1}{Cs} I_C(s) + \frac{v_C(0^-)}{s} \qquad (15\text{-}6)$$

These, and the transforms of the given sources, are all that we need!

The resistor's v–i relation remains unchanged when transformed, since it is algebraic in the time domain. For the inductor, we have the transform of the derivative, which includes the initial condition. The capacitive term includes the transform of a definite integral (division by s) and the transform of $v_C(0^-)$, a constant, being $v_C(0^-)/s$ according to Table 14-2. There is very little to memorize. For each circuit we formulate the time-domain loop equations and then transform them one by one and term by term.

EXAMPLE 15-1 _____

In the circuit shown in Figure 15-1 the initial conditions are given.

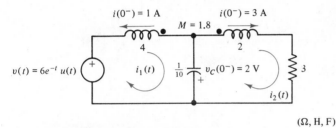

Figure 15-1   Example 15-1.

Solution.   The two loop equations are ($+$ voltage drop, $-$ voltage rise)

$$4\frac{di_1}{dt} - 1.8\frac{di_2}{dt} + 10\int_{0^-}^{t}[i_1(x) - i_2(x)]\,dx - 2 - 6e^{-t}u(t) = 0$$

$$2\frac{di_2}{dt} - 1.8\frac{di_1}{dt} + 3i_2 + 10\int_{0^-}^{t}[i_2(x) - i_1(x)]\,dx + 2 = 0\dagger$$

Transform each equation term by term. Go slowly and verify each term carefully

$$4[sI_1(s) - (-1)] - 1.8[sI_2(s) - 3] + \frac{10}{s}[I_1(s) - I_2(s)] - \frac{2}{s} - \frac{6}{s+1} = 0$$

$$2[sI_2(s) - 3] - 1.8[sI_1(s) - (-1)] + 3I_2(s) + \frac{10}{s}[I_2(s) - I_1(s)] + \frac{2}{s} = 0$$

Collect and arrange in matrix form

$$\begin{bmatrix} 4s + \dfrac{10}{s} & -1.8s - \dfrac{10}{s} \\[2mm] -1.8s - \dfrac{10}{s} & 2s + 3 + \dfrac{10}{s} \end{bmatrix} \begin{bmatrix} I_1(s) \\[2mm] I_2(s) \end{bmatrix} = \begin{bmatrix} \dfrac{6}{s+1} \\[2mm] 0 \end{bmatrix} + \begin{bmatrix} \dfrac{2}{s} - 9.4 \\[2mm] -\dfrac{2}{s} + 7.8 \end{bmatrix}$$

† Remember: Lowercase letters are functions of time, $i = i(t)$, etc.

Before actually solving, several observations deserve to be made or repeated:

1. $\mathscr{L} \, di/dt = sI(s) - i(0^-)$ always, but $i(0^-)$ can be either a positive number, when its *given* reference agrees with our choice for that current, or negative, when its reference is opposite to our choice. Here, $i_1(0^-) = -1$ and $i_2(0^-) = 3$ as shown, because of our choice of references for the two loop currents.
2. The given initial voltage across the capacitor is a rise in the first loop, and a drop in the second one.
3. In the matrix form of the equations, the right-hand side has two matrices: The first one contains the actual sources, $6/(s + 1)$ around the first loop and 0 around the second loop. This matrix causes the zero-state response in $I_1(s)$ and $I_2(s)$. The second matrix contains only terms with initial conditions. It causes the zero-input response in $I_1(s)$ and $I_2(s)$.

Suppose we want to solve just for the zero-state response of $I_2(s)$. The equations to solve are then

$$
\begin{bmatrix} 4s + \dfrac{10}{s} & -1.8s - \dfrac{10}{s} \\ -1.8s - \dfrac{10}{s} & 2s + 3 + \dfrac{10}{s} \end{bmatrix} \begin{bmatrix} I_1(s) \\ I_2(s) \end{bmatrix} = \begin{bmatrix} \dfrac{6}{s + 1} \\ 0 \end{bmatrix}
$$

according to the previous observation. The solution by Cramer's rule is†

$$
I_2(s) = \frac{\begin{vmatrix} 4s + \dfrac{10}{s} & \dfrac{6}{s + 1} \\ -1.8s - \dfrac{10}{s} & 0 \end{vmatrix}}{\begin{vmatrix} 4s + \dfrac{10}{s} & -1.8s - \dfrac{10}{s} \\ -1.8s - \dfrac{10}{s} & 2s + 3 + \dfrac{10}{s} \end{vmatrix}} = \frac{\dfrac{10.8s}{s + 1} + \dfrac{60}{s(s + 1)}}{4.76s^2 + 12s + 24 + 30/s}
$$

Here,

$$
\begin{vmatrix} 4s + \dfrac{10}{s} & -1.8s - \dfrac{10}{s} \\ -1.8s - \dfrac{10}{s} & 2s + 3 + \dfrac{10}{s} \end{vmatrix} = 4.76s^2 + 12s + 24 + \dfrac{30}{s}
$$

is the characteristic polynomial of the circuit because it is the multiplier of $I_2(s)$ in the previous step. You can see it easily by backing up one step in Cramer's solution

$$
\begin{vmatrix} 4s + \dfrac{10}{s} & -1.8s - \dfrac{10}{s} \\ -1.8s - \dfrac{10}{s} & 2s + 3 + \dfrac{10}{s} \end{vmatrix} \cdot I_2(s) = \begin{vmatrix} 4s + \dfrac{10}{s} & \dfrac{6}{s + 1} \\ -1.8s - \dfrac{10}{s} & 0 \end{vmatrix}
$$

This rule was observed in the previous chapter, and it is perfectly general. We can say, therefore, that *the characteristic polynomial of a network is the determinant of the*

† Review Appendix A, if necessary.

coefficients matrix of the loop (or node) equations. It is amazing that the characteristic polynomial obtained by *any* method of analysis is the same. Naturally, it must be so: The network does not care how we analyze it to obtain its characteristic polynomial.

Clearing fractions and collecting terms in our solution, we get

$$I_2(s) = \frac{10.8s^2 + 60}{(s + 1)(4.76s^3 + 12s^2 + 24s + 30)}$$

and this is ready for inverting, $\mathscr{L}^{-1}I_2(s) = i_2(t)$.

The final form of $I_2(s)$ is a ratio of two polynomials with integer powers of s (s^0, s^1, s^2, s^3, ...). Such a ratio is known as a *rational function*. We can think of a rational function as a generalized fraction, a fraction being a ratio of integer *numbers* ($\frac{3}{2}$ or $-\frac{11}{14}$, for example). A rational function is a ratio of *polynomials*.

A similar approach will yield, for instance, the zero-input $I_1(s)$. For this case, the right-hand side of the equation must be

$$\begin{bmatrix} \dfrac{2}{s} - 9.4 \\[2ex] -\dfrac{2}{s} + 7.8 \end{bmatrix}$$

the matrix of the terms due only to initial conditions. Again, we'll use Cramer's rule. The solution $I_1(s)$ will be another rational function. If we want the *total response* $I_1(s)$ or $I_2(s)$, we must retain on the right-hand side both matrices, that is, the sum

$$\begin{bmatrix} \dfrac{6}{s + 1} + \dfrac{2}{s} - 9.4 \\[2ex] -\dfrac{2}{s} + 7.8 \end{bmatrix}$$

In solving for the zero-input, zero-state, or the total response, we see that there are changes only in the numerator determinants in Cramer's solution. The denominator determinant remains always the characteristic polynomial.

A final glance at the characteristic polynomial will reveal the justification for its name. All its terms are characteristic of the network itself, and not of outside sources or initial conditions. Specifically, in $4s$, $4 = L_1$, the inductor in the first loop; in $10/s$, $10 = 1/C$; in $1.8s$, $M = 1.8$; in $2s$, $L_2 = 2$; and $3 = R$, the resistor. ■

Probs. 15-1, 15-2, 15-3

15-2 NODE ANALYSIS

Here the elements' *i-v* expressions are

$$i_R(t) = Gv_R(t) = \frac{1}{R}v_R(t) \tag{15-7}$$

$$i_C(t) = C\frac{dv_C(t)}{dt} \tag{15-8}$$

and

$$i_L(t) = \frac{1}{L}\int_{0^-}^{t} v_L(x)\,dx + i_L(0^-) \tag{15-9}$$

These are the duals of the v–i relations in loop analysis. We don't allow, for the moment, a mutual inductance in node equations. (This will be remedied later.)

The corresponding transform relations are

$$I_R(s) = GV_R(s) \tag{15-10}$$

$$I_C(s) = C[sV_C(s) - v_C(0^-)] \tag{15-11}$$

and

$$I_L(s) = \frac{1}{Ls} V_L(s) + \frac{i_L(0^-)}{s} \tag{15-12}$$

Again, these are the duals of Equations 15-4, 15-5, and 15-6.

EXAMPLE 15-2

The network shown in Figure 15-2 needs two node equations (or two loop equations). The unknowns are v_1 and v_2.

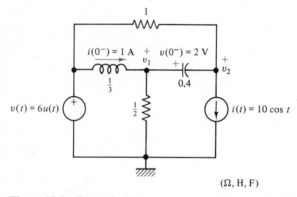

Figure 15-2   Example 15-2.

Solution.   Before even writing any equations, we establish by inspection the order of the network, $n = 2$. The characteristic equation, therefore, will be quadratic, with two characteristic values. That is a lot of useful information.

The first node equation is ($+$ current leaving, $-$ entering):

$$2v_1(t) + 0.4 \frac{d}{dt} (v_1 - v_2) + 3 \int_{0^-}^{t} [v_1(x) - 6u(x)] \, dx - 1 = 0$$

Be sure to check each term for correctness. The second node equation is

$$10 \cos t + 0.4 \frac{d}{dt} (v_2 - v_1) + \frac{v_2 - 6u(t)}{1} = 0$$

Without much delay, Laplace transform the first equation:

$$2V_1(s) + 0.4\{s[V_1(s) - V_2(s)] - 2\} + \frac{3}{s}\left[V_1(s) - \frac{6}{s}\right] - \frac{1}{s} = 0$$

here the voltage across the capacitor is $(v_1 - v_2)$, and consequently $(v_1 - v_2)_{0-} = 2$. The voltage across the inductor is $v_1(t) - 6u(t)$, and its integral is transformed as usual. The second node equation is transformed into

$$\frac{10s}{s^2 + 1} + 0.4\{s[V_2(s) - V_1(s)] - (-2)\} + V_2(s) - \frac{6}{s} = 0$$

Collect terms and arrange in matrix form

$$\begin{bmatrix} 0.4s + 2 + \dfrac{3}{s} & -0.4s \\[2ex] -0.4s & 0.4s + 1 \end{bmatrix} \begin{bmatrix} V_1(s) \\[2ex] V_2(s) \end{bmatrix} = \begin{bmatrix} \dfrac{18}{s^2} \\[2ex] \dfrac{6}{s} - \dfrac{10s}{s^2 + 1} \end{bmatrix} + \begin{bmatrix} 0.8 + \dfrac{1}{s} \\[2ex] -0.8 \end{bmatrix}$$

The right-hand side matrices contain, respectively, terms of actual sources and of initial conditions. Inspect these carefully and trace back each term; for example, what is the source (pun intended) of the term $18/s^2$?

The characteristic equation and the characteristic values can be obtained immediately from the final matrix formulation of the equations. Specifically, we get

$$\begin{vmatrix} 0.4s + 2 + \dfrac{3}{s} & -0.4s \\[2ex] -0.4s & 0.4s + 1 \end{vmatrix} = 0$$

or

$$1.2s^2 + 3.2s + 3 = 0$$

a quadratic equation, as expected. The two natural frequencies are

$$s_1 = -1.33 + j0.85 \qquad s_2 = -1.33 - j0.85 = s_1^*$$

Suppose we are asked to find only the zero-input response of $v_1(t)$. We solve therefore

$$\begin{bmatrix} 0.4s + 2 + \dfrac{3}{s} & -0.4s \\[2ex] -0.4s & 0.4s + 1 \end{bmatrix} \begin{bmatrix} V_1(s) \\[2ex] V_2(s) \end{bmatrix} = \begin{bmatrix} 0.8 + \dfrac{1}{s} \\[2ex] -0.8 \end{bmatrix}$$

writing by Cramer's rule

$$V_1(s) = \frac{\begin{vmatrix} 0.8 + \dfrac{1}{s} & -0.4s \\[2ex] -0.8 & 0.4s + 1 \end{vmatrix}}{\begin{vmatrix} 0.4s + 2 + \dfrac{3}{s} & -0.4s \\[2ex] -0.4s & 0.4s + 1 \end{vmatrix}} = \frac{1.2 + (1/s)}{1.2s + 3.2 + (3/s)}$$

The denominator is, as always, the characteristic polynomial. We clear fractions as follows:

$$V_1(s) = \frac{1.2s + 1}{1.2s^2 + 3.2s + 3}$$

$$= \frac{s + 0.833}{s^2 + 2.67s + 2.5}$$

This is the final form of $V_1(s)$ as a rational function, ready for inversion, $\mathcal{L}^{-1}V_1(s) = v_1(t)$. We took the extra step in factoring out the leading coefficient in both numerator and denominator, so that the highest powers of s have unity coefficients. Such a step is very useful later, when we'll actually invert $V_1(s)$. ■

Probs. 15-4, 15-5

We conclude these two sections with the repeated assurance, "that's all there is to it." No matter how large the linear network is, the same simple steps apply. The number of the equations may increase, but not the complexity of the terms in them. The algebra, of course, becomes more tedious; here, however, nothing is difficult or impossible. Keep in mind the original problem, case **4** in Table 14-1, and you'll realize that such a price is very fair, indeed, for the solution of this problem!

We will address several of the algebraic procedures involved in the inversion of transform answers in subsequent sections.

15-3 GENERALIZED IMPEDANCES AND ADMITTANCES

Let us consider again the transformed v–i relations for the individual elements, Equations 15-4, 15-5, and 15-6. Only now, consider just the zero-state response. We have

$$V_R(s) = RI_R(s) \tag{15-13}$$

$$V_L(s) = LsI_L(s) \tag{15-14}$$

$$V_C(s) = \frac{1}{Cs} I_C(s) \tag{15-15}$$

All of these have the same algebraic form

$$V(s) = Z(s)I(s) \tag{15-16}$$

where $Z(s)$ is called the *generalized impedance* of the element. Specifically,

$$Z_R(s) = R \qquad Z_L(s) = Ls \qquad Z_C(s) = \frac{1}{Cs} \tag{15-17}$$

with the units of ohms (Ω) for each. If we compare these with the *sinusoidal impedances* in phasor analysis (Chapters 9 to 13)

$$Z_R(j\omega) = R \qquad Z_L(j\omega) = j\omega L \qquad Z_C(j\omega) = \frac{1}{j\omega C} \tag{15-18}$$

we'll recognize, again, that we have here a special case where s, a general complex number, takes on one specific value $s = j\omega$, with ω being the single frequency of the sinusoidal source.

Equation 15-16 is sometimes called *the generalized Ohm's law*. The powerful Laplace transform provides this uniform algebraic expression for the zero-state v–i relationship in R, L, and C.

Dually, the zero-state i–v relationships are, from Equations 15-10, 15-11, and 15-12,

$$I_R(s) = \frac{1}{R}\, V_R(s) \tag{15-19}$$

$$I_L(s) = \frac{1}{Ls}\, V_L(s) \tag{15-20}$$

$$I_C(s) = Cs V_C(s) \tag{15-21}$$

Here, as before, no special memorization is needed. Quickly, recall the basics: For a resistor, Ohm's law is unchanged whether in the time domain or in the frequency domain. For an inductor without initial conditions, $i_L(t) = \phi(t)/L$ and the flux $\phi(t)$ is the integral $\int_0^t v(x)\, dx$. In the s domain, the integral becomes $V(s)$ divided by s. Similarly, for a capacitor, $i = C\, dv/dt$ becomes $I(s) = Cs V(s)$.

These three relationships are of the same form

$$I(s) = Y(s)V(s) \tag{15-22}$$

where $Y(s)$ is the *generalized admittance* (in mhos) of each element, and

$$Y(s) = \frac{1}{Z(s)} \tag{15-23}$$

for each element. In a given problem, it will be clear whether we are dealing with sinusoidal impedances and admittances or with generalized ones. So we'll use simply the terms *impedance* and *admittance*.

The rules for combining impedances in series is the same as derived in Chapter 4 for resistors only, and in Chapter 9 for sinusoidal impedances only. In general, the total driving-point impedance is

$$Z_t(s) = Z_1(s) + Z_2(s) + \cdots + Z_n(s) = \sum_{k=1}^{n} Z_k(s) \tag{15-24}$$

when n impedances are connected in series. See Figure 15-3. For admittances in parallel, we have

$$Y_t(s) = Y_1(s) + Y_2(s) + \cdots + Y_p(s) = \sum_{k=1}^{p} Y_k(s) \tag{15-25}$$

as the total equivalent admittance of p admittances connected in parallel. The rules for resistors only, or for sinusoidal phasor analysis, are special cases of Equations 15-24 and 15-25.

Probs. 15-6, 15-7

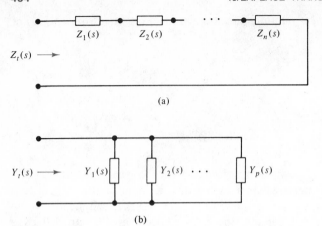

(a)

(b)

Figure 15-3   Series impedances and parallel admittances.

EXAMPLE 15-3 _____

Find the driving-point impedance of the ladder network shown in Figure 15-4. Hence, calculate the zero-state driving-point current $I(s)$ if a unit step voltage source is applied.

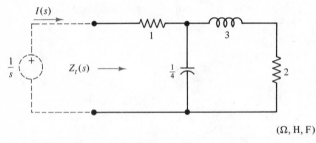

(Ω, H, F)

Figure 15-4   Example 15-3.

Solution.   We have, from right to left, a series connection of the resistor and the inductor

$$Z_{RL}(s) = 3s + 2 \qquad \therefore \quad Y_{RL}(s) = \frac{1}{3s + 2}$$

This combination is in parallel with the capacitor, therefore

$$Y_{RLC}(s) = \frac{s}{4} + \frac{1}{3s + 2} = \frac{3s^2 + 2s + 4}{12s + 8} = \frac{1}{Z_{RLC}(s)}$$

and this is in series with the 1-Ω resistor, so

$$Z_t(s) = 1 + \frac{12s + 8}{3s^2 + 2s + 4} = \frac{3s^2 + 14s + 12}{3s^2 + 2s + 4} = \frac{1}{Y_t(s)}$$

With the unit step voltage as an input, Equation 15-22 yields

$$I(s) = \frac{3s^2 + 2s + 4}{3s^2 + 14s + 12} \frac{1}{s} = \frac{3s^2 + 2s + 4}{3s^3 + 14s^2 + 12s}$$

the familiar rational function, ready for inversion.

As in the resistive (or phasor) case, we can write $Z_t(s)$ as a continued fraction

$$Z_t(s) = 1 + \frac{1}{(s/4) + [1/(3s + 2)]}$$

and clear the individual fractions. Such a step is helpful in checking your work, "$3s + 2$ are impedances in series; the reciprocal plus $s/4$ are admittances in parallel; the reciprocal plus 1 are in series." ∎

EXAMPLE 15-4

Find a one-port network for which

$$Y_t(s) = \frac{s + 1}{s + 4}$$

Solution. Here we have a problem in network design (synthesis): From a given specification, find a network. In analysis, as in the previous example, we do the opposite.

Based on our experience, we will do it slowly. First, it seems reasonable to write

$$Y_t(s) = \frac{s}{s + 4} + \frac{1}{s + 4} = Y_a(s) + Y_b(s)$$

so we have two smaller networks a and b connected in parallel. It is easy to identify network b, because

$$Z_b(s) = \frac{1}{Y_b(s)} = s + 4$$

and consequently it is a series connection of $L_b = 1$ H and $R_b = 4\ \Omega$.

To identify network a, think "backward" in the continued fraction process: Divide the numerator and the denominator of $Y_a(s)$ by s, to get

$$Y_a(s) = \frac{s}{s + 4} = \frac{1}{1 + (4/s)}$$

Therefore

$$Z_a(s) = \frac{1}{Y_a(s)} = 1 + \frac{4}{s}$$

a series connection of $R_a = 1\ \Omega$ and a capacitor $C_a = \frac{1}{4}$F. See Figure 15-5.

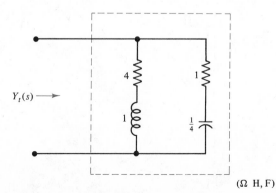

$(\Omega\ \text{H}, \text{F})$

Figure 15-5 Example 15-4.

Probs. 15-8, 15-9

15-4 TRANSFORM CIRCUIT DIAGRAMS

"O.K., so we can use impedances or admittances to calculate only the zero-state response," you argue, "but what about the *total* response?" This is a very valid question. To answer it, let us look back at the transform of the v–i relations for loop analysis (Equations 15-4, 15-5, and 15-6). They are repeated here, with additional elaborations

$$V_R(s) = RI_R(s) = Z_R(s)I_R(s) \tag{15-26}$$

$$V_L(s) = L[sI_L(s) - i_L(0^-)] = Z_L(s)I_L(s) - Li_L(0^-) \tag{15-27}$$

$$V_C(s) = \frac{1}{Cs} I_C(s) + \frac{v_C(0^-)}{s} = Z_C(s)I_C(s) + \frac{v_C(0^-)}{s} \tag{15-28}$$

In each case, the total response $V(s)$ is the sum of the zero-state drop, $Z(s)I(s)$, *plus* a zero-input term. These are shown in Figure 15-6. The associated references of $V(s)$ and $I(s)$ are the usual ones: The voltage drop is in the direction of the current arrow.

For a resistor, in Figure 15-6(a), there is only a single term $RI_R(s)$, the drop across its impedance. In an inductor, $V_L(s)$ consists of a drop across the impedance $Z_L(s) = Ls$, plus an "initial condition" voltage source, $Li_L(0^-)$, whose reference is as shown in Figure 15-6(b). In the capacitor, $V_C(s)$ consists of the drop across its impedance $Z_C(s) = 1/Cs$, plus an "initial condition" voltage source $v_C(0^-)/s$, as shown in Figure 15-6(c).

It must be stressed that (except for the resistor), the circuit diagrams in Figure 15-6 are not real-time diagrams. They are mathematical models in the frequency domain, and they serve to account for the transform relations in Equations 15-26, 15-27, and 15-28.

Given, then, a network to be analyzed by loop equations, we can skip altogether the time-domain integro-differential equations. Instead, we draw the *transform circuit diagram* using the building blocks of Figure 15-6. From that diagram, we write KVL around each loop.

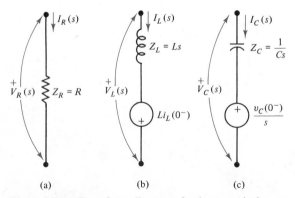

(a)   (b)   (c)

Figure 15-6 Transform diagrams for loop analysis.

EXAMPLE 15-5

To illustrate this method, consider the network in Figure 15-7(a).

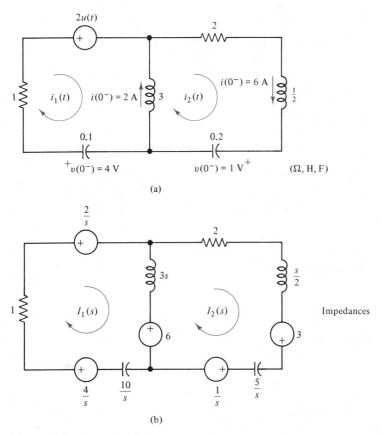

(a)

(b)

Figure 15-7 Example 15-5.

Solution. Its transform circuit diagram for loop analysis is shown in Figure 15-7(b). Be sure that you verify fully the "initial condition" sources in each inductor and in each capacitor (magnitude and reference).

The two loop equations are written by superposition: We sum up all the voltage drops around each loop caused by each loop current. Around the first loop, with $I_1(s) \neq 0$ and $I_2(s) = 0$, the total drop is $(3s + 1 + 10/s)I_1(s)$. With $I_2(s) \neq 0$ and $I_1(s) = 0$, the drop in the first loop is $-3sI_2(s)$. The total voltage drop around the first loop must be equal to the total voltage rise; therefore, the first loop equation is

$$\left(3s + 1 + \frac{10}{s}\right)I_1(s) - 3sI_2(s) = -\frac{2}{s} - 6 + \frac{4}{s}$$

Similarly, in the second loop, $I_2(s)$ alone causes the drop $[(\frac{7}{2})s + 2 + (5/s)]I_2(s)$, and $I_1(s)$ alone causes the drop $-3sI_1(s)$. Equate the total drop to the total rise, to get the second loop equation

$$-3sI_1(s) + \left(\frac{7}{2}s + 2 + \frac{5}{s}\right)I_2(s) = 9 - \frac{1}{s}$$

This method yields the transformed loop equations in their final form, with terms collected and arranged.

Probs. 15-1
15-11, 15-12

We summarize this method as follows:

1. Replace every resistor by its impedance R.
2. Replace every inductor by its impedance Ls, in series with an "initial condition" voltage source. The value of this source is $Li_L(0^-)$, and its reference is such that the given $i_L(0^-)$ leaves its $(+)$ terminal.
3. Replace every capacitor by its impedance $1/Cs$, in series with an "initial condition" voltage source. Its value is $v_C(0^-)/s$, and its reference is the given reference of $v_C(0^-)$.
4. Replace every source by its transform.
5. Draw the transform loop currents and apply superposition to each loop, summing all the drops in the impedances around the loop. On the right-hand side of each loop, write the sum of all the voltage rises around the loop (positive if a rise, negative if a drop in the direction of the loop).

The resulting equations will have the general matrix form

$$\mathbf{Z}(s)\mathbf{I}(s) = \mathbf{E}(s) \tag{15-29}$$

where

$$\mathbf{Z}(s) = \begin{bmatrix} z_{11}(s) & z_{12}(s) & \cdots & z_{1l}(s) \\ z_{21}(s) & z_{22}(s) & \cdots & z_{2l}(s) \\ \cdots\cdots\cdots\cdots\cdots\cdots\cdots\cdots\cdots \\ z_{l1}(s) & z_{l2}(s) & \cdots & z_{ll}(s) \end{bmatrix} \tag{15-30}$$

is the square $(l \times l)$ *loop impedance matrix.* The matrix $\mathbf{I}(s)$ is the column matrix of the l unknown transform loop currents. The matrix $\mathbf{E}(s)$ is the column matrix of all the inputs around each loop; it is the sum of the matrix of the actual sources (causing the zero-state responses) and of the "initial condition" sources (causing the zero-input responses).

Equation 15-29 is the generalized form of the resistive loop analysis (Equation 2-12), and of the sinusoidal phasor loop analysis (Equation 9-63).

We can also observe that, without dependent sources or current sources, a main diagonal term $z_{pp}(s)$ in Equation 15-30 is the total *self-impedance* around loop p, and it carries a $(+)$ sign. An off-diagonal term, $z_{pq}(s)$, is the common (mutual) impedance between loops p and q, positive $(+)$ if $I_p(s)$ and $I_q(s)$ flow in the same direction in $z_{pq}(s)$, and $(-)$ if opposite.

EXAMPLE 15-6 _____

In Example 15-5, we write by inspection

$$z_{11}(s) = 3s + 1 + \frac{10}{s} \qquad \text{the self-impedance around loop 1}$$

$$z_{22}(s) = 3s + \frac{s}{2} + 2 + \frac{5}{s} \qquad \text{the self-impedance around loop 2}$$

$$z_{12}(s) = z_{21}(s) = -3s \qquad \text{the mutual impedance between loop 1 and 2; it is negative because } I_1(s) \text{ and } I_2(s) \text{ are of opposite directions in it}$$

Also, by inspection,

$$E_1(s) = \frac{4}{s} + \left(-\frac{2}{s} - 6 \right) \qquad \text{the total input in loop 1, grouped into actual sources, } 4/s, \text{ and "initial condition" sources}$$

$$E_2(s) = 3 - \frac{1}{s} + 6 \qquad \text{the total input in loop 2, consisting entirely of "initial condition" sources} \qquad \blacksquare$$

With dependent sources or current sources present, it is best to proceed cautiously. The first four steps outlined above are still valid, but then it is advisable to write KVL around each loop term by term for each element, since the contributions of dependent sources and of current sources in the final form of Equation 15-29 are not obvious by inspection.

Probs. 15-13, 15-14

In a dual fashion, let us develop the transform diagrams and rules for node analysis. From Equations 15-10, 15-11, and 15-12 we have

$$I_R(s) = GV_R(s) = Y_R(s)V_R(s) \tag{15-31}$$

$$I_L(s) = \frac{1}{Ls} V_L(s) + \frac{i_L(0^-)}{s} = Y_L(s)V_L(s) + \frac{i_L(0^-)}{s} \tag{15-32}$$

$$I_C(s) = CsV_C(s) - Cv_C(0^-) = Y_C(s)V_C(s) - Cv_C(0^-) \tag{15-33}$$

displaying the admittance of the element and certain "initial condition" sources. These equations can be modeled by the transform circuits shown in Figure 15-8 for the resistor, inductor, and capacitor, respectively. Here, again, these models are not

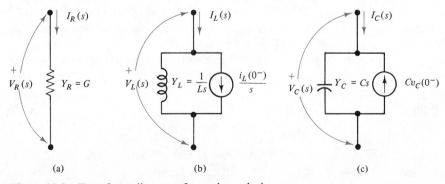

Figure 15-8 Transform diagrams for node analysis.

physical, only on paper in the s domain. The "funny" current source associated with the capacitor has the value $Cv_C(0^-) = q_C(0^-)$, the initial charge across the capacitor; it is no funnier than the previous voltage source with the inductor, $Li_L(0^-) = \phi_L(0^-)$. Be sure that you have solved Problem 15-14 by now.

The rules for writing transformed node equations from the circuit diagram are:

1. Replace every resistor by its admittance G.
2. Replace every inductor by its admittance $1/Ls$, in parallel with an "initial condition" current source. The value of this source is $i_L(0^-)/s$ and its reference is the given reference of $i_L(0^-)$.
3. Replace every capacitor by its admittance Cs, in parallel with an "initial condition" current source. Its value is $Cv_C(0^-)$ and its reference is such that its current leaves the $(+)$ terminal of the given $v_C(0^-)$.
4. Replace every source by its transform.
5. Mark the unknown transform node voltages, usually all $(+)$ with respect to a reference node. Apply superposition at each node, summing all the currents leaving that node in the admittances incident to that node. Equate this sum to all the current sources entering that node.

The resulting equations will have the general form

$$\mathbf{Y}(s)\mathbf{V}(s) = \mathbf{J}(s) \tag{15-34}$$

where

$$\mathbf{Y}(s) = \begin{bmatrix} y_{11}(s) & y_{12}(s) & \cdots & y_{1m}(s) \\ y_{21}(s) & y_{22}(s) & \cdots & y_{2m}(s) \\ \multicolumn{4}{c}{\dotfill} \\ y_{m1}(s) & y_{m2}(s) & \cdots & y_{mm}(s) \end{bmatrix} \tag{15-35}$$

is the square $(m \times m)$ *node admittance matrix*. The matrix $\mathbf{V}(s)$ is the column matrix of the m unknown node voltages. The matrix $\mathbf{J}(s)$ is the column input matrix; it is the sum of the actual sources and "initial condition" sources entering each node.

Equation 15-34 is the generalized form of the resistive node analysis (Equation 2-11), and of the sinusoidal phasor node analysis (Equation 9-91). Look them up!

Without dependent sources or voltage sources, we can write the elements of $\mathbf{Y}(s)$ by inspection: $y_{pp}(s)$ is the sum of all the admittances incident to node p, and is called the *self-admittance* of node p. The off-diagonal element $y_{pq}(s)$ is the total admittance connected between nodes p and q; this term will be negative $(-)$ if $V_p(s)$ and $V_q(s)$ are marked $(+)$ with respect to reference.

With dependent sources or voltage sources present, we proceed carefully (no "by inspection") after step 4 and write KCL at each node term by term. Only then can we collect and arrange them in the final form of Equation 15-34.

The following two examples illustrate these points. When there will be no room for misunderstanding, we'll use just capital letters for transformed variables, $V = V(s)$, $I = I(s)$, etc. Lowercase letters remain, as usual, functions of time.

EXAMPLE 15-7

In Figure 15-9(a) is the given network. Our first decision, as always, must be the choice of the method to analyze it. Three loop equations or two node equations are required.

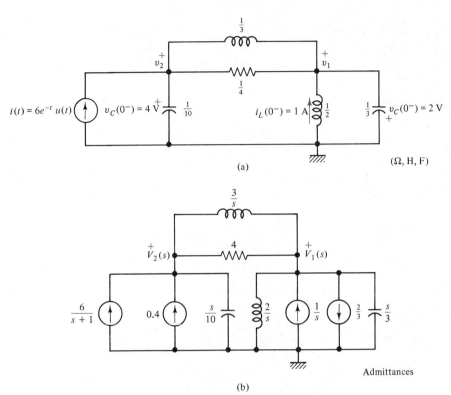

(a)

(b)

Figure 15-9 Example 15-7.

Solution. Now that we've agreed on node analysis, we draw the transform diagram as shown in Figure 15-9(b). The top inductor has no given initial current. In this circuit we can write by inspection

$$y_{11}(s) = \frac{s}{3} + \frac{2}{s} + \frac{3}{s} + 4 = \frac{s}{3} + \frac{5}{s} + 4$$

$$y_{22}(s) = \frac{s}{10} + 4 + \frac{3}{s}$$

$$y_{12}(s) = y_{21}(s) = -\left(4 + \frac{3}{s}\right)$$

Consequently, the node equations are

$$\begin{bmatrix} \dfrac{s}{3} + \dfrac{5}{s} + 4 & -\left(4 + \dfrac{3}{s}\right) \\ -\left(4 + \dfrac{3}{s}\right) & \dfrac{s}{10} + 4 + \dfrac{3}{s} \end{bmatrix} \begin{bmatrix} V_1(s) \\ V_2(s) \end{bmatrix} = \begin{bmatrix} \dfrac{1}{s} - \dfrac{2}{3} \\ \dfrac{6}{s+1} + 0.4 \end{bmatrix}$$

EXAMPLE 15-8 _____

In the network shown in Figure 15-10(a), we need two node equations (or two loop equations).

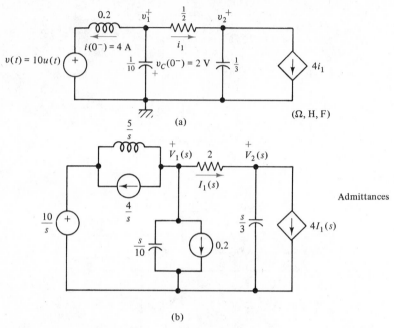

Figure 15-10   Example 15-8.

Solution.   Draw the transform diagram for node analysis, as in Figure 15-10(b). Since there is a voltage source, as well as a dependent source, we can't write $\mathbf{Y}(s)$ or $\mathbf{J}(s)$ immediately by inspection. Instead, write KCL at each node carefully (+ current leaving, −entering). At the node V_1 we have

$$\frac{s}{10} V_1 + 0.2 + \frac{4}{s} + \frac{5}{s}\left(V_1 - \frac{10}{s}\right) + 2(V_1 - V_2) = 0$$

and at the node V_2, with $I_1(s) = 2(V_1 - V_2)$,

$$4[2(V_1 - V_2)] + \frac{s}{3} V_2 - 2(V_1 - V_2) = 0$$

Collecting and rearranging terms, we obtain

$$\begin{bmatrix} \dfrac{s}{10} + 2 + \dfrac{5}{s} & -2 \\[2mm] 6 & \dfrac{s}{3} - 6 \end{bmatrix} \begin{bmatrix} V_1 \\ V_2 \end{bmatrix} = \begin{bmatrix} \dfrac{50}{s^2} - 0.2 - \dfrac{4}{s} \\[2mm] 0 \end{bmatrix}$$

as the node equations $\mathbf{Y}(s)\mathbf{V}(s) = \mathbf{J}(s)$. While $y_{11}(s)$ may have been predicted, the other entries in $\mathbf{Y}(s)$ are unpredictable—unless you have *a lot* of experience. Notice also that $y_{12}(s) \neq y_{21}(s)$, which is typical with dependent sources. The input matrix has the (unpredictable) term $50/s^2$ in $J_1(s)$. Conclusion: Do it carefully and slowly after step 4 listed above. ∎

Having reached the end of this section, you are probably asking, "Are the transform diagrams worth it?" The answer is very subjective, and will depend on your personal preference and your proficiency in handling the Laplace transform. In the early stages of your study (now), you may not wish to have yet another set of rules (Figures 15-6 and 15-8). The Laplace transform, when applied to the time-domain equations, will produce them, as well as the final answer. On the other hand, a brief mental calculation can reproduce the transform diagrams every time without too much effort. Whichever approach you prefer, you'll undoubtedly develop your own pet shortcuts and mnemonics in it.

15-5 INVERSION BY PARTIAL FRACTIONS

We are now almost at the end. The transform of the solution has been calculated, and $V(s)$ or $I(s)$ is in its final form, a rational function. If this is a ratio of quadratic polynomials or less, our table of Laplace transforms (Table 14-2, also reproduced inside the back cover) will give us the time function immediately, or after a little algebra. Let us use here $F(s)$ and $f(t)$ as generic notation for all our functions.

EXAMPLE 15-9 _____

$$F(s) = \frac{s^2 - 2s + 6}{s + 4}$$

Solution. This is an *improper* rational function, where the degree of the numerator—the highest power of s—is equal to or higher than the degree of the denominator. By comparison, an improper fraction has a numerator equal to or greater than the denominator, as in $\frac{3}{3}$, or $\frac{17}{6}$.

The first cardinal rule, *without exception*, is: If $F(s)$ is *improper*, carry out a longhand division to get a polynomial plus a *proper* remainder. This is similar to writing our improper fractions as

$$\frac{3}{3} = 1 \qquad \text{(plus zero remainder)}$$

and

$$\frac{17}{6} = 2\frac{5}{6} \qquad \text{(remainder is } \tfrac{5}{6}\text{, a proper fraction)}$$

In our case, longhand division yields (do it!)

$$F(s) = s - 6 + \frac{30}{s + 4}$$

Now, we can invert $F(s)$ term by term, using Table 14-2:

$$f(t) = \mathscr{L}^{-1}F(s) = \delta'(t) - 6\delta(t) + 30e^{-4t}u(t)$$

As we'll see in our studies, in most physical circuits, $F(s)$ will be a proper rational function, and the preliminary longhand division will not be needed. However, there will be rare cases (physical circuits, or final exams) when $F(s)$ will be improper. It is easy to spot, and then don't forget to divide first! ∎

A second useful rule is related to a quadratic denominator: If this quadratic has complex conjugate roots, it is best to complete the square in it.

EXAMPLE 15-10 _____

Find the inverse Laplace transform of

$$F(s) = \frac{3s - 1}{s^2 + 8s + 18}$$

Solution. $F(s)$ is proper, so longhand division is not needed. A quick check, $b^2 - 4ac = 64 - 72 < 0$, reveals complex conjugate roots. Therefore, we complete the square in the denominator

$$s^2 + 8s + 18 = (s + 4)^2 + 2$$

The presence of $(s + 4)$ should trigger in our mind the frequency-shifting property of the Laplace transform, listed in Table 14-2 as

$$\mathscr{L}^{-1}F(s + a) = e^{-at}f(t)u(t)$$

We need therefore to introduce $(s + 4)$ in the numerator of $F(s)$ also. This is easy. Simply replace every s by $(s + 4)$, then balance the equality.

$$3s - 1 = 3(s + 4) - 13$$

Now we have, by the linearity property,

$$\mathscr{L}^{-1}\frac{3s - 1}{s^2 + 8s + 18} = \mathscr{L}^{-1}\frac{3(s + 4) - 13}{(s + 4)^2 + 2}$$

$$= \mathscr{L}^{-1}\frac{3(s + 4)}{(s + 4)^2 + (\sqrt{2})^2} - \mathscr{L}^{-1}\frac{13}{\sqrt{2}}\frac{\sqrt{2}}{(s + 4)^2 + (\sqrt{2})^2}$$

$$= 3e^{-4t}\cos\sqrt{2}\, tu(t) - \frac{13}{\sqrt{2}}e^{-4t}\sin\sqrt{2}\, tu(t)$$

directly from Table 14-2. In the last term, we multiplied and divided by $\sqrt{2}$ to be able to recognize the entry in our table.

The process is, then, to enhance (to "doctor") the quadratic expression by legal algebraic steps in order to bring it into a form listed in Table 14-2.   ■   Prob. 15-20

Let us turn our attention to the general rational function

$$F(s) = K\frac{s^m + a_1 s^{m-1} + a_2 s^{m-2} + \cdots + a_{m-1}s + a_m}{s^n + b_1 s^{n-1} + b_2 s^{n-2} + \cdots + b_{n-1}s + b_n} = K\frac{P(s)}{Q(s)} \qquad (15\text{-}36)$$

Here, the constant multiplier K takes care of the unity coefficients of s^m and s^n. Also, since we know how to handle an improper rational function, we will assume that $F(s)$ is proper,

$$m < n \qquad (15\text{-}37)$$

In other cases, of course, you will have to make sure of it with longhand division. The numerator polynomial is conveniently denoted by $P(s)$, and the denominator by $Q(s)$.

Next, we introduce a couple of very important and common terms. A value of s which makes $F(s) = 0$ is called (obviously!) a *zero* of $F(s)$. If we denote a typical zero by z_k then

$$F(s)]_{s=z_k} = 0 \qquad (15\text{-}38)$$

A value of s which makes $F(s)$ approach infinity ("blow up") is called a *pole* of $F(s)$. If p_r is a typical pole then

$$\lim_{s \to p_r} F(s) = \infty \tag{15-39}$$

From Equation 15-36 we conclude that the roots of the equation

$$P(s) = 0 \tag{15-40}$$

are the finite zeros of $F(s)$, and the roots of

$$Q(s) = 0 \tag{15-41}$$

are the finite poles of $F(s)$. Let us look at an example.

EXAMPLE 15-11 ───

Consider $F(s)$ in Example 15-10,

$$F(s) = \frac{3s - 1}{s^2 + 8s + 18}$$

and write it in the form of Equation 15-36

$$F(s) = 3 \, \frac{s - \frac{1}{3}}{s^2 + 8s + 18} = K \frac{P(s)}{Q(s)}$$

Solution. The finite zeros of $F(s)$ are found from

$$P(s) = s - \tfrac{1}{3} = 0 \qquad \therefore \quad s = z_1 = \tfrac{1}{3}$$

and, indeed, $F(\tfrac{1}{3}) = 0$. The finite poles of $F(s)$ are given by

$$Q(s) = s^2 + 8s + 18 = 0$$

$$\therefore \quad p_1 = -4 + j\sqrt{2} \qquad p_2 = -4 - j\sqrt{2}$$

and $F(p_1) = \infty$, $F(p_2) = \infty$.

There is one additional zero of $F(s)$, at $s = z_2 \to \infty$. If you check the value of $F(s)$ as $s \to \infty$, you'll find by l'Hôpital's rule

$$\lim F(s)]_{s = z_2 = \infty} = 0$$

Be sure to distinguish between the value of s and the value of F. The infinite value of s, $s = z_2 = \infty$, makes the value of F zero. Therefore, z_2 is a zero of $F(s)$. In conclusion, $F(s)$ has one finite zero ($z_1 = \tfrac{1}{3}$), two finite poles ($p_1 = -4 + j\sqrt{2}, p_2 = p_1^* = -4 - j\sqrt{2}$), and one zero at infinity ($z_2 = \infty$). In our calculations here, we'll be interested only in the finite zeros and poles, given by Equations 15-40 and 15-41. ■ Prob. 15-21

In general, $F(s)$ will have m finite zeros, because Equation 15-40 has m roots. We can write $P(s)$ in its factored form to show them

$$P(s) = (s - z_1)(s - z_2) \cdots (s - z_m) \tag{15-42}$$

In fact, $P(z_1) = P(z_2) = \cdots = P(z_m) = 0$. Similarly, there will be n finite poles, shown by the factored $Q(s)$

$$Q(s) = (s - p_1)(s - p_2) \cdots (s - p_n) \tag{15-43}$$

A polynomial (P or Q) of order 3 or higher can be factored numerically, using a conveninent root-finding program. Some are even available on hand-held calculators.

If a zero is distinct, that is, if

$$z_1 \neq z_2 \neq \cdots \neq z_n \tag{15-44}$$

we call it *simple*. A repeated zero, say

$$z_1 = z_2 = \cdots = z_k \neq z_{k+1} \tag{15-45}$$

is called *multiple*, of multiplicity k. Similar adjectives apply to poles.

EXAMPLE 15-12 ──

$$F(s) = \frac{s^2(s + 1)(s - 4)}{(s + 3)^4(s + 2)}$$

has a multiple ($k = 2$) zero at $s = 0$, and simple zeros at $s = -1$ and $s = 4$. It has a pole at $s = -3$ of multiplicity 4, and a simple pole at $s = -2$. ■

To invert $F(s)$, we classify it as: (1) having only distinct (simple) poles, and (2) having multiple poles. We deal with each case separately.

1. Simple Poles Only   Here we have

$$F(s) = K \frac{P(s)}{(s - p_1)(s - p_2) \cdots (s - p_n)} \tag{15-46}$$

and $p_1 \neq p_2 \neq \cdots \neq p_n$. We stipulate that Equation 15-46 can be expanded in partial fractions as follows:

$$F(s) = \frac{A_1}{s - p_1} + \frac{A_2}{s - p_2} + \cdots + \frac{A_n}{s - p_n} \tag{15-47}$$

where $A_1, A_2, \ldots, A_n$, called *residues*, are constants, yet to be found. This stipulation makes a lot of sense: If we combine the right-hand side of Equation 15-47 under a common denominator with suitable values for the A's, we'll get Equation 15-46. Here, we have the reverse: Given the combined form (Equation 15-46) we want the expanded form (Equation 15-47).

There are two methods that can be used to calculate the residues. The first one is universal, since it applies to *all* cases. It can be stated briefly as follows: Equate Equations 15-46 and 15-47; then choose convenient values of s to obtain enough equations with the unknown A's. Solve these equations.

EXAMPLE 15-13 ──

$$F(s) = \frac{s - 1}{s(s + 2)(s + 3)} = \frac{A_1}{s} + \frac{A_2}{s + 2} + \frac{A_3}{s + 3}$$

Solution.   In choosing convenient values of s, stay away from the poles of $F(s)$, because you'll get ∞ on both sides—of little use.

Here, choosing $s = 1$ yields

$$0 = A_1 + \frac{A_2}{3} + \frac{A_3}{4}$$

With $s = -1$ we get

$$1 = -A_1 + A_2 + \frac{A_3}{2}$$

and $s = 2$ yields

$$\frac{1}{40} = \frac{A_1}{2} + \frac{A_2}{4} + \frac{A_3}{5}$$

These three equations can be solved by Gauss's method (Appendix A) to give

$$A_1 = -\tfrac{1}{6} \qquad A_2 = \tfrac{3}{2} \qquad A_3 = -\tfrac{4}{3}$$

Consequently

$$F(s) = \frac{-\frac{1}{6}}{s} + \frac{\frac{3}{2}}{s+2} + \frac{-\frac{4}{3}}{s+3}$$

and

$$f(t) = (-\tfrac{1}{6} + \tfrac{3}{2}e^{-2t} - \tfrac{4}{3}e^{-3t})u(t) \qquad \blacksquare$$

The second method starts also with equating the given $F(s)$ with its stipulated expansion

$$K\frac{P(s)}{(s-p_1)(s-p_2)\cdots(s-p_n)} = \frac{A_1}{s-p_1} + \frac{A_2}{s-p_2} + \cdots + \frac{A_n}{s-p_n} \qquad (15\text{-}48)$$

Multiply both sides of this equation by $(s - p_1)$, a valid algebraic step. We get

$$K\frac{P(s)}{(s-p_1)(s-p_2)\cdots(s-p_n)}(s-p_1)$$

$$= A_1 + \frac{A_2}{s-p_2}(s-p_1) + \cdots + \frac{A_n}{s-p_n}(s-p_1) \qquad (15\text{-}49)$$

where $(s - p_1)$ is canceled on the left-hand side. On the right-hand side, the residue A_1 stands free, and all the other terms are multiplied by $(s - p_1)$. Now let $s = p_1$ in Equation 15-49, with the result

$$K\frac{P(s)}{(s-p_2)(s-p_3)\cdots(s-p_n)}\bigg]_{s=p_1} = A_1 \qquad (15\text{-}50)$$

since all the other terms on the right go to zero. Equation 15-50 gives the value of A_1.
This method may be repeated for $A_2, A_3, \ldots, A_n$. It is summarized as follows:

$$A_k = F(s)(s - p_k)]_{s=p_k} \qquad (15\text{-}51)$$

In words (and in practice), we cover with our hand the factor $(s - p_k)$ in the left-hand side of Equation 15-48. This amounts to the cancellation in Equation 15-49. In what remains uncovered, we set $s = p_k$. The result is A_k. For this reason, let us dub it the "cover-up" method.

EXAMPLE 15-14 _____

For the same $F(s)$ as in Example 15-13, we have

$$A_1 = sF(s)\Big]_{s=0} = \frac{s-1}{(s+2)(s+3)}\Big]_{s=0} = -\frac{1}{6}$$

$$A_2 = (s+2)F(s)\Big]_{s=-2} = \frac{s-1}{s(s+3)}\Big]_{s=-2} = \frac{3}{2}$$

$$A_3 = (s+3)F(s)\Big]_{s=-3} = \frac{s-1}{s(s+2)}\Big]_{s=-3} = -\frac{4}{3}$$

as before.   ■

The "cover-up" method is easy to remember: In the expansion of Equation 15-47, a typical residue A_k is set "free" if we multiply throughout by its denominator $(s - p_k)$. Then all the other terms on the right will vanish for $s = p_k$, leaving only A_k. With some practice, the "cover-up" method is fast. Unlike the universal method, it applies only to simple poles and to one additional case (to be studied soon). With these limitations in mind, it is quite handy to use.

Probs. 15-22, 15-23

2. Multiple Real Poles   This case is more complicated than the previous one. Rather than derive and memorize lengthy formulas (something we never recommend), it is best to treat each problem on its own, based on some general principles that we'll outline. These are illustrated in the following example.

EXAMPLE 15-15 _____

Suppose

$$F(s) = \frac{s-2}{(s+1)(s+3)^2}$$

with a multiple pole ($k = 2$) at $s = -3$ and a simple pole at $s = -1$.

Solution.   The stipulated partial fraction expansion must be

$$F(s) = \frac{s-2}{(s+1)(s+3)^2} = \frac{A_1}{s+1} + \frac{A_2}{(s+3)^2} + \frac{A_3}{s+3}$$

where the factor $(s + 3)^2$ accounts for the last two terms. In other words, the common denominator $(s + 1)(s + 3)^2$ will accommodate the partial fraction $A_2/(s + 3)^2$ *as well as* $A_3/(s + 3)$. We cannot ignore, in advance, the last term. We must allow for it.

How do we calculate A_1, A_2, and A_3? The universal method is always valid. Three convenient values of s (but not $s = -1$ or $s = -3$) will give three equations for the A's. The "cover-up" method will work for A_1 and for A_2 only. Specifically,

$$A_1 = F(s)(s+1)\Big]_{s=-1} = -\frac{3}{4}$$

$$A_2 = F(s)(s+3)^2\Big]_{s=-3} = \frac{5}{2} \quad \text{(do it in detail!)}$$

But if we try it for A_3, we run into trouble. Normally, we would multiply through by $(s + 3)$ in order to free A_3; we get then

$$F(s)(s + 3) = \frac{s - 2}{(s + 1)(s + 3)} = \frac{A_1}{s + 1}(s + 3) + \frac{A_2}{s + 3} + A_3$$

Letting now $s = -3$ will cause trouble: Precisely because of the multiplicity of that pole, there remains a factor $(s + 3)$ in the denominators on both sides, leading to a division by zero. The "cover-up" method is good only for A_2, the partial fraction with the highest power of the multiple pole, but not for any of the partial fractions with its lower powers.

A nice algebraic method exists to handle *all* the A's in such cases. It is based on our previous study of simple poles only. Consider a *new* function

$$F_1(s) = \frac{s - 2}{(s + 1)(s + 3)}$$

which has only the simple poles of $F(s)$. We can expand $F_1(s)$ by the "cover-up" method

$$F_1(s) = \frac{\frac{5}{2}}{s + 3} + \frac{-\frac{3}{2}}{s + 1}$$

(verify it!). Next we divide $F_1(s)$ by $(s + 3)$. This restores the original $F(s)$, and we have from the previous step

$$\frac{F_1(s)}{s + 3} = F(s) = \frac{\frac{5}{2}}{(s + 3)^2} + \frac{-\frac{3}{2}}{(s + 3)(s + 1)}$$

Here the first partial fraction is good, with $A_2 = \frac{5}{2}$. The second term is expanded again by the "cover-up" method, to get the final result

$$F(s) = \frac{\frac{5}{2}}{(s + 3)^2} + \frac{\frac{3}{4}}{s + 3} + \frac{-\frac{3}{4}}{s + 1}$$

In summary, we create an auxiliary function with only simple poles, expand it, and restore the multiple pole recursively, step by step, by division.

The inversion of $F(s)$, using Table 14-2, yields

$$f(t) = (\tfrac{5}{2}te^{-3t} + \tfrac{3}{4}e^{-3t} - \tfrac{3}{4}e^{-t})u(t)$$

Probs. 15-24, 15-25, 15-26, 15-27   ∎

The last case that needs mentioning is a combination of real poles and complex conjugate poles. In principle, there is nothing new here that has not been covered already. In practice, however, there are several important points to consider.

EXAMPLE 15-16 ───────────────────────────────────

Let

$$F(s) = \frac{3s^2 + 5s + 19}{(s + 1)(s^2 + 4s + 20)}$$

having simple poles at $p_1 = -1$, $p_2 = -2 + j4$, $p_3 = p_2^* = -2 - j4$.

Solution. The most efficient expansion is

$$F(s) = \frac{A_1}{s + 1} + \frac{A_2 s + A_3}{s^2 + 4s + 20}$$

where the quadratic is left intact, as we saw earlier, ready for completing the square. Its numerator is assumed as $(A_2 s + A_3)$, a polynomial of one degree lower than the denominator.

How to find A_1, A_2, and A_3? With a combination of everything that we know. The "cover-up" method yields A_1 quickly,

$$A_1 = (s+1)F(s)\bigg]_{s=-1} = \frac{3s^2 + 5s + 19}{s^2 + 4s + 20}\bigg]_{s=-1} = 1$$

For A_2 and A_3, use the universal method. Choosing $s = 0$ we have

$$F(0) = \frac{19}{20} = \frac{1}{0+1} + \frac{A_2 \cdot 0 + A_3}{0 + 0 + 20}$$

that is,

$$A_3 = -1$$

Another choice, $s = 1$, yields

$$F(1) = \frac{27}{(2)(25)} = \frac{1}{1+1} + \frac{A_2 - 1}{1 + 4 + 20}$$

or

$$A_2 = 2$$

Therefore

$$F(s) = \frac{1}{s+1} + \frac{2s-1}{s^2 + 4s + 20}$$

This is it, as far as partial fractions are concerned. The first term is good. The second term is ready for completing the square. Let's do it.

$$F(s) = \frac{1}{s+1} + \frac{2s-1}{(s+2)^2 + 4^2} = \frac{1}{s+1} + \frac{2(s+2) - 5}{(s+2)^2 + 4^2}$$

$$= \frac{1}{s+1} + \frac{2(s+2)}{(s+2)^2 + 4^2} - \frac{5}{4}\frac{4}{(s+2)^2 + 4^2}$$

Finally,

$$\mathscr{L}^{-1}F(s) = (e^{-t} + 2e^{-2t}\cos 4t - \tfrac{5}{4}e^{-2t}\sin 4t)u(t)$$

Probs. 15-28,
15-29, 15-30

*15-6 INITIAL VALUE AND FINAL VALUE

Among the many calculations involved with $F(s)$, two are particularly useful. Without ever inverting $F(s)$, we can calculate easily the initial value of $f(t)$, $f(0^+)$, and the final value of $f(t)$, $f(\infty)$. In other words, from $F(s)$ in the s domain, we can find how the time function starts ($t = 0^+$) and how it ends ($t = \infty$).

1. The Initial Value Theorem Consider the defining equation for the Laplace transform of df/dt, repeated here

$$\mathscr{L}\frac{df}{dt} = \int_{0-}^{\infty} f'e^{-st}\,dt = sF(s) - f(0^-) \qquad (15\text{-}52)$$

Now, let $s \to \infty$ in Equation 15-52. Then $e^{-st} = 0$ and

$$\lim_{s \to \infty} \int_{0-}^{\infty} f' e^{-st} \, dt = \lim_{s \to \infty} [sF(s) - f(0^-)] \qquad (15\text{-}53)$$

Since the integrand on the left-hand side is zero, we have

$$0 = \lim_{s \to \infty} sF(s) - f(0^-) \qquad (15\text{-}54)$$

or

$$f(0^-) = \lim_{s \to \infty} sF(s) \qquad (15\text{-}55)$$

If $f(t)$ is continuous at $t = 0$, then $f(0^+) = f(0^-)$. Even if $f(t)$ has a step discontinuity (jump), Equation 15-55 holds. In summary, then,

$$f(0^+) = \lim_{s \to \infty} sF(s) \qquad (15\text{-}56)$$

gives the initial value of $f(t)$ directly from $F(s)$.

EXAMPLE 15-17 ─────────────────────────────────

Let

$$F(s) = \frac{1}{s}$$

and pretend we don't know $f(t)$.

Solution. The initial value theorem gives

$$f(0^+) = \lim_{s \to \infty} s \cdot \frac{1}{s} = 1$$

which is true, of course: The unit step $u(t)$ is 1 at $t = 0^+$. ■

EXAMPLE 15-18 ─────────────────────────────────

Take $F(s)$ from Example 15-15,

$$F(s) = \frac{s-2}{(s+1)(s+3)^2} = \frac{s-2}{s^3 + 6s^2 + 15s + 9}$$

Solution. Without ever finding $f(t)$, we have its initial value

$$f(0^+) = \lim_{s \to \infty} sF(s) = \lim_{s \to \infty} \frac{s^2 - 2s}{s^3 + 6s^2 + 15s + 9} = 0$$

This limit is zero by l'Hôpital's rule. In fact, we didn't even have to multiply out fully the denominator of $F(s)$. By inspection

$$\lim_{s \to \infty} sF(s) = \lim_{s \to \infty} \frac{s^2 + \cdots}{s^3 + \cdots} = \lim_{s \to \infty} \frac{1}{s} = 0$$

where we retained the leading terms in the numerator and denominator for $s \to \infty$.

As a verification, we found $f(t)$ in Example 15-15:

$$f(t) = (\tfrac{5}{2} te^{-3t} + \tfrac{3}{4} e^{-3t} - \tfrac{3}{4} e^{-t})u(t)$$

and so

$$f(0^+) = 0 + \tfrac{3}{4} - \tfrac{3}{4} = 0 \qquad\blacksquare$$

The initial value theorem (Equation 15-56) may be used repeatedly to find $f'(0^+)$, $f''(0^+)$, etc. If we call $f'(t)$ by $g(t)$ for a moment, then Equation 15-56 says

$$g(0^+) = \lim_{s \to \infty} sG(s) \qquad (15\text{-}57)$$

that is

$$f'(0^+) = \lim_{s \to \infty} s[sF(s) - f(0^+)] \qquad (15\text{-}58)$$

EXAMPLE 15-19

The trajectory (distance) of the DBBM (ding bat ballistic missile) has been designed and calculated as

$$X(s) = \frac{40s^2 + 900s + 1000}{s^4 + 20s^3 + 60s^2 + 100s + 200}$$

with the units of miles and seconds having been considered. Calculate its distance, velocity, and acceleration at the instant it is launched.

Solution. We have from Equation 15-56 the initial distance

$$x(0^+) = \lim_{s \to \infty} sX(s) = \lim_{s \to \infty} \frac{40s^3 + \cdots}{s^4 + \cdots} = 0$$

which is hardly surprising. The initial velocity of the DBBM is found from Equation 15-58 as

$$\left.\frac{dx(t)}{dt}\right]_{t=0^+} = x'(0^+) = \lim_{s \to \infty} s[sX(s) - x(0^+)]$$

$$= \lim_{s \to \infty} \frac{40s^4 + \cdots}{s^4 + \cdots} = 40 \text{ miles/s}$$

The initial acceleration is found from the extension of Equation 15-58, namely

$$\left.\frac{d^2x(t)}{dt^2}\right]_{t=0^+} = x''(0^+) = \lim_{s \to \infty} s[s^2 X(s) - sx(0^-) - x'(0^+)]$$

$$= \lim_{s \to \infty} s\left[s^2 \frac{40s^2 + 900s + 1000}{s^4 + 20s^3 + 60s^2 + 100s + 200} - 40 \right]$$

$$= \lim_{s \to \infty} s\left[\frac{100s^3 - 1400s^2 - 4000s - 8000}{s^4 + 20s^3 + 60s^2 + 100s + 200} \right]$$

$$= \lim_{s \to \infty} \frac{100s^4 + \cdots}{s^4 + \cdots} = 100 \text{ miles/s}^2$$

This is quite a lot of useful information about $x(t)$ without having $x(t)$! $\qquad\blacksquare$

2. The Final Value Theorem   If we take the limit of Equation 15-52 as $s \to 0$, we get

$$\lim_{s \to 0} \int_{0^-}^{\infty} f'(t)e^{-st}\, dt = \lim_{s \to 0} [sF(s) - f(0^-)] \qquad (15\text{-}59)$$

since $e^{-st} = 1$ as $s \to 0$, the left-hand side becomes

$$\int_{0^-}^{\infty} f'(t)\, dt = f(t) \Big]_{0^-}^{\infty} = \lim_{t \to \infty} f(t) - f(0^-) \qquad (15\text{-}60)$$

Equations 15-59 and 15-60 yield the result

$$\lim_{t \to \infty} f(t) = \lim_{s \to 0} sF(s) \qquad (15\text{-}61)$$

which gives the final value of $f(t)$ in terms of $F(s)$.

This result is valid *provided* all the poles of $sF(s)$ have negative real parts. This provision will be clarified in the examples that follow.

EXAMPLE 15-20 ────────────────────────────────

$$F(s) = \frac{1}{s + a} \qquad a > 0$$

$$\therefore \quad \lim_{s \to 0} sF(s) = \lim_{s \to 0} \frac{s}{s + a} = 0$$

As a verification

$$f(t) = e^{-at}u(t) \qquad a > 0$$

and its final value is indeed

$$\lim_{t \to \infty} f(t) = 0$$ ∎

EXAMPLE 15-21 ────────────────────────────────

$$F(s) = \frac{\beta}{s^2 + \beta^2}$$

if we apply Equation 15-61, it would seem that

$$\lim_{t \to \infty} f(t) = \lim_{s \to 0} s\, \frac{\beta}{s^2 + \beta^2} = 0$$

However, the function $sF(s)$ has two poles, $p_1 = j\beta$, $p_2 = -j\beta$, both with a nonnegative (here zero) real part. The final value theorem (Equation 15-61) cannot be applied. As a verification, we have

$$\lim_{t \to \infty} f(t) = \lim_{t \to \infty} \sin \beta t = ?$$

The cited restriction on the poles of $sF(s)$ guarantees the existence of

$$\lim_{t \to \infty} f(t).$$

Probs. 15-31, 15-32, 15-33, ∎ 15-34, 15-35

As a concluding remark, we notice the beautiful symmetry in these two theorems. In one, the behavior of $f(t)$ for *small* values of t is the same as the behavior of $sF(s)$ for *large* values of s. In the other, the behavior of $f(t)$ for *large* values of t is the same as the behavior of $sF(s)$ for *small* values of s.

15-7 THE LOWER LIMIT: 0^- OR 0^+?

To wrap up our study of the Laplace transform, let us consider a somewhat fine point: What if the lower limit on the integral defining the Laplace transform is 0^+ instead of 0^-? That is

$$\mathcal{L}_+ f(t) = \int_{0^+}^{\infty} f(t)e^{-st} = F(s) \tag{15-62}$$

is another legitimate definition of the Laplace transform. We would have then, for instance,

$$\mathcal{L}_+ f'(t) = sF(s) - f(0^+) \tag{15-63}$$

but if $f(0^+) \neq f(0^-)$, as we saw in several cases, then, in order to use Equation 15-63, we would have to solve first a subproblem: Given $f(0^-)$ in a circuit, calculate $f(0^+)$. We did this in Chapter 7.

As far as the Laplace transform is concerned, either lower limit (0^- or 0^+) is valid, provided we are consistent with it throughout all the steps. It is much easier, though, to work with 0^- as the lower limit, because *this choice gives automatically the correct answers* even when $f(0^-) \neq f(0^+)$. Let us consider the following example.

EXAMPLE 15-22 _____

In the circuit shown in Figure 15-11, the capacitor is initially uncharged, $v_C(0^-) = 0$. At $t = 0$, a unit impulse current excites the circuit.

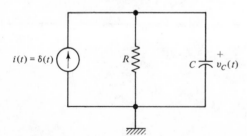

Figure 15-11   Example 15-22.

Solution.   The differential equation for this circuit is KCL

$$C\frac{dv_C(t)}{dt} + \frac{1}{R} v_C(t) = \delta(t)$$

Let us apply the Laplace transform, using 0^- as its lower limit. The result is

$$C[sV_C(s) - v_C(0^-)] + \frac{1}{R} V_C(s) = 1$$

and since $v_C(0^-) = 0$, the transform equation becomes

$$\left(Cs + \frac{1}{R}\right)V_C(s) = 1$$

and its inversion gives

$$v_C(t) = \frac{1}{C}\,e^{-t/RC}u(t)$$

If we want to use 0^+ as the lower limit, we have

$$C[sV_C(s) - v_C(0^+)] + \frac{1}{R}\,V_C(s) = 0$$

The right-hand side of this equation is zero

$$\mathscr{L}_+\,\delta(t) = \int_{0^+}^{\infty} \delta(t)e^{-st}\,dt = 0$$

because $\delta(t)$ occurs at $t = 0$, *outside* the limits of the integral. Now we have the subproblem of evaluating $v_C(0^+)$. Since $v_C(0^-) = 0$, the capacitor acts as a short circuit and the current source $i(t) = \delta(t)$ amounts to a unit step of charge (1 C) applied instantly to the capacitor. Therefore, $q_C(0^+) = 1$ and

$$v_C(0^+) = \frac{1}{C}\,q_C(0^+) = \frac{1}{C}$$

Using this value in the transformed equation, we get

$$C\left[sV_C(s) - \frac{1}{C}\right] + \frac{1}{R}\,V_C(s) = 0$$

that is,

$$\left(Cs + \frac{1}{R}\right)V_C(s) = 1$$

precisely as before! ∎

Since the evaluation of initial conditions at $t = 0^+$ is an additional task and can be quite involved (see Chapter 7), it is preferable to use the Laplace transform with 0^- as the lower limit. We will continue to do so.

Prob. 15-36

PROBLEMS

Note: When initial conditions are not given, assume them to be zero.

15-1 For the network shown:
   (a) Write the loop equations in the time domain. (How many node equations are needed?)
   (b) Transform them and arrange in final matrix form.
   (c) Solve for the total response of $I_1(s)$. Express it as a rational function.
   (d) In part (c), identify clearly the zero-input part and the zero-state part. This requires a simple but meticulous tracking of the various terms from steps (a), (b) and (c).

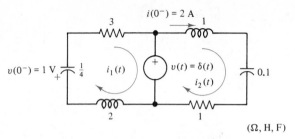

Problem 15-1

15-2 For the network shown:

(a) What is its order n?

(b) Write the loop equations.

(c) Transform them and arrange in final matrix form.

(d) Calculate the characteristic values of the network. Check with (a).

Note the current-dependent voltage source.

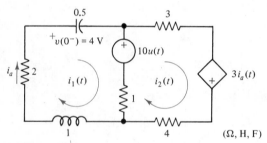

Problem 15-2

15-3 For the network shown, repeat parts (a), (b), and (c) of Problem 15-2, then find its characteristic polynomial.

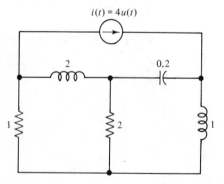

(Ω, H, F) **Problem 15-3**

15-4 Solve Example 15-2 by loop analysis, in the following steps:
  (a) Set up the two loop equations.
  (b) Transform them.
  (c) Write the expression for the zero-input response of $v_2(t)$, transform it and use part (b) to get $V_2(s)$ in its final form as a rational function. It must be the same as in Example 15-2.

15-5 Rework Example 15-2 with the following change: The voltage source is current-controlled

$$v(t) = 6i_a(t)$$

with $i_a(t)$ the current through the 0.5-Ω resistor, referenced downward. In particular, note how this dependent source affects: (1) the matrix of the coefficients and, hence, the characteristic equation; (2) the right-hand side matrices, affecting the zero-state and zero-input responses.

15-6 (a) Prove in detail Equation 15-24. *Hint*: Review the corresponding resistive case and the sinusoidal case.
  (b) Prove in detail Equation 15-25.
  (c) Prove: For *two* impedances in parallel

$$Z_t(s) = Z_1(s) \| Z_2(s) = \frac{Z_1(s)Z_2(s)}{Z_1(s) + Z_2(s)}$$

*15-7 Equations 15-24 and 15-25 reduce to the sinusoidal (phasor) case if $s = j\omega$, where ω is the single frequency of the source. For what value of s are these equations reduced to the resistive (dc) case of Chapter 4?

15-8 Obtain in its final (rational function) form the driving-point admittance $Y_t(s)$ of the network shown.

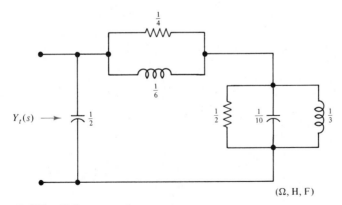

(Ω, H, F)

Problem 15-8

15-9   Calculate $Z_t(s)$ if $R = \sqrt{L/C}$. Comment on your result.

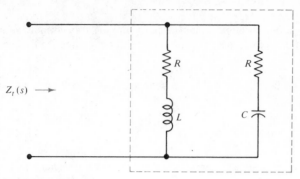

Problem 15-9

15-10   Draw the transform circuit diagram for Problem 15-1 and, from it, write the loop equations in their final form.

15-11   Draw the transform circuit diagram for Problem 15-2 and, from it, write the loop equations in their final matrix form.

15-12   Derive carefully the transform circuit diagram for two mutually coupled inductors, L_1, L_2, and M, with initial currents $i_{L_1}(0^-)$, $i_{L_2}(0^-)$. Use your results to rework Example 15-1 by the transform diagram method.

15-13   Rework Example 15-2 by loop analysis, using the transform circuit diagram. Notice the entries in its loop impedance matrix $\mathbf{Z}(s)$: Are they predictable? Is $z_{12}(s) = z_{21}(s)$? Notice also the source matrix $\mathbf{E}(s)$. Are all its entries predictable? Explain.

15-14   Check the dimensions (units) of each term in Equations 15-26, 15-27, and 15-28. *Hint*: If you run into trouble, go to the beginning and check the units of $I(s) = \mathscr{L}i(t)$, $V(s) = \mathscr{L}v(t)$. Don't take anything for granted!

15-15   Rework Example 15-2 by node analysis with the transform diagram drawn directly from Figure 15-2. Compare your results with those in Example 15-2.

15-16   Use node analysis with a transform diagram in Example 15-3, to obtain $I(s)$ and $Z_t(s)$. Without any initial conditions or dependent sources, $\mathbf{Y}(s)$ can be written by inspection from the original circuit.

*15-17   Consider a one-port (1-P) network, consisting only of R, L, M, C elements. If series-parallel combinations are not feasible, we can find its driving-point admittance $Y_{dp}(s)$ by exciting it with an arbitrary $V_{dp}(s)$, calculating the zero-state response $I_{dp}(s)$, and then

$$Y_{dp}(s) = \frac{I_{dp}(s)}{V_{dp}(s)}$$

(a) Prove that

$$Y_{dp}(s) = \frac{\det_{11}\mathbf{Z}}{\det \mathbf{Z}}$$

where $\det \mathbf{Z}$ is the determinant of the loop impedance matrix; here $I_{dp}(s)$ is in the first (driving-point) loop current; all other loop currents are inside the one-port network. The first cofactor is $\det_{11}\mathbf{Z}$ (review Appendix A). The two determinants, $\det \mathbf{Z}$ and $\det_{11}\mathbf{Z}$, can be written by inspection. *Hint*: Write the loop equations

(Equations 15-29). What is $E(s)$ here? Solve these equations for $I_{dp}(s)$. Take the ratio I_{dp}/V_{dp}.

(b) Illustrate this procedure on the one-port shown to obtain

$$Y_{dp}(s) = \frac{3s^3 + 9s^2 + 11s + 12}{21s^2 + 14s + 36}$$

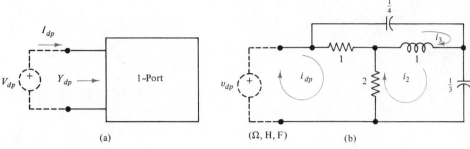

(a)

(Ω, H, F)

(b)

Problem 15-17

15-18 A capacitor C_1 is initially charged to $v_{C_1}(0^-) = V_0$ volts. At $t = 0$, it is connected through a series resistor R to an uncharged capacitor C_2.

(a) Calculate the current $i(t)$ and plot it versus t for two values of R, R_1, and $R_2 > R_1$.

(b) What is the total charge transferred to C_2?

(c) What becomes of $i(t)$ as $R \to 0$? Justify mathematically and explain physically.

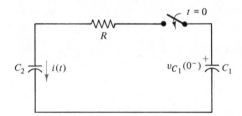

Problem 15-18

15-19 Calculate the output voltage $V_0(s)$ in the op-amp circuit shown. Leave your answer as a rational function.

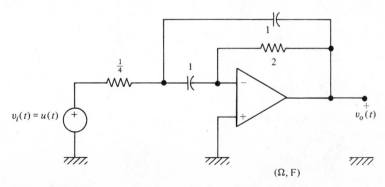

(Ω, F)

Problem 15-19

15-20 Find $\mathscr{L}^{-1}F(s)$ for each function:

(a) $\dfrac{-2s + 3}{3s + 2}$

(b) $3\,\dfrac{s^2}{s + 10}$

(c) $\dfrac{1 - 2s}{s + 3}$

(d) $\dfrac{-4}{s^2 + 4s + 4}$

(e) $\dfrac{s + 1}{s^2 + 6s + 25}$

(f) $\dfrac{-2s^2 - 10s + 4}{s^2 + s + 2}$

(g) $\dfrac{2s - 3}{(s + 1)^2}$

(h) $\dfrac{2s - 1/s}{s^2}$

(i) $\dfrac{s + 2}{s^2 + 4}$ Write $f(t)$ as $A \cos(\beta t + \theta)$

(j) $\dfrac{2s - 1}{s^2 + 100}$ Write $f(t)$ as $A \cos(\beta t + \theta)$

15-21 List the finite zeros and poles of each $F(s)$ in Problem 15-20.

15-22 Expand by partial fractions and find $f(t)$ for each $F(s)$. Write a short program to do the partial fractions here, in Problems 15-25 and 15-26, and for later use.

(a) $3.2\,\dfrac{s + 1}{s(s + 10)}$

(b) $\dfrac{1}{s(s + 1)(s + 2)}$

(c) $\dfrac{-3s + 4}{s^2 + 110s + 1000}$

(d) $\dfrac{-s - 2}{s^2 + 8s + 12}$ (carefully!)

(e) $4\,\dfrac{s - 2}{(s + 2)(s + 4)(s + 6)}$

(f) $\dfrac{-s + 4}{s^3 + 6.6s^2 + 13.31s + 8}$

(g) $\dfrac{3s - 4}{2s^3 + 12.2s^2 + 22.8s + 12.6}$

(h) $\dfrac{2s^2 - 3}{2s^2 + 4s + 10}$

15-23 Use carefully the "cover-up" method in Example 15-10 for

$$F(s) = \frac{3s - 1}{s^2 + 8s + 18} = \frac{3s - 1}{(s + 4 + j\sqrt{2})(s + 4 - j\sqrt{2})}$$

with the two simple poles $p_1 = -4 - j\sqrt{2}$, $p_2 = -4 + j\sqrt{2}$. Your $f(t)$ must finally agree with the one obtained there. Why is it better to complete the square when poles are complex numbers? *Hint*: In working this problem, you'll have to use Euler's identity when you write

$$e^{(a + jb)t} = e^{at}e^{jbt} = e^{at}(\cos bt + j \sin bt)$$

15-24 Given

$$V(s) = \frac{1}{s(s + 1)^2}$$

calculate $v(t)$ and plot it accurately. Observe the plot of each term in $v(t)$.

15-25 From

$$F(s) = \frac{2s + 3}{(s + 1)^2(s + 2)^2}$$

obtain $f(t)$.

15-26 Use the algebraic-recursive method in Example 15-15 to expand

$$F(s) = \frac{s(s - 2)}{(s + 4)^2(s + 6)}$$

Be careful: The auxiliary function $F_1(s)$ will be an improper rational function. Don't forget to divide longhand.

*15-27 There is yet another method for real multiple poles. Let us outline it using Example 15-15. We start with

$$F(s) = \frac{s - 2}{(s + 1)(s + 3)^2} = \frac{A_1}{s + 1} + \frac{A_2}{(s + 3)^2} + \frac{A_3}{s + 3}$$

and multiply by the highest power of the multiple pole, $(s + 3)^2$. We get

$$F(s)(s + 3)^2 = \frac{A_1}{s + 1}(s + 3)^2 + A_2 + A_3(s + 3)$$

If we let $s = -3$, we get A_2, as in the "cover-up" method. To free A_3, we *differentiate* both sides with respect to s, then set $s = -3$. Do this step and check your results with Example 15-15. *Note*: Differentiation of rational functions is unpleasant, at the very least. Remember

$$\left(\frac{u}{v}\right)' = \frac{vu' - uv'}{v^2}$$

and with successive differentiations it quickly becomes a mess. However, it *is* a legitimate method.

15-28 Invert

(a) $F(s) = \dfrac{1}{4} \dfrac{s^2 - 3}{(s + 2)(s^2 + 1)}$

(b) $F(s) = \dfrac{s}{(s + 1)(s^2 + s + 1)}$

15-29 Complete Example 15-1 by finding $i_2(t) = \mathscr{L}^{-1}I_2(s)$, with $I_2(s)$ as calculated there.

15-30 Complete Example 15-2 by finding $v_1(t) = \mathscr{L}^{-1}V_1(s)$ with $V_1(s)$ as calculated there.

15-31 (a) Apply the initial value theorem to $F(s)$ in Example 15-16, *then* verify by using $f(t)$ as calculated there.

   (b) Apply the initial value theorem to $F(s)$ in Example 15-16 to calculate $f'(0^+)$, *then* verify by calculating $f'(t)$ from the given $f(t)$.

   (c) Repeat, for $f''(0^+)$.

15-32 Apply the final value theorem to Examples 15-15 and 15-16. In each case, verify with $f(t)$.

15-33 Apply the final value theorem, when possible. If not possible, explain why and demonstrate by calculating $f(t)$:

(a) $F(s) = \dfrac{s - 1}{s(s + 2)}$

(b) $F(s) = \dfrac{s + 4}{s^2 - 4}$

(c) $F(s) = \dfrac{s - 2}{s^2 + 2s + 2}$

(d) $F(s) = \dfrac{s - 2}{s^2 - 2s + 2}$

15-34 Calculate the range (final distance) of the DBBM in Example 15-19. What must you check first?

15-35 You are about to plot accurately $f(t)$ for which you know

$$F(s) = \frac{1}{(s + 3)^2}$$

To help you with the plot, calculate $f(0^+)$, the initial slope $f'(0^+)$, and the initial $f''(0^+)$—some good ol' analytic geometry—without ever finding $f(t)$.

15-36 Calculate the total response $v_1(t)$ and $v_2(t)$ in Example 15-8.

Chapter 16

Generalized Theorems and Properties of Networks

As mentioned in the introduction to the Laplace transform, this method provides not only a powerful algebraic tool of analysis; several important properties of networks can be derived from the Laplace transform, as well as generalized network theorems.

16-1 THE NETWORK FUNCTION

The input-output relations in the s domain are algebraic, as we know by now. To characterize a network more specifically, we concentrate on its zero-state response only and define a *network function* $H(s)$ as follows:

$$R(s) = H(s)E(s) \qquad (16\text{-}1)$$

where $E(s)$ is the transformed input (excitation), and $R(s)$ is the transformed zero-state response.

In a block diagram, shown in Figure 16-1, we can think of the network as a signal processor (which it actually is): The input $E(s)$ is processed by the network through a multiplication by $H(s)$ to produce the zero-state response $R(s)$. It is also easy to appreciate the reason for considering only the zero-state and not the total response. Initial conditions are arbitrary, not characteristic of the network itself. Therefore, in calculating $H(s)$ we set them all to zero.

We had the resistive version of a network function in Chapter 5, Equation 5-19. There it was always a *purely real number*. A purely resistive network, we saw, can only scale (in magnitude) the input in producing the output, but it cannot change the waveform of the input.

A more general version of a network function was studied in Chapters 9 and 10. For a single-frequency pure sinusoidal input, the phasor analysis involves a network

Figure 16-1   The network function.

function which is a *complex number*. When it multiplies the phasor input, it affects both magnitude and phase angle of the output, a sinusoid of the same frequency.

Still a more general network function, $H(s)$, is a *rational function*. In a linear network, it relates an arbitrary waveform input to an arbitrary waveform output, as in Equation 16-1. These ideas are summarized in Figure 16-2. For the resistive case, no transform is needed, and we have as in Equation 5-19

$$r(t) = He(t) = Ke(t) \tag{16-2}$$

For the sinusoidal steady state, the input is $e(t) = A \cos (\omega_0 t + \alpha)$, whose phasor transform is $\mathbf{E} = A\underline{/\alpha}$. The network function is also a complex number, $\mathbf{H} = H\underline{/\theta}$. The phasor output is

$$\mathbf{R} = \mathbf{HE} = (H\underline{/\theta})(A\underline{/\alpha}) = HA\underline{/\theta + \alpha} \tag{16-3}$$

and its inverse phasor transform is the steady-state output

$$r(t) = HA \cos (\omega_0 t + \theta + \alpha) \tag{16-4}$$

For general waveforms, the Laplace transform yields the zero-state response

$$R(s) = H(s)E(s) \tag{16-5}$$

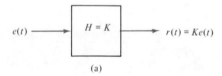

(a)

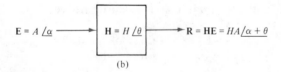

(b)

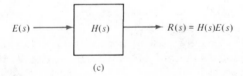

(c)

Figure 16-2   Resistive, sinudoidal, and general network functions.

and

$$r(t) = \mathcal{L}^{-1}R(s) = \mathcal{L}^{-1}H(s)E(s) \tag{16-6}$$

The inversion of the product $H(s)E(s)$ will be studied in detail later in this chapter.

EXAMPLE 16-1 _____

A driving-point impedance is a network function, because, at the two driving-point terminals of the network, we have

$$V_{dp}(s) = Z_{dp}(s)I_{dp}(s)$$

as shown in Figure 16-3(a). This equation is one specific form of Equation 16-1. Here $R(s) = V_{dp}(s)$, $E(s) = I_{dp}(s)$, and $H(s) = Z_{dp}(s)$.

Similarly, a driving-point admittance is a network function, as seen in Figure 16-3(b). Here

$$R(s) = I_{dp}(s) \qquad E(s) = V_{dp}(s) \qquad H(s) = Y_{dp}(s)$$

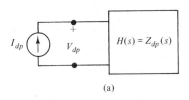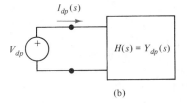

(a)   (b)

Figure 16-3 Example 16-1.   ■

In this example we have, for the same network, $Y_{dp}(s) = 1/Z_{dp}(s)$. This is a very special and restricted class of network functions. In general, the reciprocal of a network function *is not* another network function, as illustrated in the next example.

EXAMPLE 16-2 _____

For the network shown in Figure 16-4, with the indicated input and output, calculate the network function $H(s)$.

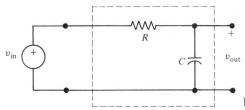

Figure 16-4 Example 16-2.

Solution. Here we identify $E(s) = V_{in}(s)$ and $R(s) = V_{out}(s)$. With zero initial conditions, we can easily see the transform diagram of the circuit in our mind's eyes. We write

$$V_{out}(s) = \frac{1}{Cs} \frac{V_{in}(s)}{R + 1/Cs}$$

which is nothing more than a voltage divider equation. Clearing fractions, we get

$$V_{out}(s) = \frac{1}{RCs + 1} V_{in}(s)$$

with the appropriate network function

$$H(s) = \frac{1}{RCs + 1}$$

Its reciprocal, $RCs + 1$, is not a network function (even if you try, in vain, to turn the network around). ■

The systematic approach to calculate a network function is apparent from these examples. The steps are the same as for the resistive and for the phasor cases:

1. Identify the input $E(s)$.
2. Identify the output $R(s)$.
3. Verify that inside the network there are no independent sources other than the given input.
4. Set all initial conditions to zero.
5. By methods of analysis studied in the previous chapter, calculate $R(s)$, the transform output. Arrange your final answer in the form of Equation 16-1 to exhibit the appropriate network function.

Depending on the nature of the input and the output, network functions can be classified as follows:

Input $E(s)$	Output $R(s)$	Network function	Symbol
Current	Voltage	Impedance	$Z(s)$
Current	Current	Current transfer	$\alpha(s)$
Voltage	Current	Admittance	$Y(s)$
Voltage	Voltage	Voltage transfer	$G(s)$

The symbols $Z(s)$, $\alpha(s)$, $Y(s)$, and $G(s)$ will be used for those specific network functions. The symbol $H(s)$ is a generic symbol for general purposes.

Two additional classifications must be made about network functions:

1. The input $E(s)$ and the output $R(s)$ are at the same port. In this case, the network function is *driving-point* by nature. We saw such examples in $Z_{dp}(s)$ and $Y_{dp}(s)$.
2. The input $E(s)$ is at a different port than the output $R(s)$. See Figure 16-5. Then we have a *transfer* network function. Quite obviously, $G(s)$ and $\alpha(s)$ are always transfer functions: It makes little sense to ask, "What is the voltage response across the terminals of a voltage input?" However, impedances and admittances may be transfer functions, as well as dp (driving-point) functions.

Figure 16-5 (a) Driving-point network functions. (b) Transfer network functions.

EXAMPLE 16-3

The network function in Example 16-2 is a voltage transfer function

$$G(s) = \frac{1}{RCs + 1}$$

Probs. 16-1, 16-2, 16-3, 16-4, 16-5

To relate our discussion here to our previous studies, consider the following example.

EXAMPLE 16-4

In the network shown in Figure 16-6(a), we are interested in the zero-state response $i(t)$. The transform circuit diagram for this case is shown in Figure 16-6(b) and you should practice seeing it directly from Figure 16-6(a).

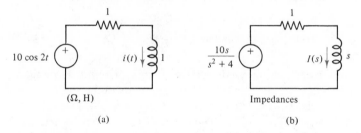

(a)

(b)

Figure 16-6 Example 16-4.

Solution. We write immediately

$$I(s) = \frac{10s}{(s + 1)(s^2 + 4)} = \frac{1}{s + 1} \cdot \frac{10s}{s^2 + 4}$$

The appropriate network function is

$$Y_{dp}(s) = \frac{1}{s + 1}$$

The partial fraction expansion of $I(s)$ yields

$$I(s) = \frac{-2}{s + 1} + \frac{2s + 8}{s^2 + 4}$$

(check it!) and its inversion gives the final answer

$$i(t) = -2e^{-t}u(t) + (2 \cos 2t + 4 \sin 2t)u(t)$$

showing the transient part of the zero-state response, $-2e^{-t}u(t)$, and the steady-state part, $(2\cos 2t + 4\sin 2t)u(t)$.

That steady-state part, *and only it*, may be also obtained by the phasor transform. We have, from Figure 16-6(a) and with $\omega = 2$

$$\mathbf{I} = \frac{10\underline{/0^\circ}}{1 + j2} = \left(\frac{1}{1 + j2}\right)10\underline{/0^\circ}$$

showing clearly the relations in Equation 16-3. Here

$$\mathbf{R} = \mathbf{I} \qquad \mathbf{E} = 10\underline{/0^\circ} \qquad \mathbf{H} = \frac{1}{1 + j2} = 0.447\underline{/-63.4^\circ}$$

Therefore

$$\mathbf{I} = (0.447\underline{/-63.4^\circ})(10\underline{/0^\circ}) = 4.47\underline{/-63.4^\circ}$$

The inverse phasor transform of $\mathbf{I}$ yields

$$i(t) = 4.47\cos(2t - 63.4^\circ) = 4.47(\cos 2t \cos 63.4^\circ + \sin 2t \sin 63.4^\circ)$$

$$= 2\cos 2t + 4\sin 2t$$

as before.   ■

One immediate and important observation to make is the following

$$H(s)\Big]_{s=j\omega_0} = H(j\omega_0) = \mathbf{H} \qquad\qquad (16\text{-}7)$$

that is, the network function $H(s)$ becomes the phasor (complex number) network function $\mathbf{H}$ when we set $s = j\omega_0$, where ω_0 is the radian frequency of the source in the circuit. In the previous example, $H(s) = 1/(s + 1)$ and $H(j2) = 1/(j2 + 1) = \mathbf{H}$.

A second important observation is: *The denominator of $H(s)$ is the characteristic polynomial of the network*. Or, the poles of $H(s)$ are the natural frequencies of the network.†

To demonstrate this statement, let us consider the general approach to finding $H(s)$, as shown in Figure 16-7. The input is, say, $V_1(s)$, and the desired zero-state output is the branch current $I_k(s)$. Let us assume that loop analysis is performed. As shown, $I_k(s) = I_p(s) - I_q(s)$, the difference of two of the unknown loop currents.

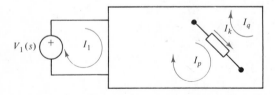

Figure 16-7   Loop analysis for $H(s)$.

† We assume that there is no cancellation of common factors between the numerator and the denominator of $H(s)$.

The loop equations are, as in Equation 15-29,

$$\begin{bmatrix} z_{11}(s) & z_{12}(s) & \cdots & z_{1l}(s) \\ z_{21}(s) & z_{22}(s) & \cdots & z_{2l}(s) \\ \cdots\cdots\cdots\cdots\cdots\cdots\cdots \\ z_{p1}(s) & z_{p2}(s) & \cdots & z_{pl}(s) \\ \cdots\cdots\cdots\cdots\cdots\cdots\cdots \\ z_{q1}(s) & z_{q2}(s) & \cdots & z_{ql}(s) \\ \cdots\cdots\cdots\cdots\cdots\cdots\cdots \\ z_{l1}(s) & z_{l2}(s) & \cdots & z_{ll}(s) \end{bmatrix} \begin{bmatrix} I_1(s) \\ I_2(s) \\ \vdots \\ I_p(s) \\ \vdots \\ I_q(s) \\ \vdots \\ I_l(s) \end{bmatrix} = \begin{bmatrix} V_1(s) \\ 0 \\ \vdots \\ 0 \\ \vdots \\ 0 \\ \vdots \\ 0 \end{bmatrix} \qquad (16\text{-}8)$$

Solution by Cramer's rule gives

$$I_p(s) = \frac{\begin{vmatrix} z_{11}(s) & \cdots & V_1(s) & \cdots & z_{1l}(s) \\ z_{21}(s) & \cdots & 0 & \cdots & z_{2l}(s) \\ \cdots\cdots\cdots\cdots\cdots\cdots\cdots\cdots\cdots \\ z_{l1}(s) & \cdots & 0 & \cdots & z_{ll}(s) \end{vmatrix}}{\det \mathbf{Z}(s)} \qquad (16\text{-}9)$$

As usual, the denominator is the determinant of the matrix of the coefficients, here the loop impedance matrix $\mathbf{Z}(s)$. In the numerator, we have the same determinant, but the pth column is replaced by the right-hand side. Expanding the numerator determinant in Equation 16-9 by the elements of its pth column we have

$$I_p(s) = \frac{1}{\det \mathbf{Z}(s)} \left[V_1(s) \cdot \det_{1p} \mathbf{Z}(s) + 0 + 0 + \cdots + 0 \right] = \frac{\det_{1p} \mathbf{Z}(s)}{\det \mathbf{Z}(s)} V_1(s) \quad (16\text{-}10)$$

where $\det_{1p} \mathbf{Z}(s)$ is the cofactor of $V_1(s)$ in that determinant. All the other cofactors are multiplied by zeros.

In a similar way, we get

$$I_q(s) = \frac{\det_{1q} \mathbf{Z}(s)}{\det \mathbf{Z}(s)} V_1(s) \qquad (16\text{-}11)$$

and so

$$I_k(s) = I_p(s) - I_q(s) = \frac{\det_{1p} \mathbf{Z}(s) - \det_{1q} \mathbf{Z}(s)}{\det \mathbf{Z}(s)} V_1(s) \qquad (16\text{-}12)$$

exhibiting clearly the network function (here a transfer admittance)

$$H(s) = \frac{\det_{1p} \mathbf{Z}(s) - \det_{1q} \mathbf{Z}(s)}{\det \mathbf{Z}(s)} \qquad (16\text{-}13)$$

There is nothing to memorize here, except to review and to draw general conclusions:

1. In finding $H(s)$, we write *any* convenient set of equations (loop, node, or state variables). The solution of the desired response will always have in its denominator the determinant of the coefficients matrix. *This is the*

denominator of $H(s)$. This denominator has coefficients that depend only on the network elements—hence its name, the characteristic polynomial.

2. The network does not care how we analyze it. Therefore, we expect to get the same characteristic polynomial by *any* method of analysis.

EXAMPLE 16-5

In Example 16-4, the loop impedance matrix, here a scalar, is by inspection $(s + 1)$. This is the characteristic polynomial and

$$s + 1 = 0$$

is the characteristic equation, yielding the natural frequency

$$s = -1$$

Suppose we wanted to analyze this network by node analysis. The node admittance matrix is $(1/1 + 1/s)$. And

$$\frac{1}{1} + \frac{1}{s} = 0$$

is the characteristic equation, the same

$$s + 1 = 0$$

as before.   ∎

EXAMPLE 16-6

Calculate the natural frequencies of the network shown in Figure 16-8.

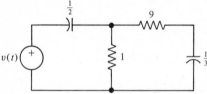

(Ω, F) **Figure 16-8**   Example 16-6.

Solution. We don't need a transform diagram—use your imagination. Two loop currents are chosen inside the two window panes, both clockwise. We write immediately

$$\det \mathbf{Z}(s) = \begin{vmatrix} \dfrac{2}{s} + 1 & -1 \\ -1 & \dfrac{3}{s} + 10 \end{vmatrix} = 9s^2 + 23s + 6 = 0$$

$$\therefore \quad s_1 = -2.26 \qquad s_2 = -0.3$$

With one loop current clockwise, the other counterclockwise, the only difference in det $\mathbf{Z}(s)$ will be the off-diagonal terms, $+1$ instead of -1. The answers s_1 and s_2 will still be the same. With node analysis (fill in details as needed) we'll have

$$\begin{vmatrix} \dfrac{s}{2} + \dfrac{10}{9} & -\dfrac{1}{9} \\ -\dfrac{1}{9} & \dfrac{s}{3} + \dfrac{1}{9} \end{vmatrix} = 9s^2 + 23s + 6 = 0$$

again. By now it may be a slightly worn observation, but the network does not know or care how we analyze it. ∎

EXAMPLE 16-7

Find the characteristic values of the network shown in Figure 16-9. The dependent source is αi_1, with α (known) but left in letter notation.

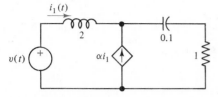

(Ω, H, F) **Figure 16-9** Example 16-7.

Solution. With dependent sources, we don't rush. One loop equation is needed (versus two node equations) and we use the outside loop. KVL reads

$$2sI_1 + \frac{10}{s}(I_1 + \alpha I_1) + 1(I_1 + \alpha I_1) = V$$

Collecting terms, we get the characteristic equation

$$2s^2 + (1 + \alpha)s + (1 + \alpha)10 = 0$$

showing clearly how the element values, including α, enter into it. ∎

Probs. 16-6, 16-7

The zero-state response (Equation 16-5) can be written more explicitly as follows

$$R(s) = H(s)E(s) = \frac{P_H(s)}{Q_H(s)} \cdot \frac{P_E(s)}{Q_E(s)} \tag{16-14}$$

where $H(s)$ and $E(s)$ are written as rational functions, each as $P(s)/Q(s)$. The subscripts H and E identify them. In the partial fraction of $R(s)$, there will be two groups of terms: Those with the poles of $H(s)$, the natural frequencies of the network, roots of $Q_H(s) = 0$, plus those with the poles of $E(s)$, the frequencies of the input, roots of $Q_E(s) = 0$

$$R(s) = \sum_n \frac{A_n}{s - p_n} + \sum_m \frac{A_n}{s - p_m} \tag{16-15}$$

Thus, the zero-state response will read

$$r(t) = \sum_n A_n e^{p_n t} + \sum_m A_m e^{p_m t} \tag{16-16}$$

where the p_n's are the poles of $H(s)$, and the p_m's the poles of $E(s)$. We assumed all poles to be simple. There are the obvious modifications for multiple poles.

It is extremely important to recognize the physical significance of Equation 16-16. A relaxed network (zero initial conditions) is excited by an input $e(t)$. The response will consist of two parts: A waveform similar to $e(t)$ *plus* a natural waveform characteristic to the network itself. Think a while about it, and see how the Laplace transform helps us in reaching this conclusion elegantly and easily.

EXAMPLE 16-8 _____

In the network of Example 16-4, Figure 16-6, we can write without any calculations

$$i(t) = A_1 \cos 2t + A_2 \sin 2t + A_3 e^{-t}$$

The first two terms reflect the input waveform. The term $A_2 \sin 2t$ must be assumed as a generalization of the input, just as we did in the classical method for the particular solution. The term $A_3 e^{-t}$ is the characteristic response of the network, due to the pole at $p_1 = -R/L = -1$.

To find A_1, A_2, and A_3 we must go through the algebra of partial fractions. Here we demonstrate only how much information can be obtained about the waveform of $i(t)$ with hardly any work at all. ■

Probs. 16-8
16-9

EXAMPLE 16-9 _____

Let $H(s)$ be given as

$$H(s) = \frac{1}{s+1}$$

with the pole $p_1 = -1$. Let the input be $e^{\alpha t}$ with α real.

Solution. The zero-state response will be

$$R(s) = \frac{1}{s+1} \cdot \frac{1}{s-\alpha}$$

For any $\alpha \neq -1$, the partial fraction of $R(s)$ is

$$R(s) = \frac{A_1}{s+1} + \frac{A_2}{s-\alpha}$$

and the time-domain response will exhibit the natural response and the input waveform

$$r(t) = A_1 e^{-t} + A_2 e^{\alpha t}$$

as expected. Let us see what happens when $\alpha = -1$. Then

$$R(s) = \frac{1}{s+1} \cdot \frac{1}{s+1} = \frac{1}{(s+1)^2}$$

and

$$\mathcal{L}^{-1} R(s) = r(t) = t e^{-t} u(t)$$

What is this? Why, resonance revisited! We excited a natural frequency of the network ($p_1 = -1$) with the source (e^{-t}). Resonance, you'll recall, is the condition when a natural frequency of the network is excited.

Two waveforms of $r(t)$ are shown in Figure 16-10. In Figure 16-10(a), we assumed $\alpha = -2$ and consequently $A_1 = 1$, $A_2 = -1$. In Figure 16-10(b), $\alpha = -1$ and resonance occurs.

In the classical method of solution this problem is

$$\frac{dr(t)}{dt} + r(t) = e^{\alpha t} \qquad r(0^-) = 0$$

The case $\alpha = -1$ is the one when the homogeneous solution is already of the same waveform as the input. Then, according to Table 7-1 and its notes, the particular solution must be multiplied by t. Notice the total correspondence of that procedure with the effortless one offered here by the Laplace transform.

As mentioned in Chapters 8 and 9, resonance may be useful or harmful. You couldn't tune to your favorite radio station without resonance; on the other hand, many a system, for example, a bridge resonating with the uniform steps of marching soldiers, can be ruined by it.

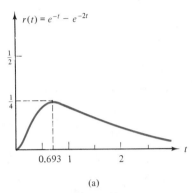

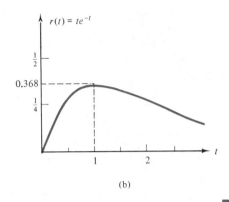

(a)

(b)

Probs. 16-10, 16-11, 16-12, 16-13

Figure 16-10 Example 16-9.

16-2 THE IMPULSE RESPONSE

At this point, a challenging question comes up. Since the characteristic (natural) response of the network is so important, as it appears together with every type of input, is there a way to obtain it just by itself, without the "contamination" of the waveform of an input? In other words, is there an input $E(s)$ that will contribute *no* poles to Equation 16-14?

A moment of thought tells us that $e(t) = \delta(t)$, the unit impulse function, is just the right input. Since $E(s) = 1$ in this case, Equation 16-14 becomes

$$R(s) = H(s) = \frac{P_H(s)}{Q_H(s)} \tag{16-17}$$

and its inverse

$$r(t) = \mathscr{L}^{-1}H(s) = h(t) \tag{16-18}$$

will contain only the terms with the natural frequencies of the network. We have labeled $\mathscr{L}^{-1}H(s) = h(t)$, and $h(t)$ is called the *impulse response* of the network. Figure 16-11 illustrates this important concept and suggests a laboratory experiment for obtaining $h(t)$: To the initially relaxed network (zero initial conditions), we apply a

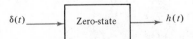

Figure 16-11 The impulse response.

unit impulse (voltage or current) as the appropriate input, and measure or record the appropriate output. This output is $h(t)$ in Equation 16-18.

What's so important about $h(t)$? We saw already that it is the natural (characteristic) response of the network. Its Laplace transform is $H(s)$, the network function that allows us, through Equation 16-14, to compute *any* zero-state response. Having $H(s)$ is just like placing an identification tag on the network. Any time that we wish to calculate the zero-state response for that network, we take $H(s)$ from the tag, the given $E(s)$, and use Equation 16-14. In the next section, we'll learn how to calculate the zero-state response $r(t)$ directly in the time domain, without the Laplace transform, using the impulse response $h(t)$.

EXAMPLE 16-10 _____

Calculate the impulse response of the network in Example 16-2. To be more specific, what will be $v_{out}(t)$ across the capacitor if $v_{in}(t) = \delta(t)$?

Solution. We have

$$h(t) = \mathcal{L}^{-1}H(s) = \mathcal{L}^{-1}\frac{1}{RCs + 1} = \frac{1}{RC}e^{-t/RC}u(t)$$

If we did not have $H(s)$, the laboratory (or paper) experiment for obtaining $h(t)$ is shown in Figure 16-12(a), and the result makes sense also intuitively, as it did in Chapter 7. At $t = 0^-$, the capacitor is uncharged and acts as a short circuit. The impulse voltage "rushes in" at $t = 0$; from $t = 0^+$, the source is zero, a short circuit, and the capacitor voltage decays in the RC circuit with the familar time constant $\tau = RC$. The plot of $h(t)$ is shown in Figure 16-12(b).

(a)   (b)

Figure 16-12   Example 16-10.   ■

EXAMPLE 16-11 _____

Calculate the impulse response in the previous example if the output voltage is across R. See Figure 16-13(a). To distinguish the two cases, let us use $\hat{H}(s)$ and $\hat{h}(t)$ here.

Solution. The relevant network function here is another voltage transfer function, obtained by the voltage divider

$$\hat{H}(s) = \frac{R}{R + 1/Cs} = \frac{RCs}{RCs + 1}$$

To find the impulse response $\hat{h}(t)$, we must first divide $\hat{H}(s)$ longhand, since it is an improper rational function (you did not forget this paramount rule, did you?). We get

$$\hat{H}(s) = 1 - \frac{1}{RCs + 1} = 1 - \frac{1/RC}{s + 1/RC}$$

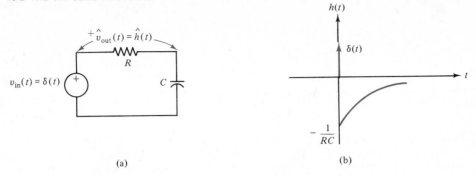

(a)

(b)

Figure 16-13 Example 16-11.

Consequently,

$$\hat{v}_{out}(t) = \hat{h}(t) = \mathscr{L}^{-1}\hat{H}(s) = \delta(t) - \frac{1}{RC}e^{-t/RC}u(t)$$

and it is shown in Figure 16-13(b). The impulse response, $\hat{h}(t)$, contains an impulse and a decaying exponential. Can we see it physically? Of course: At $t = 0$, the capacitor being still a short circuit, the entire input voltage $\delta(t)$ appears across R, according to KVL at $t = 0$. Thereafter, there is a decaying voltage component with the time constant $\tau = RC$.

   As an additional confirmation of our results in these two examples, we check KVL for the circuit for all $t \geq 0$

$$\hat{v}_{out}(t) + v_{out}(t) = v_{in}(t)$$

or

$$\delta(t) - \frac{1}{RC}e^{-t/RC} + \frac{1}{RC}e^{-t/RC} = \delta(t)$$

which is certainly true.

Probs. 16-14, 16-15, 16-16, ■ 16-17

   To conclude this discussion of network functions, let us consider the *total* response. We know that Equation 16-14 relates only the zero-state response to the network function and the input. What about the zero-input response? The answer and the conclusions are best illustrated with an example.

EXAMPLE 16-12

We are given the first-order network in Figure 16-14(a). Let us retain algebraic (letter) notation, rather than specific numerical values, so that we can trace every term in the calculations.

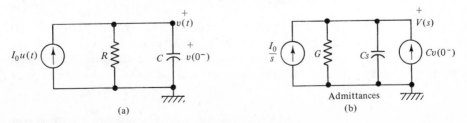

(a)

(b)

Figure 16-14 Example 16-12.

Solution. Using the transform circuit diagram of Figure 16-14(b), or transforming directly the node equation for $v(t)$, we get

$$(Cs + G)V(s) = \frac{I_0}{s} + Cv(0^-)$$

That is,

$$V(s) = \frac{1}{Cs + G}\frac{I_0}{s} + \frac{1}{Cs + G}Cv(0^-)$$

The first term on the right is the zero-state response, showing clearly the relation $R(s) = H(s)E(s)$. The second term is the zero-input response due entirely to initial conditions. In it, we see also a division by the characteristic polynomial, and this will be always true.

Inversion of these terms yields (fill in the details)

$$v(t) = RI_0(1 - e^{-t/RC})u(t) + v(0^-)e^{-t/RC}u(t)$$

where we used $R = 1/G$. The two parts, the zero-state and the zero-input, are familiar to us from our previous studies. We just want to review and see how $H(s)$ affects the total response. We see that poles of $H(s)$, the natural frequencies of the network, appear generally in both the zero-state and the zero-input parts.

Another aspect of the total response also deserves to be reviewed. Rewrite $v(t)$ as follows:

$$v(t) = RI_0u(t) + [v(0^-) - RI_0]e^{-t/RC}u(t)$$

Here, the first term is the *steady-state part* of $v(t)$, remaining nonzero for $t \to \infty$. The second term is the *transient part* of $v(t)$, decaying to zero as $t \to \infty$.

We observe that:

1. The steady-state response here is part of the zero-state response. This is reasonable to expect, since the initial conditions, causing the zero-input response, cannot affect the total response as $t \to \infty$. The initial energy stored in the dynamic elements is dissipated in the resistors of the network as $t \to \infty$.†
2. The zero-input waveform resembles the impulse response waveform; both have the natural frequencies in the exponents.
3. With a proper choice of initial conditions, we can sometimes suppress the transient part. Here, the choice $v(0^-) = RI_0$ will do so.   ■ Prob. 16-18

16-3 CONVOLUTION AND SUPERPOSITION

Let us return to the zero-state response. Its transform, $R(s)$, is given by Equation 16-14, repeated here

$$R(s) = H(s)E(s) \qquad\qquad (16\text{-}19)$$

We ask now the following question: Since the network is linear, why not obtain $r(t)$ directly in the time domain by using superposition? After all, this approach is very basic and does not require any indirect methods such as transforms. As a bonus, whatever the superposition yields in the time domain, we already know its Laplace transform from Equation 16-19: it is $H(s)E(s)$.

---

† In trivial cases, such as a single capacitor with an initial voltage $v_C(0^+)$, the total response as $t \to \infty$ is still $v_C(0^+)$.

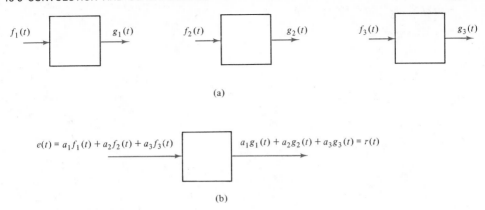

(a)

(b)

Figure 16-15 Superposition.

The idea of superposition is very simple. Assume that we know the zero-state response of the network to an elementary function. Assume that an arbitrary input $e(t)$ can be written as a sum (superposition) of such elementary functions. Then the zero-state response $r(t)$ will be the sum (superposition) of the corresponding responses to those elementary functions.

We show this idea in Figure 16-15. Let $f_1(t)$ be an elementary function that produces a known zero-state response $g_1(t)$. Similarly, let $f_2(t)$, another elementary function, produce a known $g_2(t)$, and let $f_3(t)$ produce $g_3(t)$. Let us now form the input $e(t)$ as the sum

$$e(t) = a_1 f_1(t) + a_2 f_2(t) + a_3 f_3(t) \qquad (16\text{-}20)$$

with arbitrary *constants* a_1, a_2, and a_3. The linearity of the network means that the zero-state response to this $e(t)$ will be the sum (superposition)

$$r(t) = a_1 g_1(t) + a_2 g_2(t) + a_3 g_3(t) \qquad (16\text{-}21)$$

with the *same* multipliers a_1, a_2, and a_3. Simple, right?

What do we mean by "elementary functions"? Of several possibilities, the impulse function is a promising candidate, because we know the zero-state response of the network to it: That's $h(t)$, the impulse reponse. So, to put it loosely for a moment, if an input $e(t)$ is the sum of impulse functions, the response $r(t)$ will be the sum of impulse response functions.

Our first precise job is to be able to write an arbitrary input $e(t)$ as a sum of impulse functions. Here, the sampling property of the impulse function comes to our aid. Recall it from Chapter 14 as

$$\int_{-\infty}^{\infty} f(t)\delta(t - b)\, dt = f(b) \qquad (16\text{-}22)$$

Next, consider an arbitrary input $e(x)$ plotted versus x, as shown in Figure 16-16. We use x as a dummy variable. We can write, according to Equation 16-22,

$$\int_{-\infty}^{\infty} e(x)\delta(x - t)\, dx = e(t) \qquad (16\text{-}23)$$

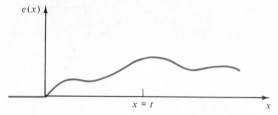

Figure 16-16   The sampling property.

where t is *any* point on the x axis. If we assume that $e(x) = 0$ for $x < 0$, as is the case for all our inputs, then the limits on the integral in Equation 16-23 can be changed as follows:

$$\int_0^\infty e(x)\delta(t - x)\,dx = e(t) \tag{16-24}$$

Here, we have used (without proof) the fact that $\delta(x - t) = \delta(t - x)$.

Our first job is done: Equation 16-24 expresses an arbitrary input $e(t)$ as the sum (integral) of shifted impulse functions $\delta(t - x)$, each of strength $e(x)$. These strengths are the multipliers a_1, a_2, and a_3 in our preliminary discussion.

Now we use linearity and superposition for the output $r(t)$. By definition, the zero-state response to an input $\delta(t)$ is $h(t)$. In a constant, time-invariant network, the response to a shifted input $\delta(t - x)$ is an identically shifted $h(t - x)$. The response to $e(x)\delta(t - x)$, a shifted impulse of strength $e(x)$, will be $e(x)h(t - x)$, where the same strength $e(x)$ multiplies the response if it multiplies the input, as in Equation 16-21. Next, a sum (integral) of these shifted impulses is the input $e(t)$, as in Equation 16-24, and a corresponding sum (integral) of impulse responses is the response $r(t)$

$$r(t) = \int_0^\infty e(x)h(t - x)\,dx \tag{16-25}$$

Finally, our physical system is *nonanticipatory* (*causal*), meaning that its response at any time $x = t$ cannot possibly depend on *future* values of the input, $x > t$. In simple words, the system does not have a crystal ball; at $x = t$, its response depends only on input values *up to* $x = t$. With this observation, the upper limit in Equation 16-25 is changed to $x = t$, and we have

$$r(t) = \int_0^t e(x)h(t - x)\,dx \tag{16-26}$$

These steps are summarized in Figure 16-17, with a brief explanation of each step. The integral in Equation 16-26 has a special name and a special symbol. It is called the *convolution* of the input $e(t)$ with the impulse response $h(t)$, and we denote convolution with a star *

$$e(t) * h(t) = \int_0^t e(x)h(t - x)\,dx \tag{16-27}$$

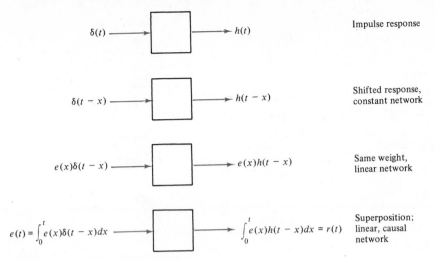

Figure 16-17 Steps in deriving convolution.

The verb is "to convolve" (*not* "convolute"), and we "convolve $e(t)$ with $h(t)$"; or else, "$e(t)$ is convolved with $h(t)$," to obtain the zero-state response $r(t)$. A quick example is needed right now.

EXAMPLE 16-13 ——

Let

$$h(t) = e^{-t}u(t)$$

be the given impulse response of a network, and let the input be†

$$e(t) = e^{-2t}u(t)$$

Solution. The zero-state response $r(t)$ is, according to Equation 16-27,

$$r(t) = e(t) * h(t) = \int_0^t e^{-2x}u(x)e^{-(t-x)}u(t-x)\,dx$$

That is, we first multiply the input as a function of x, $e(x)$, by the shifted impulse response $h(t-x)$. Then we integrate this product over x from $x = 0$ to $x = t$.

The actual evaluation of this integral, like any integral, starts with any possible simplifications of the integrand. The unit step function $u(x)$ versus x is equal to 1 over the range of integration,

$$u(x) = 1 \qquad x > 0$$

So is $u(t - x)$, because, by definition,

$$u(t - x) = \begin{cases} 1 & t - x > 0 \quad \therefore \quad x < t \\ 0 & t - x < 0 \quad \therefore \quad x > t \end{cases}$$

and we are integrating along the x axis, as shown in Figure 16-18. *We cannot stress enough the importance of drawing such sketches* when we evaluate convolution integrals.

———

† Undoubtedly an unnecessary reminder: In $e(t)$, the letter e is a generic notation for "excitation." In $e^{-2t}u(t)$, the letter $e = 2.718\ldots$ is the base of natural logarithms.

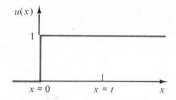 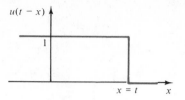

Figure 16-18   Example 16-13.

They help clarify and simplify the integrand, as well as keep tabs on x, the dummy variable of integration, and t, the running point along the x axis.

Now our integral is simplified to

$$r(t) = \int_0^t e^{-2x} e^{-(t-x)} \, dx = e^{-t} \int_0^t e^{-2x} e^x \, dx = e^{-t} \int_0^t e^{-x} \, dx$$

where e^{-t} is a constant as far as integration with respect to x is concerned, and is therefore pulled out front. Finally,

$$r(t) = e^{-t} \left[\frac{e^{-x}}{-1} \right]_{x=0}^{x=t} = e^{-t}(1 - e^{-t}) = e^{-t} - e^{-2t} \qquad t > 0$$

As a quick check, we have in the transform domain

$$H(s) = \mathscr{L}h(t) = \frac{1}{s+1}$$

$$E(s) = \mathscr{L}e(t) = \frac{1}{s+2}$$

and

$$R(s) = H(s)E(s) = \frac{1}{s+1} \cdot \frac{1}{s+2}$$

Expansion by partial fractions yields

$$R(s) = \frac{1}{s+1} + \frac{-1}{s+2}$$

and

$$r(t) = (e^{-t} - e^{-2t})u(t)$$

which checks with the convolution.   ■

This typical example raises an obvious question, "Why bother with a long, involved convolution when we can get the answer quickly and neatly by the Laplace transform?" There are several answers to this: (1) Convolution is based on a physical property of the network, superposition. (2) Superposition has a meaning in real life, in the time domain. (3) Historically, people have used convolution long before any transform methods. (4) In many practical cases, $h(t)$ is not given analytically, but is rather obtained graphically or numerically in the lab, as we saw in the previous section. Then convolution is the best way to go.

In summary, convolution stands firm on a solid foundation in the real world. It just so happens that the Laplace transform handles this operation more quickly in a nonreal world of the complex number s. That is the proper perspective to take when working with convolution.

　　　We repeat in words the result of Equation 16-27: *The zero-state response of a linear network is obtained by convolving the input with the impulse response.* The convolution consists of three steps:

1. Replace t by a dummy variable x in $e(t)$, obtaining $e(x)$.
2. Replace t by $(t - x)$ in the impulse response $h(t)$, obtaining $h(t - x)$.
3. Integrate the product $e(x)h(t - x)$ over x, between $x = 0$ and $x = t$. The result will be $r(t)$, the zero-state response to the input $e(t)$.

EXAMPLE 16-14 _____

Use the same $h(t)$ as in Example 16-13

$$h(t) = e^{-t}u(t)$$

but the input is now

$$e(t) = e^{-t}u(t)$$

Solution.　The three steps yield

$$r(t) = e(t) * h(t) = \int_0^t e^{-x}u(x)e^{-(t-x)}u(t - x)\, dx = e^{-t}\int_0^t dx = te^{-t} \qquad t > 0$$

where the simplifications of $u(x)$ and $u(t - x)$ were done as before. This result is in agreement with Example 16-9; even without referring back to it, we should recognize here immediately resonance: The impulse response $e^{-t}u(t)$ has a natural frequency at -1, and the input $e^{-t}u(t)$ excites it.　■　Prob. 16-19

　　　Two important properties of convolution are:
1. *Commutativity*, that is,

$$\int_0^t e(x)h(t - x)\, dx = e(t) * h(t) = h(t) * e(t) = \int_0^t h(x)e(t - x)\, dx \qquad (16\text{-}28)$$

which means that in setting up the convolution integral, the integrand can be $e(x)h(t - x)$ or $h(x)e(t - x)$. This is nice, since we don't need to worry which of the two functions is in terms of x and which is in terms of $(t - x)$. The choice is ours. The proof is easy: In the left-hand integral of Equation 16-28, substitute a new variable

$$t - x = y \qquad (16\text{-}29)$$

Then $dx = -dy$, the lower limit $x = 0$ becomes $y = t$, and the upper limit $x = t$ becomes $y = 0$. Therefore

$$\int_0^t e(x)h(t - x)\, dx = \int_t^0 e(t - y)h(y)(-dy) = \int_0^t e(t - y)h(y)\, dy \qquad (16\text{-}30)$$

which is precisely the right-hand integral in Equation 16-28. In both integrals, x and y are dummy variables.
　　　2. *The Laplace transform* of the convolution of two time functions is the product of their Laplace transforms, as we saw throughout this chapter

$$\mathscr{L}h(t) * e(t) = H(s)E(s) \qquad (16\text{-}31)$$

Thus, a difficult operation in the time domain (convolution) is transformed into an easier operation in the frequency domain (multiplication), as expected of a transform.

Besides serving to calculate the zero-state response, Equation 16-31 can be used to find the inverse Laplace of products. The following example illustrates it.

EXAMPLE 16-15 ───

By convolution, find

$$\mathcal{L}^{-1} \frac{1}{s(s+1)} = \mathcal{L}^{-1} \frac{1}{s} \cdot \frac{1}{s+1}$$

Solution. Therefore, identify

$$F_1(s) = \frac{1}{s} \qquad \therefore \quad f_1(t) = u(t)$$

$$F_2(s) = \frac{1}{s+1} \qquad \therefore \quad f_2(t) = e^{-t}u(t)$$

Consequently

$$\mathcal{L}^{-1}F_1(s)F_2(s) = f_1(t) * f_2(t) = \int_0^t u(t-x)e^{-x}u(x)\,dx = \int_0^t e^{-x}\,dx = (1 - e^{-t})u(t)$$

which can be quickly verified by partial fractions. Here we've used f_1, f_2, F_1, and F_2 as general functions in Equation 16-31, not necessarily input and impulse response. ■ Prob. 16-20

16-4 GRAPHICAL AND NUMERICAL CONVOLUTION

It is useful and instructive to interpret graphically the convolution integral, repeated here

$$r(t) = \int_0^t e(t-x)h(x)\,dx \qquad\qquad (16\text{-}32)$$

This expression gives $r(t)$ for any $t > 0$. Let us concentrate our attention on one specific time, $t = t_1$. Then the response at $t = t_1$ is

$$r(t_1) = \int_0^{t_1} e(t_1 - x)h(x)\,dx \qquad\qquad (16\text{-}33)$$

This value $r(t_1)$ can be interpreted as the area (the integral) under the product curve $e(t_1 - x)h(x)$, between $x = 0$ and $x = t_1$. For another value of t, $t = t_2$, the response $r(t_2)$ is the area under the product curve $e(t_2 - x)h(x)$ between $x = 0$ and $x = t_2$, etc. Proceeding in this manner we can get $r(t_1), r(t_2), r(t_3), \ldots$, in other words, the response $r(t)$ for various points in time.

EXAMPLE 16-16 ───

Let $e(t)$ and $h(t)$ be the triangular-shaped functions in Figure 16-19(a) and (b). To use the graphical approach for convolution, we must form the integrand, the product curve, as in Example 16-13. First, the curve $h(x)$ versus x is identical to the curve $h(t)$ versus t, as given. It is

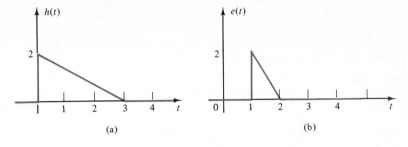

(a)   (b)

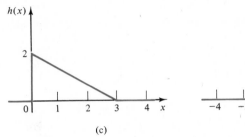

(c)

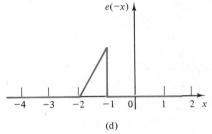

(d)

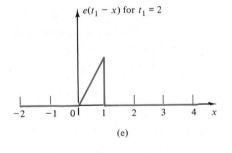

(e)

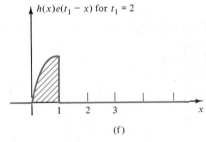

(f)

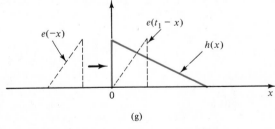

(g)

Figure 16-19   Example 16-16.

shown in Figure 16-19(c). The curve $e(t_1 - x)$ versus x is obtained in three steps: First, the plot of $e(x)$ versus x is identical to the plot of $e(t)$ versus t. Next, we get the curve $e(-x)$ versus x by reflecting (folding) the curve $e(x)$ about the vertical axis, as shown in Figure 16-19(d). Finally, we slide it to the right by t_1 units along the x axis to obtain $e(t_1 - x)$ versus x, as shown in Figure 16-19(e).

The integrand, $h(x)e(t_1 - x)$, is then the product of the curves in Figure 16-19(c) and 16-19(e). This product of curves can be done point by point, and the result is the curve shown in Figure 16-19(f). The area under this curve (shaded) is a number that is easily evaluated; it is $r(t_1)$.

The process is repeated for $t = t_2$ in the same steps:

1. Keep $h(x)$ frozen.
2. Slide the reflected $e(-x)$ by t_2 units on the x axis, and stop.
3. Multiply these two curves to get the product curve, the integrand.
4. Calculate the area under the product curve between $x = 0$ and $x = t_2$. That area, a number, is $r(t_2)$.

Repeat steps 2, 3, and 4 for $t = t_3$, etc.

In Figure 16-19(g), these steps are summarized.   ■ Prob. 16-21

This graphical method is very convenient to interpret and to execute. It is suitable when, for instance, $h(t)$ is given graphically (the display of an oscilloscope in the laboratory experiment). In addition, this method is extremely helpful in determining analytical results for some common functions, as illustrated in the next example.

EXAMPLE 16-17 _____

Use the graphical convolution to obtain the analytical expression for $r(t)$ when $e(t)$ and $h(t)$ are given in Figure 16-20(a) and (b).

Solution.   In Figure 16-20(c), we show $h(x)$ versus x. This is the stationary (frozen) curve. Figure 16-20(d) shows $e(-x)$ versus x, the reflected curve. On it, we show the point $x = t$ which changes as this curve is sliding to the right. At this instant in Figure 16-20(d) $t = 0$, and we start our clock as we slide $e(-x)$. Note also the points $x = t - 2$ and $x = t - 4$ on e.

Between $t = 0$, Figure 16-20(d), and $t = 2$, shown in Figure 16-20(e), the product of $h(x)$ and $e(t - x)$ is zero because one function is zero (though the other isn't.) Hence, the product curve, the integrand, is zero. The area under this integrand, then, is zero. Consequently

$$r(t) = 0 \qquad 0 < t < 2$$

We could have predicted this much just by looking at $e(t)$ as given in Figure 16-20(b): There is no input until $t = 2$; therefore there is no zero-state output until then also!

Things aren't so obvious for $t > 2$, but our method will work. Consider the range of $2 < t < 4$, as $e(-x)$ continues to slide to the right. A typical position is shown in Figure 16-20(f); here $h(x)$ is in solid lines and the sliding $e(t - x)$ is in dashed lines. The product curve will be zero for $x < 0$, because $h(x)$ is zero there, and for $x > t - 2$ because $e(t - x)$ is zero there. Therefore, the area under the product curve is

$$r(t) = \int_{x=0}^{x=t-2} (4)(8)\, dx = 32(t - 2) \qquad 2 < t < 4$$

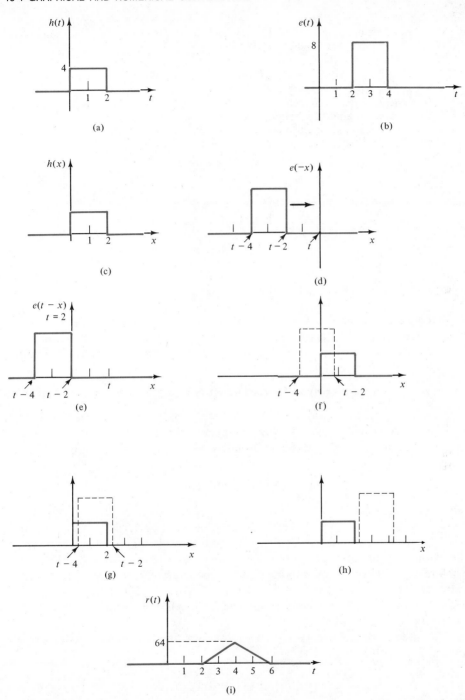

Figure 16-20 Example 16-17.

For $t > 4$ but $t < 6$, the two curves are shown in Figure 16-20(g). The product curve, and therefore the area under it, is nonzero only between $x = t - 4$ (the trailing end of e) and $x = 2$ (the fixed end of h). Therefore

$$r(t) = \int_{x=t-4}^{x=2} (8)(4)\, dx = 32(6 - t) \qquad 4 < t < 6$$

This is where the graphical method really shines: It is simple and decisive in setting the rather unexpected limits on the convolution integral.

Finally, for $t > 6$, the two curves are shown in Figure 16-20(h). Their product is obviously zero; hence the area under their product is zero. Thus

$$r(t) = 0 \qquad t > 6$$

In summary:

$$r(t) = \begin{cases} 0 & t < 2 \\ 32(t - 2) & 2 < t < 4 \\ 32(6 - t) & 4 < t < 6 \\ 0 & t > 6 \end{cases}$$

and its plot is shown in Figure 16-20(i).

Probs. 16-22
16-23

Numerical convolution is used when either $h(t)$ or $e(t)$ or both are given in a tabulated form, that is, as a sequence of numbers, in digitized form. From the lab experiment that we set up, we might, for instance, obtain $h(t)$, shown in Figure 16-21(a), as the sequence of values

$$h(n) = \{h(0), h(1), h(2), h(3), \ldots\} \tag{16-34}$$

where $h(n)$ is a simplified notation for $h(nT)$. The sample points are T seconds apart, and as T is made smaller, the accuracy is improved. As a rule, the spacing between points is the same, although this is not absolutely necessary. Similarly, let the input be given in the sequence

$$e(n) = \{e(0), e(1), e(2), e(3), \ldots\} \tag{16-35}$$

as shown in Figure 16-21(b).

We proceed as before. Reflect $e(x)$ to obtain $e(-x)$, as shown in Figure 16-21(c), and start the clock now, $t = 0$. The product curve $e(0 - x)h(x)$ is zero, and, unless h or e has impulses at the origin, the area under that product curve is zero. Therefore $r(0) = 0$. Next, slide e to the right by one interval T, as in Figure 16-21(d). The product curve is $e(1 - x)h(x)$, and is shown in Figure 16-21(e); at $x = 0$, its height is $h(0)e(1)$, and at $x = T$ ($n = 1$) its height is $h(1)e(0)$. All other points are zero in the product curve.

The area under this curve (shaded) is approximately that of a trapezoid, provided T is small enough. That is

$$r(1) = \frac{T}{2}[h(0)e(1) + h(1)e(0)] \tag{16-36}$$

where the area of a trapezoid is its height (T) times one-half the sum of its parallel sides.

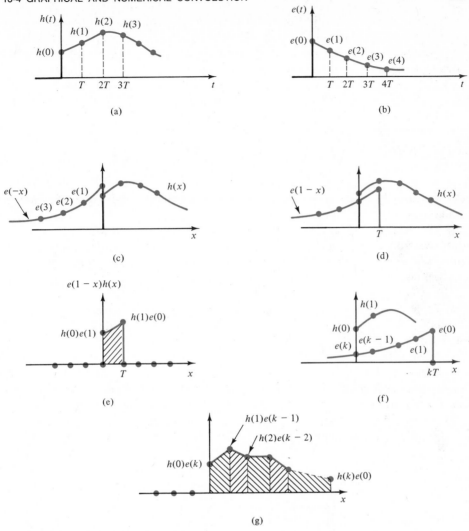

Figure 16-21   Numerical convolution.

The pattern is established now for a general time, $t = kT$, in Figure 16-21(f). The reflected curve $e(-x)$ has slid to the right k units (kT). The product curve is shown in Figure 16-21(g); pay attention to the ordinates (heights) of the various points of this curve, $h(0)e(k)$ at $x = 0$, $h(1)e(k - 1)$ at $x = T$, $h(2)e(k - 2)$ at $x = 2T$, etc. The area under the product curve is $r(k)$, the response at $t = kT$. It can be obtained as the sum of the areas of the trapezoids, that is

$$r(k) = \frac{T}{2} [h(0)e(k) + 2h(1)e(k - 1) + 2h(2)e(k - 2)$$

$$+ \cdots + 2h(k - 1)e(1) + h(k)e(0)] \qquad (16\text{-}37)$$

Inside the brackets we have the first parallel side $h(0)e(k)$, the last parallel side $h(k)e(0)$, and *twice* the internal parallel sides, since each belongs to two adjacent trapezoids.

Equation 16-37 is the numerical convolution of the two sequences $h(n)$ and $e(n)$, providing the sequence of the output $r(n) = \{r(0), r(1), r(2), \ldots\}$

Probs. 16-24
16-25, 16-26
16-27, 16-28

$$r(n) = h(n) * e(n) \qquad (16\text{-}38)$$

16-5 THÉVENIN'S AND NORTON'S THEOREMS

It is time now to generalize Thévenin's theorem. We formulated it first for resistive networks in Chapter 5, and for sinusoidal steady-state phasors in Chapter 9. It may be a good idea to leaf back and review them now.

The generalized Thévenin's theorem can be stated as follows. As far as any arbitrary load ("the observer") is concerned, as in Figure 16-22(a), the linear network can be replaced by a single voltage source $V_{Th}(s)$ in series with a single impedance, $Z_{Th}(s)$, shown in Figure 16-22(b). In both cases, the load current $I(s)$ and voltage $V(s)$ are the same.

To find $V_{Th}(s)$, the Thévenin equivalent voltage, we remove the load from terminals A–B, leaving them open-circuited (o.c.). Then we calculate the open-circuit voltage $V_{o.c.}(s)$ across those terminals caused by *all* the sources and initial conditions in the linear network. See Figure 16-22(c). This is best done with the transform circuit diagram.

To find $Z_{Th}(s)$, we set to zero all the *independent* sources (a voltage source is replaced by a short circuit and a current source by an open circuit), and we set to zero all the initial conditions. *Dependent* sources are left unchanged. Then we calculate the input impedance at terminals A–B, to get $Z_{Th}(s)$. See Figure 16-22(d).

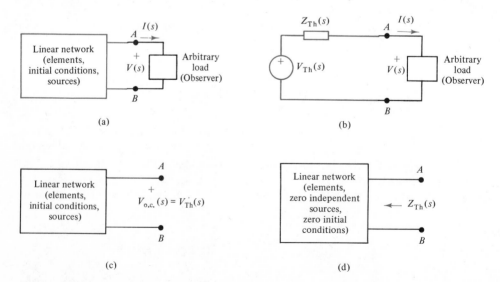

Figure 16-22   Thévenin's theorem.

The equivalent Thévenin circuit and the load are then connected as in Figure 16-22(b).

It is important to make the following remarks:

1. The load (observer) is entirely arbitrary, linear or nonlinear.
2. The network seen by the observer must be linear, and may contain independent and dependent sources.
3. There must be *no coupling* between the observer and the network, through some mutual inductance or through some dependent source.
4. The usefulness of Thévenin's theorem is twofold: First, it simplifies our calculations, particularly if we're interested only in $I(s)$ or $V(s)$ of the load; second, it gives us a conceptual simplification of a linear network as a single equivalent voltage source in series with a zero-input (no independent sources) and zero-state (no initial conditions) network, $Z_{Th}(s)$.

EXAMPLE 16-18 _____

Calculate $i_R(t)$ in the load R by Thévenin's theorem. See Figure 16-23(a).

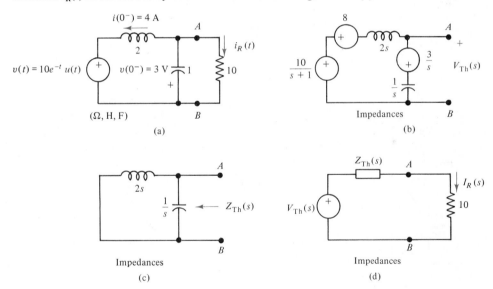

Figure 16-23   Example 16-18.

Solution.   We remove the observer and draw the transform circuit diagram of the network as shown in Figure 16-23(b). In this simplified circuit (that's a good reason to use Thévenin's theorem), we have the open-circuit voltage across A–B as

$$V_{Th}(s) = \frac{1}{s}\frac{10/(s+1) - 8 + 3/s}{2s + 1/s} - \frac{3}{s} = \frac{-6s^2 - 14s - 8s + 2}{(s+1)(2s^2+1)}$$

The Thévenin impedance is found from Figure 16-23(c) as

$$Z_{Th}(s) = 2s\left\|\frac{1}{s}\right. = \frac{2s}{2s^2 + 1}$$

Finally we connect the load to the Thévenin equivalent circuit as in Figure 16-23(d) and write from it

$$I_R(s) = \frac{V_{\text{Th}}(s)}{Z_{\text{Th}}(s) + 10}$$

There remain only the elementary algebra and inversion to be done.   ∎

Norton's equivalent circuit is merely the dual of Thévenin's. It is shown in Figure 16-24, with each step being the dual of the one in Thévenin's derivation. As far as an arbitrary load (observer) is concerned, the linear network to the left of terminals A–B can be replaced by a single current source $I_N(s)$ in parallel with a Norton admittance. The current $I_N(s)$ is the current flowing from A to B when the load is removed and terminals A and B are short-circuited. The Norton admittance is the reciprocal of the Thévenin impedance

$$Y_N(s) = \frac{1}{Z_{\text{Th}}(s)} \tag{16-39}$$

Also, the relation

$$V_{\text{Th}}(s) = Z_{\text{Th}}(s)I_N(s) \tag{16-40}$$

exists between the two circuits. This is easily verified if we replace the load in Figure 16-24(b) by an open circuit, that is, find the Thévenin equivalent of that circuit. Then $I_N(s) = Y_N(s)V_{\text{o.c.}}(s)$, and, owing to Equation 16-39, we get Equation 16-40. The result in Equation 16-40 is very useful when we have to calculate $Z_{\text{Th}}(s)$ with dependent sources.

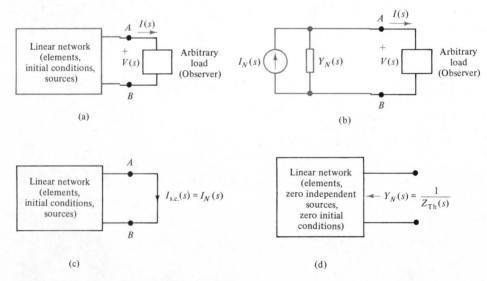

Figure 16-24   Norton's equivalent circuit.

EXAMPLE 16-19 ──

Find the Norton equivalent circuit seen by the resistive load R in the network of Figure 16-25(a). There is a voltage-controlled current source, and there are no initial conditions.

Solution. The load is removed, and in Figure 16-25(b) we calculate $V_{o.c.}(s)$ by writing KVL:

$$V_{o.c.}(s) + \frac{1}{Cs}(gV_1) - V_1 = 0$$

Note that the current gV_1 flows in the capacitor, since terminal A is open-circuited in this simplified circuit. Also,

$$V_1 = Ls(I_{in} - gV_1)$$

$$\therefore \quad V_1 = \frac{LsI_{in}}{1 + gLs}$$

$$\therefore \quad V_{o.c.} = \left(1 - \frac{g}{Cs}\right)\left(\frac{Ls}{1 + gLs}\right)I_{in}(s)$$

Now we compute $I_N(s) = I_{s.c.}(s)$ in Figure 16-25(c). We have, first,

$$V_1(s) = \frac{1}{1/Ls + Cs}I_{in}(s) = \frac{Ls}{LCs^2 + 1}I_{in}(s)$$

and therefore

$$I_{s.c.}(s) = CsV_1 - gV_1 = (Cs - g)V_1 = \frac{(Cs - g)Ls}{LCs^2 + 1}I_{in}(s)$$

Therefore, the Thévenin (or Norton) inpedance is

$$Z_{Th}(s) = \frac{V_{o.c.}(s)}{I_{s.c.}(s)} = \frac{LCs^2 + 1}{Cs(1 + gLs)}$$

With $V_{o.c.}(s) = V_{Th}(s)$, $I_{s.c.}(s) = I_N(s)$, $Z_{Th}(s) = 1/Y_N(s)$, we have all the components needed to construct *either* the Thévenin *or* the Norton equivalent circuits.

As we learned in the resistive and phasor circuits, the calculation of $Z_{Th}(s)$ can be also done by connecting a test current source $I_t(s)$ to terminals A–B in Figure 16-22(d) or 16-24(d), then calculating the resulting voltage $V_t(s)$ across it. The ratio $V_t(s)/I_t(s)$ gives $Z_{Th}(s)$. In this example, let us connect $I_t(s)$ as shown in Figure 16-25(d). Don't forget: In this step, all independent sources and initial conditions are set to zero. Node analysis here reads†

$$\begin{bmatrix} Cs & g - Cs \\ -Cs & Cs + \dfrac{1}{Ls} \end{bmatrix}\begin{bmatrix} V_t \\ V_1 \end{bmatrix} = \begin{bmatrix} I_t \\ 0 \end{bmatrix}$$

† Again, for convenience, $V_t = V_t(s)$, $I_t = I_t(s)$.

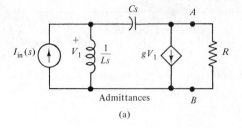

Admittances

(a)

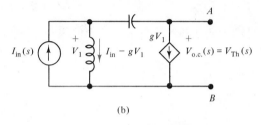

(b)

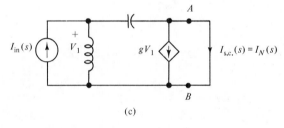

(c)

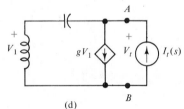

(d)

Figure 16-25   Example 16-19.

Solve by Cramer's rule

$$
V_t = \frac{\begin{vmatrix} I_t & g - Cs \\[6pt] 0 & Cs + \dfrac{1}{Ls} \end{vmatrix}}{\begin{vmatrix} Cs & g - Cs \\[6pt] -Cs & Cs + \dfrac{1}{Ls} \end{vmatrix}} = \frac{LCs^2 + 1}{Cs(1 + gLs)} I_t
$$

yielding the same answer for $Z_{\text{Th}}(s)$.

Probs. 16-29
■ 16-30

16-6 TWO-PORT NETWORKS REVISITED

In Chapter 5 we studied two-port (2-P) resistive networks and their several parameters. The extension to sinusoidal steady-state phasors was done in Chapter 9. Here, we present the generalized, Laplace transform description of such two-ports.

A linear 2-P network, shown in Figure 16-26, may contain linear resistors, inductors (including mutual coupling), capacitors, and dependent sources. The standard references for the port voltages and currents are also shown. The 2-P network is an "extended version" of a one-port (1-P) network (Figure 16-3). In a 1-P, there is only one port (two terminals) of access for one input and one output. A 2-P has two ports of access for two inputs and two outputs.

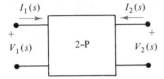

Figure 16-26   Linear two-port network.

The input-output equation for the 2-P (Equation 16-1) is still valid but must be written in *matrix form*

$$\mathbf{R}(s) = \mathbf{H}(s)\mathbf{E}(s) \tag{16-41}$$

where $\mathbf{E}(s)$ is a (2×1) column matrix of the two inputs, $\mathbf{R}(s)$ is a (2×1) column matrix of the corresponding two zero-state outputs, and $\mathbf{H}(s)$ is the appropriate matrix (2×2) network function. As always, we must be given the 2-P network and the two inputs. The two outputs are then clearly identified, and we must relate them by two linear equations to the inputs, thus getting the relationship of Equation 16-41. This informal approach is strongly recommended by comparison with the formal one where we must set up two separate experiments, some under open circuit, others under short circuit, still others under mixed conditions. The two approaches, studied in Chapter 5, will be reviewed in the examples that follow.

EXAMPLE 16-20 _____

For the 2-P network shown in Figure 16-27 (sometimes called a "T network" or a "Y network"), the three impedances in the arms are $Z_a(s)$, $Z_b(s)$, and $Z_c(s)$.

Solution.   We write KVL for the left loop, noting that the current in $Z_c(s)$ is $(I_1 + I_2)$ due to KCL at the center node:

$$-V_1 + Z_a I_1 + Z_c(I_1 + I_2) = 0$$

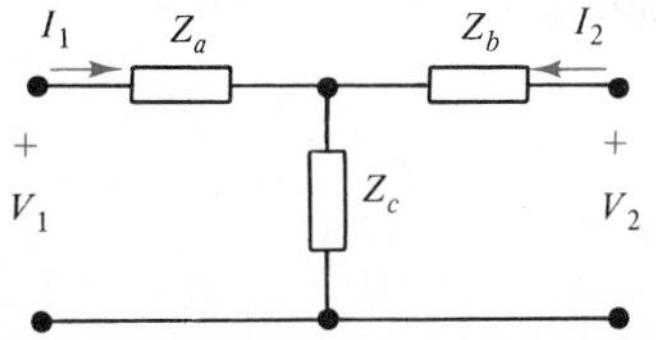

Figure 16-27   Example 16-20.

Also, KVL for the right loop is

$$-V_2 + Z_b I_2 + Z_c(I_1 + I_2) = 0$$

Rearrangement in matrix form yields

$$\begin{bmatrix} V_1 \\ V_2 \end{bmatrix} = \begin{bmatrix} Z_a + Z_c & Z_c \\ Z_c & Z_b + Z_c \end{bmatrix}\begin{bmatrix} I_1 \\ I_2 \end{bmatrix}$$

which is in the form of Equation 16-41, with

$$\mathbf{E}(s) = \begin{bmatrix} I_1(s) \\ I_2(s) \end{bmatrix} \qquad \mathbf{R}(s) = \begin{bmatrix} V_1(s) \\ V_2(s) \end{bmatrix} \qquad \mathbf{H}(s) = \begin{bmatrix} Z_a + Z_c & Z_c \\ Z_c & Z_b + Z_c \end{bmatrix}$$

By the formal approach, take one input at a time and set the other input to zero. With $I_1 \neq 0$, $I_2 = 0$, we get the partial responses

$$V_1 = (Z_a + Z_c)I_1 \qquad I_2 = 0$$

and

$$V_2 = Z_c I_1 \qquad\qquad I_2 = 0$$

Next, with $I_1 = 0$, $I_2 \neq 0$ we get

$$V_1 = Z_c I_2 \qquad\qquad I_1 = 0$$
$$V_2 = (Z_b + Z_c)I_2 \qquad I_1 = 0$$

By superposition for the linear network, the total responses add up

$$V_1 = (Z_a + Z_c)I_1 + Z_c I_2 \qquad I_1 \neq 0, I_2 \neq 0$$
$$V_2 = Z_c I_1 + (Z_b + Z_c)I_2 \qquad I_1 \neq 0, I_2 \neq 0$$

as before. In this approach, setting $I_1 = 0$, then $I_2 = 0$, means that the two experiments were conducted under open-circuit conditions at that port. The resulting 2-P parameters are called the *open-circuit impedance parameters*,

$$\mathbf{Z}_{\text{o.c.}} = \begin{bmatrix} z_{11} & z_{12} \\ z_{21} & z_{22} \end{bmatrix} = \begin{bmatrix} Z_a + Z_c & Z_c \\ Z_c & Z_b + Z_c \end{bmatrix}$$

by extension of the open-circuit resistance parameters in Chapter 5.

Be sure to recognize the distinction between these open-circuit parameters and the terms in a general loop analysis matrix (Equation 15-30 and Example 15-6). There is no relationship—except that we use the same letters and subscripts. With 26 letters and 10 digits, the choices become limited. On the other hand, there is little room for confusion between a 2-P network and a general network. ∎

EXAMPLE 16-21

Given the 2-P shown in Figure 16-28, with the dependent source and with elements denoted by their impedances.

Solution. Without memorization or formal approach, we write *any two* valid equations among V_1, I_1, V_2, and I_2. For instance, the left loop yields

$$V_1 = sI_1 + kV_2$$

The current through the capacitor is sV_2, and therefore (by KCL) the current through the resistor is $(I_2 - sV_2)$ as shown. Then the loop on the right yields

$$V_2 - kV_2 - (I_2 - sV_2) = 0$$

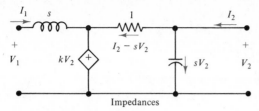

Figure 16-28   Example 16-21.

From these two informal equations, we can get *any* of the parameters of the 2-P by simply arranging them into the desired form. If, say, we want V_1 and I_2 as outputs in terms of I_1 and V_2 as inputs, we arrange the equations to read so

$$\begin{bmatrix} V_1 \\ I_2 \end{bmatrix} = \begin{bmatrix} s & k \\ 0 & \dfrac{1}{1-k+s} \end{bmatrix} \begin{bmatrix} I_1 \\ V_2 \end{bmatrix}$$

yielding the desired parameters. Don't like them? Want I_1 and I_2 as outputs, in terms of V_1 and V_2 as inputs? Again, a simple algebraic rearrangement gives

$$\begin{bmatrix} I_1 \\ I_2 \end{bmatrix} = \begin{bmatrix} \dfrac{1}{s} & -\dfrac{k}{s} \\ 0 & 1-k+s \end{bmatrix} \begin{bmatrix} V_1 \\ V_2 \end{bmatrix}$$

There is nothing to memorize, no agonizing decisions to make ("which port is open-circuited? short-circuited? and how does this affect the dependent source?").   ■

There are six sets of parameters for the 2-P network. They are the open-circuit impedance parameters, defined by

$$\begin{bmatrix} V_1 \\ V_2 \end{bmatrix} = \begin{bmatrix} z_{11} & z_{12} \\ z_{21} & z_{22} \end{bmatrix} \begin{bmatrix} I_1 \\ I_2 \end{bmatrix} \qquad \mathbf{V} = \mathbf{Z}_{\text{o.c.}} \mathbf{I} \qquad (16\text{-}42)$$

the short-circuit admittance parameters,

$$\begin{bmatrix} I_1 \\ I_2 \end{bmatrix} = \begin{bmatrix} y_{11} & y_{12} \\ y_{21} & y_{22} \end{bmatrix} \begin{bmatrix} V_1 \\ V_2 \end{bmatrix} \qquad \mathbf{I} = \mathbf{Y}_{\text{s.c.}} \mathbf{V} \qquad (16\text{-}43)$$

the hybrid parameters,

$$\begin{bmatrix} V_1 \\ I_2 \end{bmatrix} = \begin{bmatrix} h_{11} & h_{12} \\ h_{21} & h_{22} \end{bmatrix} \begin{bmatrix} I_1 \\ V_2 \end{bmatrix} = \mathbf{h} \begin{bmatrix} I_1 \\ V_2 \end{bmatrix} \qquad (16\text{-}44)$$

the inverse hybrid parameters,

$$\begin{bmatrix} I_1 \\ V_2 \end{bmatrix} = \begin{bmatrix} g_{11} & g_{12} \\ g_{21} & g_{22} \end{bmatrix} \begin{bmatrix} V_1 \\ I_2 \end{bmatrix} = \mathbf{g} \begin{bmatrix} V_1 \\ I_2 \end{bmatrix} \qquad (16\text{-}45)$$

the chain (transmission) parameters,

$$\begin{bmatrix} V_1 \\ I_1 \end{bmatrix} = \begin{bmatrix} A & B \\ C & D \end{bmatrix} \begin{bmatrix} V_2 \\ -I_2 \end{bmatrix} = \mathbf{T} \begin{bmatrix} V_2 \\ -I_2 \end{bmatrix} \qquad (16\text{-}46)$$

and the inverse chain parameters (rarely used),

$$\begin{bmatrix} V_2 \\ -I_2 \end{bmatrix} = \begin{bmatrix} E & F \\ G & H \end{bmatrix} \begin{bmatrix} V_1 \\ I_1 \end{bmatrix} \tag{16-47}$$

where all the variables and parameters are functions of s.

It does not hurt to repeat: The adjectives "open circuit" or "short circuit" *do not* mean that the 2-P network is actually operating with one port or the other open or shorted. That would be useless or disastrous. What it means is that the matrix network function may be calculated by the formal approach, setting up such an experiment on paper or in the laboratory. Once calculated, $\mathbf{Z}_{o.c.}$, or $\mathbf{Y}_{s.c.}$, or any other of the six matrix network functions, is attached to the 2-P network as its identifying tag.

The algebraic relationships among the six sets of parameters are summarized in Table 16-1. Entries in the same position are equal, so, for instance,

$$z_{21} = \frac{-y_{21}}{\Delta y} = -\frac{h_{21}}{h_{22}} = \frac{g_{21}}{g_{11}} = \frac{1}{C} \tag{16-48}$$

TABLE 16-1 2-P PARAMETERS

	$\mathbf{Z}_{o.c.}$		$\mathbf{Y}_{s.c.}$		h		g		T	
$\mathbf{Z}_{o.c.}$	z_{11}	z_{12}	$\dfrac{y_{22}}{\Delta y}$	$\dfrac{-y_{12}}{\Delta y}$	$\dfrac{\Delta h}{h_{22}}$	$\dfrac{h_{12}}{h_{22}}$	$\dfrac{1}{g_{11}}$	$\dfrac{-g_{12}}{g_{11}}$	$\dfrac{A}{C}$	$\dfrac{\Delta T}{C}$
	z_{21}	z_{22}	$\dfrac{-y_{21}}{\Delta y}$	$\dfrac{y_{11}}{\Delta y}$	$\dfrac{-h_{21}}{h_{22}}$	$\dfrac{1}{h_{22}}$	$\dfrac{g_{21}}{g_{11}}$	$\dfrac{\Delta g}{g_{11}}$	$\dfrac{1}{C}$	$\dfrac{D}{C}$
$\mathbf{Y}_{s.c.}$	$\dfrac{z_{22}}{\Delta z}$	$\dfrac{-z_{12}}{\Delta z}$	y_{11}	y_{12}	$\dfrac{1}{h_{11}}$	$\dfrac{-h_{12}}{h_{11}}$	$\dfrac{\Delta g}{g_{22}}$	$\dfrac{g_{12}}{g_{22}}$	$\dfrac{D}{B}$	$\dfrac{-\Delta T}{B}$
	$\dfrac{-z_{21}}{\Delta z}$	$\dfrac{z_{11}}{\Delta z}$	y_{21}	y_{22}	$\dfrac{h_{21}}{h_{11}}$	$\dfrac{\Delta h}{h_{11}}$	$\dfrac{-g_{21}}{g_{22}}$	$\dfrac{1}{g_{22}}$	$\dfrac{-1}{B}$	$\dfrac{A}{B}$
h	$\dfrac{\Delta z}{z_{22}}$	$\dfrac{z_{12}}{z_{22}}$	$\dfrac{1}{y_{11}}$	$\dfrac{-y_{12}}{y_{11}}$	h_{11}	h_{12}	$\dfrac{g_{22}}{\Delta g}$	$\dfrac{-g_{12}}{\Delta g}$	$\dfrac{B}{D}$	$\dfrac{\Delta T}{D}$
	$\dfrac{-z_{21}}{z_{22}}$	$\dfrac{1}{z_{22}}$	$\dfrac{y_{21}}{y_{11}}$	$\dfrac{\Delta y}{y_{11}}$	h_{21}	h_{22}	$\dfrac{-g_{21}}{\Delta g}$	$\dfrac{g_{11}}{\Delta g}$	$\dfrac{-1}{D}$	$\dfrac{C}{D}$
g	$\dfrac{1}{z_{11}}$	$\dfrac{-z_{12}}{z_{11}}$	$\dfrac{\Delta y}{y_{22}}$	$\dfrac{y_{12}}{y_{22}}$	$\dfrac{h_{22}}{\Delta h}$	$\dfrac{-h_{12}}{\Delta h}$	g_{11}	g_{12}	$\dfrac{C}{A}$	$\dfrac{-\Delta T}{A}$
	$\dfrac{z_{21}}{z_{11}}$	$\dfrac{\Delta z}{z_{11}}$	$\dfrac{-y_{21}}{y_{22}}$	$\dfrac{1}{y_{22}}$	$\dfrac{-h_{21}}{\Delta h}$	$\dfrac{h_{11}}{\Delta h}$	g_{21}	g_{22}	$\dfrac{1}{A}$	$\dfrac{B}{A}$
T	$\dfrac{z_{11}}{z_{21}}$	$\dfrac{\Delta z}{z_{21}}$	$\dfrac{-y_{22}}{y_{21}}$	$\dfrac{-1}{y_{21}}$	$\dfrac{-\Delta h}{h_{21}}$	$\dfrac{-h_{11}}{h_{21}}$	$\dfrac{1}{g_{21}}$	$\dfrac{g_{22}}{g_{21}}$	A	B
	$\dfrac{1}{z_{21}}$	$\dfrac{z_{22}}{z_{21}}$	$\dfrac{-\Delta y}{y_{21}}$	$\dfrac{-y_{11}}{y_{21}}$	$\dfrac{-h_{22}}{h_{21}}$	$\dfrac{-1}{h_{21}}$	$\dfrac{g_{11}}{g_{21}}$	$\dfrac{\Delta g}{g_{21}}$	C	D

and in each case the symbol Δ means "the determinant of...," so

$$\Delta z = z_{11}z_{22} - z_{12}z_{21} \qquad (16\text{-}49)$$

$$\Delta T = AD - BC \qquad (16\text{-}50)$$

Probs. 16-31
16-32, 16-33,
16-34

etc. Compare with Table 5-1.

PROBLEMS

16-1 Identify fully, by letter and by name, the network function in Example 15-1. Write it in its form as a rational function.

16-2 Calculate the zero-state response $V_2(s)$ in Example 15-2 with only the voltage source as input. The current source is zero (it isn't there). Identify fully the resulting network function.

16-3 Calculate the zero-state response $I_1(s)$ in Example 15-5. Identify the network function. Could you have calculated this network function quickly by another way? Do it.

16-4 Derive the voltage divider equation with generalized impedances, just as we did in the resistive and in the phasor cases. Identify and classify the network function.

16-5 Derive the current divider equation with generalized admittances. Identify and classify the network function.

16-6 Solve Example 16-7 by node analysis.

16-7 Calculate the network function in the network shown, for the output $i_L(t)$. Compute its natural frequencies if $k = 2$.

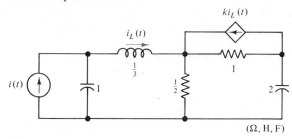

Problem 16-7

16-8 Within arbitrary multiplying constants, write the zero-state current in the 9-Ω resistor of Example 16-6 if
(a) $v(t) = 10e^{-3t}u(t)$
(b) $v(t) = 10 \cos (377t + 15°)$

16-9 Repeat Problem 16-8 for Example 16-7, with $\alpha = 2$ and
(a) $v(t) = 6u(t)$
(b) $v(t) = 10e^{-t}u(t)$

16-10 Consider a pure inductor excited by a unit step voltage. Calculate the zero-state current and plot it versus t. What is the pole of the pertinent network function? Is there resonance here? Does the answer make sense from a "primitive" point of view (how does an inductor behave for dc)?

16-11 A parallel LC circuit, $L = 1$ H, $C = \frac{1}{4}$ F, is excited by a sinusoidal current source

$$i(t) = 10 \sin \omega_0 t$$

(a) Calculate the natural frequency of the network and, consequently, the frequency of the source, ω_0, which will cause resonance in the zero-state output $v(t)$ across the circuit.

(b) Under resonance, plot $v(t)$ for $0 \le t \le 3$ s.

16-12 In Problem 16-11, a parallel resistor $R = 100\,\Omega$ is added. It accounts for the lossy inductor and capacitor.

(a) Calculate the natural frequency of this network. Will there be resonance due to the current source?

(b) Calculate the damping factor ζ for this circuit; see Chapter 8. Relate this factor and the classification of the circuit response (overdamped, underdamped, or critically damped) to part (a).

16-13 The zero-state driving-point current in a one-port network is

$$I_{dp}(s) = \frac{s}{(s^2 + 1)(s + 1)}$$

Identify completely the network and the voltage source. Is your answer unique? If not, give another solution.

16-14 With the help of a couple of examples of your choice, determine how the impulse response of a network differs from the natural response of the network with a specific input.

16-15 Calculate and plot the impulse response of each network as shown. Be sure to identify and calculate first the appropriate network function.

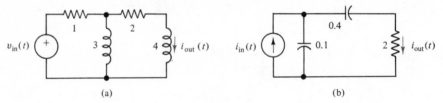

(a)   (b)

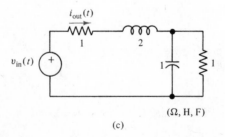

$(\Omega, \mathrm{H}, \mathrm{F})$

(c)

Problem 16-15

16-16 Use Examples 16-10 and 16-11 to develop a quick rule of thumb in deciding, at a glance of $H(s)$, whether $h(t)$ will have an impulse in it. Verify this rule on the following functions:

(a) $H_1(s) = 2.3 \dfrac{s^2 - 1}{(s + 2)(s + 10)}$ (b) $H_2(s) = \dfrac{1}{s^2 + 3s + 6}$

(c) $H_3(s) = \dfrac{1.2s^3 + 10s - 100}{s^3 + 3s^2 + 3s + 2}$ (d) $H_4(s) = 1.4 \dfrac{-s + 2}{s + 2}$

*16-17 In the lab experiments suggested for finding $h(t)$, as in Example 16-10, there is a small fly in the ointment. We don't have usually a voltage source or a current source that is truly $\delta(t)$, an impulse function. Sometimes, an approximate source will do, such as a voltage pulse of several megavolts (MV) with a very short duration of a few nanoseconds (ns). Still, we want to devise a simple lab experiment from which we can get an accurate impulse response $h(t)$.

The suggested experiment is shown in the figure. To the related network we apply as input a unit step (voltage or current), a commonly available source. We record the output, designated as $r_u(t)$, the *step response* of the network, as studied at the end of Chapter 7. Use Laplace transform calculations on this experiment, with $E(s) = 1/s$, $H(s)$ of the network, $R(s) = R_u(s)$, to prove that

$$h(t) = \frac{d}{dt} r_u(t)$$

thus providing a handy and accurate method for obtaining the impulse response from the step response.

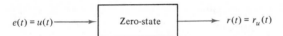

$e(t) = u(t)$ ──────→ [Zero-state] ──────→ $r(t) = r_u(t)$

Problem 16-17

16-18 An inductor L with an initial current $i_L(0^-)$ is connected at $t = 0$ across a capacitor C with an initial voltage $v_C(0^-)$. Calculate the total response $v_C(t)$ for $t > 0$. Identify clearly the zero-input response, the zero-state response, the transient response, and the steady-state response in $v_C(t)$.

16-19 For the given input and impulse response, perform convolution to find $r(t)$. Then (not earlier) check your answer by the Laplace transform.

(a) $e(t) = 10u(t)$ $h(t) = 2e^{-t}u(t)$

(b) $e(t) = 2e^{-4t}u(t)$ $h(t) = u(t).$

What name is suitable for this network?

(c) $e(t) = (4e^{-t} - e^{-2t})u(t)$ $h(t) = 2e^{-3t}u(t)$

(d) $e(t) = 10 \sin 4tu(t)$ $h(t) = \delta(t).$

Draw this network.

16-20 By convolution, find $\mathcal{L}^{-1}$ of:

(a) $\dfrac{1}{s^2} = \dfrac{1}{s} \cdot \dfrac{1}{s}$

(b) $\dfrac{1}{s^3}$ use your result from part (a)

(c) $\dfrac{10}{(s+1)(s+2)}$

*(d) $F(s)e^{-as}$

16-21 Perform graphical convolution on the two functions shown. Use the values $t_1 = 0$, $t_2 = 1$, $t_3 = 2$, $t_4 = 3$, and $t_5 = 3.1$ to calculate $r(t)$. Plot $r(t)$ versus t using these values. Extend the plot by inspection to $t_6 = 4$, $t_7 = 5$, etc.

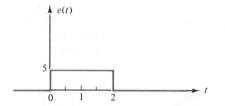 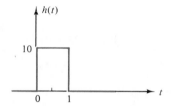

Problem 16-21

16-22 Rework Example 16-17, this time with $e(x)$ stationary and $h(t - x)$ reflected and sliding. Your final result, $r(t)$, must be the same as the one in Figure 16-20(i).

16-23 Use graphical convolution to obtain the analytical expression for $r(t)$ over the various ranges of t, given $h(t)$ and $e(t)$ as shown. Here it is easier, although not mandatory, to reflect and slide h.

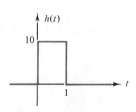 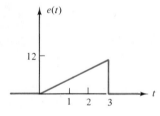

Problem 16-23

16-24 Apply the numerical convolution algorithm (Equation 16-37) to Example 16-17 by using $T = 0.5$ s. Plot $r(n)$ and compare with Figure 16-20(i).

16-25 Write a short computer program to implement numerical convolution (Equation 16-37). Run Problem 16-24 on it, with $T = 0.5$ s, then with $T = 0.01$ s. Compare the results with Figure 16-21(i), the exact response.

16-26 In Figure 16-21(c) show the present value ($x = 0$) of the input and the future values ($x > 0$) of the input. As e is reflected and sliding to the right, show to your satisfaction that only past and present values of $e(t - x)$ affect the output $r(t_1)$—not future values of e beyond t_1. This is plain common sense: Future inputs cannot possibly affect the present output. The network is *causal* and obeys the natural sequence of "cause and effect."

16-27 Use the graphical-numerical convolution to prove the sampling property of the impulse function in Equation 16-24.

16-28 By using convolution, prove the well-known property of the Laplace transform

$$\mathscr{L}\left\{\int_0^t f(x)\,dx\right\} = \frac{1}{s}F(s)$$

16-29 The network shown is in the zero state at $t = 0$. Find its Thévenin and Norton equivalent circuits at terminals A–B.

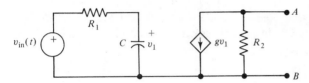

Problem 16-29

16-30 Repeat Problem 16-29 for the network shown. There is a current-controlled voltage source. (Hey, you really don't have to redraw the circuit with terminals A–B at the right!)

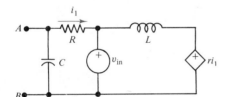

Problem 16-30

16-31 **(a)** For the 2-P network shown, calculate the short-circuit admittance parameters $\mathbf{Y}_{\text{s.c.}}$ using the informal approach.

(b) Verify by the formal approach.

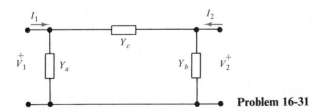

Problem 16-31

16-32 **(a)** Use the informal approach to calculate the inverse hybrid parameters **g** for the network shown.

(b) With Table 16-1, obtain $\mathbf{Z}_{\text{o.c.}}$ for this 2-P.

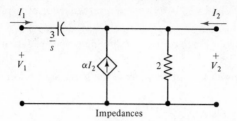

Impedances   **Problem 16-32**

16-33 A 2-P, given by its $\mathbf{Z}_{o.c.}$, is terminated by a load impedance $Z_L(s)$ as shown. Calculate the voltage transfer function $G(s)$ in

$$V_2(s) = G(s)V_1(s)$$

in terms of z_{11}, z_{12}, z_{21} and z_{22}, and Z_L. Next, let $Z_L(s) \to \infty$ (an open circuit) and compare with the result obtained directly from the defining equations of the 2-P.

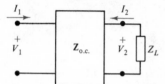

Problem 16-33

*16-34 Use Table 16-1 and $\mathbf{Y}_{s.c.}$ for the 2-P network in Problem 16-31 to obtain its $\mathbf{Z}_{o.c.}$. Compare that result with $\mathbf{Z}_{o.c.}$ of the T network in Example 16-20. What have you derived here?

Chapter 17

Frequency Response and Filters

In this chapter, we expand our preliminary discussion of frequency response (Chapter 10) and relate it to several other topics.

17-1 FREQUENCY RESPONSE, POLES AND ZEROS

In Chapters 9 and 16, we had the phasor relationship for sinusoidal steady-state input-output as

$$\mathbf{R} = \mathbf{HE} \tag{17-1}$$

where $\mathbf{E}$ is the phasor input, $\mathbf{R}$ the phasor output, and $\mathbf{H}$ the (complex number) network function. Equation 17-1 holds for the sinusoidal steady state at a *single* frequency. We also saw that if we substitute $s = j\omega$ in the general network function $H(s)$ we'll get

$$R(s)]_{s=j\omega} = H(s)E(s)]_{s=j\omega} \tag{17-2}$$

that is,

$$R(j\omega) = H(j\omega)E(j\omega) \tag{17-3}$$

This relation is valid for any zero-state sinusoidal steady state of *any* frequency ω. As such, Equation 17-3 is a generalization of Equation 17-1, and $E(j\omega)$ and $R(j\omega)$ may be considered as *generalized phasors*. For a single value of ω, Equation 17-3 becomes Equation 17-1.†

From Equation 17-3, we obtain the magnitude of the response as

$$|R(j\omega)| = |H(j\omega)||E(j\omega)| \tag{17-4}$$

---

† It is a curious paradox in semantics that a *real* frequency ω is denoted as the *imaginary* part of s, $s = \sigma + j\omega$. Equally curious is the fact that the real part of the complex frequency, σ, is not a real frequency in the common sense: The waveform $e^{\sigma t}$ does not oscillate! But, after all, it's only a matter of semantics.

and the phase of the response

$$\theta_R(j\omega) = \theta_H(j\omega) + \theta_E(j\omega) \qquad (17\text{-}5)$$

Compare these results with Equation 16-3. Therefore, the magnitude $|H(j\omega)|$ and phase $\theta_H(j\omega)$ of a network function are very important in determining the zero-state output. They can be obtained either analytically (on paper) or in the laboratory. Together, they are called the *frequency response*. In specific cases, we'll refer to the *magnitude response* or the *phase response* as the behavior of $|H(j\omega)|$ versus ω and of $\theta_H(j\omega)$ versus ω, respectively. Our first example will cover all the important points.

EXAMPLE 17-1 ———————————————————————————

For the network shown in Figure 17-1(a), calculate the voltage transfer function and plot its frequency response curves.

> *Solution.* Although we did some of the work in Example 16-2, let us do it here. We write a voltage divider equation
>
> $$V_{out}(s) = \frac{1/Cs}{R + 1/Cs} V_{in}(s) = \frac{1/RC}{s + 1/RC} V_{in}(s)$$
>
> and identify the network function as
>
> $$H(s) = G(s) = \frac{1/RC}{s + 1/RC}$$
>
> For $s = j\omega$, we have
>
> $$G(j\omega) = \frac{1/RC}{j\omega + 1/RC}$$
>
> The magnitude of $G(j\omega)$ is the ratio of the magnitudes of its numerator and denominator
>
> $$|G(j\omega)| = \frac{|1/RC|}{|j\omega + 1/RC|} = \frac{1/RC}{\sqrt{\omega^2 + (1/RC)^2}}$$
>
> Its phase is the difference between the angles of the numerator and the denominator,
>
> $$\theta_G(j\omega) = 0° - \tan^{-1}\omega RC$$
>
> Here, the numerator is a real, positive number whose angle is zero. The denominator is of the general form $a + jb$ whose angle is $\tan^{-1}(b/a)$.
>
> The plots of $|G(j\omega)|$ and $\theta_G(j\omega)$ are obtained by calculating a few significant points for each:
>
> 1. For very low frequencies ($\omega \to 0$), we get $|G| \to 1$. This may be obtained immediately from the original $G(s)$, because if $\omega \to 0$ then $s = j\omega \to 0$. Setting $s = 0$ in $G(s)$ gives $G(0) = 1$. The angle $\theta_G(j\omega)$ is calculated as $0°$ when $\omega \to 0$.
>
>   These two points, $|G(0)| = 1$ and $\theta_G(0) = 0°$, are plotted. They show how the two curves begin. We call them the *low-frequency asymptotes*. Do these values make sense physically? (Never fail to check it out!) Yes, they do. As $\omega \to 0$, we get dc (constant) input and output. For dc, the capacitor is an open circuit and since there is no current through R, then $v_{out}(t) = v_{in}(t)$. The magnitude and phase of the dc output are identical to those of the dc input, and so $|G| = 1$ and $\theta_G = 0°$.

2. The *high-frequency asymptotes* (how the two curves end) are also easy to compute. As $\omega \to \infty$, the magnitude of $G(j\omega)$ approaches zero

$$\lim_{\omega \to \infty} \frac{1/RC}{\sqrt{\omega^2 + (1/RC)^2}} = 0$$

while the phase approaches $-90°$

$$\lim_{\omega \to \infty} (-\tan^{-1}\omega RC) = -90°$$

Physically, as $\omega \to \infty$, the capacitor becomes a short circuit, and therefore the output voltage across it approaches zero in magnitude, and a phase of $-90°$ due to $1/(j\omega C)$, the angle of $1/j = -j$ being $-90°$.

3. A couple of points in the middle range of ω will complete the plots. At $\omega = 1/RC$, the magnitude of $G(j\omega)$ is

$$\left| G\left(j\frac{1}{RC}\right) \right| = \frac{1/RC}{\sqrt{2/(RC)^2}} = \frac{1}{\sqrt{2}} \approx 0.707$$

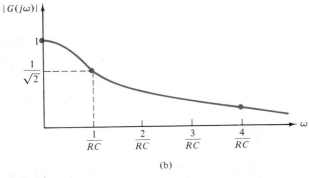

(a)

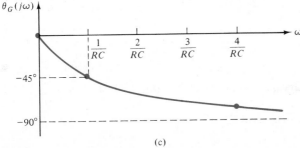

(b)

(c)

Figure 17-1 Example 17-1.

and the phase

$$\theta_G\left(j\frac{1}{RC}\right) = -\tan^{-1}1 = -45°$$

At $\omega = 4/RC$ we calculate

$$\left|G\left(j\frac{4}{RC}\right)\right| = 0.243$$

and

$$\theta_G\left(j\frac{4}{RC}\right) = -76°$$

The frequency plots are shown in Figure 17-1(b) and (c).

In Chapter 9 we classified such a network as a *lowpass filter*. Specifically, inputs in the frequency range near $\omega = 0$ will pass through the network, $|G| \approx 1$, while signals of high frequencies will not show at the output because $|G| \approx 0$.

A convenient *cutoff frequency*, ω_c, separating the *passband* from the *stopband*, is one where the magnitude response is down from its maximum by a factor of $\sqrt{2}$. Here, then $\omega_c = 1/\sqrt{2}$. The passband is then

$$0 \leq \omega \leq \omega_c$$

and the stopband is

$$\omega_c \leq \omega \leq \infty$$

■ Prob. 17-1

The general network function is a rational function, as in Equation 15-36

$$H(s) = K\frac{s^m + a_1 s^{m-1} + a_2 s^{m-2} + \cdots + a_{m-1}s + a_m}{s^n + b_1 s^{n-1} + b_2 s^{n-2} + \cdots + b_{n-1}s + b_n} = K\frac{P(s)}{Q(s)} \qquad (17\text{-}6)$$

Let $s = j\omega$ to get

$$H(j\omega) = K\frac{(j\omega)^m + a_1(j\omega)^{m-1} + \cdots + a_{m-1}(j\omega) + a_m}{(j\omega)^n + b_1(j\omega)^{n-1} + \cdots + b_{n-1}(j\omega) + b_n} = K\frac{P(j\omega)}{Q(j\omega)} \qquad (17\text{-}7)$$

and we can write the numerator and the denominator in their rectangular forms as

$$H(j\omega) = K\frac{A + jB}{C + jD} \qquad (17\text{-}8)$$

where A is the real part of $P(j\omega)$, and B its imaginary part. Similarly, C is the real part of $Q(j\omega)$ and D its imaginary part. They are all functions of ω, but we want to keep the notation simple.

From Equation 17-8 we have

$$|H(j\omega)| = |K|\frac{|P|}{|Q|} = |K|\sqrt{\frac{A^2 + B^2}{C^2 + D^2}} \qquad (17\text{-}9)$$

and

$$\theta_H(j\omega) = \theta_K + \theta_P - \theta_Q = \begin{cases} \tan^{-1}\dfrac{B}{A} - \tan^{-1}\dfrac{D}{C} \\[2ex] 180° + \tan^{-1}\dfrac{B}{A} - \tan^{-1}\dfrac{D}{C} \end{cases} \qquad (17\text{-}10)$$

where $\theta_K = 0°$ if $K > 0$ and $\theta_K = 180°$ if $K < 0$.

These equations look formidable. Actually, they need not be memorized. Their derivation is presented only to show the general approach. It is easier to work out each case on its own, following these guidelines.

EXAMPLE 17-2 ———————————————————————————————————————

Given

$$H(s) = -2\,\frac{s-3}{2s^2 + 3s + 8}$$

Solution. Set $s = j\omega$ to get

$$H(j\omega) = -2\,\frac{j\omega - 3}{2(j\omega)^2 + 3j\omega + 8} = -2\,\frac{-3 + j\omega}{(8 - 2\omega^2) + j3\omega}$$

where we collected real terms and imaginary terms in both numerator and denominator and wrote them as in Equation 17-8. The magnitude of $H(j\omega)$ is then

$$|H(j\omega)| = 2\,\sqrt{\frac{(-3)^2 + \omega^2}{(8 - 2\omega^2)^2 + (3\omega)^2}} = 2\,\sqrt{\frac{\omega^2 + 9}{(8 - 2\omega^2)^2 + 9\omega^2}}$$

and the phase is (here $K = -2$ and it contributes 180°)

$$\theta_H(j\omega) = 180° + \tan^{-1}\frac{\omega}{-3} - \tan^{-1}\frac{3\omega}{8 - 2\omega^2} \qquad \blacksquare$$

It would be helpful to develop general guidelines for the high-frequency and low-frequency asymptotes for the frequency response in Equation 17-7. As $\omega \to \infty$, we have

$$\lim_{\omega \to \infty} H(j\omega) = \lim_{\omega \to \infty} K\,\frac{(j\omega)^m}{(j\omega)^n} = \lim_{\omega \to \infty} K(j\omega)^{m-n} \qquad (17\text{-}11)$$

because then the highest powers of ω prevail over the lower powers in P and in Q. There are three possibilities to consider:

1. $m > n$. The magnitude of $H(j\omega)$ is then very large

$$\lim_{\omega \to \infty} |H(j\omega)| = \lim_{\omega \to \infty} |K|\omega^{m-n} \to \infty \qquad (17\text{-}12)$$

Don't forget that $|j|^{m-n} = 1$. The phase of $H(j\omega)$ is

$$\lim_{\omega \to \infty} \theta_H(j\omega) = (m - n)90° \qquad (17\text{-}13)$$

because every j contributes 90°. There may be an additional 180° in Equation 17-13 if $K < 0$.

2. $m = n$. Then the magnitude is

$$\lim_{\omega \to \infty} |H(j\omega)| = \lim_{\omega \to \infty} |K| = |K| \qquad (17\text{-}14)$$

and

$$\lim_{\omega \to \infty} \theta_H(j\omega) = \begin{cases} 0° & K > 0 \\ 180° & K < 0 \end{cases} \qquad (17\text{-}15)$$

3. $m < n$. Then

$$\lim_{\omega \to \infty} |H(j\omega)| = \lim_{\omega \to \infty} |K|\omega^{m-n} = 0 \qquad (17\text{-}16)$$

and

$$\lim_{\omega \to \infty} \theta_H(j\omega) = (m - n)90° \quad (+180°) \qquad (17\text{-}17)$$

These are shown in Figure 17-2. In a similar way, the low-frequency asymptotes are dictated by the *lowest* powers in P and Q, and they can be determined quickly.

In summary, let us repeat that no memorization is needed of Equations 17-11 to 17-17. High-frequency behavior is governed by the highest powers of ω (or of s), and low-frequency behavior by the lowest powers of ω (or s). Each case is then treated on its own.

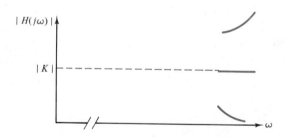

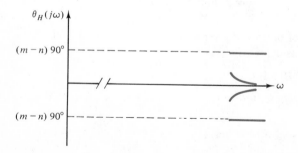

Figure 17-2   High-frequency asymptotes.

EXAMPLE 17-3 ——————————————————————————————

Investigate the asymptotic behavior of

$$H(s) = -2 \frac{s - 3}{2s^2 + 3s + 8}$$

from Example 17-2.

Solution. For high frequencies, we have

$$\lim_{s \to \infty} H(s) = -2 \frac{s}{2s^2} = -\frac{1}{s} \xrightarrow{s = j\omega} -\frac{1}{j\omega} = j\frac{1}{\omega}$$

Therefore,

$$\lim_{\omega \to \infty} |H(j\omega)| = \lim_{\omega \to \infty} \frac{1}{\omega} = 0$$

and

$$\lim_{\omega \to \infty} \theta_H(j\omega) = 90°$$

For low frequencies

$$\lim_{s \to 0} H(s) = \lim_{s \to 0} \left[-2 \left(\frac{-3}{8} \right) \right] = \frac{3}{4}$$

and so

$$\lim_{\omega \to 0} |H(j\omega)| = \frac{3}{4}$$

and

$$\lim_{\omega \to 0} \theta_H(j\omega) = 0°$$

Several intermediate points must be calculated by the expressions developed in Example 17-2 before the complete plots can be drawn. ■   Probs. 17-2, 17-3, 17-4

A graphical method for evaluating points in the frequency response will be a most welcome addition to our knowledge. After all, engineers love graphical methods. The graphical method will depend (not surprisingly) on the location of the poles and zeros of $H(s)$ in the s plane.

We write $H(s)$ in its factored form, taken from Equation 17-6, as

$$H(s) = K \frac{P(s)}{Q(s)} = K \frac{(s - z_1)(s - z_2) \cdots (s - z_m)}{(s - p_1)(s - p_2) \cdots (s - p_n)} \qquad (17\text{-}18)$$

showing the finite zeros $z_1, z_2, \ldots, z_m$, the roots of $P(s) = 0$, and the finite poles $p_1, p_2, \ldots, p_n$, the natural frequencies, the roots of $Q(s) = 0$. They may be real, complex, simple, or multiple. We show them on the complex s plane with a small circle **O** for a zero, and a small **x** for a pole.

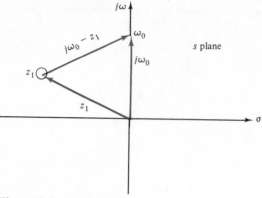

Figure 17-3 Geometry of $(j\omega_0 - z_1)$.

To calculate $H(j\omega_0)$, at a specific $\omega = \omega_0$, we write Equation 17-18 as

$$H(j\omega_0) = K \frac{(j\omega_0 - z_1)(j\omega_0 - z_2)\cdots(j\omega_0 - z_m)}{(j\omega_0 - p_1)(j\omega_0 - p_2)\cdots(j\omega_0 - p_n)} \qquad (17\text{-}19)$$

and consider the geometry of the term $(j\omega_0 - z_1)$. We let z_1 be a complex number, as shown in Figure 17-3. Therefore, it is given graphically by the directed line from the origin to z_1. In a similar way, the number $j\omega_0$ is a directed line from the origin to $j\omega_0$. The subtraction $(j\omega_0 - z_1)$ yields geometrically a directed line going *from z_1 to $j\omega_0$*. As a quick check, add vectorially to get $z_1 + (j\omega_0 - z_1) = j\omega_0$. In a similar way $(j\omega_0 - z_2)$ is a directed line *from z_2 to $j\omega_0$*, etc. These are called *zero lines*, from each zero to the point $j\omega_0$. Likewise, $(j\omega_0 - p_1)$, $(j\omega_0 - p_2)$, etc., are *pole lines*, from each pole to the point $j\omega_0$. Therefore, Equation 17-19 can be written, for $\omega = \omega_0$, as

$$H(j\omega_0) = K \frac{\text{product of all zero lines to } j\omega_0}{\text{product of all pole lines to } j\omega_0} \qquad (17\text{-}20)$$

We'll write every zero line and pole line in its polar form, in magnitude and angle, then calculate $H(j\omega_0)$ according to Equation 17-20.

EXAMPLE 17-4 _____

Let

$$H(s) = 10 \frac{s - 2}{s^2 + 2s + 10}$$

with a finite zero at $z_1 = 2$ and two complex conjugate poles at $p_1 = -1 + j3$, $p_2 = p_1^* = -1 - j3$. See Figure 17-4(a). Let us evaluate $H(j1)$.

Solution. We draw the zero line from $z_1 = 2$ to $j1$, and the two pole lines, from p_1 to $j1$ and from p_2 to $j1$. The lengths and angles of the lines can be measured directly from an accurate sketch, with only a ruler and a protractor; it's that easy! Or else, lengths and angles are calculated easily from the geometry of the sketch. Here we have

$$H(j1) = 10 \frac{\sqrt{5}\underline{/153.4°}}{(\sqrt{17}\underline{/76°})(\sqrt{5}\underline{/-63.4°})} = \frac{10\sqrt{5}}{\sqrt{17}\sqrt{5}}\underline{/153.4° - (76° - 63.4°)}$$

$$= 2.43\underline{/140.8°} = |H(j1)|\underline{/\theta_H(j1)}$$

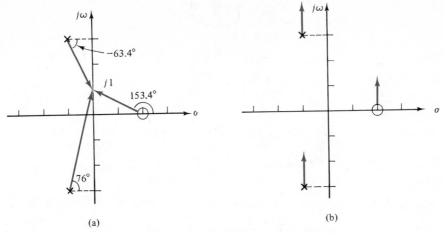

Figure 17-4 Example 17-4.

The same approach is repeated for as many points ω_0 as needed, to get the magnitude and the angle plots.

In Figure 17-4(b) we show this construction for the high-frequency asymptote $\omega_0 \to \infty$. Then we get a zero line of infinite length (M) at 90°, and two pole lines, each of infinite length (M) at 90°. Therefore

$$\lim_{\omega \to \infty} H(j\omega) = \lim_{M \to \infty} 10 \frac{M \underline{/90°}}{(M \underline{/90°})(M \underline{/90°})} = \lim_{M \to \infty} \frac{10}{M} \underline{/-90°}$$

Consequently, $|H(j\infty)| \to 0$ and $\theta_H(j\infty) = -90°$. ∎

Probs. 17-5, 17-6

A concluding remark is in order here. For a multiple zero, the corresponding factor is $(s - z_i)^r$, the zero z_i being of multiplicity $r = 2, 3, \ldots$. There will be r zero lines sitting on top of each other in the graphical method. Length and angle will be counted r times. The same is true for a multiple pole.

17-2 FREQUENCY RESPONSE—BODE PLOTS

It is common in engineering to measure and plot the *gain function* $\alpha(j\omega)$ which is related to the magnitude $|H(j\omega)|$ as follows:

$$\alpha(j\omega) = 20 \log |H(j\omega)| \qquad \text{dB} \qquad (17\text{-}21)$$

with the unit of decibel (dB). The log is to the base 10. Also, because of the wide range of frequencies, we compress the horizontal axis from a linear scale in ω to a logarithmic scale. In Figure 17-5 we see a linear scale of x, with equal distances between the points $x = 0, 1, 2, \ldots$. Below it we have a logarithmic scale of ω, with

$$x = \log \omega \qquad \omega = 10^x \qquad (17\text{-}22)$$

allowing us to squeeze a big range of frequencies on a reasonable plot. We speak of two frequencies ω_1 and ω_2 being one *decade* apart if

$$\omega_1 = 10\omega_2 \qquad (17\text{-}23)$$

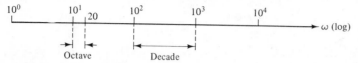

Figure 17-5 Logarithmic scale.

so, for example, $\omega_1 = 2000$ is one decade above $\omega_2 = 200$. Another common interval is the *octave*, defined by

$$\omega_1 = 2\omega_2 \qquad (17\text{-}24)$$

The term octave is borrowed from music, where the frequencies of two notes that are one octave apart obey Equation 17-24.

Frequency response plots of gain and phase versus a logarithmic scale of ω are called Bode plots, named after H. W. Bode (1905–1982), an American engineer and educator.† To appreciate fully the advantages of Bode plots, let us consider the general form of $H(s)$ as in Equation 17-18. To obtain the gain function, we write

$$\alpha(j\omega) = 20 \log |H(j\omega)| = 20 \log |K| \frac{|j\omega - z_1||j\omega - z_2| \cdots |j\omega - z_m|}{|j\omega - p_1||j\omega - p_2| \cdots |j\omega - p_n|} \quad (17\text{-}25)$$

that is,

$$\alpha(j\omega) = 20 \log |K| + \sum_i 20 \log |j\omega - z_i| - \sum_k 20 \log |j\omega - p_k| \qquad (17\text{-}26)$$

The advantage is clear now: Instead of *multiplying* and *dividing* terms in $|H(j\omega)|$, as in Equation 17-19, we *add* and *subtract* terms in $\alpha(j\omega)$. The phase plot is, as before,

$$\theta_H(j\omega) = \theta_K + \sum \theta_z - \sum \theta_p \qquad (17\text{-}27)$$

where θ_K is the angle contributed by K, θ_z is the angle of a zero factor $(j\omega - z_i)$, and θ_p is the angle of a pole factor $(j\omega - p_k)$. All that we have to do now is calculate the individual contributions in Equations 17-26 and 17-27. We do it step by step:

1. The Constant K   The gain is

$$\alpha = 20 \log |K| \qquad \text{dB} \qquad (17\text{-}28)$$

which is a constant. It can be added to the total gain in Equation 17-26 by raising or lowering the plot. In practice, such a "dc gain" is often ignored in the plots, since it does not affect the shape of the plot. The angle contribution of K is

$$\theta_K = \begin{cases} 0° & K > 0 \\ 180° & K < 0 \end{cases} \qquad (17\text{-}29)$$

† The name rhymes with *Jody*, not with *node*.

2. A First-Order Factor $(j\omega - z_i)$ Let us write it in a standard form as

$$j\omega - z_i = \frac{1}{T_i}(1 + j\omega T_i) \quad -z_i = \frac{1}{T_i} \tag{17-30}$$

The factor $1/T_i$ can be absorbed in K. The gain of the first-order factor is then

$$\alpha = 20 \log |1 + j\omega T_i| = 20 \log (1 + \omega^2 T_i^2)^{1/2} \quad dB \tag{17-31}$$

and the angle

$$\theta = \tan^{-1} \omega T_i \tag{17-32}$$

Another advantage of the Bode plot is the asymptotic plot of the gain. For very small frequencies, $\omega \ll 1/T_i$, $\omega^2 T_i^2 \ll 1$, and Equation 17-31 yields

$$\alpha \approx 20 \log 1 = 0 \, dB \tag{17-33}$$

and for large frequencies, $\omega \gg 1/T_i$, $\omega^2 T_i^2 \gg 1$, and

$$\alpha \approx 20 \log \omega T_i \quad dB \tag{17-34}$$

As we increase ω by a decade, from ωT_i to $10\omega T_i$, the value of α will increase by $20 \log 10 = 20 \, dB$. Therefore, the plot of Equation 17-34 is a straight line with a slope of 20 dB/decade ($= 6$ dB/octave).

The low-frequency asymptote and the high-frequency asymptote meet when

$$0 = 20 \log \omega T_i \tag{17-35}$$

or at

$$\omega = \frac{1}{T_i} \tag{17-36}$$

which is called the *break frequency* or the *corner frequency*. This is an old friend, the half-power frequency in the resonant *RLC* circuit (Chapters 8 and 9), and the reciprocal of the time constant, as in Figure 17-1(b).

At the break frequency, the exact gain is

$$\alpha = 20 \log \sqrt{2} \approx 3 \, dB \tag{17-37}$$

A few additional points are listed in Table 17-1. At the corner frequency, the exact curve of $\alpha(j\omega)$ is 3 dB away from the asymptote; one octave above and below the corner frequency, the curve is 1 dB away. Figure 17-6(a) summarizes these results. They apply also to a pole factor $(s - p_k)$, with negative signs.

TABLE 17-1   GAIN OF $(1 + j\omega T_i)$

ω	Exact α	Asymptote
$1/2T_i$	$20 \log \sqrt{\frac{5}{4}} \approx 1$ dB	0 dB
$1/T_i$	$20 \log \sqrt{2} \approx 3$ dB	0 dB
$2/T_i$	$20 \log \sqrt{5} \approx 7$ dB	6 dB

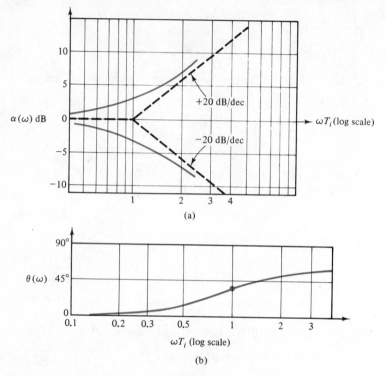

Figure 17-6 Bode plots of a first-order factor.

The phase, according to Equation 17-32, has 0° as the low-frequency asymptote and 90° as the high-frequency asymptote. At the break frequency it is 45°. Its plot is shown in Figure 17-6(b).

EXAMPLE 17-5 _____

$$H(s) = 2.3 \frac{(s + 10)(s + 3200)}{(s + 80)(s + 600)}$$

The break frequencies are at $\omega = 10$, $\omega = 3200$ (zero factors), and $\omega = 80$ and $\omega = 600$ (pole factors). Draw the gain asymptote for each factor, as in Figure 17-7(a), each at 0 dB below its break frequency, and a straight line of ±6 dB/octave beyond its break frequency. Each slope is marked beside its asymptote. Now add these individual asymptotes, to get the total asymptote, as shown in Figure 17-7(b):

Up to $\omega = 10$, the sum of the asymptotes is $0 + 0 + 0 + 0 = 0$.

Between $\omega = 10$ and $\omega = 80$, it is $0 + 6 = 6$ dB/octave.

Between $\omega = 80$ and $\omega = 600$, it is $6 - 6 = 0$ dB/octave.

Between $\omega = 600$ and $\omega = 3200$, it is $0 - 6 = -6$ dB/octave and beyond $\omega = 3200$ it is $-6 + 6 = 0$ dB/octave.

In most cases, this total asymptote is good enough! We add the dc gain = $20 \log\{2.3[(10)(3200)/(80)(600)]\} = 3.71$ dB, and note the asymptotic value for large ω,

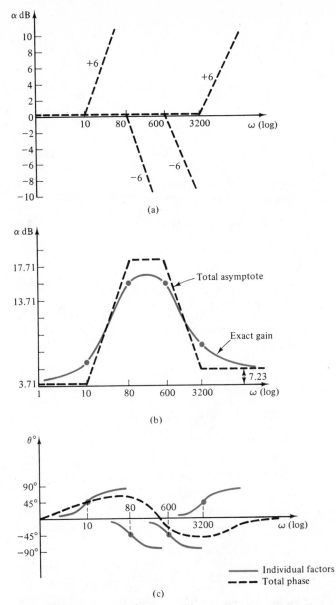

Figure 17-7 Example 17-5.

$20 \log 2.3 = 7.23$ dB. If we need more accuracy, we add the exact points at the break frequencies and draw the exact gain curve. The individual phase plots and the total phase are shown in Figure 17-7(c). ■

3. A Second-Order Factor A second-order factor arises from a pair of complex conjugate zeros or poles,

$$(s + b + jc)(s + b - jc) = s^2 + 2bs + b^2 + c^2 \qquad (17\text{-}38)$$

If we let

$$\frac{1}{T^2} = \omega_0^2 = b^2 + c^2 \tag{17-39}$$

$$\zeta = \frac{b}{\omega_0} \tag{17-40}$$

Then

$$s^2 + 2bs + b^2 + c^2 = \frac{1}{T^2}(1 + 2\zeta sT + s^2 T^2) \tag{17-41}$$

where $\zeta \leq 1$ is the *damping factor* from Chapter 8. Again, the constant $1/T^2$ is absorbed in K, and so we have the gain function

$$\alpha(j\omega) = \pm 10 \log [(1 - \omega^2 T^2)^2 + (2\zeta\omega T)^2] \tag{17-42}$$

and the phase function

$$\theta(j\omega) = \tan^{-1}\frac{2\zeta\omega T}{1 - \omega^2 T^2} \tag{17-43}$$

Several plots of α and θ, for common values of ζ, are given in Figure 17-8 for a second-order pole factor. For a zero factor, use the negative values of α and θ. Here we note that the low-frequency asymptote for the gain is

$$\alpha \approx 10 \log 1 = 0 \text{ dB} \qquad \omega \ll \omega_0 \tag{17-44}$$

while, for $\omega \gg \omega_0$, the high-frequency asymptote is

$$\alpha \approx \pm 10 \log (-\omega^2 T^2)^2 = \pm 40 \log \omega T \qquad \text{dB} \tag{17-45}$$

that is, a straight line with a slope of 40 dB/decade. This is twice the slope of a first-order factor, and it makes good sense: A quadratic is, in principle, a squared first-order factor. Similarly, from Equation 17-43, we get the low-frequency and the high-frequency asymptotes for θ

$$\theta \approx 0° \qquad \omega \ll \omega_0 \tag{17-46}$$

and

$$\theta \approx \pm 180° \qquad \omega \gg \omega_0 \tag{17-47}$$

At the break frequency, $\theta = \pm 90°$ for all ζ, as seen from Equation 17-43.

In summary, these are the steps for plotting the Bode frequency response curves:

1. Write $H(s)$ in its factored form, in both numerator $P(s)$ and denominator $Q(s)$.
2. Rewrite all first-order factors in their standard form (Equation 17-30). Identify and mark the break frequencies on the $\omega(\log)$ axis. Draw the first-order gain asymptotes, flat 0 dB up to the break frequency, a straight line ± 6 dB/octave ($= \pm 20$ dB/decade) beyond. The $+6$ goes with zero factors, -6 with pole factors.

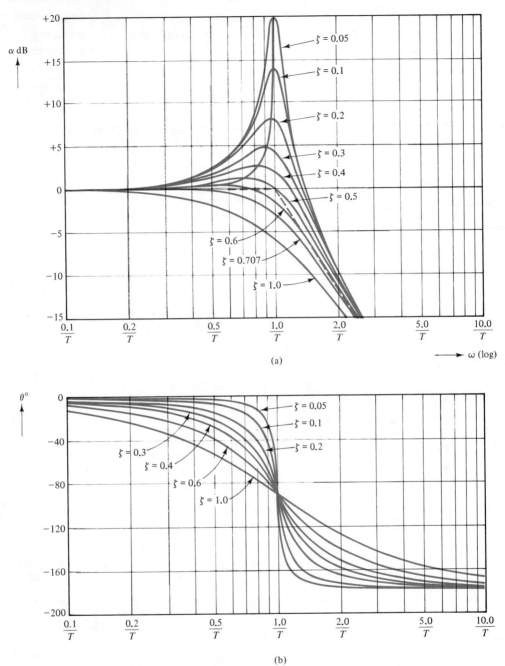

Figure 17-8 Bode plots for a second-order factor.

3. Rewrite all second-order factors in their standard form (Equation 17-41). Identify and mark the break frequencies. Identify the damping factor ζ for each case. Draw the individual gain asymptotes, 0 dB up to the break frequency, ± 12 dB/octave ($= \pm 40$ dB/decade) beyond.
4. Add all the asymptotes to get the total asymptote.
5. Draw the necessary correction values for the first-order factors from Table 17-1, and for second-order factors from Figure 17-8.
6. Draw a smooth curve through these points, guided by the asymptotes.
7. Follow the same steps for the phase plot.

EXAMPLE 17-6 ───

$$H(s) = 1.6 \, \frac{s(s-2)}{s^2 + 4s + 100}$$

Solution. Rewrite each factor in its standard form to get

$$H(s) = \frac{(1.6)(2)}{100} \, \frac{s(s/2 - 1)}{(s/10)^2 + \frac{4}{10}(s/10) + 1}$$

Therefore, we have:

1. A constant multiplier $K = 0.032$
2. A first-order zero factor, s
3. A first-order zero factor $(s/2) - 1$
4. A second-order pole factor $(s/10)^2 + 2(0.2)(s/10) + 1$

The gain plot proceeds as follows:

1. The gain of K is $20 \log 0.032 = -30$ dB, and will be added at the end.
2. The zero factor s has a break frequency $\omega = 1$ and the line rises at $+6$ dB/octave (see Problem 17-8, which you did earlier, right?).
3. The zero factor $(s/2) - 1$ has a break frequency $\omega = 2$, and the line rises at $+6$ dB/octave.
4. The second-order pole has a break frequency at $\omega = 10$, and a damping factor $\zeta = 0.2$. The high-frequency asymptote goes down at -12 dB/octave.
      These individual asymptotes are shown in Figure 17-9(a). The total asymptote is shown in Figure 17-9(b) in dashed lines. A correction of $+3$ dB is added at the break frequency $\omega = 2$ for the factor $(s - 2)$. At $\omega = 10$, a correction of $+7$ dB is read from Figure 17-8(a) for $\zeta = 0.2$, and added to the curve; the corrections of s and of $(s - 2)$ are negligible at $\omega = 10$. The solid line in Figure 17-9(b) is the true gain curve $\alpha(\omega)$. Finally, the entire curve must be shifted down by 30 dB, due to the factor K. Or else, relabel the vertical axis of α, making the old 0-dB point a new point of -30 dB, etc. This was actually done in the figure.

The phase plot goes along these lines:

1. K contributes $0°$.
2. The factor $s = j\omega$ contributes $90°$ for all ω.

3. The angle of $(j\omega - 2)$ is 180° for $\omega \ll 2$, it is 135° at $\omega = 2$, and it is 90° for $\omega \gg 2$. Don't rely on lengthy memorizations; rather, think quickly $\theta = \tan^{-1}(\omega/-2)$ and take it from there, or use the zero line.
4. The angle contribution of the second-order pole is 0° for $\omega \ll 10$, $-180°$ for $\omega \gg 10$, and $-90°$ at $\omega = 10$, according to Figure 17-8(b).

These individual phase plots are shown in dotted lines in Figure 17-9(c). Their sum is the total phase plot, shown in a solid line.

Probs. 17-10, 17-11, 17-12

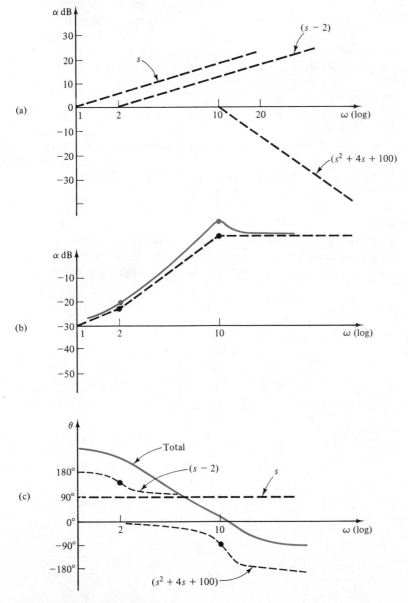

Figure 17-9 Example 17-6.

17-3 FILTERS—AGAIN

In our previous encounter with filters in Chapters 8, 10, and 16, we defined a filter as a frequency-selective two-port network. Its transfer function, $H(j\omega)$, has specific characteristics which let inputs at certain frequencies pass, others not to pass to the output. In Figure 17-10 we show the *ideal* magnitude characteristics of a lowpass, bandpass, bandstop, and highpass filter. The ideal lowpass filter, for example, has $|H(j\omega)| = 1$† for $0 < \omega < \omega_c$, and therefore all inputs of frequencies in the passband $0 < \omega < \omega_c$ will produce outputs according to $|R(j\omega)| = |H(j\omega)||E(j\omega)| = |E(j\omega)|$. In the stopband $\omega_c < \omega < \infty$, $|H(j\omega)| = 0$, and consequently $R(j\omega)$ is zero. Similar remarks apply to the passband filter, where only inputs at frequencies $\omega_{c_1} < \omega < \omega_{c_2}$ are passed. The bandstop (or band-elimination, or band-rejection) filter does not pass signals of frequencies $\omega_{c_1} < \omega < \omega_{c_2}$. The highpass filter passes all signals of frequencies greater than the cutoff frequency ω_c.

It is obvious why these are ideal characteristics. In practice and in design we make certain allowances, typical in all engineering. For example, a common bandpass filter specification is shown in Figure 17-11. There is a *transition band* between the stopbands and the passband, at $\omega_a < \omega < \omega_{c_1}$ and $\omega_{c_2} < \omega < \omega_b$. In the stopband, we allow a deviation of ε_2 from the ideal zero, and in the passband ε_1 from the ideal 1. A sample practical curve is shown also. The parameters ω_a, ω_{c_1}, ω_{c_2}, ω_b, ε_1, and ε_2 are typical engineering specifications.

From our studies so far, we appreciate also the facts that at (or near) a zero of $H(s)$, the function is zero, $H(z_i) = 0$, and therefore its magnitude is zero. At (or near) a pole, $H(s)$ becomes very large, $H(p_k) \to \infty$, as does its magnitude. An elementary rule for design is therefore *to place zeros of $H(s)$ in the desired stopband and poles in the*

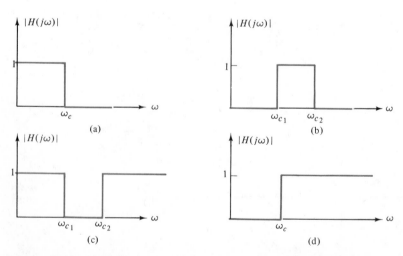

Figure 17-10 Ideal filter characteristics: (a) Lowpass. (b) Bandpass. (c) Bandstop. (d) Highpass.

† We use magnitude scaling for this convenient value. See Appendix C.

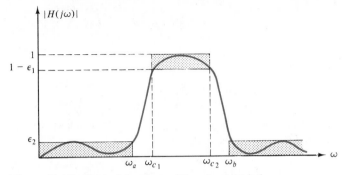

Figure 17-11 Typical bandpass filter characteristic.

desired passband. In the following examples, we explore this relationship between locations of poles and zeros of $H(s)$ and the corresponding frequency response. We do it for some common first-order factors.

EXAMPLE 17-7

Let

$$H(s) = \frac{1}{s}$$

with a pole at $s = 0$ ($\omega = 0$) and a zero at $s = \infty$ ($\omega = \infty$).

Solution. The magnitude response is

$$|H(j\omega)| = \frac{1}{\omega}$$

and the phase is $-90°$. See Figure 17-12(a) for the pole-zero configuration in the s plane. The magnitude and phase are shown in Figure 17-12(b) and (c). We have here a very elementary (primitive?) lowpass filter, essentially a single capacitor or inductor.

Prob. 17-13

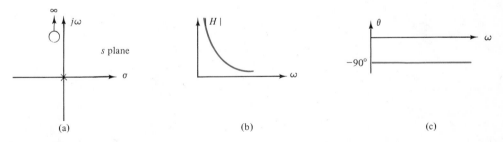

Figure 17-12 Example 17-7.

■

EXAMPLE 17-8

$$H(s) = \frac{1}{s + p_1} \qquad p_1 > 0$$

Solution. Here, there is a simple pole on the $-\sigma$ axis, and a simple zero at infinity. This is exactly the network function of Example 17-1 (with a suitable scaling.) We have

$$|H(j\omega)| = \frac{1}{\sqrt{\omega^2 + p_1^2}}$$

$$\theta(j\omega) = -\tan^{-1}\frac{\omega}{p_1}$$

See Figure 17-13. How do these characteristics change for $p_2 > p_1$ (a pole farther away from $\omega = 0$ on the $-\sigma$ axis)? For $p_2 < p_1$? For $p_2 \to 0$?

Prob. 17-14

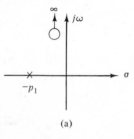

(a)

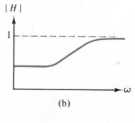

(b)

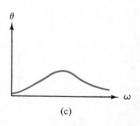

(c)

Figure 17-13   Example 17-8.   ■

EXAMPLE 17-9

$$H(s) = \frac{s + z_1}{s + p_1} \qquad \begin{aligned} z_1 &> 0 \\ p_1 &> 0 \end{aligned}$$

with a simple zero on the $-\sigma$ axis and a simple pole on the $-\sigma$ axis. Also, let $z_1 < p_1$. See Figure 17-14(a).

Solution.   Here

$$|H(j\omega)| = \sqrt{\frac{\omega^2 + z_1^2}{\omega^2 + p_1^2}}$$

and

$$\theta(j\omega) = \tan^{-1}\frac{\omega}{z_1} - \tan^{-1}\frac{\omega}{p_1}$$

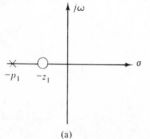

(a)

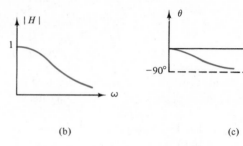

(b)

(c)

Figure 17-14   Example 17-9.

The frequency plots are shown in Figure 17-14(b) and (c), and we have a highpass (sort of…) filter, where the low frequencies are not quite attenuated. Where would you place z_1 to improve the attenuation?

Probs. 17-15, 17-16

■

A classical magnitude response for filters is known as a *Butterworth response*. If

$$H(s) = \frac{1}{s+1} \tag{17-48}$$

then

$$|H(j\omega)| = \frac{1}{\sqrt{\omega^2 + 1}} \tag{17-49}$$

and we speak of a Butterworth of order $n = 1$. A second-order ($n = 2$) Butterworth response is given by

$$H(s) = \frac{1}{s^2 + \sqrt{2}s + 1} \tag{17-50}$$

that is,

$$|H(j\omega)| = \frac{1}{\sqrt{(\omega^2)^2 + 1}} \tag{17-51}$$

The third-order ($n = 3$) Butterworth response is given by

$$H(s) = \frac{1}{s^3 + 2s^2 + 2s + 1} \tag{17-52}$$

that is,†

$$|H(j\omega)| = \frac{1}{\sqrt{(\omega^2)^3 + 1}} \tag{17-53}$$

and, in general, the nth-order Butterworth magnitude response is

$$|H(j\omega)| = \frac{1}{\sqrt{(\omega^2)^n + 1}} \tag{17-54}$$

Several magnitude plots are shown in Figure 17-15. Here the cutoff frequency is scaled, $\omega_c = 1$. It is also the -3-dB break frequency. The Bode plot has a low-frequency asymptote of

$$\alpha = 20 \log \frac{1}{\sqrt{(\omega^2)^n + 1}} \approx 0 \text{ dB} \qquad \omega \ll 1 \tag{17-55}$$

† Don't take it on blind faith; derive Equation 17-51 from 17-50, and 17-53 from 17-52. It's easy.

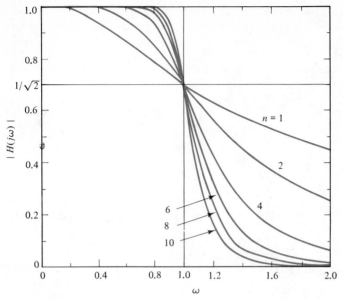

Figure 17-15 Butterworth response.

and a high-frequency asymptote

$$\alpha \approx 20 \log \frac{1}{\sqrt{(\omega^2)^n}} = -20n \log \omega \qquad \omega \gg 1 \qquad (17\text{-}56)$$

that is, a rolloff of $20n$ dB/decade $= 6n$ dB/octave in the stopband.

EXAMPLE 17-10 _____

A certain lowpass filter must satisfy the following specification: At $4\omega_c$, its attenuation must be at least -100 dB.

> *Solution.* The Butterworth filter that will do the job is found as follows: $4\omega_c$ is 2 octaves beyond $\omega_c = 1$, and the rolloff is $6n$ dB/octave. So
>
> $$12n \geq 100$$
> $$\therefore \quad n \geq 8.3$$

and so $n = 9$, a ninth-order Butterworth response, will do:

$$|H(j\omega)| = \frac{1}{\sqrt{(\omega^2)^9 + 1}}$$

Standard tables are available with listings of $H(s)$ for various values of n. In Equations 17-48, 17-50, and 17-52 we have these listings for $n = 1$, 2, and 3.   ■   Prob. 17-17

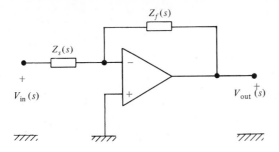

Figure 17-16 Inverting op amp circuit.

As examples of *active filters* (with active elements such as dependent sources), let us consider two op amp circuits, studied for resistive circuits in Chapter 4. The *inverting* op amp circuit, shown in Figure 17-16, has the voltage transfer function

$$G(s) = \frac{V_{out}(s)}{V_{in}(s)} = -\frac{Z_f(s)}{Z_s(s)} \tag{17-57}$$

with f standing for "feedback" and s for "source." Equation 17-57 is a generalization of the resistive case, Equation 4-33 and Example 5-17. A judicious choice of $Z_f(s)$ and $Z_s(s)$ provides the design of such a filter.

EXAMPLE 17-11 _____

Design an RC inverting op amp circuit for the same voltage transfer function as in Example 17-9,

$$G(s) = \frac{s + z_1}{s + p_1} \qquad z_1 > 0, p_1 > 0$$

Solution. We have

$$\frac{Z_f(s)}{Z_s(s)} = \frac{s + z_1}{s + p_1}$$

but we cannot identify $Z_f(s) = s + z_1$, $Z_s(s) = s + p_1$ because this will require inductors. Instead, write

$$\frac{Z_f(s)}{Z_s(s)} = \frac{1/(s + p_1)}{1/(s + z_1)}$$

and therefore

$$Y_f(s) = s + p_1$$
$$Y_s(s) = s + z_1$$

with the complete circuit shown in Figure 17-17(a).

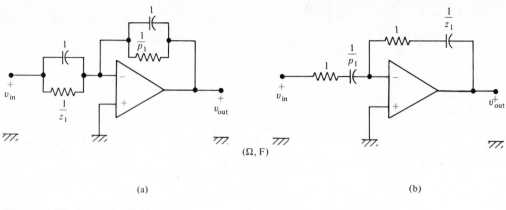

(a)   (b)

Figure 17-17   Example 17-11.

Another possible realization may be obtained if we write

$$\frac{Z_f(s)}{Z_s(s)} = \frac{(s + z_1)/s}{(s + p_1)/s} = \frac{1 + z_1/s}{1 + p_1/s}$$

with the circuit shown in Figure 17-17(b). *Question*: What happened to the minus sign in Equation 17-57 in these circuits? *Answer*: Think magnitude and phase!   ∎

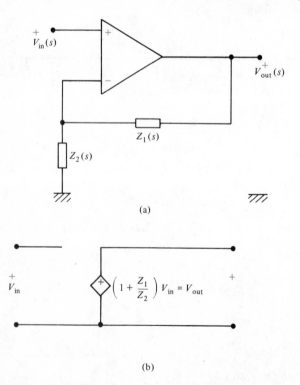

Figure 17-18   Noninverting op amp circuit.

The *noninverting* op amp circuit, shown in Figure 17-18(a), has the voltage transfer function

$$G(s) = \frac{V_{out}(s)}{V_{in}(s)} = 1 + \frac{Z_1(s)}{Z_2(s)} \qquad (17\text{-}58)$$

Compare it with the resistive circuit in Figure 4-17 and Example 5-18. Just to refresh our memory about the nature of the op amp, the equivalent circuit in Figure 17-18(b) shows the voltage-controlled voltage source which obeys Equation 17-58, as well as the infinite input impedance of the op amp itself.

Here, as before, a wise identification of $Z_1(s)$ and $Z_2(s)$ from the given $G(s)$ will realize the circuit.

EXAMPLE 17-12 _____

With the same $G(s)$ as in Example 17-11, we have here

$$1 + \frac{Z_1(s)}{Z_2(s)} = \frac{s + z_1}{s + p_1}$$

or,

$$\frac{Z_1(s)}{Z_2(s)} = \frac{z_1 - p_1}{s + p_1}$$

Solution. To realize the circuit with R and C elements, we must have $z_1 > p_1$. If we write

$$\frac{Z_1(s)}{Z_2(s)} = \frac{1/(s + p_1)}{1/(z_1 - p_1)}$$

we can select

$$Z_1(s) = \frac{1}{s + p_1} \qquad Z_2(s) = \frac{1}{z_1 - p_1}$$

as shown in Figure 17-19(a). An alternate realization may be obtained if we divide by s,

$$\frac{Z_1(s)}{Z_2(s)} = \frac{(z_1 - p_1)/s}{1 + p_1/s}$$

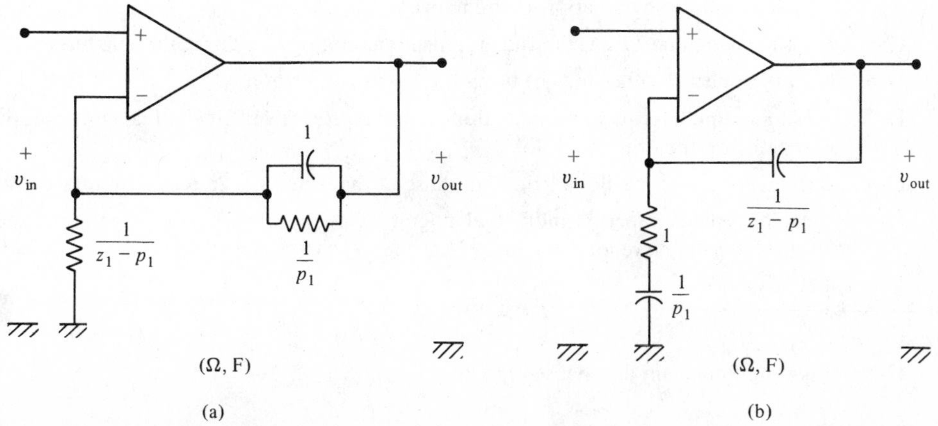

(a)

(b)

Figure 17-19 Example 17-12.

and identify in Figure 17-19(b)

$$Z_1(s) = \frac{z_1 - p_1}{s} \qquad \text{a single } C$$

$$Z_2(s) = 1 + \frac{p_1}{s} \qquad \text{a series } RC$$

■ Probs. 17-1
  17-19

In this chapter we studied in detail the concepts of frequency response and their relations to poles and zeros; it would not be a bad idea to review now Chapters 8 and 10 and the preliminary discussion of resonance, frequency response, bandwidth, and damping factor as these relate to our present discussion.

The introduction (that's all it is) of some ideas from filter design is, we hope, exciting enough to make you pursue this topic further. It is fun!

PROBLEMS

17-1 Rework Example 17-1 with the resistor R and the capacitor C exchanging places. Calculate the magnitude and phase functions, low and high asymptotes, and cutoff frequency. Plot the frequency response curves and classify this filter.

17-2 Develop the low-frequency asymptote values for the general network function to complement Equations 17-11 through 17-17. Draw the corresponding figure to Figure 17-2.

17-3 Complete Examples 17-2 and 17-3 by drawing the frequency response curves. (If you have access to a plotter on your computer, be honest about it. Use it only to confirm your work, not *instead of* your work. There are no shortcuts in the learning process.)

17-4 Plot completely the frequency response curves for

(a)
$$H(s) = \frac{s - 1}{s + 1}$$

(b)
$$H(s) = \frac{s^2 - s + 1}{s^2 + s + 1}$$

Classify each filter by an appropriate name.

17-5 Work out Example 17-2 using the graphical method of zero lines and pole lines.

17-6 Repeat Problem 17-4(a) and (b) using the graphical method.

17-7 Finish Example 17-1 by plotting its Bode phase plot. Draw the individual contribution of each factor, then the total plot.

17-8 Calculate and plot the Bode gain and phase of the factor s, a zero or a pole at $s = 0$.

17-9 Sketch on semilog paper the individual magnitude asymptotes, the total asymptote, and the smooth gain curve for

$$H(s) = 8.4 \frac{s}{(s + 10)(s + 100)}$$

17-10 Draw the Bode gain and phase plots for

$$H(s) = -10 \frac{s - 4}{(s + 1)(s^2 + 20s + 400)}$$

17-11 By comparing with Example 17-6, draw the Bode gain and phase plots for

$$H(s) = 1.6 \frac{s(s + 2)}{s^2 + 4s + 100}$$

and draw conclusions on the contributions of a zero factor of the form $(s - a)$ and $(s + a)$, $a > 0$.

17-12 From each given total gain asymptote, find the network function $H(s)$, including the constant multiplier K.

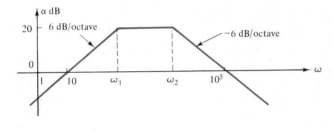

(a)

(b)

Problem 17-12

17-13 Prepare, as in Example 17-7, the pole-zero configuration and the frequency response plots for $H(s) = s$.

17-14 Repeat Problem 17-13 for $H(s) = s + z_1$, $z_1 > 0$, and $z_1 < 0$. Compare with Example 17-8.

17-15 Repeat Example 17-9 with $z_1 > p_1$; that is, the pole is nearer to the origin ($s = 0$) than the zero.

17-16 Consider the voltage transfer function

$$G(s) = \frac{Z_1(s)}{Z_1(s) + Z_2(s)}$$

of the class of filters shown.

(a) Design such a filter for the given specification

$$G(s) = \frac{s}{s + 10}$$

as: (1) an RL network, (2) an RC network. That's when the real fun begins, when you have more than one solution to a design problem. Factors of desirable elements (C) versus bulky nonlinear elements (L) come into consideration.

(b) Plot the pole-zero configuration of $G(s)$ and its frequency response curves.

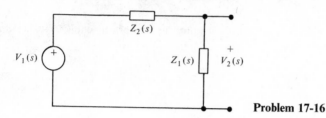

Problem 17-16

*17-17 (a) Show on the s plane the location of the poles of the Butterworth responses (Equations 17-48, 17-50, and 17-52).

(b) The general rule for finding these poles is as follows: Solve for the $2n$ roots of the equation

$$(-1)^n s^{2n} = -1$$

and retain as poles only the n roots in the left-hand s plane, i.e., those with a negative real part. Then write the nth Butterworth network function as

$$H(s) = \frac{1}{(s - p_1)(s - p_2)\cdots(s - p_n)}$$

Check this rule for $n = 1, 2,$ and 3; then obtain the Butterworth transfer function for $n = 4$ and 5.

17-18 Try to realize with passive R, L, C elements a noninverting op amp filter (Figure 17-18) for

(a)
$$G(s) = \frac{1}{s}$$

(b)
$$G(s) = \frac{1}{s + p_1} \qquad p_1 > 0$$

Comment on your results.

*17-19 Another popular method of filter design is the *cascade* realization. Say we wish to realize the voltage transfer function

$$G(s) = \frac{2}{(s + 1)(s + 2)}$$

We write it as

$$G(s) = \frac{1}{s + 1} \cdot \frac{2}{s + 2} = G_1(s) \cdot G_2(s)$$

and realize $G_1(s)$ and $G_2(s)$, then cascade them with a voltage follower (buffer) between them, as shown. The buffer, as you recall, is essentially a noninverting op amp circuit used to isolate the two RC sections.

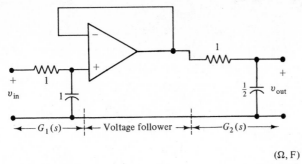

(Ω, F)

Problem 17-19

(a) Prove by actual calculations that for the two RC sections cascaded without the op amp,

$$G(s) \neq \frac{1}{s+1} \cdot \frac{2}{s+2}$$

(b) Realize by cascade RC sections

$$G(s) = \frac{s}{(s+1)^2}$$

then sketch the Bode gain plot.

Chapter 18

Stability

The stability of a network is extremely important in analysis and particularly in design (synthesis). Intuitively, we expect in a stable network to have all the voltages and currents finite (*bounded*) as time goes on, so that there will be no burning of components or, more likely, catastrophic distortions of waveforms. Let us elaborate on this intuition more precisely, and discuss three related concepts of stability.

18-1 IMPULSE RESPONSE STABILITY

A network is *impulse response stable* if

$$\lim_{t \to \infty} |h(t)| \neq \infty \tag{18-1}$$

or, in plain words, if its impulse response doesn't "blow up" as time increases; more succinctly, if its impulse response is bounded. We place magnitude signs on $h(t)$ because it is equally bad if $h(t)$ becomes unbounded with large negative or with large positive values.

   This concept is very sensible: A relaxed network is excited by $\delta(t)$, and a good measure of its stability is the boundedness of $h(t)$. A stricter requirement is sometimes imposed as follows:

$$\lim_{t \to \infty} |h(t)| = 0 \tag{18-2}$$

meaning that the impulse response must approach zero as t increases. We will be able to distinguish this requirement easily.

In Chapter 16, we learned that the impulse response is

$$h(t) = \mathcal{L}^{-1}H(s) = \mathcal{L}^{-1}\frac{P_H(s)}{Q_H(s)} \tag{18-3}$$

the inverse of the network function given as the rational function $P_H(s)/Q_H(s)$. We also learned that $Q_H(s)$ is the characteristic polynomial of the network, and, in factored form,

$$\frac{P_H(s)}{Q_H(s)} = \frac{P_H(s)}{(s - p_1)(s - p_2)\cdots(s - p_k)^r\cdots} \tag{18-4}$$

where the roots of $Q_H(s) = 0$ are the poles of $H(s)$. To be more specific, we show simple poles at $s = p_1, s = p_2$, and a multiple pole of order r at $s = p_k$. By partial fractions, we get $h(t)$ as

$$h(t) = \sum_{i=1}^{k} A_i e^{p_i t} u(t) + \sum_{r=2}^{k} K_r t^{r-1} e^{p_k t} u(t) + \cdots \tag{18-5}$$

Here, the first summation is due to all the simple poles and to the first power of the multiple pole, and the second summation is due to the multiple pole.

The decay rates and the frequencies of $h(t)$ are entirely dependent on the poles, the characteristic values. The zeros of $H(s)$, $P_H(s) = 0$, affect only the *finite* multipliers A_i and K_r as obtained in the partial fraction expansion of Equation 18-4. Thus, zeros do not affect the boundedness of $h(t)$.

As far as their location in the s plane, all poles of $H(s)$ fall into three categories:†

1. Real Poles on the σ Axis   A simple, real pole on the σ axis is shown in Figure 18-1(a), $p_1 = -\sigma_1$. It yields the waveform

$$\mathcal{L}^{-1}\frac{A_1}{s - p_1} = \mathcal{L}^{-1}\frac{A_1}{s + \sigma_1} = A_1 e^{-\sigma_1 t} \tag{18-6}$$

which is bounded if

$$\sigma_1 \geq 0 \tag{18-7}$$

that is, if the pole is in the left half of the s plane (LHP), to the left of the $j\omega$ axis, or at the origin. The inequality sign satisfies Equation 18-2, and the equal sign satisfies Equation 18-1 because then $A_1 e^{0t} = A_1$ remains bounded. The waveform of Equation 18-6 is shown in Figure 18-1 (b), assuming $A_1 > 0$. We also recognize that the farther σ_1 is to the left in the LHP, the faster the decay of this waveform.

A simple pole on the positive σ axis, in the right half of the s plane (RHP) yields a waveform

$$\mathcal{L}^{-1}\frac{A_1}{s - \sigma_1} = A_1 e^{\sigma_1 t} \tag{18-8}$$

which is unbounded since it increases with time.

† See your Problem 8-20!

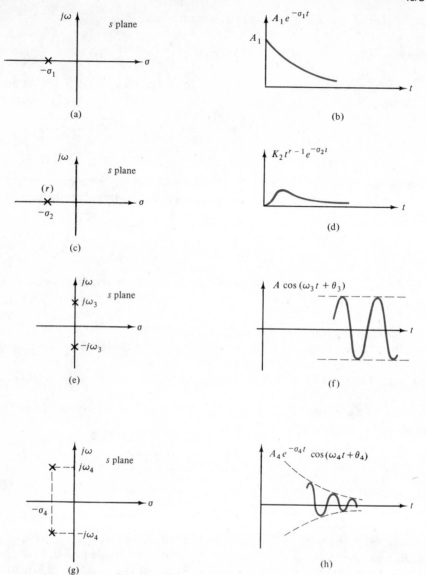

Figure 18-1 Pole locations and related waveforms.

A multiple pole on the negative real axis, shown in Figure 18-1(c), has

$$\mathscr{L}^{-1} \frac{A_2}{(s + \sigma_2)^r} = K_2 t^{r-1} e^{-\sigma_2 t} \tag{18-9}$$

and this waveform is shown in Figure 18-1(d). Here $K_2 = A_2/(r - 1)!$, a finite number. Therefore, we are concerned with

$$\lim_{t \to \infty} t^{r-1} e^{-\sigma_2 t} = \lim_{t \to \infty} \frac{t^{r-1}}{e^{\sigma_2 t}} = 0 \qquad \sigma_2 > 0 \tag{18-10}$$

where the last limit is evaluated by l'Hôpital's rule. If $\sigma_2 = 0$, a multiple pole at the origin, Equation 18-9 shows an unbounded response. A multiple pole in the RHP on the $+\sigma$ axis will obviously cause an unbounded waveform, $t^{r-1}e^{+\sigma_2 t}$.

2. Imaginary Poles on the $j\omega$ Axis A simple pole and its complex conjugate are shown in Figure 18-1(e), $p_3 = j\omega_3$, $p_3^* = -j\omega_3$. They appear as a pair of conjugates because the coefficients of $Q_H(s)$ must be real numbers

$$(s - p_3)(s - p_3^*) = s^2 + \omega_3^2 \tag{18-11}$$

This pair yields

$$\mathscr{L}^{-1} \frac{A_3 s + A_4}{s^2 + \omega_3^2} = A \cos(\omega_3 t + \theta_3) \tag{18-12}$$

or a similar sine waveform. See Figure 18-1(f). This is a bounded response according to Equation 18-1, since its magnitude never exceeds its amplitude A. We also observe that the larger ω_3 (farther on the $j\omega$ axis from the origin), the larger the frequency of oscillation of this waveform.

A multiple pole of the $j\omega$ axis, $(s^2 + \omega_3^2)^2$ or $(s^2 + \omega_3^2)^3$, etc., will cause an unbounded response of the form $A \cos(\omega_3 t + \alpha) + Bt \cos(\omega_3 t + \beta) + Ct^2 \cos(\omega_3 t + \gamma)$.

3. Complex Conjugate Poles A pair of simple, complex conjugate poles inside the LHP is shown in Figure 18-1(g):

$$\begin{aligned} p_4 &= -\sigma_4 + j\omega_4 \\ p_4^* &= -\sigma_4 - j\omega_4 \end{aligned} \qquad \sigma_4 > 0 \tag{18-13}$$

They contribute the factor $(s - p_4)(s - p_4^*) = (s + \sigma_4)^2 + \omega_4^2$ in $Q_H(s)$, and

$$\mathscr{L}^{-1} \frac{K_4 s + K_5}{(s + \sigma_4)^2 + \omega_4^2} = A_4 e^{-\sigma_4 t} \cos(\omega_4 t + \theta_4) \tag{18-14}$$

a damped sinusoid shown in Figure 18-1(h). The amplitude of this sinusoid decreases as $e^{-\sigma_4 t}$, and its frequency is ω_4. Obviously, this is a bounded response satisfying Equation 18-1.

Multiple poles inside the LHP will give the waveform

$$\sum_{r=2}^{k} A_{r-1} t^{r-1} e^{-\sigma_4 t} \cos(\omega_4 t + \theta_{r-1}) \tag{18-15}$$

which is bounded provided $\sigma_4 > 0$.

A pair of complex conjugate poles inside the RHP

$$\begin{aligned} p_5 &= \sigma_5 + j\omega_5 \\ p_5^* &= \sigma_5 - j\omega_5 \end{aligned} \qquad \sigma_5 > 0 \tag{18-16}$$

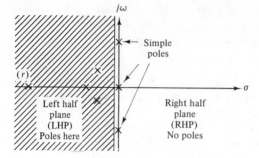

Figure 18-2 Pole locations for impulse response stability.

will yield an unbounded waveform

$$\lim_{t \to \infty} e^{\sigma_s t} \cos (\omega_s t + \theta_s) \to \infty \qquad (18\text{-}17)$$

and so will such poles if they are multiple.

We have exhausted all the possibilities and can summarize our conclusions: *For impulse response stability, the poles of H(s) must be inside the LHP ($\sigma < 0$), where they can be of any multiplicity r; or, at most, on the jω axis ($\sigma = 0$), where they must be simple, r = 1. No poles are allowed in the RHP ($\sigma > 0$).* This is illustrated in Figure 18-2.

EXAMPLE 18-1

All these network functions represent networks which are impulse response stable:

(a)
$$H(s) = 2.1 \frac{s - 1}{s^2 + 2s + 10}$$

(b)
$$H(s) = \frac{s + 4}{(s^2 + 1)(s^2 + s + 1)^2}$$

(c)
$$H(s) = -3 \frac{s}{s^4 + 2s^3 + 2s^2 + s + \frac{1}{4}}$$

Solution. In (a), $H(s)$ has two complex conjugate poles in the LHP at $p_1 = -1 + j3$, $p_2 = p_1^*$. It has a zero in the RHP (!) at $z_1 = 1$ but that's O.K.—there are no restrictions on zeros, in general. In (b), $H(s)$ has *simple* poles *on the jω axis* at $\pm j1$, and multiple poles in the LHP. In (c), $H(s)$ has all four poles in the LHP. How do we know so fast? This will be discussed in Section 18-4.   ■

EXAMPLE 18-2

Comment on the following network functions:

(a)
$$H(s) = \frac{4 - 3s}{s^2(s + 10)}$$

(b)
$$H(s) = 4 \frac{s - 1}{s + 3 + j2}$$

(c)
$$H(s) = - \frac{2s}{s^2 - 10}$$

Solution. In (a), $H(s)$ has a multiple pole ($r = 2$) at the origin, and the origin is part of the $j\omega$ axis; the impulse response will be unbounded since $\mathscr{L}^{-1}(1/s^2) = tu(t)$.† In (b), this is not a network function at all, since its coefficients are not all real; the pole at $p_1 = -3 - j2$ must be accompanied by its conjugate $p_1^* = -3 + j2$ to make these coefficients real. In (c), $H(s)$ has a pole in the RHP at $p_1 = \sqrt{10}$; therefore, the impulse response is unstable.†   ■

Let us look at this situation from the point of view of energy. With the impulse input, we impart to the network a certain amount of energy. If the network consists entirely of passive elements (R, L, M, C), the impulse response will go to zero as $t \to \infty$ because of the heat energy dissipated in the resistors, and Equation 18-2 is satisfied. With only lossless elements (L, M, C) but without dependent sources, the initial energy will oscillate between the elements, as we saw in Chapter 8, and Equation 18-1 is satisfied. Therefore, *passive* networks are *stable*. Dependent sources that deliver energy may sometimes add to the original imparted energy and cause instability through a characteristic value in the RHP.

One class of network functions deserves additional attention. These are driving-point impedances and admittances. Since

$$Z_{dp}(s) = \frac{1}{Y_{dp}(s)} \qquad Y_{dp}(s) = \frac{1}{Z_{dp}(s)} \qquad (18\text{-}18)$$

we see that the reciprocal of one network function is another network function. Consequently, the zeros of $Y_{dp}(s)$ are the poles of $Z_{dp}(s)$, and the zeros of $Z_{dp}(s)$ are the poles of $Y_{dp}(s)$. Therefore, *zeros of driving-point functions are restricted in the same way as poles* for impulse response stability. In transfer functions, no such restriction holds, because the reciprocal of a transfer function is not a network function, in general.   Probs. 18-1, 18-2, 18-3

18-2 ZERO-INPUT RESPONSE STABILITY

As its name indicates, this is the stability of the zero-input response. The network is excited by a given initial state, consisting of n initial capacitive voltages and inductive currents, $v_{C_1}(0^-), v_{C_2}(0^-), \ldots, i_{L_1}(0^-), i_{L_2}(0^-), \ldots$. Here, as before, we consider three cases:

1. The poles of $H(s)$, the characteristic values of the network, are in the LHP; that is, they all have negative real parts. In this case, the zero-input response will approach zero as $t \to \infty$; in particular, the final state of the network—those n capacitive voltages and inductive currents—will approach zero. This type of stability is also called *asymptotic stability*.

2. The poles of $H(s)$ are on the $j\omega$ axis and are simple. Here, an initial state will cause sinusoidal oscillations in the lossless (L, M, C) networks.

3. The poles of $H(s)$ are in the RHP, with positive real parts. Here the zero-input response becomes unbounded as $t \to \infty$ and the network is unstable.

---

† Actually, unbounded responses are allowed—in fact, are essential—in oscillator circuits (to start up) and in computer flip-flops. Such responses, of course, are limited to very short durations of time.

EXAMPLE 18-3

In a parallel RLC circuit, the initial state is given, $v_C(0^-)$ and $i_L(0^-)$. The zero-input response $v_C(t)$ is obtained from KCL as

$$\left(Cs + \frac{1}{R} + \frac{1}{Ls}\right)V_C(s) = Cv_C(0^-) - \frac{i_L(0^-)}{s}$$

that is,

$$V_C(s) = \frac{LCv_C(0^-)s - Li_L(0^-)}{LCs^2 + L/Rs + 1}$$

Solution. The characteristic values (natural frequencies) are the roots of

$$LCs^2 + \frac{L}{R}s + 1 = 0$$

and they fall into the three familiar categories of overdamped, underdamped, or critically damped. In all cases they are in the LHP, meaning that the network is asymptotically stable: The two state variables, $v_C(t)$ and $i_L(t)$, will decay to zero as $t \to \infty$. ∎

EXAMPLE 18-4

In the previous example, let $R \to \infty$ (open circuit).

Solution. The network becomes lossless, and we get

$$V_C(s) = \frac{LCv_C(0^-)s - Li_L(0^-)}{LCs^2 + 1}$$

yielding two poles on the $j\omega$ axis, at $\omega_0 = 1/(LC)^{1/2}$. The state variable $v_C(t)$ will be of the form

$$\mathcal{L}^{-1}V_C(s) = v_C(t) = A\cos(\omega_0 t + \theta)$$

and so will be $i_L(t)$. This is a bounded oscillatory zero-input response. ∎

EXAMPLE 18-5

In the network shown in Figure 18-3, the initial state is given, $v_C(0^-) = 0$, $i_L(0^-) \neq 0$. The zero-input response of I_L is calculated as

$$I_L(s) = \frac{LCi_L(0^-)s}{LCs^2 + Cs + (1 - \alpha)}$$

You should fill in the details here. The characteristic equation is a quadratic, and if $\alpha > 1$, the characteristic values will be in the RHP; the $i_L(t)$ will grow unbounded from its initial finite value, and the network will be zero-input unstable.

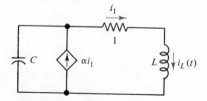

Figure 18-3 Example 18-5.

Prob. 18-4

18-3 BOUNDED-INPUT, BOUNDED-OUTPUT (BIBO) STABILITY

The third type of stability is concerned with input-output relations. Specifically, it requires that for every bounded input, the zero-state output be also bounded. Hence its brief name "bounded input, bounded output" and the acronym BIBO.

We resort to the general expression of the zero-state response

$$R(s) = H(s)E(s) \tag{18-19}$$

and, as rational functions,

$$\frac{P_R(s)}{Q_R(s)} = \frac{P_H(s)}{Q_H(s)} \cdot \frac{P_E(s)}{Q_E(s)} \tag{18-20}$$

We see therefore that the poles of $R(s)$, which determine the waveform of the output, are the poles of $H(s)$ *and* the poles of $E(s)$.† By assumption, $e(t)$ is bounded, and so the poles of $E(s)$ are either in the LHP or (simple) on the $j\omega$ axis. From our previous discussion, we have the same restrictions on the poles of $H(s)$. However, for BIBO stability, we require that all the poles of $H(s)$ be confined inside the LHP, but none on the $j\omega$ (and certainly not in the RHP). To illustrate this point, consider the following example.

EXAMPLE 18-6 ——————————————————————————————

A parallel *LC* circuit has the driving-point impedance

$$Z_{dp}(s) = \frac{Ls}{LCs^2 + 1} = \frac{1}{C}\frac{s}{s^2 + \omega_0^2}$$

where

$$\omega_0^2 = \frac{1}{LC}$$

This network function has two poles in the $j\omega$ axis at $p_1 = j\omega_0$ and $p_2 = p_1^* = -j\omega_0$. If we excite the network with the *bounded* sinusoidal input

$$i(t) = I_m \sin \omega_0 t \qquad I(s) = \frac{I_m \omega_0}{s^2 + \omega_0^2}$$

the output voltage will be

$$V(s) = Z_{dp}(s)I(s) = \frac{I_m \omega_0}{C} \frac{s}{(s^2 + \omega_0^2)^2}$$

that is, $v(t)$ will be *unbounded*

$$v(t) = \mathscr{L}^{-1}V(s) = \frac{I_m}{2C} t \sin \omega_0 t$$

Hello, again! we have excited here a natural frequency of the network at $j\omega_0$ with an outside source, causing resonance—and BIBO instability. It is for this reason that we exclude the characteristic values of the network even from the $j\omega$ axis and allow them only inside the LHP when BIBO stability is important. ■ Prob. 18-5

———

† We assume that there is no cancellation of terms in forming the product of Equation 18-20.

18-4 STABILITY TESTS

An obvious way to check the location of the characteristic values is to actually find them by solving the characteristic equation $Q_H(s) = 0$. First-order and quadratic characteristic equations are no problem. Higher-order ones need a numerical root-finding computer program. In many cases, however, we only want to make sure that there are no poles in the RHP, without actually calculating them.

Let us see what conclusions we can reach about a given polynomial

$$Q_H(s) = b_0 s^n + b_1 s^{n-1} + \cdots + b_{n-1}s + b_n \qquad (18\text{-}21)$$

if we insist that all its zeros† be in the LHP or, at the most, on the $j\omega$ axis (where they must be simple) for impulse response stability. From our studies and Figure 18-1, the only possible factors of $Q_H(s)$ are:

1. $(s + \alpha)^k$, $\alpha > 0$, $k = 1, 2, \ldots$ accounting for real, negative poles of any multiplicity
2. $(s + \beta + j\gamma)^r(s + \beta - j\gamma)^r = (s^2 + 2\beta s + \beta^2 + \gamma^2)^r$, $\beta > 0, \gamma > 0$, $r = 1, 2, \ldots$ accounting for complex conjugate poles inside the LHP, of any multiplicity
3. $(s^2 + \omega_i^2)$, $\omega_i > 0$, accounting for simple poles on the $j\omega$ axis
4. s, a simple pole at the origin, on the $j\omega$ axis ($\omega = 0$)

That's all! The polynomial $Q_H(s)$ can therefore be written as a product of these factors

$$Q_H(s) = s \prod_k (s + \alpha_k) \prod_r (s^2 + 2\beta_r s + \beta_r^2 + \gamma_r^2) \prod_i (s^2 + \omega_i^2) \qquad (18\text{-}22)$$

where the symbol $\prod$ means "the product of ...." A polynomial of the form in Equation 18-22 is called a *Hurwitz* polynomial.

When we multiply out fully the factors in Equation 18-22 we reach the following conclusions:

1. All the coefficients b_i in $Q_H(s)$, Equation 18-21, must be *real* and *positive*, because no negative sign appears in any of the factors in Equation 18-22.
2. No power of s can be missing in $Q_H(s)$ below its highest (leading) power. Reason? In carrying out the products of these factors, a power of s can disappear only by subtraction; for example, $20s^3$ will cancel $-20s^3$. However, subtraction cannot happen without negative signs in Equation 18-22.

There are two exceptions to this observation: If all the poles are on the $j\omega$ axis, $Q_H(s)$ will consist only of the product $s(s^2 + \omega_1^2)(s^2 + \omega_2^2)\ldots$. Then $Q_H(s)$ will have only odd powers of s, and all the even powers will be absent. If there is no pole at $s = 0$, $Q_H(s)$ will consist of product $(s^2 + \omega_1^2)(s^2 + \omega_2^2)\ldots$, and will have only even powers.

---

† A reminder of nomenclature: The zeros of $Q_H(s) = 0$ are the poles of $H(s)$.

The two observations, 1 and 2, are necessary but not sufficient. In other words, if $Q_H(s)$ does not satisfy them, then it is definitely not Hurwitz; however, if this inspection test is satisfied, $Q_H(s)$ may or may not be Hurwitz.

EXAMPLE 18-7 ———————————————————————————

Consider

$$Q(s) = 2s^4 + 10s^3 + 24s + 8$$

Solution. This is not a Hurwitz polynomial. The term in s^2 is missing, and therefore one or more poles are in the RHP. A network with this $Q(s)$ as a characteristic polynomial is unstable. ∎

EXAMPLE 18-8 ———————————————————————————

The polynomial

$$Q(s) = 3s^5 + 4s^4 + 2s^3 + 5s^2 + 3s + 1$$

passes the test by inspection, and therefore may or may not be a Hurwitz polynomial. Further investigation is needed. ∎

EXAMPLE 18-9 ———————————————————————————

The polynomial

$$Q(s) = s^6 + 6s^4 + 11s^2 + 6$$

may, or may not, be a Hurwitz polynomial.

EXAMPLE 18-10 ———————————————————————————

The polynomial

$$Q(s) = s^5 + s^4 + 6s^3 + 6s^2 + 25s + 25$$

looks O.K., right? However, two of its roots are in fact in the RHP,

$$p_1 = 1 + j2 \qquad p_2 = p_1^* = 1 - j2$$

This example illustrates the fact that the two observations by inspection are only necessary conditions but not sufficient. Clearly, this polynomial is not Hurwitz. ∎

We are ready now for a conclusive test that will tell us definitely whether a polynomial is Hurwitz or not. Two such tests will be explained.

The Routh Test This test is named after E. J. Routh (1831–1907), a British scientist. After $Q_H(s)$ passes the inspection test, we form an array of numbers as follows. The first two rows of the array are the given coefficients of $Q_H(s)$ in Equation 18-21, arranged as

$$\begin{array}{c|ccccc} s^n & b_0 & b_2 & b_4 & b_6 & \cdots \\ s^{n-1} & b_1 & b_3 & b_5 & b_7 & \cdots \end{array} \qquad (18\text{-}23)$$

that is, the first row contains the first, third, fifth, ..., coefficients, and the second row contains the second, fourth, sixth, ..., coefficients. The notation in the margin s^n and

s^{n-1}, is for "bookkeeping" purposes and will be useful later. The third row of the array is generated from the two rows preceding it, as follows

$$s^{n-2} \left| \frac{b_1 b_2 - b_0 b_3}{b_1} \quad \frac{b_1 b_4 - b_0 b_5}{b_1} \quad \frac{b_1 b_6 - b_0 b_7}{b_1} \right. \tag{18-24}$$

This pattern is shown in Figure 18-4, and can be stated as, "the first terms in the two preceding rows interact with every two terms in succession." For convenience, denote this row as

$$s^{n-2} \mid c_1 \quad c_2 \quad c_3 \quad \dots \tag{18-25}$$

The fourth row is generated in the same pattern from its two preceding rows, rows two and three

$$s^{n-3} \left| \frac{c_1 b_3 - b_1 c_2}{c_1} \quad \frac{c_1 b_5 - b_1 c_3}{c_1} \quad \dots \right. \tag{18-26}$$

and is denoted conveniently as

$$s^{n-3} \mid d_1 \quad d_2 \quad d_3 \quad \dots \tag{18-27}$$

Don't be scared by the "complicated" notation in Equation 18-26. The pattern is easy to learn, and instead of letters you'll be working with real numbers, so the rows will look a lot simpler. The fifth row is formed similarly, using the third and the fourth rows. Every row is generated by this algorithm, using the two preceding rows. The array is finished when the marginal power is zero, s^0, and we have the *Routh array*:

$$
\begin{array}{c|cccc}
s^n & b_0 & b_2 & b_4 & b_6 & \cdots \\
s^{n-1} & b_1 & b_3 & b_5 & b_7 & \cdots \\
s^{n-2} & c_1 & c_2 & c_3 & \cdots \\
s^{n-3} & d_1 & d_2 & d_3 & \cdots \\
\vdots & \\
s^0 & \cdots
\end{array}
$$

The Routh criterion says: "Count the number of sign changes going down the first column of this array, from b_0 to b_1, to c_1, to d_1, etc. This number is the number of zeros of $Q_H(s)$ in the RHP." Amazingly simple and powerful, isn't it? It is too bad that we have to present these results without proof, which is rather long. But, after all, we are mainly interested in the working results.

As a result, $Q_H(s)$ will have all its zeros inside the left half plane (LHP) if, and only if, there are no sign changes in the first column of the Routh array.

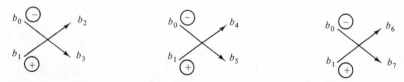

Figure 18-4   Pattern for the Routh array.

EXAMPLE 18-11 _____

Consider part (c) of Example 18-1. There we had

$$Q_H(s) = s^4 + 2s^3 + 2s^2 + s + \tfrac{1}{4}$$

Solution. After it passes the inspection test, we set up its Routh array as follows. The first two rows are

$$
\begin{array}{c|ccc}
s^4 & 1 & 2 & \tfrac{1}{4} \\
s^3 & 2 & 1 &
\end{array}
$$

The third row is formed according to Equation 18-24

$$
\begin{array}{c|cc}
s^2 & \dfrac{(2)(2) - (1)(1)}{2} & \dfrac{(2)(\tfrac{1}{4}) - (1)(0)}{2}
\end{array}
$$

that is

$$
\begin{array}{c|cc}
s^2 & \tfrac{3}{2} & \tfrac{1}{4}
\end{array}
$$

The fourth row is generated along the same lines, with Equation 18-26

$$
\begin{array}{c|c}
s^1 & \dfrac{(\tfrac{3}{2})(1) - (2)(\tfrac{1}{4})}{\tfrac{3}{2}}
\end{array}
$$

that is

$$
\begin{array}{c|c}
s^1 & \tfrac{2}{3}
\end{array}
$$

The last row is

$$
\begin{array}{c|c}
s^0 & \dfrac{(\tfrac{2}{3})(\tfrac{1}{4}) - (\tfrac{3}{2})(0)}{\tfrac{2}{3}}
\end{array}
$$

or

$$
\begin{array}{c|c}
s^0 & \tfrac{1}{4}
\end{array}
$$

The complete Routh array is therefore

$$
\begin{array}{c|ccc}
s^4 & 1 & 2 & \tfrac{1}{4} \\
s^3 & 2 & 1 & \\
s^2 & \tfrac{3}{2} & \tfrac{1}{4} & \\
s^1 & \tfrac{2}{3} & & \\
s^0 & \tfrac{1}{4} & &
\end{array}
$$

There are no sign changes down the first column. Therefore, all four roots of $Q_H(s)$ are in the LHP. (That's how we predicted it in Example 18-1c.) ∎

EXAMPLE 18-12 _____

Check $Q(s) = s^5 + 2s^4 + 3s^3 + 4s^2 + 3s + 1$.

Solution. It passes the inspection test. Its Routh array is (do it step by step!)

$$
\begin{array}{c|ccc}
s^5 & 1 & 3 & 3 \\
s^4 & 2 & 4 & 1 \\
s^3 & 1 & 2.5 & \\
s^2 & -1 & 1 & \\
s^1 & 3.5 & & \\
s^0 & 1 & &
\end{array}
$$

There are two sign changes in the first column, from $+1$ to -1 and from -1 to $+3.5$. Therefore, there are two roots of $Q(s) = 0$ in the RHP and the network is unstable. ■ Prob. 18-6

There are two special cases of the Routh array that need consideration. We do them by two specific examples.

EXAMPLE 18-13 _____

$$Q(s) = s^4 + 2s^3 + 3s^2 + 6s + 1$$

passes inspection.

Solution. Its Routh array starts with

$$
\begin{array}{c|ccc}
s^4 & 1 & 3 & 1 \\
s^3 & 2 & 6 \\
s^2 & 0 & 1
\end{array}
$$

The next row is in trouble because, to form it, we must divide by zero. What's the cure? A favorite of mathematicians (and engineers): Replace the zero by $\varepsilon > 0$, carry on bravely, then let $\varepsilon \to 0$.

Following this approach we get the complete array as

$$
\begin{array}{c|ccc}
s^4 & 1 & 3 & 1 \\
s^3 & 2 & 6 \\
s^2 & \varepsilon & 1 \\
s^1 & \dfrac{6\varepsilon - 2}{\varepsilon} \\
s^0 & 1
\end{array}
$$

For a very small $\varepsilon > 0$, the entry in the fourth row is negative. There are two sign changes in the first column, and therefore $Q(s)$ has two zeros in the RHP (by now it may be called the wrong, not the right, half plane). ■

EXAMPLE 18-14 _____

Take $Q(s)$ of Example 18-10

$$Q(s) = s^5 + s^4 + 6s^3 + 6s^2 + 25s + 25$$

Solution. Here the Routh array starts as

$$
\begin{array}{c|ccc}
s^5 & 1 & 6 & 25 \\
s^4 & 1 & 6 & 25 \\
s^3 & 0 & 0
\end{array}
$$

and an entire row (not just the first entry as before) is full of zeros. The trick with ε won't work here, if we try. (Do it!)

What we have here is a possibility to factor the original polynomial (again, this is in the general proof of the Routh method). We form an auxiliary polynomial whose coefficients are those of the last nonzero row, with powers starting with the one in the margin and going down by two. Here the auxiliary polynomial is

$$1s^4 + 6s^2 + 25s^0 = s^4 + 6s^2 + 25$$

Now we can continue in either of two ways:

1. Knowing that the auxiliary polynomial is a factor of $Q(s)$, divide it into $Q(s)$ to get a partial (or total) factorization. Here we get

$$Q(s) = (s + 1)(s^4 + 6s^2 + 25)$$

and, furthermore, the auxiliary polynomial is a quadratic in s^2; if we let $s^2 = x$ then

$$s^4 + 6s^2 + 25 = x^2 + 6x + 25$$
$$\therefore \quad x = -3 \pm j4 = 5\underline{/\pm 126.8°}$$
$$\therefore \quad s = \pm\sqrt{x} = \pm\sqrt{5\underline{/\pm 63.4°}} = \pm 1 \pm j2$$

Therefore

$$Q(s) = (s + 1)(s + 1 + j2)(s + 1 - j2)(s - 1 + j2)(s - 1 - j2)$$

and we have factored completely a fifth-order polynomial! Sit back for a moment and be properly impressed: It is no small feat to find all the zeros of a fifth-order polynomial, in the first place. That it can be done with so little work is doubly amazing. Two of these zeros are in the RHP and the network is unstable.

Finally (again, without proof), a row full of zeros occurs in the Routh array whenever the roots of the auxiliary polynomial are in *quadrant symmetry* about the origin, as shown in Figure 18-5.

2. Suppose you don't want to factor $Q(s)$, but just to complete the Routh array and to draw the conclusions from it. In that case, replace the row full of zeros with the derivative of the auxiliary polynomial with respect to s. Here

$$\frac{d}{ds}(s^4 + 6s^2 + 25) = 4s^3 + 12s$$

and the Routh array is completed as follows

s^5	1	6	25
s^4	1	6	25
s^3	4	12	
s^2	3	25	
s^1	-21.33		
s^0	25		

The two sign changes in the first column indicate the two roots in the RHP.

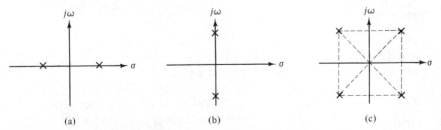

(a)　　　　　　　　(b)　　　　　　　　(c)

Figure 18-5　Quadrant symmetry.

EXAMPLE 18-15 _____

$$Q(s) = s^6 + 7s^4 + 14s^2 + 8$$

Solution. The test by inspection is still good here. Setting up the Routh array

$$
\begin{array}{c|ccc}
s^6 & 1 & 7 & 14 & 8 \\
s^5 & 0 & 0 & 0
\end{array}
$$

we run immediately into a row full of zeros. The auxiliary polynomial here is $Q(s)$ itself, so factorization is not possible, $Q(s) \cdot 1 = Q(s)$. We must, therefore, use the derivative. Then the Routh array is

$$
\begin{array}{c|cccc}
s^6 & 1 & 7 & 14 & 8 \\
s^5 & 6 & 28 & 28 \\
s^4 & 2.33 & 9.33 & 8 \\
s^3 & 4 & 7.43 \\
s^2 & 5 & 8 \\
s^1 & 1.033 \\
s^0 & 8
\end{array}
$$

and there are no sign changes down the first column.   ■

The Routh array, by itself, will not detect multiple roots on the $j\omega$ axis, as shown in the next example.

EXAMPLE 18-16 _____

Let

$$Q(s) = s^6 + 3s^4 + 3s^2 + 1 = (s^2 + 1)^3$$

and pretend that we don't know in advance that there is a multiple root at $j\omega = j1$.

Solution. The Routh array, without the option of partial factorization, is

$$
\begin{array}{c|cccc}
s^6 & 1 & 3 & 3 & 1 \\
s^5 & 6 & 12 & 6 \\
s^4 & 1 & 2 & 1 & \quad (s^4 + 2s^2 + 1) \\
s^3 & 4 & 4 \\
s^2 & 1 & 1 & \quad (s^2 + 1) \\
s^1 & 2 \\
s^0 & 1
\end{array}
$$

and the first column simply tells us correctly that there are no zeros in the RHP. On the right, we show the two auxiliary polynomials that appear in the process. The first one will reveal immediately a multiple zero,

$$s^4 + 2s^2 + 1 = (s^2 + 1)^2$$

and the second one offers a "last chance" at factoring. However, using the derivative and the straight array does not reveal the multiple zero.   ■   Probs. 18-7, 18-8

As a concluding example, let us consider the stability of *active networks*, such as those with dependent sources. In particular, we'll see that variations of a parameter in a dependent source may cause instability. See also Example 18-5 earlier.

EXAMPLE 18-17 _____

The network shown in Figure 18-6(a) is a second-order active RC filter, containing a voltage-controlled voltage source. The constant μ is positive and can vary, $0 < \mu < \infty$. Calculate the transfer function and discuss the stability of this network in terms of μ.

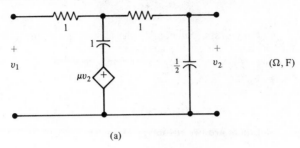

(a)

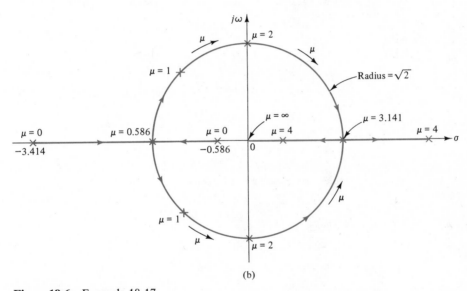

(b)

Figure 18-6 Example 18-17.

Solution. A straightforward analysis yields the voltage transfer function. (Check it in detail!)

$$G(s) = \frac{V_2(s)}{V_1(s)} = \frac{2}{s^2 + (4 - 2\mu)s + 2}$$

Here

$$Q(s) = s^2 + (4 - 2\mu)s + 2$$

is a quadratic, and the location of the poles clearly depends on μ. The Routh array here is

$$
\begin{array}{c|cc}
s^2 & 1 & 2 \\
s^1 & 4 - 2\mu & \\
s^0 & 2 &
\end{array}
$$

and, therefore, for stability, we must have

$$4 - 2\mu > 0$$

or

$$0 < \mu < 2$$

(*Note*: For a quadratic polynomial, there is no formal need for the Routh array; the quadratic formula will get the same results.)

Let us explore the variations of μ and the resulting locations of the poles:

1. *For $\mu = 0$:* By assumption, this is the lowest value of μ. Then

$$Q(s) = s^2 + 4s + 2 = (s + 2 + \sqrt{2})(s + 2 - \sqrt{2})$$

and there are two distinct, negative, real poles,

$$p_1 = -3.414 \qquad p_2 = -0.586$$

2. *For $\mu = 2 - \sqrt{2} = 0.586$:* Then $Q(s)$ is a perfect square (that's the reason for choosing this μ!):

$$Q(s) = s^2 + 2\sqrt{2}s + 2 = (s + \sqrt{2})^2$$

yielding two equal poles on the negative real axis

$$p_1 = p_2 = -\sqrt{2} = -1.414$$

3. *For an increasing value of μ, $\mu = 1$:* Then

$$Q(s) = s^2 + 2s + 2$$

with two complex conjugate poles inside the LHP

$$p_{1,2} = -1 \pm j$$

4. *For $\mu = 2$:* Then

$$Q(s) = s^2 + 2$$

and the two poles are on the $j\omega$ axis

$$p_1 = j\sqrt{2} \qquad p_2 = -j\sqrt{2}$$

5. *For $\mu > 2$:* Now the poles migrate into the RHP, causing instability. For instance, when $\mu = 3$, we get

$$Q(s) = s^2 - 2s + 2$$

yielding two complex conjugate poles in the RHP

$$p_1 = 1 + j1 \qquad p_2 = 1 - j1$$

When $\mu = 2 + \sqrt{2}$, the two poles merge on the real positive axis

$$p_1 = p_2 = \sqrt{2}$$

and, as μ continues to increase, one pole moves towards infinity while the other moves towards zero on the $+\sigma$ axis. See Figure 18-6(b), called a *root locus plot*. It shows the locus of the poles, roots of $Q(s) = 0$, as they vary with μ. We see how μ can affect the output of this network from stable, to oscillatory, to unstable.

Root locus plots are very useful in both the analysis and the design of networks. ∎

The Hurwitz Test This test, named after the German mathematician A. Hurwitz (1862–1909), is totally equivalent to the Routh test. It starts, in fact, from the first two rows of the Routh array. Two auxiliary polynomials are formed, each using the coefficients of one row. Specifically,

$$Q_1(s) = b_0 s^n + b_2 s^{n-2} + b_4 s^{n-4} + \cdots \tag{18-28}$$

and

$$Q_2(s) = b_1 s^{n-1} + b_3 s^{n-3} + b_5 s^{n-5} + \cdots \tag{18-29}$$

Obviously, the given polynomial $Q(s)$ in their sum

$$Q(s) = Q_1(s) + Q_2(s) \tag{18-30}$$

and $Q_1(s)$ contains only the even (or odd) powers of $Q(s)$, while $Q_2(s)$ contains only the odd (or even) powers of $Q(s)$.

The Hurwitz algorithm states simply, "divide longhand $Q_1(s)/Q_2(s)$ once. This yields a quotient of the form $\alpha_1 s$ plus a remainder. Invert the remainder and repeat a longhand division. Continue to divide once, then invert the remainder." The definitive criterion is: $Q(s)$ is Hurwitz if, and only if, all the coefficients of the quotients, $\alpha_1, \alpha_2, \ldots$, are *positive*.

EXAMPLE 18-18 _____

Take

$$Q(s) = s^4 + 2s^3 + 2s^2 + s + \tfrac{1}{4}$$

as in Example 18-11.

Solution. Here we start as follows

$$\frac{Q_1(s)}{Q_2(s)} = \frac{s^4 + 2s^2 + \tfrac{1}{4}}{2s^3 + s} = \underset{\alpha_1}{\left(\tfrac{1}{2}\right)s} + \frac{\tfrac{3}{2}s^2 + \tfrac{1}{4}}{2s^3 + s}$$

Invert the remainder, then divide once:

$$\frac{2s^3 + s}{\tfrac{3}{2}s^2 + \tfrac{1}{4}} = \underset{\alpha_2}{\left(\tfrac{4}{3}\right)s} + \frac{\tfrac{2}{3}s}{\tfrac{3}{2}s^2 + \tfrac{1}{4}}$$

Continue inverting and dividing

$$\frac{\tfrac{3}{2}s^2 + \tfrac{1}{4}}{\tfrac{2}{3}s} = \underset{\alpha_3}{\left(\tfrac{9}{4}\right)s} + \frac{\tfrac{1}{4}}{\tfrac{2}{3}s}$$

$$\frac{\tfrac{2}{3}s}{\tfrac{1}{4}} = \underset{\alpha_4}{\left(\tfrac{8}{3}\right)s}$$

and the test is finished. All the partial quotients are positive $\alpha_1 > 0$, $\alpha_2 > 0$, $\alpha_3 > 0$, $\alpha_4 > 0$. Hence $Q(s)$ has no roots in the RHP. ∎

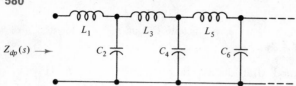

Figure 18-7 Ladder LC network.

The divide-invert process of the Hurwitz test can be summarized as follows

$$\frac{Q_1(s)}{Q_2(s)} = \alpha_1 s + \cfrac{1}{\alpha_2 s + \cfrac{1}{\alpha_3 s + \cfrac{1}{\alpha_4 s + \cdots}}} \qquad (18\text{-}31)$$

that is, a *continued fraction*. In this connection, there is a very interesting relationship between the Hurwitz test—a pure mathematical operation—and certain ladder networks. The driving-point impedance of the *LC* ladder network in Figure 18-7 is

$$Z_{dp}(s) = L_1 s + \cfrac{1}{C_2 s + \cfrac{1}{L_3 s + \cfrac{1}{C_4 s + \cdots}}} \qquad (18\text{-}32)$$

and with all L's and C's positive, we see the equivalence between Equations 18-31 and 18-32.

Let us explore the two special cases of the Routh array here, with the same polynomials.

EXAMPLE 18-19 _____

$$Q(s) = s^4 + 2s^3 + 3s^2 + 6s + 1$$

as in Example 18-13, with the first entry in a row being zero.

Solution. The Hurwitz test reads

$$\frac{Q_1(s)}{Q_2(s)} = \frac{s^4 + 3s^2 + 1}{2s^3 + 6s} = \underset{\underset{\alpha_1}{\nearrow}}{\left(\tfrac{1}{2}\right)} s + \frac{1}{2s^3 + 6s}$$

and the "defect" becomes obvious immediately: In the remainder, the numerator is *not* one degree less than the denominator. The next inversion and division will *not* yield $\alpha_2 s$ plus a remainder. ■

EXAMPLE 18-20 _____

$$Q(s) = s^5 + s^4 + 6s^3 + 6s^2 + 25s + 25$$

as in Example 18-14, with a row full of zeros in the Routh array.

Solution. Here we get

$$\frac{Q_1(s)}{Q_2(s)} = \frac{s^5 + 6s^3 + 25s}{s^4 + 6s^2 + 25} = \underbrace{\textcircled{1}}_{\alpha_1} s + 0$$

a zero remainder and a premature ending of the process of "divide-invert." Here, the last divisor is the same auxiliary polynomial found in the Routh array.

It is easy to explain the presence of such an auxiliary polynomial by the Hurwitz test. Let us write

$$Q_1(s) = p_x(s)q_1(s)$$

and

$$Q_2(s) = p_x(s)q_2(s)$$

where, by assumption, $p_x(s)$ is such an auxiliary polynomial. Then obviously Hurwitz's algorithm will end prematurely

$$\frac{Q_1(s)}{Q_2(s)} = \frac{p_x(s)q_1(s)}{p_x(s)q_2(s)}$$

through the cancellation of $p_x(s)$. On the other hand, the original polynomial is

$$Q(s) = Q_1(s) + Q_2(s) = p_x(s)[q_1(s) + q_2(s)]$$

showing that $p_x(s)$ is indeed a factor of $Q(s)$. ∎

Probs. 18-9, 18-10, 18-11, 18-12, 18-13

PROBLEMS

18-1 Classify the location of poles in the s plane in terms of the damping factor ζ and the undamped natural frequency ω_n as studied in Chapter 8. Consider the four cases (a) $\zeta < 1$, (b) $\zeta > 1$, (c) $\zeta = 1$, (d) $\zeta = 0$.

18-2 For each network function given, check by inspection its impulse response stability. Write $h(t)$ by inspection, within arbitrary constant multipliers.

(a)
$$H(s) = 30.4 \frac{2 - s}{s + 2}$$

(b)
$$H(s) = \frac{1}{(s^2 + 1)(s^2 + 4s + 20)}$$

(c)
$$H(s) = \frac{4s - 1}{s(s + 1)(s + 2)}$$

18-3 Someone insists that the following network function provides a stable impulse response. Check it carefully, and draw conclusions.

$$H(s) = 2.4 \frac{s - 1}{s^2 - 1}$$

18-4 Consider the parallel *RLC* zero-input circuit in Example 18-3. This time, let $R < 0$ be a negative resistor. Such an active element is a model of several devices. Calculate the natural frequencies and determine the zero-input stability or instability of this circuit.

18-5 **(a)** A series *LC* circuit is excited by a unit step voltage. Verify that the zero-state response does not contain a unit step. Note the pole-zero cancellation.

(b) Repeat with a series *RL* circuit. In both cases, explain the result by physical (electrical) reasoning. This is a good place to remind ourselves that the Laplace transform, or any other tool, is no substitute for basic thinking.

18-6 From the results of Examples 18-11 and 18-12, show on the *s* plane the possible locations of *all* the zeros of each polynomial. *Hint*: If complex, zeros must be conjugate.

18-7 Write a short program for the Routh test, and check for stability each network whose characteristic polynomial is:

(a) $$Q(s) = s^3 + s^2 + 2s + 24$$

(b) $$Q(s) = s^5 + 2s^4 + 4s^2 + 3s + 8$$

(c) $$Q(s) = 2s^5 + 3s^4 + 7s^3 + 7s^2 + 6s + 2$$

(d) $$Q(s) = s^4 + 2s^3 + s^2 + 3s + 3$$

(e) $$Q(s) = 2s^3 + 3s^2 + 10s + 1$$

(f) $$Q(s) = 2s^6 + 4.1s^5 + s^4 + 1.3s^3 + 2s^2 + 1.4s + 0.3$$

(g) $$Q(s) = s^5 + 2s^4 + 3s^3 + s^2 + 4s + 6$$

(h) $$Q(s) = s^8 + 7s^6 + 17s^4 + 17s^2 + 6$$

(i) $$Q(s) = s^7 + 4s^5 + 2s^3 + s$$

18-8 In the design of a certain active circuit, the following network function is proposed

$$H(s) = \frac{1}{s^3 + Ks^2 + 10s + 1}$$

where *K* is an adjustable parameter of an active element (say, the gain of a voltage-controlled voltage source).

(a) Determine the value, or values, of *K* to ensure the stability of this network.

(b) Is there a value of *K* that will place a natural frequency on the *jω* axis and thus cause a problem for BIBO stability? If so, at what frequency will this happen?

(c) Sketch the root locus as *K* varies, $0 < K < \infty$.

18-9 Run the Hurwitz test on each polynomial in Problem 18-7.

18-10 Try the Hurwitz test on Problem 18-8. Recognize how much easier the Routh test is here. That's one good reason to know both methods, don't you agree?

18-11 Show the dual ladder network to Figure 18-6; that is, $Y_{dp}(s)$ is given by Equation 18-31.

18-12 Check BIBO stability of the network for which

$$H(s) = \frac{1}{s^7 + 14s^5 + 49s^3 + 36s}$$

18-13 The characteristic equation of a network is

$$s^3 + 2Ks^2 + 10s + (K + 19) = 0$$

Investigate the zero-input stability of the network in terms of the adjustable constant *K*.

Chapter 19

State Variable Analysis

We have had some exposure to state variables, starting in Chapter 7. Here we will study this topic in detail, including the systematic formulation of the state equations and their solution.

As a partial review, and as further motivation, let us list the advantages of state variable analysis:

1. It is yet another general method of analysis, in addition to loop analysis and node analysis.
2. It provides a better understanding of the physical (electrical) aspects of the network. State variables are easily measured or displayed, unlike some loop currents, for example. State variables are associated directly with the energy in the circuit.
3. It generates always a set of simultaneous first-order differential equations, unlike the integro-differential equations of loop or node analysis.
4. The extension to time-varying and nonlinear networks is very easy, which is not the case with loop or node analysis.
5. These equations are particularly suitable for solution by analog or digital computers.

We will emphasize these points as we progress in our study.

19-1 SYSTEMATIC FORMULATION OF STATE EQUATIONS

A formal definition of the state of a network will be helpful, particularly if we recall the several examples, starting in Chapter 7. The *state of a network* is the minimal set of data satisfying the following two conditions:

1. Given the state at any time, $t = t_0$, and the inputs to the network for $t > t_0$, the future state can be determined uniquely for any time $t > t_0$.
2. The state at any time t_0 and the inputs at t_0 determine uniquely any variable in the network.

EXAMPLE 19-1 ───

In the series RC circuit excited by a voltage source $v_s(t)$, studied in Chapter 7, the capacitor's voltage $v_C(t)$ qualifies as the state of the network. Given $v_C(t_0)$, particularly if $t_0 = 0$, but not necessarily, and given $v_s(t)$ for $t > t_0$, we can—and did—calculate $v_C(t)$ for any $t > t_0$. Also given $v_C(t)$ and $v_s(t)$, we can calculate, for example, $i_R(t)$.

These two conditions qualify $v_C(t)$ as the state of this network; however, this is not a unique choice: We can choose the capacitor's charge $q_C(t)$ as the state. As we'll see later, such a choice is particularly useful for time-varying or nonlinear networks.   ■

The systematic formulation of state variable equations begins with drawing a *proper tree* for the network. (Review quickly the idea of a tree in Chapter 3.) First, the graph of the network is drawn, with a branch representation for each element; no series or parallel combinations must be made. In the proper tree we choose as tree branches the following elements, in order of preference:

1. All the voltage sources
2. All the capacitors†
3. Some resistors, as needed
4. None of the inductors†
5. No current sources

To begin with, note the beautiful duality here: All the voltage sources in the tree, and all the current sources in the co-tree; capacitors in the tree, inductors in the co-tree; resistors, being dual of themselves, go either way. The reasons for these choices will become clear as we go on; for now, let us comment that:

Choice 1 is possible because voltage sources by themselves cannot form a closed loop; if they did, one of them is redundant or else they violate KVL. Since they can't form a loop, they will fit into the tree which, by definition, has no closed loops.

Similarly, all the capacitors will fit into the tree if they don't form any closed loops. Their initial voltages, $v_{C_1}(0^-)$, $v_{C_2}(0^-)$, ..., are part of the initial state of the network, and $v_{C_1}(t)$, $v_{C_2}(t)$, ..., are state variables.

A resistor may be needed in the proper tree if, for example, there is a node to which are incident only resistors. This node *must* be connected to the tree via one resistor.

By a dual argument to the capacitors, all the inductors will be in the co-tree as links. The given $i_{L_1}(0^-)$, $i_{L_2}(0^-)$, ... form part of the initial state of the network. These currents for $t > 0$, $i_{L_1}(t)$, $i_{L_2}(t)$, ..., are state variables.

───────────

† A small modification to this rule will be explained later.

All these somewhat abstract notions will become clear as we work out our examples.

EXAMPLE 19-2 _____

For the network shown in Figure 19-1(a), we draw the graph in Figure 19-1(b)

Solution. The proper tree, drawn according to the list above, is shown in Figure 19-1(c). The orientation of the branches is either given, as, for example, the voltage of the source and the initial conditions, or arbitrarily assigned, as in both resistors.

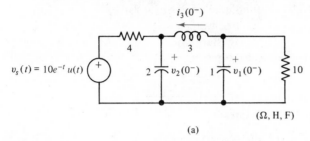

(a)

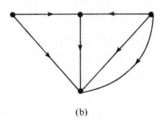

(b)

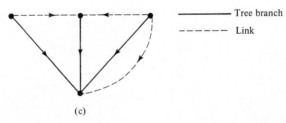

——— Tree branch

------ Link

(c)

Figure 19-1 Example 19-2.

As mentioned, *the state variables for a network are the capacitive voltages (or charges) of the tree branches and the inductive currents (or fluxes) of the links.* In a carefully presented problem, the initial state will consist of the initial values of those variables, thereby helping us to confirm the choice of state variables.

The next step is the formulation of the state equations. Here, too, there is little to memorize. Instead, we are guided by the attractive promise of only first-order derivatives. The first derivative of a capacitive voltage, dv_C/dt, triggers in our mind

$C\, dv_C/dt$, the current through the capacitor. Dually, di_L/dt reminds us of $L\, di_L/dt$, the voltage across the inductor. The two rules are very simple, then:

1. Write KCL for every fundamental cut set in the network formed by each capacitor in the tree.
2. Write KVL for every fundamental loop in the network formed by each inductor in the co-tree.

We remember that a fundamental cut set is formed by *one* tree branch (a capacitor in our proper tree) and some links. A fundamental loop is formed by *one* link (an inductor in the proper co-tree) and some tree branches.

This is it. The algorithm for writing state equations is that simple. Let us continue to illustrate with the previous example.

EXAMPLE 19-3 —————————————————————————————

The fundamental cut set for the 1-F capacitor reads

$$i_1(t) + i_3(t) + i_{10}(t) = 0$$

where the subscripts correspond to the element values. This fundamental cut set consists of the capacitor that creates it (tree branch), the inductor (link), and the 10-Ω resistor (link).

The second fundamental cut set consists of the 2-F capacitor which creates it (tree branch), the inductor (link), and the 4-Ω resistor (link). KCL for it reads

$$i_2(t) - i_3(t) - i_4(t) = 0$$

The fundamental loop consists of the inductor that creates it (link) and the two capacitors (tree branches). KVL for it is

$$v_3(t) + v_2(t) - v_1(t) = 0$$

In the first state equation $i_1(t) = 1(dv_1/dt)$, so

$$\frac{dv_1(t)}{dt} = -i_3(t) - i_{10}(t)$$

Similarly, in the second state equation $i_2(t) = 2(dv_2/dt)$, and therefore

$$\frac{dv_2(t)}{dt} = \tfrac{1}{2}i_3(t) + \tfrac{1}{2}i_4(t)$$

In the third state equation, $v_3(t) = 3(di_3/dt)$, so

$$\frac{di_3(t)}{dt} = -\tfrac{1}{3}v_2(t) + \tfrac{1}{3}v_1(t)$$

We are almost there! The left-hand sides of these equations, as promised, show only the first derivatives of the state variables. On the right-hand side we have "desired" state variables, $i_3(t)$, $v_1(t)$, and $v_2(t)$, and "undesired" variables $i_{10}(t)$ and $i_4(t)$. To substitute a desired variable for $i_{10}(t)$, write a fundamental loop equation for it, because the 10-Ω resistor is a link. We get

$$10i_{10}(t) = v_1(t)$$
$$\therefore \quad i_{10}(t) = 0.1v_1(t)$$

To replace $i_4(t)$, also in a link, write a fundamental loop for it

$$4i_4(t) + v_2(t) = v_s(t)$$

$$\therefore \quad i_4(t) = \tfrac{1}{4}v_s(t) - \tfrac{1}{4}v_2(t)$$

where $v_s(t)$, a known function, is a desired variable.

With these substitutions, the three state equations are

$$\frac{dv_1(t)}{dt} = -0.1v_1(t) - i_3(t)$$

$$\frac{dv_2(t)}{dt} = -\tfrac{1}{8}v_2(t) + \tfrac{1}{2}i_3(t) + \tfrac{1}{8}v_s(t)$$

$$\frac{di_3(t)}{dt} = \tfrac{1}{3}v_1(t) - \tfrac{1}{3}v_2(t)$$

In matrix form they read

$$\begin{bmatrix} \dfrac{dv_1(t)}{dt} \\[2mm] \dfrac{dv_2(t)}{dt} \\[2mm] \dfrac{di_3(t)}{dt} \end{bmatrix} = \begin{bmatrix} -0.1 & 0 & -1 \\ 0 & -\tfrac{1}{8} & \tfrac{1}{2} \\ \tfrac{1}{3} & -\tfrac{1}{3} & 0 \end{bmatrix} \begin{bmatrix} v_1(t) \\ v_2(t) \\ i_3(t) \end{bmatrix} + \begin{bmatrix} 0 \\ \tfrac{1}{8} \\ 0 \end{bmatrix} v_s(t)$$

∎

The *standard*, or *normal*, form of the state equations is illustrated in the previous example. It is given by

$$\frac{d}{dt}\mathbf{x}(t) = \mathbf{A}\mathbf{x}(t) + \mathbf{B}\mathbf{e}(t) \tag{19-1}$$

Here $\mathbf{x}(t)$ is the column matrix of the n unknown state variables

$$\mathbf{x}(t) = \begin{bmatrix} v_{C_1}(t) \\ v_{C_2}(t) \\ \vdots \\ i_{L_1}(t) \\ i_{L_2}(t) \\ \vdots \end{bmatrix} \tag{19-2}$$

and n, as usual, is the order of the network. $\mathbf{A}$ is a square $(n \times n)$ matrix. The p independent sources in the network are in the column matrix $\mathbf{e}(t)$, and $\mathbf{B}$ is $(p \times n)$. In Equation 19-1 we have n simultaneous first-order differential equations, or a single first-order *matrix differential equation*, subject to the n given initial conditions

$$\mathbf{x}(0^-) = \begin{bmatrix} v_{C_1}(0^-) \\ v_{C_2}(0^-) \\ \vdots \\ i_{L_1}(0^-) \\ i_{L_2}(0^-) \\ \vdots \end{bmatrix} \tag{19-3}$$

the initial state of the network.

Once the state is known, that is, after Equation 19-1 is solved for $\mathbf{x}(t)$, we ought to be able to express any output in terms of the state and the inputs (see the second condition of the definition of state). In a linear network, this will be done by superposition, as follows:

$$\mathbf{r}(t) = \mathbf{C}\mathbf{x}(t) + \mathbf{D}\mathbf{e}(t) \tag{19-4}$$

where $\mathbf{r}(t)$ is the column matrix of those outputs, and $\mathbf{C}$ and $\mathbf{D}$ are appropriate matrices. Equation 19-4 is called the *output equation*.

EXAMPLE 19-4 _____

In Figure 19-1, let the desired outputs be $i_{10}(t)$ and $i_4(t)$.

Solution.   Then we have

$$i_{10}(t) = \tfrac{1}{10} v_1(t)$$
$$i_4(t) = \tfrac{1}{4}(v_s - v_2)$$

and Equation 19-4 reads

$$\begin{bmatrix} i_{10}(t) \\ i_4(t) \end{bmatrix} = \begin{bmatrix} \tfrac{1}{10} & 0 & 0 \\ 0 & -\tfrac{1}{4} & 0 \end{bmatrix} \begin{bmatrix} v_1(t) \\ v_2(t) \\ i_3(t) \end{bmatrix} + \begin{bmatrix} 0 \\ \tfrac{1}{4} \end{bmatrix} v_s(t) \qquad ■$$

We should recognize that the state equation (Equation 19-1) is a differential equation with the unknown $\mathbf{x}(t)$. It must be *solved*, and we'll do it. The output equation (Equation 19-4) does not need any solution: The output $\mathbf{r}(t)$ is merely a linear combination of $\mathbf{x}(t)$—presumably solved already—and of $\mathbf{e}(t)$, the known input.

We summarize the algorithm for formulating the state and output equations:

1. Draw the proper tree. Capacitive voltages (or charges) of the tree branches and inductive currents (or fluxes) of the links are the state variables. There are n state variables, determining the order n of the network.
2. For each capacitive tree branch, write its fundamental cut set KCL equation.
3. For each inductive link, write its fundamental loop KVL equation.
4. In steps 2 and 3, substitute for undesired variables. If such a variable is in a link, write a fundamental loop for it; if a tree branch—a fundamental cut set.
5. Arrange in final form, as in Equation 19-1.
6. Write the output equation in its form (Equation 19-4).

Let us do another example, this time with a dependent source. Be sure to track this algorithm step by step, as developed.

EXAMPLE 19-5 _____

In Figure 19-2(a) we are given the initial state

$$\mathbf{x}(0^-) = \begin{bmatrix} v_3(0^-) \\ i_2(0^-) \end{bmatrix}$$

therefore, the state variables are

$$\mathbf{x}(t) = \begin{bmatrix} v_3(t) \\ i_2(t) \end{bmatrix} \qquad n = 2$$

and the independent input matrix is

$$e(t) = i_5(t) = 6u(t)$$

a scalar in this case.

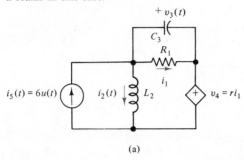

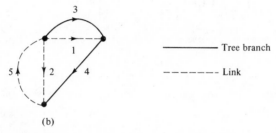

(b)

Figure 19-2 Example 19-5.

Solution.

1. The proper tree and co-tree are shown in Figure 19-2(b), where the branches are labeled as the elements.
2. The fundamental cut set equation for the capacitor is†

$$C_3 \frac{dv_3}{dt} = -i_1 - i_2 + i_5$$

3. The fundamental loop equation for the inductor is

$$L_2 \frac{di_2}{dt} = ri_1 + v_3$$

4. The undesired variables are those that do not belong in $x(t)$ and $e(t)$; here it is only $i_1(t)$. Since R_1 is a link, write the fundamental loop for it

$$R_1 i_1 = v_3$$

$$\therefore \quad i_1 = \frac{1}{R_1} v_3$$

† Don't forget: Lowercase letters are functions of time, $v_3 = v_3(t)$, etc.

5. Substitute this $i_1(t)$ into the two state equations and arrange in final form

$$\frac{dv_3}{dt} = -\frac{1}{R_1 C_3} v_3 - i_2 + i_5$$

$$\frac{di_2}{dt} = \frac{r}{R_1 L_2} v_3 + \frac{1}{L_2} v_3$$

or in matrix form

$$\begin{bmatrix} \dfrac{dv_3}{dt} \\ \dfrac{di_2}{dt} \end{bmatrix} = \begin{bmatrix} -\dfrac{1}{R_1 C_3} & -1 \\ \dfrac{r}{R_1 L_2} + \dfrac{1}{L_2} & 0 \end{bmatrix} \begin{bmatrix} v_3 \\ i_2 \end{bmatrix} + \begin{bmatrix} 1 \\ 0 \end{bmatrix} 6u(t)$$

6. Let the desired output be $i_1(t)$. Then Equation 19-4 reads

$$i_1(t) = \frac{1}{R_1} v_3(t) = \begin{bmatrix} \dfrac{1}{R_1} & 0 \end{bmatrix} \begin{bmatrix} v_3(t) \\ i_2(t) \end{bmatrix} + [0] i_5(t)$$

Probs. 19-1
19-2, 19-3,
19-4

To get just a taste of nonlinear networks, consider the following example.

EXAMPLE 19-6 ——————————————————————

In the network in Figure 19-3(a), the inductor is nonlinear, defined by a functional relation between its current and flux

$$i_L = f(\phi_L)$$

and shown in Figure 19-3(b).

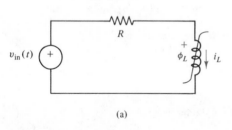

(a)   (b)

Figure 19-3 Example 19-6.

Solution. The algorithmic formulation of the state equations is still valid, and this universality is one of the most attractive features of this method. Using the proper tree, and with ϕ_L as the state variable, we write the fundamental loop equation as

$$v_L + v_R = v_{in}$$

that is

$$v_L = \frac{d\phi_L}{dt} = v_{in} - v_R = v_{in} - R i_R = v_{in} - R i_L$$

$$= v_{in} - R f(\phi_L)$$

In other words,

$$\frac{d\phi_L}{dt} = F(\phi_L, v_{in}) = F[\phi_L, e(t)]$$

The first derivative of the state variable is a nonlinear function of the state variable and the input. In that sense, state variable formulation is universal, always relating the first derivative of the state to the state and the input.   ■

Let us get back to our main discussion and consider the cases when not all capacitors can fit into the proper tree or when not all inductors can fit into the co-tree. In the capacitors' case it is very easy to spot: You begin to build the proper tree, having first used all the voltage sources, by adding capacitors as required. Suddenly you add a capacitor to the tree and you find that you have closed a loop! A tree (proper or otherwise) must not contain closed loops.

What's happened? In the network there is an *all-capacitive loop*, consisting of only capacitors and possibly voltage sources. In this loop, one capacitive voltage does not qualify as an independent variable since its value depends on KVL for that loop. To illustrate, consider the network shown in Figure 19-4. To build the proper tree, we start with the voltage source, then add C_1. By doing so, C_2 must be excluded from the tree, because KVL around this all-capacitive loop says

$$v_{C_1}(t) + v_{C_2}(t) = v_{in}(t) \tag{19-5}$$

With $v_{in}(t)$ known, only $v_{C_1}(t)$ *or* $v_{C_2}(t)$ is independent. Hence, if $v_{C_1}(t)$ is chosen as an unknown state variable, then $v_{C_2}(t)$ is not an unknown any longer—and it does not belong in the tree. We could, of course, choose C_2 in the tree; then $v_{C_2}(t)$ is a state variable and $v_{C_1}(t)$ is not. Thus, an all-capacitive loop reduces the number of capacitive state variables, and therefore the order of the network n, by one.†

As we continue to build our proper tree, we may find the dual of an all-capacitive loop. It is an *all-inductive cut set*, consisting only of inductors and possibly current sources. One inductor must be a tree branch because a tree, by definition, is connected. In Figure 19-4, we see such a cut set, and either L_1 or L_2 must be a tree branch. As such, it is no longer an independent state variable, because it depends on the link current in KCL for that cut set

$$i_{L_1}(t) - i_{L_2}(t) = 0 \tag{19-6}$$

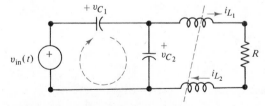

Figure 19-4   All-C loop and all-L cut set.

---

† The initial capacitive voltages at $t = 0^-$ are generally independent: $v_{C_1}(0^-), v_{C_2}(0^-), \ldots$ can be specified arbitrarily if they don't form a loop at $t = 0^-$. However, for $t \geq 0$, they are dependent and tied by KVL around the all-capacitive loop.

An all-inductive cut set reduces the number of inductive state variables, and therefore the order of the network by one.† Only inductive link currents qualify as independent state variables.

The simplest and straightforward way to detect all-C loops and all-L cut sets is via the actual construction of the proper tree. Of course, you may be able to detect some of them just by looking at the network. However, it is safer to be systematic about it.

EXAMPLE 19-7 _____

Formulate the state equations for the network in Figure 19-4.

Solution. The proper tree is shown in Figure 19-5. From it, we write immediately

$$\mathbf{x}(t) = \begin{bmatrix} v_{C_1}(t) \\ i_{L_2}(t) \end{bmatrix} \qquad n = 2 \quad \mathbf{e}(t) = v_{in}(t)$$

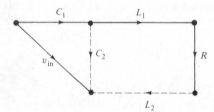

Figure 19-5 Example 19-7.

The fundamental cut set equation for C_1 reads

$$C_1 \frac{dv_{C_1}}{dt} - C_2 \frac{dv_{C_2}}{dt} - i_{L_2} = 0$$

The fundamental loop equation for L_2 is

$$L_2 \frac{di_{L_2}}{dt} - v_{in} + v_{C_1} + L_1 \frac{di_{L_1}}{dt} + v_R = 0$$

The undesirable variables are v_{C_2}, i_{L_1}, and v_R. The all-C loop yields for v_{C_2}, as in Equation 19-5,

$$v_{C_2} = v_{in} - v_{C_1}$$

The all-L cut set (Equation 19-6) gives

$$\frac{di_{L_1}}{dt} = \frac{di_{L_2}}{dt}$$

The resistor is a tree branch; therefore, a fundamental cut set equation for it yields

$$i_R = i_{L_2} \qquad \therefore \quad v_R = R i_{L_2}$$

† The initial inductive currents at $t = 0^-$ are generally independent: $i_{L_1}(0^-)$, $i_{L_2}(0^-)$, ... can be specified arbitrarily if they don't form a cut set at $t = 0^-$. However, for $t \geq 0$, they are dependent and tied by KCL at the all-inductive cut set.

With these substitutions, the two state equations become

$$C_1 \frac{dv_{C_1}}{dt} - C_2 \frac{d}{dt}(v_{\text{in}} - v_{C_1}) - i_{L_2} = 0$$

$$L_2 \frac{di_{L_2}}{dt} - v_{\text{in}} + v_{C_1} + L_1 \frac{di_{L_2}}{dt} + Ri_{L_2} = 0$$

Collecting terms and arranging in final form, we get

$$\frac{dv_{C_1}}{dt} = \frac{1}{C_1 + C_2} i_{L_2} + \frac{C_2}{C_1 + C_2} \frac{dv_{\text{in}}}{dt}$$

$$\frac{di_{L_2}}{dt} = -\frac{1}{L_1 + L_2} v_{C_1} - \frac{R}{L_1 + L_2} i_{L_2} + \frac{1}{L_1 + L_2} v_{\text{in}}$$

In passing, we note that the *derivative* of the input appears here. Nothing to worry about: Since $v_{\text{in}}(t)$ is a *known* function, so is its derivative. ∎

Probs. 19-5, 19-6

*19-2 FORMULATION BY SUPERPOSITION

This method is very instructive in bringing out certain basic ideas in circuit analysis. It can be reasonably fast, with some practice. As its name implies, this method is valid only for a linear network, since we'll be using superposition.

The first idea here is to recognize that, as far as the entire network is concerned, a capacitor may be replaced by a voltage source whose waveform is $v_C(t)$, the voltage across that capacitor. In fact, we can replace it by anything, as long as the voltage across this anything is $v_C(t)$ and the current through this anything is $C\, dv_C/dt$. For the formulation of the state equations, $v_C(t)$ is, of course, an unknown. It will be found during the solution. Dually, an inductor can be replaced by a current source $i_L(t)$ whose voltage is $L\, di_L(t)/dt$. Figure 19-6 illustrates these ideas.

To formulate the state equations for a linear, constant network, we:

1. Replace every capacitor by a voltage source and every inductor by a current source, as in Figure 19-6.
2. Use superposition in the resulting network to calculate the partial contribution of *each* source (real, inductive and capacitive) to the current in each capacitor and to the voltage across each inductor. Add them up to give the total $C\, dv_C/dt$ or $L\, di_L/dt$. These are the state equations.

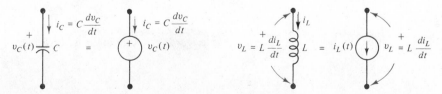

Figure 19-6 Equivalence of state variables and sources.

EXAMPLE 19-8 _____

The network shown in Figure 19-7(a) is replaced by the one in Figure 19-7(b).

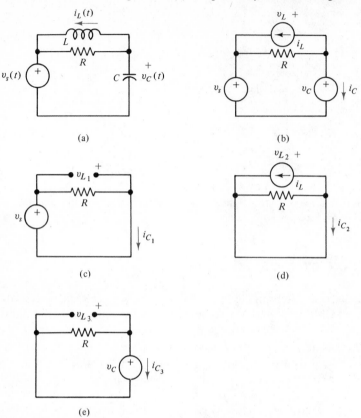

Figure 19-7   Example 19-8.

Solution.   By superposition, we consider v_s alone, with $i_L = 0$ and $v_C = 0$, as in Figure 19-7(c). The partial response i_{C_1} is, by inspection,

$$i_{C_1} = \frac{1}{R} v_s$$

and the partial response v_{L_1} is

$$v_{L_1} = -v_s$$

Next, with i_L acting alone, $v_s = 0$ and $v_C = 0$, as in Figure 19-7(d), we get the partial responses

$$i_{C_2} = -i_L$$
$$v_{L_2} = 0$$

since R is shorted out. Finally, with v_C acting alone, $i_L = 0$, $v_s = 0$, as in Figure 19-7(e), we have

$$i_{C_3} = -\frac{1}{R} v_C$$

$$v_{L_3} = v_C$$

By superposition, then,

$$C\frac{dv_C}{dt} = i_C = i_{C_1} + i_{C_2} + i_{C_3} = \frac{1}{R}v_s - i_L - \frac{1}{R}v_C$$

$$L\frac{di_L}{dt} = v_L = v_{L_1} + v_{L_2} + v_{L_3} = -v_s + 0 + v_C$$

These are the state equations. Dividing the first one by C and the second one by L, we arrange them in matrix form as

$$\begin{bmatrix} \dfrac{dv_C}{dt} \\[2ex] \dfrac{di_L}{dt} \end{bmatrix} = \begin{bmatrix} -\dfrac{1}{RC} & -\dfrac{1}{C} \\[2ex] \dfrac{1}{L} & 0 \end{bmatrix} \begin{bmatrix} v_C \\[1ex] i_L \end{bmatrix} + \begin{bmatrix} \dfrac{1}{R} \\[2ex] -1 \end{bmatrix} v_s$$

Do it the "conventional" way and compare. Such learning practice is good! ∎

This method can be very fast, because each of the subnetworks is purely resistive with only one source. (*Exception*: A controlled source always stays with its controller.) Partial answers can be written quickly, using such tools as current dividers or voltage dividers, as necessary. The summation of the partial answers yields the state equations in their final form.

Probs. 19-7, 19-8, 19-9

19-3 SOLUTION BY THE LAPLACE TRANSFORM: THE INFORMAL WAY

The solution of state equations (Equation 19-1) is a breeze with the Laplace transform. All we need is the transform of the first derivative

$$\mathscr{L}\frac{dx(t)}{dt} = sX(s) - x(0^-) \tag{19-7}$$

and this, surely, has become second nature to us by now.

To get a solid feel for it, let us start with a first-order network ($n = 1$), as in Problem 19-1. Its state equation is

$$\frac{dx(t)}{dt} = ax(t) + be(t) \tag{19-8}$$

a single (scalar) first-order differential equation, of the form of Equation 19-1. Its Laplace transform is

$$sX(s) - x(0^-) = aX(s) + bE(s) \tag{19-9}$$

Collect terms and rearrange

$$(s - a)X(s) = x(0^-) + bE(s) \tag{19-10}$$

On the left side, we recognize immediately the characteristic equation and value

$$s - a = 0 \qquad \therefore \quad s_1 = a \tag{19-11}$$

and on the right side is the initial state $x(0^-)$ which will account for the zero-input response, and the input $bE(s)$ which will account for the zero-state response. Specifically, we divide Equation 19-10 by the characteristic polynomial to get

$$X(s) = \frac{1}{s-a} x(0^-) + \frac{1}{s-a} bE(s) \tag{19-12}$$

Here, the first term on the right is the zero-input solution (more precisely, it is the zero-input state response), and the second one is the zero-state solution (the zero-state state response, which is a mouthful to say!).

The inversion of Equation 19-12 is easy

$$x(t) = e^{at}x(0^-) + e^{at} * be(t) \tag{19-13}$$

In the first term, $x(0^-)$ is a constant, so we are really dealing with $\mathscr{L}^{-1}[K/(s-a)] = Ke^{at}$. Why do we write it in a "funny" way in Equation 19-13, with the constant $x(0^-)$ postmultiplying e^{at}? Just to get used to such a term. When we'll deal with higher-order networks $(n > 1)$, matrix manipulation will require us to keep these matrices in a certain order, so it's nice to get acquainted with such terms at an early stage.

In the second term, we have used the convolution theorem and applied it to the product of two transformed functions

$$\mathscr{L}^{-1}\left(\frac{1}{s-a}\right)(bE(s)) = e^{at} * be(t) \tag{19-14}$$

as studied in Chapter 16. This result also fits our general conclusion there: The zero-state response is the convolution of the impulse response with the input. See Equation 16-27. So, we have found, as a nice "by-product," the state impulse response of a first-order network

$$h(t) = e^{at} \tag{19-15}$$

In words: An initially relaxed first-order network is excited by an impulse $\delta(t)$. The single capacitive voltage *or* the single inductive current response will be of the form e^{at}. As far as the single C or L is concerned, the rest of the network was replaced by a Thévenin or a Norton equivalent circuit.

Probs. 19-10 19-11

EXAMPLE 19-9 ───

Let us consider a second-order network described by the state equation

$$\begin{bmatrix} \dfrac{dv_C(t)}{dt} \\[2ex] \dfrac{di_L(t)}{dt} \end{bmatrix} = \begin{bmatrix} -1 & 6 \\ -1 & -6 \end{bmatrix} \begin{bmatrix} v_C(t) \\ i_L(t) \end{bmatrix} + \begin{bmatrix} 1 \\ 0 \end{bmatrix} u(t) \qquad \begin{bmatrix} v_C(0^-) \\ i_L(0^-) \end{bmatrix} \text{ given}$$

Solution. Take the Laplace transform of each equation separately:

$$sV_C(s) - v_C(0^-) = -V_C(s) + 6I_L(s) + \frac{1}{s}$$

and

$$sI_L(s) - i_L(0^-) = -V_C(s) - 6I_L(s)$$

From here on, algebra takes over, as expected in the Laplace transform method. This is the general rule, and we realize with relief, again, how little (or no) memorization is required.

Collect terms and rearrange for algebraic solution

$$(s + 1)V_C(s) - 6I_L(s) = v_C(0^-) + \frac{1}{s}$$

$$V_C(s) + (s + 6)I_L(s) = i_L(0^-)$$

These are the two simultaneous algebraic equations for the unknown $V_C(s)$ and $I_L(s)$. Let us solve just for $V_C(s)$ by determinants, using Cramer's rule

$$V_C(s) = \frac{\begin{vmatrix} v_C(0^-) + \dfrac{1}{s} & -6 \\ i_L(0^-) & s + 6 \end{vmatrix}}{\begin{vmatrix} s + 1 & -6 \\ 1 & s + 6 \end{vmatrix}} = \underbrace{\frac{(s + 6)v_C(0^-) + 6i_L(0^-)}{s^2 + 7s + 12}}_{\substack{\text{zero-input} \\ \text{response}}} + \underbrace{\frac{(1/s)(s + 6)}{s^2 + 7s + 12}}_{\substack{\text{zero-state} \\ \text{response}}}$$

Here we recognize the determinant in the denominator as the characteristic polynomial. This, again, follows the general rule established earlier: The polynomial that multiplies the transform unknown, just prior to solving algebraically for that unknown, is the characteristic polynomial. In the first-order case, it is $(s - a)$ in Equation 19-10. In the case of $n = 2, 3, \ldots$, it is the determinant of the coefficients of the n unknowns, when arranged in their final form, ready for solution.

The characteristic equation and the natural frequencies are therefore

$$s^2 + 7s + 12 = 0$$

$$\therefore \quad s_1 = -3 \qquad s_2 = -4$$

In the solution for $V_C(s)$, we arranged the terms in two groups. The first one contains only the initial state, $v_C(0^-)$ and $i_L(0^-)$; that is clearly the zero-input part of $V_C(s)$. The second one contains only $1/s$, the input, and is therefore the zero-state part of $V_C(s)$. From here, the inversion of $V_C(s)$ is done by partial fractions, as usual.

In a typical case, we will have numerical values for $v_C(0^-)$ and $i_L(0^-)$, of course. Here we used letter notation simply to show how easy and orderly it is to keep track of the parts of the solution—another outstanding feature of the Laplace transform and the state equations. ∎

Probs. 19-12, 19-13

The previous example shows the general pattern for solving n state equations for an nth-order network. The Laplace transform requires only the use of an old friend, Equation 19-7; after that, we keep track of each step in the algebraic process. The determinant of the coefficients of the unknown transform state variables *must* yield an nth-order characteristic polynomial, as expected. That is good to know in advance. (By contrast, you'll recall, the number of loop equations or node equations has no relationship to the order of the network.) Finally, the zero-state part and the zero-input part are easily tractable in the process of solution.

The output equation (Equation 19-4), as noted, requires no solution. Its Laplace transform is

$$\mathbf{R}(s) = \mathbf{C}\mathbf{X}(s) + \mathbf{D}\mathbf{E}(s) \tag{19-16}$$

Having just found $\mathbf{X}(s)$, we merely substitute it into this equation to obtain $\mathbf{R}(s)$.

*19-4 FORMAL SOLUTION BY THE LAPLACE TRANSFORM

Let us extend the formal solution of a first-order circuit (Equations 19-9 through 19-13), to higher-order circuits. Specifically, let us take $n = 2$, which is enough to obtain the most general results.

The state equations, in matrix form, are

$$\begin{bmatrix} \dfrac{dx_1}{dt} \\ \dfrac{dx_2}{dt} \end{bmatrix} = \begin{bmatrix} a_{11} & a_{12} \\ a_{21} & a_{22} \end{bmatrix} \begin{bmatrix} x_1(t) \\ x_2(t) \end{bmatrix} + \mathbf{B}e(t) \tag{19-17}$$

Here we keep on purpose the generic notation for the state variables $x_1(t)$ and $x_2(t)$. These can be capacitive voltages and/or inductive currents. Similarly, letter notation is retained for the elements of the matrix $\mathbf{A}$, as well as for the input term $\mathbf{B}e(t)$. Take the Laplace transform of Equation 19-17 to get

$$\begin{bmatrix} sX_1(s) - x_1(0^-) \\ sX_2(s) - x_2(0^-) \end{bmatrix} = \begin{bmatrix} a_{11} & a_{12} \\ a_{21} & a_{22} \end{bmatrix} \begin{bmatrix} X_1(s) \\ X_2(s) \end{bmatrix} + \mathbf{B}E(s) \tag{19-18}$$

Collect terms and rearrange as

$$\begin{bmatrix} (s - a_{11}) & -a_{12} \\ -a_{21} & (s - a_{22}) \end{bmatrix} \begin{bmatrix} X_1(s) \\ X_2(s) \end{bmatrix} = \begin{bmatrix} x_1(0^-) \\ x_2(0^-) \end{bmatrix} + \mathbf{B}E(s) \tag{19-19}$$

The matrix of the coefficients on the left-hand side is recognized as

$$\begin{bmatrix} s & 0 \\ 0 & s \end{bmatrix} - \begin{bmatrix} a_{11} & a_{12} \\ a_{21} & a_{22} \end{bmatrix} = s \begin{bmatrix} 1 & 0 \\ 0 & 1 \end{bmatrix} - \mathbf{A} = s\mathbf{U} - \mathbf{A} \tag{19-20}$$

where $\mathbf{U}$ is the unit matrix, a square matrix of order $(n \times n)$, with 1's on the main diagonal and zeros elsewhere. Therefore, Equation 19-19 is rewritten as

$$(s\mathbf{U} - \mathbf{A})\mathbf{X}(s) = \mathbf{x}(0^-) + \mathbf{B}E(s) \tag{19-21}$$

This is the set of n simultaneous algebraic equations in the unknowns $\mathbf{X}(s)$. The characteristic equation is

$$\det(s\mathbf{U} - \mathbf{A}) = 0 \tag{19-22}$$

where, as usual, "det..." means "the determinant of." It will be an nth-order algebraic equation, as expected, with n roots as the characteristic values.

The formal solution of Equation 19-21 continues by premultiplying by $(s\mathbf{U} - \mathbf{A})^{-1}$ to get the transform answer

$$\mathbf{X}(s) = (s\mathbf{U} - \mathbf{A})^{-1}\mathbf{x}(0^-) + (s\mathbf{U} - \mathbf{A})^{-1}\mathbf{B}E(s) \tag{19-23}$$

Here you are strongly encouraged to compare this result with the scalar case ($n = 1$) in Equation 19-12, repeated here:

$$X(s) = \frac{1}{s - a} x(0^-) + \frac{1}{s - a} bE(s) \tag{19-12}$$

Division by a matrix is not defined: Instead, we use the inverse of that matrix. In the scalar case, this is the same

$$(s - a)^{-1} = \frac{1}{s - a} \tag{19-24}$$

Also, the order of premultiplication by $(s\mathbf{U} - \mathbf{A})^{-1}$ is important in Equation 19-23. In the scalar case, such an order is not important; however, in anticipation of Equation 19-23 and for comparison purposes, we kept the same order in the scalar case also.

If we are bold, we can take the final step in this formal derivation, writing the inverse transform of Equation 19-23 with a sharp eye on the scalar case (Equation 19-13), also repeated here

$$x(t) = e^{at}x(0^-) + e^{at} * be(t) \tag{19-13}$$

The result is the final solution

$$\mathbf{x}(t) = e^{\mathbf{A}t}\mathbf{x}(0^-) + e^{\mathbf{A}t} * \mathbf{B}e(t) \tag{19-25}$$

showing clearly the zero-input and the zero-state parts.

Also, by comparison, we write

$$e^{\mathbf{A}t} = \mathscr{L}^{-1}(s\mathbf{U} - \mathbf{A})^{-1} \tag{19-26}$$

We here have a totally new (gasp!) time function, $e^{\mathbf{A}t}$, a *matrix exponential function*: It is $e = 2.718\ldots$ raised to the power of the matrix $\mathbf{A}t$.

EXAMPLE 19-10

For the matrix $\mathbf{A}$ in Example 19-9,

$$\mathbf{A} = \begin{bmatrix} -1 & 6 \\ -1 & -6 \end{bmatrix}$$

we have

$$s\mathbf{U} - \mathbf{A} = \begin{bmatrix} s+1 & -6 \\ 1 & s+6 \end{bmatrix}$$

Next†

$$(s\mathbf{U} - \mathbf{A})^{-1} = \frac{1}{\det(s\mathbf{U} - \mathbf{A})} \begin{bmatrix} s+6 & 6 \\ -1 & s+1 \end{bmatrix}$$

$$= \begin{bmatrix} \dfrac{s+6}{s^2 + 7s + 12} & \dfrac{6}{s^2 + 7s + 12} \\ \dfrac{-1}{s^2 + 7s + 12} & \dfrac{s+1}{s^2 + 7s + 12} \end{bmatrix}$$

$$= \begin{bmatrix} \dfrac{s+6}{(s+3)(s+4)} & \dfrac{6}{(s+3)(s+4)} \\ \dfrac{-1}{(s+3)(s+4)} & \dfrac{s+1}{(s+3)(s+4)} \end{bmatrix}$$

† See Appendix A.

Expansion by partial fractions yields

$$(s\mathbf{U} - \mathbf{A})^{-1} = \begin{bmatrix} \dfrac{3}{s+3} + \dfrac{-2}{s+4} & \dfrac{6}{s+3} + \dfrac{-6}{s+4} \\ \dfrac{-1}{s+3} + \dfrac{1}{s+4} & \dfrac{-2}{s+3} + \dfrac{3}{s+4} \end{bmatrix}$$

and so its inverse Laplace is

$$e^{\mathbf{A}t} = \begin{bmatrix} 3e^{-3t} - 2e^{-4t} & 6e^{-3t} - 6e^{-4t} \\ -e^{-3t} + e^{-4t} & -2e^{-3t} + 3e^{-4t} \end{bmatrix}$$

This $e^{\mathbf{A}t}$ is then used in Equation 19-25 to get the answer $\mathbf{x}(t)$.   ■ Prob. 19-14

Without going further into details (some things must be left for other courses!), let us just enjoy the complete solution and the parallel steps between the scalar case and the matrix case. We are reassured in knowing that, for all practical purposes, the informal and straightforward method is totally satisfactory; it gives the same final results as the formal one.

*19-5 STATE TRAJECTORY IN THE STATE SPACE

Let us consider, again, a second-order ($n = 2$) network and its zero-input response only. In other words, the network starts with the initial state

$$\mathbf{x}(0^-) = \begin{bmatrix} x_1(0^-) \\ x_2(0^-) \end{bmatrix} \tag{19-27}$$

and no other input. We are interested in the state $\mathbf{x}(t)$ at any time $t > 0$.

At every instant, say $t = t_1$, $t = t_2, \ldots$, the state of the network will be given by $\mathbf{x}(t_1)$, $\mathbf{x}(t_2), \ldots$—a pair of numbers for each instant. Let us show each such pair $[x_1(t_k), x_2(t_k)]$ as the coordinates of a point in the x_1-x_2 plane. Such a plane is called the *state space* for this network, and, as t varies from $t = 0$ to $t = \infty$, the points $[x_1(t_k), x_2(t_k)]$ trace a curve called the *state trajectory*.

EXAMPLE 19-11 _____

For the circuit in Example 19-9, let the initial state be given as

$$\mathbf{x}(0^-) = \begin{bmatrix} v_C(0^-) \\ i_L(0^-) \end{bmatrix} = \begin{bmatrix} -4 \\ 2 \end{bmatrix}$$

The zero-input response is therefore (fill in the details!)

$$v_C(t) = -4e^{-4t} \qquad t \geq 0$$
$$i_L(t) = 2e^{-4t} \qquad t \geq 0$$

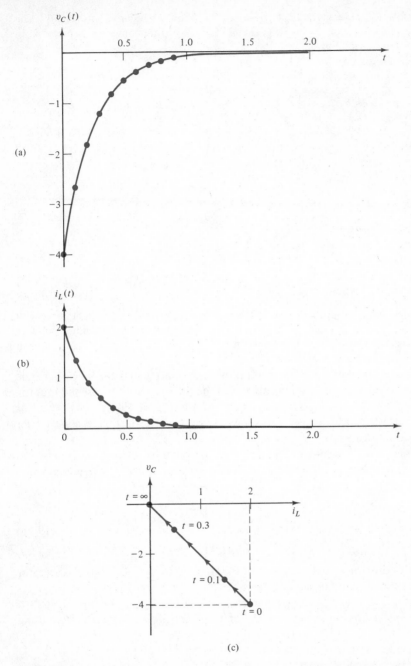

Figure 19-8   State trajectory.

The separate plots of $v_C(t)$ versus t and of $i_L(t)$ versus t are shown in Figure 19-8(a) and (b). This is clearly the *overdamped* case in this *RLC* circuit. These plots are conveniently prepared by using the following calculations:

t	v_C	i_L
0	−4.00	2.00
0.1	−2.68	1.34
0.2	−1.80	0.90
0.3	−1.20	0.60
0.4	−0.80	0.40
0.5	−0.54	0.27
0.6	−0.36	0.18
0.7	−0.24	0.12
0.8	−0.16	0.08
0.9	−0.11	0.05
1.0	−0.07	0.04
2.0	−0.001	0.0006
∞	0.00	0.00

The state trajectory is shown in Figure 19-8(c). It just happens to be a straight line here. It starts at $(-4, 2)$ for $t = 0$, and ends at $(0, 0)$ for $t = \infty$. For any t, the corresponding point on the trajectory specifies the state (v_C, i_L) at that time. ∎

Probs. 19-15, 19-16

An alternate and equivalent interpretation of the state $\mathbf{x}(t)$ is the following: The two numbers $[x_1(t), x_2(t)]$ are the coordinates of the tip of a vector whose tail is at the origin $(0, 0)$ of the state space. Such a vector is shown in Figure 19-9(a) for $t = t_1$. For different times, the tip of this vector traces the state trajectory. For this reason we refer to $\mathbf{x}(t)$ as the *state vector*.

A state vector $\mathbf{x}(t_1)$ is shown in Figure 19-9(b) for a third-order $(n = 3)$ network. The state space here has x_1, x_2, and x_3 as coordinate axes; at $t = t_1$, the components of $\mathbf{x}(t_1)$ are $x_1(t_1)$, $x_2(t_1)$, and $x_3(t_1)$

$$\mathbf{x}(t_1) = \begin{bmatrix} x_1(t_1) \\ x_2(t_1) \\ x_3(t_1) \end{bmatrix} \tag{19-28}$$

For networks of order $n \geq 4$, the concept of the state vector is still valid. It is a vector in the n-dimensional state space with coordinate axes $x_1, x_2, \ldots, x_n$, whose tip has the coordinates $x_1(t), x_2(t), \ldots, x_n(t)$ for every t. The actual drawing of such a multidimensional vector is another matter, but it certainly does not stop us from using the concept.

To conclude this discussion, let us tie the ideas of the state trajectory to the study of stability (Chapter 18). We can define *zero-input stability* as follows: A network is zero-input stable if, for any given initial state $\mathbf{x}(0^-)$, the state trajectory remains bounded for all $0 < t < \infty$. In other words, every component of the state vector, $x_p(t)$, must be bounded

$$|x_p(t)| < K \qquad p = 1, 2, \ldots, n \tag{19-29}$$

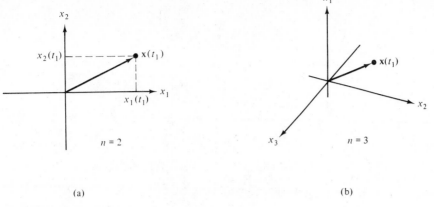

Figure 19-9   State vectors.

where K is some *fixed* finite constant. Thus, none of the capacitive voltages and inductive currents must grow without bound as $t \to \infty$. This, in turn, puts the familiar restrictions on the characteristic values of the network: They must be in the left half plane (LHP) or, at the most, on the $j\omega$ axis where they must be simple.

19-6 STATE VARIABLES AND THE NETWORK FUNCTION

The network function $H(s)$ relates the zero-state response to the input. How do we do it with the state variables? The answer is fairly simple: Eliminate (mathematically) the state variables between the state equation and the output equation. Before doing it for the general case, let us consider a scalar case, $n = 1$.

EXAMPLE 19-12 _____

In the *RL* network shown in Figure 19-10, the voltage transfer function $H(s)$ is found immediately by the voltage divider

$$V_{\text{out}}(s) = \frac{R}{Ls + R} V_{\text{in}}(s)$$

$$\therefore H(s) = \frac{R}{Ls + R}$$

Let us obtain it via our state variable analysis.

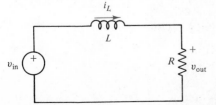

Figure 19-10   Example 19-12.

Solution. The state equation in $i_L(t)$ is

$$\frac{di_L}{dt} = -\frac{R}{L}i_L + \frac{1}{L}v_{in}$$

and the output equation is

$$v_{out} = Ri_L + 0 \cdot v_{in}$$

exhibiting the state matrices (here scalars)

$$\mathbf{A} = -\frac{R}{L} \qquad \mathbf{B} = \frac{1}{L} \qquad \mathbf{C} = R \qquad \mathbf{D} = 0$$

Laplace transform these two equations, with zero initial conditions. [Remember: $H(s)$ is calculated for the zero-state response!] We get

$$sI_L(s) = -\frac{R}{L}I_L(s) + \frac{1}{L}V_{in}(s)$$

and

$$V_{out}(s) = RI_L(s)$$

Now eliminate $I_L(s)$—the "intermediate" state variable—between these two equations. Specifically, from the first one we get

$$\left(s + \frac{R}{L}\right)I_L(s) = \frac{1}{L}V_{in}(s)$$

$$\therefore \quad I_L(s) = \frac{V_{in}(s)}{Ls + R}$$

and substitute this expression into the output equation

$$V_{out}(s) = \frac{R}{Ls + R}V_{in}(s)$$

with the network function clearly identified. No big deal, right? Yet, this process contains all the steps that we'll be using for networks of higher order. ∎

To review this case and to prepare us for the general derivation, let us rework it using the standard generic notation in state variable analysis. The state equation (Equation 19-1) is here a scalar equation

$$\frac{dx(t)}{dt} = ax(t) + be(t) \tag{19-30}$$

and the output equation (Equation 19-4), is here

$$r(t) = cx(t) + de(t) \tag{19-31}$$

Laplace transform both equations, under zero-state conditions:

$$sX(s) = aX(s) + bE(s) \tag{19-32}$$

and

$$R(s) = cX(s) + dE(s) \tag{19-33}$$

The first one (Equation 19-32) is familiar to us from Section 19-3, where the *total* solution of $x(t)$ was obtained. Compare with Equations 19-9 and 19-10 there.

To eliminate $X(s)$, we solve it from Equation 19-32 as

$$(s - a)X(s) = bE(s) \tag{19-34}$$

that is,

$$X(s) = \frac{1}{s - a} bE(s) \tag{19-35}$$

and substitute into Equation 19-33 to get

$$R(s) = c \frac{1}{s - a} bE(s) + dE(s)$$

$$= \left(c \frac{1}{s - a} b + d \right) E(s) \tag{19-36}$$

Therefore

$$H(s) = c \frac{1}{s - a} b + d \tag{19-37}$$

is the desired network function whose pole is the characteristic value $s = a$. Here, again, we use some "funny" algebraic notation, but it is perfectly correct. The reason? As earlier, we are preparing to plunge into the general matrix notation.

The general state equation (Equation 19-1) is repeated here

$$\frac{d\mathbf{x}(t)}{dt} = \mathbf{A}\mathbf{x}(t) + \mathbf{B}e(t) \tag{19-1}$$

and the general output equation is Equation 19-4:

$$\mathbf{r}(t) = \mathbf{C}\mathbf{x}(t) + \mathbf{D}e(t) \tag{19-4}$$

Laplace transform these two equations with zero initial conditions, to get

$$s\mathbf{X}(s) = \mathbf{A}\mathbf{X}(s) + \mathbf{B}E(s) \tag{19-38}$$

and

$$\mathbf{R}(s) = \mathbf{C}\mathbf{X}(s) + \mathbf{D}E(s) \tag{19-39}$$

Solve Equation 19-39 for $\mathbf{X}(s)$, as follows

$$(s\mathbf{U} - \mathbf{A})\mathbf{X}(s) = \mathbf{B}E(s) \tag{19-40}$$

$$\therefore \quad \mathbf{X}(s) = (s\mathbf{U} - \mathbf{A})^{-1}\mathbf{B}E(s) \tag{19-41}$$

where $\mathbf{U}$ is the $(n \times n)$ unit matrix. Please review here the similar steps, detailed in Equations 19-18 through 19-23, but with $\mathbf{x}(0^-) = \mathbf{0}$ here.

Substitute Equation 19-41 into Equation 19-39 to get

$$\mathbf{R}(s) = \mathbf{C}(s\mathbf{U} - \mathbf{A})^{-1}\mathbf{B}E(s) + \mathbf{D}E(s)$$

$$= [\mathbf{C}(s\mathbf{U} - \mathbf{A})^{-1}\mathbf{B} + \mathbf{D}]E(s) \tag{19-42}$$

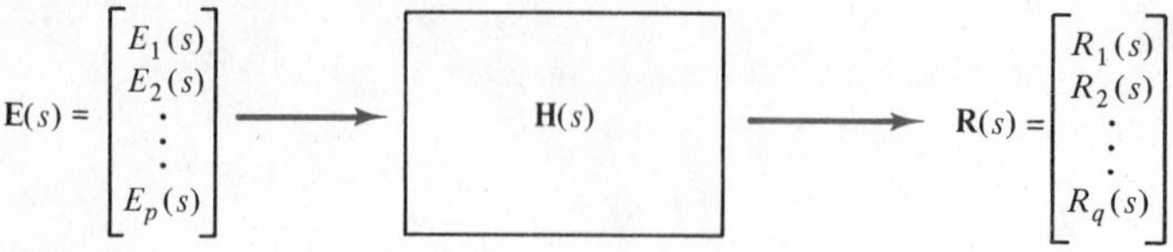

$$E(s) = \begin{bmatrix} E_1(s) \\ E_2(s) \\ \cdot \\ \cdot \\ E_p(s) \end{bmatrix} \longrightarrow \boxed{\quad H(s) \quad} \longrightarrow R(s) = \begin{bmatrix} R_1(s) \\ R_2(s) \\ \cdot \\ \cdot \\ R_q(s) \end{bmatrix}$$

Figure 19-11   Matrix network function.

where the order of multiplication of the matrices is carefully preserved; in the last step of Equation 19-42, the input matrix $E(s)$ is factored out as a postmultiplier.

In Equation 19-42, we recognize the desired input-output relationship

$$\mathbf{R}(s) = \mathbf{H}(s)\mathbf{E}(s) \tag{19-43}$$

with

$$\mathbf{H}(s) = \mathbf{C}(s\mathbf{U} - \mathbf{A})^{-1}\mathbf{B} + \mathbf{D} \tag{19-44}$$

as the *matrix* network function, relating any number, p, of given inputs to any number, q, of specified zero-state outputs. This is illustrated in Figure 19-11. The single-input ($p = 1$) single-output ($q = 1$) case yields, of course, a scalar network function $H(s)$. In our studies of 2-P networks (Chapter 16), we had two inputs, $p = 2$, and two outputs, $q = 2$. The various network functions, $\mathbf{Z}_{\text{o.c.}}$, $\mathbf{Y}_{\text{s.c.}}$, $\mathbf{h}$, etc., were matrix network functions.

EXAMPLE 19-13 _____

Consider the second-order ($n = 2$) network in Figure 19-12. The order $n = 2$ is big enough to illustrate fully all the matrix operations, yet small enough to make them manageable.

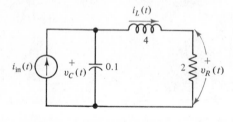

$(\Omega, \text{H}, \text{F})$

Figure 19-12   Example 19-13.

Solution.   The input here is $E(s) = I_{\text{in}}(s)$, a single one ($p = 1$). The desired outputs are $I_L(s)$ and $V_R(s)$, so $q = 2$ and

$$\mathbf{R}(s) = \begin{bmatrix} I_L(s) \\ V_R(s) \end{bmatrix}$$

The state equations are (do it in detail!)

$$\frac{dv_C(t)}{dt} = -10i_L(t) + 10i_{\text{in}}(t)$$

$$\frac{di_L(t)}{dt} = \frac{v_C(t)}{4} - \tfrac{1}{2}i_L(t)$$

That is,

$$\frac{d}{dt}\begin{bmatrix} v_C(t) \\ i_L(t) \end{bmatrix} = \begin{bmatrix} 0 & -10 \\ \frac{1}{4} & -\frac{1}{2} \end{bmatrix}\begin{bmatrix} v_C(t) \\ i_L(t) \end{bmatrix} + \begin{bmatrix} 10 \\ 0 \end{bmatrix}i_{in}(t)$$

exhibiting the matrices **A** and **B**

$$\mathbf{A} = \begin{bmatrix} 0 & -10 \\ \frac{1}{4} & -\frac{1}{2} \end{bmatrix} \qquad \mathbf{B} = \begin{bmatrix} 10 \\ 0 \end{bmatrix}$$

The output equations are

$$i_L(t) = i_L(t)$$
$$v_R(t) = 2i_R(t) = 2i_L(t)$$

or

$$\begin{bmatrix} i_L \\ v_R \end{bmatrix} = \begin{bmatrix} 0 & 1 \\ 0 & 2 \end{bmatrix}\begin{bmatrix} v_C \\ i_L \end{bmatrix} + [0]i_{in}$$

Therefore

$$\mathbf{C} = \begin{bmatrix} 0 & 1 \\ 0 & 2 \end{bmatrix} \qquad \mathbf{D} = 0$$

With the four matrices fully identified, we use Equation 19-44 to calculate **H**(s)

$$\mathbf{H}(s) = \begin{bmatrix} 0 & 1 \\ 0 & 2 \end{bmatrix}\begin{bmatrix} s & 10 \\ -\frac{1}{4} & s+\frac{1}{2} \end{bmatrix}^{-1}\begin{bmatrix} 10 \\ 0 \end{bmatrix}$$

$$= \begin{bmatrix} 0 & 1 \\ 0 & 2 \end{bmatrix}\frac{1}{s^2 + \frac{1}{2}s + 2.5}\begin{bmatrix} s+\frac{1}{2} & -10 \\ \frac{1}{4} & s \end{bmatrix}\begin{bmatrix} 10 \\ 0 \end{bmatrix}$$

$$= \frac{1}{s^2 + \frac{1}{2}s + 2.5}\begin{bmatrix} 2.5 \\ 5 \end{bmatrix} = \begin{bmatrix} \dfrac{5}{2s^2 + s + 5} \\ \dfrac{10}{2s^2 + s + 5} \end{bmatrix}$$

Consequently, Equation 19-43 reads

$$\begin{bmatrix} I_L(s) \\ V_R(s) \end{bmatrix} = \begin{bmatrix} \dfrac{5}{2s^2 + s + 5} \\ \dfrac{10}{2s^2 + s + 5} \end{bmatrix}I_{in}(s) = \mathbf{H}(s)I_{in}(s) = \begin{bmatrix} H_{11}(s) \\ H_{21}(s) \end{bmatrix}I_{in}(s)$$

showing the (2 × 1) matrix network function **H**(s). Its denominator is the expected characteristic polynomial, det (s**U** − **A**).

As a quick independent check, use the current divider equation to calculate $I_L(s)$ in Figure 19-12. We have there

$$I_L(s) = \frac{\dfrac{1}{4s+2}}{\dfrac{1}{4s+2} + 0.1s}I_{in}(s) = \frac{5}{2s^2 + s + 5}I_{in}(s)$$

which confirms $H_{11}(s)$ in $\mathbf{H}(s)$. The second output is

$$V_R(s) = 2I_L(s) = 2 \frac{5}{2s^2 + s + 5} I_{in}(s) = \frac{10}{2s^2 + s + 5} I_{in}(s)$$

confirming $H_{21}(s)$.

Probs. 19-17
19-18, 19-19

This concludes our discussion of relating the state variables description $(\mathbf{A}, \mathbf{B}, \mathbf{C}, \mathbf{D})$ to the network function $\mathbf{H}(s)$. In a sense, we have completed the circle. We started with the network function via loop or node analysis. Next, we took a big detour to explore the new method of state variable analysis. Finally, we returned to the network function via the state variables.

A fascinating subject is the reverse problem: From a given $\mathbf{H}(s)$, find the matrices $\mathbf{A}, \mathbf{B}, \mathbf{C}, \mathbf{D}$. This problem of network identification (or synthesis) is considerably more difficult and is reserved for advanced courses.

19-7 NUMERICAL CALCULATIONS

To get an idea of the numerical methods available in the calculations of state variables, let us go back and consider the state trajectory of Section 19-4. What we did there was to calculate the zero-input state response from the state equation

$$\frac{d}{dt} \mathbf{x}(t) = \mathbf{A}\mathbf{x}(t) \qquad \mathbf{x}(0^-) \text{ given} \tag{19-45}$$

with $\mathbf{e}(t) = \mathbf{0}$ as required for zero input. The exact analytical solution is given in Equation 19-25, namely,

$$\mathbf{x}(t) = e^{\mathbf{A}t}\mathbf{x}(0^-) \tag{19-46}$$

with $e^{\mathbf{A}t}$ defined in Equation 19-26.

To develop an algorithmic step-by-step numerical approximate solution to Equation 19-45, we divide the time axis into intervals, $t = 0, \Delta t, 2\,\Delta t, 3\,\Delta t, \ldots, k\,\Delta t, \ldots$. If Δt is taken sufficiently small, the derivative in Equation 19-45 can be approximated as follows, for $t = 0$:

$$\left. \frac{d\mathbf{x}(t)}{dt} \right]_0 \approx \frac{\mathbf{x}(\Delta t) - \mathbf{x}(0)}{\Delta t} = \mathbf{A}\mathbf{x}(0) \tag{19-47}$$

that is, the derivative of $\mathbf{x}(t)$ at $t = 0$ is approximately the difference between $\mathbf{x}(\Delta t)$ and $\mathbf{x}(0)$ divided by Δt. If you recall your calculus, this is the basic idea in defining the first derivative of a function. From Equation 19-47 we get $\mathbf{x}(\Delta t)$ as

$$\mathbf{x}(\Delta t) \approx \mathbf{x}(0) + \mathbf{A}\mathbf{x}(0)\Delta t \tag{19-48}$$

We repeat this step, to calculate $\mathbf{x}(2\,\Delta t)$ from $\mathbf{x}(\Delta t)$:

$$\frac{d\mathbf{x}(t)}{dt}\bigg]_{\Delta t} \approx \frac{\mathbf{x}(2\,\Delta t) - \mathbf{x}(\Delta t)}{\Delta t} = \mathbf{A}\mathbf{x}(\Delta t) \qquad (19\text{-}49)$$

or

$$\mathbf{x}(2\,\Delta t) \approx \mathbf{x}(\Delta t) + \mathbf{A}\mathbf{x}(\Delta t)\Delta t \qquad (19\text{-}50)$$

In general, we obtain

$$\mathbf{x}(k\Delta t) \approx \mathbf{x}[(k-1)\Delta t] + \mathbf{A}\mathbf{x}[(k-1)\Delta t]\Delta t \qquad (19\text{-}51)$$

This expression gives the state $\mathbf{x}(t)$ at any time $k\,\Delta t$ as the sum of the previous state plus a "correction term." Equation 19-51 is easy to implement on a digital computer, and, the smaller Δt, the closer will the numerical solution be to the exact one.

Let us illustrate with the same equation as in Examples 19-9 and 19-11.

EXAMPLE 19-14 _____

We have

$$\frac{d}{dt}\begin{bmatrix} x_1(t) \\ x_2(t) \end{bmatrix} = \begin{bmatrix} -1 & 6 \\ -1 & -6 \end{bmatrix}\begin{bmatrix} x_1(t) \\ x_2(t) \end{bmatrix} \qquad \mathbf{x}(0^-) = \begin{bmatrix} -4 \\ 2 \end{bmatrix}$$

Solution. For comparison purposes, take $\Delta t = 0.1\,\text{s}$, as in the exact calculations of Example 19-11. Here we get from Equation 19-48

$$\mathbf{x}(0.1) = \begin{bmatrix} -4 \\ 2 \end{bmatrix} + \begin{bmatrix} -1 & 6 \\ -1 & -6 \end{bmatrix}\begin{bmatrix} -4 \\ 2 \end{bmatrix}(0.1) = \begin{bmatrix} -2.4 \\ 1.2 \end{bmatrix}$$

Compare it with the exact (two decimal) result there

$$\mathbf{x}(0.1) = \begin{bmatrix} -2.68 \\ 1.34 \end{bmatrix}$$

We continue the numerical calculations and comparison

$$\mathbf{x}(0.2) = \begin{bmatrix} -2.4 \\ 1.2 \end{bmatrix} + \begin{bmatrix} -1 & 6 \\ -1 & -6 \end{bmatrix}\begin{bmatrix} -2.4 \\ 1.2 \end{bmatrix}(0.1) = \begin{bmatrix} -1.44 \\ 0.72 \end{bmatrix}$$

and the exact one is

$$\mathbf{x}(0.2) = \begin{bmatrix} -1.8 \\ 0.90 \end{bmatrix}$$

The approximation is not very good. It can be improved by taking a smaller Δt, say, $\Delta t = 0.01$; at the same time, the exact (analytical) values should be calculated to 3 or 4 decimal places. More sophisticated numerical methods may be used. ■ Prob. 19-20

PROBLEMS

19-1 Obtain the state equation for each network shown. Assume initial conditions in letter notation, $i_L(0^-)$ and $v_C(0^-)$.

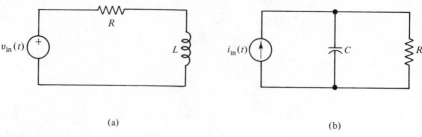

(a)   (b)

Problem 19-1

19-2 Obtain the standard matrix state equation for the network shown. The initial state is zero.

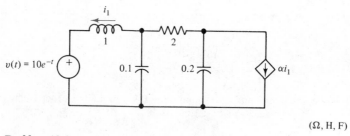

(Ω, H, F)

Problem 19-2

19-3 Obtain the state and the output equations for the network shown.

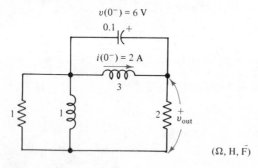

$(\Omega, H, \bar{F})$

Problem 19-3

*19-4 In Example 19-2, let all the elements be linear, time-varying. That is, the two resistors are defined by

$$v_4(t) = R_4(t)i_4(t)$$
$$v_{10}(t) = R_{10}(t)i_{10}(t)$$

The inductor is defined by

$$\phi_3(t) = L_3(t)i_3(t)$$

and the two capacitors by

$$q_1(t) = C_1(t)v_1(t)$$
$$q_2(t) = C_2(t)v_2(t)$$

Choose $q_1(t)$, $q_2(t)$, and $\phi_3(t)$ as the state variables, and formulate the state equations. Compare with Equation 19-1 for a linear constant network.

19-5 For each network shown, determine its order n, draw its proper tree, and list its state variables.

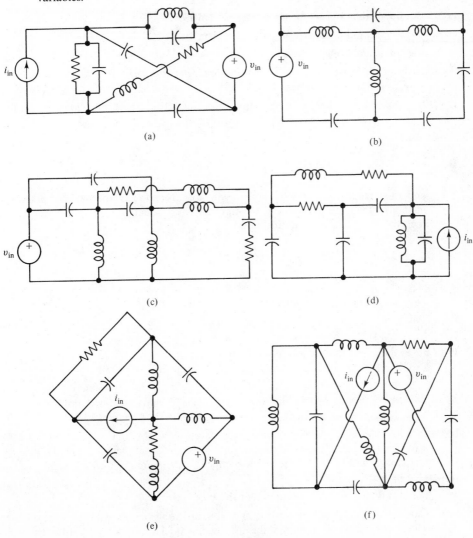

(a)

(b)

(c)

(d)

(e)

(f)

Problem 19-5

19-6 Formulate the state equations for the network shown. Do you need the initial state $\mathbf{x}(0^-)$ for the formulation? Explain.

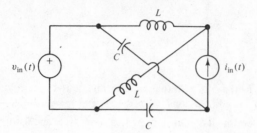

Problem 19-6

19-7 Use the method of superposition for Problem 19-1.

19-8 Use the method of superposition for Problem 19-3.

*19-9 Use the method of superposition for Example 19-7.

19-10 In the characteristic equation of the first-order network (Equation 19-11), what is the obvious stability requirement on a?

19-11 For old times' sake, solve Equation 19-8 by the classical time-domain method (homogeneous, particular, etc.). Start with its form

$$\frac{dx(t)}{dt} - ax(t) = be(t)$$

and go as far as you can. Where must you stop? Next, as an alternate route, try the method of the integrating factor: Multiply the entire equation by e^{-at}; then recognize the left-hand side as a total differential,

$$\frac{d}{dt}[x(t)e^{-at}]$$

Now you can integrate both sides and continue.

19-12 Calculate the zero-state solution for $v_3(t)$ in Example 19-5, Figure 19-2, with the following values: $R_1 = 1\,\Omega$, $r = 0.5\,\Omega$, $L_2 = \frac{1}{4}$H, $C_3 = 0.1$ F.

19-13 In Problem 19-12, let the response (output) be $r(t) = i_1(t)$. With $X(s)$ as obtained in Problem 19-12, calculate $R(s) = I_1(s)$. Confirm this answer by either loop analysis or node analysis.

19-14 Rework Problem 19-12 by the formal approach of matrix manipulation (Equations 19-18 through 19-23). Calculate $e^{\mathbf{A}t}$ for this circuit, as in Equation 19-26.

19-15 In a parallel RLC circuit we are given $R = 1\,\Omega$, $L = 1$ H, $C = 1$ F, $i_L(0^-) = 1$ A, $v_C(0^-) = 1$ V. Calculate the zero-input state response for this network and plot its state trajectory. Classify this circuit as overdamped, underdamped, or critically damped.

19-16 In a lossless parallel circuit, $L = 1$ H, $C = \frac{1}{4}$ F, $i_L(0^-) = 1$ A, $v_C(0^-) = 1$ V. Calculate the zero-input state response and plot the state trajectory.

19-17 Calculate $\mathbf{H}(s)$, the matrix network function, in Problem 19-6, with the given inputs. The desired output is the current through the top inductor, referenced positive from left to right.

19-18 Calculate $\mathbf{H}(s)$ for Example 19-5, Figure 19-2, with

$$\mathbf{r}(t) = \begin{bmatrix} i_1(t) \\ v_3(t) \\ i_2(t) \end{bmatrix}$$

as the desired outputs. Element values are: $R_1 = 1\ \Omega$, $L_2 = 2\ \mathrm{H}$, $C_3 = 0.1\ \mathrm{F}$, $r = 10$.

19-19 If the input vector $\mathbf{E}(s)$ is $(p \times 1)$, the output vector $\mathbf{R}(s)$ is $(q \times 1)$, and the order of the network is n, what are the dimensions of $\mathbf{A}$, $\mathbf{B}$, $\mathbf{C}$, $\mathbf{D}$, and $\mathbf{H}$ in Equation 19-44?

19-20 Rework Example 19-14 by using $\Delta t = 0.01$ for $0 \le t \le 1$. Write a short program to help you here. Plot and compare with Figure 19-8.

Chapter 20

The Fourier Series

Periodic waveforms are very common in electronic circuits, communication networks, computer circuits, and power generation and distribution. In Chapters 9 to 13, we studied in detail the steady-state response due to one special kind of a periodic input, a pure sinusoid. In this chapter, we extend this study to other periodic waveforms. A few such waveforms are shown in Figure 20-1, with their common names. We will be interested in finding the zero-state periodic response of a linear constant circuit excited by such a periodic input, the related power calculations, etc.

Why not use the Laplace transform? There are several reasons: First, these periodic functions are defined for *all* t, $-\infty < t < \infty$, but our Laplace transform is defined for functions that are zero for $t < 0$. This difficulty may be overcome, but even then the Laplace transform approach is very complicated; in addition, it masks the frequency response aspects of the circuit, and, as we saw in previous chapters, such aspects are basic to understanding the behavior of the circuit. With phasors, we were never too far from the physical aspects of the waveform—magnitude and phase.

For these reasons, we will study and apply the ideas of the Fourier series.

20-1 THE FOURIER SERIES

A *periodic* function $f(t)$ is one that repeats itself every T seconds. More precisely

$$f(t \pm nT) = f(t) \qquad n = 1, 2, 3, \ldots \tag{20-1}$$

for $-\infty < t < \infty$. The *period* of $f(t)$ is T, and one *cycle* of $f(t)$ is the portion of $f(t)$ over one period. In Figure 20-1 we show a cycle of $f(t)$ as $f_1(t)$ in color. The *frequency*

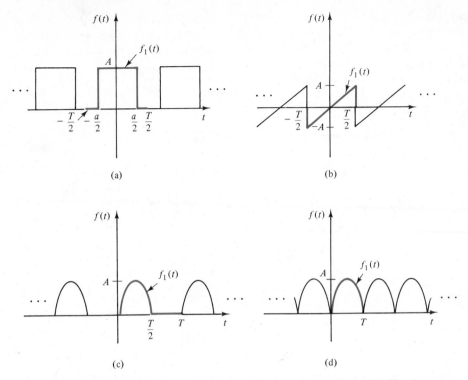

Figure 20-1 Various periodic waveforms. (a) The square waveform. (b) The "saw-tooth" waveform. (c) The half-wave rectified waveform. (d) The full-wave rectified waveform.

f of $f(t)$ is the number of cycles per second (cps) and is measured in hertz (Hz)

$$f = \frac{1}{T} \qquad (20\text{-}2)$$

Given, then, one cycle $f_1(t)$ and the frequency (or the period), we know the periodic function $f(t)$. These notions are familar to us from pure sinusoidal waveforms; here, we review them in preparation for extending them to other periodic functions.

The French mathematician Jean B. J. Fourier (1768–1830), while studying problems of heat flow, discovered that a periodic function can be written as a sum of pure sine and cosine functions of different frequencies. More specifically, if we define the *fundamental* radian frequency ω_0 as

$$\omega_0 = 2\pi f = \frac{2\pi}{T} \qquad \text{rad/s} \qquad (20\text{-}3)$$

then we can write the Fourier series for $f(t)$ as

$$f(t) = \frac{a_0}{2} + a_1 \cos \omega_0 t + a_2 \cos 2\omega_0 t + \cdots + a_n \cos n\omega_0 t + \cdots$$

$$+ \, b_1 \sin \omega_0 t + b_2 \sin 2\omega_0 t + \cdots + b_n \sin n\omega_0 t + \cdots \qquad (20\text{-}4)$$

Here, the first term $a_0/2$ is a constant. The sine and cosine terms are of integer multiples of ω_0, and they are called the *harmonics*. For example, the terms $a_4 \cos 4\omega_0 t + b_4 \sin 4\omega_0 t$ are the fourth harmonic. In general, the series is infinite, and a common practice in engineering is to truncate it after a certain finite number of terms with an acceptable error of approximation.

EXAMPLE 20-1 _____

The Fourier series for the "sawtooth" waveform in Figure 20-2(a) is given by

$$f(t) = 2 \sin t - \sin 2t + \tfrac{2}{3} \sin 3t + \cdots$$

So here $a_0/2 = 0$, $a_1 = a_2 = \cdots = a_n = 0$, $b_1 = 2$, $b_2 = -1$, $b_3 = \tfrac{2}{3}, \ldots$. In Figure 20-2(b) we plotted the three functions

$$f_a(t) = 2 \sin t$$

$$f_b(t) = 2 \sin t - \sin 2t$$

$$f_c(t) = 2 \sin t - \sin 2t + \tfrac{2}{3} \sin 3t$$

showing how the approximation improves with more and more terms.   Prob. 20-1

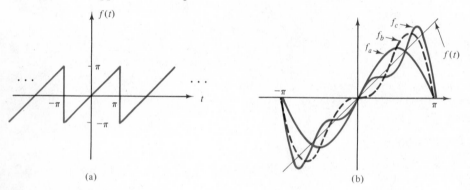

Figure 20-2   Example 20-1.   ■

Our main problem is: *For a given $f(t)$, such as in Figure 20-1,* calculate its Fourier series, *that is,* the a's and the b's in Equation 20-4. As an interesting sidelight we mention the known sufficient conditions for $f(t)$ to have a converging Fourier series:

1. $f(t)$ is single-valued.
2. $f(t)$ has, at most, a finite number of maxima and minima over the period T.
3. $f(t)$ has, at most, a finite number of discontinuities ("jumps") over the period T.
4. The integral

$$\int_{t_0}^{t_0+T} |f(t)| \, dt$$

is finite.

These conditions (known as Dirichlet's conditions) are only *sufficient*; in other words, if $f(t)$ satisfies them, then it has a Fourier series. However, they are not necessary: If $f(t)$ does not satisfy them, it may still have a Fourier series. To this day, the necessary *and* sufficient conditions are not known. Of more interest to us is the comforting fact that most of the common periodic waveforms in engineering *have* a Fourier series.

20-2 CALCULATION OF THE FOURIER COEFFICIENTS

In order to calculate the a's and b's in Equation 20-4, we'll need the following results:

$$\int^{T} \sin k\omega_0 t \, dt = 0 \qquad k \text{ integer} \tag{20-5}$$

$$\int^{T} \cos k\omega_0 t \, dt = 0 \qquad k \text{ integer} \tag{20-6}$$

$$\int^{T} (\sin k\omega_0 t)(\cos p\omega_0 t) \, dt = 0 \qquad k \text{ and } p \text{ integers} \tag{20-7}$$

$$\int^{T} (\sin k\omega_0 t)(\sin p\omega_0 t) \, dt = \begin{cases} 0 & k \neq p \\ \dfrac{T}{2} & k = p \end{cases} \tag{20-8}$$

$$\int^{T} (\cos k\omega_0 t)(\cos p\omega_0 t) \, dt = \begin{cases} 0 & k \neq p \\ \dfrac{T}{2} & k = p \end{cases} \tag{20-9}$$

Here we use the convenient shorthand notation $\int^{T}$ to indicate the integral over one period, from some t_0 to $t_0 + T$. The choice of t_0 is ours to make, and we'll do so often. Equations 20-5 and 20-6 are quickly established (think geometrically: What is the net area, over one period, under a pure cosine or sine?). To prove Equations 20-7 to 20-9, we use such trigonometric identities as

$$\frac{\sin x \pm \sin y}{2} = \sin \frac{x \pm y}{2} \cos \frac{x \mp y}{2} \tag{20-10}$$

$$\frac{\cos x - \cos y}{2} = -\sin \frac{x + y}{2} \sin \frac{x - y}{2} \tag{20-11}$$

$$\frac{\cos x + \cos y}{2} = \cos \frac{x + y}{2} \cos \frac{x - y}{2} \tag{20-12}$$

to express the integrands as sums or differences of sines and cosines.

Functions that obey relations like Equations 20-7, 20-8, and 20-9 are named *orthogonal*; the sinusoidal functions are orthogonal over the period T, and it is precisely this orthogonality that allows us to calculate the Fourier coefficients.

Ready? Let's start with $a_0/2$, the constant term in the Fourier series. To calculate it, we integrate both sides of Equation 20-4 over one period:

$$\int^T f(t)\, dt = \int^T \frac{a_0}{2}\, dt + \int^T a_1 \cos \omega_0 t\, dt + \cdots + \int^T a_n \cos n\omega_0 t\, dt + \cdots$$

$$+ \int^T b_1 \sin \omega_0 t\, dt + \cdots + \int^T b_n \sin n\omega_0 t\, dt + \cdots \quad (20\text{-}13)$$

which reduces to

$$\int^T f(t)\, dt = \frac{a_0}{2} T \quad\quad (20\text{-}14)$$

since all the other integrals on the right-hand side vanish because of Equations 20-5 and 20-6. From Equation 20-14 we get

$$\frac{a_0}{2} = \frac{1}{T} \int^T f(t)\, dt \quad\quad (20\text{-}15)$$

This is the *average value* of $f(t)$, sometimes called the *dc term*. In many cases, it can be determined by inspection.

To calculate a_1, multiply both sides of Equation 20-4 by $\cos \omega_0 t$ (certainly a valid operation!), then integrate over one period. We get

$$\int^T f(t) \cos \omega_0 t\, dt = \int^T \frac{a_0}{2} \cos \omega_0 t\, dt + \int^T a_1 (\cos \omega_0 t)^2\, dt$$

$$+ \int^T a_2 (\cos 2\omega_0 t)(\cos \omega_0 t)\, dt + \cdots$$

$$+ \int^T b_1 (\sin \omega_0 t)(\cos \omega_0 t)\, dt + \cdots \quad\quad (20\text{-}16)$$

On the right-hand side, every integral, except the second one, vanishes because of the orthogonality property. Therefore

$$\int^T f(t) \cos \omega_0 t\, dt = a_1 \frac{T}{2} \quad\quad (20\text{-}17)$$

that is,

$$a_1 = \frac{2}{T} \int^T f(t) \cos \omega_0 t\, dt \quad\quad (20\text{-}18)$$

Now we know why we multiplied Equation 20-4 by $\cos \omega_0 t$ before integrating. It was precisely to save only a_1 on the right-hand side.

The other a's are calculated in the same way. In general, to find a_n, we multiply Equation 20-4 by $\cos n\omega_0 t$, then integrate over a period. The result is

$$a_n = \frac{2}{T} \int^T f(t) \cos n\omega_0 t\, dt \quad\quad n = 0, 1, 2, \ldots \quad\quad (20\text{-}19)$$

This equation is also good for $n = 0$, if we call the first (dc) term $a_0/2$. Compare it with Equation 20-15.

In a similar way, we get the general expression for the b's

$$b_n = \frac{2}{T} \int^T f(t) \sin n\omega_0 t \, dt \qquad n = 1, 2, \ldots \qquad (20\text{-}20)$$

In Equations 20-19 and 20-20 we have the required Fourier coefficients for Equation 20-4. For emphasis and review, we repeat them: The periodic function $f(t)$ is expressed in its Fourier series

$$f(t) = \frac{a_0}{2} + \sum_{n=1}^{\infty} a_n \cos n\omega_0 t + \sum_{n=1}^{\infty} b_n \sin n\omega_0 t \qquad (20\text{-}4)$$

where

$$a_n = \frac{2}{T} \int^T f(t) \cos n\omega_0 t \, dt \qquad (20\text{-}19)$$

$$b_n = \frac{2}{T} \int^T f(t) \sin n\omega_0 t \, dt \qquad (20\text{-}20)$$

EXAMPLE 20-2

Let us confirm the Fourier series for Figure 20-2(a), as given in Example 20-1.

Solution. Here $T = 2\pi$ and $\omega_0 = 2\pi/T = 1$. Also, over a period

$$f_1(t) = t \qquad -\pi < t < \pi$$

where our choice $t_0 = -\pi$ and $t_0 + T = \pi$ is rather obvious. Any other choice, though valid, would require *different* expressions for $f(t)$ over the various ranges of T.

Now calculate the coefficients. The dc (average) value of $f(t)$ is zero by inspection: The net area under $f(t)$ over $-\pi < t < \pi$ is zero. Therefore

$$\frac{a_0}{2} = 0$$

Next, use Equation 20-19

$$a_n = \frac{2}{2\pi} \int_{-\pi}^{\pi} t \cos nt \, dt = \frac{1}{\pi} \left[\frac{\cos nt}{n^2} + \frac{t \sin nt}{n} \right]_{-\pi}^{\pi}$$

The integration can be done by parts or by looking it up in integral tables. The result is

$$a_n = \frac{1}{\pi} \left[\frac{\cos n\pi}{n^2} + \frac{\pi \sin n\pi}{n} - \frac{\cos(-n\pi)}{n^2} + \frac{\pi \sin(-n\pi)}{n} \right] = 0$$

Now calculate b_n from Equation 20-20

$$b_n = \frac{2}{2\pi} \int_{-\pi}^{\pi} t \sin nt \, dt = \frac{1}{\pi} \left[\frac{\sin nt}{n^2} - \frac{t \cos nt}{n} \right]_{-\pi}^{\pi}$$

For $n = 1$ we get

$$b_1 = \frac{1}{\pi}\left[\frac{\pi}{1} + \frac{\pi}{1}\right] = 2$$

For $n = 2$

$$b_2 = \frac{1}{\pi}\left[\frac{-\pi}{2} + \frac{-\pi}{2}\right] = -1$$

For $n = 3$

$$b_3 = \frac{1}{\pi}\left[\frac{\pi}{3} + \frac{\pi}{3}\right] = \frac{2}{3}$$

etc., which agrees with the given series in Example 20-1.   ■

As mentioned, $f(t)$ may have different expressions over one period. Then we must write the integral in Equations 20-19 and 20-20 as a sum of integrals, each one with the appropriate expression over its range.

EXAMPLE 20-3 ⎯⎯⎯⎯⎯⎯⎯⎯⎯⎯⎯⎯⎯⎯⎯⎯⎯⎯⎯⎯⎯⎯⎯

For the square waveform of Figure 20-1(a), we calculate a_n as follows:

$$a_n = \frac{2}{T}\left[\int_{-T/2}^{-a/2} 0 \cdot \cos n\omega_0 t\, dt + \int_{-a/2}^{a/2} A \cdot \cos n\omega_0 t\, dt + \int_{a/2}^{T/2} 0 \cdot \cos n\omega_0 t\, dt\right] = \cdots$$

and similarly for b_n.   ■   Probs. 20-2,
20-3, 20-4

EXAMPLE 20-4 ⎯⎯⎯⎯⎯⎯⎯⎯⎯⎯⎯⎯⎯⎯⎯⎯⎯⎯⎯⎯⎯⎯⎯

A "staircase" periodic $v(t)$ is shown in Figure 20-3(a), and its first cycle is given by

$$v_1(t) = \begin{cases} 1 & 0 < t < 0.25 \\ 2 & 0.25 < t < 0.5 \\ 0 & 0.5 < t < 1 \end{cases}$$

with $T = 1$. Therefore, $\omega_0 = 2\pi/T = 2\pi$.

Solution. The Fourier coefficients are calculated as

$$\frac{a_0}{2} = \frac{1}{1}\left[\int_0^{0.25} 1\, dt + \int_{0.25}^{0.5} 2\, dt\right] = \frac{3}{4}$$

$$a_1 = \frac{2}{1}\left[\int_0^{0.25} 1 \cos 2\pi t\, dt + \int_{0.25}^{0.5} 2 \cos 2\pi t\, dt\right] = -\frac{1}{\pi}$$

$$b_1 = \frac{2}{1}\left[\int_0^{0.25} 1 \sin 2\pi t\, dt + \int_{0.25}^{0.5} 2 \sin 2\pi t\, dt\right] = \frac{3}{\pi}$$

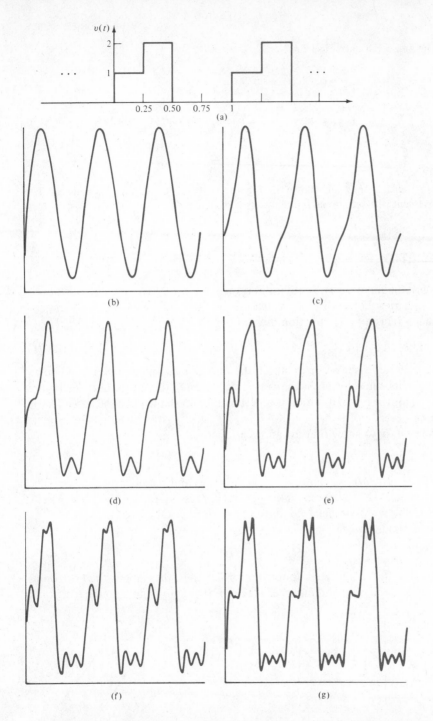

Figure 20-3 Example 20-3. (a) "Staircase" $v(t)$. (b) $v_1(t) = \frac{3}{4} - (1/\pi)\cos 2\pi t$. (c) $v_2(t) = v_1(t) - (1/\pi)\sin 4\pi t$. (d) $v_3(t) = v_2(t) + (1/3\pi)\cos 6\pi t + (1/\pi)\sin 6\pi t$. (e) $v_5(t) = v_3(t) - (1/5\pi)\cos 10\pi t + (3/5\pi)\sin 10\pi t$. (f) $v_6(t) = v_5(t) - (1/3\pi)\sin 12\pi t$. (g) $v_7(t) = v_6(t) + (1/7\pi)\cos 14\pi t + (3/7\pi)\sin 14\pi t$.

etc. The first seven harmonics are

$$v(t) = \frac{3}{4} - \frac{1}{\pi}\cos 2\pi t + \frac{3}{4}\sin 2\pi t - \frac{1}{\pi}\sin 4\pi t + \frac{1}{3\pi}\cos 6\pi t$$

$$+ \frac{1}{\pi}\sin 6\pi t - \frac{1}{5\pi}\cos 10\pi t + \frac{3}{5\pi}\sin 10\pi t - \frac{1}{3\pi}\sin 12\pi t$$

$$+ \frac{1}{7\pi}\cos 14\pi t + \frac{3}{7\pi}\sin 14\pi t + \cdots$$

Plots of partial sums are shown in Figure 20-3(b), (c), (d), (e), (f), and (g). The improvement with additional harmonics is very clear.   ■

20-3 SYMMETRY OF f(t)

When the periodic function $f(t)$ exhibits certain symmetry, there are simplifications for its Fourier series. Let us discuss two cases:

1. When $f(t)$ is an *even* function defined by

$$f(-t) = f(t) \tag{20-21}$$

that is, $f(t)$ is reflected, as in a mirror, about the vertical axis. In other words, if you go to the right t_1 units, the value of f is the same as if you go to the left t_1 units. See Figure 20-4(a), where, for the fun of it, we show an even function shaped like the letter M, the letter M being "even."

It is fairly easy to show that for an even function

$$b_n = 0 \qquad n = 1, 2, 3, \ldots \tag{20-22}$$

Therefore, the *Fourier series contains only cosine terms* and possibly the dc term. The proof of Equation 20-22 follows from the general expression for b_n (Equation 20-20), in which we invoke the symmetry condition (Equation 20-21).

2. When $f(t)$ is an *odd* function satisfying

$$f(-t) = -f(t) \tag{20-23}$$

This is a function that is reflected about the vertical axis, then about the horizontal axis. Go t_1 units to the right and find the value of f. If you go t_1 units to the left, you'll

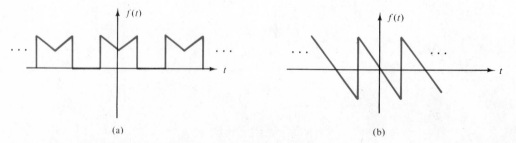

(a)   (b)

Figure 20-4   Even and odd functions.

find the negative of that value. See, for example, Figure 20-4(b), where an N-shaped function is odd.

For an odd function

$$a_n = 0 \qquad n = 0, 1, 2, 3, \ldots \tag{20-24}$$

and its *Fourier series consists only of sine terms.*

These results are also acceptable intuitively: A pure cosine waveform is even, so the sum of only cosines yields an even $f(t)$. A pure sine is odd, so the sum of only sines yields an odd $f(t)$. This, however, is not a proof, just an easy mnemonic.

EXAMPLE 20-5 _____

The square waveform in Figure 20-1(a) is even, so it contains only cosine terms, and a dc term ($= aA/T$ by inspection). Check your answer to Problem 20-2.

The sawtooth waveform in Figure 20-1(b) is odd, and its Fourier series consists of sine terms only.

The full-wave rectified waveform of Figure 20-1(d) is even and contains cosine terms and a dc term.   ■   Prob. 20-5

There are other symmetry conditions (such as half-wave and quarter-wave symmetries), but their "simplifications" in deriving the Fourier coefficients are not worth the trouble (and the memorization!) It is better to work with the general expressions, Equations 20-19 and 20-20.

20-4 THE COSINE FORM OF THE FOURIER SERIES

In many cases, it is more convenient to combine the two terms of the nth harmonic $(a_n \cos \omega_0 t + b_n \sin n\omega_0 t)$ into a single equivalent cosine term. We did it in Chapter 9, but let us repeat it here. Divide and multiply by $(a_n^2 + b_n^2)^{1/2}$ to get

$$a_n \cos n\omega_0 t + b_n \sin n\omega_0 t = \sqrt{a_n^2 + b_n^2} \left[\frac{a_n}{\sqrt{a_n^2 + b_n^2}} \cos n\omega_0 t + \frac{b_n}{\sqrt{a_n^2 + b_n^2}} \sin n\omega_0 t \right]$$

$$\tag{20-25}$$

and with the help of the triangle in Figure 20-5, we recognize that

$$a_n \cos n\omega_0 t + b_n \sin n\omega_0 t = A_n \cos (n\omega_0 t - \theta_n) \tag{20-26}$$

with

$$A_n = +\sqrt{a_n^2 + b_n^2} \tag{20-27}$$

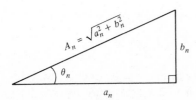

Figure 20-5   Amplitude and phase of the nth harmonic.

the *amplitude* of the *n*th harmonic, and

$$\theta_n = \tan^{-1} \frac{b_n}{a_n} \tag{20-28}$$

the *phase* of the *n*th harmonic. Thus, the cosine Fourier series of a periodic function is

$$f(t) = A_0 + A_1 \cos(\omega_0 t - \theta_1) + A_2 \cos(2\omega_0 t - \theta_2) + \cdots \tag{20-29}$$

where $A_0 = a_0/2$, the dc term.

EXAMPLE 20-6

A periodic voltage, with $\omega_0 = 2$, is given by its Fourier series

$$v(t) = 4.2 - 3.1 \cos 2t + 1.3 \cos 4t + 0.4 \cos 6t + \cdots$$
$$+ 2.8 \sin 2t + 1.1 \sin 4t - 0.3 \sin 6t + \cdots$$

Solution.   The dc term is

$$A_0 = 4.2$$

For the first harmonic we have

$$A_1 = \sqrt{(-3.1)^2 + (2.8)^2} = 4.18$$

and

$$\theta_1 = \tan^{-1} \frac{2.8}{-3.1} = 137.9°$$

For the second harmonic

$$A_2 = \sqrt{(1.3)^2 + (1.1)^2} = 1.7$$

$$\theta_2 = \tan^{-1} \frac{1.1}{1.3} = 40.2°$$

For the third harmonic

$$A_3 = \sqrt{(0.4)^2 + (-0.3)^2} = 0.5$$

$$\theta_3 = \tan^{-1} \frac{-0.3}{0.4} = -36.9°$$

Be sure to get the correct quadrant for θ_n, as given by the signs of b_n and a_n. The cosine Fourier series for $v(t)$ is therefore

$$v(t) = 4.2 + 4.18 \cos(2t - 137.9°) + 1.7 \cos(4t - 40.2°)$$
$$+ 0.5 \cos(6t + 36.9°) + \cdots \qquad \blacksquare \quad \text{Prob. 20-6}$$

20-5 FOURIER SERIES, SUPERPOSITION, AND PHASORS

By now you must have seen the reason for the cosine form of the Fourier series: If an input to a *linear* network is periodic, as in Equation 20-29, we can use *superposition and phasors* to calculate the steady-state response. This is a major highlight of the Fourier series, as mentioned in the introduction to this chapter.

EXAMPLE 20-7 ————————————————————————————————

Let a periodic voltage source, given by $v(t)$ in the previous example, be applied to a series RL circuit, as shown in Figure 20-6(a). Calculate the steady-state current $i(t)$.

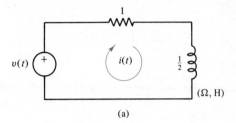

(a)

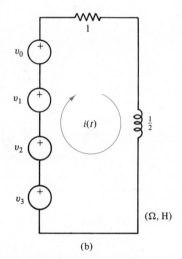

(b)

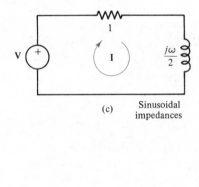

(c)   Sinusoidal impedances

Figure 20-6   Example 20-7.

Solution.   Since

$$v(t) = 4.2 + 4.18 \cos (2t - 137.9°) + 1.7 \cos (4t - 40.2°) + 0.5 \cos (6t + 36.9°)$$

it can be considered as a series connection of four sources

$$v(t) = v_0(t) + v_1(t) + v_2(t) + v_3(t)$$

as in Figure 20-6(b). Using superposition, let us take one source at a time. Since each source is purely sinusoidal, we can use phasors. In particular here

$$\mathbf{I} = \mathbf{YV}$$

is the appropriate input-output phasor relationship, with $\mathbf{Y}$ being the driving-point admittance. See Figure 20-6(c), where, for any ω,

$$\mathbf{Y} = \frac{1}{1 + j(\omega/2)}$$

The dc source ($\omega = 0$) $v_0(t) = 4.2$ yields the phasor response

$$\mathbf{I}_0 = \frac{1}{1 + j0}\, 4.2 = 4.2$$

$$\therefore \quad i_0(t) = 4.2$$

which could be written by inspection, since the inductor acts as a short circuit for dc. The first harmonic ($\omega = \omega_0 = 2$) yields

$$\mathbf{I}_1 = \frac{1}{1 + j}\, 4.18\underline{/-137.9°} = 2.96\underline{/-182.9°}$$

$$\therefore \quad i_1(t) = 2.96 \cos(2t - 182.9°)$$

The contribution of the second harmonic ($\omega = 4$) is

$$\mathbf{I}_2 = \frac{1}{1 + j2}\, 1.7\underline{/-40.2°} = 0.76\underline{/-103.6°}$$

$$\therefore \quad i_2(t) = 0.76 \cos(4t - 103.6°)$$

The third harmonic ($\omega = 6$) response is

$$\mathbf{I}_3 = \frac{1}{1 + j3}\, 0.5\underline{/36.9°} = 0.16\underline{/-34.7°}$$

$$\therefore \quad i_3(t) = 0.16 \cos(6t - 34.7°)$$

Therefore, the total steady-state reponse is

$$i(t) = 4.2 + 2.96 \cos(2t - 182.9°) + 0.76 \cos(4t - 103.6°) + 0.16 \cos(6t - 34.7°) + \cdots$$

and it is periodic, with the same ω_0 and T as the input, and given by its Fourier cosine series. ■

Probs. 20-7, 20-8, 20-9

20-6 THE EXPONENTIAL FORM OF THE FOURIER SERIES

Another useful form of the Fourier series can be obtained by using Euler's identities

$$\cos x = \frac{e^{jx} + e^{-jx}}{2} \tag{20-30}$$

and

$$\sin x = \frac{e^{jx} - e^{-jx}}{2j} \tag{20-31}$$

in Equation 20-4. We get

$$f(t) = \frac{a_0}{2} + \sum_{n=1}^{\infty} a_n \frac{e^{jn\omega_0 t} + e^{-jn\omega_0 t}}{2} + \sum_{n=1}^{\infty} b_n \frac{e^{jn\omega_0 t} - e^{-jn\omega_0 t}}{2j} \tag{20-32}$$

Collect together the positive exponentials and the negative ones:

$$f(t) = \frac{a_0}{2} + \sum_{n=1}^{\infty} \left(\frac{a_n - jb_n}{2} \right) e^{jn\omega_0 t} + \sum_{n=1}^{\infty} \left(\frac{a_n + jb_n}{2} \right) e^{-jn\omega_0 t} \tag{20-33}$$

Now let us define a new coefficient, complex in general

$$c_k = \frac{a_k - jb_k}{2} \qquad c_0 = \frac{a_0}{2} \tag{20-34}$$

Then Equation 20-33 is written as

$$f(t) = c_0 + \sum_{n=1}^{\infty} c_n e^{jn\omega_0 t} + \sum_{n=1}^{\infty} c_n^* e^{-jn\omega_0 t} \tag{20-35}$$

where, as usual

$$c_n^* = \left(\frac{a_n - jb_n}{2}\right)^* = \frac{a_n + jb_n}{2} \tag{20-36}$$

is the complex conjugate of c_n. One final simplification in notation will be introduced; let

$$c_n^* = c_{-n} \tag{20-37}$$

Then Equation 20-35 has the compact form

$$f(t) = \sum_{n=-\infty}^{\infty} c_n e^{jn\omega_0 t} \tag{20-38}$$

Be sure to recognize that here the summation index n goes from $-\infty$ to $+\infty$ through all the integers. For negative integers, say, $n = -6$, we get the term

$$c_{-6} e^{-j6\omega_0 t} = c_6^* e^{-j6\omega_0 t} \tag{20-39}$$

because of Equation 20-37. Thus, for negative integers, we get the second summation in Equation 20-35. Next, for $n = 0$, we get c_0, the dc term. Finally, for positive integers, we get the first summation in Equation 20-35. The *exponential form* of the Fourier series, Equation 20-38, is very compact and pleasing to the eye. Moreover, it has certain advantages that we'll explore.

First, let us obtain the explicit form for c_n, the complex coefficients. From Equations 20-34, 20-19, and 20-20, we write

$$c_n = \frac{1}{2}(a_n - jb_n) = \frac{1}{2}\left[\frac{2}{T}\int^T f(t)\cos n\omega_0 t \, dt - j\frac{2}{T}\int^T f(t)\sin n\omega_0 t \, dt\right]$$

$$= \frac{1}{T}\int^T f(t)(\cos n\omega_0 t - j\sin n\omega_0 t) \, dt$$

$$= \frac{1}{T}\int^T f(t)e^{-jn\omega_0 t} \, dt \tag{20-40}$$

where Euler's identity, $e^{-jx} = \cos x - j\sin x$, was used in the last step. Equation 20-40 is valid for all integers n, negative, zero, and positive, due to Equation 20-37.

To summarize and bring them together, we repeat our results: A periodic $f(t)$ is written as

$$f(t) = \sum_{n=-\infty}^{\infty} c_n e^{jn\omega_0 t} \tag{20-38}$$

where the coefficients c_n are given by

$$c_n = \frac{1}{T} \int^T f(t) e^{-jn\omega_0 t} \, dt \qquad (20\text{-}40)$$

Note the symmetry between these equations: $f(t)$ in one is replaced by c_n in the other, the summation and the integral, the exponent in one and its negative in the other. These observations are not just skin deep. Their full significance and extensions are the subject of the following discussions. However, an example is in order now.

EXAMPLE 20-8 _____

For the square waveform of Figure 20-1(a), let $a = T/2$.

Solution. We evaluate c_n according to Equation 20-40

$$c_n = \frac{1}{T} \int_{-T/4}^{T/4} A e^{-jn\omega_0 t} \, dt$$

since over the remaining range of T the integrand is zero. Carry out the integration

$$c_n = \frac{A}{T} \frac{e^{-jn\omega_0 t}}{-jn\omega_0}\Bigg]_{-T/4}^{T/4} = \frac{A}{T} \frac{e^{jn\omega_0 T/4} - e^{-jn\omega_0 T/4}}{jn\omega_0}$$

and since $\omega_0 T = 2\pi$ we get

$$c_n = \frac{A}{T} \frac{e^{jn\pi/2} - e^{-jn\pi/2}}{jn\omega_0} = \frac{A}{2} \frac{1}{n\omega_0 T/4} \frac{e^{jn\pi/2} - e^{-jn\pi/2}}{2j} = \frac{A}{2} \frac{\sin(n\pi/2)}{n\pi/2}$$

So here c_n is real, not complex. But this could have been predicted from the beginning: $f(t)$ is an even function, with $b_n = 0$, and so $c_n = (a_n - jb_n)/2 = a_n/2$ here.

Let us calculate the first few coefficients. For $n = 0$ we have

$$c_0 = \frac{A}{2} \frac{\sin(n\pi/2)}{n\pi/2}\Bigg]_{n=0} = \frac{A}{2} \frac{(\pi/2)\cos(n\pi/2)}{\pi/2}\Bigg]_{n=0} = \frac{A}{2}$$

where we had to use l'Hôpital's rule on the indeterminate form $0/0$. Next, for $n = 1$

$$c_1 = \frac{A}{2} \frac{\sin \pi/2}{\pi/2} = \frac{A}{\pi}$$

and immediately for $n = -1$

$$c_{-1} = c_1^* = \frac{A}{\pi}$$

For $n = 2$,

$$c_2 = \frac{A}{2} \frac{\sin \pi}{\pi} = 0$$

and therefore

$$c_{-2} = c_2^* = 0$$

For $n = 3$,

$$c_3 = \frac{A}{2} \frac{\sin(3\pi/2)}{3\pi/2} = -\frac{A}{3\pi}$$

and

$$c_{-3} = c_3^* = -\frac{A}{3\pi}$$

The Fourier series is then, according to Equation 20-38,

$$f(t) = \cdots \frac{-A}{3\pi} e^{-j3\omega_0 t} + 0e^{-j2\omega_0 t} + \frac{A}{\pi} e^{-j\omega_0 t} + \frac{A}{2} + \frac{A}{\pi} e^{j\omega_0 t} + 0e^{j2\omega_0 t} + \frac{-A}{3\pi} e^{j3\omega_0 t} + \cdots$$

To bring it into a trigonometric form, combine exponentials

$$f(t) = \frac{A \cdot}{2} + \frac{A}{\pi}(e^{j\omega_0 t} + e^{-j\omega_0 t}) + \frac{-A}{3\pi}(e^{j3\omega_0 t} + e^{-j3\omega_0 t}) + \cdots$$

$$= \frac{A}{2} + \frac{2A}{\pi} \cos \omega_0 t - \frac{2A}{3\pi} \cos 3\omega_0 t + \cdots$$

where, again, Euler comes to the rescue in the last step.

Probs. 20-10, 20-11

20-7 POWER CALCULATIONS AND EFFECTIVE VALUES

Consider a one-port linear network, where the voltage and current at the port are periodic, expressed in their cosine Fourier series

$$v(t) = V_0 + \sqrt{2}\, V_{1\text{eff}} \cos(\omega_0 t + \alpha_1) + \sqrt{2}\, V_{2\text{eff}} \cos(2\omega_0 t + \alpha_2) + \cdots \quad (20\text{-}41)$$

and

$$i(t) = I_0 + \sqrt{2}\, I_{1\text{eff}} \cos(\omega_0 t + \beta_1) + \sqrt{2}\, I_{2\text{eff}} \cos(2\omega_0 t + \beta_2) + \cdots \quad (20\text{-}42)$$

Here we remember that both functions have the same ω_0. Also, we expressed the amplitude of each individual harmonic (a pure sinusoid) as $\sqrt{2}$ times its effective value, as studied in Chapter 11.

With the usual associated references for $v(t)$ and $i(t)$, as in Figure 20-7, the instantaneous power in the one-port is $p(t) = v(t)i(t)$. Of practical interest is the *average* power, given as always by

$$P_{\text{av}} = \frac{1}{T}\int^T p(t)\, dt = \frac{1}{T}\int^T v(t)i(t)\, dt \quad (20\text{-}43)$$

If we substitute the expressions for $v(t)$ and $i(t)$, and recall the orthogonality equations of the cosine function (Equation 20-9), we get

$$P_{\text{av}} = V_0 I_0 + V_{1\text{eff}} I_{1\text{eff}} \cos(\alpha_1 - \beta_1) + V_{2\text{eff}} I_{2\text{eff}} \cos(\alpha_2 - \beta_2) + \cdots \quad (20\text{-}44)$$

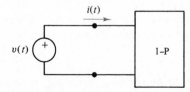

Figure 20-7   Power calculations with periodic waveforms.

that is,

$$P_{av} = V_0 I_0 + \sum_{n=1}^{\infty} V_{n\,eff} I_{n\,eff} \cos \theta_n = P_0 + P_1 + P_2 + P_3 + \cdots \quad (20\text{-}45)$$

In words, *the average power is the sum of the average powers of all the harmonics,* including dc. The average power of the kth harmonic is, as in Chapter 11,

$$P_k = V_{k\,eff} I_{k\,eff} \cos \theta_k \quad (20\text{-}46)$$

with $\cos \theta_k = \cos (\alpha_k - \beta_k)$, the power factor of that harmonic.

Different harmonics of voltage and current, such as the third harmonic of $v(t)$ and the second harmonic of $i(t)$, do not contribute to the average power.

EXAMPLE 20-9 _____

In Example 20-7, we had for the RL circuit

$v(t) = 4.2 + 4.18 \cos (2t - 137.9°) + 1.7 \cos (4t - 40.2°) + 0.5 \cos (6t + 36.9°) + \cdots$

$i(t) = 4.2 + 2.96 \cos (2t - 182.9°) + 0.76 \cos (4t - 103.6°) + 0.16 \cos (6t - 34.7°) + \cdots$

Solution. The average power dissipated is

$$P_{av} = (4.2)(4.2) + \left(\frac{4.18}{\sqrt{2}}\right)\left(\frac{2.96}{\sqrt{2}}\right) \cos 45° + \left(\frac{1.7}{\sqrt{2}}\right)\left(\frac{0.76}{\sqrt{2}}\right) \cos 63.4°$$

$$+ \left(\frac{0.5}{\sqrt{2}}\right)\left(\frac{0.16}{\sqrt{2}}\right) \cos 71.6° = 17.64 + 4.374 + 0.289 + 0.013$$

$$= 22.3 \text{ W}$$

where we've used terms through the third harmonic. ∎

Probs. 20-12, 20-13

While we are at it, let us calculate the effective (RMS) value of a periodic waveform in terms of its Fourier coefficients. To be specific, let us look at $i(t)$, as given in Equation 20-42. By definition,

$$I_{eff} = \sqrt{\frac{1}{T} \int^{T} [i(t)]^2 \, dt} \quad (20\text{-}47)$$

the effective (RMS) value being the square *root* of the (average) *mean* of the *square* of $i(t)$, as obtained in Section 11-2. With Equation 20-42 we write

$$I_{eff}^2 = \frac{1}{T} \int^{T} [I_0 + \sqrt{2}\, I_{1\,eff} \cos (\omega_0 t + \beta_1) + \sqrt{2}\, I_{2\,eff} \cos (2\omega_0 t + \beta_2) + \cdots]^2 \, dt$$

$$(20\text{-}48)$$

Here, again, because of the orthogonality of the cosine functions, all the "cross-product" terms involving different harmonics will integrate to zero. The only surviving terms will be products of the same frequency. Therefore (fill in the details!)

$$I_{eff}^2 = I_0^2 + I_{1\,eff}^2 + I_{2\,eff}^2 + \cdots \quad (20\text{-}49)$$

the sum of the squares of the effective values of all the harmonics, including the dc term.

EXAMPLE 20-10 _____

For $v(t)$ and $i(t)$ of Example 20-7, we have

$$V_{eff} = \left[(4.2)^2 + \left(\frac{4.18}{\sqrt{2}}\right)^2 + \left(\frac{1.7}{\sqrt{2}}\right)^2 + \left(\frac{0.5}{\sqrt{2}}\right)^2 \right]^{1/2} = 5.29 \text{ V}$$

and

$$I_{eff} = \left[(4.2)^2 + \left(\frac{2.96}{\sqrt{2}}\right)^2 + \left(\frac{0.76}{\sqrt{2}}\right)^2 + \left(\frac{0.16}{\sqrt{2}}\right)^2 \right]^{1/2} = 4.73 \text{ A}$$

As a quick check, the average power in the resistor there is

$$P_{av} = I_{eff}^2 R = (4.73)^2 1 = 22.3 \text{ W}$$

as before. *A quickie*: Why can't we calculate here P_{av} as V_{eff}^2/R? ■

Probs. 20-14, 20-15

As a concluding note, we give a geometrical interpretation to Equation 20-49. With only the dc term and the first harmonic present, the effective value is given by

$$I_{eff}^2 = I_0^2 + I_{1eff}^2 \tag{20-50}$$

This relation is shown in Figure 20-8(a), where the two sides of the right-angle triangle are I_0 and I_{1eff}, and I_{eff} is the hypotenuse. With the second harmonic also present, we have

$$I_{eff}^2 = I_0^2 + I_{1eff}^2 + I_{2eff}^2 \tag{20-51}$$

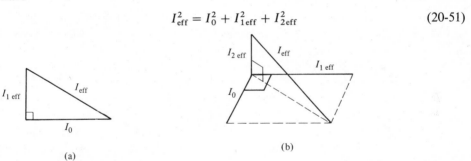

(a)

(b)

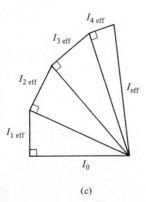

(c)

Figure 20-8 Geometry of effective values.

and I_{eff} is the hypotenuse of a three-dimensional right-angle triangle. See Figure 20-8(b). Continuing in this manner, we can interpret I_{eff} in Equation 20-49 as the hypotenuse of a multidimensional right-angle triangle, each side of which is the effective value of each harmonic. See Figure 20-8(c). When drawn to scale, such a sketch provides a handy way for finding the effective value.

Or, you can think of I_{eff} as the length (norm) of a vector whose components along each axis are the effective values of each harmonic, and the axes are orthogonal. In this sense, it is quite similar to the length of the state vector discussed in the previous chapter.

EXAMPLE 20-11

Let us consider the sawtooth waveform shown in Figure 20-9. It is the same as in Example 11-6, where we found directly and exactly

$$V_{\text{eff}} = \sqrt{\frac{1}{T}\int_0^T \left(\frac{A}{T}t\right)^2 dt} = \frac{A}{\sqrt{3}} \approx 0.5773A$$

Let us do the same, using harmonic analysis. The complex Fourier coefficients are given by

$$c_n = \frac{1}{T}\int_0^T \frac{A}{T} t e^{-jn\omega_0 t} \, dt = \frac{Ae^{-j2\pi n}}{-jn\omega_0 T} = j\frac{A}{2\pi n} \qquad n \neq 0$$

The dc term is $c_0 = (1/T)(AT/2) = A/2$. Therefore

$$v(t) = \cdots - j\frac{A}{6\pi}e^{-j3\omega_0 t} - j\frac{A}{4\pi}e^{-j2\omega_0 t} - j\frac{A}{2\pi}e^{-j\omega_0 t} + \frac{A}{2}$$

$$+ j\frac{A}{2\pi}e^{j\omega_0 t} + j\frac{A}{4\pi}e^{j2\omega_0 t} + j\frac{A}{6\pi}e^{j3\omega_0 t} + \cdots$$

Collect exponentials and use Euler's identity to get

$$v(t) = \frac{A}{2} - \frac{A}{\pi}\sin \omega_0 t - \frac{A}{2\pi}\sin 2\omega_0 t - \frac{A}{3\pi}\sin 3\omega_0 t + \cdots$$

Therefore

$$V_{\text{eff}} = \sqrt{\left(\frac{A}{2}\right)^2 + \left(\frac{A}{\sqrt{2}\,\pi}\right)^2 + \left(\frac{A}{\sqrt{2}\,2\pi}\right)^2 + \left(\frac{A}{\sqrt{2}\,3\pi}\right)^2 + \cdots} = 0.5725A$$

where we included terms through the tenth harmonic. Compared with the *exact* value $A/\sqrt{3}$, the error is 0.83 percent.

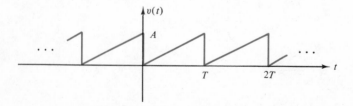

Figure 20-9   Example 20-11.

*20-8 CONVERGENCE AND ERROR

We cited earlier the sufficient Dirichlet conditions for the convergence of the Fourier series. In the previous example, we saw the error introduced when the Fourier series is truncated after a finite number of terms—a step that is inevitable in all practical applications.

Consider, then, a *finite* Fourier series

$$S_n(t) = \frac{a_0}{2} + \sum_{k=1}^{n} (a_k \cos k\omega_0 t + b_k \sin k\omega_0 t) \qquad (20\text{-}52)$$

with $(2n + 1)$ terms, which approximates $f(t)$. The error in this approximation is

$$\varepsilon_n(t) = f(t) - S_n(t) \qquad (20\text{-}53)$$

where the subscript n reminds us that the error is dependent on the number of terms taken in S_n. In many theoretical and applied cases, we do not deal with $\varepsilon_n(t)$, but with the *mean square error*, given by

$$E_n = \frac{1}{T} \int^{T} [\varepsilon_n(t)]^2 \, dt \qquad (20\text{-}54)$$

The usefulness of E_n, rather than $\varepsilon_n(t)$, is twofold: (1) we are not interested in instantaneous positive or negative errors, so we square $\varepsilon_n(t)$ and get a positive quantity $[\varepsilon_n(t)]^2$; (2) we average $[\varepsilon_n(t)]^2$ to get E_n, a constant that depends only on the number of terms taken in S_n.

Specifically here we have

$$E_n = \frac{1}{T} \int^{T} [f(t) - S_n(t)]^2 \, dt \qquad (20\text{-}55)$$

If we minimize E_n, the following conclusion is reached: Of all possible finite trigonometric series that approximate $f(t)$

$$f(t) = \alpha_0 + \sum_{k=1}^{n} (\alpha_k \cos k\omega_0 t + \beta_k \sin k\omega_0 t) \qquad (20\text{-}56)$$

the *minimum* E_n will be obtained if the series is a Fourier series, that is, if

$$\alpha_0 = \frac{a_0}{2}$$

$$\alpha_1 = a_1 \qquad \beta_1 = b_1 \qquad (20\text{-}57)$$

$$\vdots \qquad\qquad \vdots$$

$$\alpha_n = a_n \qquad \beta_n = b_n$$

Stated in other words, *the finite Fourier series is the best trigonometric series approximation with the least mean square error.*†

---

† There are, of course, other error criteria. For example, the minimization of the maximum possible value of $|\varepsilon_n(t)|$ is one such criterion. It does *not* lead to Equation 20-57.

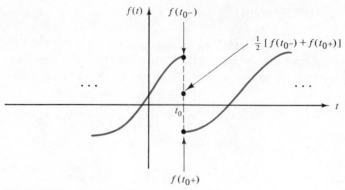

Figure 20-10   A point of discontinuity.

At a point of discontinuity (allowed by Dirichlet's condition), the Fourier series converges to the average value

$$f(t_0) = \tfrac{1}{2}[f(t_0-) + f(t_0+)] \qquad (20\text{-}58)$$

as shown in Figure 20-10. Also, the following properties of the Fourier coefficients are useful:

1. When $f(t)$ has discontinuities, the coefficients a_n and b_n decrease in magnitude as $1/n$. For example, the coefficients b_n in the sawtooth wave (Example 20-11) are proportional to $1/n$. There are no a_n's there. Another example is the square waveform of Figure 20-1(a).

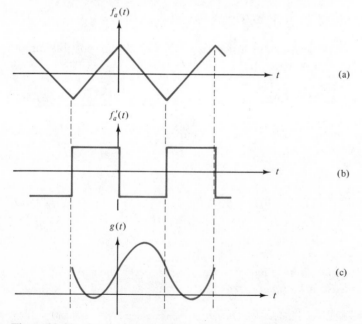

(a)

(b)

(c)

Figure 20-11   Continuous and discontinuous waveforms.

2. When $f(t)$ is continuous but $df/dt = f'(t)$ has discontinuities, a_n and b_n decrease in magnitude as $1/n^2$. An example is shown in Figure 20-11(a) and (b). The waveform $f_a(t)$ is continuous, but $f'_a(t)$ has discontinuities.

3. When $f(t)$ and $f'(t)$ are continuous, but $f''(t) = d^2f/dt^2$ is discontinuous, a_n and b_n decrease in magnitude as $1/n^3$. In Figure 20-11(c), the function $g(t)$ is made up of parabolic sections (t^2). It is continuous, as is its first derivative given by $f_a(t)$. Its second derivative $g''(t) = f'_a(t)$ is discontinuous.

4. In general, when $f(t)$, $f'(t)$, $f''(t)$, ..., $f^{(p-1)}(t)$ are continuous, but $f^{(p)}(t) = d^p f(t)/dt^p$ is discontinuous, a_n and b_n will decrease in magnitude as $1/n^{p+1}$.

Probs. 20-16, 20-17

20-9 DISCRETE SPECTRA

We have developed so far several equivalent ways of representing a periodic $f(t)$ by its Fourier series. From a given $f(t)$, we can obtain its ω_0, a_n, and b_n (Equations 20-4, 20-19, and 20-20); or else its A_n and θ_n (Equations 20-27 and 20-28); or its c_n (Equations 20-38 and 20-40). Conversely, from a_n and b_n, or from A_n and θ_n, or from c_n, we can obtain $f(t)$. What have we got here? Why, a transform! A function $f(t)$ in the time domain is transformed into the frequency (ω) domain. From the frequency domain, the inverse transform (here: the explicit writing of the series) yields back $f(t)$. This idea is illustrated in Table 20-1.

The major advantage of any transform is also clear here, namely, the ease of calculations in the transform (frequency) domain. The amplitudes and the phase angles of the various harmonics are handled easily, for example, by superposition. The effects of higher-order harmonics can be easily calculated or estimated, as was done in the previous section. Such insight into the frequency domain is the main reason for the use of the Fourier series.

To illustrate these advantages further, electrical engineers resort to graphical displays. Consider again the Fourier cosine series (Equation 20-29), repeated for convenience

$$f(t) = A_0 + \sum_{n=1}^{\infty} A_n \cos(n\omega_0 t - \theta_n) \tag{20-59}$$

The amplitudes A_0, A_1, ..., and the phase angles θ_0, θ_1, ... of the harmonics can be plotted versus the harmonics $n\omega_0$. These plots are called, respectively, the *discrete amplitude spectrum* and the *discrete phase spectrum* of $f(t)$. They provide graphically all the information on the frequency contents of $f(t)$ in Equation 20-59.

TABLE 20-1 FOURIER SERIES TRANSFORM

Time domain	Frequency domain (ω_0, $2\omega_0$, $3\omega_0$, ...)
$f(t)$	ω_0, a_n, b_n
Period $= T$	or: ω_0, A_n, θ_n
	or: ω_0, c_n

EXAMPLE 20-12 _____

In Example 20-7, we had a (hypothetical) periodic voltage

$$v(t) = 4.2 + 4.18 \cos (2t - 137.9°) + 1.7 \cos (4t - 40.2°) + 0.5 \cos (6t + 36.9°) + \cdots$$

Solution. Its two discrete spectra are plotted in Figure 20-12(a) and (b). These plots are discrete because A_n and θ_n exist only for integer values of $\omega = 0, \omega_0, 2\omega_0, \ldots$. There is nothing, for instance, at $\omega = 1.3\omega_0$ or at $\omega = 2.01\omega_0$. Only integer harmonics, by definition and by derivation, enter into the Fourier series.

Note: In the amplitude spectrum, we always plot positive amplitudes. If a harmonic has a negative sign in front of it, we add 180° to the angle,

$$-A_k \cos (k\omega_0 t - \theta_k) = + A_k \cos (k\omega_0 t + 180° - \theta_k)$$

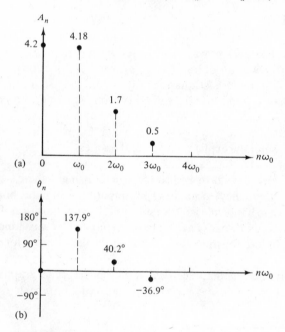

Figure 20-12   Example 20-12.

In a related way, we treat the exponential form of the Fourier series. Here

$$c_n = \frac{a_n - jb_n}{2} = |c_n| e^{j\theta_n} \tag{20-60}$$

and therefore the amplitude spectrum is

$$|c_n| = \tfrac{1}{2}\sqrt{a_n^2 + b_n^2} = \tfrac{1}{2}A_n \tag{20-61}$$

a scaled (halved) version of A_n. Furthermore, since

$$|c_{-n}| = |c_n^*| = |c_n| \tag{20-62}$$

the discrete plot versus $\pm n\omega_0$ extends from $-\infty$ to $+\infty$ and is _even._

EXAMPLE 20-13

The square waveform of Figure 20-1(a) has the complex Fourier coefficients

$$c_n = \frac{1}{T} \int^T f(t)e^{-jn\omega_0 t} \, dt = \frac{A}{T} \int_{-a/2}^{a/2} e^{-jn\omega_0 t} \, dt = \frac{Aa}{T} \frac{\sin{(n\pi a/T)}}{n\pi a/T}$$

which is derived in the same way as in Example 20-8. There we chose $a = T/2$, but here we left it arbitrary, $a < T$. Since $f(t)$ is an even function, $b_n = 0$ and c_n is real, positive or negative. If $c_p > 0$, then $\theta_p = 0°$, and if $c_q < 0$ then $\theta_p = 180°$, as agreed earlier.

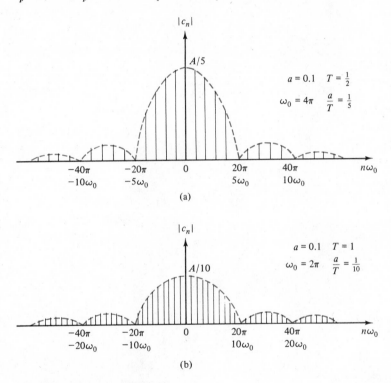

Figure 20-13 Example 20-13.

Solution. In Figure 20-13, we show two plots of discrete spectra $|c_n|$ versus $n\omega_0$. In (a), we chose $T = \frac{1}{2}$, $a = 0.1$, so that $\omega_0 = 4\pi$ and $a/T = \frac{1}{5}$. In (b) we took $T = 1$, $a = 0.1$, $\omega_0 = 2\pi$, $a/T = \frac{1}{10}$.

We make the following observations. As the period T increases, the fundamental frequency ω_0 decreases, since $\omega_0 = 2\pi/T$. Consequently, with a larger T, there are more harmonics in the Fourier series. This is also reasonable: To approximate $f(t) = 0$ from $-T/2$ to $-a/2$ and $a/2$ to $T/2$, for a large T, more sines and cosines of many harmonics are needed so that their oscillations (+ and − values) can cancel to give the required zero value.

Finally, the dashed line joining the tips of the amplitude spectra is shown only because it is the plot of the function $|\sin x/x|$, a very common curve which we'll study later. ∎

Probs. 20-18, 20-19, 20-20 20-21

EXAMPLE 20-14 _____

The attenuation and the phase response plots of a certain network are shown in Figure 20-14(a) and (b). Calculate the first three terms of its voltage output if the input $v_{in}(t)$ is given as

$$v_{in}(t) = 10 + 4.6 \cos (t - 15°) - 2.1 \cos (2t + 40°) + \cdots$$

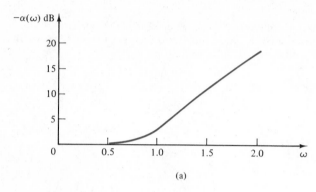

(a)

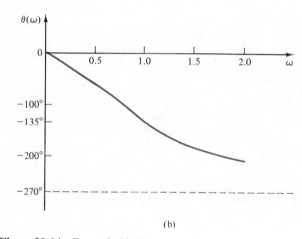

(b)

Figure 20-14   Example 20-14.

The attenuation is the negative logarithmic gain (Chapter 17)

$$-\alpha_H(\omega) = -20 \log |H(j\omega)|$$

From the given continuous frequency response plots, we read the required discrete points and calculate, in a similar way to Example 20-7,

   1. At dc ($\omega = 0$):

$$-\alpha_H = 0$$
$$\therefore \quad |H(j0)| = 10^{0/20} = 1$$
$$\therefore \quad |V_{out}(j0)| = (1)(10) = 10$$
$$\theta_H = 0° \quad \therefore \quad \theta_{out} = 0° + \theta_{in} = 0°$$

2. At $\omega = 1$:

$$-\alpha_H = 3$$
$$\therefore \quad |\mathbf{H}(j1)| = 10^{-3/20} = 0.707$$
$$\therefore \quad |\mathbf{V}_{\text{out}}(j1)| = (0.707)(4.6) = 3.25$$
$$\theta_H = -135° \quad \therefore \quad \theta_{\text{out}} = -135° - 15° = -150°$$

3. At $\omega = 2$:

$$-\alpha_H = 18$$
$$\therefore \quad |\mathbf{H}(j2)| = 10^{-18/20} = 0.126$$
$$\therefore \quad |\mathbf{V}_{\text{out}}(j2)| = (0.126)(2.1) = 0.26$$
$$\theta_H = -210° \quad \therefore \quad \theta_{\text{out}} = -210° + 40° = -170°$$

Therefore

$$v_{\text{out}}(t) = 10 + 3.25 \cos (t - 150°) - 0.26 \cos (2t - 170°) + \cdots$$

This example illustrates the use of frequency response plots together with discrete spectra for calculating a periodic output. This is one instance where the Laplace transform enters the picture, since $H(j\omega) = H(s)$ for $s = j\omega$. ∎

The *discrete power spectrum* in Equation 20-45 can also be plotted versus $n\omega_0$. See Figure 20-15. We may do it in either of two ways: A single-sided plot of P_0, P_1, $P_2,\ldots$ versus $\omega = 0, \omega_0, 2\omega_0, \ldots$, similar to the amplitude plot of A_n in Figure 20-11, or a double-sided even plot versus $\omega = 0, \pm\omega_0, \pm 2\omega_0, \ldots$, similar to the plot of $|c_n|$ in Figure 20-12. In the double-sided plot, it is customary to assign one half of P_k to $\omega = +k\omega_0$ and the other half to $\omega = -k\omega_0$. The same happened in the plot of $|c_n|$ in Equation 20-61. Remember: Physically, there are only *positive* frequencies, $+k\omega_0$. The negative frequencies are a mathematical convenience related to the exponential form of the Fourier series and Euler's identity for cosines and exponentials.

Prob. 20-22

The idea of larger and larger T, briefly mentioned earlier, is the subject of the next chapter. We'll let the period $T \to \infty$, thereby obtaining a nonperiodic waveform in the time domain. The corresponding Fourier series, a sum, becomes then a Fourier integral.

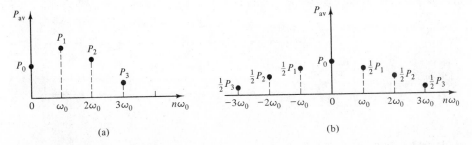

Figure 20-15 Power spectra.

PROBLEMS

20-1 **(a)** Plot $\cos^2 t$ versus t and determine its period T.
   (b) Use Euler's identity

$$\cos x = \frac{e^{jx} + e^{-jx}}{2}$$

   to find the finite (and exact) Fourier series of $\cos^2 t$.

20-2 Calculate the Fourier series for the square waveform in Example 20-3 with $T = 8$, $a = 2$, $A = 10$. Write the series through the fifth harmonic.

20-3 Repeat Problem 20-2 for the half-wave rectified waveform in Figure 20-1(c), with $T = 2\pi$, $A = 1$. Plot successively the partial sum of: (1) the dc term and the first harmonic; (2) add the next harmonic; (3) add the next harmonic.

20-4 Repeat Problem 20-3 for the full-wave rectified waveform of Figure 20-1(d).

20-5 The half-wave rectified waveform in Figure 20-1(c) is neither even nor odd, so it contains sines and cosines in its Fourier series. See Problem 20-3. Nevertheless, we can make it even by shifting the origin to the right by $T/4$. That is, with τ as the new (shifted) time variable

$$\tau = t - \frac{T}{4}$$

the function $f(\tau)$ is even. Use this property to expand $f(\tau)$ in its Fourier series containing only $\cos n\omega_0\tau$ terms. Then substitute for τ to get the original Fourier series in t. Use the same numerical values as in Problem 20-3, and compare results (they must be the same, of course!).

20-6 Write the cosine Fourier series for the half-wave rectified waveform of Problem 20-3.

20-7 A square-wave current source, of Example 20-3 and Problem 20-2, is applied to a parallel RC circuit, $R = 0.25\ \Omega$, $C = 3$ F. Calculate the steady-state voltage response up to and including the third harmonic. Plot it together with $i(t)$.

20-8 A half-wave rectifier circuit is connected to the common household outlet, as shown. Its output is $v_r(t)$ and is given by (compare your answer with Problem 20-3)

$$v_r(t) = \frac{120\sqrt{2}}{\pi} + \frac{120\sqrt{2}}{2}\sin 377t - \frac{2}{\pi}120\sqrt{2} \cdot \sum_{n=2,4,6,\ldots} \frac{1}{n^2 - 1}\cos 377nt$$

where $\omega_0 = 2\pi(60) = 377$ rad/s. This waveform must be smoothed out before it is applied to a resistive load, for example, a car battery that needs to be charged. For this purpose, an LC filter is inserted as shown.

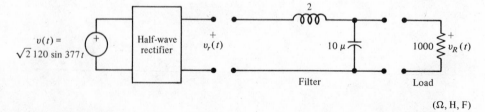

$(\Omega, \text{H}, \text{F})$

Problem 20-8

Calculate the first three terms (dc, the first, and the second harmonic) of the output voltage $v_R(t)$.

Notice the reduced amplitudes of the harmonics in $v_R(t)$, causing a smoother waveform. Plot it together with the half-wave rectified voltage $v_r(t)$.

*20-9 As we saw in Chapter 9 and in Example 20-5 here, a constant linear network does not change the frequency of the periodic input, and the output is of the same frequency. This is not true, in general, in a nonlinear network. Consider a nonlinear resistor, defined by

$$v = i^2$$

over a certain range of currents and voltages. Let the input current be

$$i(t) = 2 + 1.1 \cos t$$

Calculate the output $v(t)$. What is its period? What harmonics are present in $v(t)$ that are not in $i(t)$? Plot $i(t)$ and $v(t)$.

20-10 Calculate the exponential Fourier series for the sawtooth waveform of Figure 20-1(b), with $A = 10$ and $T = 4$. Get the first five coefficients c_n, $n = 0, \pm 1, \pm 2, \pm 3, \pm 4$, write the series as in Equation 20-38, then reduce it to its trigonometric form. *Hint*: In your calculations you'll need the following integral

$$\int xe^{ax}\, dx = \frac{xe^{ax}}{a} - \frac{e^{ax}}{a^2}$$

(you're welcome).

20-11 A periodic voltage is given by its first cycle

$$v_1(t) = 20e^{-t} \qquad 0 < t < 1, \; T = 1$$

Calculate, through $n = \pm 4$, its exponential Fourier series and reduce it to its trigonometric form.

20-12 Let the exponential Fourier series for $v(t)$ and $i(t)$ be

$$v(t) = \sum_{-\infty}^{\infty} c_n e^{jn\omega_0 t}$$

$$i(t) = \sum_{-\infty}^{\infty} d_n e^{jn\omega_0 t}$$

Prove that the average power is given by

$$P_{av} = \sum_{-\infty}^{\infty} c_n d_n^*$$

20-13 For the 1-P network in Figure 20-6, we are given

$$v(t) = 2.4 + 1.6 \cos 2t - 0.9 \cos 4t$$

and

$$i(t) = 1.3 \sin 2t + 0.8 \sin 4t$$

Without any calculations, determine what type of passive elements (R, L, M, C) are inside the 1-P.

20-14 Express the effective value of $f(t)$, Equation 20-4, in terms of the a's and the b's of its Fourier series. In this form, it is known as *Parseval's theorem*.

20-15 Express the effective value of $f(t)$, Equation 20-38, in terms of the c's in its exponential Fourier series.

20-16 Use the properties of a_n and b_n, as related to $1/n$, $1/n^2$, etc., to check your results in Problems 20-2, 20-3, and 20-10.

20-17 Plot the periodic function

$$f(t) = |t| \qquad -\pi < t < \pi$$

with $T = 2\pi$. Expand it into its sine and cosine Fourier series, making full use of such considerations as evenness or oddness, the discontinuity of its derivative, etc. Write out fully its series up to and including the fourth harmonic.

20-18 In Example 20-13, calculate for both cases (a) and (b) the values of $|c_n|$ up to and including $n = 10$. Prove that, in (a), $c_n = 0$ for $n = \pm 5, \pm 10, \pm 15, \dots$ and that, in (b), $c_n = 0$ for $n = \pm 10, \pm 20, \pm 30, \dots$

20-19 Plot the discrete spectra for Problem 20-3.

*20-20 How do the discrete spectra of $f(t)$ change if $f(t)$ is shifted by t_0 units to the right? *Hint*: Write $f(t)$ in its exponential Fourier series, then replace t by $(t - t_0)$.

20-21 A periodic $f(t)$ is given by its first cycle

$$f_1(t) = e^{-t} \qquad 0 < t < 1, T = 1$$

Calculate and plot $|c_n|$ and θ_n for $n = 0, \pm 1, \pm 2, \pm 3, \pm 4$.

20-22 In the circuit shown, the current source is periodic and given by

$$i(t) = 10 + 15 \cos (100t - 30°) + 8 \cos (200t + 45°)$$

(a) Calculate the cosine Fourier series of $v(t)$.
(b) Plot the discrete magnitude and phase spectra of $i(t)$ and of $v(t)$.
(c) Calculate and plot the power spectra for the circuit.
Hint: Think basics!

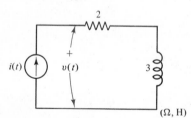

(Ω, H)   **Problem 20-22**

Chapter 21

The Fourier Transform

21-1 THE SQUARE PULSE REVISITED

Consider again the periodic square waveform, as discussed in Chapter 20. It is repeated here in Figure 21-1(a). Its exponential Fourier coefficients are

$$c_n = \frac{Aa}{T} \frac{\sin (n\pi a/T)}{n\pi a/T} = \frac{Aa}{T} \frac{\sin (n\omega_0 a/2)}{n\omega_0 a/2} \tag{21-1}$$

as derived previously. They are all real (positive or negative), because $f(t)$ is even. Two discrete amplitude spectra are shown in Figure 21-1(b) and (c), for $a/T = \frac{1}{5}$ and for $a/T = \frac{1}{10}$. We repeat several important observations:

1. If we keep a fixed and increase T (here, from $T = 5a$ to $T = 10a$), the fundamental frequency decreases, since $\omega_0 = 2\pi/T$.
2. The discrete frequencies $k\omega_0$ become closer to each other with an increase in T; that is, more and more harmonics are present, and their separation

$$(n + 1)\omega_0 - n\omega_0 = \omega_0 \tag{21-2}$$

   gets smaller and smaller.
3. The *envelope* of the plot is a continuous function given by

$$(c_n) = \frac{Aa}{T} \frac{\sin (\omega a/2)}{\omega a/2} \tag{21-3}$$

where (c_n) means the continuous envelope plot. The plot of Equation 21-3 is the standard mathematical curve of $\sin x/x$, with $x = \omega a/2$, shown in Figure 21-2. Because we plot $|c_n|$, there are only positive lobes in Figure 21-1.

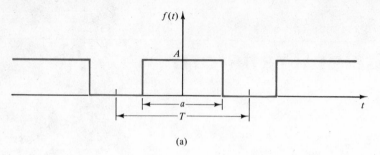

(a)

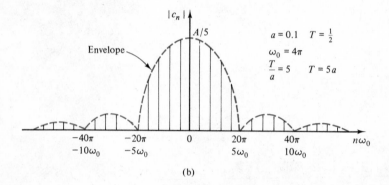

(b)

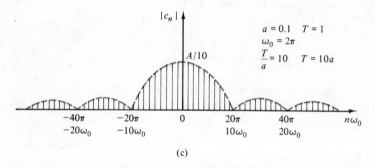

(c)

Figure 21-1 The square pulse.

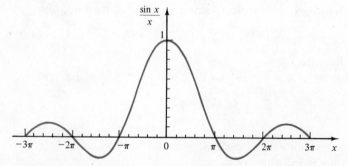

Figure 21-2 Plot of $\sin x/x$.

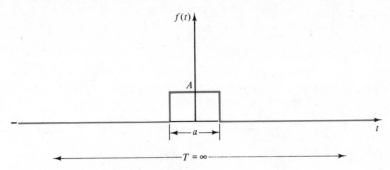

Figure 21-3 A nonperiodic function.

4. The amplitude c_n is zero when

$$\sin \frac{n\pi a}{T} = 0 \qquad (21\text{-}4)$$

that is, when

$$\frac{na}{T} = \pm 1, \ \pm 2, \ \pm 3, \ldots \qquad (21\text{-}5)$$

For $a/T = \frac{1}{5}$, these zeros occur when

$$n = \pm 5, \ \pm 10, \ \pm 15, \ldots \qquad (21\text{-}6)$$

that is, at

$$\omega = \pm 5\omega_0, \ \pm 10\omega_0, \ \pm 15\omega_0, \ldots \qquad (21\text{-}7)$$

For $a/T = \frac{1}{10}$, they occur at

$$\omega = \pm 10\omega_0, \ \pm 20\omega_0, \ \pm 30\omega_0, \ldots \qquad (21\text{-}8)$$

In general, for $a/T = 1/N$, they occur at

$$\omega = \pm N\omega_0, \ \pm 2N\omega_0, \ \pm 3N\omega_0, \ldots \qquad (21\text{-}9)$$

and, as T increases (N increases), there are more and more harmonics between these points.

5. As $T \to \infty$, the function becomes *nonperiodic*, as shown in Figure 21-3.

21-2 THE FOURIER INTEGRAL

The exponential Fourier series is

$$f(t) = \sum_{n=-\infty}^{\infty} c_n e^{jn\omega_0 t} \qquad \omega_0 = \frac{2\pi}{T} \qquad (21\text{-}10)$$

with

$$c_n = \frac{1}{T} \int_{-T/2}^{T/2} f(t) e^{-jn\omega_0 t} \, dt \qquad (21\text{-}11)$$

From Equation 21-11 we have

$$Tc_n = \int_{-T/2}^{T/2} f(t)e^{-jn\omega_0 t}\, dt \qquad (21\text{-}12)$$

Let us see what happens when $T \to \infty$. First, the separation between successive harmonics (Equation 21-2) becomes infinitesimally small

$$\omega_0 \to \Delta\omega \qquad (21\text{-}13)$$

and the discrete frequency variable $n\omega_0$ becomes a continuous variable

$$n\omega_0 \to \omega \qquad (21\text{-}14)$$

Thus, the integral in Equation 21-12 becomes

$$\int_{-\infty}^{\infty} f(t)e^{-j\omega t}\, dt = F(\omega) \qquad (21\text{-}15)$$

and Equation 21-10 is rewritten as

$$f(t) = \frac{1}{2\pi} \sum_{n=-\infty}^{\infty} c_n T e^{jn\omega_0 t}\, \Delta\omega \qquad (21\text{-}16)$$

using Equation 21-13. In the limit, the summation becomes an integral, and we get

$$f(t) = \frac{1}{2\pi} \int_{-\infty}^{\infty} F(\omega)e^{j\omega t}\, d\omega \qquad (21\text{-}17)$$

due to Equation 21-15.

Equations 21-15 and 21-17 are the *Fourier transform* for $f(t)$. More specifically, the direct Fourier transform of $f(t)$ is given by Equation 21-15, transforming a time-domain function into the frequency domain. Symbolically we write it as

$$\mathscr{F}f(t) = F(\omega) \qquad (21\text{-}18)$$

where $\mathscr{F}$ is the symbol of the Fourier transform. The inverse Fourier transform of $F(\omega)$ is given by Equation 21-17, retrieving the time function from its frequency transform. Symbolically,

$$\mathscr{F}^{-1}F(\omega) = f(t) \qquad (21\text{-}19)$$

EXAMPLE 21-1 ─────────────────────────────

Let us calculate the Fourier transform of the nonperiodic square pulse of Figure 21-3.

Solution. From the defining equation, (Equation 21-15) we have

$$\mathscr{F}f(t) = \int_{-a/2}^{a/2} Ae^{-j\omega t}\, dt = A\, \frac{e^{-j\omega t}}{-j\omega}\bigg]_{-a/2}^{a/2} = \frac{A}{\omega}\, 2\sin\frac{\omega a}{2}$$

$$= Aa\, \frac{\sin(\omega a/2)}{\omega a/2}$$

Compare it with Equation 21-1 for the discrete spectrum of a periodic square pulse, and to the envelope of this spectrum, Equation 21-3. We recognize the transition as $T \to \infty$: The periodic $f(t)$ becomes nonperiodic; its discrete frequency spectrum becomes a *continuous* spectrum containing *all* the frequencies, $F(\omega)$. ∎

Our derivation, while not totally rigorous, is more than adequate for our engineering applications. Just as in the Fourier series, there are sufficient Dirichlet conditions for the Fourier integral to exist. They are similar to those for the Fourier series: (1) over a finite interval, $f(t)$ may have, at most, a finite number of finite discontinuities; (2) $f(t)$ may have, at most, a finite number of maxima and minima; (3) $f(t)$ is absolutely integrable, that is, $\int_{-\infty}^{\infty} |f(t)|\,dt = M$, a finite number. However, several important functions that do not satisfy these conditions still have Fourier transforms, because those conditions are only sufficient, not necessary.

The function $F(\omega)$, as an extension of c_n, can be complex, in general. (Why is it real in Example 21-1?) We write it either in its rectangular form

$$F(\omega) = R(\omega) + jX(\omega) \tag{21-20}$$

showing the real and imaginary parts of $F(\omega)$; or else in its polar (exponential) form

$$F(\omega) = |F(\omega)|e^{j\phi(\omega)} \tag{21-21}$$

showing its magnitude and angle. We call $|F(\omega)|$ the *continuous amplitude (magnitude) spectrum* and $\phi(\omega)$ the *continuous phase spectrum* of $f(t)$. Quite obviously, these are the extensions of the discrete spectra of a periodic function.

EXAMPLE 21-2 _____

In Example 21-1, the continuous amplitude spectrum is

$$|F(\omega)| = Aa \left| \frac{\sin (\omega a/2)}{\omega a/2} \right| = \left| \frac{\sin \omega}{\omega} \right|$$

for $A = 0.5$ and $a = 2$.

Solution. Its plot will be the same as in Figure 21-2, but with the negative lobes made positive. The continuous phase spectrum $\phi(\omega)$ will be

$$\phi(\omega) = \begin{cases} 0° & -\pi < \omega < \pi, \quad 2\pi < \omega < 3\pi, \dots \\ 180° & -2\pi < \omega < -\pi, \quad \pi < \omega < 2\pi, \dots \end{cases}$$

The earlier plot of $\theta(\omega)$ is not continuous; the adjective "continuous" in "continuous phase spectrum" reminds us that it is computed and plotted for the continuous frequency variable ω. ∎

EXAMPLE 21-3 _____

Let us calculate the Fourier transform of

$$f(t) = e^{-at}u(t) \qquad a > 0$$

as shown in Figure 21-4(a).

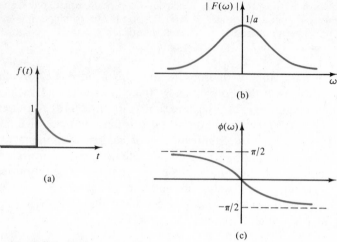

Figure 21-4   Example 21-3.

Solution.   We write

$$\mathscr{F}e^{-at}u(t) = \int_{-\infty}^{\infty} e^{-at}u(t)e^{-j\omega t}\,dt = \int_{0}^{\infty} e^{-(a+j\omega)t}\,dt = \frac{1}{a+j\omega} = F(\omega)$$

which we recognize as $F(s) = 1/(s + a)$ with $s = j\omega$. We'll return to this topic later in this chapter. The magnitude spectrum is

$$|F(\omega)| = \left| \frac{1}{a+j\omega} \right| = \frac{1}{\sqrt{a^2 + \omega^2}}$$

and the phase spectrum is

$$\phi(\omega) = -\tan^{-1}\frac{\omega}{a}$$

Their plots are shown in Figure 21-4(b) and (c).   Probs. 21-1,
                                                                                        ■ 21-2

EXAMPLE 21-4 _____

To illustrate the use of the impulse function in the Fourier transform, let us consider, first, the direct evaluation of

$$\mathscr{F}1 = ?$$

Solution.   Clearly, the function $f(t) = 1$ does not satisfy one of the Dirichlet conditions, since

$$\int_{-\infty}^{\infty} 1\,dt \to \infty$$

and is not a finite number M. If we attempt to evaluate the direct transform

$$\mathscr{F}1 = \int_{-\infty}^{\infty} 1e^{-j\omega t}\,dt$$

we run into a similar trouble: The integral does not converge.

Nevertheless, we can find this Fourier transform with the help of the impulse function, coming in through the back door, so to speak. Suppose we calculate the inverse Fourier transform of

$$F(\omega) = 2\pi\delta(\omega)$$

According to Equation 21-17, it is

$$f(t) = \mathscr{F}^{-1}2\pi\delta(\omega) = \frac{1}{2\pi}\int_{-\infty}^{\infty} 2\pi\delta(\omega)e^{j\omega t}\,d\omega = e^{j\omega t}\bigg]_{\omega=0} = 1$$

where we've used the sampling property of $\delta(\omega)$ to evaluate the integral. Consequently,

$$\mathscr{F}1 = 2\pi\delta(\omega)$$

as shown in Figure 21-5.

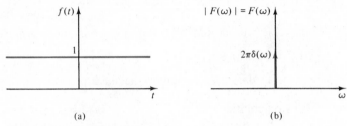

(a)   (b)

Figure 21-5 Example 21-4.

What does it mean? If you stretch your imagination somewhat, you might think of a Fourier series of the "periodic" constant 1 as having only one coefficient at $\omega = 0$, a dc term—which makes a lot of sense, because, after all, the constant 1 *is* a dc function. It is quite a strain, though, to visualize this Fourier coefficient as an impulse, of strength 2π yet! It is better to treat the Fourier transform on its own merits, as representing the frequency contents (magnitude and phase) of a time function. ■

EXAMPLE 21-5 ───────────────────────────

Find the Fourier transform of the unit impulse function $\delta(t)$, Figure 21-6(a).

Solution. We write

$$F(\omega) = \mathscr{F}\delta(t) = \int_{-\infty}^{\infty} \delta(t)e^{-j\omega t}\,dt = e^{-j\omega t}\bigg]_{t=0} = 1$$

with the help of the sampling property. This is a most interesting result. The magnitude spectrum of $\delta(t)$ is a *constant*, meaning that *all* frequencies are present, and all have unity amplitude and zero phase.

A little thought will confirm this startling discovery: The unit impulse $\delta(t)$ can be considered the limiting case of the square pulse in Example 21-1, Figure 21-3, with $a \to 0$, $A \to \infty$, and $Aa = 1$. Then its Fourier transform may be obtained from Example 21-1 as

$$\lim_{a\to 0}\frac{\sin(\omega a/2)}{\omega a/2} = \lim_{a\to 0}\frac{(\omega/2)\cos(\omega a/2)}{\omega/2} = 1$$

by l'Hôpital's rule. See Figure 21-6.

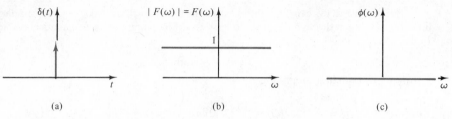

Figure 21-6   Example 21-5.   ■

21-3 PROPERTIES OF THE FOURIER TRANSFORM

The Fourier transform has several properties similar to those of the Laplace transform. Let us derive them and list them for future use:

Linearity   For two (or more) functions, and with a_1 and a_2 constants, we have

$$\mathscr{F}[a_1 f_1(t) + a_2 f_2(t)] = a_1 F_1(\omega) + a_2 F_2(\omega) \tag{21-22}$$

The proof is direct, relying on Equation 21-15 and on the linear property of integrals.

Time Differentiation   Let us differentiate Equation 21-17 with respect to t. Since the integral on the right is with respect to ω, a variable independent of t, we can do the differentiation under the integral sign. Then

$$\frac{df}{dt} = \frac{1}{2\pi} \int_{-\infty}^{\infty} F(\omega) \frac{d}{dt} e^{j\omega t} \, d\omega = \frac{1}{2\pi} \int_{-\infty}^{\infty} j\omega F(\omega) e^{j\omega t} \, d\omega$$

$$= \mathscr{F}^{-1}[j\omega F(\omega)] \tag{21-23}$$

Consequently,

$$\mathscr{F} \frac{df(t)}{dt} = j\omega F(\omega) \tag{21-24}$$

a result very similar to the Laplace transform, where multiplication by s corresponds to differentiation in the time domain. *A quickie*: Why is there no initial condition, $f(0^-)$, in Equation 21-24?

Time Shifting   The time-shifted function $f(t - a)$ has a Fourier transform

$$\mathscr{F} f(t - a) = e^{-j\omega a} F(\omega) \tag{21-25}$$

where $F(\omega) = \mathscr{F} f(t)$, the Fourier transform of the original, unshifted function. The proof of Equation 21-25 is direct, using Equation 21-15 and a change of variables, $t - a = y$, say. Again, the similarity to the Laplace transform, with $s = j\omega$, is evident.

EXAMPLE 21-6 _____

The magnitude spectrum of the time-shifted function in Equation 21-25 is

$$|e^{-j\omega a} F(\omega)| = |e^{-j\omega a}| |F(\omega)| = |F(\omega)|$$

because $|e^{-j\omega a}| = 1$. Therefore, the amplitude spectrum is identical to the one of the nonshifted function; the phase spectrum, on the other hand, is affected by the time shift

$$\phi_s(\omega) = -\omega a + \phi(\omega)$$

where $\phi_s(\omega)$ is the phase spectrum of the time-shifted function and $\phi(\omega)$ is that of the original, nonshifted function.

∎

Frequency Shifting

$$\mathscr{F}[f(t)e^{j\omega_0 t}] = F(\omega - \omega_0) \qquad (21\text{-}26)$$

where $F(\omega) = \mathscr{F}f(t)$. Here, again, the proof is direct, from either Equation 21-15 or 21-17, with a change of variables, $x = \omega - \omega_0$.

Probs. 21-3 through 21-7

These and other useful properties are listed in Table 21-1.

TABLE 21-1 THE FOURIER TRANSFORM

Properties and operations		
	$f(t)$	$F(\omega)$
Linearity	$a_1 f_1(t) + a_2 f_2(t)$	$a_1 F_1(\omega) + a_2 F_2(\omega)$
Time differentiation	$\dfrac{d^n f(t)}{dt^n}$	$(j\omega)^n F(\omega)$
Time integration	$\displaystyle\int_{-\infty}^{t} f(x)\,dx$	$\dfrac{1}{j\omega}F(\omega) + \pi F(0)\delta(\omega)$
Time shifting	$f(t-a)$	$F(\omega)e^{-j\omega a}$
Modulation (frequency shifting)	$f(t)e^{j\omega_0 t}$	$F(\omega - \omega_0)$
Time scaling	$f(at)$	$\dfrac{1}{a}F\!\left(\dfrac{\omega}{a}\right) \quad a > 0$
Time convolution	$\displaystyle\int_{-\infty}^{\infty} h(\lambda)x(t-\lambda)\,d\lambda$	$H(\omega)\cdot X(\omega)$
Frequency convolution	$h(t)x(t)$	$\dfrac{1}{2\pi}\displaystyle\int_{-\infty}^{\infty} H(\omega-\lambda)X(\lambda)\,d\lambda$
Symmetry	$F(t)$	$2\pi f(-\omega)$
Frequency differentiation	$(-jt)^n f(t)$	$\dfrac{d^n F(\omega)}{d\omega^n}$

Some elementary functions	
$\delta(t)$	1
1	$2\pi\delta(\omega)$
$e^{-at}u(t) \quad a > 0$	$\dfrac{1}{a + j\omega}$
$e^{-a|t|}$	$\dfrac{2a}{a^2 + \omega^2}$
$e^{j\omega_0 t}$	$2\pi\delta(\omega - \omega_0)$
$\cos \omega_0 t$	$\pi[\delta(\omega + \omega_0) + \delta(\omega - \omega_0)]$
$\sin \omega_0 t$	$j\pi[\delta(\omega + \omega_0) - \delta(\omega - \omega_0)]$
$u(t)$	$\pi\delta(\omega) + \dfrac{1}{j\omega}$
$\operatorname{sgn}(t)$	$\dfrac{2}{j\omega}$

Time Integration   The Fourier transform of the integral of $f(t)$ has the expected division by $j\omega$ in the frequency domain, plus an additional impulse term. Unlike in the Laplace transform, where the integral goes from 0^- to t, here the time integration includes the entire negative axis, from $-\infty$ to t. The additional term $\pi F(0)\delta(\omega)$ accounts for this.

EXAMPLE 21-7 _____

Let us find $\mathscr{F}u(t)$, the Fourier transform of the unit step function.

Solution.   Since

$$\frac{du(t)}{dt} = \delta(t)$$

we have

$$u(t) = \int_{-\infty}^{t} \delta(x)\,dx$$

Therefore

$$\mathscr{F}u(t) = \frac{1}{j\omega} + \pi\delta(\omega)$$

because $F(\omega) = 1$, and $F(0) = 1$ also.   ■   Prob. 21-8

EXAMPLE 21-8 _____

Let us derive the same result through a limiting process.

Solution.   Consider first the function

$$f(t) = \begin{cases} e^{-at}u(t) & t > 0 \quad a > 0 \\ -e^{at}u(-t) & t < 0 \quad a > 0 \end{cases}$$

as shown in Figure 21-7(a). Its Fourier transform is

$$F(\omega) = \int_{-\infty}^{0} -e^{at}e^{-j\omega t}\,dt + \int_{0}^{\infty} e^{-at}e^{-j\omega t}\,dt = \frac{-j2\omega}{\omega^2 + a^2}$$

Now let $a \to 0$. The function $f(t)$ becomes

$$\mathrm{sgn}\,(t) = -u(-t) + u(t) = \begin{cases} 1 & t > 0 \\ -1 & t < 0 \end{cases}$$

called "signum t" or "sign t," as shown in Figure 21-7(b). Its Fourier transform is the limit of $F(\omega)$ as $a \to 0$, that is

$$\mathscr{F}\,\mathrm{sgn}\,(t) = \frac{2}{j\omega}$$

Now, the unit step function $u(t)$ can be written as

$$u(t) = \frac{1 + \mathrm{sgn}\,(t)}{2}$$

and consequently

$$\mathscr{F}u(t) = \frac{1}{2}\left[2\pi\delta(\omega) + \frac{2}{j\omega}\right] = \pi\delta(\omega) + \frac{1}{j\omega}$$

as obtained before.

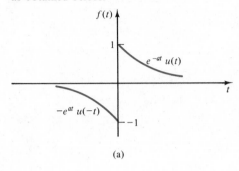

(a)

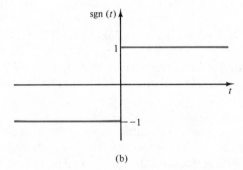

(b)   **Figure 21-7**   Example 21-8.   ■

Of the other entries in the table, let us comment on the following ones:

Time Scaling   A similar result holds in the Laplace transform, but there we had no physical feel for expanding time and compressing frequency. In view of the entire process of reaching the Fourier integral, and since t and ω are inversely related, it is not surprising that when we multiply t by a, the frequency ω gets divided by a. Of course, a may be less than or greater than unity. The factor $1/a$ in front of $F(\omega/a)$ has only a mathematical justification—it pops up in the derivation.

Frequency Shifting and Modulation   The property

$$\mathscr{F}[f(t)e^{j\omega_0 t}] = F(\omega - \omega_0) \tag{21-26}$$

is symmetric to the shifting property in Equation 21-25. There is, however, more to it. From Equation 21-26 we get

$$\mathscr{F}[f(t)e^{j\omega_0 t} + f(t)e^{-j\omega_0 t}] = F(\omega - \omega_0) + F(\omega + \omega_0) \tag{21-27}$$

that is,

$$\mathscr{F}[f(t)\cos\omega_0 t] = \tfrac{1}{2}F(\omega - \omega_0) + \tfrac{1}{2}F(\omega + \omega_0) \tag{21-28}$$

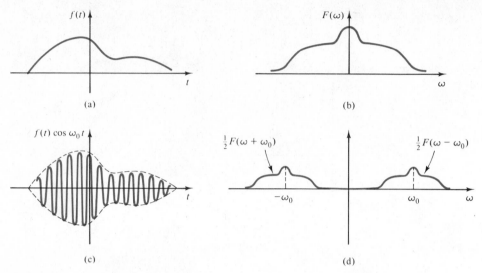

Figure 21-8   AM signal and spectra.

In Equation 21-28, the waveform $f(t) \cos \omega_0 t$ is an *amplitude-modulated (AM)* signal, with the pure cosine being the *carrier* and $f(t)$ the *envelope* of the signal. Such modulation is common in commercial broadcasting (540 to 1600 kHz on your radio dial). A typical AM signal with the corresponding magnitude spectra are shown in Figure 21-8.

Time Convolution   As in the case of the Laplace transform, time convolution is transformed into multiplication in the frequency domain. Here, however, the limits on the time waveforms and on the convolution integral extend over the entire time axis, $-\infty < t < \infty$. The product of two transform functions will be discussed in the next section.

21-4 APPLICATIONS TO CIRCUITS

Let us consider several examples of the application of the Fourier transform in circuit analysis.

EXAMPLE 21-9 _____

In the *RL* network shown in Figure 21-9(a), the input is $v(t) = 10e^{-t}u(t)$. The state equation, or KVL, reads

$$\frac{di(t)}{dt} + 2i(t) = 10e^{-t}u(t)$$

Solution.   The Fourier transform of this differential equation is

$$j\omega I(\omega) + 2I(\omega) = \frac{10}{1 + j\omega}$$

that is,

$$I(\omega) = \frac{10}{(2 + j\omega)(1 + j\omega)}$$

Expanding in partial fractions yields

$$I(\omega) = \frac{-10}{2 + j\omega} + \frac{10}{1 + j\omega}$$

and inversion (easy with Table 21-1, very difficult with Equation 21-17!) gives us

$$i(t) = \begin{cases} -10e^{-2t} + 10e^{-t} & t > 0 \\ 0 & t < 0 \end{cases}$$

as shown in Figure 21-9(b).

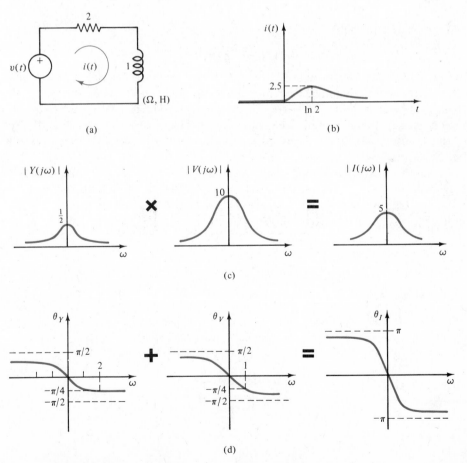

Figure 21-9   Example 21-9.

True, we could have solved this problem by the Laplace transform. Why do we bother with the Fourier transform? Because it provides a direct insight into the frequency response of the circuit. Specifically, we recognize in the expression of $I(\omega)$

$$I(\omega) = \frac{1}{2 + j\omega} \cdot \frac{10}{1 + j\omega} = Y(j\omega)V(j\omega) \qquad (21\text{-}29)$$

the familiar input-output relation

$$R(j\omega) = H(j\omega)E(j\omega) \qquad (21\text{-}30)$$

with $E(j\omega)$ the Fourier transform of the excitation (input), $H(j\omega)$ the appropriate network function, and $R(j\omega)$ the Fourier transform of the response (output).

Note: Throughout the study of Fourier transforms, the notation $I(\omega)$, $Y(\omega)$, etc., is simpler than $I(j\omega)$, $Y(j\omega)$, etc. The latter notation serves as a reminder of its connection to network functions studied earlier.

From Equation 21-29 we get, as usual, the magnitude spectrum

$$|I(\omega)| = |Y(\omega)||V(\omega)| \qquad (21\text{-}31)$$

and the phase spectrum

$$\theta_I(\omega) = \theta_Y(\omega) + \theta_V(\omega) \qquad (21\text{-}32)$$

as shown in Figure 21-9(c) and (d). We see here that magnitude spectra are *multiplied* to produce the resulting magnitude spectrum, and phase spectra are *added* to get the resulting phase spectrum. This is a complementary point of view to the frequency response results that we got in Chapters 15 and 16. Here we see them through the eyes of the Fourier transform and as an extension of the discrete spectra of periodic functions.

Another attractive feature of the Fourier transform is that, for sinusoidal sources, it provides the steady-state response, as illustrated in the next example.

EXAMPLE 21-10 ───

In the previous example, let

$$v(t) = 10 \cos 2t \qquad -\infty < t < \infty$$

and we wish to calculate $i(t)$.

Solution. Because $v(t)$ is applied very early $(t > -\infty)$, all transients have died, and we are dealing with the steady-state response. Here,

$$V(\omega) = \mathscr{F}v(t) = 10\pi[\delta(\omega + 2) + \delta(\omega - 2)]$$

from Table 21-1. The network function is, as before,

$$Y(\omega) = \frac{1}{2 + j\omega}$$

Consequently,

$$I(\omega) = \frac{1}{2 + j\omega} \cdot 10\pi[\delta(\omega + 2) + \delta(\omega - 2)]$$

is the Fourier transform of the response. We invert it using Equation 21-17

$$i(t) = \mathscr{F}^{-1}I(\omega) = \frac{1}{2\pi} \int_{-\infty}^{\infty} \frac{10\pi}{2 + j\omega} [\delta(\omega + 2) + \delta(\omega - 2)]e^{j\omega t} \, d\omega$$

$$= 5\left[\frac{e^{-j2t}}{2 - j2} + \frac{e^{j2t}}{2 + j2}\right]$$

where the evaluation of the integrals

$$\int_{-\infty}^{\infty} \delta(\omega + 2) \frac{e^{j\omega t}}{2 + j\omega} \, d\omega \qquad \int_{-\infty}^{\infty} \delta(\omega - 2) \frac{e^{j\omega t}}{2 - j\omega} \, d\omega$$

is done *by inspection* (no small feat!) with the help of the sampling property of the impulse function, bless its heart. It is in such formidable integrals that this property comes through, shining brightly!

A final rearrangement of $i(t)$ yields

$$i(t) = 5\left[\left(\frac{e^{-j2t}}{2 - j2}\right) + \left(\frac{e^{-j2t}}{2 - j2}\right)^*\right] = \frac{5}{\sqrt{2}} \cos(2t - 45°)$$

where we've used the fact that the sum of a complex number and its conjugate equals twice its real part (see Appendix B).

The same result may be obtained by using phasors,

$$\mathbf{V} = 10\underline{/0°} \quad (\omega = 2) \qquad \mathbf{Z} = 2 + j2$$

$$\therefore \quad \mathbf{I} = \frac{10\underline{/0°}}{2 + j2} = \frac{5}{\sqrt{2}} \underline{/-45°} \qquad (\omega = 2)$$

∎

While there is no doubt that phasors are faster here, the main idea is to illustrate the applications of the Fourier transform and its relationship to phasors. As with all our tools, we must learn their broad use, their most efficient applications, and their interrelationships. The last examples put in such a perspective the Fourier transform, the frequency response, and the phasor transform. Probs. 21-10, 21-11

21-5 FILTERS—AGAIN!

We have studied filters in Chapters 10 and 17. Let us tie those studies with our present discussion of the Fourier transform. In this context, we can define a *filter* as a linear network with certain specific amplitude spectrum characteristics. In most filters, we are interested in amplitude, not in phase. In others, a phase-correcting filter may be required, without regard to amplitude.

Consider, for instance, a *distortionless* filter. Such a filter has an output that is identical in shape to the input, but may be delayed

$$r(t) = Ke(t - a) \qquad a > 0 \tag{21-33}$$

as shown in Figure 21-10(a). The constant K is a real number, a scale factor. The frequency spectra are therefore (compare with Problem 20-20)

$$R(\omega) = Ke^{-j\omega a}E(\omega) \tag{21-34}$$

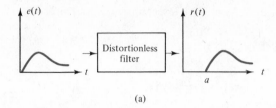

(a)

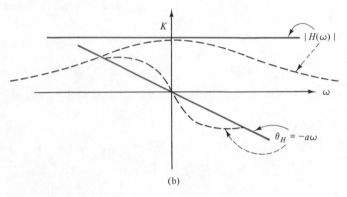

(b)

Figure 21-10   Distortionless filter.

From this result, the network function of this filter is

$$H(\omega) = Ke^{-j\omega a} \tag{21-35}$$

with a constant amplitude (magnitude) and with a linear phase

$$|H(\omega)| = K \qquad \theta_H(\omega) = -\omega a \tag{21-36}$$

as shown in Figure 21-10(b).

The *ideal* characteristics of amplitude and phase are only approximated by real filters, as shown in dotted lines. As a result, some distortion is introduced, and Equation 21-33 is correct only approximately.

The ideal *lowpass* filter, as studied in Chapter 17, has a magnitude spectrum

$$|H(\omega)| = \begin{cases} K & |\omega| < \omega_c \\ 0 & |\omega| > \omega_c \end{cases} \tag{21-37}$$

where ω_c is the *cutoff frequency* (often scaled to $\omega_c = 1$). The *bandwidth* of this filter is $0 < \omega < \omega_c$. With a linear phase, the network function of this filter is

$$H(\omega) = \begin{cases} Ke^{-j\omega a} & |\omega| < \omega_c \\ 0 & |\omega| > \omega_c \end{cases} \tag{21-38}$$

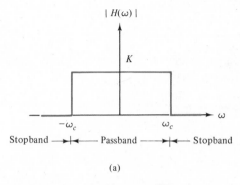

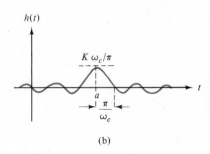

(a)

(b)

Figure 21-11   Ideal lowpass filter.

as shown in Figure 21-11(a). It would be instructive to find the impulse response of this filter. The inversion integral (Equation 21-17) is applied to Equation 21-38 to yield

$$h(t) = \frac{1}{2\pi} \int_{-\omega_c}^{\omega_c} K e^{-j\omega a} e^{j\omega t} \, d\omega$$

$$= \frac{K}{2\pi} \int_{-\omega_c}^{\omega_c} e^{j\omega(t-a)} \, d\omega$$

$$= \frac{K\omega_c}{\pi} \frac{\sin \omega_c(t-a)}{\omega_c(t-a)} \tag{21-39}$$

and the plot of $h(t)$ is shown in Figure 21-11(b). This response shows another reason why this filter is ideal: There is an output at $t < 0$, *before* the application of the input—clearly a physical impossibility.

Prob. 21-12

Let us review (again!) the sinusoidal steady-state response. Since

$$R(\omega) = H(\omega)E(\omega) \tag{21-40}$$

let us use an input $e(t) = e^{j\omega_0 t}$, thus allowing for $\cos \omega_0 t$, its real part, or $\sin \omega_0 t$, its imaginary part. From Table 21-1 and Equation 21-40 we have

$$R(\omega) = H(\omega)2\pi\delta(\omega - \omega_0) \tag{21-41}$$

and inversion of this result gives us

$$r(t) = \frac{1}{2\pi} \int_{-\infty}^{\infty} R(\omega)e^{j\omega t} \, d\omega = H(\omega_0)e^{j\omega_0 t} \tag{21-42}$$

This result is familiar, but worth repeating: In a constant, linear network, the steady-state response to a sinusoidal input $e^{j\omega_0 t}$ is another sinusoid at the *same* frequency, $e^{j\omega_0 t}$, scaled by the multiplier $H(\omega_0)$. We've used this idea in the study of phasors and in the techniques of frequency response.

*21-6 ENERGY SPECTRUM

Let us prove Parseval's theorem for the Fourier transform, as we did in the previous chapter for the Fourier series. Parseval's theorem here is

$$\int_{-\infty}^{\infty} [f(t)]^2 \, dt = \frac{1}{2\pi} \int_{-\infty}^{\infty} |F(\omega)|^2 \, d\omega \tag{21-43}$$

Its proof is based on the frequency convolution in Table 21-1, with $h(t) = x(t) = f(t)$. Thus

$$\mathscr{F}[f(t)]^2 = F(\omega) * F(\omega) \tag{21-44}$$

that is,

$$\int_{-\infty}^{\infty} [f(t)]^2 e^{-j\omega t} \, dt = \frac{1}{2\pi} \int_{-\infty}^{\infty} F(\lambda) F(\omega - \lambda) \, d\lambda \tag{21-45}$$

Since this is true for all ω, set $\omega = 0$, to get

$$\int_{-\infty}^{\infty} [f(t)]^2 \, dt = \frac{1}{2\pi} \int_{-\infty}^{\infty} F(\lambda) F(-\lambda) \, d\lambda \tag{21-46}$$

Also, for $f(t)$ real we have

$$F(-\lambda) = F^*(\lambda) \tag{21-47}$$

and Equation 21-43 follows.

What does it mean? Assume that $f(t)$ is a current source driving a 1-Ω resistor. Then the left-hand side of Equation 21-43 is the total energy supplied by the signal $f(t)$, being the sum (integral) over all time of the power, or of the *energy per unit time*, $[f(t)]^2$. According to Parseval's theorem (Equation 21-43) the same total energy can be obtained by integrating (summing) the *energy per unit frequency*, $|F(\omega)|^2/2\pi$, in joules per hertz, over all frequencies. Thus, $|F(\omega)|^2$ can be considered the frequency contents of energy, and it is called the *energy density spectrum* of the signal $f(t)$.

Parseval's theorem thus provides a close tie between time-domain and frequency-domain calculations. It is a very interesting basic property that we can calculate power and energy directly in the frequency domain.

EXAMPLE 21-11 ⎯⎯⎯⎯⎯⎯⎯⎯⎯⎯⎯⎯⎯⎯⎯⎯⎯⎯⎯⎯⎯⎯⎯⎯⎯⎯⎯⎯

Calculate the energy spectral density of

$$f(t) = 10e^{-t}u(t)$$

and confirm it by Parseval's theorem.

Solution.   In the time domain, we have

$$\int_{-\infty}^{\infty} [f(t)]^2 \, dt = \int_{0}^{\infty} 100e^{-2t} \, dt = 50 \text{ J}$$

In the frequency domain, we have $F(\omega) = 10/(j\omega + 1)$, and

$$\frac{1}{2\pi} \int_{-\infty}^{\infty} \left| \frac{10}{j\omega + 1} \right|^2 \, d\omega = \frac{1}{\pi} \int_{0}^{\infty} \frac{100}{\omega^2 + 1} \, d\omega = \frac{100}{\pi} \tan^{-1} \omega \Big]_{0}^{\infty} = 50 \text{ J}$$

Here, we have used the fact that $|F(\omega)|^2$ is an even function; therefore, its integral over $-\infty < \omega < \infty$ is twice its integral over $0 < \omega < \infty$. ■

Probs. 21-13
through 21-17

21-7 FOURIER AND LAPLACE†

Similarities between the Laplace transform and the Fourier transform are apparent by now. In order to review and summarize them, we write again the direct transforms

$$F_L(s) = \mathscr{L} f(t)u(t) = \int_{0^-}^{\infty} f(t)e^{-st} \, dt \qquad (21\text{-}48)$$

$$F_F(\omega) = \mathscr{F} f(t) = \int_{-\infty}^{\infty} f(t)e^{-j\omega t} \, dt \qquad (21\text{-}49)$$

where the subscripts L and F are added for clarity and distinction. A comparison between these two transforms requires, first, that $f(t)$ be *causal*, that is,

$$f(t) = 0 \qquad t < 0 \qquad (21\text{-}50)$$

because of the different lower limits on the two integrals. Our Laplace transform handles only causal functions, while the Fourier transform can handle also noncausal functions over the entire time axis, $-\infty < t < \infty$.‡

With Equation 21-50 in mind, we consider three cases:

1. All the poles of $F_L(s)$ are *inside* the left half plane (LHP), $\sigma < 0$. Then Equation 21-49 is just a special case of Equation 21-48, with $s = j\omega$,

$$F_F(\omega) = F_L(s) \Big]_{s=j\omega} \qquad (21\text{-}51)$$

---

† An interesting historical vignette: In 1807, Fourier presented his studies on heat flow, including his trigonometric series, to the prestigious Institut de France. One of the examiners appointed to review his paper was Laplace. Another examiner, Lagrange, objected to the acceptance of Fourier's work. Undaunted, Fourier had it published elsewhere in 1822.

Here is a lesson in tenacity (Fourier) and fallibility (Lagrange) among giants.

‡ There is also a two-sided (bilateral) Laplace transform, defined for all t, $-\infty < t < \infty$. Our usual Laplace transform is one-sided (unilateral), for $0^- < t < \infty$.

The requirement for causality, again, is the lower limit on the integral in Equation 21-48 by comparison with Equation 21-49. The restriction on the poles of $F_L(s)$ guarantees that $f(t)$ is absolutely integrable

$$\int_0^\infty |f(t)|\, dt < \infty \qquad (21\text{-}52)$$

and, consequently, the Fourier integral exists.

EXAMPLE 21-12 _____

For

$$f(t) = e^{-at}u(t) \qquad a > 0$$

we have

$$F_L(s) = \frac{1}{s + a}$$

Solution. Therefore

$$F_F(\omega) = \frac{1}{s + a}\Bigg]_{s = j\omega} = \frac{1}{j\omega + a} \qquad \blacksquare$$

As mentioned, the Fourier transform is defined for $-\infty < t < \infty$ and therefore it can handle also noncausal functions

$$f(t) \neq 0 \qquad t < 0 \qquad (21\text{-}53)$$

In fact, this is one of its strengths, as we saw in Section 21-4. To relate $F_F(\omega)$ to $F_L(s)$ here, we write such a noncausal function as

$$f(t) = f_1(-t)u(-t) + f_2(t)u(t) \qquad (21\text{-}54)$$

where $f_1(-t)u(-t)$ is the negative-time part of $f(t)$ and $f_2(t)u(t)$ is its positive-time (causal) part. See Figure 21-12.

The Fourier transform of $f_1(-t)u(-t)$ is obtained by, first, "flipping" this function over to the positive time axis. Then we take the usual Laplace transform of this reflected function and set $s = -j\omega$. That is,

$$\mathscr{F}f(t) = \mathscr{L}[f_1(t)u(t)]_{s = -j\omega} + \mathscr{L}[f_2(t)u(t)]_{s = j\omega} \qquad (21\text{-}55)$$

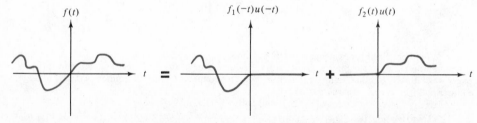

Figure 21-12 Decomposition of $f(t)$ into its negative-time and positive-time parts.

EXAMPLE 21-13 _____

See Example 21-8 and Figure 21-7, where

$$f_1(-t)u(-t) = -e^{at}u(-t)$$

$$f_2(t)u(t) = e^{-at}u(t) \qquad a > 0$$

$$\therefore \quad f_1(t)u(t) = -e^{-at}u(t)$$

$$\therefore \quad F_F(\omega) = \frac{-1}{s+a}\bigg]_{s=-j\omega} + \frac{1}{s+a}\bigg]_{s=j\omega} = \frac{-j2\omega}{\omega^2 + a^2}$$

as before. We actually derived this result in Example 21-8 by a direct evaluation of the Fourier integral. Here we illustrated the application of Equation 21-55. ■ Prob. 21-18

2. If $F_L(s)$ has poles inside the LHP ($\sigma < 0$) as well as *simple* poles *on* the $j\omega$ axis ($\sigma = 0$), we can write

$$F_L(s) = \hat{F}_L(s) + \sum_p \frac{K_p}{s - j\omega_p} \qquad (21\text{-}56)$$

where $\hat{F}_L(s)$ has only poles inside the LHP, and the summation accounts for the simple poles on the $j\omega$ axis. To get a "feel" for this case, consider a preliminary example.

EXAMPLE 21-14 _____

For

$$f(t) = e^{j\omega_0 t}u(t)$$

we have

$$F_L(s) = \frac{1}{s - j\omega_0}$$

Solution. On the other hand, the Fourier transform of $u(t)$ is, as in Example 21-7,

$$\mathscr{F}u(t) = \frac{1}{j\omega} + \pi\delta(\omega) \qquad (21\text{-}57)$$

and, using the frequency shifting property of Table 21-1, we obtain

$$F_F(\omega) = \mathscr{F}e^{j\omega_0 t}u(t) = \frac{1}{j(\omega - \omega_0)} + \pi\delta(\omega - \omega_0) \qquad ■$$

Comparing $F_L(s)$ and $F_F(\omega)$, we reach the following result here

$$F_F(\omega) = F_L(s)]_{s=j\omega} + \sum_p \pi K_p \delta(\omega - \omega_p) \qquad (21\text{-}58)$$

The first term is in accordance with Equation 21-51, and the second one has the Fourier transform of terms like $K_p e^{j\omega_p t}u(t)$. This, in turn, is found in Table 21-1, with one significant change: There, the time function is $e^{j\omega_p t}$ for $-\infty < t < \infty$, but here $e^{j\omega_p t}u(t)$ is causal, $0 < t < \infty$. Therefore, instead of 2π, we multiply only by π.

EXAMPLE 21-15 _____

For $u(t)$, the unit step function, we have

$$F_L(s) = \frac{1}{s}$$

Solution. Equation 21-58 then yields immediately

$$F_F(\omega) = \frac{1}{s}\bigg]_{s=j\omega} + \pi\delta(\omega - 0) = \frac{1}{j\omega} + \pi\delta(\omega)$$

as obtained earlier.

Prob. 21-19

3. If $F_L(s)$ has poles in the right half plane (RHP), $\sigma > 0$, then the Fourier transform of $f(t)$ does not exist because that integral does not converge.

EXAMPLE 21-16 _____

Consider

$$f(t) = e^{at}u(t) \qquad a > 0$$

Solution. If we set up its Fourier integral

$$\int_0^\infty e^{at}e^{-j\omega t}\, dt = ?$$

it fails to converge: At the upper limit, even though $|e^{-j\omega t}| = 1$, the term e^{at} becomes unbounded (it "blows up").

The Laplace transform, on the other hand, *exists*

$$\mathscr{L}e^{at}u(t) = \int_{0^-}^\infty e^{at}e^{-(\sigma+j\omega)t}\, dt = \int_{0^-}^\infty e^{(a-\sigma)t}e^{-j\omega t}\, dt$$

and the integral *converges* for $\sigma > a$. Therefore

$$F_L(s) = \frac{1}{s-a} \qquad \sigma > a$$

The Laplace transform has a "built-in" convergence factor, $e^{-\sigma t}$, which the Fourier transform lacks.

Probs. 21-20, 21-21, 21-22

These three cases, 1, 2, and 3 are shown in Figure 21-13.

By way of a summary, let us repeat several differences between the two transforms:

1. The Laplace transform is useful for causal functions and initial conditions. The Fourier transform can handle noncausal functions (initial conditions are meaningless in such cases).
2. The steady-state response obtained by the Fourier transform is also available as the zero-state response to a periodic, causal input using the Laplace transform.

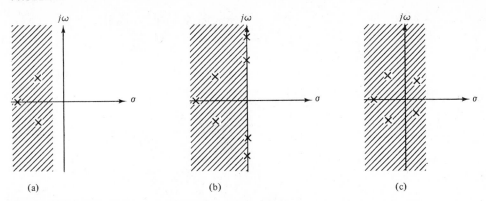

(a)　　　　　　　　　(b)　　　　　　　　　(c)

Figure 21-13 From $F_L(s)$ to $F_F(\omega)$: (a) $F_L(s)$ has poles only inside LHP. Use Equation 21-51. (b) $F_L(s)$ has also simple poles on the $j\omega$ axis. Use Equation 21-58. (c) $F_L(s)$ has poles inside RHP. No $F_F(\omega)$.

3. The Fourier transform is based on the frequency contents of waveforms. The Laplace transform masks this physical aspect, but yields the network function $H(j\omega) = H(s)$ with $s = j\omega$, so essential in the Fourier transform analysis.

4. The Fourier transform contains important information about the energy spectrum of signals—a feature totally unavailable in the Laplace transform.

PROBLEMS

21-1　Calculate the Fourier transform, and plot the two continuous spectra, for

$$f(t) = \begin{cases} A & 0 < t < a \\ 0 & \text{elsewhere} \end{cases}$$

Compare with Example 21-1 and with Problem 20-2.

21-2　Calculate the Fourier transform of

$$f(t) = e^{-a|t|} \qquad a > 0 \qquad -\infty < t < \infty$$

Plot $f(t)$, $|F(\omega)|$, and $\phi(\omega)$. Next, let $a \to 0$. What happens to $f(t)$ and to $F(\omega)$? *Hint:* First calculate the area under $F(\omega)$.

21-3　Prove Equation 21-25 in detail.

21-4　Prove Equation 21-26 in detail.

21-5　Calculate

$$f(t) = \mathscr{F}^{-1} 2\pi\delta(\omega - \omega_0)$$

21-6　From your result in Problem 21-5, find $\mathscr{F} \cos \omega_0 t$ and $\mathscr{F} \sin \omega_0 t$. *Hint:* Use Euler's identities for $\cos \omega_0 t$, $\sin \omega_0 t$, and $e^{\pm j\omega_0 t}$. Plot both Fourier transforms, in magnitude and phase.

21-7 Calculate the Fourier transform of the triangular pulse shown. Relate it and its transform to Problem 21-1.

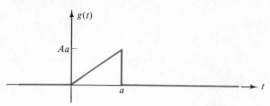

Problem 21-7

21-8 Prove

$$\mathscr{F}\left[\int_{-\infty}^{t} f(x)\, dx\right] = \frac{1}{j\omega} F(\omega)$$

if

$$\int_{-\infty}^{\infty} f(t)\, dt = 0$$

21-9 Prove the scaling property of the Fourier transform

$$\mathscr{F}f(at) = \frac{1}{|a|} F\left(\frac{\omega}{a}\right)$$

Hint: In setting up the integral, let $at = x$.

21-10 Let the triangular pulse of Problem 21-7 ($A = 2$, $a = 1$) be a current source exciting a parallel RC circuit, with $R = 1\,\Omega$, $C = \frac{1}{2}$ F, and with zero initial conditions. Calculate and plot the magnitude and phase spectra of the voltage $v(t)$ across C.

21-11 Let the triangular pulse of Problem 21-7 be a voltage source ($A = 2$, $a = 1$) exciting a series RC circuit, $R = 1\,\Omega$, $C = 0.1$ F, with zero initial state. Calculate and plot the magnitude and phase spectra of the capacitor's voltage $v_C(t)$.

21-12 Let the input to an ideal lowpass filter be periodic,

$$e(t) = \sum_{n=-\infty}^{\infty} c_n e^{jn\omega_0 t}$$

Calculate the response by considering, first, the response to $e^{jn\omega_0 t}$ for $n\omega_0 < \omega_c$ and for $n\omega_0 > \omega_c$. Then use superposition. Your answer should make sense intuitively—state it in words.

21-13 Prove that if $f(t)$ is real (the usual case), then

$$F(-\omega) = F^*(\omega)$$

Hint: Write Equation 21-15 with $e^{-j\omega t} = \cos \omega t - j \sin \omega t$, and separate $F(\omega)$ into its rectangular form

$$F(\omega) = R(\omega) + jX(\omega)$$

21-14 Prove that for a periodic $f(t)$, Parseval's theorem (Equation 21-43) has the form

$$\frac{1}{T} \int^T [f(t)]^2 \, dt = \sum_{n=-\infty}^{\infty} |c_n|^2$$

See also Problem 20-14. Interpret this result in terms of energy per period and energy per harmonic.

21-15 Calculate the energy density spectrum of the impulse function $\delta(t)$.

21-16 Calculate the energy density spectrum of the triangular pulse in Problem 21-7.

21-17 The input $e(t)$ to an ideal lowpass filter, with $K = 1$, $\omega_c = 20$ rad/s, is given by

$$e(t) = 100e^{-10t}u(t)$$

(a) Plot $|E(\omega)|^2$ for this input.
(b) Plot $|R(\omega)|^2$ for the output.
(c) Using Parseval's theorem, calculate the percentage of the total input energy available at the output.

21-18 Prove Equation 21-55.

21-19 Given

$$F_L(s) = \frac{1}{s^2 + 1}$$

Find $F_F(\omega)$. Plot carefully the time function.

21-20 From your table of the Laplace transform, obtain the Fourier transform of each function:

(a) $f(t) = e^{-at} \cos \beta t \, u(t)$ $a > 0$
(b) $f(t) = e^{-at} \sin \beta t \, u(t)$ $a > 0$
(c) $f(t) = \cos \beta t \, u(t)$
(d) $f(t) = t^n e^{-at} u(t)$ $a > 0, n = 1, 2, 3, \ldots$

*21-21 It is claimed that

$$F_L(s) = \mathscr{L} f(t)u(t) = \mathscr{F}[f(t)e^{-at}u(t)]$$

Derive this relationship and, based on the convergence of the Fourier integral, state the limitations on the location of the poles of $F_L(s)$.

*21-22 Extend Equation 21-58 to *multiple* poles on the $j\omega$ axis. Specifically, start with

$$F_L(s) = \frac{1}{(s - j\omega_0)^2} \qquad f(t) = te^{j\omega_0 t}u(t)$$

and use the frequency differentiation property in Table 21-1 to get

$$F_F(\omega) = j\pi\delta'(\omega - \omega_0) + \frac{1}{(j\omega - j\omega_0)^2}$$

where δ' is the first derivative of δ.

Appendix A

Matrices and Determinants

A-1 DEFINITIONS

A *matrix* is a rectangular array of scalars, called *elements*. These scalars may be real numbers, complex numbers, or functions of some parameter. Square brackets are used around a matrix.

EXAMPLE A-1 ————————————————————————————————————

$$\mathbf{A} = \begin{bmatrix} -1 & 2.3 & 0 \\ 1 & 1 & -6 \end{bmatrix} \quad \mathbf{B} = \begin{bmatrix} 2+j & -j \\ -1 & 3-j \end{bmatrix} \quad \mathbf{C} = \begin{bmatrix} \dfrac{1}{s} \\ \dfrac{s}{s^2+1} \\ \dfrac{-2}{s+4} \end{bmatrix}$$

Here matrix $\mathbf{A}$ is *real* (its elements are real); matrix $\mathbf{B}$ is *complex*, and in matrix $\mathbf{C}$ the elements are functions of s. ∎

The *order* of a matrix is $m \times n$ ("m by n"), where m is its number of rows and n the number of columns. So, the previous matrices are of order (2×3), (2×2), and (3×1), respectively. A shorthand notation for a matrix is

$$\mathbf{A} = [a_{ij}] \qquad \begin{matrix} 1 \le i \le m \\ 1 \le j \le n \end{matrix} \qquad \text{(A-1)}$$

where a_{ij} is the ijth element of $\mathbf{A}$, located in row i and column j.

A *square* matrix has an equal number of rows and columns

$$m = n \qquad \text{(A-2)}$$

A *column matrix* (also called a *column vector*) has m rows and one column

$$n = 1 \tag{A-3}$$

A *row matrix* (a *row vector*) has one row and n columns

$$m = 1 \tag{A-4}$$

A *diagonal* matrix is square, with all the elements off the main diagonal being zero

$$a_{ij} = 0 \qquad i \neq j \tag{A-5}$$

where the main diagonal is $a_{11}, a_{22}, \ldots, a_{nn}$, from the upper left to the lower right.

The *unit matrix* (also called the *identity matrix*) is a diagonal matrix with 1's on its main diagonal. It is usually denoted by $\mathbf{U}$

$$\mathbf{U} = [u_{ii}] \qquad u_{ii} = 1 \tag{A-6}$$

It plays the same role in matrix algebra as the scalar 1 in scalar algebra.

The *zero matrix* (*null matrix*) has all its elements zero

$$\mathbf{0} = [0_{ij}] \tag{A-7}$$

The *transpose* of a matrix, denoted by a superscript T, is obtained by interchanging rows and columns

$$\mathbf{A}^T = [a_{ji}] \qquad \begin{array}{l} 1 \leq i \leq m \\ 1 \leq j \leq n \end{array} \tag{A-8}$$

EXAMPLE A-2 _____

The transposes of the matrices given in Example A-1 are, respectively,

$$\mathbf{A}^T = \begin{bmatrix} -1 & 1 \\ 2.3 & 1 \\ 0 & -6 \end{bmatrix}$$

$$\mathbf{B}^T = \begin{bmatrix} 2+j & -1 \\ -j & 3-j \end{bmatrix}$$

$$\mathbf{C}^T = \begin{bmatrix} \frac{1}{s} & s & -2 \\ \frac{1}{s} & s^2+1 & s+4 \end{bmatrix} \qquad \blacksquare$$

A-2 BASIC OPERATIONS

Two matrices $\mathbf{A}$ and $\mathbf{B}$ are *equal* if and only if they are of the same order and every a_{ij} is equal to the corresponding b_{ij}

$$a_{ij} = b_{ij} \qquad \begin{array}{l} 1 \leq i \leq m \\ 1 \leq j \leq n \end{array} \tag{A-9}$$

A matrix is *symmetric* if it is equal to its transpose

$$\mathbf{A} = \mathbf{A}^T \tag{A-10}$$

that is, it must be square, and its off diagonal elements must be equal

$$a_{ij} = a_{ji} \qquad \begin{array}{l} 1 \le i \le m \\ 1 \le j \le n \end{array} \qquad \text{(A-11)}$$

Matrices can be *added* (or *subtracted*) only if they have the same order; then their sum (or difference) is done element by element, that is

$$\mathbf{A} \pm \mathbf{B} = \mathbf{C} \qquad \text{(A-12)}$$

means

$$a_{ij} \pm b_{ij} = c_{ij} \qquad \text{(A-13)}$$

In particular

$$\mathbf{A} - \mathbf{A} = \mathbf{0} \qquad \text{(A-14)}$$

the zero matrix. The commutative law is valid for addition

$$\mathbf{A} + \mathbf{B} = \mathbf{B} + \mathbf{A} \qquad \text{(A-15)}$$

and so is the associative law

$$\mathbf{A} + (\mathbf{B} + \mathbf{C}) = (\mathbf{A} + \mathbf{B}) + \mathbf{C} \qquad \text{(A-16)}$$

A matrix $\mathbf{A}$, when multiplied by a scalar α, yields the matrix

$$\alpha\mathbf{A} = [\alpha a_{ij}] \qquad \text{(A-17)}$$

where *every* element of $\mathbf{A}$ is multiplied by α.

The *product* of two matrices, $\mathbf{AB}$, is defined only if $\mathbf{A}$ is of order $(m \times r)$ and $\mathbf{B}$ of order $(r \times n)$; that is, the number of columns of $\mathbf{A}$ must equal the number of rows of $\mathbf{B}$. Such matrices are called *conformable* or *compatible*. Then the product $\mathbf{AB}$ is a matrix $\mathbf{C}$, of order $m \times n$, with its elements given by

$$c_{ij} = \sum_{k=1}^{k=r} a_{ik}b_{kj} \qquad \begin{array}{l} 1 \le i \le m \\ 1 \le j \le n \end{array} \qquad \text{(A-18)}$$

In words, we multiply the elements of row i of $\mathbf{A}$ by the corresponding elements of column j of $\mathbf{B}$, then add these products to get the ijth element of $\mathbf{C}$.

EXAMPLE A-3 —————————————————————————————————

The matrix equation $\mathbf{Ax} = \mathbf{B}$ reads fully

$$\begin{bmatrix} a_{11} & a_{12} & a_{13} \\ a_{21} & a_{22} & a_{23} \\ a_{31} & a_{32} & a_{33} \end{bmatrix} \begin{bmatrix} x_1 \\ x_2 \\ x_3 \end{bmatrix} = \begin{bmatrix} b_{11} \\ b_{21} \\ b_{31} \end{bmatrix}$$

and when multiplied out gives

$$a_{11}x_1 + a_{12}x_2 + a_{13}x_3 = b_{11}$$
$$a_{21}x_1 + a_{22}x_2 + a_{23}x_3 = b_{21}$$
$$a_{31}x_1 + a_{32}x_2 + a_{33}x_3 = b_{31}$$

∎

In general, the commutative law of multiplication *does not* hold

$$\mathbf{AB} \ne \mathbf{BA} \qquad \text{(A-19)}$$

since the product $\mathbf{BA}$ may not even be defined! Even if it is defined, it may not be equal to the product $\mathbf{AB}$.

EXAMPLE A-4 _____

In Example A-3, the product $\mathbf{xA}$ is not defined, because the number of columns of $\mathbf{x}$ is not equal to the number of rows of $\mathbf{A}$.   ■

EXAMPLE A-5 _____

Let

$$\mathbf{A} = \begin{bmatrix} 1 & -1 \\ 0 & 4 \end{bmatrix} \qquad \mathbf{B} = \begin{bmatrix} 1 & -2 \\ 3 & 6 \end{bmatrix}$$

Then

$$\mathbf{AB} = \begin{bmatrix} 1 & -1 \\ 0 & 4 \end{bmatrix}\begin{bmatrix} 1 & -2 \\ 3 & 6 \end{bmatrix} = \begin{bmatrix} -2 & -8 \\ 12 & 24 \end{bmatrix}$$

but

$$\mathbf{BA} = \begin{bmatrix} 1 & -2 \\ 3 & 6 \end{bmatrix}\begin{bmatrix} 1 & -1 \\ 0 & 4 \end{bmatrix} = \begin{bmatrix} 1 & -9 \\ 3 & 21 \end{bmatrix} \neq \mathbf{AB} \qquad\qquad ■$$

For this reason, we must be careful in stating the order of multiplication. We say that, in the product $\mathbf{AB}$, matrix $\mathbf{A}$ *premultiplies* $\mathbf{B}$, or, alternately, matrix $\mathbf{A}$ is *postmultiplied* by $\mathbf{B}$. The distributive law is valid in multiplication

$$(\mathbf{A} + \mathbf{B})\mathbf{C} = \mathbf{AC} + \mathbf{BC} \tag{A-20}$$

and so is the associative law

$$(\mathbf{AB})\mathbf{C} = \mathbf{A}(\mathbf{BC}) = \mathbf{ABC} \tag{A-21}$$

provided the order of multiplication is preserved in each case.
    The unit matrix commutes in multiplication

$$\mathbf{UA} = \mathbf{AU} = \mathbf{A} \tag{A-22}$$

where the order of $\mathbf{U}$ is $(n \times n)$, the same as that of $\mathbf{A}$.
    Positive integer powers of a square matrix are defined by

$$\mathbf{A}^2 = \mathbf{AA} \tag{A-23}$$

$$\mathbf{A}^n = \mathbf{A}^{n-1}\mathbf{A} \tag{A-24}$$

Also

$$\mathbf{A}^0 = \mathbf{U} \tag{A-25}$$

    Often, it is convenient to partition a matrix into *submatrices*.

EXAMPLE A-6 _____

The matrix

$$\mathbf{A} = \begin{bmatrix} 2 & 0 & 1 \\ -1 & 3 & 2 \end{bmatrix}$$

may be partitioned as follows:

$$\mathbf{A} = \left[\begin{array}{cc:c} 2 & 0 & 1 \\ -1 & 3 & 2 \end{array} \right] = [\mathbf{A}_{11} \vdots \mathbf{A}_{12}]$$

where A_{11} is the (1, 1) submatrix, taken as a single element in A, occupying the first row and column; the submatrix A_{12} is in the (1, 2) position. ∎

Partitioning is not unique, and may be chosen for convenience. However, in a product, any partitioning must be done so that the submatrices remain compatible.

EXAMPLE A-7

Let A of the previous example be postmultiplied by B, given as

$$B = \begin{bmatrix} -3 \\ 1 \\ -1 \end{bmatrix}$$

Then

$$AB = \begin{bmatrix} 2 & 0 & 1 \\ -1 & 3 & 2 \end{bmatrix} \begin{bmatrix} -3 \\ 1 \\ -1 \end{bmatrix} = \begin{bmatrix} -7 \\ 4 \end{bmatrix}$$

Now, if A is partitioned as before (two columns), then B *must* be partitioned to have two rows, that is

$$B = \begin{bmatrix} -3 \\ 1 \\ \hline -1 \end{bmatrix} = \begin{bmatrix} B_{11} \\ B_{21} \end{bmatrix}$$

and the product AB is, in partitioned form,

$$AB = [A_{11} \mathrel{\vdots} A_{12}] \begin{bmatrix} B_{11} \\ B_{21} \end{bmatrix} = A_{11}B_{11} + A_{12}B_{21}$$

$$= \begin{bmatrix} 2 & 0 \\ -1 & 3 \end{bmatrix} \begin{bmatrix} -3 \\ 1 \end{bmatrix} + \begin{bmatrix} 1 \\ 2 \end{bmatrix}[-1] = \begin{bmatrix} -7 \\ 4 \end{bmatrix}$$

as before. ∎

The *inverse* of a square matrix A is denoted by A^{-1} and satisfies the relation

$$AA^{-1} = A^{-1}A = U \tag{A-26}$$

(Just as a mnemonic, remember that for a scalar a we have $aa^{-1} = a^{-1}a = 1$, provided $a \neq 0$). If A has an inverse, we call A *nonsingular*. Otherwise, matrix A is *singular*. The inverse of A exists if and only if its determinant is nonzero

$$\det A \neq 0 \tag{A-27}$$

The determinant of a square matrix can be evaluated by familiar methods to be reviewed soon. It is important to remember that a matrix has no value—it is just an array of elements. The determinant of a square matrix, on the other hand, has a value.

EXAMPLE A-8

The matrix

$$A = \begin{bmatrix} 1 & -1 \\ 2 & 4 \end{bmatrix}$$

is nonsingular because its determinant is nonzero

$$\det A = \begin{vmatrix} 1 & -1 \\ 2 & 4 \end{vmatrix} = (1)(4) - (-1)(2) = 6$$

Therefore, A has an inverse (yet to be found).

The matrix

$$\mathbf{B} = \begin{bmatrix} 2 & -1 \\ -4 & 2 \end{bmatrix}$$

is singular and has no inverse because

$$\det \mathbf{B} = \begin{vmatrix} 2 & -1 \\ -4 & 2 \end{vmatrix} = (2)(2) - (-1)(-4) = 0$$

The standard notation for a determinant is two vertical lines (square brackets are for a matrix). ∎

A-3 DETERMINANTS

The determinant of a (1×1) matrix is the element itself, that is, if

$$\mathbf{A} = [a] \tag{A-28}$$

then

$$\det \mathbf{A} = a \tag{A-29}$$

The determinant of a (2×2) matrix

$$\mathbf{A} = \begin{bmatrix} a_{11} & a_{12} \\ a_{21} & a_{22} \end{bmatrix} \tag{A-30}$$

is given by

$$\det \mathbf{A} = a_{11}a_{22} - a_{12}a_{21} \tag{A-31}$$

and it follows the general procedure for evaluating a determinant of order $(n \times n)$, to be discussed below.

The determinant of a submatrix of $\mathbf{A}$, obtained by deleting from $\mathbf{A}$ the ith row and the jth column, is called the *minor* of the element a_{ij} or, briefly, the ijth minor; it is denoted by m_{ij}.

EXAMPLE A-9 _____

In Example A-8, the four minors in A are

$$m_{11} = 4 \qquad m_{12} = 2 \qquad m_{21} = -1 \qquad m_{22} = 1$$

∎

The *cofactor* c_{ij} is a "signed" minor given by

$$c_{ij} = (-1)^{i+j} m_{ij} \tag{A-32}$$

EXAMPLE A-10 _____

The four cofactors in Example A-9 are

$$c_{11} = (-1)^{1+1} 4 = 4 \qquad c_{12} = (-1)^{1+2} 2 = -2$$
$$c_{21} = (-1)^{2+1}(-1) = 1 \qquad c_{22} = (-1)^{2+2} 1 = 1$$

∎

The recursive rule for evaluating the determinant of a square $(n \times n)$ matrix $\mathbf{A}$ is

$$\det \mathbf{A} = \sum_{j=1}^{n} a_{ij} c_{ij} \qquad \text{any } i = 1, 2, \ldots, n \tag{A-33}$$

This is called the *expansion* of the determinant *about the ith row*. Alternately, we can *expand* the determinant of **A** *about the jth column* to get

$$\det \mathbf{A} = \sum_{i=1}^{n} a_{ij} c_{ij} \qquad \text{any } j = 1, 2, \dots, n \qquad \text{(A-34)}$$

EXAMPLE A-11 ——

Expand det **A** from Example A-8 about the second row, according to Equation A-33. With $i = 2$, we have there

$$\det \mathbf{A} = 2c_{21} + 4c_{22} = 2(-1)^{2+1}(-1) + 4(-1)^{2+2}(1) = 6$$

Or, if we want, we may expand about the first column, with $j = 1$ in Equation A-34, as follows:

$$\det \mathbf{A} = 1c_{11} + 2c_{21} = 1(-1)^{1+1}(4) + 2(-1)^{2+1}(-1) = 6 \qquad \blacksquare$$

EXAMPLE A-12 ——

Evaluate

$$\det \mathbf{A} = \begin{vmatrix} 1 & -2 & 3 \\ 0 & 4 & -2 \\ 6 & -1 & -1 \end{vmatrix}$$

Since the expansion can be done about any one row or one column, it is convenient to choose a row or a column with as many zeros in it as possible. Here the first column has a zero, and expanding about this column yields

$$\det \mathbf{A} = 1c_{11} + 0c_{21} + 6c_{31} = 1(-1)^{1+1} \begin{vmatrix} 4 & -2 \\ -1 & -1 \end{vmatrix}$$

$$+ 6(-1)^{3+1} \begin{vmatrix} -2 & 3 \\ 4 & -2 \end{vmatrix} = -6 + 6(-8) = -54 \qquad \blacksquare$$

The expansion of an $(n \times n)$ determinant (Equation A-33 or A-34) is in terms of $(n-1) \times (n-1)$ determinants; therefore, it is recursive, as illustrated in Example A-12 above. Some useful properties of determinants are:

1. $$\det \mathbf{A}^T = \det \mathbf{A} \qquad \text{(A-35)}$$

2. If all the elements of a row (or of a column) in **A** are zero then

$$\det \mathbf{A} = 0 \qquad \text{(A-36)}$$

and matrix **A** is singular then.

3. If a *single* row (or column) of **A** is multiplied by a scalar α, then the determinant of the new matrix is multiplied by α. This is distinct from multiplying the matrix **A** by α, in accordance with Equation A-17. In fact, a matrix multiplied by a scalar has *every* element multiplied by this scalar. Consequently, we have

$$\det (\alpha \mathbf{A}) = \alpha^n \det \mathbf{A} \qquad \text{(A-37)}$$

4. The value det **A** remains unchanged if a *k*th row is multiplied by α, then added to another row.

EXAMPLE A-13 ——

In Example A-12, multiply the first row by -6 and add to the third row. The result is

$$\det \mathbf{A} = \begin{vmatrix} 1 & -2 & 3 \\ 0 & 4 & -2 \\ 0 & 11 & -19 \end{vmatrix} = 1 \begin{vmatrix} 4 & -2 \\ 11 & -19 \end{vmatrix} = -54 \qquad \blacksquare$$

This property is useful in generating zeros in a particular row in order to expand the determinant about that row later. The same property holds if "column" is substituted for "row."

A-4 CRAMER'S RULE

Consider the system of n linear algebraic equations with n unknowns

$$
\begin{aligned}
a_{11}x_1 + a_{12}x_2 + \cdots + a_{1n}x_n &= b_1 \\
a_{21}x_1 + a_{22}x_2 + \cdots + a_{2n}x_n &= b_2 \\
&\cdots\cdots\cdots\cdots\cdots\cdots\cdots\cdots \\
a_{n1}x_1 + a_{n2}x_2 + \cdots + a_{nn}x_n &= b_n
\end{aligned}
\tag{A-38}
$$

or, in matrix form,

$$
\mathbf{Ax} = \mathbf{B} \tag{A-39}
$$

Example A-3 is such a system with $n = 3$. Such a system has a solution provided det $\mathbf{A} \neq 0$, that is, if matrix $\mathbf{A}$ is nonsingular. Then any unknown x_k is found as

$$
x_k = \frac{N_k}{\det \mathbf{A}} \qquad k = 1, 2, \ldots, n \tag{A-40}
$$

where N_k is the determinant of $\mathbf{A}$ with the kth column replaced by the column matrix $\mathbf{B}$. Equation A-40 is commonly called *Cramer's rule* for solving the system.

EXAMPLE A-14 _____

The solution for the system

$$
\begin{bmatrix} 2 & 3 & -4 \\ 4 & -1 & 3 \\ 3 & 2 & 1 \end{bmatrix}
\begin{bmatrix} x_1 \\ x_2 \\ x_3 \end{bmatrix}
=
\begin{bmatrix} 12 \\ -1 \\ 6 \end{bmatrix}
$$

is, according to Cramer's rule,

$$
x_1 = \frac{\begin{vmatrix} 12 & 3 & -4 \\ -1 & -1 & 3 \\ 6 & 2 & 1 \end{vmatrix}}{\begin{vmatrix} 2 & 3 & -4 \\ 4 & -1 & 3 \\ 3 & 2 & 1 \end{vmatrix}} = 1
$$

$$
x_2 = \frac{\begin{vmatrix} 2 & 12 & -4 \\ 4 & -1 & 3 \\ 3 & 6 & 1 \end{vmatrix}}{\begin{vmatrix} 2 & 3 & -4 \\ 4 & -1 & 3 \\ 3 & 2 & 1 \end{vmatrix}} = 2
$$

and

$$x_3 = \frac{\begin{vmatrix} 2 & 3 & 12 \\ 4 & -1 & -1 \\ 3 & 2 & 6 \end{vmatrix}}{\begin{vmatrix} 2 & 3 & -4 \\ 4 & -1 & 3 \\ 3 & 2 & 1 \end{vmatrix}} = -1$$

Here we have shown the column matrix **B** in dotted lines just for emphasis. ■

Note that in Cramer's rule the denominator is the same for all the unknowns and needs to be evaluated only once. The numerators are different and unique for each unknown.

EXAMPLE A-15 ───────────────────────────────

Solve the system

$$\begin{bmatrix} s + 2 + \dfrac{1}{s} & -\dfrac{1}{s} \\ -\dfrac{1}{s} & 4 + \dfrac{1}{s} \end{bmatrix} \begin{bmatrix} I_1 \\ I_2 \end{bmatrix} = \begin{bmatrix} -10 \\ 0 \end{bmatrix}$$

Solution. We have

$$I_1 = \frac{\begin{vmatrix} -10 & -\dfrac{1}{s} \\ 0 & 4 + \dfrac{1}{s} \end{vmatrix}}{\begin{vmatrix} s + 2 + \dfrac{1}{s} & -\dfrac{1}{s} \\ -\dfrac{1}{s} & 4 + \dfrac{1}{s} \end{vmatrix}} = \frac{-40s - 10}{4s^2 + 9s + 6}$$

$$I_2 = \frac{\begin{vmatrix} s + 2 + \dfrac{1}{s} & -10 \\ -\dfrac{1}{s} & 0 \end{vmatrix}}{4s^2 + 9s + 6} = \frac{-10}{4s^2 + 9s + 6}$$ ■

The denominator in Cramer's rule, det **A**, is called the *characteristic determinant* if it is a number, as in Example A-14; it is the *characteristic polynomial* if it is a polynomial in some parameter, as in Example A-15.

EXAMPLE A-16 ───────────────────────────────

The system

$$\begin{bmatrix} R & -R \\ -R & R \end{bmatrix} \begin{bmatrix} i_1 \\ i_2 \end{bmatrix} = \begin{bmatrix} v_1 \\ v_2 \end{bmatrix}$$

cannot be solved for the i's in terms of the v's because the coefficient matrix

$$\mathbf{R} = \begin{bmatrix} R & -R \\ -R & R \end{bmatrix}$$

is singular, det $\mathbf{R} = 0$. We also say then that the original system of equations cannot be inverted.   ■

A-5 MATRIX INVERSION

With Equation A-39, let us write the relationship that defines the inverse of $\mathbf{A}$, $\mathbf{A}^{-1}$, that is

$$\mathbf{AA}^{-1} = \mathbf{U} \tag{A-41}$$

and consider it as n systems of n linear equations, with the unknowns being the elements of $\mathbf{A}^{-1}$. Cramer's rule yields then

$$p_{ij} = \frac{c_{ji}}{\det \mathbf{A}} \tag{A-42}$$

where p_{ij} is the ijth element of $\mathbf{A}^{-1}$ and c_{ji} is the cofactor of a_{ji} in the matrix $\mathbf{A}$. Note the reversed order of the subscripts in c_{ji}. In other words, we form, first, the *adjoint matrix* of $\mathbf{A}$ as follows

$$\text{Adj } \mathbf{A} = [c_{ij}]^T \tag{A-43}$$

that is, every element of $\mathbf{A}$ is replaced by its cofactor, and the resulting matrix is transposed. Then the inverse of $\mathbf{A}$ is given by

$$\mathbf{A}^{-1} = \frac{1}{\det \mathbf{A}} \text{Adj } \mathbf{A} \tag{A-44}$$

and this inverse exists only if det $\mathbf{A} \neq 0$.

EXAMPLE A-17 ————————————————————————————

For the (2×2) matrix

$$\mathbf{A} = \begin{bmatrix} a_{11} & a_{12} \\ a_{21} & a_{22} \end{bmatrix}$$

and provided det $\mathbf{A} \neq 0$, the inverse is

$$\mathbf{A}^{-1} = \frac{1}{a_{11}a_{22} - a_{12}a_{21}} \begin{bmatrix} a_{22} & -a_{12} \\ -a_{21} & a_{11} \end{bmatrix} \qquad ■$$

With $\mathbf{A}^{-1}$ found, the system of equations

$$\mathbf{Ax} = \mathbf{B} \tag{A-39}$$

can be solved (inverted) formally: We premultiply Equation A-39 by $\mathbf{A}^{-1}$

$$\mathbf{A}^{-1}\mathbf{Ax} = \mathbf{A}^{-1}\mathbf{B} \tag{A-45}$$

that is,

$$\mathbf{Ux} = \mathbf{A}^{-1}\mathbf{B} \tag{A-46}$$

or, finally,

$$\mathbf{x} = \mathbf{A}^{-1}\mathbf{B} \tag{A-47}$$

EXAMPLE A-18 _____

For the matrix **A** in Example A-14, we have

$$c_{11} = \begin{vmatrix} -1 & 3 \\ 2 & 1 \end{vmatrix} = -7 \qquad c_{12} = -\begin{vmatrix} 4 & 3 \\ 3 & 1 \end{vmatrix} = 5 \qquad c_{13} = \begin{vmatrix} 4 & -1 \\ 3 & 2 \end{vmatrix} = 11$$

$$c_{21} = -\begin{vmatrix} 3 & -4 \\ 2 & 1 \end{vmatrix} = -11 \qquad c_{22} = \begin{vmatrix} 2 & -4 \\ 3 & 1 \end{vmatrix} = 14 \qquad c_{23} = -\begin{vmatrix} 2 & 3 \\ 3 & 2 \end{vmatrix} = 5$$

$$c_{31} = \begin{vmatrix} 3 & -4 \\ -1 & 3 \end{vmatrix} = 5 \qquad c_{32} = -\begin{vmatrix} 2 & -4 \\ 4 & 3 \end{vmatrix} = -22 \qquad c_{33} = \begin{vmatrix} 2 & 3 \\ 4 & -1 \end{vmatrix} = -14$$

and

$$\det \mathbf{A} = -43$$

Therefore, the inverse of **A** is

$$\mathbf{A}^{-1} = \frac{1}{-43} \begin{bmatrix} -7 & -11 & 5 \\ 5 & 14 & -22 \\ 11 & 5 & -14 \end{bmatrix} = \begin{bmatrix} \dfrac{7}{43} & \dfrac{11}{43} & \dfrac{-5}{43} \\ \dfrac{-5}{43} & \dfrac{-14}{43} & \dfrac{22}{43} \\ \dfrac{-11}{43} & \dfrac{-5}{43} & \dfrac{14}{43} \end{bmatrix}$$

As a check, we calculate

$$\mathbf{A}^{-1}\mathbf{A} = \begin{bmatrix} \dfrac{7}{43} & \dfrac{11}{43} & \dfrac{-5}{43} \\ \dfrac{-5}{43} & \dfrac{-14}{43} & \dfrac{22}{43} \\ \dfrac{-11}{43} & \dfrac{-5}{43} & \dfrac{14}{43} \end{bmatrix} \begin{bmatrix} 2 & 3 & -4 \\ 4 & -1 & 3 \\ 3 & 2 & 1 \end{bmatrix} = \begin{bmatrix} 1 & 0 & 0 \\ 0 & 1 & 0 \\ 0 & 0 & 1 \end{bmatrix} = \mathbf{U}$$

The solution to Example A-14 by matrix inversion is

$$\mathbf{x} = \begin{bmatrix} x_1 \\ x_2 \\ x_3 \end{bmatrix} = \mathbf{A}^{-1}\mathbf{B} = \begin{bmatrix} \dfrac{7}{43} & \dfrac{11}{43} & \dfrac{-5}{43} \\ \dfrac{-5}{43} & \dfrac{-14}{43} & \dfrac{22}{43} \\ \dfrac{-11}{43} & \dfrac{-5}{43} & \dfrac{14}{43} \end{bmatrix} \begin{bmatrix} 12 \\ -1 \\ 6 \end{bmatrix} = \begin{bmatrix} 1 \\ 2 \\ -1 \end{bmatrix}$$

as before. ∎

The two methods, Cramer's rule and matrix inversion, are totally equivalent for the solution of $\mathbf{Ax} = \mathbf{B}$. Your particular choice may depend on taste, dexterity, and availability of computing facilities.

A useful property of the inverse is the following: Let **A** and **B** be nonsingular, each of order $(n \times n)$. Then

$$(\mathbf{AB})^{-1} = \mathbf{B}^{-1}\mathbf{A}^{-1} \tag{A-48}$$

that is, the inverse of a product is the product of the inverses in opposite order. The proof is quite simple, because

$$(\mathbf{AB})(\mathbf{B}^{-1}\mathbf{A}^{-1}) = \mathbf{A}(\mathbf{BB}^{-1})\mathbf{A}^{-1} = \mathbf{AUA}^{-1} = \mathbf{AA}^{-1} = \mathbf{U} \qquad (A\text{-}49)$$

showing that the inverse of $\mathbf{AB}$ is, indeed, $\mathbf{B}^{-1}\mathbf{A}^{-1}$.

A-6 GAUSS–JORDAN ELIMINATION

Matrix inversion or Cramer's rule, while totally correct, are not very efficient for solving large systems. Instead, we resort to any one of several equivalent methods.

An effective algorithm to solve numerically the system $\mathbf{Ax} = \mathbf{B}$ is based on the properties of determinants studied earlier. It can be summarized as follows: Use suitable multiples of the kth equation and add them to the $(k + 1)$st, $(k + 2)$nd, ..., nth equation in order to eliminate x_k from them. Do so in n cycles, starting with $k = 1$ and ending with $k = n$. An example will illustrate it.

EXAMPLE A-19 —————————————————————————————————————

Take again the system in Example A-14.

$$2x_1 + 3x_2 - 4x_3 = 12$$
$$4x_1 - x_2 + 3x_3 = -1$$
$$3x_1 + 2x_2 + x_3 = 6$$

and perform on it the following cycles:

First cycle ($k = 1$): Divide the first equation by 2, getting

$$x_1 + 1.5x_2 - 2x_3 = 6$$

and call it the "slave equation" for this cycle, because it will do all the work. Multiply it by -4 and add to the second equation, to get

$$0 \cdot x_1 - 7x_2 + 11x_3 = -25$$

the idea being to eliminate x_1 from the second equation. Next, multiply the slave equation by -3 and add to the third one, to eliminate x_1 from it:

$$0 \cdot x_1 - 2.5x_2 + 7x_3 = -12$$

If there were more equations, we'd continue in this fashion to eliminate x_1 from each one, with the help of the slave equation. At the end of the first cycle, then, we have the new equivalent system

$$x_1 + 1.5x_2 - 2x_3 = 6$$
$$- 7x_2 + 11x_3 = -25$$
$$- 2.5x_2 + 7x_3 = -12$$

Second cycle ($k = 2$): Divide the (new) second equation by -7 to get the second slave equation

$$x_2 - 1.571x_3 = 3.571$$

and use it to eliminate x_2 from all the next equations, here the third one. Specifically, multiply this slave equation by 2.5 and add to the third equation of the first cycle to get

$$3.073x_3 = -3.073$$

If there were more equations, we'd continue to eliminate x_2 from each one, using the slave equation of this cycle.

Third cycle ($k = 3$): Divide the new third equation by 3.073, to get the new slave equation for this cycle

$$x_3 = -1$$

which is here the answer for x_3.

Now use back substitution. With x_3 known, go back to the slave equation of cycle 2, to get x_2

$$x_2 - 1.571(-1) = 3.571$$
$$\therefore \quad x_2 = 2$$

With x_3 and x_2 known, go to the first slave equation to get x_1

$$x_1 + (1.5)(2) - 2(-1) = 6$$
$$\therefore \quad x_1 = 1 \qquad\qquad ■$$

The essence of this method, called the *Gauss elimination*, is to reduce the original system of equations

$$\mathbf{Ax = B} \tag{A-50}$$

to an equivalent system of equations, having the same solution, but of the form

$$\mathbf{Tx = \hat{B}} \tag{A-51}$$

where $\mathbf{T}$ is a *triangular* matrix, that is

$$\mathbf{T} = \begin{bmatrix} 1 & t_{12} & t_{13} & \cdots & t_{1n} \\ 0 & 1 & t_{23} & \cdots & t_{2n} \\ 0 & 0 & 1 & \cdots & t_{3n} \\ & \cdots & & \cdots & \\ 0 & 0 & 0 & \cdots & 1 \end{bmatrix} \tag{A-52}$$

Here, in the kth row, the kth element is 1, being the coefficient of x_k in the kth slave equation. The t's are the results of the various operations done on the original a's. Similarly, the new matrix $\hat{\mathbf{B}}$ is the result of these operations on the original b's in $\mathbf{B}$.

A variation on this approach is the *Gauss–Jordan elimination* method, where the kth slave equation is used to eliminate x_k from *all* the other $(n - 1)$ equations, not just from the equations below the kth one. Here the final result is

$$\begin{bmatrix} 1 & 0 & 0 & \cdots & 0 \\ 0 & 1 & 0 & \cdots & 0 \\ 0 & 0 & 1 & \cdots & 0 \\ & \cdots & & \cdots & \\ 0 & 0 & 0 & \cdots & 1 \end{bmatrix} \mathbf{x} = \hat{\mathbf{B}} \tag{A-53}$$

that is,

$$\mathbf{Ux = x = \hat{B}} \tag{A-54}$$

and all the solutions are immediately ready without the need of back substitutions, $x_k = \hat{b}_k$.

Appendix B

Complex Numbers

B-1 RECTANGULAR, EXPONENTIAL, AND POLAR FORMS

While the origin of imaginary numbers was in the search of solutions to equations like

$$x^2 + 1 = 0 \tag{B-1}$$

they soon found wide applications in other disciplines, including engineering in general and electrical network analysis in particular.

We denote the solution to Equation B-1 by j, where†

$$j^2 = -1 \qquad j = \sqrt{-1} \tag{B-2}$$

A *complex number* is given in general as

$$c = a + jb \tag{B-3}$$

where a is the real part of c, and written as

$$\mathrm{Re}\,\{c\} = a \tag{B-4}$$

and b is the imaginary part of c,

$$\mathrm{Im}\,\{c\} = b \tag{B-5}$$

It is important to recognize that the imaginary part of c is b, not jb. In other words, the j in jb simply identifies b as the imaginary (jmaginary?) part.

The expression in Equation B-3 is known as the *rectangular form* of the complex number c, and is shown in Figure B-1. The complex plane has the real axis and the imaginary axis. Along the real axis, we go a units, and along the imaginary axis, b units. The coordinates (a, b) define the complex number c. Alternately, the directed line from the origin to c is the complex number c.

† In mathematics, the letter i is used. To avoid confusion with the notation for current, electrical engineers (and others) use the letter j.

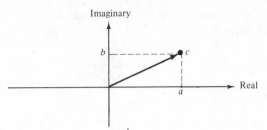

Figure B-1   Rectangular form.

The geometry of c in the complex plane is extremely useful—and easy. We cannot overemphasize the importance of such a quick sketch; it can save serious mistakes.

An alternate way to specify the complex number c is by its distance r from the origin and by the angle θ formed by r with the positive real axis. See Figure B-2. The angle θ is considered positive in the usual mathematical sense, counterclockwise from the positive real axis; it is negative clockwise.

From the geometry of Figures B-1 and B-2 we have

$$r = +\sqrt{a^2 + b^2} \tag{B-6}$$

distance being always a positive number, and

$$\theta = \tan^{-1}\frac{b}{a} \tag{B-7}$$

The number r is also called the *magnitude* of c, denoted as $|c|$. We have also

$$a = r \cos \theta = |c| \cos \theta \tag{B-8}$$
$$b = r \sin \theta = |c| \sin \theta \tag{B-9}$$

As a result, we have

$$c = a + jb = r(\cos \theta + j \sin \theta) \tag{B-10}$$

Because of Euler's identity

$$e^{\pm j\theta} = \cos \theta \pm j \sin \theta \tag{B-11}$$

the complex number in Equation B-10 can be written as

$$c = re^{j\theta} = |c|e^{j\theta} \tag{B-12}$$

which is its *exponential form*. A convenient shorthand notation for the exponential form is

$$c = r\underline{/\theta} \tag{B-12}$$

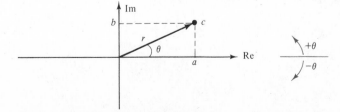

Figure B-2   Exponential and polar forms.

(read as "r at θ degrees"), called the *polar form* of c. We must stress that the polar form is just a notation for the exponential form. All mathematical operations with c will obey the usual rules when c is written in exponential form.

While Equations B-6 to B-10 are nice and simple, *don't memorize them*! Better yet, don't use them blindly. The reason? Those familiar quadrants in the plane. As shown in Figure B-3, the angle θ falls in one of four quadrants, depending on the individual signs of a and b. *Always draw a picture*!

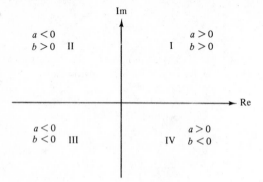

Figure B-3   Quandrants and tangents.

EXAMPLE B-1 _____

Show the following complex numbers in the complex plane, and convert them into their polar form:

$$
\begin{array}{ll}
(1)\ c_1 = 6 + j8 & (2)\ c_2 = 6 - j8 \\
(3)\ c_3 = -6 + j8 & (4)\ c_4 = -6 - j8
\end{array}
$$

(1) Here

$$r_1 = \sqrt{6^2 + 8^2} = 10$$

$$\theta_1 = \tan^{-1}\frac{8}{6} = 53.13°$$

as shown in Figure B-4 (a). Therefore,

$$c_1 = 10\underline{/53.13°}$$

(2) $r_2 = 10$ and $\theta_2 = \tan^{-1}(\frac{-8}{6}) = -53.13°$, from Figure B-4(b). Therefore,

$$c_2 = 10\underline{/-53.13°}$$

(3) $r_3 = 10$ and $\theta_3 = 180° - \tan^{-1}\frac{8}{6} = 126.87°$, from Figure B-4(c). Therefore,

$$c_3 = 10\underline{/126.87°}$$

(4) $r_4 = 10$ and $\theta_4 = -180° + \tan^{-1}\frac{8}{6} = -126.87°$

$$\therefore\quad c_4 = 10\underline{/-126.87°}$$

The point worth repeating is: A simple sketch on paper or in your mind's eyes is essential. It gives you the assurance that your calculator is right—don't *ever* relegate your thinking to it!

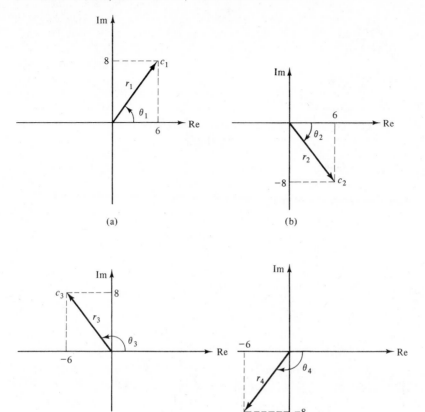

Figure B-4 Example B-1.   ■

EXAMPLE B-2

Without any calculator, decide if the following conversions are reasonable. The left side is given correctly:

(a) $$3\underline{/-35°} \overset{?}{=} 2.46 + j1.72$$

(b) $$-4 + j6 \overset{?}{=} 7.2\underline{/56.3°}$$

(c) $$10\underline{/214°} \overset{?}{=} -5.6 - j8.3$$

(d) $$3\underline{/78°} \overset{?}{=} 1.1 + j5.175$$

In (a), the given angle is in the fourth quadrant. There, the imaginary part must be negative. Therefore, the rectangular form is wrong. (In fact, the correct one is $2.46 - j1.72$.)

In (b), the given rectangular form is in the second quadrant, where $90° < \theta < 180°$. Obviously, while the magnitude r seems reasonable, the angle is wrong. (The correct calculated angle is $180° - 56.3° = 123.7°$.)

In (c), the angle puts the complex number in the third quadrant. Both real and imaginary parts must be negative, as they are. However, a little thought (still no calculator, just a picture!) tells us that

$$\tan^{-1}\frac{8.3}{5.6} > 45°$$

since $\tan^{-1} 1 = 45°$. But the given angle shows that we have $214° - 180° = 34°$. So the answer in rectangular form is wrong. The correct answer is $-8.3 - j5.6$, with the real and imaginary parts switched—a common mistake, sometimes even with a calculator. Here, we were able to spot it by inspection.

In (d), the imaginary part is greater than the magnitude r—an impossibility according to Pythagoras. [The angle just happens to be correct, since $78° = \tan^{-1}(5.175/1.1)$.] ■

EXAMPLE B-3 ——————————————————————————————

Find the rectangular form of the complex number

$$c = 2/\tan^{-1}(-0.7)$$

This is an ill-posed problem, because the angle is ambiguous. It can be in the second quadrant

$$\tan^{-1} \frac{0.7}{-1} = 180° - 35° = 145°$$

or in the fourth quadrant

$$\tan^{-1} \frac{-0.7}{1} = -35°$$ ■

The *conjugate* of a complex number $c = a + jb = re^{j\theta}$ is

$$c^* = a - jb = re^{-j\theta} \tag{B-13}$$

and is formed by changing the sign of the imaginary part of c, or by changing the sign of the angle θ of c. In graphical form, c^* is the reflection of c about the real axis. See Figure B-5.

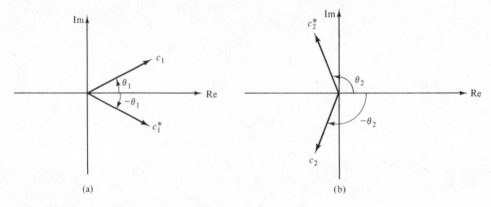

Figure B-5 Complex conjugate numbers.

Important relations among c and c^* are

$$c + c^* = 2a = 2\,\text{Re}\,\{c\} = 2\,\text{Re}\,\{c^*\} \tag{B-14}$$
$$c - c^* = j2b = j2\,\text{Im}\,\{c\} = -j2\,\text{Im}\,\{c^*\} \tag{B-15}$$

The magnitude of c is the same as of c^*

$$|c| = |c^*| = r \tag{B-16}$$

and

$$cc^* = r^2 \tag{B-17}$$

These equations are not to be memorized. When needed, derive them quickly with the help of a sketch.

In particular, if $c = 1e^{j\theta} = 1\underline{/\theta}$, then Equation B-14 yields

$$\cos \theta = \tfrac{1}{2}(e^{j\theta} + e^{-j\theta}) \tag{B-18}$$

and Equation B-15 gives

$$\sin \theta = \frac{1}{2j}(e^{j\theta} - e^{-j\theta}) \tag{B-19}$$

These two relations are sometimes also called *Euler's equalities*, together with Equation B-11.

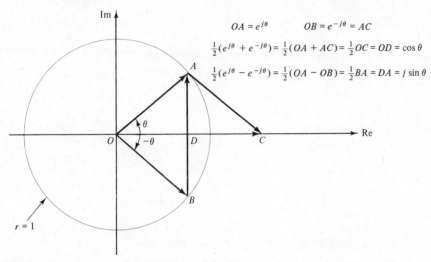

$$OA = e^{j\theta} \qquad OB = e^{-j\theta} = AC$$

$$\tfrac{1}{2}(e^{j\theta} + e^{-j\theta}) = \tfrac{1}{2}(OA + AC) = \tfrac{1}{2}OC = OD = \cos \theta$$

$$\tfrac{1}{2}(e^{j\theta} - e^{-j\theta}) = \tfrac{1}{2}(OA - OB) = \tfrac{1}{2}BA = DA = j \sin \theta$$

Figure B-6 Geometry of Euler's equalities.

B-2 MATHEMATICAL OPERATIONS

Two complex numbers are equal if, and only if, their real parts are equal *and* their imaginary parts are equal; that is, given $c_1 = a_1 + jb_1$ and $c_2 = a_2 + jb_2$, then the equality

$$c_1 = c_2 \tag{B-20}$$

implies

$$a_1 = a_2 \quad \text{and} \quad b_1 = b_2 \tag{B-21}$$

Alternately, in polar form, the equality of complex numbers means that their magnitudes are equal *and* their angles are equal; that is, with

$$c_1 = r_1 e^{j\theta_1} \qquad c_2 = r_2 e^{j\theta_2} \tag{B-22}$$

the equality of Equation B-20 implies

$$r_1 = r_2 \quad \text{and} \quad \theta_1 = \theta_2 \tag{B-23}$$

Addition (or subtraction) of complex numbers *must* be done in rectangular form. Then, if $c_1 = a_1 + jb_1$ and $c_2 = a_2 + jb_2$, their sum is

$$c_1 + c_2 = (a_1 + a_2) + j(b_1 + b_2) \tag{B-24}$$

and their difference is

$$c_1 - c_2 = (a_1 - a_2) + j(b_1 - b_2) \tag{B-25}$$

In words, real parts are added (or subtracted) to yield the resulting real part, and imaginary parts are added (subtracted) to yield the resulting imaginary part.

EXAMPLE B-4 ————————————————————————————————

Given

$$c_1 = 2.2\underline{/20°}$$
$$c_2 = 1.29 - j1.25$$
$$c_3 = 0.6\underline{/-110°}$$

Calculate

$$c_4 = c_1 - c_2 + c_3$$

First, we must convert c_1 and c_3 into their rectangular form. We get (check it!)

$$c_1 = 2.07 + j0.75$$
$$c_3 = -0.21 - j0.56$$

We have then

$$c_4 = 2.07 + j0.75 - (1.29 - j1.25) + (-0.21 - j0.56) = 0.57 + j1.44$$

or, when converted into its polar form,

$$c_4 = 1.55\underline{/68.4°}$$ ■

The graphical methods of addition and subtraction are simple, quick, and reasonably accurate with the help of only a ruler (to measure lengths) and a protractor (to measure angles). These methods are invaluable in many areas of electrical engineering.

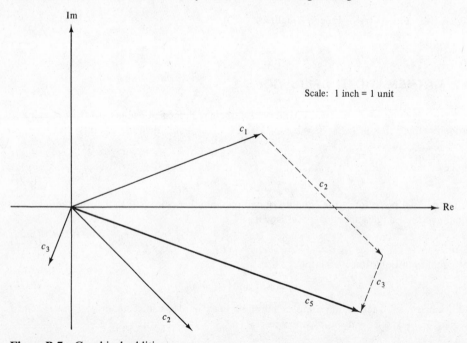

Figure B-7   Graphical addition.

To add graphically complex numbers, we follow the rule of "head-to-tail": From the tip of the first complex number (the head of the arrow), draw the second complex number; from the tip of that one, draw the third one, etc. Their sum is the complex number whose tail is at the tail of the first one and whose head is at the head of the last one. In Figure B-7, we show this construction for $c_5 = c_1 + c_2 + c_3$. From it, with a ruler and a protractor, we read directly, $c_5 = 3.3\underline{/-18.5°}$. This method is recognized as a simple extension of the parallelogram addition of vectors in mechanics and physics.

Graphical subtraction of two complex numbers follows a similar procedure if we recognize that

$$c_6 - c_7 = c_6 + (-c_7) \tag{B-26}$$

Therefore, to subtract c_7 from c_6, we reverse c_7 and add to c_6 by the head-to-tail method. This result is shown in Figure B-8(a). From it, we see that $(c_6 - c_7)$ is a complex number directed from c_7 to c_6, as shown in the simplified construction of Figure B-8(b).

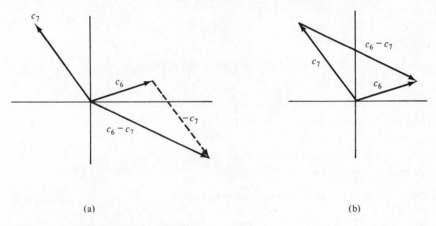

(a)   (b)

Figure B-8   Graphical subtraction.

Multiplication or division of complex numbers may be done either in rectangular or in polar form, unlike addition or subtraction, which must be done in rectangular form. For multiplication (or division), the polar form is much easier.

EXAMPLE B-5 ────────────────────────────────

In the previous example, calculate the product $c_2 c_3$.

In polar form, we have

$$c_2 c_3 = (1.8\underline{/-44°})(0.6\underline{/-110°})$$

Remember that this notation really stands for the exponential form. The usual rules of exponents in multiplication yield

$$c_2 c_3 = (1.8e^{-j44°})(0.6e^{-j110°}) = (1.8)(0.6)e^{j(-44°-110°)} = 1.08e^{-j154°}$$

that is,

$$c_2 c_3 = 1.08\underline{/-154°}$$

which, when converted into rectangular form, gives

$$c_2 c_3 = -0.97 - j0.47$$

Alternately, the product can be done in rectangular form as

$$c_2 c_3 = (1.29 - j1.25)(-0.21 - j0.56) = (1.29)(-0.21) + (1.29)(-j0.56)$$
$$+ (j1.25)(0.21) - (1.25)(0.56) = -0.97 - j0.46$$

correct to two significant figures. Here we used the equality $(-j)(-j) = -1$, a common one in such calculations. ∎

We see that, in polar form, magnitudes are *multiplied* to give the resulting magnitude, while angles are *added* to give the resulting angle. In polar and exponential form

$$(r_1 \underline{/\theta_1})(r_2 \underline{/\theta_2}) = r_1 e^{j\theta_1} r_2 e^{j\theta_2} = r_1 r_2 e^{j(\theta_1 + \theta_2)} = r_1 r_2 \underline{/\theta_1 + \theta_2} \qquad \text{(B-27)}$$

Division in polar form is equally easy. Here, again, we use the usual rules of exponents

$$\frac{r_1 \underline{/\theta_1}}{r_2 \underline{/\theta_2}} = \frac{r_1 e^{j\theta_1}}{r_2 e^{j\theta_2}} = \frac{r_1}{r_2} e^{j(\theta_1 - \theta_2)} = \frac{r_1}{r_2} \underline{/\theta_1 - \theta_2} \qquad \text{(B-28)}$$

showing that we must *divide* the magnitudes and *subtract* the angles.

EXAMPLE B-6 _____

From the previous example, calculate c_2/c_3. In polar form we write

$$\frac{c_2}{c_3} = \frac{1.8 \underline{/-44°}}{0.6 \underline{/-110°}} = 3 \underline{/66°} = 1.22 + j2.74$$

where the answer in polar form was converted into its rectangular form in the last step. ∎

Division in rectangular form requires, first, that we multiply numerator and denominator by the conjugate of the denominator—a valid operation, multiplying a given ratio by 1. Thus

$$\frac{c_2}{c_3} = \frac{c_2}{c_3} \cdot \frac{c_3^*}{c_3^*} = \frac{c_2 c_3^*}{|c_3|^2} \qquad \text{(B-29)}$$

The idea here is to get a *real number*, $|c_3|^2$, in the denominator of the answer, as illustrated in the following example.

EXAMPLE B-7 _____

From Example B-4, we use this approach to get

$$\frac{c_2}{c_3} = \frac{1.29 - j1.25}{-0.21 - j0.56} = \frac{1.29 - j1.25}{-0.21 - j0.56} \cdot \frac{-0.21 + j0.56}{-0.21 + j0.56}$$

$$= \frac{(1.29)(-0.21) + (1.29)(j0.56) + (j1.25)(0.21) + (1.25)(0.56)}{(-0.21)^2 + (0.56)^2}$$

$$= \frac{0.43 + j0.98}{0.36} = 1.21 + j2.73$$

as before. ∎

A few useful relations here are listed now:†

$$\frac{1}{j} = -j \qquad \text{(B-30)}$$

$$(-j)(+j) = 1 \qquad \text{(B-31)}$$

$$e^{j\pi/2} = j \qquad \text{(B-32)}$$

$$e^{-j\pi/2} = -j \qquad \text{(B-33)}$$

$$e^{\pm j\pi} = -1\ddagger \qquad \text{(B-34)}$$

$$e^{\pm j2\pi} = e^{\pm j4\pi} = \cdots = e^{\pm j2k\pi} = 1 \qquad k = 0, 1, 2, 3, \ldots \qquad \text{(B-35)}$$

To raise a complex number to an integer power, use the polar form

$$c^k = (re^{j\theta})^k = r^k e^{jk\theta} = r^k \underline{/k\theta} \qquad \text{(B-36)}$$

EXAMPLE B-8 _____

Given

$$c = -\frac{1}{2} + j\frac{\sqrt{3}}{2}$$

calculate c^2 and c^{-3}. Give the answers in rectangular form.

We convert c into its polar form

$$c = 1\underline{/120°}$$

Therefore

$$c^2 = (1\underline{/120°})^2 = 1\underline{/240°}$$

In rectangular form, it is

$$1\underline{/240°} = -\frac{1}{2} - j\frac{\sqrt{3}}{2}$$

Similarly,

$$c^{-3} = (1\underline{/120°})^{-3} = 1\underline{/-360°} = 1\underline{/0°}$$

that is,

$$c^{-3} = 1 + j0$$

in rectangular form. (*Note*: This particular complex number, $c = 1\underline{/120°}$, is very useful in the studies of electric power systems.) ∎

The same approach holds for fractional powers, that is, roots of complex numbers. Let us illustrate with an example.

EXAMPLE B-9 _____

Find the five roots of the equation

$$x^5 + 2 = 0$$

† Remember that $j = 0 + j1$.

‡ A fondly remembered professor used to extol the sheer beauty of this equation which combines transcendental numbers (e and π), the unit imaginary number (j), and the unit real number (1) in such an elegant fashion.

We know that there are five roots, in accordance with the fundamental theorem of algebra which states that an nth-order polynomial equation has n roots.

Write the equation as

$$x^5 = -2$$
$$\therefore \quad x = (-2)^{1/5}$$

We express -2 in its exponential form, adding multiples of 360° to the angle:†

$$-2 = 2e^{j180°} = 2e^{j(180° + 360°)} = 2e^{j(180° + 720°)} = \cdots$$

Therefore,

$$x_1 = (2e^{j180°})^{1/5} = 2^{1/5}e^{j36°} = 1.149\underline{/36°}$$
$$x_2 = (2e^{j540°})^{1/5} = 2^{1/5}e^{j108°} = 1.149\underline{/108°}$$
$$x_3 = (2e^{j900°})^{1/5} = 2^{1/5}e^{j180°} = 1.149\underline{/180°} = -1.149$$
$$x_4 = (2e^{j1260°})^{1/5} = 2^{1/5}e^{j252°} = 1.149\underline{/252°}$$
$$x_5 = (2e^{j1620°})^{1/5} = 2^{1/5}e^{j324°} = 1.149\underline{/324°}$$

If we attempt to continue and add another 360°, we get back to x_1

$$x_6 = (2e^{j1980°})^{1/5} = 2^{1/5}e^{j396°} = 2^{1/5}e^{j36°} = x_1$$

These five roots are shown in Figure B-9.

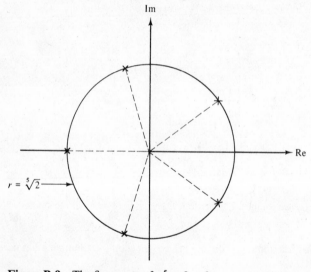

Figure B-9 The five roots of $x^5 + 2 = 0$.

Exponentiation with complex numbers is defined as follows:

$$e^{(a + jb)} = e^a e^{jb} = e^a(\cos b + j \sin b) \tag{B-37}$$

where the last equality follows Euler's identity, Equation B-11.

† *Note:* $\theta = \theta + 360° = \theta + 720° = \theta + 1080° = \cdots$ (in degrees)
    $\theta = \theta + 2\pi = \theta + 4\pi = \theta + 6\pi = \cdots$ (in radians)

EXAMPLE B-10 ──

Calculate the real part of the complex number

$$c = 100e^{-(2+j4)t}$$

where t is time, a real number.

We have

$$c = 100e^{-2t}(\cos 4t + j \sin 4t)$$

and therefore

$$\text{Re}\{c\} = 100e^{-2t} \cos 4t$$

Similarly,

$$\text{Im}\{c\} = 100e^{-2t} \sin 4t$$

Such functions of time are very common in circuit analysis. ∎

EXAMPLE B-11 ──

Just for the fun of it, let us calculate j^j.

From Equation B-32, we have

$$j = e^{j\pi/2} = e^{j5\pi/2} = \cdots = e^{j(\pi/2 + 2k\pi)} \qquad k = 0, \pm 1, \pm 2, \ldots$$

since we can add multiples of 2π to the angle of a complex number. Therefore,

$$j^j = (e^{j\pi/2})^j = e^{j^2\pi/2} = e^{-\pi/2} = 0.20788$$

or

$$j^j = (e^{j5\pi/2})^j = e^{j^25\pi/2} = e^{-5\pi/2} = 3.8820 \times 10^{-4}$$

or, in general, the *real* number

$$j^j = e^{-(\pi/2 + 2k\pi)} \qquad k = 0, \pm 1, \pm 2, \ldots \qquad ∎$$

In summary, complex numbers obey the same laws of algebra as do real numbers. Furthermore, complex numbers reduce to real numbers when their imaginary part vanishes, $\text{Im}\{c\} = 0$.

In conclusion, here are a few practical suggestions for working with complex numbers:

1. Don't memorize.
2. Draw a sketch.
3. Don't rely blindly on your calculator. Always think!
4. Convert every complex number into its other form, rectangular to polar and polar to rectangular. Then you'll have it when needed in either form.

Appendix C

Scaling

"Why are we dealing, throughout this book (and, in fact, in most other books, articles, etc.) with such 'unrealistic' values as $R = 1\,\Omega, C = 0.5\,F, L = 2\,H, t = 1, 2, 3\,s, \omega = 1\,rad/s$?" The answer to this justified question is found in *scaling* (or *normalization*). Just think of scaled distances on a map, making it easy to draw the maps and to calculate real (unscaled) distances. Or think of a scale model of a newly designed airplane being tested in a wind tunnel to calculate and evaluate many variables and parameters of the real, unscaled airplane.

In circuit analysis and circuit design, we often have a very wide range of numbers, such as $10\,k\Omega, 20\,pF\ (= 20 \times 10^{-12}\,F), 100\,ns\ (= 10^{-7}\,s), 5\,MHz\ (= 5 \times 10^6\,Hz)$, etc. Scaling is useful for several reasons. For one, it reduces human error in computations, or possible over- or underflow in machine calculations. In circuit design, normalized (scaled) configurations of prototype filters, with their element values, are readily available; for a particular problem, such a filter is easily unscaled to provide the final answer.

Two common scalings are for *magnitude* and for *frequency*, so that, for instance, a scaled resistor is $1\,\Omega$ and a scaled radian frequency is $1\,rad/s$. Let k_m, a positive constant, be the magnitude scaling factor; then the normalized (scaled) magnitude of an impedance is

$$Z_m = \frac{Z}{k_m} \tag{C-1}$$

where the subscript m reminds us of *magnitude* scaling. Letters without subscripts indicate here the actual, unscaled values. Specifically, for a resistor, we have

$$Z_m = \frac{R}{k_m} \tag{C-2}$$

for an inductor we have

$$Z_m = \frac{\omega L}{k_m} \tag{C-3}$$

and for a capacitor

$$Z_m = \frac{1}{\omega C k_m} \qquad\qquad \text{(C-4)}$$

Here we are dealing with magnitudes of impedances in the sinusoidal steady-state case; hence the j is missing. Similar results hold for the generalized (Laplace transform) impedances.

For the magnitude-scaled elements, the impedances are

$$Z_m = R_m \qquad Z_m = \omega L_m \qquad Z_m = \frac{1}{\omega C_m} \qquad\qquad \text{(C-5)}$$

respectively. Comparing these equations, we get the relationships between actual and magnitude-scaled element values:

$$R = k_m R_m \qquad\qquad \text{(C-6)}$$

$$L = k_m L_m \qquad\qquad \text{(C-7)}$$

$$C = \frac{C_m}{k_m} \qquad\qquad \text{(C-8)}$$

Equations C-1, C-6, C-7, and C-8 are the basis for magnitude scaling and unscaling. If $k_m > 1$, magnitude is scaled down, and if $k_m < 1$, it is scaled up, according to Equation C-1.

Next, let us scale the frequency by a positive constant k_f, such that the scaled frequency f_f is

$$f_f = \frac{f}{k_f} \qquad\qquad \text{(C-9)}$$

and the scaled radian frequency is

$$\omega_f = \frac{\omega}{k_f} \qquad\qquad \text{(C-10)}$$

Similarly, the complex (Laplace transform) frequency is scaled

$$s_f = \frac{s}{k_f} \qquad\qquad \text{(C-11)}$$

Here, again, the subscript f denotes $frequency$-scaled variables.

If frequency scaling is not to affect the magnitude of impedances, we must have for a resistor

$$Z_f = R \qquad\qquad \text{(C-12)}$$

since resistance is frequency-independent. For an inductor, we have

$$Z_f = \omega L = \left(\frac{\omega}{k_f}\right) k_f L \qquad\qquad \text{(C-13)}$$

and for a capacitor

$$Z_f = \frac{1}{\omega C} = \frac{1}{\left(\frac{\omega}{k_f}\right) k_f C} \qquad\qquad \text{(C-14)}$$

These frequency-scaled impedances are, respectively,

$$Z_f = R_f \qquad Z_f = \omega_f L_f \qquad Z_f = \frac{1}{\omega_f C_f} \tag{C-15}$$

Therefore, frequency-scaled and actual element values are related by

$$R = R_f \tag{C-16}$$

$$L = \frac{L_f}{k_f} \tag{C-17}$$

$$C = \frac{C_f}{k_f} \tag{C-18}$$

For both magnitude and frequency scalings, we combine these results as follows

$$R = k_m R_{m,f} \tag{C-19}$$

$$L = \frac{k_m}{k_f} L_{m,f} \tag{C-20}$$

$$C = \frac{1}{k_m k_f} C_{m,f} \tag{C-21}$$

where the subscripts m, f indicate *magnitude* and *frequency* scalings.

EXAMPLE C-1 _____

In Figure C-1(a) we have a real, unscaled network with its element values. The driving-point impedance of this network is given as

$$Z(s) = 100 + \cfrac{1}{1.6 \times 10^{-9}s + \cfrac{1}{16 \times 10^{-6}s}}$$

which is rather clumsy and prone to errors. After clearing fractions (correctly!), we get

$$Z(s) = \frac{25.6 \times 10^{-13}s^2 + 16 \times 10^{-6}s + 100}{25.6 \times 10^{-15}s^2 + 1}$$

which is hardly an improvement. On the other hand, if we choose

$$k_m = 100 \qquad k_f = \frac{10^7}{1.6}$$

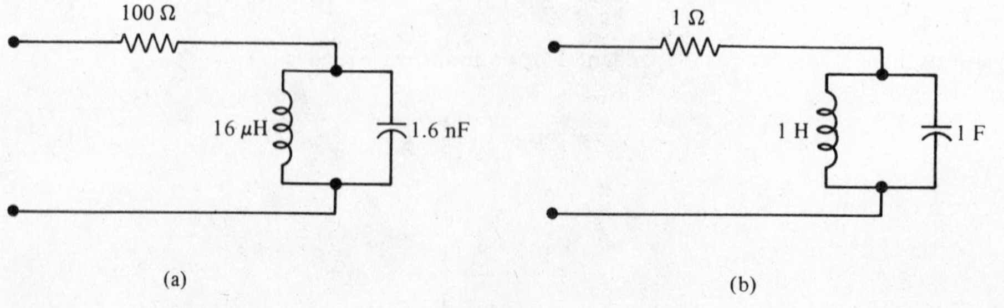

(a)   (b)

Figure C-1   Example C-1.

we get the scaled network in Figure C-1(b). Verify all these scaled element values, according to Equations C-19, C-20, and C-21. For this scaled network, we write neatly

$$Z_{m,f} = 1 + \cfrac{1}{s + \cfrac{1}{s}} = \frac{s^2 + s + 1}{s^2 + 1}$$

What a relief!

◼

EXAMPLE C-2

A certain filter is designed using magnitude and frequency scalings. The scaled cutoff frequency is $\omega_f = 1$ rad/s. The filter's structure and element values were obtained from standard tables, as shown in Figure C-2(a). The actual (unscaled) filter, we are told, must be terminated in a resistor of 600 Ω, and the actual cutoff frequency is 1 MHz. (By the way, this is a common way to select these scale factors in circuit design.) Thus,

$$k_m = 600 \qquad k_f = 2\pi \times 10^6$$

and the actual, unscaled filter is shown in Figure C-2(b).

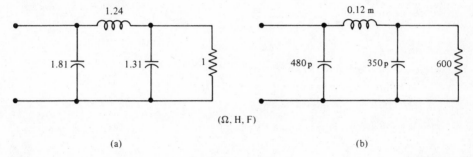

(Ω, H, F)

(a)   (b)

Figure C-2 Example C-2.

◼

Scaling of time is not independent of frequency scaling. In fact, if we write a typical sinusoidal voltage as

$$v(t) = V_0 \cos \omega t = V_0 \cos \left(\frac{\omega}{k_f}\right)(k_f t) \qquad (C\text{-}22)$$

we recognize that the scaled frequency

$$\omega_f = \frac{\omega}{k_f} \qquad (C\text{-}23)$$

must be accompanied by the inverse scaling of time

$$t_f = k_f t \qquad (C\text{-}24)$$

This reciprocal relation is also obvious intuitively: A compressed time means an expanded frequency, and vice versa. The same holds true for the generalized (Laplace transform) frequency since, in that case, we have

$$e^{-st} = e^{-(s/k_f)(k_f t)} = e^{-s_f t_f} \qquad (C\text{-}25)$$

Finally, it can be shown that an entire network function whose units are ohms (such as a driving-point or a transfer impedance) is affected by magnitude scaling of all its elements as follows:

$$Z_m(s_f) = \frac{Z(s_f)}{k_m} \tag{C-26}$$

as in Example C-1. On the other hand, a dimensionless network function (such as a voltage transfer function or a current transfer function) is *not* affected by the magnitude scaling of the elements in the network; that is, for a voltage transfer function we have

$$G_m(s_f) = G(s_f) \tag{C-27}$$

and for a current transfer function we have

$$\alpha_m(s_f) = \alpha(s_f) \tag{C-28}$$

In both Equations C-26 and C-27, frequency scaling is done independently of magnitude scaling.

EXAMPLE C-3 _____

In the network shown in Figure C-3, the output current is

$$I_{out}(s) = \frac{1/R}{1/R + 1/Ls + Cs} I_{in}(s)$$

with the current transfer function

$$\alpha(s) = \frac{1/R}{1/R + 1/Ls + Cs}$$

Apply to this network magnitude and frequency scalings to get

$$\alpha(s_f) = \frac{1/k_m R_{m,f}}{\dfrac{1}{k_m R_{m,f}} + \dfrac{1}{\left(\dfrac{k_m}{k_f} L_{m,f}\right)(s_f k_f)} + \left(\dfrac{1}{k_m k_f} C_{m,f}\right)(s_f k_f)} = \frac{1/R_{m,f}}{\dfrac{1}{R_{m,f}} + \dfrac{1}{L_{m,f} s_f} + C_{m,f} s_f}$$

showing how k_m does no' affect a dimensionless network function.

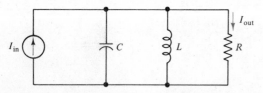

Figure C-3 Example C-3.   ■

In summary: Scaling reduces the complexity of calculations, improves accuracy, saves time, and allows us to use available normalized tables for circuits. These reasons are more than sufficient to justify the wide use of scaling, both in analysis and in design. We've done it throughout this book.

Appendix D

Selected Hints and Answers

CHAPTER 1

1-2 **(a)** Lumped valid
(d) Distributed

1-3 2.56×10^{-11} c/m³

1-5 3.75 kV

1-10 6.242×10^{18} eV

1-14 Receives power $0 < t < \pi/200$, $\pi/100 < t < 3\pi/200$, etc.

CHAPTER 2

2-1 0.022 Ω

2-5 **(b)**

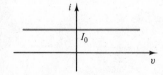

2-6 $\dfrac{di}{dv} < 0$

2-7 $I_{\text{eff}} = 6.93$ A

2-11 $v_1 = -5.6$ V

2-12 $i_6 = 18.57$ A

2-15 $l - n_i$

2-18 $i_C = 60 \sin t - 36$

2-23 $i_0 = 0$

CHAPTER 3

3-2   30 ($\frac{1}{2}$ the sum of all incidences)

3-7   (2) Not necessarily; see Figure 3-6(a)

3-8   (c) $\sum = 0$

3-10 $v_1 = \frac{63}{27}$ V

3-16 $i_e = 15$ A

3-17 Three node equations, two loop equations, $i_a = 2$ A

3-20 Source delivers 2.95 W

CHAPTER 4

4-1   (b) 3.834 Ω

4-2   Check dimensions (units)!

4-4 $R = \dfrac{1 + \sqrt{5}}{2}$

4-8 $-R_0$, negative resistor

4-14 2.542 Ω

4-17 For 5 V, $R = 490\,\Omega$

4-21 $R_x = 1000$ V

4-26 $i_0 = 12$ mA

CHAPTER 5

5-2   (a) Nonlinear $(r_1 + r_2) \neq m(e_1 + e_2) + b$

5-4 $v_0 = -24$ V

5-5 $v_3 = -0.64$ V

5-7 $i_4 = 2.57$ A

5-8 $p = p_1 + p_2$ because $v = $ const.

5-9 $p_{2A} = 1.30$ W   delivering

5-10 $i_2 = i_1 = 0.5$ A $i_3 = 0.2$ A

5-11 $i_1 = -0.028$ $i_2 = -0.113$ $i_3 = -0.0083$

5-12 $v_1 = 0.536$ $v_2 = 0.893$ $v_3 = 0.536$

5-13 $v_1 = -8$ $v_2 = -4.67$

5-14 $v_{Th} = 18$ $R_{Th} = 3$

5-16 $R_1 = \dfrac{R_2 R_3}{R_4}$

5-17 $v_2 = -4.667\ \text{V}$

5-19 $\alpha = -2 \qquad R = -R_0 \qquad$ negative resistance converter

5-24 $R_N = (R_1 \| R_3) + (R_2 \| R_4)$

5-25 $R \to 0 \qquad \therefore p = 0$
$R \to \infty \qquad \therefore p = 0$
but $p \geq 0 \qquad \therefore$ plot must be

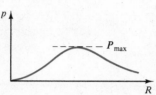

5-27 $p_R = \dfrac{3600R}{(R + 6)^2} \qquad \dfrac{dp_R}{dR} = 0 \to R = 6\ \Omega$

5-28 $p = \dfrac{V_{\text{Th}}^2}{4R}$

5-31 $\mathbf{H} = \begin{bmatrix} 0.083 & 0.024 \\ 0.16 & 0.02 \\ 0.02 & 0.087 \end{bmatrix}$

5-33 $\mathbf{R}_{\text{o.c.}} = \begin{bmatrix} r_e - k_1 k_2 r_c & k_1 r_c \\ -k_2 r_c & r_c \end{bmatrix}$

5-35 $v_0 = \tfrac{11}{10} v_1$

$\mathbf{G}_{\text{s.c.}} = \begin{bmatrix} 0 & 0 \\ -11 & 10 \end{bmatrix}$ is singular

5-39 $\begin{bmatrix} v_1 \\ i_2 \end{bmatrix} = \begin{bmatrix} 6 & 0 \\ 2 & 0.5 \end{bmatrix}\begin{bmatrix} i_1 \\ v_2 \end{bmatrix}$

5-42 $\begin{bmatrix} i_1 \\ v_2 \end{bmatrix} = \begin{bmatrix} 0 & 0 \\ 1.1 & 0.1 \end{bmatrix}\begin{bmatrix} v_1 \\ i_2 \end{bmatrix}$

5-45 $\begin{bmatrix} v_1 \\ i_1 \end{bmatrix}_{\text{total}} = \begin{bmatrix} A & B \\ C & D \end{bmatrix}_{\text{I}} \begin{bmatrix} A & B \\ C & D \end{bmatrix}_{\text{II}} \begin{bmatrix} v_2 \\ -i_2 \end{bmatrix}_{\text{total}}$

5-48 $v_{\text{Th}} = \dfrac{r_{21}}{r_{11} + R_s} v_s \qquad$ see Equation 5-58

5-49 $i_2 = \dfrac{-r_{21}}{r_{22} + R_L} i_s$

5-50 $\mathbf{R}_{\text{o.c.}} = \begin{bmatrix} r_b + r_e & r_e \\ r_a - \alpha r_c & r_e + r_c(1 - \alpha) \end{bmatrix}$

CHAPTER 6

6-2 $q(t) = \begin{cases} 0 & t < 0 \\ 0.3 t e^{-4t} & t > 0 \end{cases}$
$t = \tfrac{1}{4}$ for max q

6-4 $\displaystyle\int_{-\infty}^{t} i(x)\,dx = \int_{-\infty}^{t_0} i(x)\,dx + \int_{t_0}^{t} i(x)\,dx = q(t_0) - q(-\infty) + \int_{t_0}^{t} i(x)\,dx$

Assume $q(-\infty) = 0$ (reasonable!) to get Equation 6-7

6-5 $v(t) = 100e^{-2t}$

6-8 $p = (377)(50)\sin 754t$

6-9 $p(a) = -p(b)$

6-10 Parallel $\quad C = \sum C_i \qquad$ Series $\quad \dfrac{1}{C} = \sum \dfrac{1}{C_i}$

6-11 $t = 2$ for i_{max}

$v(t) = 0.2te^{-t}(2 - t)$

6-14 Spring $w\Big]_{t_1}^{t_2} = \dfrac{1}{2K}\left[f_2^2 - f_1^2\right]$

6-16 (b) $M = 2.2$ H

6-19 $v_2 = L_2 \dfrac{di_2}{dt} + M\dfrac{di_1}{dt}$

6-23 $w_a = w_b = \dfrac{L}{2}\left[2^2 - \left(\dfrac{2}{3}\right)^2\right]$

CHAPTER 7

7-2 $v_o = v_i + \dfrac{1}{RC}\displaystyle\int_0^t v_i(x)\,dx$

7-5 $v_C(t) = 1.2e^{-t/0.2}$

7-8 $i_1(t) = i - 2 \times 10^{-4}e^{-3t}$

7-9 $v_s = 0 \quad$ zero input $\qquad \tau = 0.65$ s

7-10 $\tau = 1.5$ s

7-11 $w_R\Big]_0^{\infty} = \displaystyle\int_0^{\infty} R(i_L)^2\,dt = \cdots$

7-14 $M\ddot{x} = -D\dot{x} \qquad$ Newton's law

$\therefore M\dot{v} + Dv = 0 \qquad \tau = M/D$

Analogous to $C\dot{v} + Gv = 0$ in RC circuit

7-16 Tangent intersects asymptote at $t = \tau$.

7-18 (a1) $v = 0.2e^{-1.5t} - 1.2e^{-4t}$

(a5) $v_H = Ke^t \quad$ physically impossible for $t \to \infty$

7-19 $i_L = -9e^{-(2/3)t} + 5$

7-20 $i_L = \frac{1}{2}[1 - e^{-(2/3)t}]$

7-21 $i_L(0.1) = 0.236 =$ initial value for b

7-22 $s^2 + 2s + 1 = 0 \qquad s_1 = s_2 = -1$

$i_H = K_1 e^{-t} + K_2 t e^{-t}$

$i_P = Bt^2 e^{-t} \qquad$ etc.

7-23 $i_P = Bt^3 e^{-t}$

7-24 $i_L = 5[1 - e^{-(2/3)(t-2)}]$

7-26 (c) Area $= \dfrac{1}{\varepsilon} \cdot C \cdot \varepsilon \equiv C$ for any ε

Area $= \displaystyle\int_0^\varepsilon i_C(t)\, dt = q(t) =$ finite

7-31 (a) $i_C(0^+) = 0 \qquad v_{L_1}(0^+) = 0 \qquad v_{L_2}(0^+) = 0$
 (b) $v_{L_2}(0^+) = -24\,\text{V}$

7-33 $v_C''(0^+) = \dfrac{1}{LC}(2R + 1)$

CHAPTER 8

8-1 $v_{1H} = K_b e^{-30t} \qquad 0 + 30B = K_a$

$$\therefore B = \frac{K_a}{30} = K_c$$

8-2 (d) $v_1(\infty) = v_2(\infty) \qquad$ since $i_R = 0 \qquad \therefore v_R = 0$

 (f) $w_C(\infty) = \frac{4}{3}\,\text{J} - w_R = \displaystyle\int_0^\infty R i^2\, dt = \cdots$

8-3 (a) No current $\qquad v_{C_1}(t) = v_{C_2}(t) = 2$

8-6 $A = C \cdot 1$ (coulombs)

8-10 $\zeta = \dfrac{R}{R_{cr}} \qquad R_{cr} = 2\sqrt{\dfrac{L}{C}}$

8-11 $\sigma^2 + \omega^2 = \omega_0^2 \qquad$ a circle of radius $= \omega_0$

8-13 (c) $C > \frac{1}{2000}\,\text{F}$

8-14 No, because if $K_4 = 0$ then $i_L = K_3 e^{-\omega_0 t}$, a *first-order* circuit???

8-16 Find first the maximum of each $i_L(t)$, with $t_{\max}$, then $i_L(t_{\max})$, etc.

8-17 $i_L(0^-) = 10 \qquad v_C(0^-) = 0$

8-18 $i_L(0^-) = 6 \qquad v_C(0^-) = \pm 6$ (depends on your reference)

8-19 $v_C(0^-) = -10 \qquad i_L(0^-) = 0$

8-20 (b) Decays faster than **(a)**
 (c) Critically damped
 (f) Oscillates faster than **(e)**

8-21 $v_C(0^-) = 0 \qquad i_L(0^-) = 2$
 After $t = 0$ convert current source and R into a Thévenin circuit. Then use Section 8-5.

8-22 $i_L(0^+) = -1.2\,\text{A} \qquad v_C(0^+) = 0$

8-23 4.9 joules

8-24 $0.3\dfrac{di}{dt} + v_C = 0 \qquad$ zero input!

$i = 0.1\dfrac{dv_C}{dt} + \frac{1}{2}v_C$

etc.

8-25 All transient

8-31 *One* node equation

8-32 $\dfrac{d^6 v_1}{dt^6} + a_1 \dfrac{d^5 v_1}{dt^5} + \cdots + a_6 v_1 = \cdots$

CHAPTER 9

9-5 1320 Hz

9-6 *i* lags behind *v* by 45°

9-9 Divide and multiply by $\sqrt{A^2 + B^2}$

9-12 KCL at the inverting terminals

9-16 $\sqrt{24} = \sqrt{25 - 1} = \cdots$

9-17 (5) $\mathbf{I}_5 = 10.6\underline{/-120°}$

9-20 Different ω's. Phasors require one ω

9-22 $\mathbf{V}_0 = 21\underline{/-110.4°}$

9-24 (b) $\mathbf{I} = 0.0007\underline{/-56°}$

9-26 $f(t) = 24.64 \cos (377t + 150°)$

9-28 $0.06(j50)\mathbf{V}_C + \dfrac{3}{(2 \times 10^{-3})(j50)} \mathbf{V}_C + 4\mathbf{V}_C = 80\underline{/40°}$

etc.

9-29 Resistor

9-30 $v_R = -20i_R$ as shown

9-33 $i_L = 4.2 \cos (2\pi 10^4 t - 120°)$

9-39 $v_1(t) = 2.546 \cos (377t - 28.83°)$

9-40 $v_0 = 4V_m \sin 100t$ integrator

9-44 If $|\mathbf{V}_L| = |\mathbf{V}_C|$ they cancel in KVL. No unique solution. $|\mathbf{V}_L| = |\mathbf{V}_C|$ is arbitrary

9-45 No, everything is in phase

9-51 Wrong dimensions (units)

9-53 $\mathbf{V}_3 = -3.33$ $\mathbf{Y} = 0.474\underline{/71.6°}$

9-58 $Z_{Th} = \dfrac{-1 + g_m/j\omega C_2}{g_m}$

9-62 Superposition! one circuit with $\omega = 50$ (current source only), the other with $\omega = 100$ (voltage source only)

CHAPTER 10

10-1 (b) $i_P = Bte^{-\alpha t}$ critically damped, resonant

10-2 $G^2 + B^2 = \dfrac{G}{R}$. Circle centered at $\left(\dfrac{1}{2R}, 0\right)$, $r = \dfrac{1}{2R}$

10-8 $v_R \neq f(R)$!

10-13 $|\mathbf{H}| \equiv 1$, all-pass filter

CHAPTER 11

11-1 **(d)** $\mathbf{Z} = 0.632\underline{/71.56°}$ 0.316 lagging

11-3 **(d)** Series $R = 0.4\,\Omega$ $X_L = 0.7\,\Omega$

11-5 $P = 180$ watts

11-7 $\cos\theta < 0$ delivering power

11-8 $V_{\text{eff}} = \dfrac{V_m}{2}$

11-10 5.84 W

11-13 $\cos\theta_t = 0.981$ leading

11-14 $L = 6.35$ mH

11-16 $P_t = 66$ kW $Q_t = 39$ kVAR
      wanted: $\hat{P}_t = 66$ kW $\hat{Q}_t = 13.4$ kVAR
      $C = 1.678$ mF

11-17 **(a)** $Q_C = 0.662$ MVAR   **(b)** $Q_C = 0.605$ MVAR

11-20 $R_0 = R_{\text{Th}}$

11-22 $P_0 = \dfrac{|V_0|^2}{10}$

CHAPTER 12

12-1 **(b)** No state variables, $i_{L_1} \equiv i_1$ $i_{L_2} \equiv i_2$

12-5 $a = a_1 a_2$

12-7 $L_1 L_2 - M^2 > 0$ (not $=$)
      Remember that $\dfrac{1}{j\omega}$ corresponds to integration

12-16 5.4 percent regulation

12-18 **(a)** $P_{\text{in}} = W_c$ **(b)** $P_{\text{in}} = |I_2|^2 r_{\text{eq}}$, since $|V_1| \approx 0$

12-19 $\eta = 91.6$ percent

12-20 $\eta_{\text{ad}} = 91.9$ percent

CHAPTER 13

13-2 $v_C(t) = 120\sqrt{2}\cos(377t + 180°)$

13-3 $\sum v = 0$ around this loop!

13-6 $\mathbf{I}_{CN} = 21\underline{/117.2°}$

13-7 $\sum \mathbf{I} = 0$ across this cut set in *all* cases

13-8 $\mathbf{I}_{CA} = 14.7\underline{/173.1°}$

13-10 $\mathbf{I}_{aA} = 25.48\underline{/-23.1°}$

13-11 $\mathbf{V}_{AB}$ leads $\mathbf{V}_{AN}$ by 30°

13-12 Start with the given $\mathbf{I}_{aA}$ and try two possibilities of decomposing it into $(\mathbf{I}_{ba} + \mathbf{I}_{ca})$.

13-16 Two loop equations or, better yet, one node equation at N, with n as reference.
$\mathbf{V}_N = 103.5\underline{/15.6°}$ etc.

13-18 $I_l = 26.6$ A

13-19 **(a)** $C = 0.58 \ \mu\text{F}$ 13.8 kV
(b) $C = 1.76 \ \mu\text{F}$ 8 kV

13-20 By power triangles

13-25 $Q_t = \sqrt{3}\,(W_1 - W_2)$

13-29 0.866

CHAPTER 14

14-1   **(d)** $f_d(t) = 100 \sin 377t[u(t) - u(t - \frac{1}{60})]$

14-6 $i_L(t) = u(t)$ $v_L(t) = \dfrac{d}{dt}\,(Lu(t)) = L\delta(t)$

14-7   Area $\equiv 1$ independent of ε

14-9 $\tan^{-1} 2 = 63.44°$

14-10 $K(t - a)^2 u(t - a)$ the unit parabola

14-11 **(c)** 0

14-12 **(a)** $v(0)$ is finite, no impulse, $\phi(0^+) - \phi(0^-) = \displaystyle\int_{0^-}^{0^+} v(t)\,dt = 0$

14-14 **(d)** Yes, $c > -a$
(e) Yes, $c > 0$
(h) No. There is no c

14-19 $s_1 = -39.875$ $s_2 = -0.125$

14-24 **(a)** $v(t) = e^{-t} \cos \sqrt{14}\,tu(t)$
(b) $i(t) = \frac{3}{4}e^{-2t} \sin 4tu(t)$

14-27 $\mathscr{L}t \sin \beta t\, u(t) = \dfrac{2\beta s}{(s^2 + \beta^2)^2}$

14-28 $\dfrac{\partial}{\partial b} \sin bt\, u(t) = t \cos bt\, u(t)$

$\dfrac{\partial}{\partial b}\dfrac{b}{s^2 + b^2} = \dfrac{s^2 - b^2}{(s^2 + b^2)^2}$

14-29 **(c)** $\displaystyle\int_0^\infty t^3 e^{-4t}dt = \mathscr{L}[t^n e^{-at}]_{\substack{n=3\\a=4\\s=0}} = \dfrac{3!}{4^3} = \dfrac{6}{64} = \dfrac{3}{32}$

CHAPTER 15

15-1 **(d)** $I_1(s) = \underbrace{\dfrac{-s}{2s^2 + 3s + 4}}_{\text{Zero-state}} + \underbrace{\dfrac{-1}{2s^2 + 3s + 4}}_{\text{Zero-input}}$

15-2 $8s^2 + 20s + 16 = 0$

15-3 $2s^3 + 7s^2 + 12s + 15 = 0$

15-7 $s = 0$

15-9 $Z_t(s) \equiv R$ always

15-14 $[I(s)] = [\text{amperes}][\text{seconds}] = [\text{coulombs}]$

15-17 $Y_{dp} = \dfrac{I_{dp}}{V_{dp}} = \dfrac{\Delta_{11}}{\Delta}$ $\Delta = $ loop impedance determinant

15-18 $R = 0$ $\rightarrow i(t) = C_t V_0 \delta(t)$

15-19 $V_0 = \dfrac{-4}{s^2 + s + 2}$

15-20 **(b)** $3\delta'(t) - 30\delta(t) + 300e^{-10t}u(t)$

15-22 See below. In **(d)**, the factor $(s + 2)$ cancels

15-25 $\dfrac{1}{(s + 1)^2} - \dfrac{1}{(s + 2)^2}$

15-33 **(a)** $-\frac{1}{2}$ **(b)** Not applicable   **(c)** 1   **(d)** NA

15-34 Check first the locations of poles of $sF(s)$ (Routh!)

15-35 $f(0^+) = 0$ $f'(0^+) = 1$ $f''(0^+) = -6$

15-22

```
c     subroutine pfexp (n,p,m,coef,c)


c     partial fraction expansion

c     This subroutine was designed to perform partial fraction
c     expansion of proper fractions. It will accommodate
c     any number of poles and multiplicities and is bounded
c     only by the dimension statement. The calling sequence
c     is:
c         call pfexp (n,p,m,coeff,c)
c     where
c         n = the number of distinct poles
c         p = a one-dimensional array containing
c           the poles
c         m = a one-dimensional array containing
c           the multiplicities
c         (note that there is a one to one
c         correspondence between the arrays
c         p and m. Thus m(1) is the multiplicity
c         of the pole p(1).)
```

```
c          coef = a one-dimensional array containing
c                 the coefficients of the numerator
c                 polynomial. For n distinct poles
c                 there must be n coefficients. They
c                 must be specified from order zero
c                 to order (n − 1). For example, if the
c                 numerator for a polynomial of n = 5 is
c                 (s**4)+(2*(s**3)) − 6*s+12
c                 then coef(1)=12,coef(2)= −6,coef(3)=0,
c                 coef(4)=2,coef(5)=1.
c          c = a (6 × n) array used for output from
c                 the subroutine. The row designates the
c                 order of the constant and the column
c                 designates the particular pole. For
c                 example, if c(3,6)=4.33 and the sixth
c                 pole is −2, then the expansion includes
c                          4.33/((s+2)**3)

       dimension a(6,6),c(6,6),p(6),m(6),coef(6)
c    zero fill the work array
       do 10 i=1,6
       do 10 j=1,6
       c(i,j)=0.0
   10  a(i,j)=0.0
c    compute first order coefficients
       do 20 jj=1,n
       hold=1.0
       cnumer=0.0
       s=p(jj)
       do 15 i=1,n
       if(jj.ne.i)hold=hold*(s − p(i))
       if(s.ne.0.0)cnumer=cnumer+coef(i)*(s**(i − 1))
   15  continue
       c(1,jj)=cnumer/hold
   20  continue
c    set order pointer: korder
       do 900 korder=2,6
c    set pole pointer: npole
       do 800 npole=1,n
c    if multiplicity of npole < korder try next pole
       if(m(npole).lt.korder) go to 800
c    if multiplicity of npole not < korder then
c      (1) shift npole column of the coefficient array of
c             output values down one element
c      (2) calculate contribution of npole with all other terms
c             and add contributions to corresponding locations in
c             the coefficient array of output values
       do 50 iwork=1,5
       iw=7 − iwork
       jw=iw − 1
       a(iw,npole)=c(jw,npole)
```

```
  50   c(iw,npole)=c(jw,npole)
       a(1,npole)=0.0
       c(1,npole)=0.0
       do 700 jpole=1,n
       if (jpole.eq.npole)go to 700
       kwork=korder
       a(kwork,jpole)=0.0
 200   continue
       if (c(kwork,jpole).ne.0.0)go to 300
       kwork=kwork − 1
       if(kwork.eq.0) go to 700
       go to 200
 300   khold=kwork
 400   con=c(kwork,jpole)
 500   a1=con/(p(npole) − p(jpole))
       b=con/(p(jpole) − p(npole))
       a(kwork,jpole)=a(kwork,jpole)+b
       if(kwork.eq.1)go to 600
       kwork=kwork − 1
       con=a1
       go to 500
 600   a(1,npole)=a(1,npole)+a1
       khold=khold − 1
       if(khold.eq.0)go to 700
       kwork=khold
       go to 400
 700   continue

c   replace coefficient array for output values with work array
       do 750 i=1,6
       do 750 j=1,n
       c(i,j)=a(i,j)
 750   a(i,j)=0.0
 800   continue
 900   continue
c   print the coefficient array of output values in matrix form
       write(6,955)
 955   format('0','coefficient array for partial fraction expansion')
       do 950 i=1,6
 950   write(6,960)(c(i,j),j=1,n)
 960   format(6f10.3)
       return
       end
```

CHAPTER 16

16-1 $Y_{tr}(s) = \dfrac{1.8s^2 + 10}{4.76s^3 + 12s^2 + 24s}$

16-7 $k = 2 \to 6s^3 + 20s^2 + 21s + 66 = 0$

16-9   **(a)** $i_L(t) = K_1 u(t) + K_2 e^{-0.75t} \cos(3.8t + K_3)$

16-10 $I(s) = \dfrac{1}{Ls^2} \rightarrow \dfrac{1}{L} tu(t)$ resonance at dc

16-12 $s_{1,2} = -0.02 \pm j2$

       No resonance with $s = j2$ of source

16-13 **(a)** $H(s) = \dfrac{s}{s^2 + 1} = Y_{dp} = \dfrac{1}{s + 1/s}$

          $E(s) = \dfrac{1}{s + 1}$

      (b) $H(s) = \dfrac{1}{s + 1} = Y_{dp}$

          $E(s) = \dfrac{s}{s^2 + 1}$

16-14 Within the arbitrary multipliers

16-15 **(c)** $i_{out}(t) = \frac{1}{2} e^{-(3/4)t} \left(\cos \dfrac{\sqrt{7}}{4} t + \dfrac{1}{\sqrt{7}} \sin \dfrac{\sqrt{7}}{4} t \right)$

16-16 **(a)** $h(t) = 2.3\delta(t) + \cdots$
      (b) No
      (c) $h(t) = 1.2\delta(t) + \cdots$

16-17 Experiment yields $R_u(s) = H(s) \cdot \dfrac{1}{s}$

       $\therefore sR_u(s) = H(s)$

       $\therefore sR_u(s) - r_u(0^-) = H(s)$

              $= 0$

       $\mathscr{L}^{-1} \rightarrow \dfrac{d}{dt} r_u(t) = h(t)$

16-18 All zero-input, all steady state

16-19 **(d)** $r(t) = 10 \sin 4tu(t)$ $1\text{-}\Omega$ resistor

16-25

```
c    this program performs the numerical convolution of two functions,
c    h and x, defined in the function statements at the end of the program.
c    the program assumes the functions to equal 0 for t < 0, therefore the
c    convolution begins for t > 0. the user should define the desired functions
c    properly, using fortran 77 statements.

c    the program also requires that the user input two variables when
c    prompted:
c       1) the sampling interval, t; and
c       2) the number of intervals over which the convolution
c            should take place, k.
```

```
c    program convolution

     integer l,k,n
     real   y,t,sum,h,x
     common t

     print*,'enter an integer value for k.'
     read*,k
     print*,'enter a value for t.'
     read*,t
     print*
     print*,' t = ',t
     print*

c    perform the convolution

         do 40 l = 1 , k
            sum = 0
            do 30 n = 1 , 1 − 1
               sum = sum + 2*h(n)*x(l − n)
30          continue
            y = t*(sum + h(0)*x(l) + h(l)*x(0))/2.0
            print*,'y( ',l,'t) = ',y
40       continue
     end

     function h(arg)
     common t
     integer arg
         h = exp(−2*arg*t)
     return
     end

     function x(arg)
     common t
     integer arg
         x = 10
     return
     end
c    this program performs the numerical convolution of two functions,
c    x and h. the program assumes the functions to equal 0 for
c    t < 0, therefore the convolution begins for t > 0.
c    the functions must be given at equally spaced intervals, where t is
c    the spacing. these data must be stored in a file called convol.dat.
c    the format is one data entry per line. if the convolution is to take
c    place over k intervals, x(0),x(1t),x(2t),...,x(kt),h(0),h(1t),h(2t),...
c    h(kt) must be given in that order in the file.
```

```
c    the program also requires that the user input 2 variables when
c    prompted:
c         1) the sampling interval, t; and
c         2) the number of intervals over which the convolution
c             should take place, k.

     program convolution

     integer l,k,n,j
     real   y,t,sum,h(0:20),x(0:20)

     print*,'enter an integer value for k.'
     read*,k
     print*,'enter a value for t.'
     read*,t
     print*
     print*,' t = ',t
     print*

     open(unit = 15,file = 'convol.dat',status = 'old')
     rewind(unit = 15)
c    read in the data

     do 10 j=0,k
         read(15,*)x(j)
10   continue
     do 20 j=0,k
         read(15,*)h(j)
20   continue

c    perform the convolution

     do 40 l = 1 , k
         sum = 0
         do 30 n = 1 , l − 1
             sum = sum + 2*h(n)*x(l − n)
30       continue
         y = t*(sum + h(0)*x(l) + h(l)*x(0))/2.0
         print*,'y( ',l,'t) = ',y
40   continue
     end
```

16-26

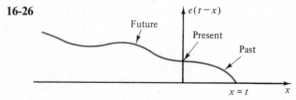

16-31 $Y_{s.c.} = \begin{bmatrix} Y_a + Y_c & -Y_c \\ -Y_c & Y_b + Y_c \end{bmatrix}$

16-33 $G = -\dfrac{z_{21}Z_L}{\Delta_z + z_{11}Z_L}$

16-34 The Δ-Y equivalence!

CHAPTER 17

17-1 $G(s) = \dfrac{s}{s + 1/RC} \rightarrow \dfrac{j\omega}{j\omega + 1/RC}$

$\omega = 0 \rightarrow |G| = 0 \qquad \theta = 0°$

$\omega \rightarrow \infty \quad |G| = 1 \qquad \theta = 0°$

17-2 $m = n \qquad |H| = |K| \left| \dfrac{a_m}{b_n} \right|$

17-4 All-pass

17-6 **(b)** $|H| \equiv 1$

$\theta = 0° \rightarrow -360°$

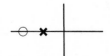

17-8 $\alpha = \pm 20 \log |j\omega| = \pm 20 \log \omega \qquad$ dB

$\theta = \pm 90°$

17-12 **(a)** $H(s) = K \dfrac{s}{(s + 100)(s + 10^4)}$

at $\omega = 10 \qquad \alpha = 0 \qquad \therefore |H(j10)| = 1$

$\therefore 1 = K \dfrac{10}{\sqrt{10^2 + 10^4}\sqrt{10^2 + 10^8}} \approx K \dfrac{10}{10^6} \qquad \therefore K = 10^5$

17-13

17-15

17-17 $n = 5 \qquad H(s) = \dfrac{1}{s^5 + 3.236s^4 + 5.236s^3 + 5.236s^2 + 3.236s + 1}$

All poles (for all n) are located on the unit circle

17-18 Can't do because of $-s$

17-19 (b)

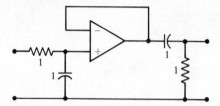

CHAPTER 18

18-1 $\zeta < 1$ complex conjugate
          $\zeta > 1$ real, distinct
          $\zeta = 1$ real, multiple
          $\zeta = 0$ pure imaginary

18-2   (c) $h(t) = Fu(t) + Ge^{-t} + Ke^{-2t}$

18-4   Unstable

18-7
```
c    this program generates the routh array for a given polynomial.
c    the program asks for the order of the polynomial and then requires
c    that the coefficients be entered when prompted in descending order,
c    i.e. first enter the coefficient of the highest power, then the
c    coefficient of the next highest power, etc. until the coefficient of
c    the constant term is entered.  note: a coefficient for each power
c    must be entered.

     program routharray

     integer order, horder,q,x,i,j,k,z,r,current,l,b,m,h
     real routh(21,11),sum
     print*,'enter order'
     read(*,*)order
     data routh/231*0.0/
     horder=order+1
     l=nint(float(horder/2.0))

c    l is the length of rows 1 and 2 of the routh array.

     q=l-1
```

```
c     q is the length of the row to be generated.

      x=1
      i=1
      j=-1
      m=(-1)**order
      if (m.lt.0) then
        r=1
      else
        r=2
      endif

c     horder, as a row name, refers to the first row of the routh array.
c     order, as a row name, refers to the second row of the routh array.

c     read in the given polynomial.

      do 15 k=1,l
          b=2
          if ((r.eq.2).and.(k.eq.l)) b=1
        do 18 z=1,b
          print*,'enter coefficient'
          read(*,*)routh(horder,i)
          x=j*x
          horder=horder+x
18        continue
          i=i+1
15    continue
      horder=order+1
      print*,(routh(horder,j), =1,q+1)
      print*
      sum=0
      do 20 j=1,l
          sum=sum+routh(order,j)
20    continue
      h=order

c     check to see if every other power of the given polynomial is missing.
c     if it is, replace the zero row with the derivative of the row above it.

      if(sum.eq.0.0) then
          do 22 j=1,l
              routh(order,j)=routh(horder,j)*h
              h=h-2
22        continue
      endif
      print*,(routh(order,j),j=1,q+1)
      if(routh(order,1).eq.0) routh(order,1)=0.000001
      current=horder-2
```

c   current, as a row name, refers to the row of the routh array currently
c   being generated.

c   generate the elements of the routh array.

```
      do 80 w=1,order − 1
        do 30 j=1,q
          routh(current,j)=(routh(current+1,1)*rough(current+2,j+1) −
    1       routh(current+2,1)*routh(current+1,j+1))/routh(current+1,1)
30      continue
        sum=0
        do 35 j=1,q
        sum=sum+routh(current,j)
35      continue
```

c   check to see if an auxiliary polynomial is generated.

```
        if (sum.eq.0) then
          print*,'the degree of the auxiliary polynomial is ',current
          print*,'the coefficients in descending order with every other'
          print*,'power missing are '
          print*,(routh(current+1,j),j=1,l)
          go to 100
        elseif (routh(current,1).eq.0) then
          routh(current,1)=0.000001
        endif
        print*
        print*,(routh(current,j),j=1,q)
        current=current−1
        r=r+1
        if(mod(r,3).eq.0)then
            q=q − 1
            r=r+1
        endif
80      continue
100 end
```

18-8   **(a)** $0 < K \le 0.1$

       (b) $K = 0.1$ $\omega = \sqrt{10}$

18-13 Avoid sinusoids at $\omega = 0, 1, 2, 3$

18-14 $K > -19$

CHAPTER 19

19-3 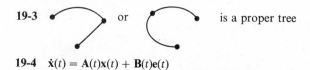   or   is a proper tree

19-4 $\dot{\mathbf{x}}(t) = \mathbf{A}(t)\mathbf{x}(t) + \mathbf{B}(t)\mathbf{e}(t)$

19-5 **(c)** $n = 3$ **(e)** $n = 4$

19-6 Formulation does not require initial conditions. Solution does.

19-10 $a \leq 0$ (real, of course)

19-11 $\dfrac{dx_H}{dt} - ax_H = 0$ $\quad \therefore x_H = Ke^{at}$

By integrating factor

$$e^{-at}\dot{x} - ae^{-at}x = e^{-at}be(t)$$

$$\frac{d}{dt}[e^{-at}x(t)] = e^{-at}be(t)$$

Integrate both sides, etc.

$$x(t) = Ke^{at} + e^{at}*be(t)$$

19-12 $s_{1,2} = -5 \pm j5.92$

19-13 $I_1(s) = \dfrac{60}{s^2 + 10s + 60}$ by node analysis

19-15

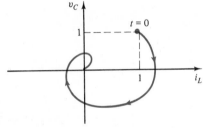

19-16 An ellipse

CHAPTER 20

20-1 $\cos^2 t = [\frac{1}{2}(e^{jt} + e^{-jt})]^2 = \cdots = \frac{1}{2} + \frac{1}{2}\cos 2t$

20-3 Half-wave rectified

$$f(t) = \frac{A}{\pi} + \frac{A}{2}\sin \omega_0 t - \frac{2A}{\pi}\sum_{n=2,4,6,\ldots}^{\infty}\frac{\cos n\omega_0 t}{n^2 - 1}$$

20-4 Full-wave rectified

$$f(t) = \frac{2A}{\pi} - \frac{4A}{\pi}\sum_{n=1}^{\infty}\frac{\cos n\omega_0 t}{4n^2 - 1}$$

20-7 $\mathbf{V} = \mathbf{ZI} = \dfrac{1}{4 + j3\omega}\mathbf{I}$ for $\omega = 0, \omega_0, 2\omega_0, 3\omega_0$

20-8 $\mathbf{V}_R = \dfrac{\mathbf{Z}_{RC}}{\mathbf{Z}_{RC} + \mathbf{Z}_L}\mathbf{V}_r$

$$v_R(t) = \frac{156}{\pi} + 39.2\cos(377t + 112.3°) - 3.16\cos(754t - 171.7°) + \cdots$$

20-10 c_n is pure imaginary since $f(t)$ is odd $\quad c_0 = 0$

20-13 L, M, C but no R since $P_{av} = 0$

20-15 $F_{eff}^2 = c_0^2 + 2 \sum_1^\infty |c_n|^2$

20-20 $|c_n|$ remains unchanged
θ_n has $-n\omega t_0$ added to it

20-21 $c_n = \dfrac{0.632}{1 + j2\pi n}$

CHAPTER 21

21-2 $F(\omega) = \dfrac{2a}{a^2 + \omega^2}$ $\phi(\omega) = 0°$

Area $= \displaystyle\int_{-\infty}^{\infty} F(\omega)\, d\omega = 2\pi$

As $a \to 0$ $f(t) \to 1$ and $F(\omega) \to 2\pi\delta(\omega)$

21-5 $\mathscr{F}^{-1} 2\pi\delta(\omega - \omega_0) = e^{j\omega_0 t}$

21-7 $\dfrac{dg}{dt} = f(t)$ of Problem 21-1

$j\omega G(\omega) = F(\omega)$ $\therefore G(\omega) = \dfrac{F(\omega)}{j\omega}$

21-8 $F(0) = \left[\displaystyle\int_{-\infty}^{\infty} f(t)e^{-j\omega t} dt \right]_{\omega=0} = \displaystyle\int_{-\infty}^{\infty} f(t)\, dt$

$\therefore$ the term $\pi F(0)\delta(\omega) = 0$

21-12 $e^{jn\omega t} \to K e^{jn\omega_0(t-t_0)}$ if $n\omega_0 < \omega_c$

$\to 0$ if $n\omega_0 > \omega_c$

$\therefore g(t) = K \displaystyle\sum_{n=-N}^{+N} c_n e^{jn\omega_0(t-t_0)}$

where N is the largest harmonic contained in the passband $N\omega_0 < \omega_c < (N+1)\omega_0$

21-17 $F(\omega) = \dfrac{100}{10 + j\omega}$

Energy input $= \dfrac{1}{2\pi} \displaystyle\int_{-\infty}^{\infty} |F|^2 d\omega = \cdots = 500$ J

Energy output $= \dfrac{1}{2\pi} \displaystyle\int_{-10}^{10} |G|^2 d\omega = \cdots = 250$ J

Index

Table 14-2   The Laplace Transform

Definition: $F(s) = \mathscr{L}f(t)u(t) = \int_{0^-}^{\infty} f(t)e^{-st}\,dt$

Functions	
$f(t)$	$F(s)$
$\delta(t)$	1
$u(t)$	$\dfrac{1}{s}$
$tu(t)$	$\dfrac{1}{s^2}$
A, constant	$\dfrac{A}{s}$
$t^n u(t) \quad n = 0, 1, 2, \ldots$	$\dfrac{n!}{s^{n+1}}$
$e^{-at}u(t)$	$\dfrac{1}{s+a}$
$t^n e^{-at}u(t)$	$\dfrac{n!}{(s+a)^{n+1}}$
$\sin \beta t u(t)$	$\dfrac{\beta}{s^2 + \beta^2}$
$\cos \beta t u(t)$	$\dfrac{s}{s^2 + \beta^2}$
$e^{-at} \sin \beta t u(t)$	$\dfrac{\beta}{(s+a)^2 + \beta^2}$
$e^{-at} \cos \beta t u(t)$	$\dfrac{s+a}{(s+a)^2 + \beta^2}$
$\sinh \beta t u(t)$	$\dfrac{\beta}{s^2 - \beta^2}$
$\cosh \beta t u(t)$	$\dfrac{s}{s^2 - \beta^2}$